D1233218

Organic Reactions

ADVISORY BOARD

John E. Baldwin
Virgil Boekelheide
George A. Boswell, Jr.
Donald J. Cram
David Y. Curtin
Samuel Danishefsky
William G. Dauben
Heinz W. Gschwend
Stephen Hanessian
Ralph F. Hirschmann
Herbert O. House

Andrew S. Kende
Steven V. Ley
Blaine C. McKusick
James A. Marshall
Jerrold Meinwald
Gary H. Posner
Hans J. Reich
Charles Sih
Barry M. Trost
Milán Uskokovic

FORMER MEMBERS OF THE BOARD
NOW DECEASED

Roger Adams
Homer Adkins
Werner E. Bachmann
A. H. Blatt
Theodore L. Cairns
Arthur C. Cope
Louis F. Fieser

John R. Johnson
Willy Leimgruber
Frank C. McGrew
Carl Niemann
Harold R. Snyder
Boris Weinstein

Organic Reactions

VOLUME 49

EDITORIAL BOARD

LEO A. PAQUETTE, *Editor-in-Chief*

PETER BEAK
ENGELBERT CIGANEK
DENNIS CURRAN
ANTHONY W. CZARNIK
SCOTT E. DENMARK
LOUIS HEGEDUS

ROBERT C. KELLY
LARRY E. OVERMAN
WILLIAM ROUSH
AMOS B. SMITH, III
JAMES D. WHITE

ROBERT BITTMAN, *Secretary*
Queens College of The City University of New York, Flushing, New York

JEFFERY B. PRESS, *Secretary*
Brewster, New York

EDITORIAL COORDINATOR

ROBERT M. JOYCE
Sun City Center, Florida

ASSOCIATE EDITORS

GURNOS JONES
GAGIK G. MELIKYAN

JAMES H. RIGBY
STEPHEN P. STANFORTH

JOHN WILEY & SONS, INC.

York   •   Chichester   •   Brisbane   •   Toronto   •   Singapore

547
0680

This text is printed on acid-free paper.

Published by John Wiley & Sons, Inc.

Copyright © 1997 by Organic Reactions, Inc.

All rights reserved. Published simultaneously in Canada.

Reproduction or translation of any part of this work beyond
that permitted by Section 107 or 108 of the 1976 United
States Copyright Act without the permission of the copyright
owner is unlawful. Requests for permission or further
information should be addressed to the Permissions Department,
John Wiley & Sons, Inc., 605 Third Avenue, New York, NY
10158-0012

Library of Congress Catalog Card Number 42-20265

ISBN 0-471-15655-8

Printed in the United States of America

10 9 8 7 6 5 4 3 2 1

PREFACE TO THE SERIES

In the course of nearly every program of research in organic chemistry the investigator finds it necessary to use several of the better-known synthetic reactions. To discover the optimum conditions for the application of even the most familiar one to a compound not previously subjected to the reaction often requires an extensive search of the literature; even then a series of experiments may be necessary. When the results of the investigation are published, the synthesis, which may have required months of work, is usually described without comment. The background of knowledge and experience gained in the literature search and experimentation is thus lost to those who subsequently have occasion to apply the general method. The student of preparative organic chemistry faces similar difficulties. The textbooks and laboratory manuals furnish numerous examples of the application of various syntheses, but only rarely do they convey an accurate conception of the scope and usefulness of the processes.

For many years American organic chemists have discussed these problems. The plan of compiling critical discussions of the more important reactions thus was evolved. The volumes of *Organic Reactions* are collections of chapters each devoted to a single reaction, or a definite phase of a reaction, of wide applicability. The authors have had experience with the processes surveyed. The subjects are presented from the preparative viewpoint, and particular attention is given to limitations, interfering influences, effects of structure, and the selection of experimental techniques. Each chapter includes several detailed procedures illustrating the significant modifications of the method. Most of these procedures have been found satisfactory by the author or one of the editors, but unlike those in *Organic Syntheses* they have not been subjected to careful testing in two or more laboratories.

Each chapter contains tables that include all the examples of the reaction under consideration that the author has been able to find. It is inevitable, however, that in the search of the literature some examples will be missed, especially when the reaction is used as one step in an extended synthesis. Nevertheless, the investigator will be able to use the tables and their accompanying bibliographies in place of most or all of the literature search so often required.

Because of the systematic arrangement of the material in the chapters and the entries in the tables, users of the books will be able to find information desired by reference to the table of contents of the appropriate chapter. In the interest of economy the entries in the indices have been kept to a minimum, and, in particular, the compounds listed in the tables are not repeated in the indices.

The success of this publication, which will appear periodically, depends upon the cooperation of organic chemists and their willingness to devote time and effort to the preparation of the chapters. They have manifested their interest already by the almost unanimous acceptance of invitations to contribute to the work. The editors will welcome their continued interest and their suggestions for improvements in *Organic Reactions*.

Chemists who are considering the preparation of a manuscript for submission to *Organic Reactions* are urged to write either secretary before they begin work.

CONTENTS

Organic Reactions

CHAPTER 1

THE VILSMEIER REACTION OF FULLY CONJUGATED CARBOCYCLES AND HETEROCYCLES

GURNOS JONES

Department of Chemistry, University of Keele,
Keele, England

STEPHEN P. STANFORTH

Department of Chemical and Life Sciences,
University of Northumbria at Newcastle,
Newcastle upon Tyne, England

CONTENTS

Organic Reactions, Vol. 49, Edited by Leo A. Paquette et al.
ISBN 0-471-15655-8 © 1997 Organic Reactions, Inc. Published by John Wiley & Sons, Inc.

INTRODUCTION

In 1925 Fischer, Müller, and Vilsmeier[1] published a paper describing the reaction between phosphoryl chloride and N-methylacetanilide, giving a number of products, including the quinolinium salt **1** and the salt **2** (Eq. 1). The probable course of the reaction was given in a paper by Vilsmeier and Haack in 1927,[2] and they made the important discovery that the reagent obtained from N-methylformanilide and phosphoryl chloride, represented as the salt **3**, would react with N,N-dimethylaniline, giving 4-N,N-dimethylaminobenzaldehyde (**4**)(Eq. 2). No

(Eq. 1)

(Eq. 2)

2-substituted products were observed in this reaction. Other N,N-dialkylaniline derivatives, including 3,N,N-trimethylaniline and 1-N,N-dimethylaminonaphthalene were also successfully used as substrates to prepare aromatic aldehyde derivatives.

The gradual development of the reagent for synthesis was accompanied by interest in the nature of the reagent. It was discovered that other acid chlorides (e.g., thionyl chloride, carbonyl chloride, and oxalyl chloride) could be used in the reaction and that substituted amides other than formamides gave ketones, although in generally poorer yields. Thionyl chloride frequently gives sulfur-containing products. The most commonly used amide is dimethylformamide (DMF) and there is now a consensus that the reagent formed from DMF and most acid chlorides, other than phosphoryl chloride, can be represented by the structure **5**, and this is illustrated for the reaction between DMF and carbonyl chloride (Eq. 3). Salt **5** is a stable compound and is often isolated before being reacted

(Eq. 3)

with a substrate. It seems likely that the most commonly used reagent, that made from DMF and phosphoryl chloride, is an equilibrium mixture of the iminium salts **6** and **7** (Eq. 4). Recent unpublished spectroscopic studies[3] have indicated that in DMF solution there is an equilibrium mixture of iminium compounds, including the dipolar structure **8** (Eq. 5).

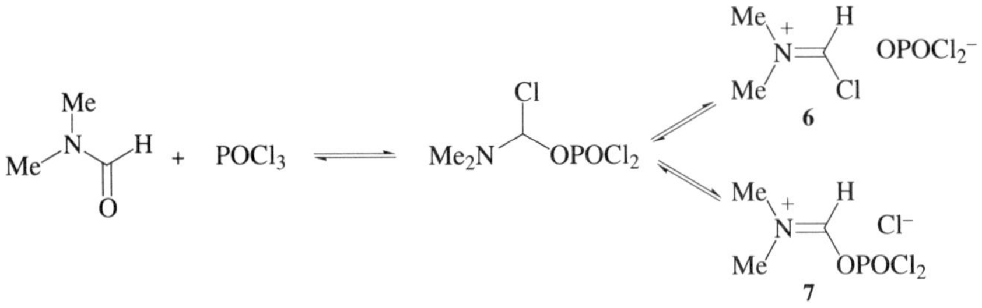

(Eq. 4)

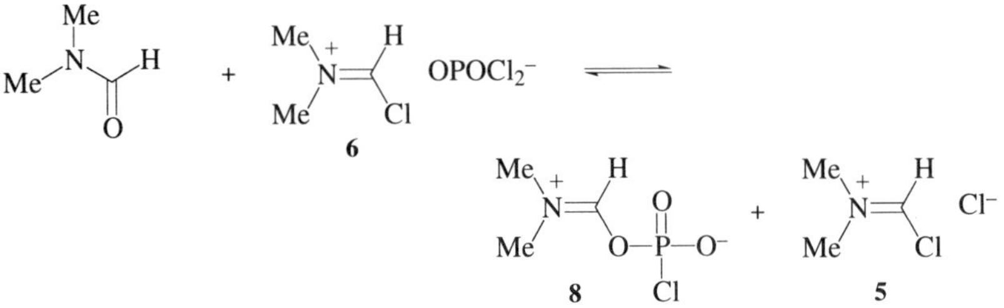

(Eq. 5)

The electrophilic chloroiminium salt **6** or salt **7** then reacts with a substrate in an electrophilic substitution process yielding an iminium salt **9**, which is usually hydrolyzed to the aromatic aldehyde **10** (Eq. 6). Vinylogous chloroiminium salts

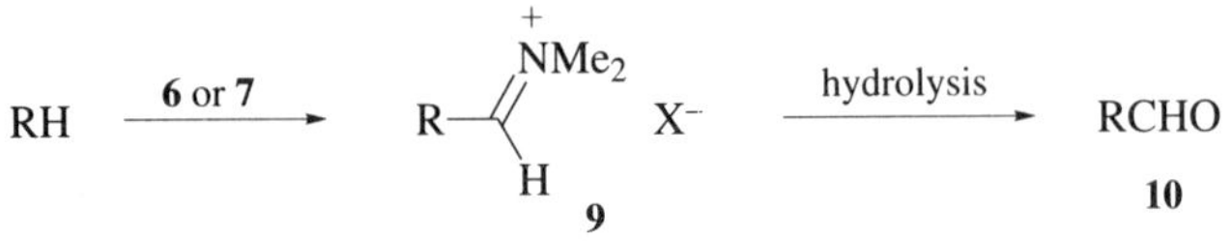

such as **11** can be prepared from the corresponding vinylogous formamide derivatives and these yield, after hydrolysis, α,β-unsaturated products **12** (Eq. 7). This particular reaction is generally limited to more reactive substrates.

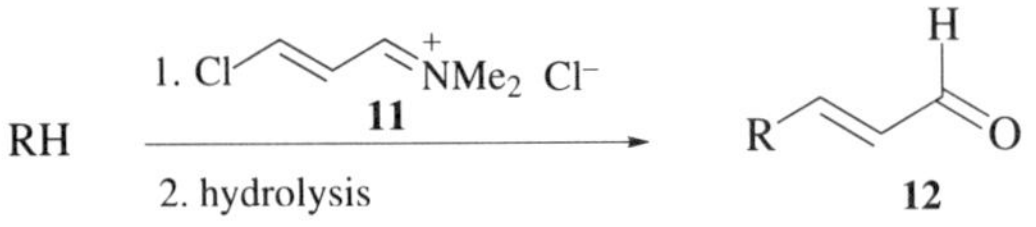

The formation of carbon–carbon bonds to fully conjugated carbocycles and heterocycles is the subject of this chapter; a subsequent chapter considers carbon–carbon bond-formation reactions in alkenes (including heterosubstituted alkenes such as enamines and enol ethers), alkynes, and activated methyl and methylene compounds (aldehydes, ketones, carboxylic acid derivatives, and nitriles).

It is not surprising that the Vilsmeier reaction has been the subject of many review articles of varying scope and length. The author(s) and dates are, in chronological order: Vilsmeier (1951),[4] Bayer (1954),[5] Bredereck et. al. (1959),[6] Eilingsfeld, Seefelder, and Weidinger (1960),[7] Minkin and Dorofeenko (1960),[8] Oda and Yamamoto (1960),[9] de Maheas (1962),[10] Hafner et al. (1963),[11] Gore (1964),[12] Hazebroucq (1966),[13] Jutz (1968),[14] Ulrich (1968),[15] Kuehne (1969),[16] Seshadri (1973),[17] Jutz (1976),[18] Meth-Cohn and Tarnowski (1982),[19] Simchen (1983),[20] Marson (1992),[21] Meth-Cohn and Stanforth (1991),[22] and Meth-Cohn (1993).[23]

Kantlehner (1976)[24] has reviewed adducts from acid amides and acylation reagents and also the preparation and reaction of chloromethyliminium salts with nucleophiles. Liebscher and Hartmann (1979)[25] have published an article relating to vinylogous chloroiminium salts. With so many excellent reviews dealing with the Vilsmeier reaction, its mechanism, and the structure of the various electrophilic reagents, we have restricted our coverage to important concepts rather than reiterate all the literature material.

MECHANISM AND REGIOCHEMISTRY

A brief description of the mechanism and regiochemistry of the Vilsmeier reaction is now presented and this is elaborated with appropriate cases in the following sections which deal with specific compound types.

From the early stages it was obvious that the Vilsmeier reagent is a relatively weak electrophile, generally requiring an activated aromatic nucleus. For this reason, it has been particularly successful with π-excessive heterocycles, such as furan, thiophene, and pyrrole. A recent discussion of the mechanism[26] has indicated that in the equilibrium leading to the formation of the reactive species, the use of an electrophilic anhydride (e.g., trifluoromethanesulfonic anhydride) should lead to a cationic species that cannot react with a nucleophile to give a lower energy and less reactive species, and the action of pyrophosphoryl chloride on DMF indeed produces an intermediate **13** (Eq. 8) for which the pathway to the

$$\text{(Eq. 8)}$$

iminium chloride of the type **5** is precluded. Indeed, the reagent **13** thus produced is associated with a high reactivity and increased steric demand, which can increase regioselectivity.

Early reports of the Vilsmeier reaction, particularly on benzenoid aromatics, suggested regiospecificity, but careful examination has indicated that regioisomers are frequently formed, although quantities of minor isomers were small. Thus, the most recent study of the Vilsmeier formylation of anisole[26] reports a mixture of 2-methoxybenzaldehyde (4.5%) and 4-methoxybenzaldehyde (70.5%).

Commonly used solvents are chlorinated hydrocarbons, excess dimethylformamide, or excess phosphoryl chloride. When DMF is used as the solvent, a further equilibrium involving the chloroiminium salt **5** and DMF can produce a less reactive electrophilic species **14** (Eq. 9).[27-29] The choice of solvent and reaction temperature is an important consideration in many Vilsmeier reactions.

$$(Eq. 9)$$

In all cases, the initial product of the Vilsmeier reaction is an iminium salt **9** (Eq. 6), which is often isolated as its chloride, perchlorate, or tetrafluoroborate, but is more commonly hydrolyzed to the aldehyde **10**. Iminium salt **9** can also be converted into groups other than aldehydes (e.g., thioaldehydes) by treatment with hydrogen sulfide, nitriles by treatment with hydoxylamine, and amines by reduction. In some cases, iminium salts can react intramolecularly with appositely located functional groups to generate fused-ring systems.

SCOPE AND LIMITATIONS

In this section we consider first carbocyclic substrates and then heterocyclic substrates.

Monosubstituted Benzenes

A powerful activating group is normally required to achieve a successful Vilsmeier reaction on monosubstituted benzene derivatives. Most reactions have been performed on *O*-alkylated phenols and *N,N*-dialkylated aniline derivatives. Thioanisole has been reported to give 4-formylthioanisole in low yield.[30] *N,N*-Dimethylaniline offers representative examples of most of the reagents used for simple formylation and acylation reactions (Eq. 10). The reagents range from the

$$(Eq. 10)$$

most commonly used mixture of DMF and phosphoryl chloride (maximum reported yield 85%);[31,32] to the recent pyrophosphoryl chloride and DMF mixture
reported to give 4-formyl-*N,N*-dimethylaniline (**4**) in 99% yield.[26] Diformylation
of *N,N*-dimethylaniline has been reported with an excess of DMF and phosphoryl
chloride to give a low yield of 2,4-diformyl-*N,N*-dimethylaniline together with
the monoformylated product **4**.[33] Other acid chlorides which have been used in
this formylation reaction include carbonyl chloride,[31] thionyl chloride (with[34] or
more commonly without[31] aluminum trichloride), triphenylphosphine dibro-
mide,[35] and the acid chloride equivalent, 2,4,6-trichloro-1,3,5-triazine.[36] Reaction
with dimethylacetamide (DMA)[33] gives 4-acetyl-*N,N*-dimethylaniline (**15**) in
poor yield, but moderate to good yields of benzophenone derivatives **16** have been
obtained with benzamides (Eq. 10).[37] The synthesis of cinnamaldehyde derivative
17[38] and dienal derivative **18**[39] from aldehyde **19** (cf. Eq. 7) and phosphoryl chloride illustrates an extension of the Vilsmeier reaction to the synthesis of α,β-
unsaturated aldehydes and their homologues (Eq. 11). The aza-analogue **20** of
aldehyde **19** yields iminium salt **21** (Eq. 11).[40]

17, n = 1 (70-80%)
18, n = 2 (19%)   (Eq. 11)

21 (50%)

The iminium salt **22** formed by the action of DMF and phosphoryl chloride on
N,N-dimethylaniline has been reacted with *N,N*-dimethylaniline to give the substituted triphenylmethane derivative **23** (Eq. 12).[41]

(Eq. 12)

23 (93%)

No authors quote any lack of regioselectivity in the reactions of *N,N*-dimethylaniline. Recent studies of the reaction of anisole (Eq. 13) have shown

A, DMF, (Cl₂PO)₂O
B, MFA, (Cl₂PO)₂O
C, DMF, POCl₃

A (75%) 94:6
B (72%) >98:2
C (58%) 89:11

(Eq. 13)

that in this less-hindered substrate, perceptible amounts of the 2-formyl isomer can be formed. Not surprisingly, the steric demands of the reagent influence the ratio of 4- and 2-substituted products[26] so that pyrophosphoryl chloride and DMF give a 94:6 ratio, whereas *N*-methylformanilide (MFA) gives a ratio greater than 98:2.[26] The conventional DMF and phosphoryl chloride gave, in the hands of these authors, a ratio of 89:11, but with a much poorer yield. Thionyl chloride and DMF give the sulfur-containing product **24** with anisole, and other ethers react similarly (Eq. 14).[42]

SOCl₂
DMF

24
(34%) based on SOCl₂

(Eq. 14)

Di- and Polysubstituted Benzenes

Regioselectivity becomes more important with di- or polysubstitution. A second factor which can give rise to abnormal products is the presence of an appositely located group which can react intramolecularly with the primary product to give a new ring system, as illustrated for the production of benzo[*b*]furans **25** (Eq. 15).[43] Benzo[*b*]thiophenes can be formed in a similar process.[44] Other prod-

DMF
POCl₃

R	
OMe	(25%)
OEt	(31%)
NEt₂	(78%)

25

(Eq. 15)

ucts formed by intramolecular cyclizations include isoquinolines **26** (Eq. 16)[45] and salt **27** (Eq. 17).[46] The formation of isoquinolines **26** provides an interesting

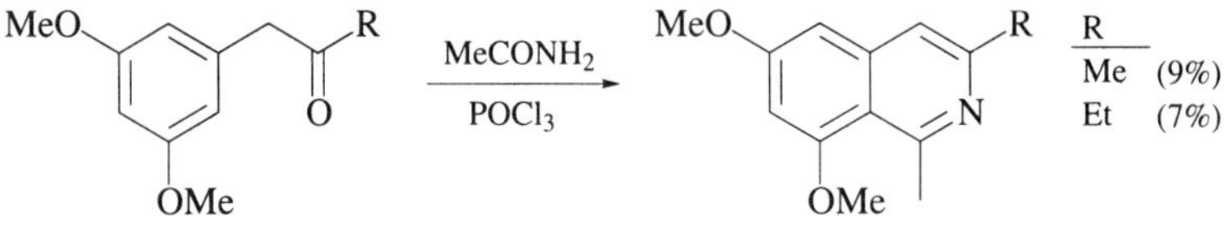

R	
Me	(9%)
Et	(7%)

26

(Eq. 16)

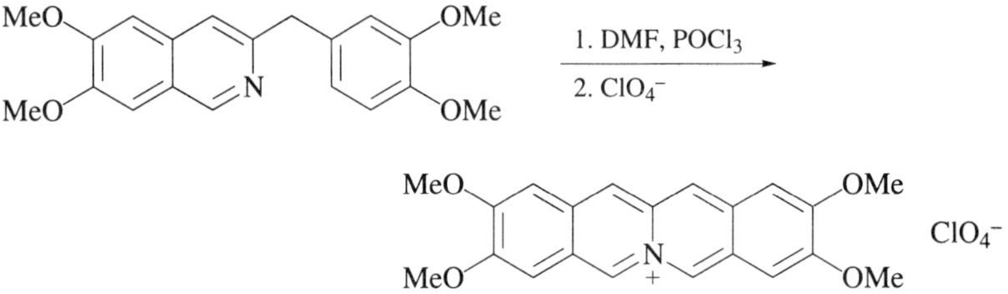

1. DMF, POCl₃

2. ClO₄⁻

27

ClO₄⁻

(Eq. 17)

application of a cyclization reaction because the nitrogen of the Vilsmeier reagent is retained in the product. However, the yield in these cyclizations is often quite low.

The lowest degree of activation reported to give successful formylation is shown by 1,3,5-trimethylbenzene (Eq. 18), which is formylated by a mixture of

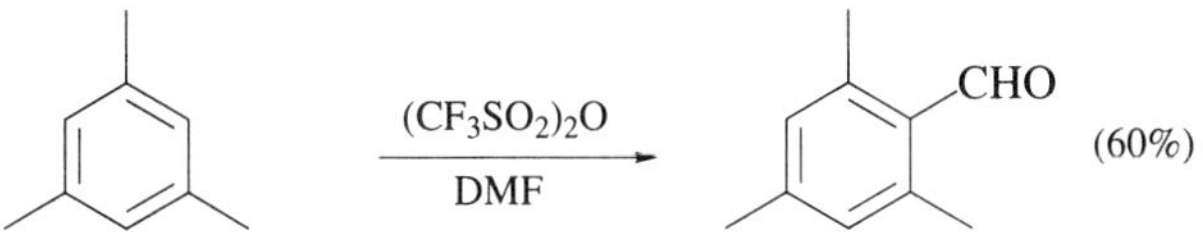

(CF₃SO₂)₂O

DMF

(60%)   (Eq. 18)

DMF and trifluoromethanesulfonic anhydride.[47] The position of substitution is normally predictable by considering the relative directing power of the substituents, as illustrated for the trisubstituted benzene derivatives **28**[48] (Eq. 19) and

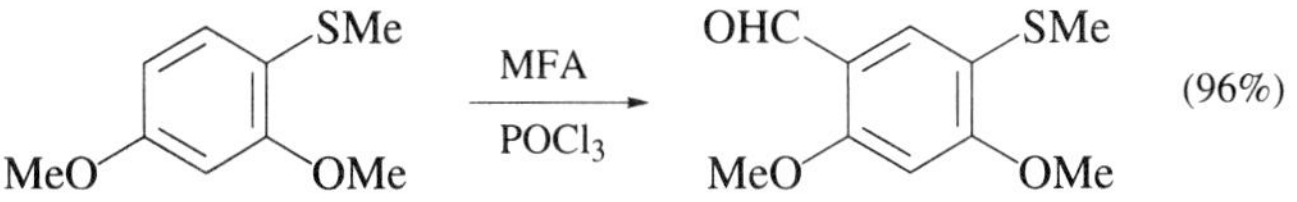

28

MFA

POCl₃

(96%)   (Eq. 19)

29[49] (Eq. 20), but can be modified by steric factors, as shown by the formation of the 6-formyl derivative **30** from 3-*tert*-butylanisole (Eq. 21).[50] In a number of

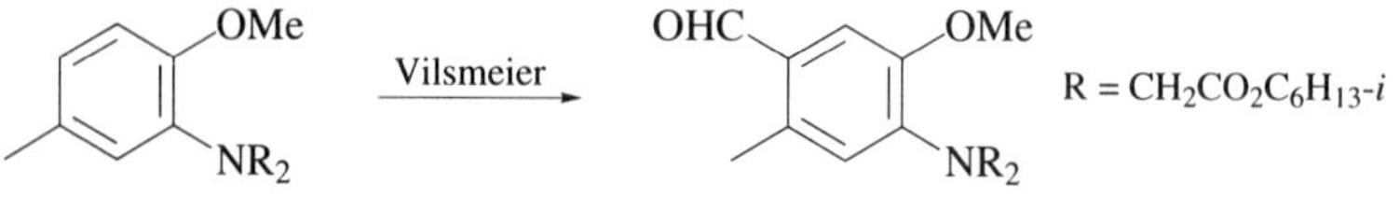

$$R = CH_2CO_2C_6H_{13}\text{-}i$$

(Eq. 20)

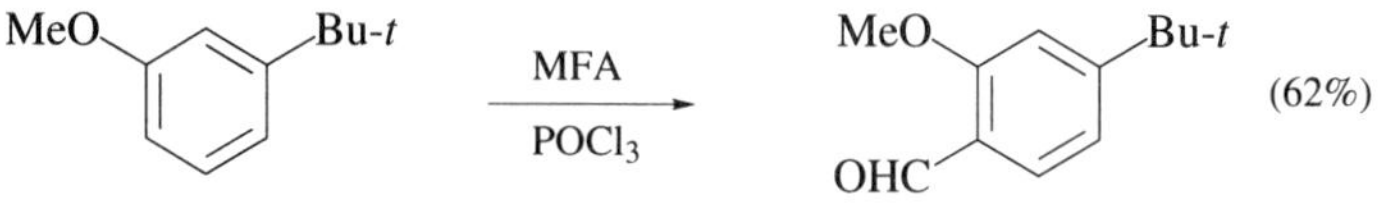

(62%)

(Eq. 21)

cases, a free phenolic hydroxy group has been reported to exercise direction over a methyl ether, but yields were low and other products may have been formed.[51,52] High regioselectivity can be achieved when the Vilsmeier reagent replaces the tributylstannyl group (Eq. 22).[53]

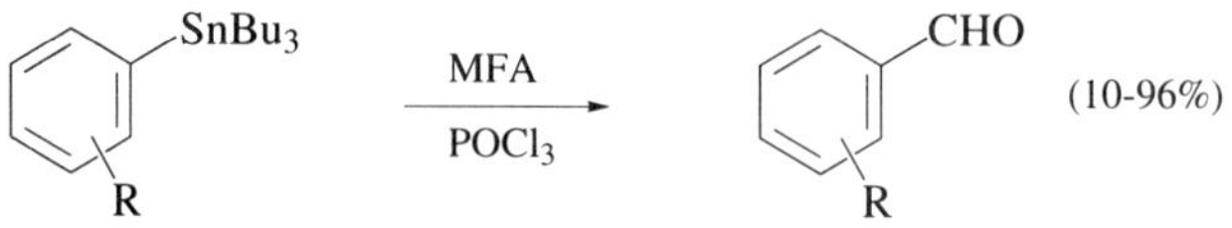

(10-96%)   (Eq. 22)

An interesting reaction of substituted *N*-benzyl-4-methylaniline derivatives **31** has been reported (Eq. 23).[54, 55] If the benzyl substituent is relatively electron

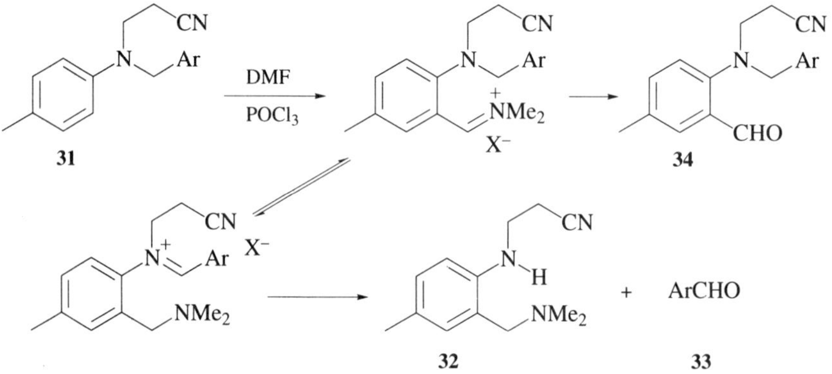

(Eq. 23)

rich, hydride transfer is the major reaction, yielding amine **32** and aldehyde **33**. When the benzyl group is relatively electron deficient, aldehydes **34** are the major products.

Naphthalenes and Polycyclic Benzenoid Hydrocarbons

Naphthalene has not been reported to undergo formylation with the usual DMF and phosphoryl chloride mixture, but with the more potent combination of DMF and trifluoromethanesulfonic anhydride, naphthalene-1-carbaldehyde is produced in 50% yield.[47] The presence of a single activating substituent, as in N,N-dimethylaminonaphthalene, has been reported to give up to 70% of diformylated product as well as the monoformylated product (Eq. 24).[56]

(Eq. 24)

The pattern of entry of the reagent is that normally found in naphthalenes possessing electron-rich substituents (i.e., 1-substituents direct to the 4-position and 2-substituents direct to the 1-position). It has been reported that 1,6-dimethoxynaphthalene gives 4-formylation with DMF and phosphoryl chloride;[57] previous work had indicated 5-formylation with MFA and phosphoryl chloride (Eq. 25).[58]

(61%)   (Eq. 25)

Pericyclization has also been reported (Eq. 26).[59]

(47%)   (Eq. 26)

Anthracene gives anthracene-9-carbaldehyde in the Vilsmeier reaction[60] with MFA and phosphoryl chloride, and many other formylating mixtures also yield this product. A methoxy group provides sufficient activation to allow substitution to occur at other positions (Eq. 27).[50] Phenanthrene gives a poor yield of phenanthrene-3-carbaldehyde with the potent trifluoromethanesulfonic anhy-

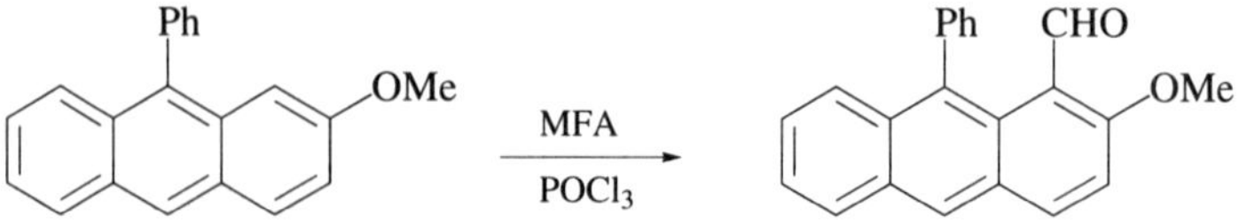

$$(Eq.\ 27)$$

dride and DMF mixture,[47] and 3-methoxyphenanthrene gives a moderate yield of 3-methoxyphenanthrene-9-carbaldehyde.[61]

Nonbenzenoid Aromatic Hydrocarbons

Derivatives of the cyclopentadienyl anion and its benzologues react with the simple Vilsmeier reagent (Eq. 28)[62] or with various vinylogous equivalents such as the polymethyleniminium salt **35** (Eq. 29)[63] to give fulvene derivatives. By

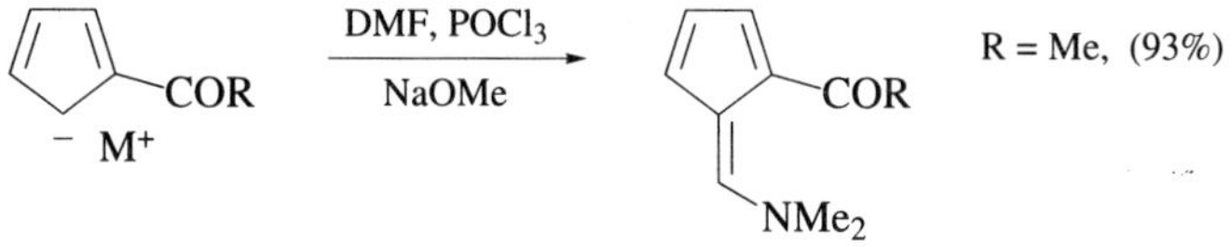

$$(Eq.\ 28)$$

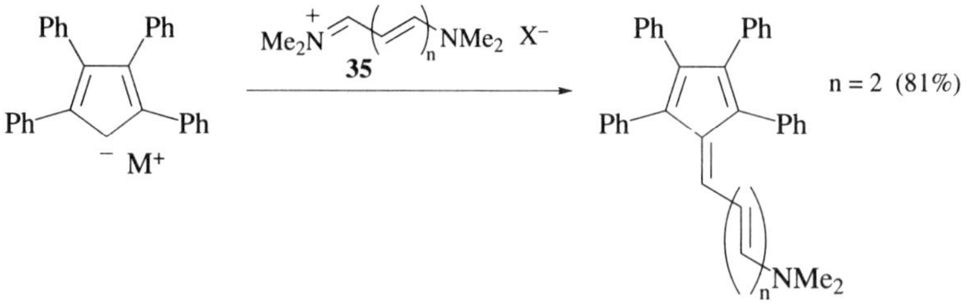

$$(Eq.\ 29)$$

using the cross-conjugated reagent **36**, it is possible to prepare a variety of linked products such as compound **37** (Eq. 30).[63] The presence of a second potential

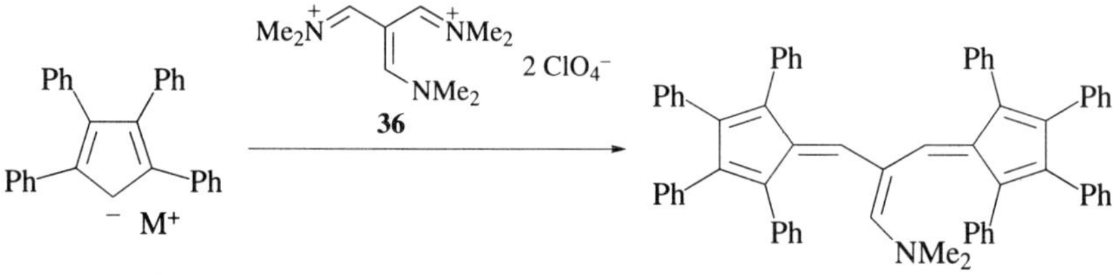

$$(Eq.\ 30)$$

electrophile in the reagent allows cyclization to occur, resulting in the formation of a new ring, as exemplified by the production of the azulene derivative **38** (Eq. 31)[64] and pyrene (Eq. 32).[65]

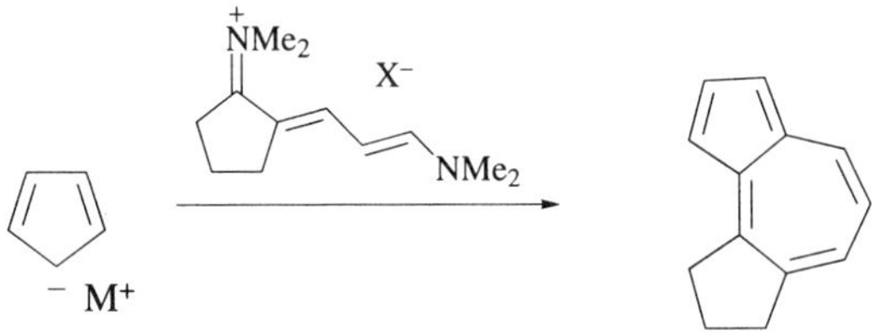

(Eq. 31)

38 (28%)

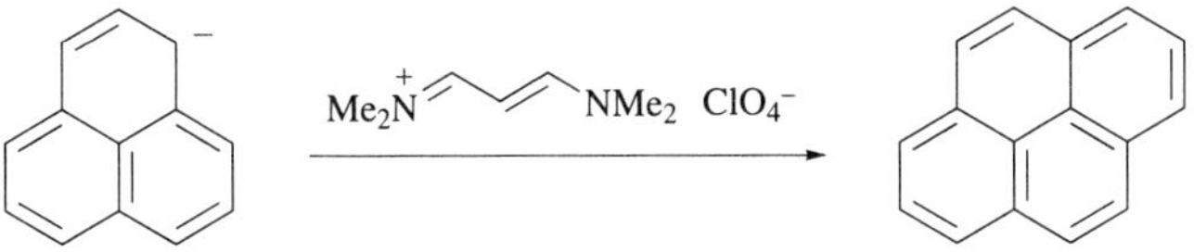

(91%)   (Eq. 32)

Azulenes are the most comprehensively studied bicyclic nonbenzenoid aromatic hydrocarbons and yield 1-substituted products, often in excellent yields (Eq. 33).[66] The reactivity of the azulene ring is sufficient to allow the preparation of ketone derivatives (Eq. 34),[67] acraldehydes, or pentadienaldehydes (Eq. 35).[68]

(91%)   (Eq. 33)

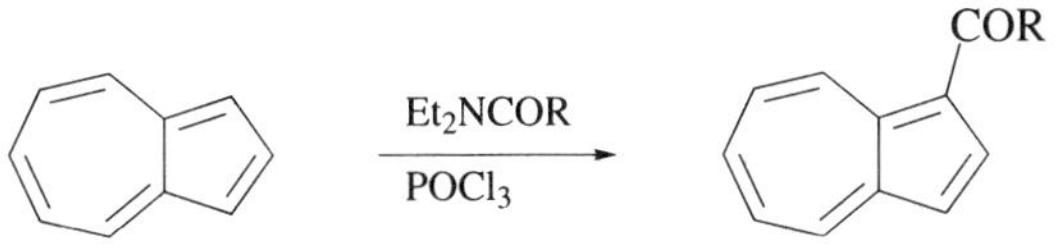

R = (CH$_2$)$_3$CO$_2$Et (86%)

(Eq. 34)

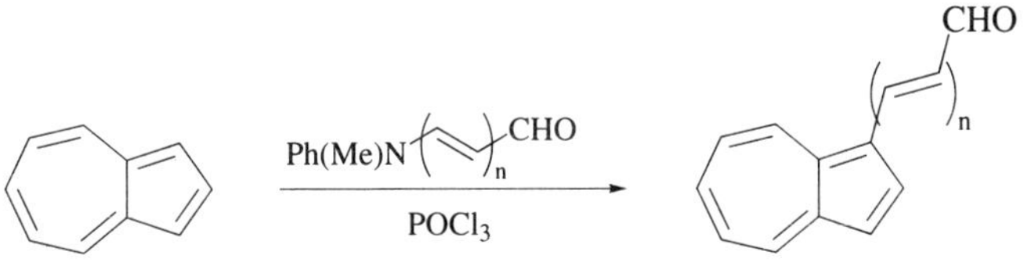

n = 1, 2
(95-97%)

(Eq. 35)

The bridged 10 π-annulene derivative **39** gave a modest yield of the 4-formyl derivative (Eq. 36)[69] and the polycyclic compound **40** afforded a mixture of formylated products (Eq. 37).[70]

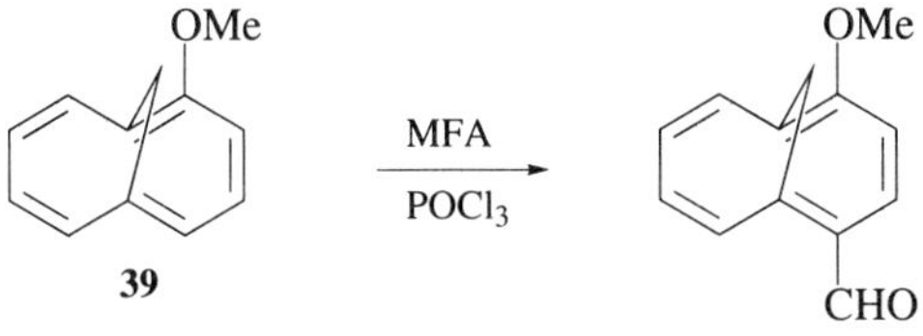

39

(38%)   (Eq. 36)

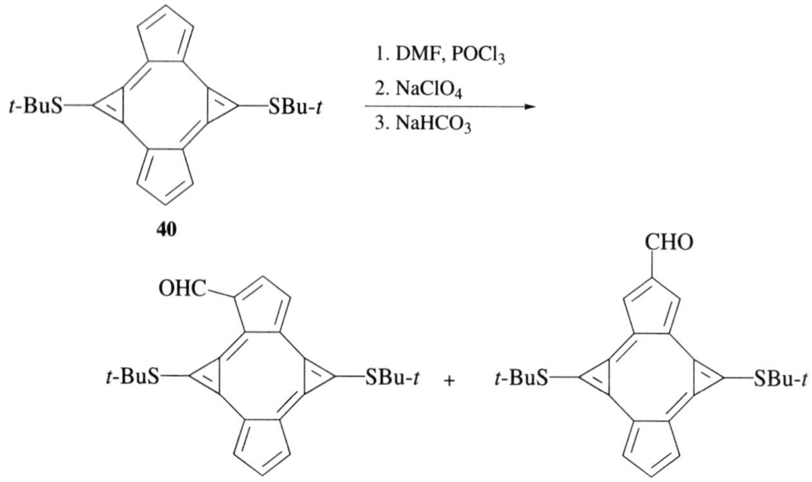

(13%)   (69%)   (Eq. 37)

Organometallic compounds react normally, and examples include mono- and diformylation of ferrocene (Eq. 38),[71] monoformylation of the cycloheptatrienyl

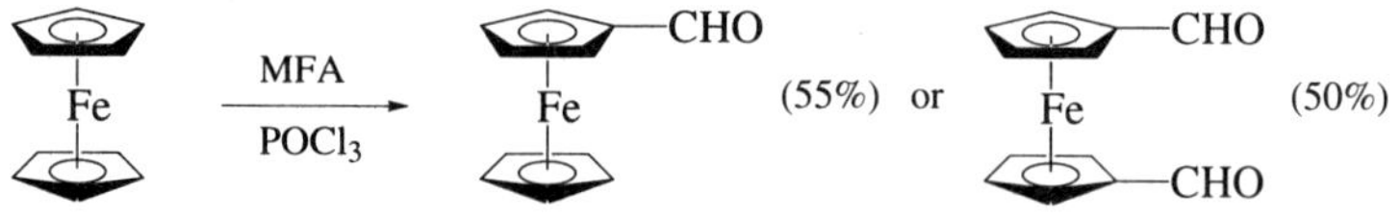

(55%) or   (50%)

(Eq. 38)

iron tricarbonyl compound **41** (Eq. 39),[72] diformylation of the biscyclobutadiene iron tricarbonyl compound **42** (Eq. 40), and monoformylation of the cobalt derivative **43** (Eq. 41).[73]

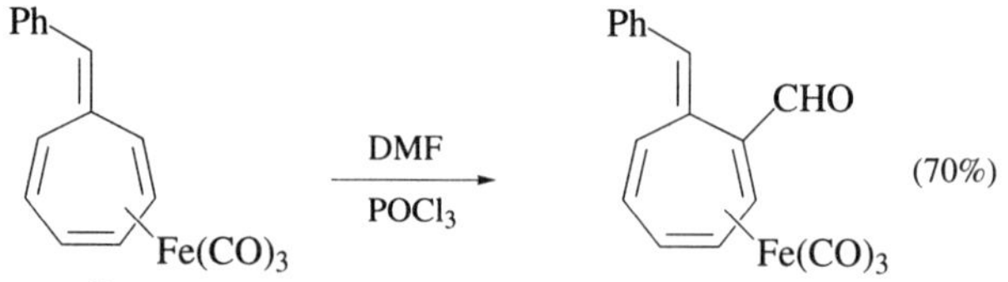

(70%)   (Eq. 39)

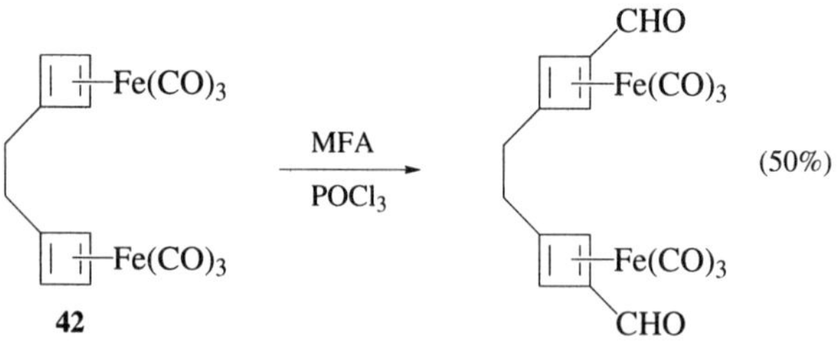

(50%)   (Eq. 40)

(Eq. 41)

Furans, Thiophenes, Selenophenes, and Pyrroles

The extension of the Vilsmeier reaction from activated benzene derivatives to the electron-rich heterocycles furan, thiophene, selenophene, and pyrrole is well documented and the main interest is associated with the relative reactivities of these heterocycles and the regioselectivity within these systems, which is generally good.

The relative reactivity of these electron-rich heterocycles is well established as pyrrole > furan > thiophene, and this has been confirmed (Eq. 42) by the

(Eq. 42)

44, X = O, 2- or 3-substituted   X = O, 2-substituted (62%)
45, X = S, 2- or 3-substituted   X = O, 3-substituted (56%)

formylation of the pyrrolylfurans **44**[74] and pyrrolylthiophenes **45**.[74,75] The pyrrolylfurans **44** also give a small quantity of diformylated products where formylation has occurred in both rings.[74] The presence of a deactivating ester group in the pyrroles **46** and **47** provides a more precise picture of relative reactivity where the deactivated pyrrole ring competes successfully against a thiophene ring but not a furan ring (Eq. 43).[74]

(Eq. 43)

Regioselectivity within each group of heterocycles can be influenced by electronic or steric effects. Furans that possess a single substituent at the 2-position give uniformly 5-formyl derivatives, whereas substituents at the 3-position, such

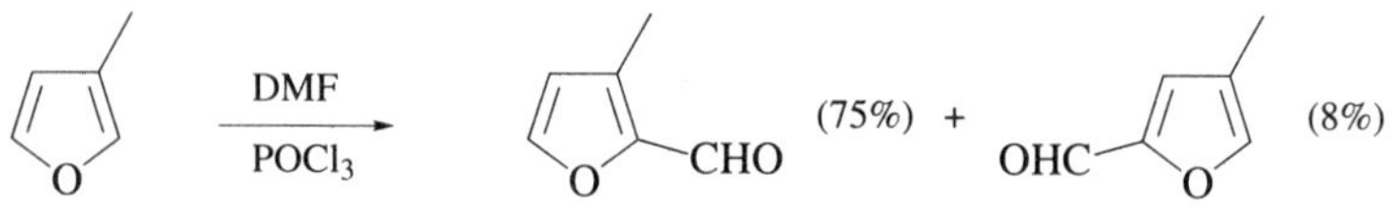

(Eq. 44)

as compound **48**, produce, rather surprisingly, 2-formyl derivatives (Eq. 44).[76] Thiophene shows a similar pattern of substitution; 3-substituted compounds such as **49** give mainly products of formylation at the 2-position (Eq. 45).[77] In the

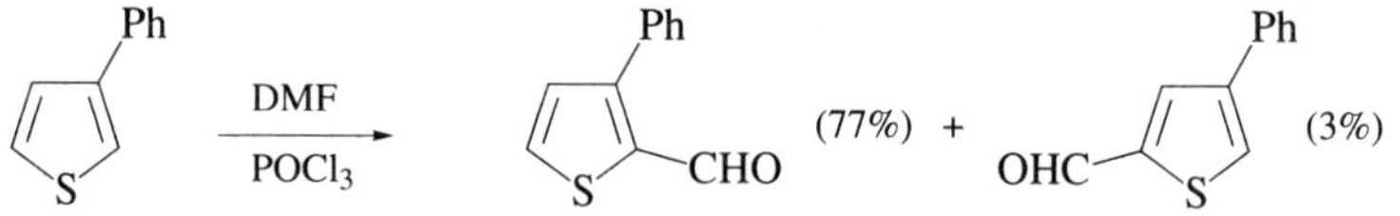

(Eq. 45)

cases of 3,4- or 2,5-disubstituted thiophenes, the most strongly electron-releasing substituent determines the point of attack, as illustrated for compounds **50** (Eq. 46)[76] and **51** (Eq. 47).[78] The replacement of a 2-bromo substituent in com-

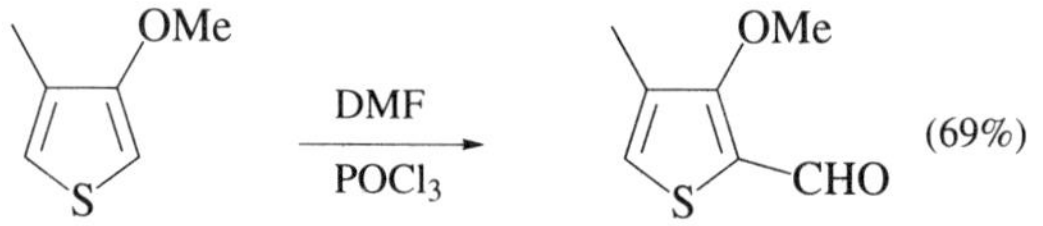

(Eq. 46)

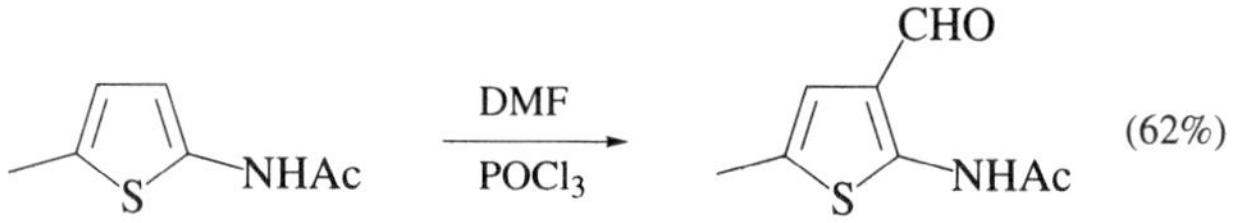

(Eq. 47)

pound **52** by chlorine has been reported[79] when DMF/phosphorus oxychloride was the formylating reagent, but this problem can be overcome[80] by using phosphorus oxybromide (Eq. 48); 3-bromothiophene appears stable to DMF and phosphorus oxychloride.[76]

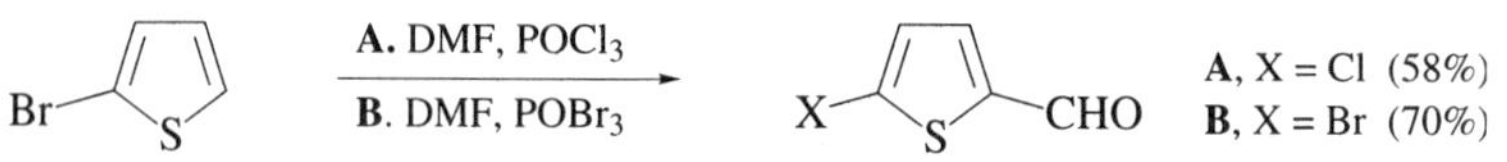

(Eq. 48)

A hydroxy group has also been reported to be replaced by chlorine in the formylation of compound **53** (Eq. 49),[81] although not during benzoylation.[82] This is analogous to the reactions of acyclic and alicyclic ketones, and hence may indi-

(Eq. 49)

cate the intermediacy of the tautomeric dihydrothiophene-3-one **54**, which is be-having as an activated methylene compound. Dealkylation of 2-alkoxythiophenes **55** has been observed to give products **56** (Eq. 50).[83]

Selenophene reacts similarly to thiophene, with the same regioselectivity.[84]

Pyrroles have provided numerous substrates for Vilsmeier formylations or acy-lations with appropriate amide derivatives, and lactams have also been exten-sively used because of the importance of the products in the synthesis of pyrrole pigments. Included in a very large range of *N*-substituted pyrroles **57**, the iso-propyl and *tert*-butyl derivatives provide good examples of steric hindrance forc-ing formylation to the normally unfavored 3-position (Eq. 51).[85]

R	I	II
Me	(95%)	(0%)
i-Pr	(8%)	(71%)
t-Bu	(5%)	(64%)

(Eq. 51)

The greater reactivity of pyrrole has allowed the production of a wide range of product types, including ketones such as compounds **58**[86] and products such as compound **59**[87] (Eq. 52) (the latter example illustrating the production of bicyclic heterocycles), imines such as compound **60** (Eq. 53),[88] the bipyrrolyl derivative **61** (Eq. 54),[89] and the acraldehyde **62** (Eq. 55).[38]

Pyrroles that are not substituted on nitrogen can be formylated, and the result-ing iminium salts can then be deprotonated, giving azafulvene derivatives **63** (Eq. 56).[90] Azafulvene derivative **64** has been reacted with a variety of alkyl-lithium reagents to give pyrroles **65** (Eq. 57).[91] Some illustrative examples of the

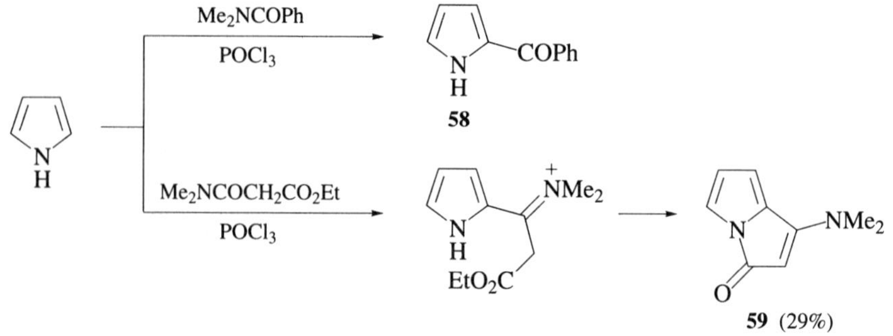

(Eq. 52)

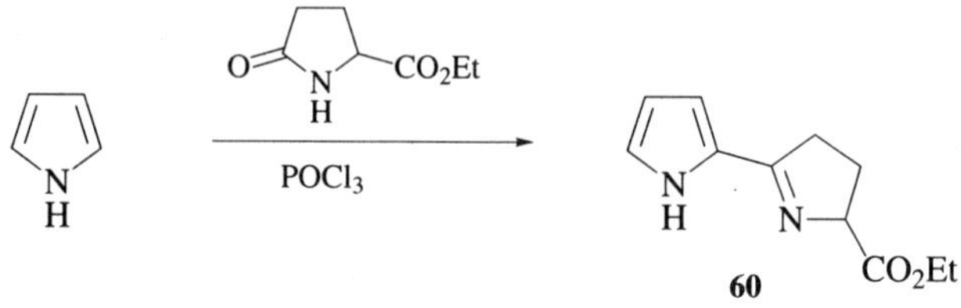

(46%)   (Eq. 53)

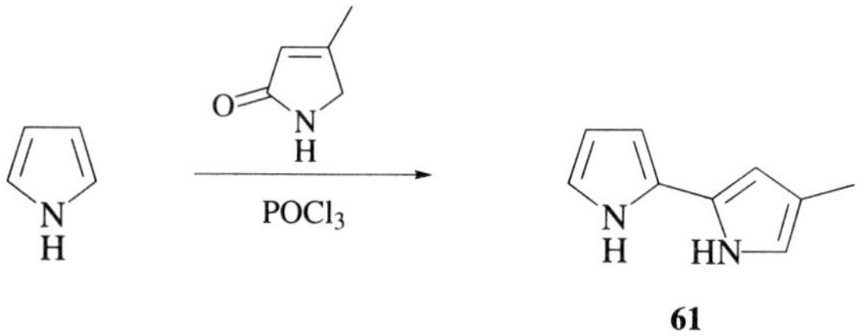

(79%)   (Eq. 54)

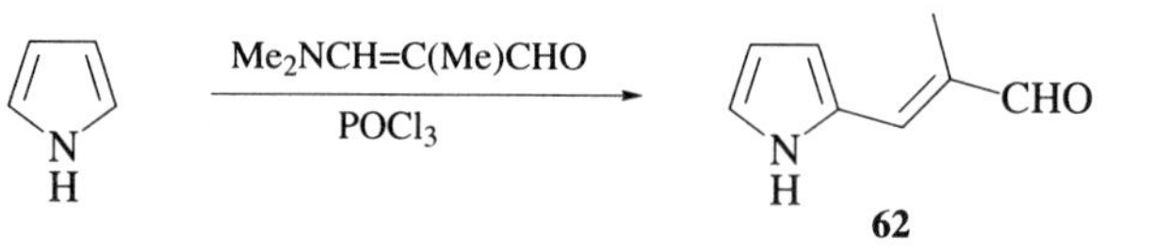

(51%)   (Eq. 55)

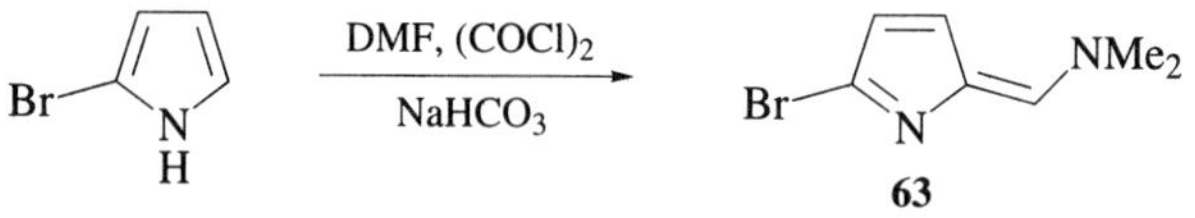

(80%)   (Eq. 56)

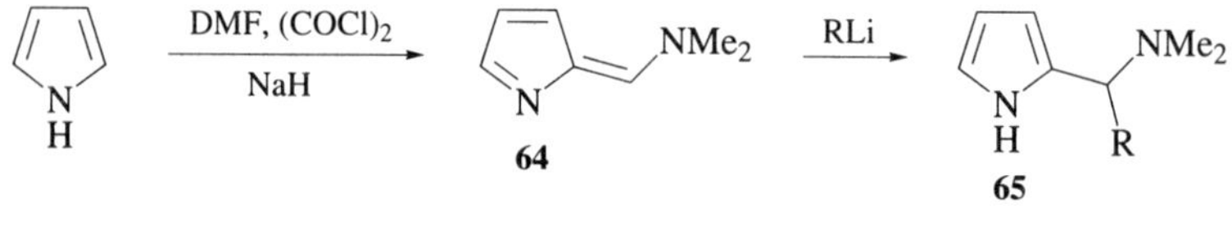

(Eq. 57)

application of the Vilsmeier reaction to give intermediates for pyrrole pigments are provided by compounds **66** (Eq. 58)[92] and **67** (Eq. 59).[93]

(75%)

(Eq. 58)

(100%)

(Eq. 59)

The presence of a 3-substituent in pyrrole gives less selectivity toward formylation at the 2-position than is observed with furan and thiophene. An interesting observation reported for the pyrrole derivative **68** is that the presence of 1 equivalent of hexamethylphosphoric triamide (HMPT) changes the ratio of 2- and 5-substitution (Eq. 60), and the reaction can be conducted at lower temperature.[94]

A 60:40
B 40:60

(Eq. 60)

Annulated Furans, Thiophenes, and Pyrroles. Benzo[*b*]furan participates in the Vilsmeier reaction giving benzo[*b*]furan-2-carbaldehyde[95] (Eq. 61),

(62%)   (Eq. 61)

whereas benzo[*b*]thiophene[96] and indole[97] undergo substitution at the 3-position (Eq. 62). The presence of appropriate substituents can allow changes in regioselectivity and provides an approximate guide to relative reactivities. Thus, in benzo[*b*]furan, a methoxy substituent in the benzene ring of compound **69** is not

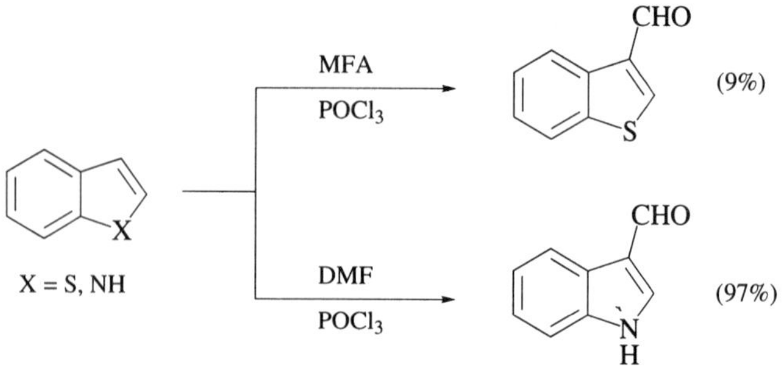

(Eq. 62)

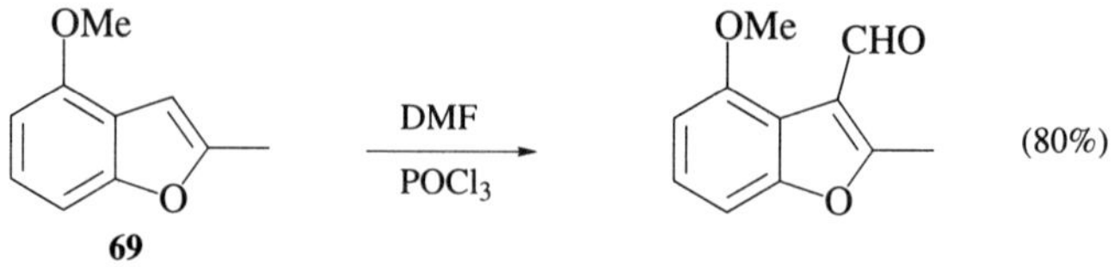

(Eq. 63)

sufficiently activating to direct formylation away from the furan ring (Eq. 63),[98] whereas in the related benzo[*b*]thiophene derivatives **70** and **71**, formylation does occur in the benzene-ring fragment of the molecule (Eq. 64).[99] For

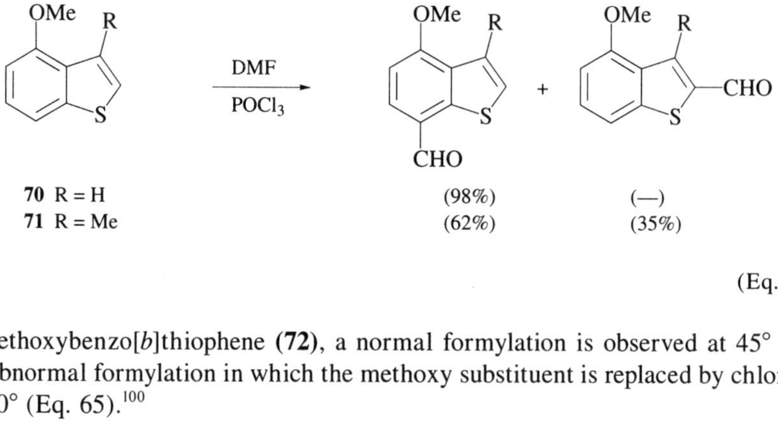

(Eq. 64)

3-methoxybenzo[*b*]thiophene (**72**), a normal formylation is observed at 45° and an abnormal formylation in which the methoxy substituent is replaced by chlorine at 90° (Eq. 65).[100]

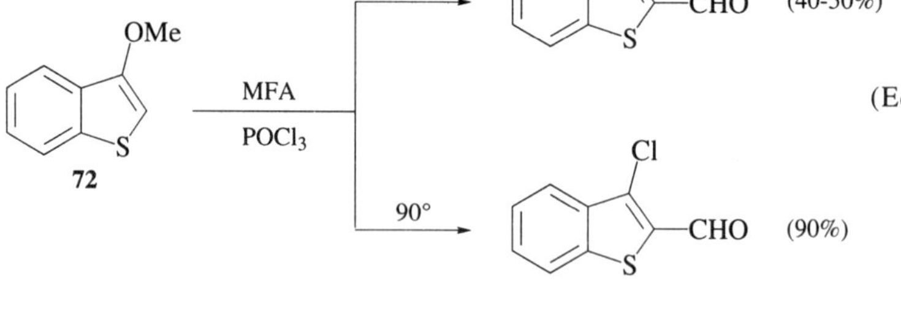

(Eq. 65)

Indole has the highest reactivity, and a large number of substituents have been introduced into the 3-position, as exemplified by the preparation of compounds **73**,[101] **74**,[40] and **75**;[102] the last example illustrates that indolenines can be isolated (Eq. 66). There is a report[103] that a trace of 1-acetylindole is formed when indole

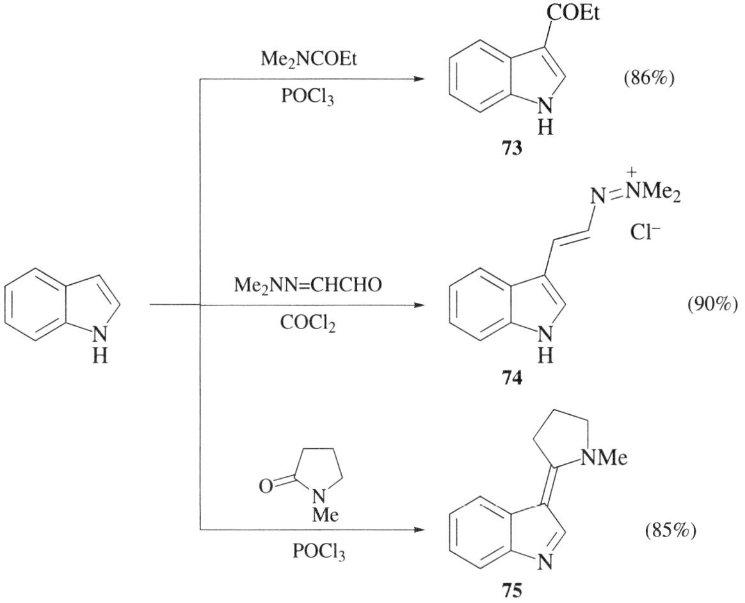

(Eq. 66)

is treated with DMA and carbonyl chloride, but only in the ratio 2:98 against the expected 3-acetylindole. Substituents at the 2-position do not affect the normal 3-substitution, but a substituent in the 3-position generally produces predominantly *N*-formylation (or acylation), as illustrated for skatole **76** (Eq. 67),[104] unless the substituent has a strong activating effect, as shown for 3-

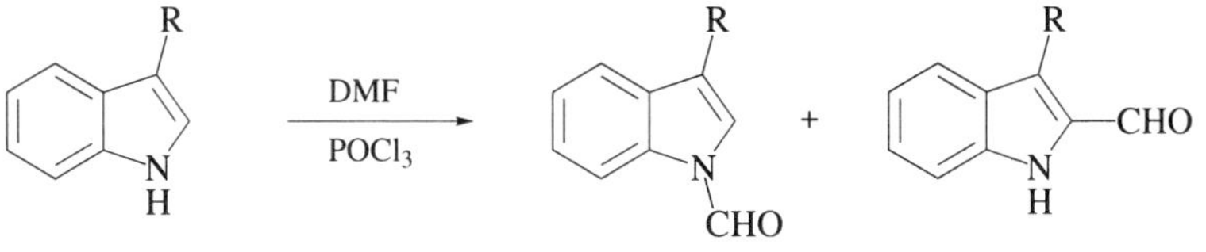

| **76** R = Me | (71%) | (22%) |
| **77** R = NHAc | (—) | (31%) |

(Eq. 67)

acetylaminoindole (**77**), in which case substitution at the 2-position occurs (Eq. 67).[104,105] Compound **78** gave heterocycle **79** (Eq. 68).[106]

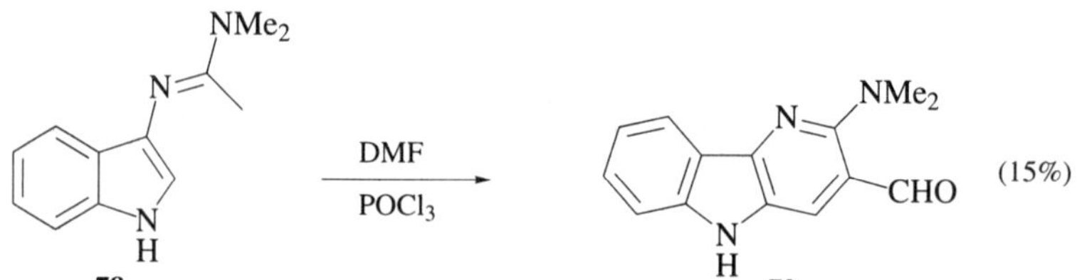

(Eq. 68)

There are no reports of single substituents in the benzene ring that can override the normal 3-substitution in indole. High yields of 2-formyl indole derivatives are obtained from 1,3-disubstituted indoles,[107,108] but if the two substituents produce a large degree of steric obstruction, a mixture of formylated products results (Eq. 69).[108] With two reinforcing methoxy substituents in the benzene ring

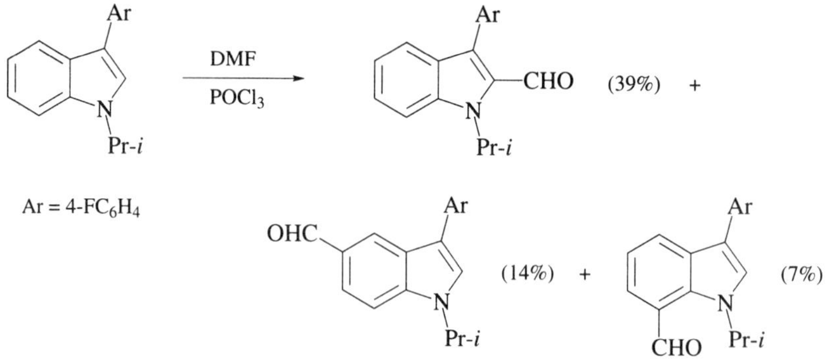

(Eq. 69)

and a degree of steric obstruction at the 2 position, a mixture of products derived from substitution in both the benzene and pyrrole rings is observed (Eq. 70).[109]

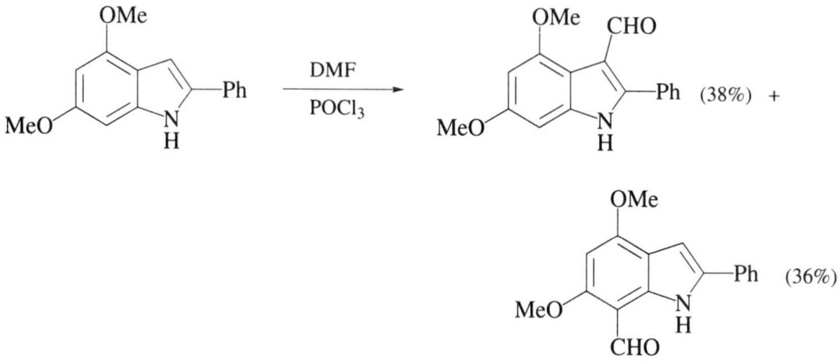

(Eq. 70)

Tetrahydrocarbazole derivatives undergo a wealth of unusual transformations in which the product distribution is very sensitive to the reaction conditions, and this is illustrated for 1-methyltetrahydrocarbazole (**80**) (Eq. 71).[110]

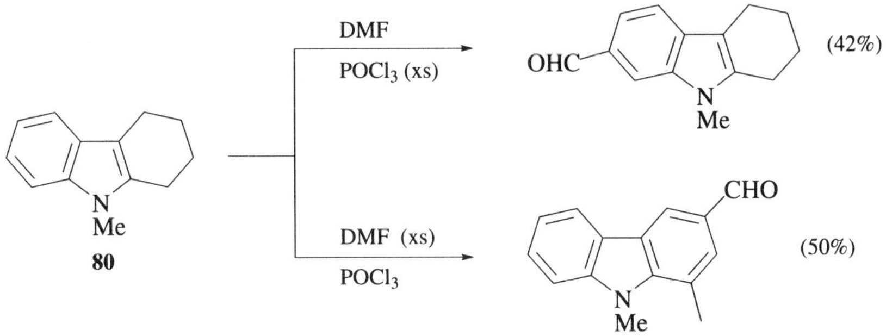

(Eq. 71)

Of the dibenzo derivatives, carbazole has attracted some interest. Benzoylation is reported to give the product of *N*-substitution **81** (Eq. 72),[111] but the normal substitution position is 3 or 6 (Eq. 73),[112,113] unless an activating substituent directs to other sites, as illustrated for hydroxy derivatives **82** (Eq. 74).[114]

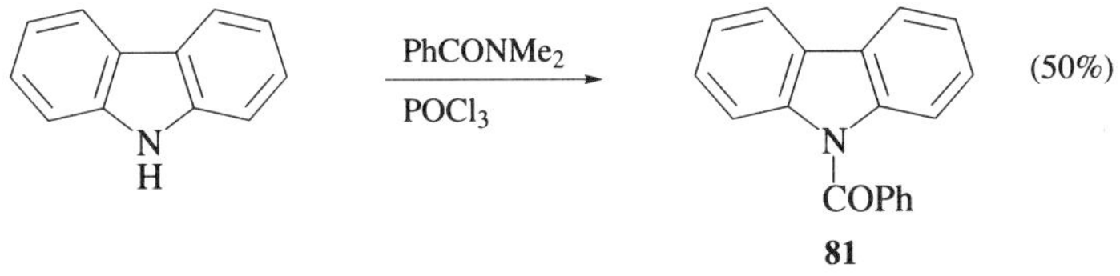

(Eq. 72)

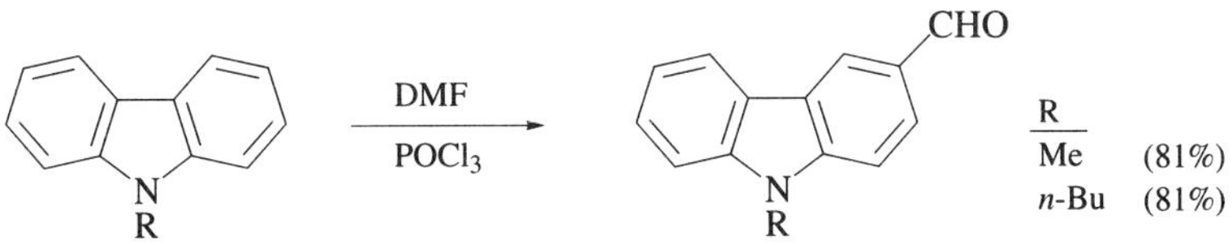

	R	
	Me	(81%)
	n-Bu	(81%)

(Eq. 73)

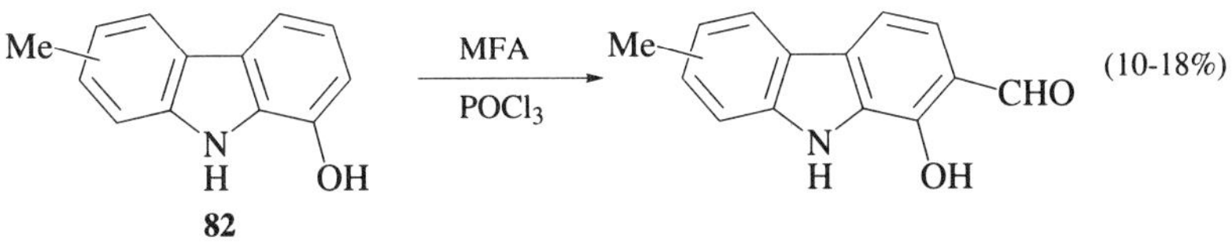

(Eq. 74)

Other Heterocycles with One Fully Conjugated Ring

There are few examples of Vilsmeier formylation of common five-membered heterocycles with two heteroatoms. A range of 1-substituted pyrazoles has, however, been formylated, the simplest examples giving 4-formyl derivatives (Eq. 75).[115] A series of 1-aryl-4-phenyl-5-aminopyrazoles **83** gave products of ring formylation and amino substitution (Eq. 76).[116] Pyrazolone derivative **84** reacts normally, giving the expected aldehyde (Eq. 77).[117] Pyrazol-5-ones

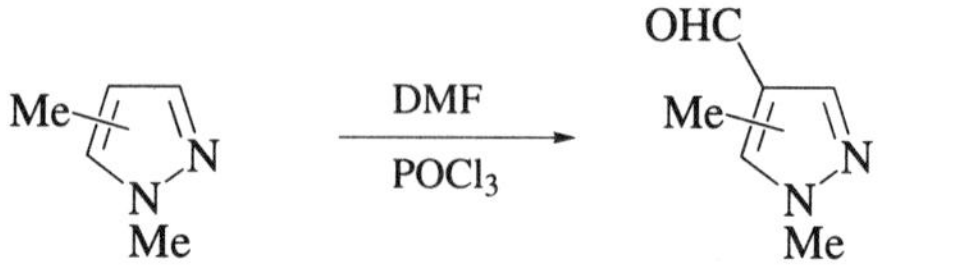

$$\text{3-Me (58\%)} \atop \text{5-Me (57\%)} \qquad \text{(Eq. 75)}$$

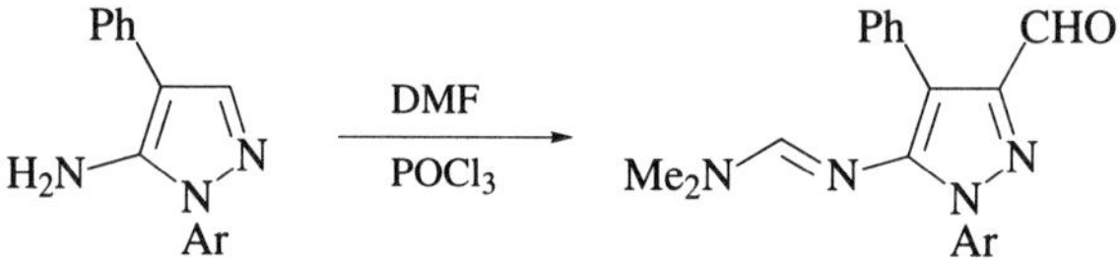

Ar	
Ph	(83%)
4-ClC$_6$H$_4$	(94%)
4-O$_2$NC$_6$H$_4$	(99%)
4-MeC$_6$H$_4$	(91%)
4-MeOC$_6$H$_4$	(74%)

(Eq. 76)

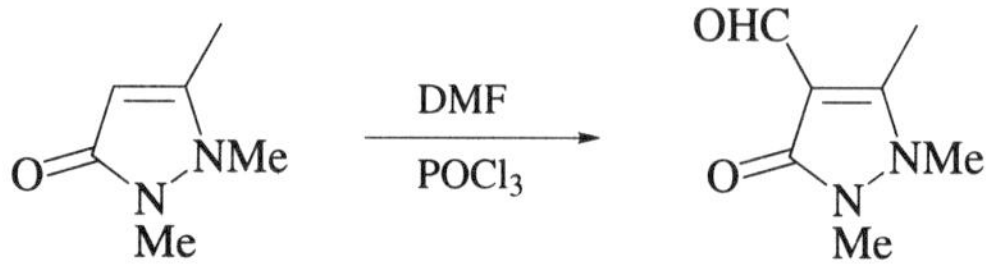

(40%)   (Eq. 77)

85 can give 5-chloro-4-formyl derivatives such as compound **87**,[118] or 4-dimethylaminomethylene derivatives such as compound **88**[119] (Eq. 78). Although

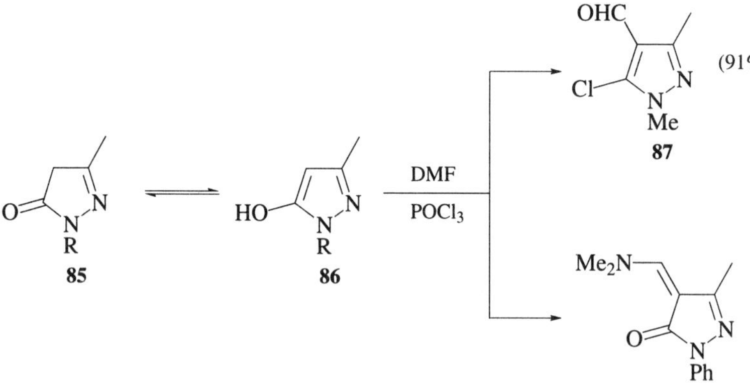

(Eq. 78)

compound **85** is not fully conjugated, and its reactivity is similar to that of cyclic ketones and lactams, this system is conveniently dealt with in this chapter because its fully conjugated tautomer **86** may well be a reaction intermediate. Other potentially tautomeric systems are similarly considered in this chapter.

The reaction of imidazole derivative **89**[120] indicates that formylation of this ring system is possible (Eq. 79), and the imidazol-4-one derivative **90** gives a chlorocarbaldehyde product **91** (Eq. 80).[121]

$$\text{89} \xrightarrow[\text{POCl}_3]{\text{DMF}} \qquad (64\%) \qquad \text{(Eq. 79)}$$

Ar = 2,4-Cl$_2$C$_6$H$_3$

$$\text{90} \xrightarrow[\text{POCl}_3]{\text{DMF}} \text{91} \qquad (90\%) \qquad \text{(Eq. 80)}$$

Isoxazol-5-ones **92** can give many types of products in the Vilsmeier reaction depending upon the reaction conditions (Eq. 81).[122] Besides the expected products

95, X = H
96, X = CHO

97, X =

(Eq. 81)

93 and **94** of the Vilsmeier reaction, ring-expansion products **95–97** and fragmentation products **98** can also be isolated. Normal formylation is observed for 2-phenyloxazole (Eq. 82),[123] and oxazole derivative **99** underwent an interesting conversion giving pyrrole derivative **100** in the Vilsmeier reaction (Eq. 83).[124]

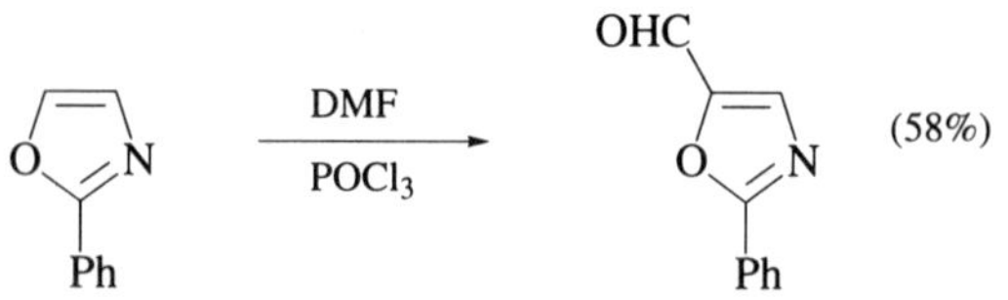

(Eq. 82)

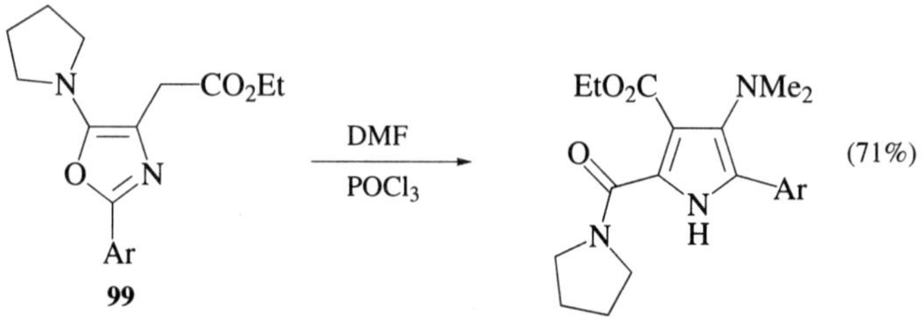

(Eq. 83)

A low-yield formylation is reported for 2-amino-4-phenylthiazole (Eq. 84),[125] and thiazol-4-one derivative **101** gave the chloroaldehyde derivative **102** (Eq. 85).[121] Tetrazoles fragment with elimination of nitrogen (Eq. 86),[126] whereas sydnones give variable yields of formylated products (Eq. 87).[127]

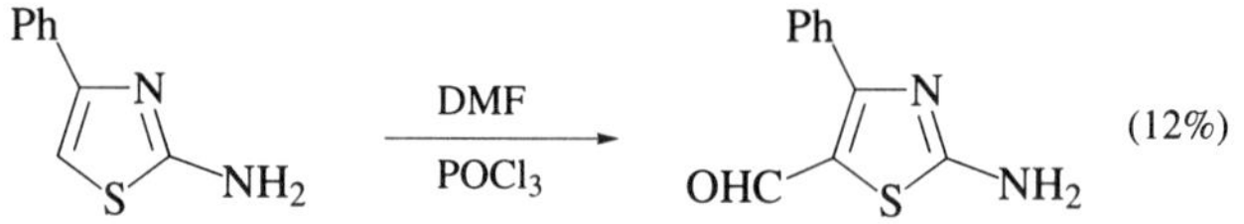

(12%)   (Eq. 84)

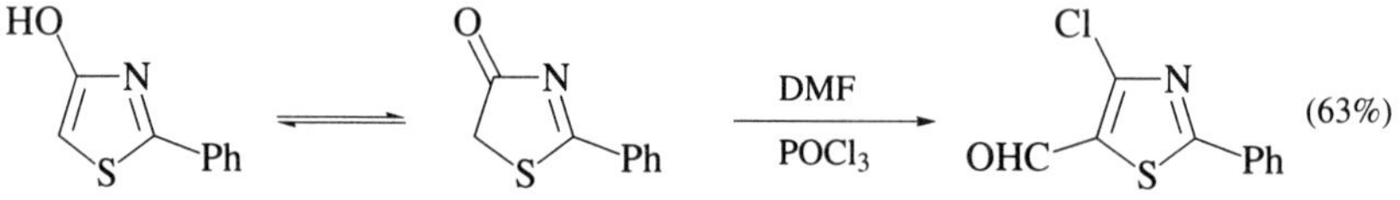

(63%)

(Eq. 85)

(58-83%)

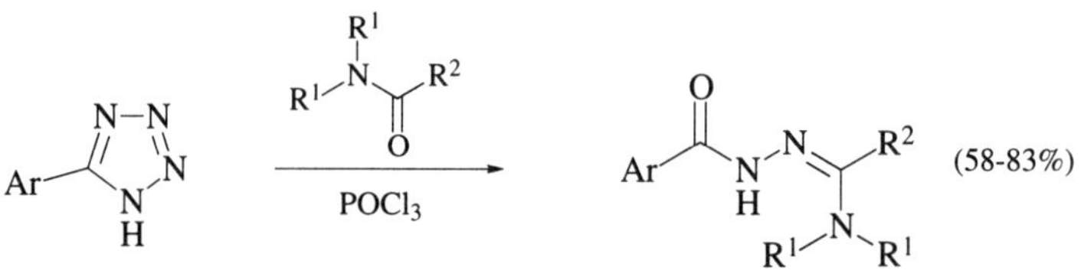

(Eq. 86)

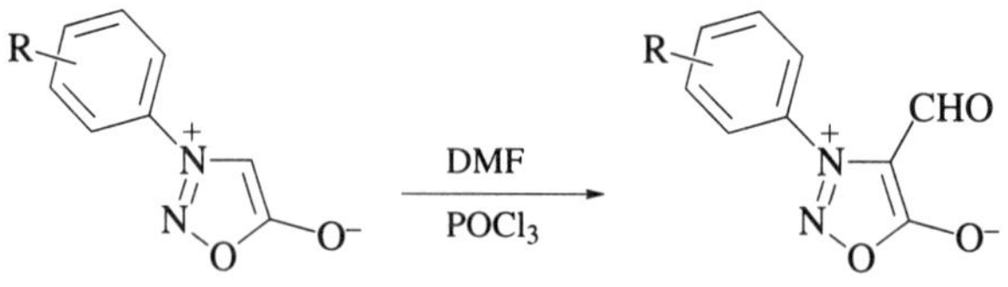

R	
4-MeO	(76%)
2-Me	(30%)
4-Me	(61%)
4-Cl	(34%)
4-Br	(40%)

(Eq. 87)

Six-membered heterocycles such as pyridine and pyrimidine require strongly electron-donating groups to be present for successful formylation. Most of the reported reactions involve pyridinones and pyrimidinones, and in substrates that have an electron-donating hydroxy group, conversion of this group to a chloro substituent is frequently observed. Examples are provided by the pyridinone **103** (Eq. 88)[128] and the pyrimidinone **104** (Eq. 89).[129] Cyclization is possible to a suitably placed phenyl ring, as illustrated for compound **105** (Eq. 90).[130]

(73%) (Eq. 88)

(80%) (Eq. 89)

R	
Me	(96%)
Et	(98%)
n-Pr	(93%)

(Eq. 90)

Other Heterocycles with Two Fully Conjugated Rings

A wide variety of bicyclic heterocycles react with Vilsmeier reagents, particularly those with one or two five-membered rings. The simplest examples have two fused five-membered heterocyclic rings, and in these compounds substitution occurs predominantly as predicted by the parent monocyclic systems. Thus, thieno[2,3-c]thiophene **(106)** (Eq. 91),[131] its selenium analogue **107** (Eq. 92),[132]

(17%) + (39%)

(Eq. 91)

(37%) (Eq. 92)

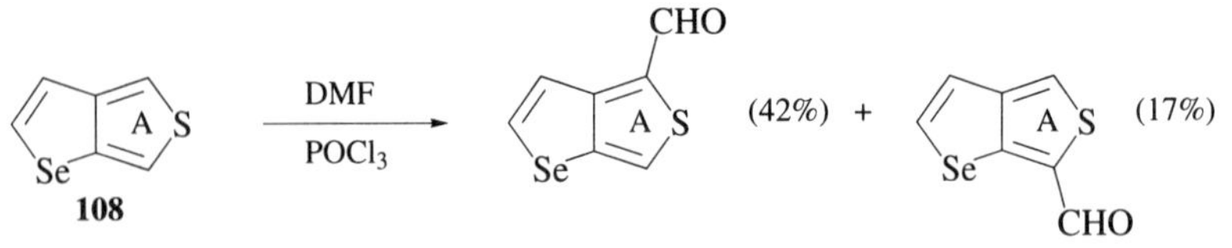

(Eq. 93)

and thieno[3,4-*b*]selenophene (**108**) (Eq. 93)[133] all react to give mixtures of aldehyde products where substitution has occurred in ring A. Some ring-opening and
degradation products are also reported with compound **107**. Thieno[2,3-*b*]pyrrole
(**109**) undergoes formylation mainly in the pyrrole ring as expected (Eq. 94),[134]

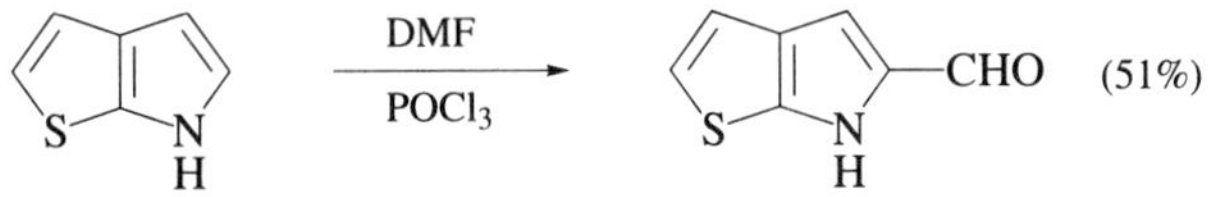

(Eq. 94)

and traces of other formylated isomers are also obtained. The furo[3,2-*b*]pyrrole
derivative **110** is formylated in the furan ring (Eq. 95),[135] and with the phenyl de-

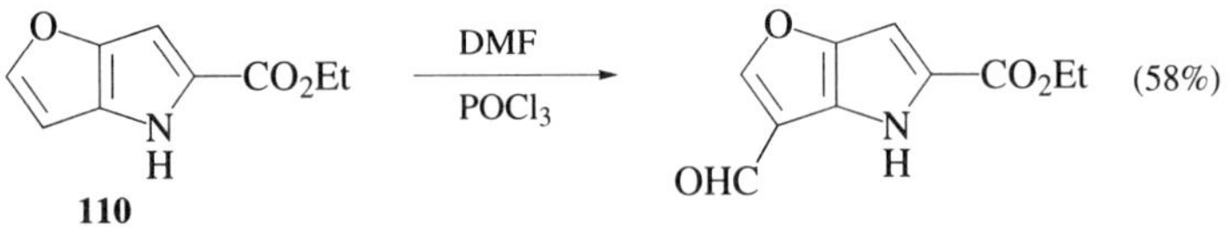

(Eq. 95)

rivative **111** of compound **110** where the furan ring is blocked, formylation occurs
first on the pyrrole nitrogen giving compound **112**, but eventually the thermodynamically more stable product **113** is obtained after a protracted reaction time
(Eq. 96).[135]

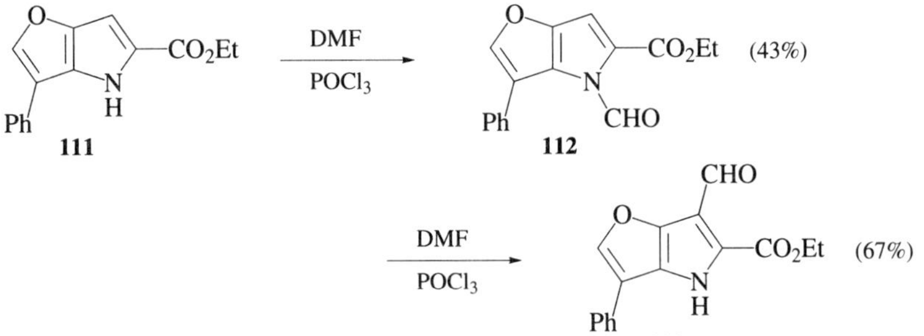

(Eq. 96)

Bicycles with bridgehead nitrogen react well, as illustrated for compounds **114**
(Eq. 97)[136] and **115** (Eq. 98).[137] Compound **114** also provides an example of the

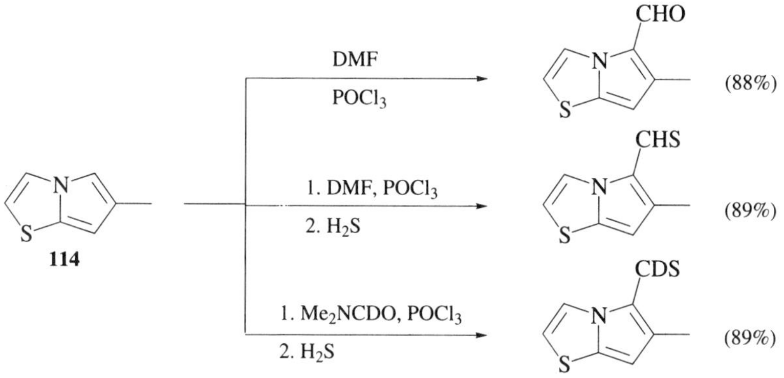

(Eq. 97)

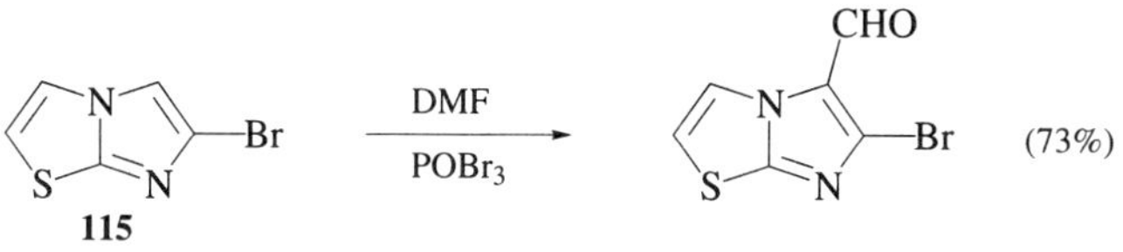

(73%)   (Eq. 98)

synthesis of a thioaldehyde and a deuterated thioaldehyde. Phosphoryl bromide was used in the reaction of compound **115** because of the potentially replaceable bromine substituent. Compounds with a bridgehead sulfur atom such as heterocycle **116** can also be formylated, although yields are commonly lower than that shown (Eq. 99).[138]

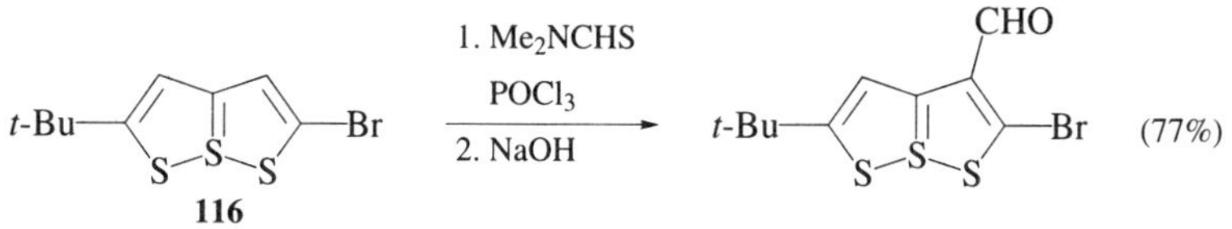

(Eq. 99)

Fusion of a six-membered ring to a five-membered ring (excluding those systems with a bridgehead nitrogen) leads predictably to substitution in the five-membered ring. Yields are usually lower than those for indole; the pyrrolo[2,3-b]pyridines **117** illustrate the range of yields achieved (Eq. 100).[139]

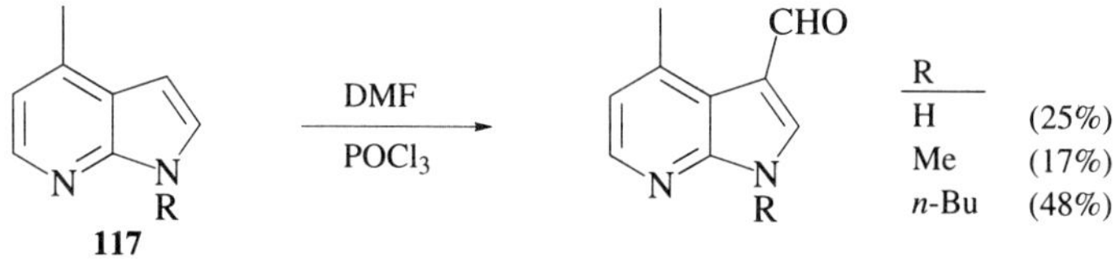

R	
H	(25%)
Me	(17%)
n-Bu	(48%)

(Eq. 100)

The furo[2,3-*b*]pyridine derivative **118** (Eq. 101) and the furo[3,2-*b*]pyridine derivative **119** (Eq. 102) can undergo dealkylation and subsequently behave as hydroxyfurans, and chloroformylation as well as normal substitution is observed.[140]

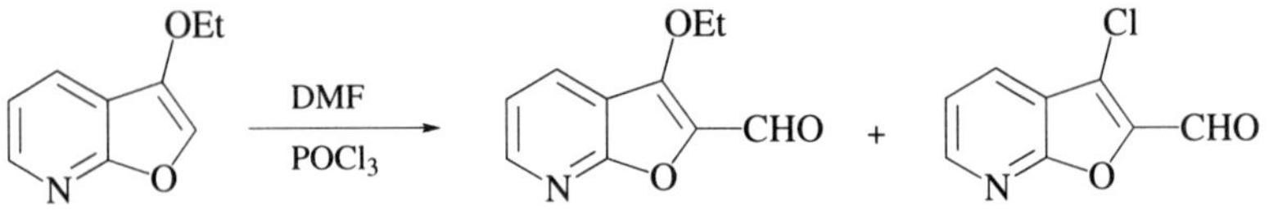

118   (22%)   (43%)

(Eq. 101)

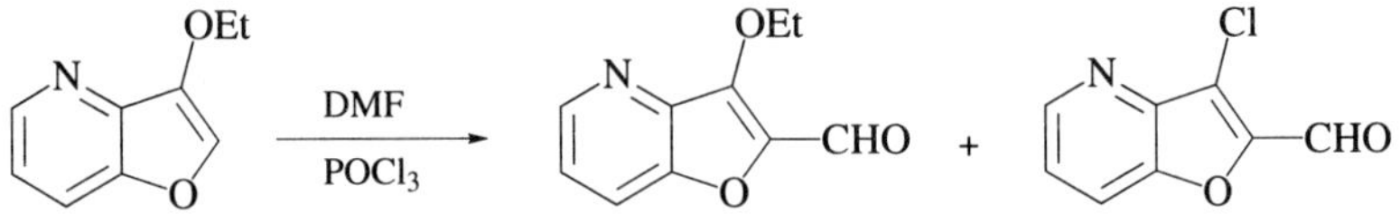

119   (30%)   (35%)

(Eq. 102)

Bicycles with bridgehead nitrogen are readily formylated. Indolizines such as compound **120** (Eq. 103) are substituted in high yield in the 3-position, or if this

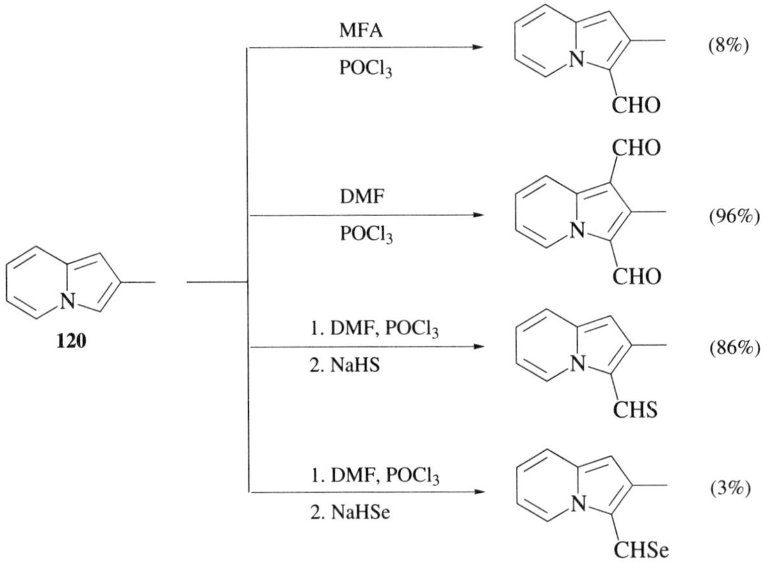

(Eq. 103)

is blocked, in the 1-position. Compound **120** also illustrates the production of a thioaldehyde,[141,142] a selenoaldehyde,[143] and a dialdehyde under forcing conditions.[141,142] Azaindolizines also undergo Vilsmeier formylation whether the second nitrogen is located in the five-membered ring (Eq. 104)[144] or in the six-membered ring (Eq. 105).[145] The poor yield of formyl derivative from triazolopyridine **121** probably reflects the ease of ring opening with loss of nitrogen (Eq. 106).[146]

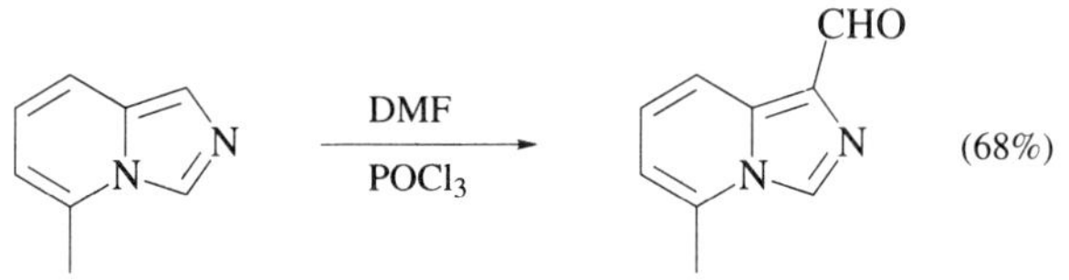

(68%) (Eq. 104)

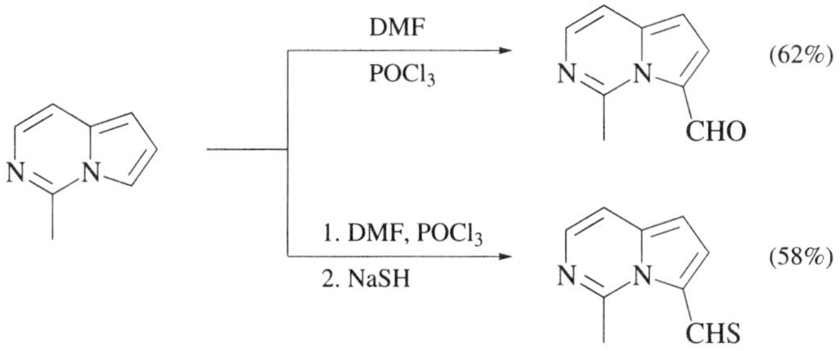

(62%)

(Eq. 105)

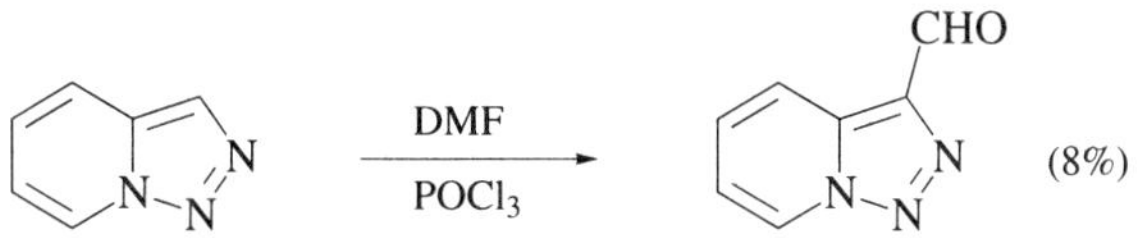

(58%)

121

(8%) (Eq. 106)

The pyrroloazepinone **122** has been formylated in good yield (Eq. 107).[147] Coumarins can give various products if the benzenoid ring is deactivated as shown for the formylation of compound **123** (Eq. 108).[148] Activation in the ben-

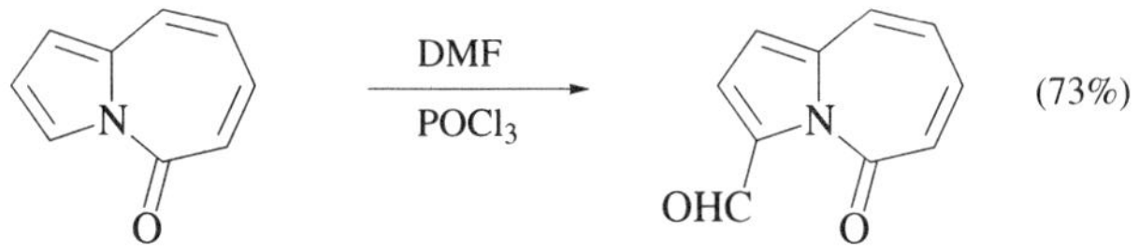

122

(73%)

(Eq. 107)

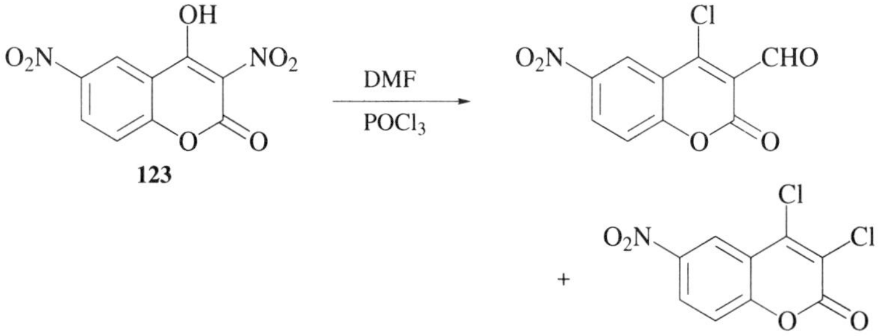

123

(Eq. 108)

zenoid ring gives formylation in the predicted position (Eq. 109)[149] and activation
in the heterocyclic ring gives formylation which can be followed by cyclization
(Eq. 110).[150]

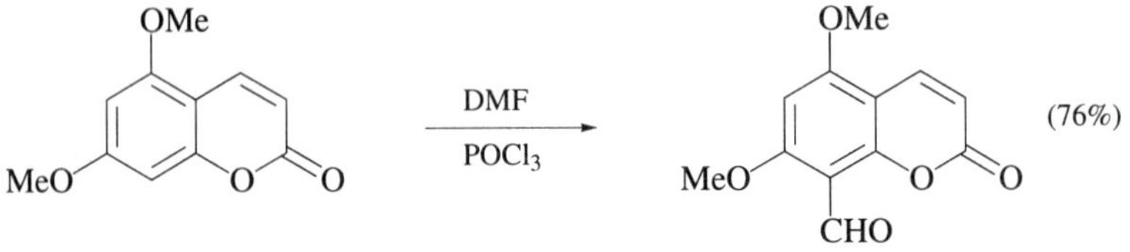

(Eq. 109)

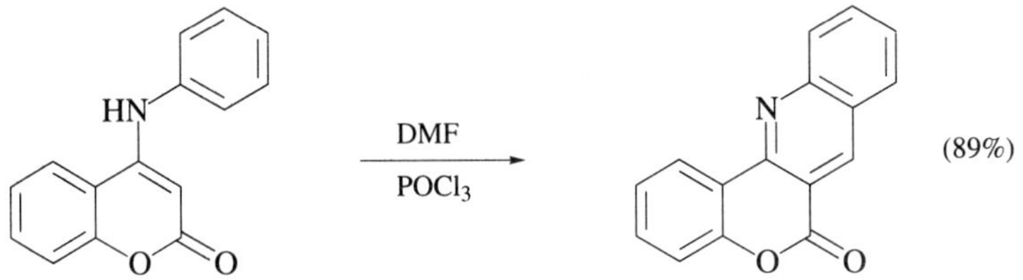

(Eq. 110)

Pyrimidopyridines need electron-donating substituents for successful formylation (Eq. 111),[151] and when the activation is provided by an *N*-arylamino group,
polycyclic products can be formed (Eq. 112).[152]

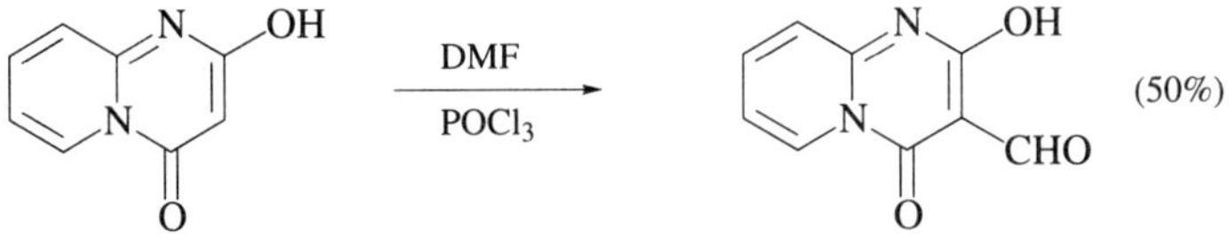

(Eq. 111)

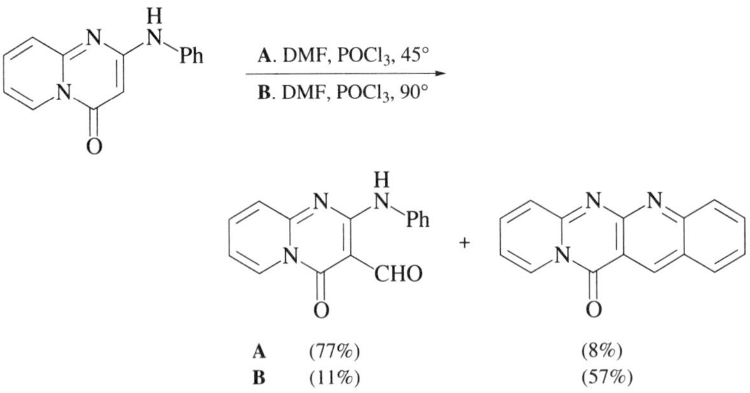

A	(77%)	(8%)
B	(11%)	(57%)

(Eq. 112)

Other Heterocycles with Three or More Fully Conjugated Rings

The simplest examples of the many systems with three or more fully conjugated rings are the benzindolizine **124** (Eq. 113)[153] and the benzindoles **125**

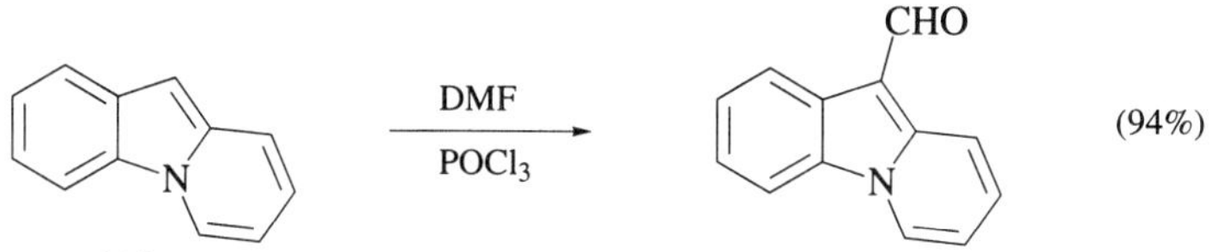

(94%)

124

(Eq. 113)

(Eq. 114)[154,155] and **126** (Eq. 115).[154] The benzo[e]indoles **127** (Eq. 116) and **128** (Eq. 117) may illustrate some similarities with phenanthrene since formylation can be achieved in the central ring.[154,155]

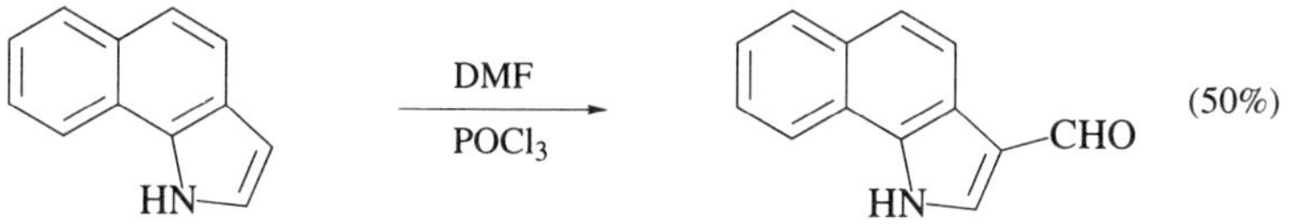

(50%)

125

(Eq. 114)

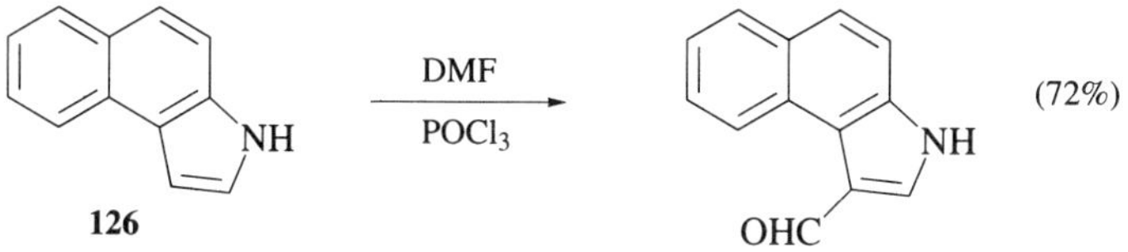

(72%)

126

(Eq. 115)

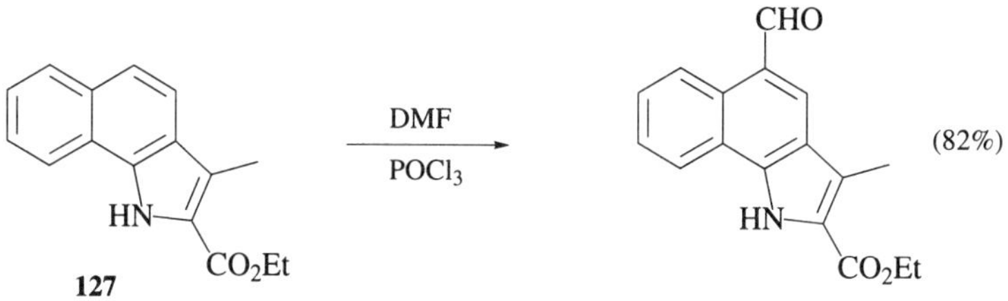

(82%)

127

(Eq. 116)

(80%)

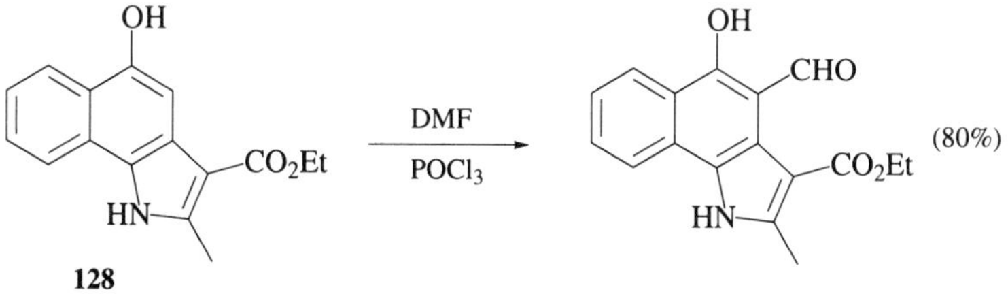

128

(Eq. 117)

The tricyclic furan derivative **129** was originally reported to undergo formylation in the furan ring,[156] but a reinvestigation of this reaction concluded that substitution had occurred in the naphthalene ring (Eq. 118).[157]

$$\xrightarrow[\text{POCl}_3]{\text{DMF}}$$

(65%)   (Eq. 118)

In larger polycycles based on indole, the phenanthrenopyrrole **130** (Eq. 119)[158] and the pyrrolocarbazole **131** (Eq. 120)[159] give the expected products in formylation and acylation reactions, but benzo[e]pyrrolo[3,2-g]indole **132** undergoes an unprecedented chlorination when excess phosphoryl chloride is used (Eq. 121).[160]

$$\xrightarrow{\text{Formylation}}$$

(85%)   (Eq. 119)

$$\xrightarrow[\text{POCl}_3]{\text{Me}_2\text{NCOR}}$$

R	
Me	(10%)
CH$_2$Cl	(40%)

(Eq. 120)

$$\xrightarrow[\text{B. DMA, POCl}_3 \text{ (xs)}]{\text{A. DMA, POCl}_3}$$

A, R = H (50%)
B, R = Cl (62%)

(Eq. 121)

Examples of 5:5:6 tricycles which undergo the Vilsmeier reaction include the benzthienothiophenes **133** (Eq. 122) and **134** (Eq. 123)[161] and the furo[3,2-b]indoles **135** (Eq. 124).[162] Heterocycle **136** yielded a benzthienothiophene derivative (Eq. 125).[163]

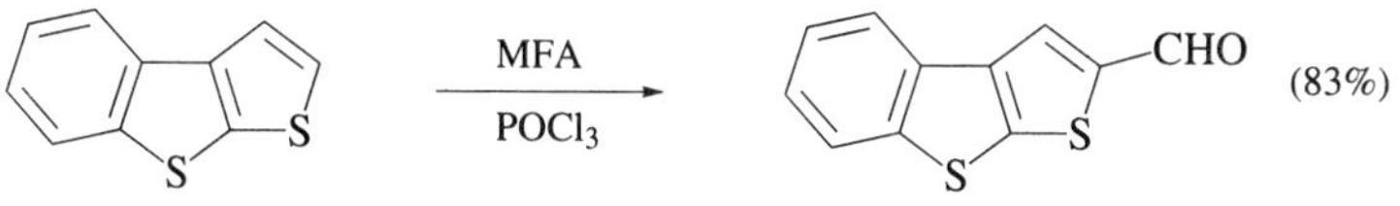

(Eq. 122)

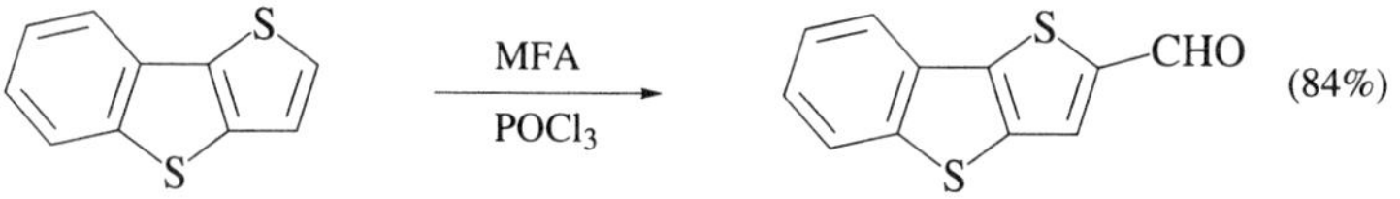

(Eq. 123)

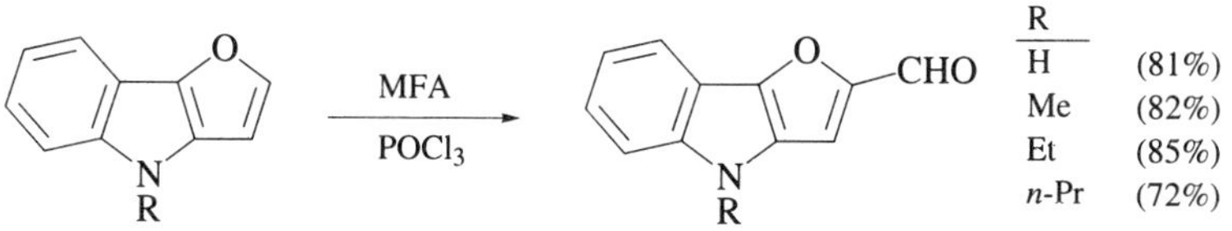

R	
H	(81%)
Me	(82%)
Et	(85%)
n-Pr	(72%)

(Eq. 124)

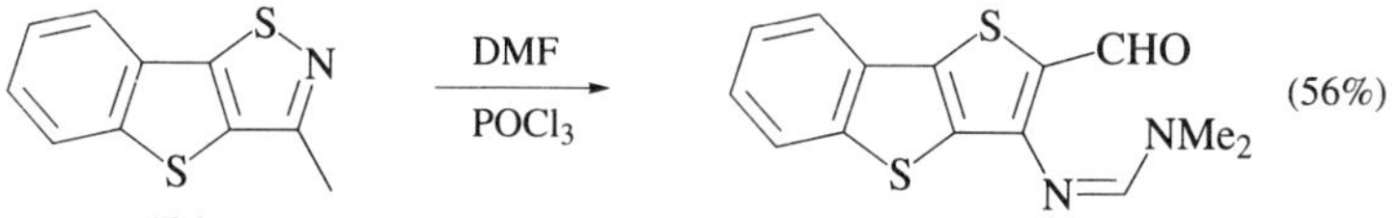

(Eq. 125)

With three heteroatoms distributed between the two five-membered rings in 5:5:6 systems, the yields of aldehydes remain excellent, as illustrated for heterocycles **137** (Eq. 126),[164] **138** (Eq. 127),[165] and **139** (Eq. 128).[166]

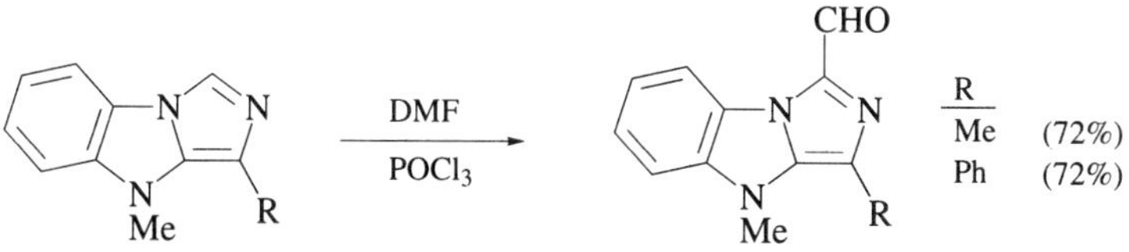

R	
Me	(72%)
Ph	(72%)

(Eq. 126)

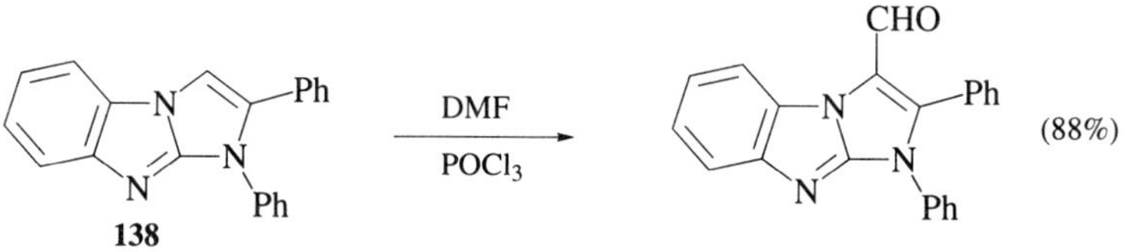

(Eq. 127)

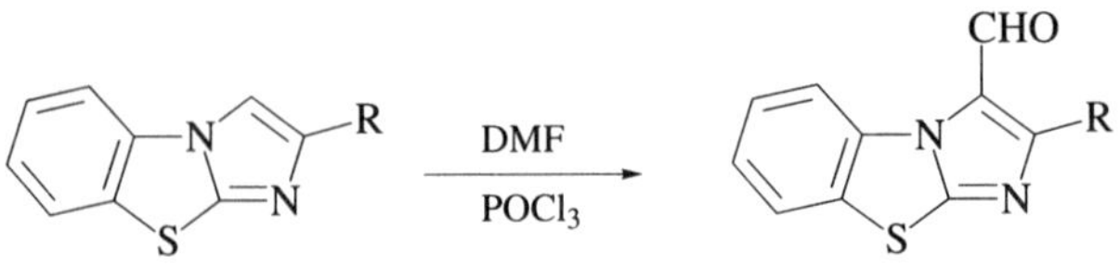

R	
Ph	(89%)
4-ClC$_6$H$_4$	(91%)

139

(Eq. 128)

The azacyclazine **140** gives a mixture of two products (Eq. 129).[167] The isomers **141** (Eq. 130)[168] and **142** (Eq. 131)[169] give contrasting results, the former

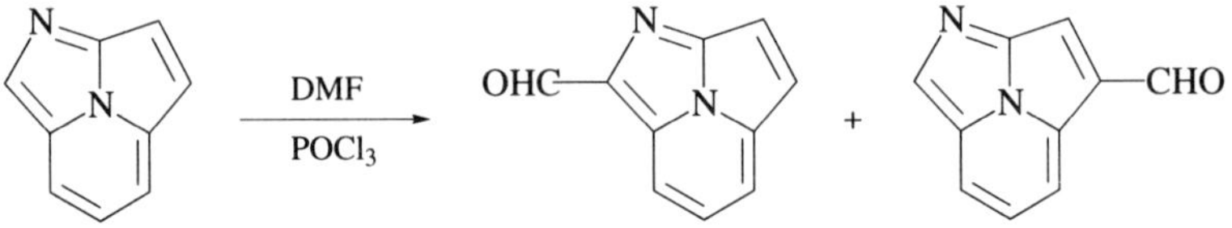

140 (37%) (25%)

(Eq. 129)

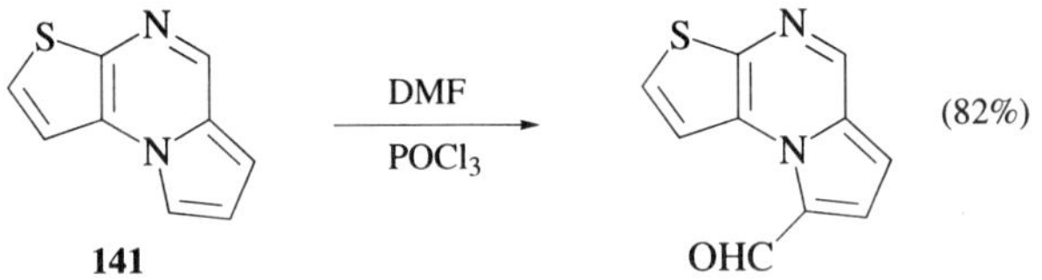

141 (82%) (Eq. 130)

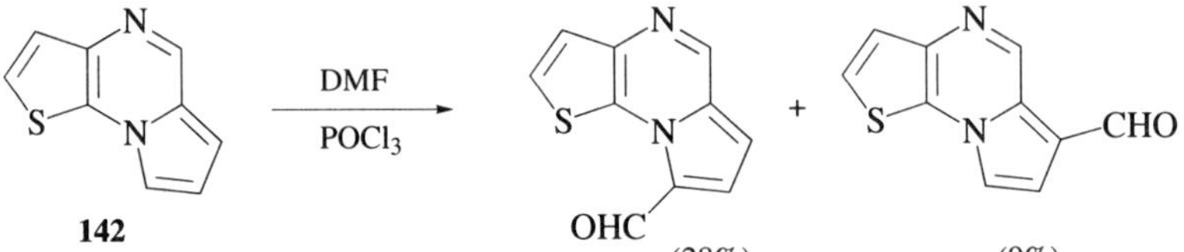

142 (28%) (9%)

(Eq. 131)

with a single high-yield site for formylation and the latter giving poor yields of two products.

Protected sugars are tolerant of the Vilsmeier reagent, as illustrated for heterocycle **143** (Eq. 132),[170,171] although an unusually low temperature was used in one

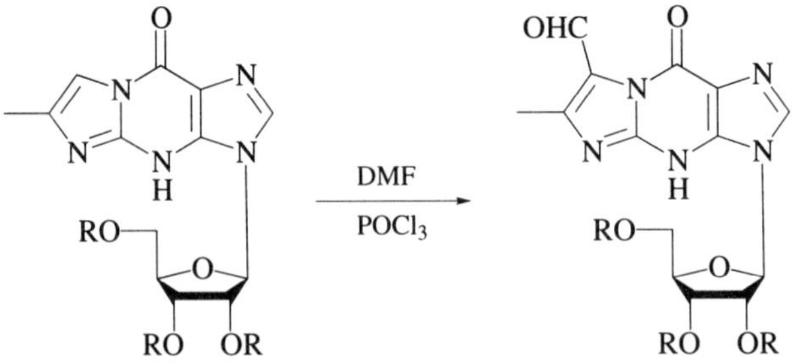

R	
COMe	(92%)
Bn	(63%)

143

(Eq. 132)

case. Acridine requires electron-donating substituents for successful formylation, and with two powerful groups diformylation is possible (Eq. 133).[172] The tricycle

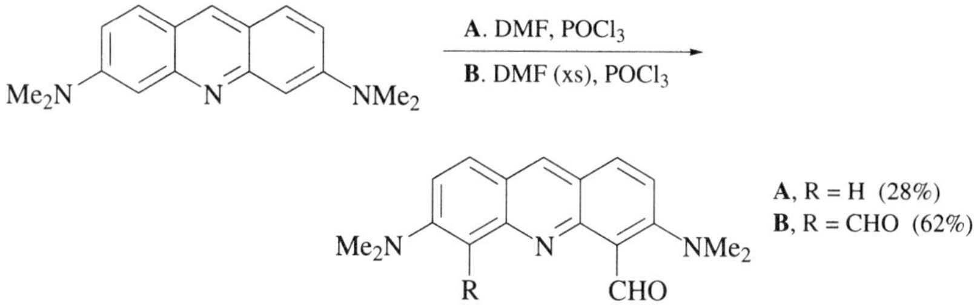

A. DMF, $POCl_3$

B. DMF (xs), $POCl_3$

A, R = H (28%)
B, R = CHO (62%)

(Eq. 133)

144 behaves as a vinylogous enamine in the Vilsmeier reaction and aldehyde **146** is obtained via salt **145** (Eq. 134).[173]

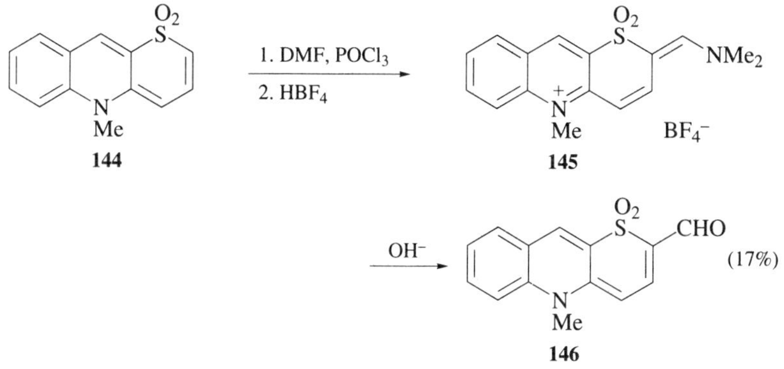

1. DMF, $POCl_3$

2. HBF_4

144

145

BF_4^-

OH^-

146 (17%)

(Eq. 134)

Porphyrins.   There are many examples of porphyrins that participate in the Vilsmeier reaction. The general rule is that substitution occurs at one or more of the *meso* positions (Eq. 135)[174,175] and mixtures result from unsymmetrical por-

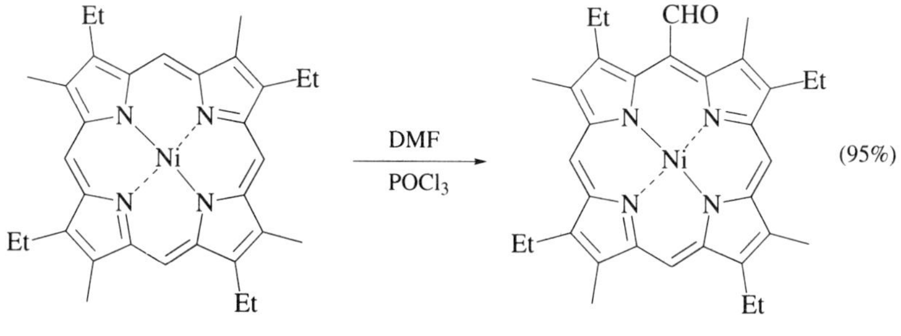

DMF

$POCl_3$

(95%)

(Eq. 135)

phyrins. The use of a vinylogous amide can allow annulation to an adjacent pyrrole ring (Eq. 136),[176] although in other cases the *meso* acraldehyde is

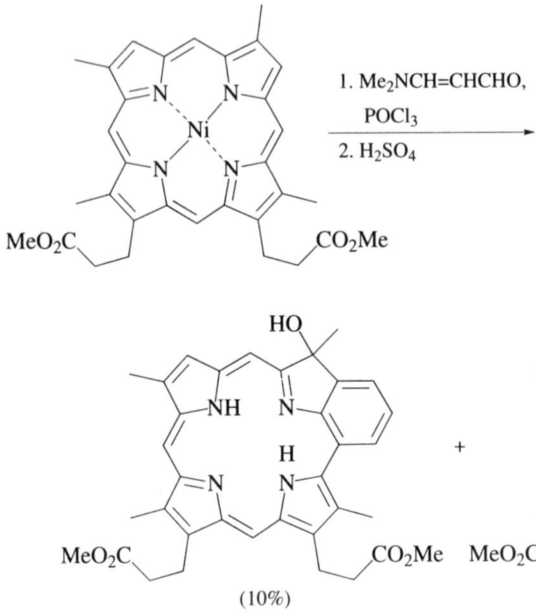

obtained.[176] Substitution can occur at a pyrrole ring if a hindered amide is used (Eq. 137).[177]

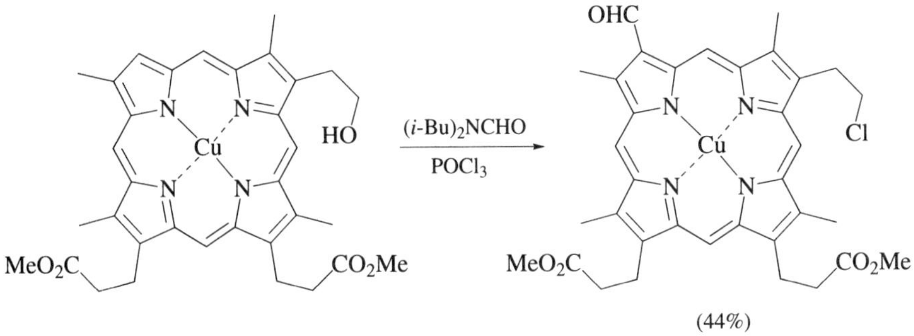

COMPARISON WITH OTHER METHODS

There are several methods for converting an aromatic derivative ArH into its corresponding aldehyde derivative ArCHO using an electrophilic substitution reaction and the scope and limitations of these reactions have been summarized.[20,178] The Gattermann–Koch reaction uses carbon monoxide in the presence of hydrogen chloride and a Lewis-acid catalyst as the formylating reagent and is applicable to simple alkyl- and haloaromatics and polycyclic aromatic hydrocarbons, but fails for phenols. In contrast, the Gattermann reaction, which uses hydrogen cyanide, hydrogen chloride, and a Lewis-acid catalyst, will formylate phenols. Related to the Gattermann–Koch reaction is the direct formylation of aromatic hydrocarbons with formyl fluoride in the presence of boron trifluoride. Dichloromethyl alkyl ethers also formylate aromatic hydrocarbons in the presence of a Lewis-acid catalyst to give α-alkoxybenzyl chlorides from which aldehydes can be obtained by heating or hydrolysis. Electron-rich aromatics such as phenols and aromatic amines can be formylated with substantial *ortho* selectivity using hexamethylenetetramine in acetic acid (the Duff reaction). With trifluoroacetic acid, simple aromatics such as toluene can be formylated but the reaction is *para* selective in these cases. The Reimer–Tiemann reaction of phenols uses chloroform and alkali and gives a high proportion of *ortho* products.[178]

EXPERIMENTAL PROCEDURES

4-*N,N*-Dimethylaminobenzaldehyde (Formylation of an Activated Benzene Derivative with Pyrophosphoryl Chloride and DMF).[26] Pyrophosphoryl chloride (2.27 g, 9.0 mmol) was added dropwise to stirred, cold (ice bath) DMF (1.10 g, 15.0 mmol) and *N,N*-dimethylaniline (0.91 g, 7.5 mmol) to give a thick green syrup. The mixture was heated at 65° for 15 hours and allowed to cool, and the resulting green solid was dissolved in water and basified with 2 M sodium hydroxide solution giving a yellow solid. The solid was sublimed at 140–150° (0.6 mm) to give 1.11 g (99%) of 4-*N,N*-dimethylaminobenzaldehyde, mp 73–75°; IR 1656 cm^{-1}; ^{1}H NMR (CDCl$_3$) δ 3.06 (s, 6 H), 6.67 (d, $J = 9$ Hz, 2 H), 7.71 (d, $J = 9$ Hz, 2 H), 9.72 (s, 1 H).

4-*N,N*-Dimethylaminobenzaldehyde (Formylation of an Activated Benzene Derivative with Phosphoryl Chloride and DMF). A detailed procedure for this reaction is described in *Organic Syntheses.*[32] The yield of 4-*N,N*-dimethylaminobenzaldehyde was 84%.

Anthracene-9-carbaldehyde (Formylation of a Polycyclic Hydrocarbon with Phosphoryl Chloride and *N*-Methylformanilide). A detailed procedure for this reaction is described in *Organic Syntheses.*[179] The yield of anthracene-9-carbaldehyde was 84%.

Thiophene-2-carbaldehyde (Formylation of Thiophene with Phosphoryl Chloride and N-Methylformanilide). A detailed procedure for this reaction is described in *Organic Syntheses.*[180] The yield of thiophene-2-carbaldehyde was 74%.

Pyrrole-2-carbaldehyde (Formylation of Pyrrole with Phosphoryl Chloride and DMF). A detailed procedure for this reaction is described in *Organic Syntheses.*[181] The yield of pyrrole-2-carbaldehyde was 95%.

1-Benzylpyrrole-2-carbaldehyde and 1-Benzylpyrrole-3-carbaldehyde (Formylation of a Pyrrole Derivative with Pyrophosphoryl Chloride and MFA).[26] Pyrophosphoryl chloride (2.14 g, 8.5 mmol) was added dropwise to a stirred mixture of cold (ice bath) MFA (2.03 g, 15 mmol) and 1-benzylpyrrole (1.18 g, 7.5 mmol) and the resulting syrup was stirred at 20° for 19 hours. The mixture was basified with 2 M sodium hydroxide solution, and extracted with dichloromethane, and the combined organic extracts were washed with dilute hydrochloric acid, dried (MgSO$_4$), and evaporated. The unreacted MFA was removed by distillation (Kugelrohr) at 70° (1 mm) and the residue was fractionated by column chromatography on silica gel (eluting with petroleum ether–ether 5:1) to give 1.03 g (75%) of 1-benzylpyrrole-2-carbaldehyde; IR 1658 cm^{-1}; ^{1}H NMR (CDCl$_3$) δ 5.54 (s, 2 H), 6.24 (m, 1 H), 6.94 (m, 2 H), 7.14–7.10 (m, 2 H), 7.31–7.18 (m, 3 H), 9.54 (s, 1 H), and 0.303 g (22%) of 1-benzylpyrrole-3-carbaldehyde; IR 1660 cm^{-1}; ^{1}H NMR (CDCl$_3$) δ 5.08 (s, 2 H), 6.65 (dd, J = 2.8 and 1.8 Hz, 1 H), 6.70 (t, J = 2.8 Hz, 1 H), 7.18–7.13 (m, 2 H), 7.37–7.29 (m, 4 H), 9.73 (s, 1 H).

2-Acetyl-1-phenylpyrrole and 3-Acetyl-1-phenylpyrrole (Acylation of a Pyrrole Derivative with Phosphoryl Chloride and DMA).[182] DMA (2.87 g, 33 mmol) was added to phosphoryl chloride (6.0 g, 39 mmol) cooled to 5° and the mixture was then stirred at room temperature for 6 hours. A solution of 1-phenylpyrrole (4.0 g, 27 mmol) in dichloroethane (25 mL) was added and the mixture was stirred at 60° for 24 hours under a nitrogen atmosphere, poured into 30% aqueous sodium carbonate solution (100 mL), extracted with dichloromethane, and evaporated. The residue was fractionated by column chromatography on silica gel (eluting with toluene–ethyl acetate 9:1) to give 3.0 g (60%) of 2-acetyl-1-phenylpyrrole, mp 56–57°, and 0.8 g (15%) of 3-acetyl-1-phenylpyrrole, bp 165° (1 mm). 2-Acetyl-1-phenylpyrrole: ^{1}H NMR (CDCl$_3$) δ 2.3 (s, 3 H), 6.65 (d, J = 3 Hz, 1 H), 7.0 (d, J = 3 Hz, 1 H), 7.3 (s, 5 H), 7.55 (s, 1 H).

2-Benzoylpyrrole (Benzoylation of Pyrrole with Phosphoryl Chloride and N-Benzoylmorpholine).[183] A mixture of N-benzoylmorpholine (2.96 g, 20 mmol) and phosphoryl chloride (4.0 mL, 20 mmol) was kept at 25° for 6 hours. A solution of pyrrole (1.38 g, 20 mmol) in anhydrous 1,2-dichloroethane (100 mL) was added and, after swirling, the reaction mixture was left at 25° for 14 hours. After hydrolysis with 10% aqueous sodium carbonate solution (100 mL), the organic layer was separated and the aqueous layer washed with 1,2-dichloroethane (2 × 20 mL). The combined organic layers were dried (Na$_2$CO$_3$), the solvent removed, and the residue recrystallized (charcoal) from petroleum ether giving 2-benzoylpyrrole, 2.95 g (86%) as colorless needles, mp 77.5–78°.

4,6-Dimethoxyindole-7-carbaldehyde (Formylation of an Indole Derivative in the Benzene Ring with Phosphoryl Chloride and DMF).[109] To a

stirred solution of 4,6-dimethoxyindole (0.18 g, 4 mmol) in DMF (1 mL) at 0°
was added dropwise an ice-cold solution of phosphoryl chloride (0.37 mL, 4.05
mmol) in DMF (1 mL). The mixture was kept at 0° for 1 hour and then allowed
to warm to room temperature, added to ice water (10 mL), and made strongly al-
kaline with 10% sodium hydroxide solution. The resulting solid was collected,
washed with water, dried, and recrystallized from chloroform–petroleum ether
to give 0.11 g (56%) of 4,6-dimethoxyindole-7-carbaldehyde, mp 201–202°; IR
1639 cm^{-1}; ^{1}H NMR (CDCl$_3$) δ 3.92 (s, 3 H), 3.98 (s, 3 H), 6.4 (m, 2 H), 7.05 (t,
$J = 2.5$ Hz, 1 H), 10.26 (s, 1 H), 11.20 (broad s, 1 H); mass spectrum, m/z (rel.
intensity) 205 (100).

**5-Chloro-3-*tert*-butyl-1-phenylpyrazole-4-carbaldehyde (Chloroformyla-
tion of a Pyrazolone Derivative with Phosphoryl Chloride and DMF).[184]**
Phosphoryl chloride (14.85 mL, 0.162 mol) was slowly added with stirring to ice-
cold DMF (5.37 mL, 0.069 mol) over 0.75 hour. 3-*tert*-Butyl-1-phenyl-5(1*H*)-
pyrazolone (5.0 g, 0.023 mol) was added and the mixture was heated at reflux for
0.25 hour. The cooled mixture was poured into water (100 mL, 0°) and extracted
continuously with ether. The organic extact was dried (Na$_2$SO$_4$), evaporated, and
distilled under reduced pressure at 86° (2 mm) to give 4.8 g (79%) of 5-chloro-3-
tert-butyl-1-phenylpyrazole-4-carbaldehyde; IR (neat) 1689 cm^{-1}; ^{1}H NMR
(CDCl$_3$) δ 1.45 (s, 9 H), 7.30–7.90 (m, 5 H); mass spectrum, m/z (rel. intensity)
262 (52), 77 (100).

**1-(*N*,*N*-Dimethylaminomethylidene)-3-phenyl-1*H*-isoindole and 1-
Phenylisoindole-3-carbaldehyde (Dimethylaminomethylenation of an Isoin-
dole Derivative with Phosphoryl Chloride and DMF and its Conversion to
the Corresponding Aldehyde).[185]** Phosphoryl chloride (2.5 g, 0.016 mol) was
added to DMF (20 mL, 0.26 mol) with stirring at −5 to −10° under an argon atmosphere. 1-Phenylisoindole (1.5 g, 7.77 mmol) in DMF (20 mL) was added
dropwise and the mixture was stirred at −5° for 1 hour, then at room temperature
for 5 hours. The mixture was poured into saturated sodium hydrogen carbonate
solution (500 mL) with stirring, and the resulting brownish yellow solid (1.84 g)

was washed with water and dried. The solid was purified by column chromatography on silica gel (eluting with petroleum ether–acetone 2:1) and the product was recrystallized from ethyl acetate–petroleum ether to give 1.31 g (68%) of 1-(*N,N*-dimethylaminomethylidene)-3-phenyl-1*H*-isoindole as brilliant yellow-green needles, mp 152–153°; ^{1}H NMR (CDCl$_3$) δ 3.57 (broad s, 6 H) and 7.0–8.2 (m, 10 H); mass spectrum, m/z (rel. intensity) 248 (100).

The above compound (0.218 g, 0.93 mmol), ethanol (20 mL), and 4% sodium hydroxide solution (1.3 mL) were heated under reflux under an argon atmosphere for 3.25 hours. The mixture was evaporated and the residue was dissolved in dichloromethane. The organic layer was washed with water and the aqueous washings were back-extracted with dichloromethane. The combined organic extracts were dried (Na$_2$SO$_4$) and evaporated to give a residue which was triturated with ether (2 mL). The resulting solid was recrystallized from ethyl acetate to give 93 mg (48%) of 1-phenylisoindole-3-carbaldehyde as bright-yellow needles, mp 183–184°; IR (Nujol) 3210, 1630, 1615, 1290, 1275, 1230, 745 cm^{-1}; ^{1}H NMR (CD$_3$SOCD$_3$) δ 7.17–8.25 (m, 9 H), 9.96 (s, 1 H), 13.34 (broad s, 1 H); mass spectrum, m/z (rel. intensity) 221 (100).

Nickel(II) *meso*-(2-Formylvinyl)octaethylporphyrin [Vinylogous Formylation of a Porphyrin Derivative with Phosphoryl Chloride and 3-(Dimethylamino)acrolein].[176] Phosphoryl chloride (1.0 mL, 10 mmol) was added dropwise to a solution of 3-(dimethylamino)acrolein (1.0 mL, 10 mmol) in dichloromethane (4.0 mL) and the mixture was kept at 0° for 0.25 hour. The mixture was added to nickel(II) octaethylporphyrin (200 mg, 0.338 mmol) at 0° and then stirred at room temperature for 8 hours. After basic hydrolysis, the final neutralized residue was purified by column chromatography over silica gel (eluting with dichloromethane) and was further purified by preparative TLC on silica plates (eluting with dichloromethane–petroleum ether 8:2). The less polar compound was collected and crystallized from dichloromethane–methanol to give 186 mg (85%) of nickel(II) *meso*-(2-formylvinyl)octaethylporphyrin, mp 245–246°; ^{1}H NMR (CDCl$_3$) δ 1.66–1.79 (overlapping t, 24 H), 3.65–3.86 (overlapping q, 16 H), 5.53 (dd, $J = 7.7$ and 15.2 Hz, 1 H), 9.36 (s, 3 H), 9.69 (d, $J = 15.2$ Hz, 1 H), 9.84 (d, $J = 7.7$ Hz, 1 H).

TABULAR SURVEY

We have attempted to cover thoroughly the literature until the end of 1991, but some additional references after this date have also been included. The principal exclusions from this chapter are intramolecular cyclization reactions of the Bischler–Napieralski type. Only carbon–carbon bond formation reactions are included in the tables. Where a reaction has been performed by different workers, the yield in the table corresponds to the first reference.

To assist the reader who might be searching for specific substitution patterns, Tables I, II, V, and VIII–XV have been subdivided according to substitution pattern with the unsubstituted parent compounds being considered before monosubstituted compounds. Within subdivisions which are associated with multiple substitution, for example, disubstituted compounds, the 1,2-substitution pattern is considered before the 1,3-disubstituted pattern.

Tables III, IV, and VII have been subdivided according to the number of component rings, with monocyclic systems having precedence over bicyclic systems. Rings containing the least number of carbon atoms are considered first within this subdivision. Similarly, in Tables XVI–XVIII, rings containing the least number of carbon atoms have precedence. To give additional structure to Tables XVI–XVIII, the parent heterocyclic ring system (e.g., pyrimidine) is considered before derivatives possessing a conjugated exocyclic substituent (e.g., pyrimidinediones) which have precedence over derivatives possessing two conjugated exocyclic substituents (e.g., pyrimidiones). Heterocycles which can become fully conjugated through tautomerism are considered in these tables.

Table XIX is not subdivided, and entries are ordered by increasing number of carbon atoms.

The following abbreviations are used in the tables:

DMF   Dimethylformamide
MFA   *N*-Methylformanilide
DMA   *N,N*-Dimethylacetamide

TABLE I. BENZENES

R³ R⁴ on benzene ring, R⁵, R², R⁶, R¹

$R^1 - R^6 = H$ except as indicated

Substrate	Reagents	Product(s) and Yield(s) (%)	Refs.
		A. Monosubstituted Benzenes	
C_6 R^1 = OH	DMF, POCl₃	R^4 = CHO (5)	186
	DMF, SOCl₂, AlCl₃	R^4 = CHO (60)	34
C_7 R^1 = OMe	DMF, POCl₃	R^4 = CHO (~70)	186
	DMF, POCl₃	R^2 = CHO (4) + R^4 = CHO (34)	26
	Me₂N¹³CHO, POCl₃	R^4 = ¹³CHO (—)	187
	DMF, SOCl₂, AlCl₃	R^2 = CHO (4) + R^4 = CHO (69)	34
	DMF, (Cl₂PO)₂O	R^2 = CHO (5) + R^4 = CHO (70)	26, 188
	MFA, (Cl₂PO)₂O	R^2 = CHO (<1) + R^4 = CHO (72)	26, 188
	MFA, POCl₃	R^4 = CHO (—)	189
	—	R^4 = CHO (70)	190
	DMF, (CF₃SO₂)₂O	R^2 = CHO (22) + R^4 = CHO (78)	47
	[Me₂N=CHCl]⁺ Cl⁻	R^4 = CHO (<4)	26
R^1 = SMe	DMF, POCl₃	R^4 = CHO (14)	30, 186
C_{7-11} R^1 = OR	DMF, SOCl₂	(structure: RO–C₆H₄–S–C₆H₄–OR)	42

R	
Me	(34)
Et	(30)
n-Pr	(29)

TABLE I. BENZENES (Continued)

	Substrate	Reagents	Product(s) and Yield(s) (%)	Refs.
	R			
	n-Bu		(30)	
	i-Bu		(29)	
	C₅H₁₁		(33)	
C₈				
	R¹ = OEt	DMF, POCl₃	R⁴ = CHO (—)	186
	R¹ = NMe₂	DMF, POCl₃	R⁴ = CHO (85)	31, 32, 191, 192
		DMF, POCl₃ (2-3 eq)	R⁴ = CHO (—) + R² = R⁴ = CHO (15–20)	33
		DMF, COCl₂	R⁴ = CHO (50)	31
		DMF, SOCl₂	R⁴ = CHO (60)	31
		DMF, [2,4,6-trichloro-1,3,5-triazine]	R⁴ = CHO (41)	36
		(CD₃)₂NCXO, COCl₂ or POCl₃	R⁴ = CXO (—), X = H, D	29
		DMF, SOCl₂, AlCl₃	R⁴ = CHO (70)	34
		DMF, SO₂Cl₂	R⁴ = CHO (—)	193
		DMF, (Cl₂PO)₂O	R⁴ = CHO (99)	188, 26
		DMF, Ph₃PBr₂	R⁴ = CHO (72)	35
		DMF, P(NCl₂)₃	R⁴ = CHO (21)	194
		DMF, P(NCl₂)₃, AlCl₃	R⁴ = CHO (51)	194
		1. DMF, POCl₃ 2. Me₂NC₆H₅	(Me₂NC₆H₄)₃CH (93)	41
		MeNHCOMe, POCl₃	R⁴ = COMe (25)	31
		DMA, POCl₃	R⁴ = COMe (15)	31
		MeNHCOPh, POCl₃	R⁴ = COPh (65)	31
		PhNHCOPh, POCl₃	R⁴ = COPh (85)	31
		Me₂NCOPh, POCl₃	R⁴ = COPh (80)	31, 111

46

TABLE I. BENZENES (*Continued*)

Substrate	Reagents	Product(s) and Yield(s) (%)	Refs.
PhNHCOAr		R^4 = COAr	37
	$\underline{\text{Ar}}$		
	Ph	(65)	
	4-BrC$_6$H$_4$	(75)	
	3-HOC$_6$H$_4$	(50)	
	4-HOC$_6$H$_4$	(60)	
	2-O$_2$NC$_6$H$_4$	(—)	
	3-O$_2$NC$_6$H$_4$	(40)	
	4-O$_2$NC$_6$H$_4$	(45)	
	2-MeOC$_6$H$_4$	(55)	
	3-MeOC$_6$H$_4$	(40)	
	4-MeOC$_6$H$_4$	(50)	
	1-C$_{10}$H$_7$	(35)	
	2-C$_{10}$H$_7$	(35)	
PhN(Me)CH=CHCHO, POCl$_3$		R^4 = CH=CHCHO (70-80)	39
PhN(Me)CH=CHCHO, MeCOBr or PhCOBr		R^4 = CH=CHCHO (—)	195
Me$_2$NCH=C(R)CHO, POCl$_3$		R^4 = CH=C(R)CHO	38
	$\underline{\text{R}}$		
	H	(42)	
	Me	(61)	
	Et	(37)	
	n-Pr	(32)	
	n-C$_5$H$_{11}$	(21)	
PhN(Me)(CH=CH)$_2$CHO, POCl$_3$		R^4 = (CH=CH)$_2$CHO (19)	39
Me$_2$NN=CHCHO, POCl$_3$		R^4 = CH=CHN=NMe$_2$$^+$ Cl$^-$ (—)	40

TABLE I. BENZENES (*Continued*)

Substrate	Reagents	Product(s) and Yield(s) (%)	Refs.
C$_{10}$			
R^1 = NEt$_2$	H$_2$NCHO, POCl$_3$, Py	R^4 = CHO (94)	196
	DMF, POCl$_3$	R^4 = CHO (—)	191
	MFA, POCl$_3$	R^4 = CHO (—)	2
	1. MFA, POCl$_3$	R^4 = CN (30)	197
	2. H$_2$NOH		
	3. Dehydration		
	PhNHCOAr, POCl$_3$	R^4 = COAr	137
	$\underline{\text{Ar}}$		
	Ph	(70)	
	2-ClC$_6$H$_4$	(50)	
	3-O$_2$NC$_6$H$_4$	(20)	
	4-O$_2$NC$_6$H$_4$	(60)	
	4-MeOC$_6$H$_4$	(40)	
C$_{10\text{-}16}$			
R^1 = N(CH$_2$CH$_2$)$_2$X	PhN(Me)CH=CHCHO, POCl$_3$	R^4 = CH=CHCHO (84)	39
$\underline{\text{X}}$			
O	DMF, POCl$_3$	R^4 = CHO (97)	198, 191
NCOCF$_3$	DMF, POCl$_3$	R^4 = CHO (—)	199
NPh	DMF, POCl$_3$	R^4 = R$^{4'}$ = CHO (—)	191, 200
C$_{12\text{-}25}$			
R^1 = N(Me)R			
$\underline{\text{R}}$			
(CH$_2$)$_4$CO$_2$H	DMF, POCl$_3$	R^4 = CHO (~60)	201
Ph	DMF, POCl$_3$	R^4 = R$^{4'}$ = CHO (—)	191, 202
(CH$_2$)$_4$CO$_2$Me	DMF, POCl$_3$	R^4 = CHO (~60)	201
(CH$_2$)$_2$NEt$_2$	MFA, POCl$_3$	R^4 = CHO (—)	203
Bn	DMF, POCl$_3$	R^4 = CHO (80)	204
	MFA, POCl$_3$	R^4 = CHO (82)	2

48

TABLE I. BENZENES (Continued)

Substrate	Reagents	Product(s) and Yield(s) (%)	Refs.
R			
(CH$_2$)$_2$(N-piperidyl)	—	R^4 = CHO (—)	203
(CH$_2$)$_2$N(Me)Ph	DMF, POCl$_3$	R^4 = R$^{4'}$ = CHO (—)	191
n-C$_9$H$_{19}$	DMF, POCl$_3$	R^4 = CHO (~60)	201
(CH$_2$)$_3$N(Me)Ph	DMF, POCl$_3$	R^4 = R$^{4'}$ = CHO (50)	205
n-C$_{18}$H$_{37}$	DMF, POCl$_3$	R^4 = CHO (~60)	201
C$_{14}$			
R^1 = NBu$_2$	DMF, POCl$_3$	R^4 = CHO (58)	206
C$_{14-16}$			
R^1 = N(R)(CH$_2$)$_2$CN			
R			
CH$_2$C$_6$H$_4$Cl-4	DMF, POCl$_3$	R^4 = CHO (91)	55
Bn	DMF, POCl$_3$	R^4 = CHO (100)	55
(CH$_2$)$_2$OCO$_2$Et	—	R^4 = CHO (—)	207
C$_{14-18}$			
R^1 = N(Et)R			
R			
(CH$_2$)$_2$NEt$_2$	—	R^4 = CHO (—)	203
CH$_2$C$_6$H$_4$(SO$_2$NMe$_2$-3)	—	R^4 = CHO (—)	208
(CH$_2$)$_2$N(Et)Ph	DMF, POCl$_3$	R^4 = R$^{4'}$ = CHO (—)	191
C$_{16}$			
R^1 = CH=CHC$_6$H$_4$(NMe$_2$-4)	DMF, POCl$_3$ (1 eq)	R^4 = CHO (33)	209
	DMF, POCl$_3$ (4 eq)	R^4 = R$^{2'}$ = CHO (60)	209
C$_{16-18}$			

R^1 =

49

TABLE I. BENZENES (*Continued*)

Substrate	Reagents	Product(s) and Yield(s) (%)	Refs.
$\dfrac{n}{1}$	DMF, POCl$_3$	R^4 = CHO (75)	210
2	DMF, POCl$_3$	R^4 = CHO (65)	210
C$_{18}$			
R^1 = NPh$_2$	DMF, POCl$_3$	R^4 = R$^{4'}$ = R$^{4''}$ = CHO (—)	191
R^1 = SnBu$_3$	MFA, POCl$_3$	R^1 = CHO (56)	53
C$_{20}$ R^1 = N(Bn)$_2$	MFA, POCl$_3$	R^4 = CHO (100)	2
C$_{21}$ 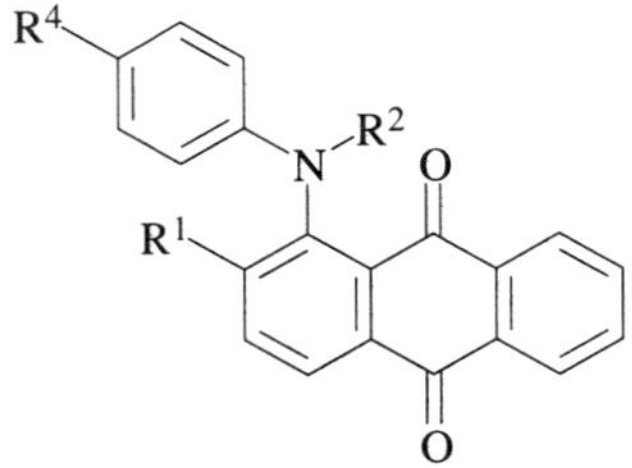	DMF, POCl$_3$	R^4 = CHO	211
C$_{21\text{-}36}$		(100)	
		(100)	

R^1 = NPh$_2$
R^1 = SnBu$_3$

R^1	R^2	R^4
H	Me	H
Me	H	H

R^1 = Ar^1

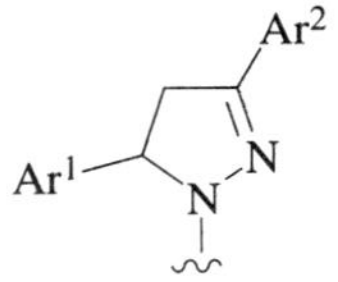

TABLE I. BENZENES (*Continued*)

Substrate		Reagents	Product(s) and Yield(s) (%)	Refs.
Ar¹	Ar²			
Ph	4-O₂NC₆H₄	DMF, POCl₃	R⁴ = CHO (60)	212
Ph	(naphthalene-1,8-dicarboxylic anhydride structure)	DMF, POCl₃	R⁴ = CHO (50)	212
	4-MeOC₆H₄ (naphthalimide–benzimidazole structure)	DMF, POCl₃	R⁴ = CHO (83)	212
C₂₂	(structure, Et–N–OAr; Ar = 4-(C₆H₁₁)C₆H₄)	1. — 2. CH₂(CN)₂	(structure) (—)	213
	R¹ = N(C₈H₁₇)₂	1. — 2. CH₂(CN)₂	R⁴ = CH=C(CN)₂ (—)	214

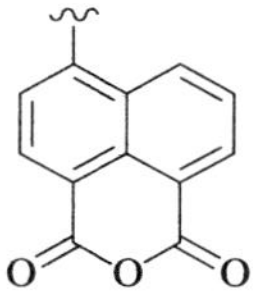

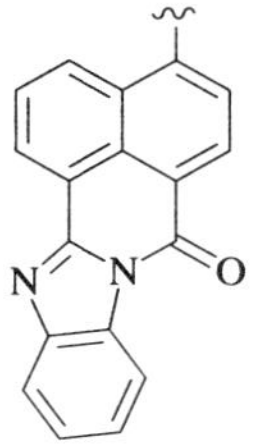

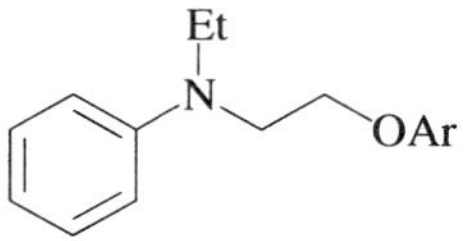

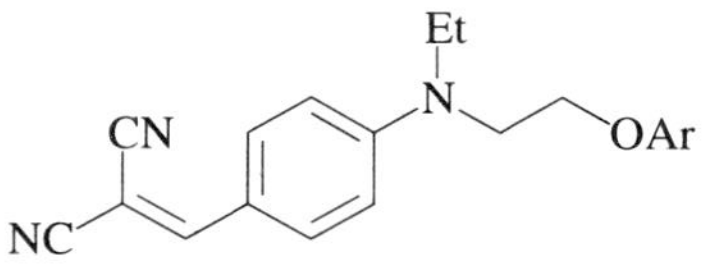

TABLE I. BENZENES (*Continued*)

Substrate	Reagents	Product(s) and Yield(s) (%)	Refs.
B. Disubstituted Benzenes			
B1. R^1, R^2 Substituents			
C_7			
$R^1 + R^2 = -OCH_2O-$	DMF, $POCl_3$	$R^4 = CHO$ (70)	215
C_8			
$R^1 + R^2 = -NH(CH_2)_2-$	—	$R^4 = CHO$ (36)	216
$R^1 = OMe, R^2 = Me$	DMF, $POCl_3$	$R^4 = CHO$ (70)	217, 186
$R^1 = R^2 = OMe$	DMF, $POCl_3$	$R^4 = CHO$ (30–40)	186
	DMF, $(Cl_2PO)_2O$	$R^4 = CHO$ (54)	26
	MFA, $(Cl_2PO)_2O$	$R^4 = CHO$ (83)	26
	MFA, $POCl_3$	$R^4 = CHO$ (38)	218, 219
C_9 $R^1 + R^2 = -N(Me)CONMe-$	DMF, $POCl_3$	$R^4 = CHO$ (47)	220
C_{10} $R^1 + R^2 = -N(Me)(CH_2)_3-$	DMF, $POCl_3$	$R^4 = CHO$ (90) + $R^4 = R^6 = CHO$ (4)	221
$R^1 = OH, R^2 = t$-Bu	MFA, $POCl_3$	$R^4 = CHO$ (18)	222
C_{11} $R^1 + R^2 = -OC(Me)_2(CH_2)_2-$	DMF, $POCl_3$	$R^4 = CHO$ (71)	223
$R^1 = NEt_2, R^2 = Me$	DMF, $POCl_3$	$R^4 = CHO$ (—)	191
C_{12}	DMF, $POCl_3$	(80)	224
C_{13} $R^1 = Ph, R^2 = OMe$	DMF, $POCl_3$	$R^5 = CHO$ (—)	186

52

TABLE I. BENZENES (*Continued*)

Substrate	Reagents	Product(s) and Yield(s) (%)	Refs.
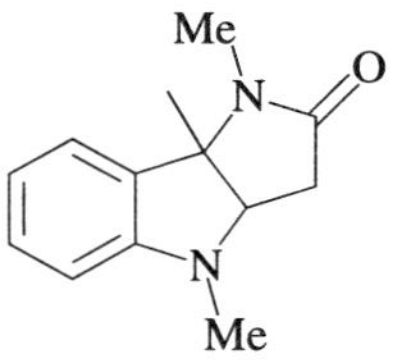 C$_{15}$	DMF, POCl$_3$	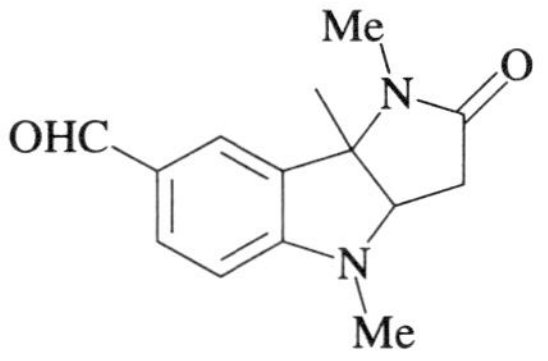 (80)	224
R^1 + R^2 = N(Bn)(CH$_2$)$_2$	DMF, POCl$_3$	R^4 = CHO (88)	225
C$_{15-19}$ 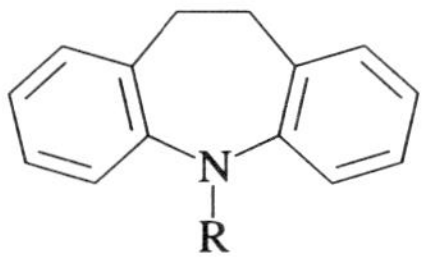	DMF, POCl$_3$	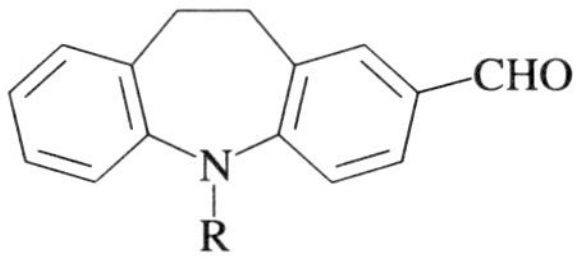	
$\dfrac{\text{R}}{\text{Me}}$		(—)	226
C$_{16}$ R^1 = (CH$_2$)$_3$NMe$_2$,	DMF, POCl$_3$	R^4 = CHO (—)	227
R^1 = NEt$_2$, R^2 = Ph			
C$_{19}$ R^1 = OMe, R^2 = SnBu$_3$	MFA, POCl$_3$	R^2 = CHO (79)	191
			53

TABLE I. BENZENES (Continued)

Substrate	Reagents	Product(s) and Yield(s) (%)	Refs.
C$_{20}$ 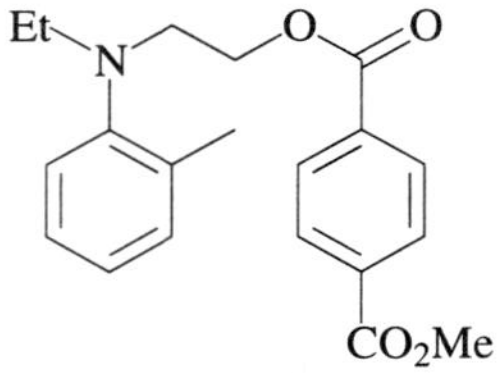 (structure: Et–N attached to 2-methylphenyl and to CH$_2$CH$_2$O–C(=O)–C$_6$H$_4$–CO$_2$Me)	1. DMF, POCl$_3$ 2. CH$_2$(CN)$_2$	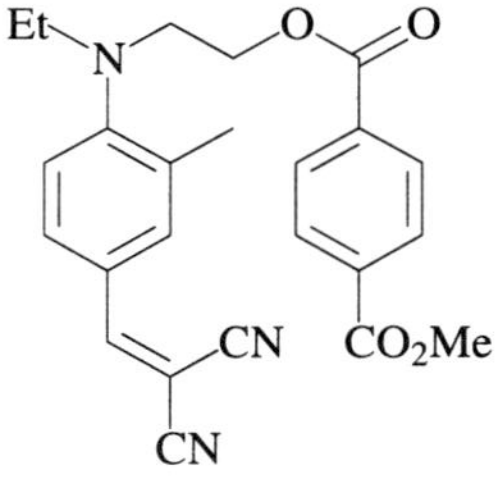 (structure with CO$_2$Me, CN, CN vinyl) (85)	228
C$_{21}$ 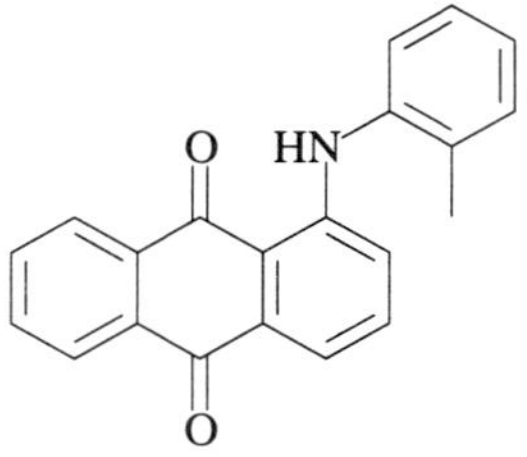 (anthraquinone with HN–(2-methylphenyl))	DMF, POCl$_3$	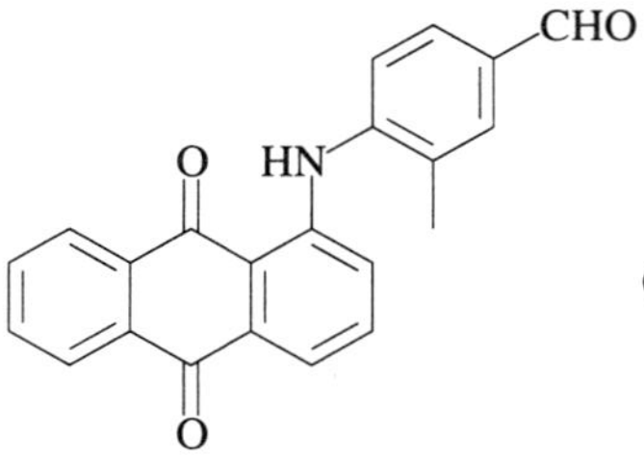 (anthraquinone with HN–aryl–CHO) (6) 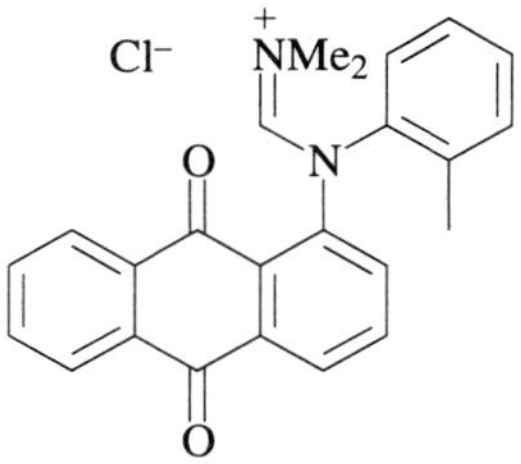 + (anthraquinone with N–aryl–NMe$_2$, Cl$^-$) (17)	211

TABLE I. BENZENES (*Continued*)

Substrate	Reagents	Product(s) and Yield(s) (%)	Refs.

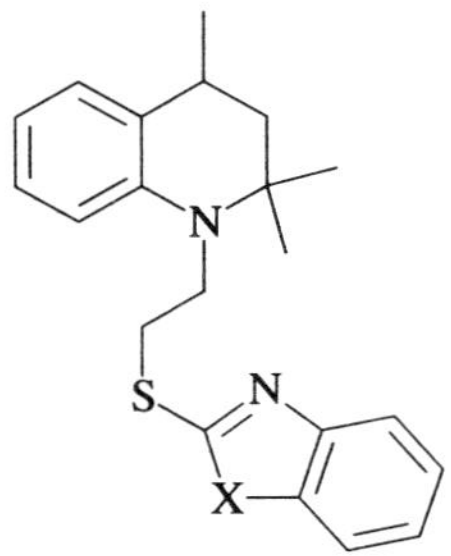

1. DMF, POCl₃ → $1.\ \text{DMF, POCl}_3$
2. $\text{CH}_2(\text{CN})_2$

229

$\dfrac{X}{S}$
O

(—)
(—)

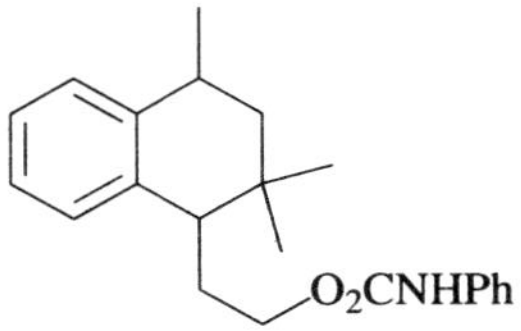

1. DMF, POCl_3
2. $\text{CH}_2(\text{CN})_2$

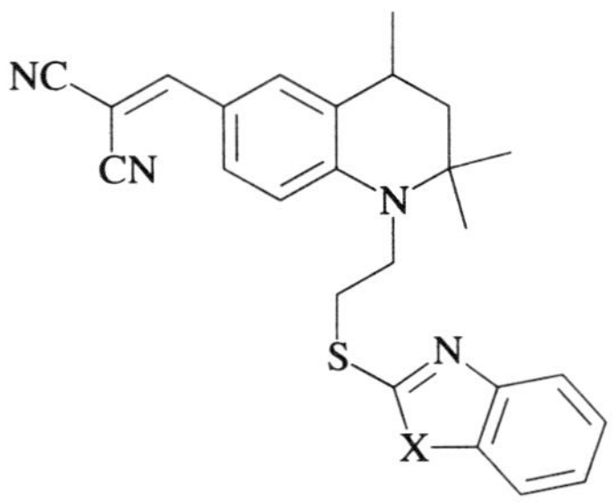

(83) 230

— 231

O_2CNHPh

OAc, OAc, Et, Me, OHC

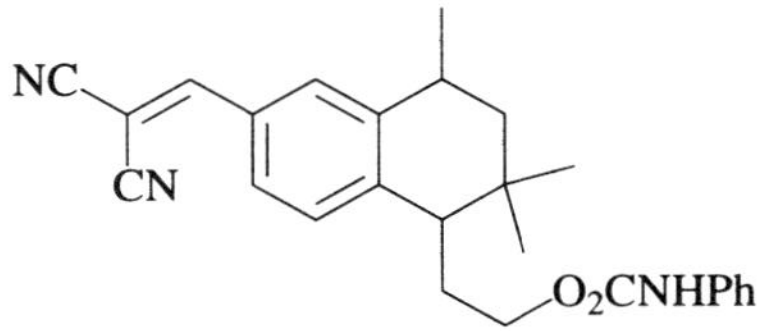

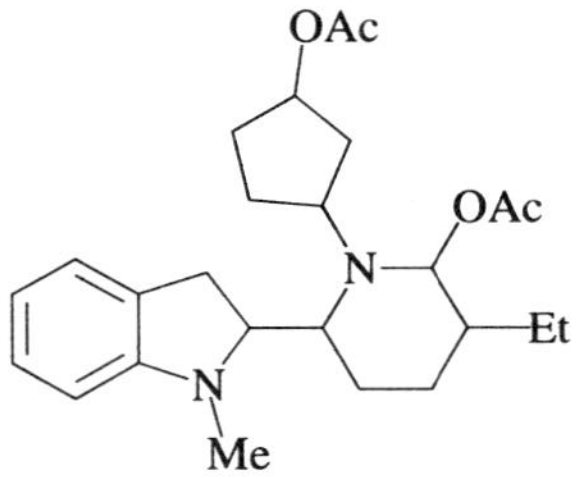

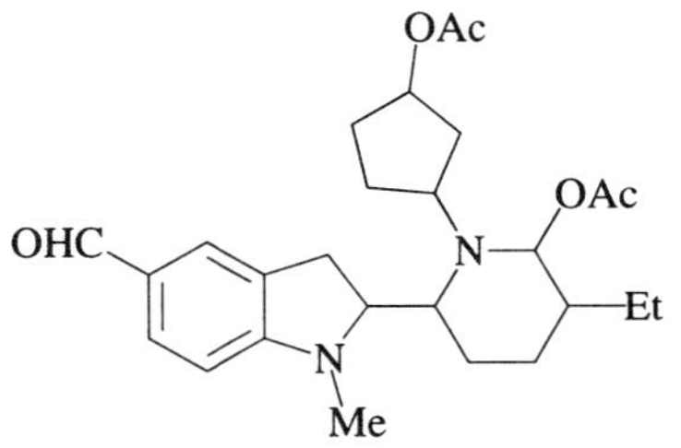

C₂₅

TABLE I. BENZENES (Continued)

Substrate	Reagents	Product(s) and Yield(s) (%)	Refs.
C$_{26}$ 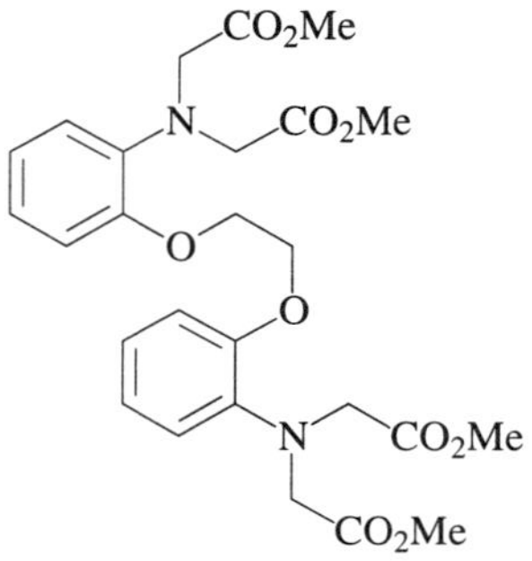	—	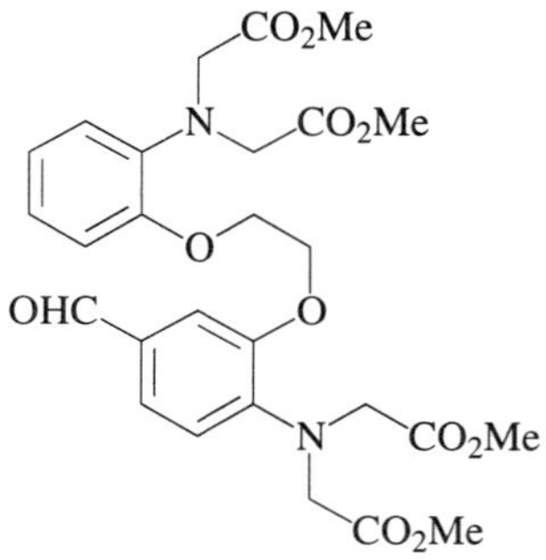 (—)	232

B 2. R^1, R^3 Substituents

Substrate	Reagents	Product(s) and Yield(s) (%)	Refs.
C$_6$ R^1 = R^3 = OH	DMF, POCl$_3$	R^4 = CHO (68)	52
	H$_2$NCHO, POCl$_3$	R^4 = CHO (75)	233
	DMF, (Cl$_2$PO)$_2$O	R^4 = CHO (88)	26
C$_7$ R^1 = OH, R^3 = Me	DMF, POCl$_3$	R^4 = CHO (—)	186
R^1 = OH, R^3 = OMe	DMF, POCl$_3$	R^4 = CHO (25)	51, 52
C$_8$ R^1 = OMe, R^3 = Me	DMF, POCl$_3$	R^4 = CHO (72)	217
R^1 = R^3 = OMe	DMF, POCl$_3$	R^4 = CHO (8998)	234, 52
	MFA, POCl$_3$	R^4 = CHO (85)	218, 50
	DMF, 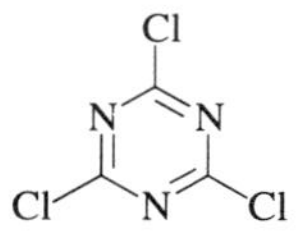	R^4 = CHO (36)	36

TABLE I. BENZENES (Continued)

Substrate	Reagents	Product(s) and Yield(s) (%)	Refs.
	1. DMF, POCl₃ 2. NH₂OH 3. Dehydrate	R⁴ = CN (43)	197, 235
	PhN(Me)CH=CHCHO, POCl₃	R⁴ = CH=CHCHO (90)	39
	DMF, SOCl₂	(18) + R⁴ = CHO (59)	42
C₈₋₁₂ R¹ = OR, R³ = Me $\frac{R}{\text{Me / Et / }n\text{-Pr / }n\text{-Bu / }n\text{-C}_5\text{H}_{11}}$	DMF, SOCl₂	R⁴ = CHO (42) (52) (44) (46) (37)	42
C₉ R¹ = NMe₂, R³ = Me	MFA, POCl₃	R⁴ = CHO (—)	2
C₁₀ R¹ = OCH₂CN, R³ = OEt	DMF, POCl₃	R⁴ = CHO (11) + (7)	236

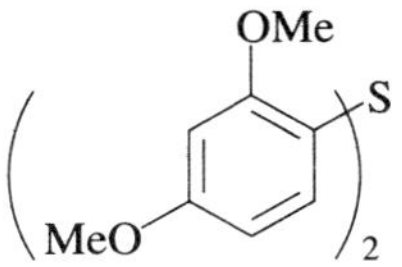

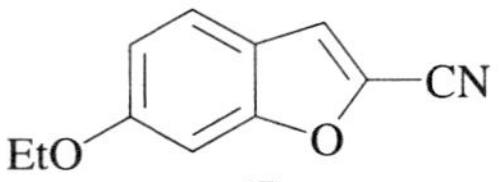

TABLE I. BENZENES (Continued)

Substrate	Reagents	Product(s) and Yield(s) (%)	Refs.
$R^1 = OMe$, $R^3 = CH_2CH=CH_2$	MFA, $POCl_3$	$R^4 = CHO$ (40) 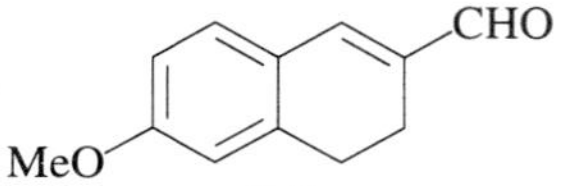 + [structure: CHO-substituted tetrahydronaphthalene with MeO] (11)	237
$R^1 = R^3 = OEt$	DMF, $SOCl_2$	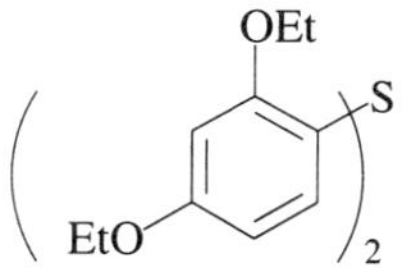 [structure with OEt, EtO, S]$_2$ (33) + $R^4 = CHO$ (20)	42
C_{11} $R^1 = OMe$, $R^3 = t$-Bu	MFA, $POCl_3$	$R^6 = CHO$ (62)	50
$R^1 = NEt_2$, $R^3 = Me$	DMF, $POCl_3$	$R^4 = CHO$ (—)	191
$R^1 = N(Et)(CH_2)_2OH$, $R^3 = Me$	1. DMF, $POCl_3$ 2. 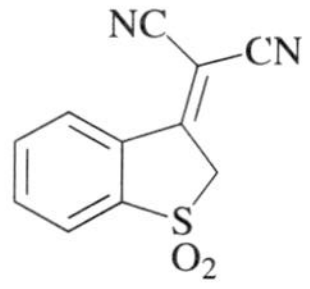 [structure NC, CN, S, O_2]	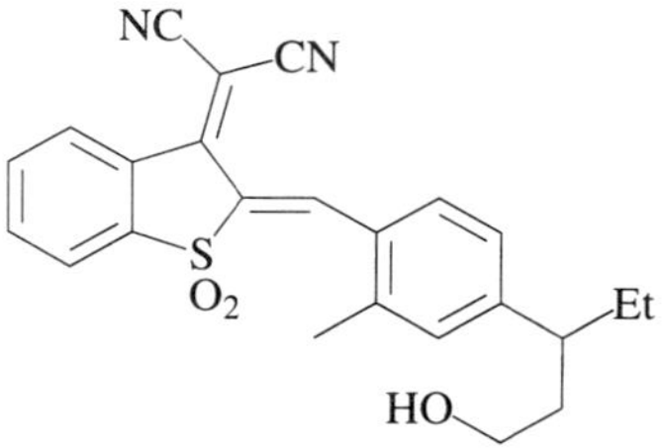 [structure with NC, CN, Et, HO, O_2] (—)	238
C_{12} $R^1 = NEt_2$, $R^3 = OCH_2CN$	DMF, $POCl_3$	[benzofuran structure with Et_2N, CN] (4)	236

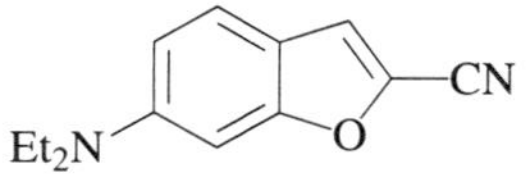

58

TABLE I. BENZENES (*Continued*)

Substrate	Reagents	Product(s) and Yield(s) (%)	Refs.
C$_{13}$			
R^1 = NEt$_2$, R^3 = OTMS	1. DMF, POCl$_3$ 2. hydrolysis 3. 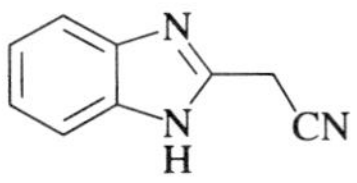	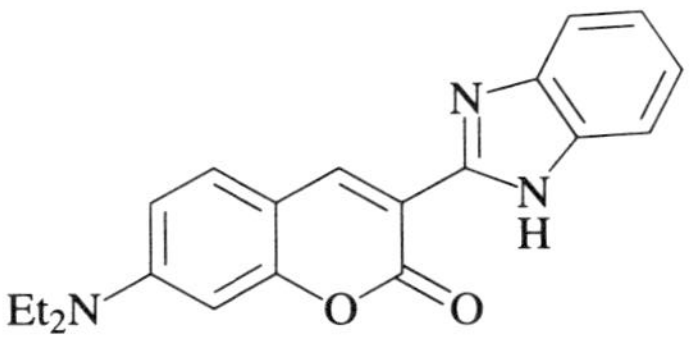 (—)	239
C$_{13\text{-}16}$			
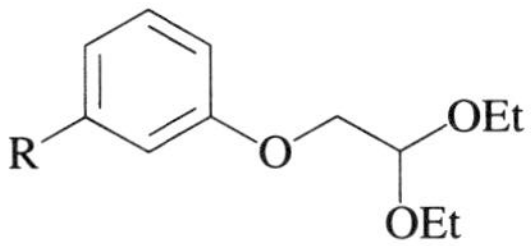	DMF, POCl$_3$	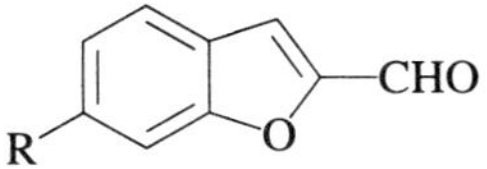	240
$\dfrac{R}{\begin{array}{l}\text{OMe}\\\text{OEt}\\\text{NEt}_2\end{array}}$		(27) (29) (33)	
C$_{14}$			
R^1 = R^3 = NEt$_2$ R^1 = N(Me)(CH$_2$)$_2$NEt$_2$, R^3 = Me	DMF, POCl$_3$ —	An aldehyde (—) R^4 = CHO (—)	191 203
C$_{15\text{-}18}$			
R^1 = R, R^3 = OCH$_2$COPh R = OMe	DMF, POCl$_3$	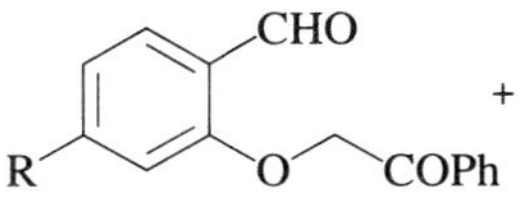 (8) + (19) + (25)	43

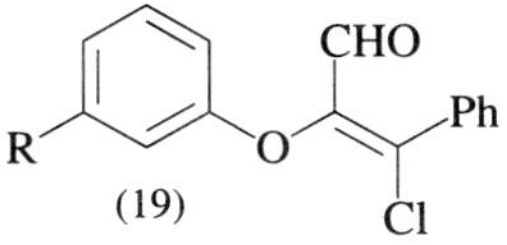

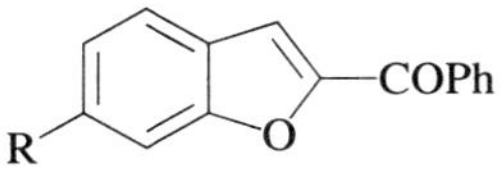

59

TABLE I. BENZENES (Continued)

Substrate	Reagents	Product(s) and Yield(s) (%)	Refs.
R = OEt		I (31)	
R = NEt$_2$		I (78)	
C$_{18}$			
R^1 = N(Et)(CH$_2$)$_2$O$_2$CPh, R^3 = Me	DMF, POCl$_3$	R^4 = CHO (—)	241
C$_{18-21}$			
R^1 = N(Et)(CH$_2$)$_2$R, R^3 = Me	DMF, POCl$_3$	R^4 = CHO	242
R			
SCH$_2$C$_6$H$_3$Cl$_2$-2,5		(—)	
SCH(OAc)CH$_2$C$_6$H$_3$Cl$_2$-2,5		(—)	
C$_{18}$			
C$_{19}$			
R^1 = Me, R^3 = SnBu$_3$	MFA, POCl$_3$	R^3 = CHO (55)	53
C$_{20}$	1. DMF, POCl$_3$ 2. CH$_2$(CN)$_2$	(85)	243
C$_{20-22}$			
R^1 = N(Bn)$_2$, R^3 = R	DMF, POCl$_3$	R^4 = CHO	244

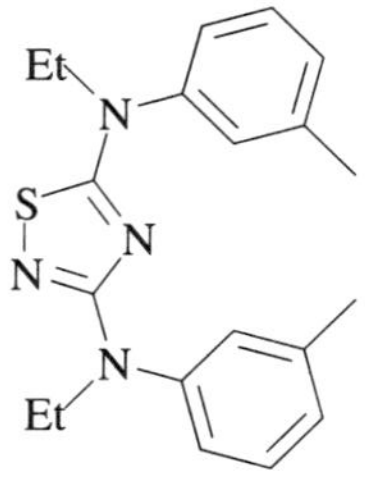

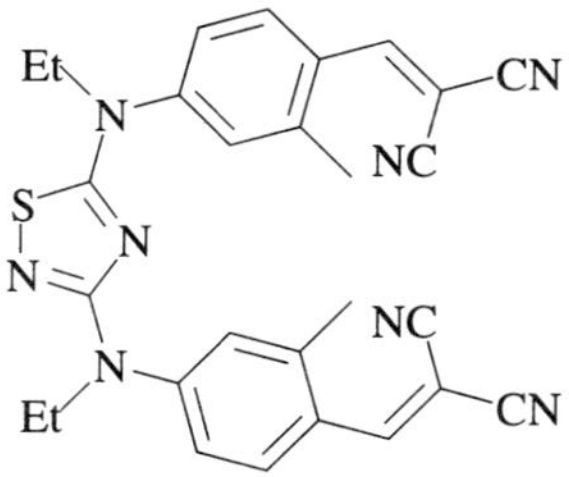

R	
F	(93)
Cl	(83)
Me	(81)
Et	(80)

TABLE I. BENZENES (*Continued*)

Substrate	Reagents	Product(s) and Yield(s) (%)	Refs.
C$_{21}$ $\dfrac{R}{Me}$ OMe	DMF, POCl$_3$	**I** + **II** **I** (4) + **II** (84) **I** (13) + **II** (83)	211
C$_{24}$ R^1 = N[(CH$_2$)$_2$SAr][(CH$_2$)$_3$SAr], R^3 = Me $\dfrac{Ar}{\text{2,5-Cl}_2\text{C}_6\text{H}_3}$ 4-ClC$_6$H$_4$	DMF, POCl$_3$	R^4 = CHO (—) (—)	242

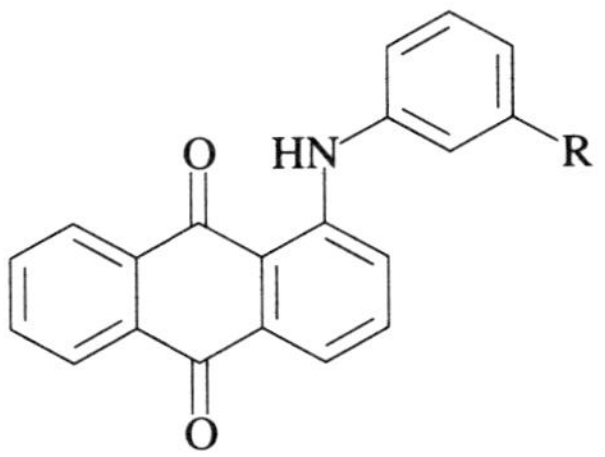

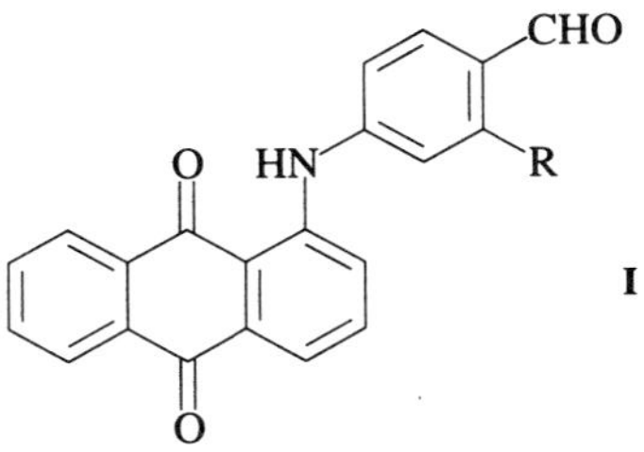

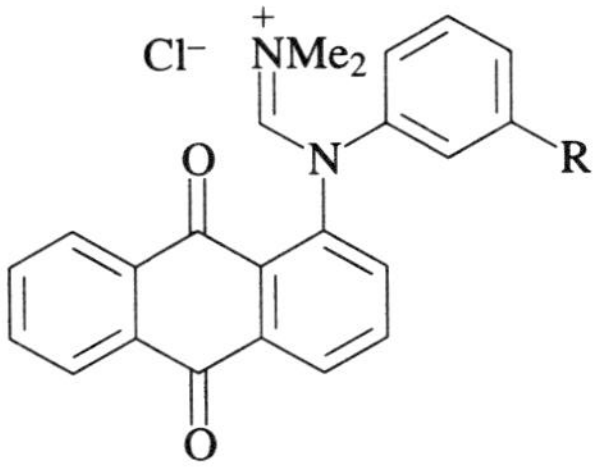

61

TABLE I. BENZENES (*Continued*)

Substrate	Reagents	Product(s) and Yield(s) (%)	Refs.
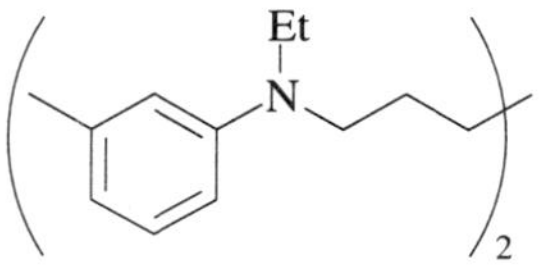 ![Et–N, methyl-substituted aniline structure]	—	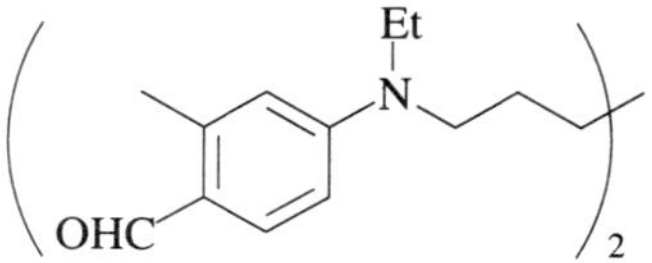 ![OHC, methyl, Et–N product structure] (—)	245

B3. R¹, R⁴ Substituents

Substrate	Reagents	Product(s) and Yield(s) (%)	Refs.
C$_7$ R^1 = OH, R^4 = Me	DMF, POCl$_3$	R^2 = CHO (—)	186
C$_8$ R^1 = R^4 = OMe	MFA, POCl$_3$	R^2 = CHO (16)	218
	DMF, (Cl$_2$PO)$_2$O	R^2 = CHO (40)	26, 188
C$_{10}$ R^1 = OMe, R^4 = CH$_2$CH=CH$_2$	MFA, POCl$_3$	R^2 = CHO (16)	237
R^1 = NMe$_2$, R^4 = COMe	(Me$_2$N=CHCl)$^+$ Cl$^-$	R^2 = CHO, R^4 = C(Cl=C(CHO)$_2$ (25) + R^4 = CCl=C(CHO)$_2$ (23)	246
C$_{13}$ R^1 = N[(CH$_2$)$_2$CN]$_2$, R^4 = Me	DMF, POCl$_3$	R^2 = CHO (95)	55
C$_{15}$ R^1 = NMe$_2$, R^4 = C((NMe$_2$)=CHCH=NMe$_2$$^+$ ClO$_4$$^-$	1. DMF, POCl$_3$ 2. NaOH	R^2 = CHO, R^4 = COC(CHO)=CHOH (73)	246

62

TABLE I. BENZENES (Continued)

Substrate	Reagents	Product(s) and Yield(s) (%)	Refs.

C$_{15-19}$

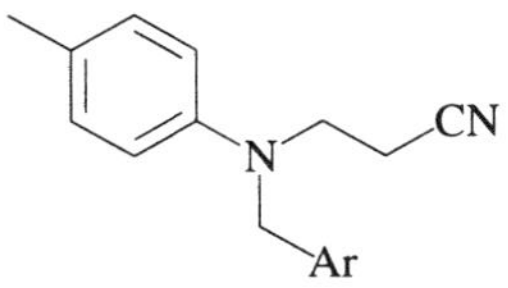

Reagents: ArCHO (I) +

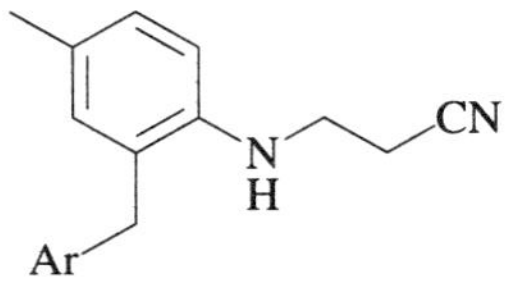

Products II and III

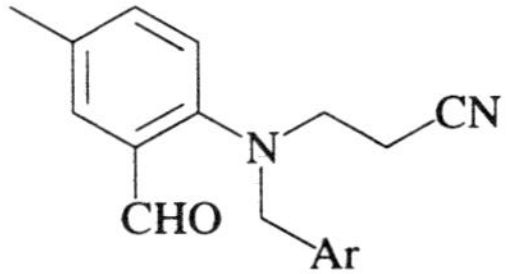

Ar	I	II	III	Refs.
2,3,4,5,6-F$_5$C$_6$	(0)	(0)	(67)	54
2,4-F$_2$C$_6$H$_3$	(30)	(39)	(43)	54
3,5-F$_2$C$_6$H$_3$	(32)	(39)	(35)	54
2-Thienyl	(47)	(59)	(0)	54
3-FC$_6$H$_4$	(40)	(49)	(13)	54
4-FC$_6$H$_4$	(71)	(82)	(0)	54
4-ClC$_6$H$_4$	(80)	(67)	(0)	55
3-BrC$_6$H$_4$	(71)	(58)	(0)	54
4-BrC$_6$H$_4$	(82)	(62)	(0)	54
4-O$_2$NC$_6$H$_4$	(30)	(0)	(0)	54
Ph	(94)	(71)	(0)	55
3-MeC$_6$H$_4$	(92)	(79)	(0)	54
4-MeC$_6$H$_4$	(88)	(76)	(0)	54
3-MeOC$_6$H$_4$	(89)	(80)	(0)	54
4-MeOC$_6$H$_4$	(76)	(74)	(0)	54
Cinnamyl	(40)	(24)	(0)	54

TABLE I. BENZENES (Continued)

Substrate	Reagents	Product(s) and Yield(s) (%)	Refs.
C_{16}	DMF, POCl$_3$	(40) + (35)	247
C_{18} R^1 = Cl, R^4 = SnBu$_3$	MFA, POCl$_3$	R^4 = CHO (10)	53
R^1 = N[(CH$_2$)$_2$CN]CH$_2$Ar, R^4 = Me	1. DMF, POCl$_3$ 2. NaBH$_4$	$\underline{Ar}$ 4-MeC$_6$H$_4$ R^2 = CH$_2$NMe$_2$ (39); 4-MeOC$_6$H$_4$ R^2 = CH$_2$NMe$_2$ (76)	54
C_{19} R^1 = Me, R^4 = SnBu$_3$	MFA, POCl$_3$	R^4 = CHO (70)	53
R^1 = OMe, R^4 = SnBu$_3$	MFA, POCl$_3$	R^4 = CHO (96)	53
C_{19-22} R^1 = N(Bn)CH$_2$Ar, R^4 = Me	DMF, POCl$_3$	PhCHO (I) + ArCHO (II)	54

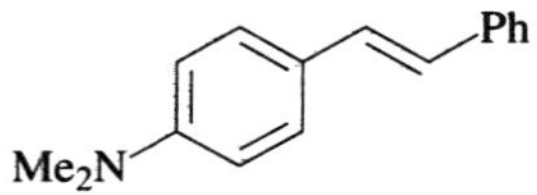

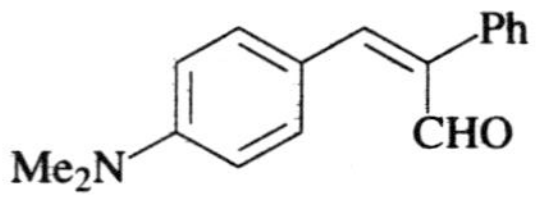

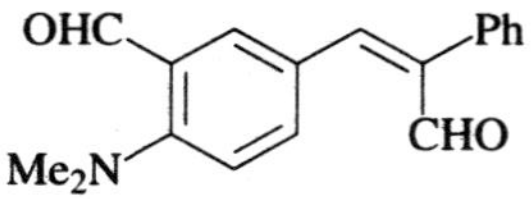

$\underline{Ar}$	$\underline{I:II}$
2-Thienyl	1:1.7
2,3,4,5,6-F$_5$C$_6$	1:0
4-FC$_6$H$_4$	1.3:1
4-ClC$_6$H$_4$	1.6:1
4-BrC$_6$H$_4$	1.6:1
4-MeOC$_6$H$_4$	1:8.5

64

TABLE I. BENZENES (Continued)

Substrate	Reagents	Product(s) and Yield(s) (%)	Refs.
C_{32}	DMF, POCl$_3$	(67)	248
	—	(75)	248

C. Trisubstituted Benzenes
C1. R^1, R^2, R^3 Substituents

Substrate	Reagents	Product(s) and Yield(s) (%)	Refs.
C_8 $R^1 = OMe$, $R^2 + R^3 = OCH_2O$	—	$R^6 = CHO$ (50)	250
	DMF, POCl$_3$, KI	$R^4 = CHO$ (21) + $R^6 = CHO$ (59)	251
	DMF, POCl$_3$	$R^4 = CHO$ (30) + $R^6 = CHO$ (53)	251
	DMF, POCl$_3$	$R^4 = CHO + R^6 = CHO$ (—) 70:30	252
	MFA, POCl$_3$	$R^6 = CHO$ (63)	253, 254
	MFA, POCl$_3$	$R^4 = CHO$ (10) + $R^6 = CHO$ (34)	255
C_9 $R^1 = i\text{-}Pr$, $R^2 = R^3 = OH$	DMF, POCl$_3$	$R^4 = CHO$ (50)	256
$R^1 = R^3 = OMe$, $R^2 = Me$	DMF, POCl$_3$	$R^4 = CHO$ (83)	234, 257
$R^1 = R^2 = OMe$	DMF, POCl$_3$	$R^4 = CHO$ (70–80)	234
C_{10} $R^1 + R^2 = OCH_2O$, $R^3 = i\text{-}Pr$	—	$R^5 = CHO$ (—)	258

TABLE I. BENZENES (*Continued*)

Substrate	Reagents	Product(s) and Yield(s) (%)	Refs.
C_{11}			
$R^1 = R^2 = OMe$, $R^3 = CH_2CH=CH_2$	MFA, POCl$_3$	$R^5 = CHO$ (10) 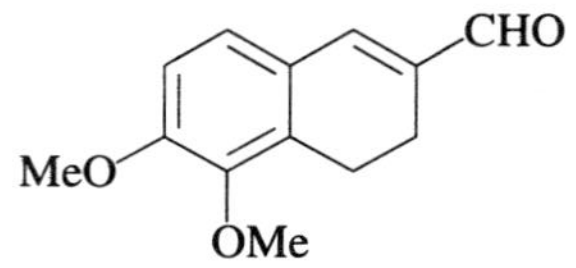 + (8)	237
$R^1 + R^2 = (CH_2)_4$, $R^3 = OMe$	MFA, POCl$_3$	$R^6 = CHO$ (—)	189
$R^1 + R^2 = N(Et)(CH_2)_2$, $R^3 = Me$	DMF, POCl$_3$	$R^4 = CHO$ (65)	225
$R^1 + R^2 = N(Et)(CH_2)_2$, $R^3 = Me$	1. DMF, POCl$_3$ 2. EtNO$_2$ 3. LiAlH$_4$ 4. Pd, H$_2$	$R^4 = CH_2CH(Me)NH_2$ (65)	225
C_{12}			
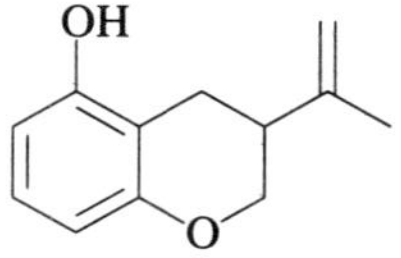	—	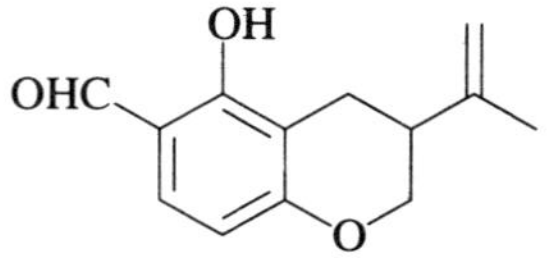 (—)	259
C_{13}			
$R^1 + R^2 = CH(Me)(CH_2)_2CH(Me)$, $R^3 = OMe$	DMF, POCl$_3$	$R^6 = CHO$ (60)	260
C_{15}			
$R^1 + R^2 = N(Bn)(CH_2)_2$, $R^3 = Cl$	DMF, POCl$_3$	$R^4 = CHO$ (98)	225

66

TABLE I. BENZENES (Continued)

Substrate	Reagents	Product(s) and Yield(s) (%)	Refs.
C2. R^1, R^2, R^4 Substituents			
C_7			
R^1 + R^2 = OCH$_2$O, R^4 = OH	—	R^5 = CHO (—)	261
C_8			
R^1 + R^2 = OCH$_2$O, R^4 = SeMe	MFA, POCl$_3$	R^5 = CHO (65)	262
C_{8-11}			
R^1 + R^2 = OCH$_2$O, R^4 = SR		R^5 = CHO	262

$\dfrac{\text{R}}{\text{Me}}$			
Me	DMF, POCl$_3$	(75)	
Me	MFA, POCl$_3$	(86)	
	4-ClC$_6$H$_4$N(Me)CHO, POCl$_3$	(70)	
	4-O$_2$NC$_6$H$_4$N(Me)CHO, POCl$_3$	(50)	
	4-MeC$_6$H$_4$N(Me)CHO, POCl$_3$	(91)	
	4-MeOC$_6$H$_4$N(Me)CHO, POCl$_3$	(93)	
	4-i-PrC$_6$H$_4$N(Me)CHO, POCl$_3$	(90)	
	1,3,5-Me$_3$C$_6$H$_2$N(Me)CHO, POCl$_3$	(78)	
	4-t-BuC$_6$H$_4$N(Me)CHO, POCl$_3$	(89)	
Et	MFA, POCl$_3$	(80)	
n-Pr	MFA, POCl$_3$	(81)	
i-Pr	MFA, POCl$_3$	(79)	
n-Bu	MFA, POCl$_3$	(80)	
i-Bu	MFA, POCl$_3$	(80)	
s-Bu	MFA, POCl$_3$	(79)	
C_9			
R^1 + R^2 = OCH$_2$CH$_2$, R^4 = OMe	MFA, POCl$_3$	R^5 = CHO (56)	263
R^1 = R^4 = Me, R^2 = OMe	DMF, POCl$_3$	R^5 = CHO (80)	217, 234
R^1 = Me, R^2 = R^4 = OMe	DMF, POCl$_3$	R^5 = CHO (87)	234, 264
R^1 = R^2 = OMe, R^4 = Me	DMF, POCl$_3$	R^5 = CHO (35-40)	186
R^1 = R^2 = R^4 = OMe	DMF, POCl$_3$	R^5 = CHO (88)	234
R^1 = R^2 = OMe, R^4 = SMe	MFA, POCl$_3$	R^5 = CHC (85)	48

TABLE I. BENZENES (*Continued*)

Substrate	Reagents	Product(s) and Yield(s) (%)	Refs.
$R^1 = R^4 = OMe$, $R^2 = SMe$	MFA, POCl$_3$	$R^5 = CHO$ (53)	48, 265
$R^1 = SMe$, $R^2 = R^4 = OMe$	MFA, POCl$_3$	$R^5 = CHO$ (96)	48
$R^1 + R^2 = O(CH_2)O$, $R^4 = OMe$	DMF, POCl$_3$	$R^5 = CHO$ (77)	266, 267
C$_{10}$			
$R^1 = R^4 = OMe$, $R^2 = NMe_2$	DMF, POCl$_3$	$R^5 = CHO$ (47)	204
$R^1 + R^2 = OCH(Me)CH_2$, $R^4 = OMe$	MFA, POCl$_3$	$R^5 = CHO$ (78)	263
C$_{11}$			
$R^1 = R^2 = OMe$, $R^4 = CH_2CH=CH_2$	MFA, POCl$_3$, 8 h	$R^5 = CHO$ (5) + **I** (3) + **II** (15)	237
$R^1 = CH_2CH=CH_2$, $R^2 = R^4 = OMe$	MFA, POCl$_3$, 72 h	**I** (8) + **II** (39)	237
$R^1 + R^2 = (CH_2)_4$, $R^4 = OMe$	MFA, POCl$_3$	$R^5 = CHO$ (63)	237
	DMF, POCl$_3$	$R^5 = CHO$ (—)	186
	MFA, POCl$_3$	$R^5 = CHO$ (—)	189
$R^1 + R^2 = (CH_2)_2N(Et)$, $R^4 = Me$	DMF, POCl$_3$	$R^5 = CHO$ (51)	225
$R^1 = Me$, $R^2 = OMe$, $R^4 = i\text{-}Pr$	DMF, POCl$_3$	$R^5 = CHO$ (76)	186
$R^1 = i\text{-}Pr$, $R^2 = R^4 = OMe$	DMF, POCl$_3$	$R^5 = CHO$ (76)	264
C$_{11\text{-}12}$	MeCONH$_2$, POCl$_3$	**I** + **II**	45

TABLE I. BENZENES (*Continued*)

Substrate	Reagents	Product(s) and Yield(s) (%)	Refs.
$\dfrac{R}{Me}$ Et		**I** (9) **I** (7) + **II** (8)	
C$_{12}$			
R^1 + R^2 = (CH$_2$)$_2$C(Me)$_2$O, R^4 = OMe	DMF, POCl$_3$	R^5 = CHO (79)	268
R^1 + R^2 = CH=CHC(Me)$_2$, R^4 = OMe	DMF, POCl$_3$	R^5 = CHO (89)	269
R^1 = R^4 = OMe, R^2 = (CH$_2$)$_2$CO$_2$Me	—	R^5 = CHO (85)	270
R^1 + R^2 = (CH$_2$)$_4$, R^4 = OEt	MFA, POCl$_3$	R^3 = CHO (—)	189
C$_{12-14}$ $\dfrac{R}{Me}$ Et Pr	DMF, POCl$_3$	 (78) (80) (78)	247
C$_{13}$			
R^1 = R^2 = OMe, R^4 = (CH$_2$)$_2$N(Me)COCF$_3$	DMF, POCl$_3$	R^5 = CHO (45)	271
R^1 + R^2 = OCH$_2$O, R^4 = OCH$_2$CH(OEt)$_2$	DMF, POCl$_3$	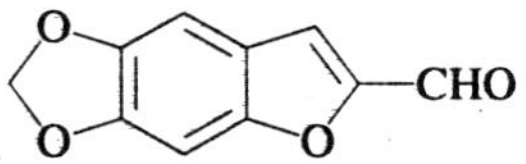 CHO (15)	240
R^1 + R^2 = CH(Me)(CH$_2$)$_2$CH(Me), R^4 = OMe	DMF, POCl$_3$	R^5 = CHO (69)	272

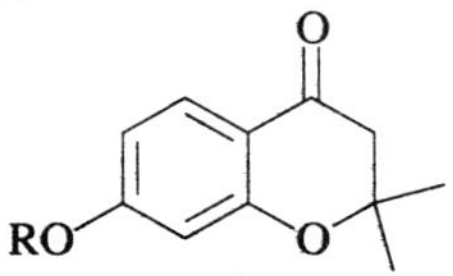

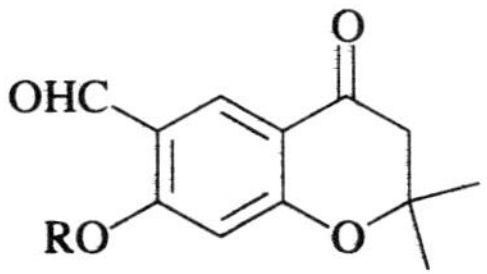

TABLE I. BENZENES (*Continued*)

Substrate	Reagents	Product(s) and Yield(s) (%)	Refs.
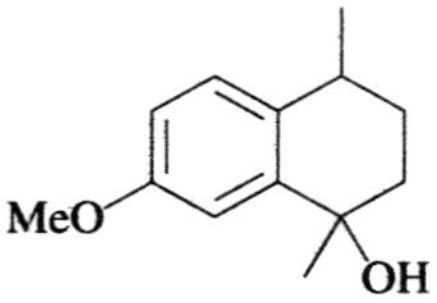 (MeO-tetralin with OH, Me substituents)	DMF, POCl$_3$	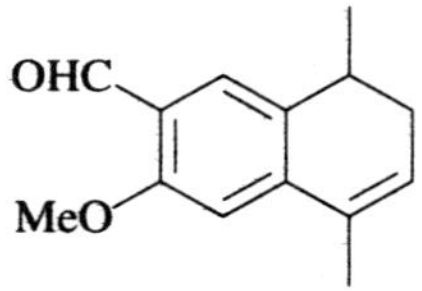 (28) 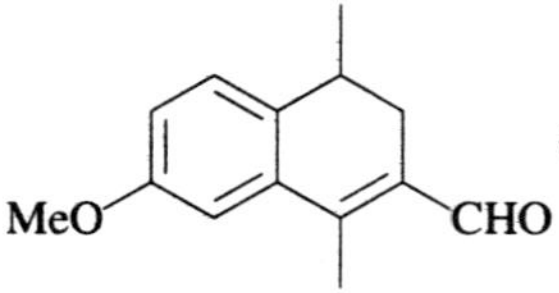 + (28) structures (OHC..., MeO...; MeO..., CHO)	260
R^1 = i-C$_5$H$_{11}$, R^2 = R^4 = OMe	DMF, POCl$_3$	R^5 = CHO (84)	268
R^1 + R^2 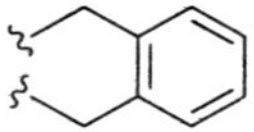 = (indane); R^4 = OMe	MFA, POCl$_3$	R^5 = CHO (82)	50
C$_{16}$			
R^1 = COCH$_2$C$_6$H$_4$Me-4, R^2 = R^4 = OMe	DMF, POCl$_3$	R^5 = CHO (60)	273
R^1 = CH=CHPh, R^2 = R^4 = OMe	DMF, POCl$_3$	R^5 = CHO (39)	209
R^1 + R^2 = (CH$_2$)$_2$N(Bn), R^4 = Me	1. DMF, POCl$_3$ 2. MeNO$_2$ 3. LiAlH$_4$ 4. H$_2$, Pd	R^5 = (CH$_2$)$_2$NH$_2$ (70)	255

TABLE I. BENZENES (*Continued*)

Substrate	Reagents	Product(s) and Yield(s) (%)	Refs.

C$_{17}$

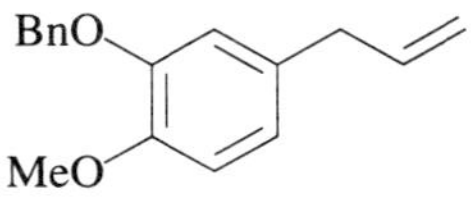

MFA, POCl$_3$

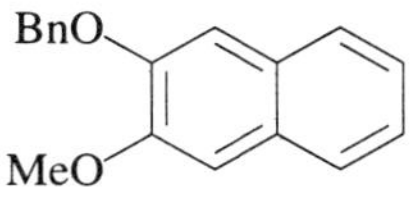

BnO, MeO — allyl substituted benzene

BnO, MeO — naphthalene (3) +

HO, MeO — naphthalene-CHO (24) +

BnO, MeO — naphthalene-CHO (9)

237

C$_{20}$

MeO, MeO — isoquinoline with OMe, MeO substituted benzyl group

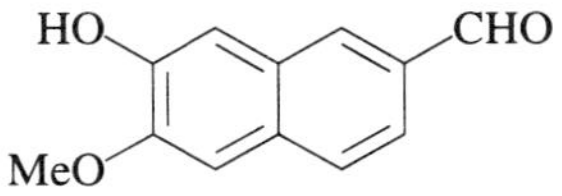

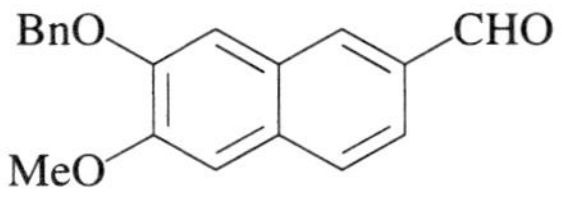

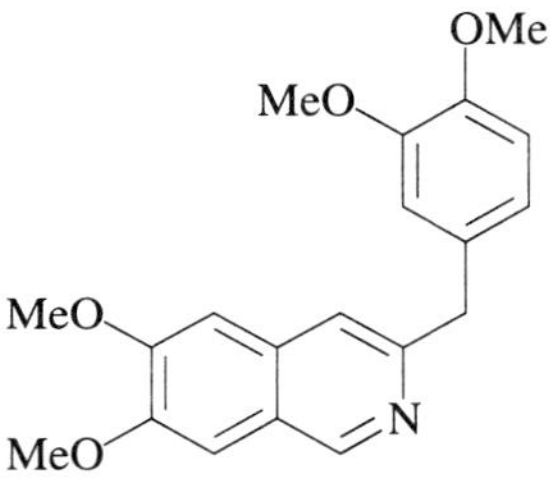

1. DMF, POCl$_3$
2. ClO$_4^-$

MeO, MeO / OMe, OMe — fused ring system N$^+$ ClO$_4^-$ (—)

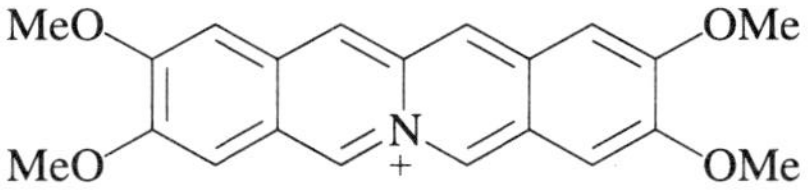

46

71

TABLE I. BENZENES (Continued)

Substrate	Reagents	Product(s) and Yield(s) (%)	Refs.
C$_{21-31}$ $\dfrac{R}{H}$	MFA, POCl$_3$	(82) (79)	50
C$_{22}$ $\dfrac{R}{Me}$ OMe	DMF, POCl$_3$	**I** + **II** **I** (37) + **II** (8) **I** (100)	211
C$_{24}$ — R = OMe, R^4 = Me, R^2 = N(CH$_2$CO$_2$C$_6$H$_{13}$)$_2$	—	R^5 = CHO (—)	49

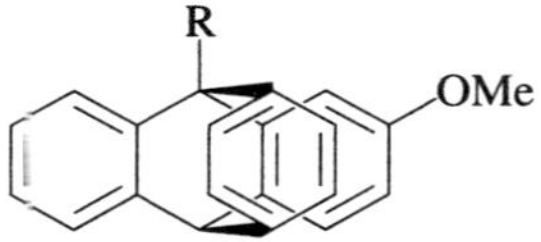

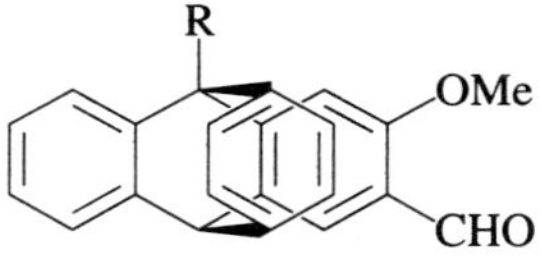

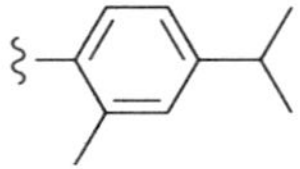

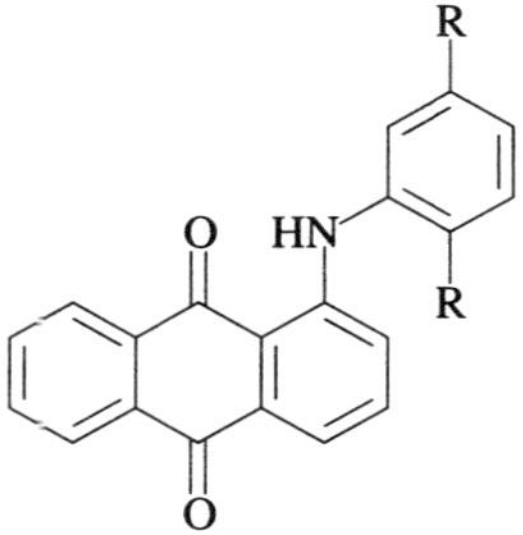

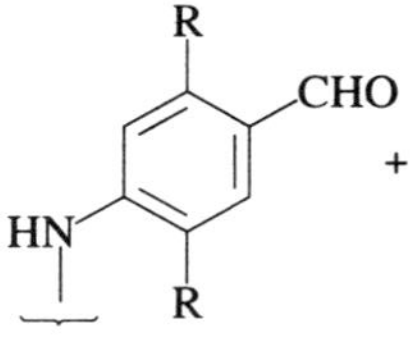

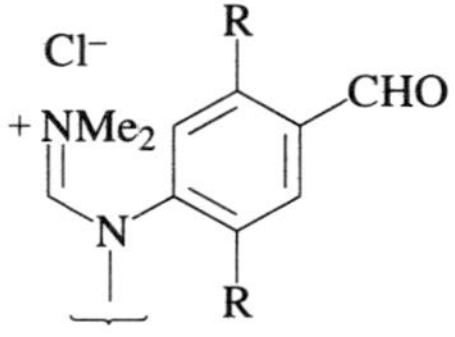

72

TABLE I. BENZENES (*Continued*)

Substrate	Reagents	Product(s) and Yield(s) (%)	Refs.
C3. R¹, R³, R⁵ Substituents			
C_8			
$R^1 = Br, R^3 = R^5 = OMe$	DMF, POCl₃	$R^2 = CHO$ (84)	274
$R^1 = OH, R^3 = R^5 = OMe$	DMF, POCl₃	$R^2 = CHO$ (35) + $R^4 = CHO$ (45)	275
	DMF, POCl₃	$R^2 = CHO$ (11) + $R^4 = CHO$ (52) + $R^6 = CHO$ (1)	276
C_9			
$R^1 = R^3 = R^5 = Me$	DMF, (CF₃SO₂)₂O	$R^2 = CHO$ (60)	47
$R^1 = R^3 = Me, R^5 = OMe$	DMF, POCl₃	$R^2 = CHO$ (70)	217
$R^1 = Me, R^3 = R^5 = OMe$	DMF, POCl₃	$R^2 = CHO$ (84)	234
$R^1 = R^3 = R^5 = OMe$	DMF, POCl₃	$R^2 = CHO$ (94)	274, 186
C_{10}			
$R^1 = R^3 = OMe, R^5 = OCH_2CN$	DMF, POCl₃	$R^2 = CHO$ (7-13) + $R^6 = CHO$ (29-36) 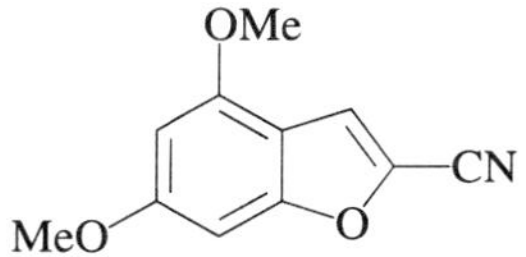 + (0-61)	236
C_{10-16}			
$R^1 = NMe_2, R^3 = R^5 = Me$	DMF, POCl₃ (2 eq)	$R^4 = CHO$ (69) + $R^2 = R^4 = CHO$ (—)	33
	DMF, POCl₃ (>2 eq)	$R^2 = R^4 = CHO$ (81)	33
	DMF, POCl₃	(44)	44

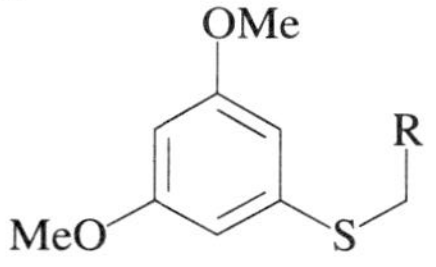

73

TABLE I. BENZENES (Continued)

Substrate	Reagents	Product(s) and Yield(s) (%)	Refs.

R / CN

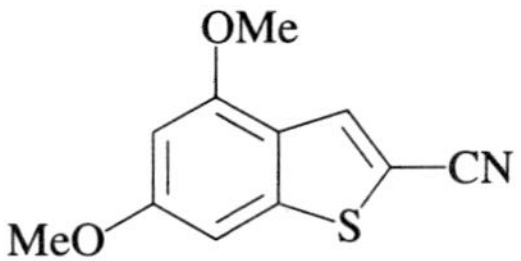

(12) + (3)

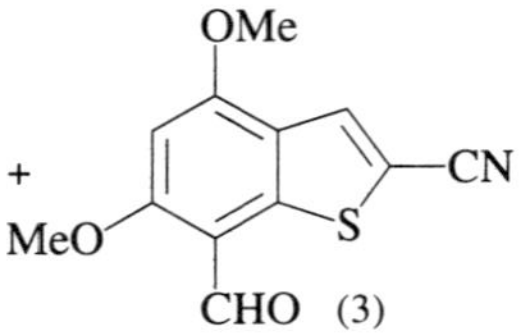

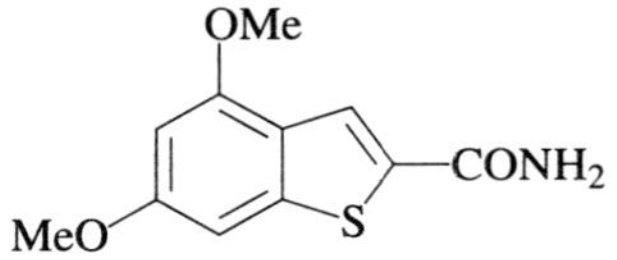

(22)

(56)

CO₂H

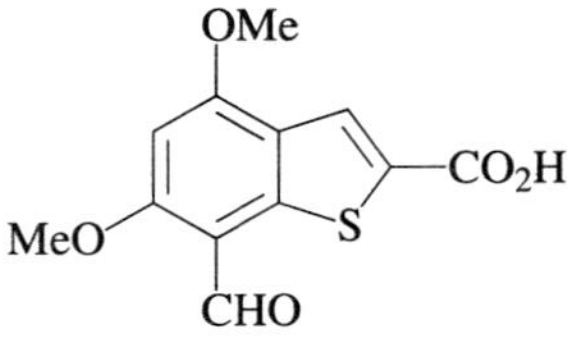

(35)

CO₂Et

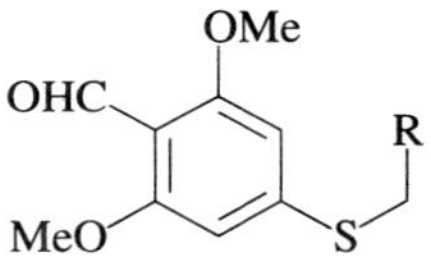

(25)

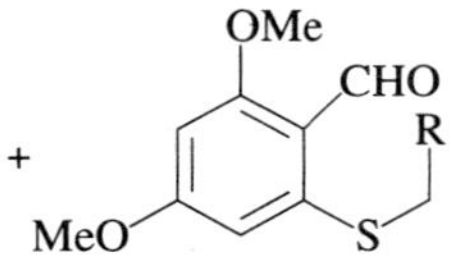

74

TABLE I. BENZENES (*Continued*)

Substrate	Reagents	Product(s) and Yield(s) (%)	Refs.

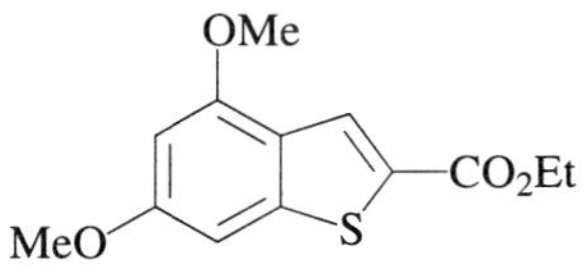

COPh

Products: benzothiophene structures with OMe, MeO, CO₂Et substituents — (4) + (16); Bz-substituted benzothiophene with OMe, MeO, OHC — (2), (52)

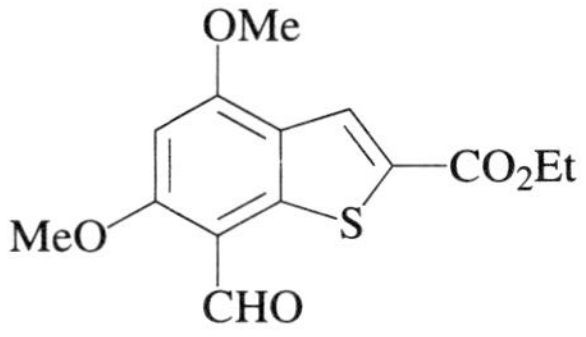

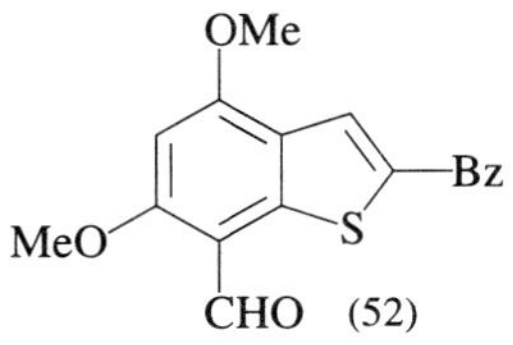

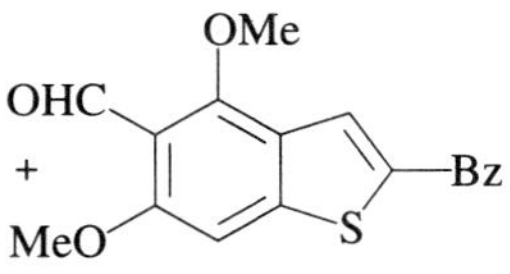

CH(OEt)₂

Products: (12) + (18)

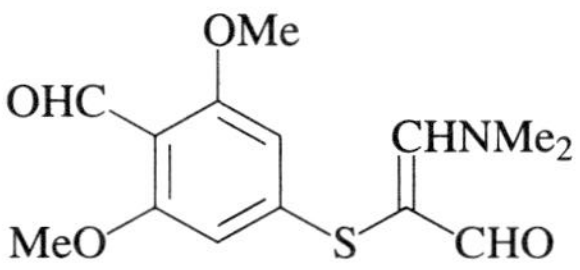

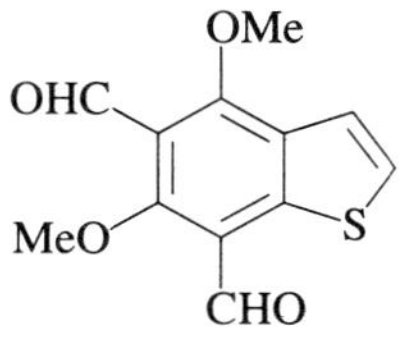

75

TABLE I. BENZENES (Continued)

Substrate	Reagents	Product(s) and Yield(s) (%)	Refs.
C_{12} $R^1 = CH_2CO_2Et$, $R^3 = R^5 = OMe$	—	$R^2 = CHO$ (—)	277
C_{14} $R^1 = OH$, $R^3 = R^5 = N(CH_2CH_2)_2O$	DMF, POCl₃	$R^2 = CHO$ (—)	278
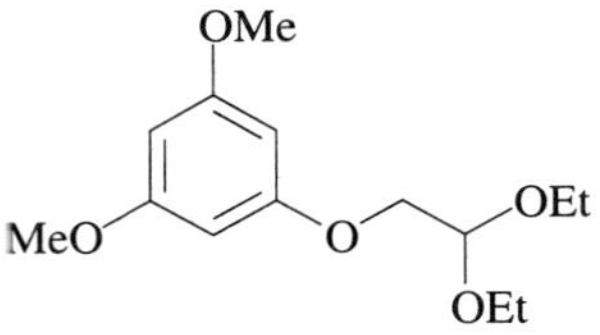 C_{14} (structure: MeO, OMe substituted benzene with OCH(OEt)₂ chain)	DMF, POCl₃	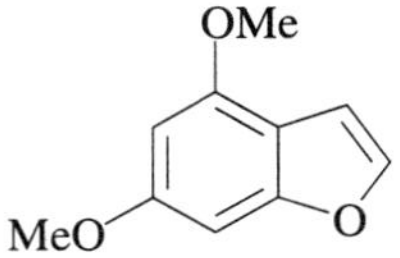 (benzofuran: MeO, OMe) (58)	240
C_{15} (structure: diphenylamine with OMe, CO₂H, OMe)	DMF, POCl₃	(acridone: OMe, OMe, CHO) (15)	279
C_{16} (structure: MeO, OMe benzene with OCH₂COPh)	DMF, POCl₃	(14) + benzofuran (2-Bz, OMe, MeO) (39) + benzofuran (2-Bz, OMe, MeO, CHO) (39)	43

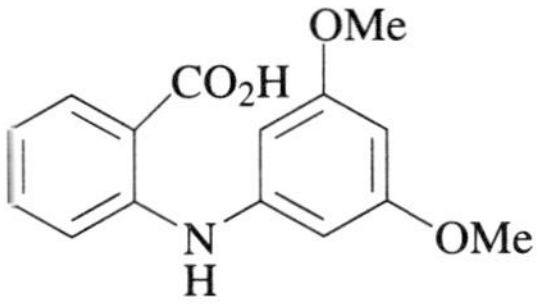

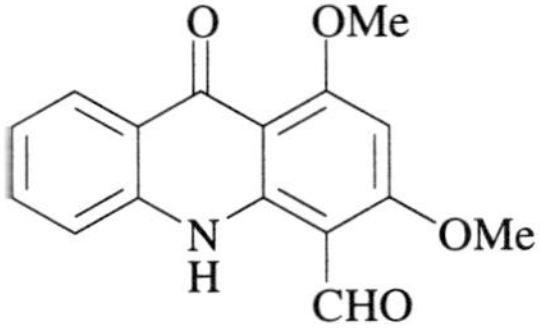

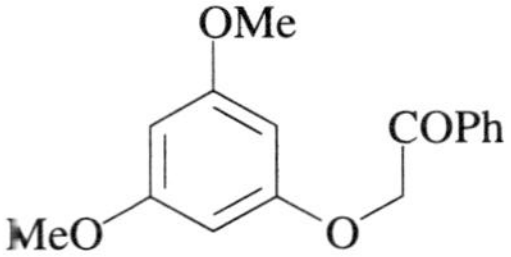

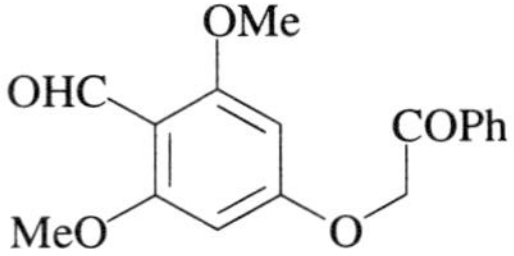

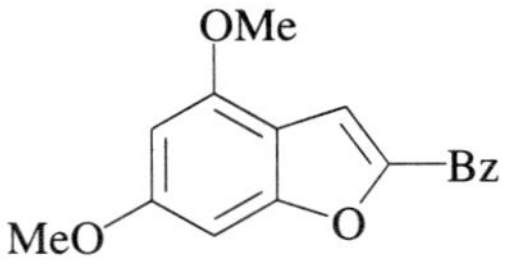

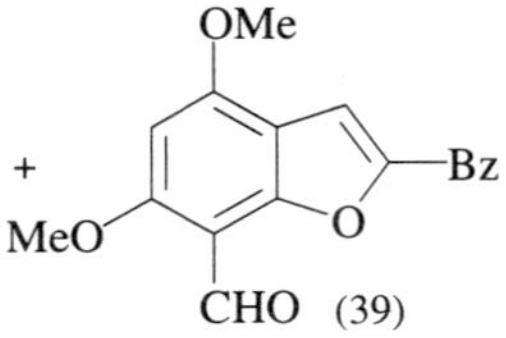

76

TABLE I. BENZENES (*Continued*)

Substrate	Reagents	Product(s) and Yield(s) (%)	Refs.
C_{20}			
$R^1 = R^3 = Cl, R^5 = N(Bn)_2$	DMF, POCl$_3$	$R^2 = CHO$ (66)	244
C_{28}			
$R^1 = CH_2O_2CPh, R^3 = R^5 = OBn$	DMF, POCl$_3$	$R^2 = CHO$ (66)	280
D. Tetrasubstituted Benzenes			
D1. R^1, R^2, R^3, R^4 Substituents			
C_{10}			
$R^1 = R^2 = R^3 = R^4 = OMe$	MFA, POCl$_3$	$R^5 = CHO$ (94)	281–283
C_{16}			
$R^1 = R^2 = R^3 = OMe, R^4 = OBn$	MFA, POCl$_3$	$R^5 = CHO$ (23) + $R^6 = CHO$ (33)	284
D2. R^1, R^2, R^3, R^5 Substituents			
C_{10}			
$R^1 + R^2 = OCH_2O, R^3 = OMe, R^5 = (CH_2)_2Cl$	DMF, POCl$_3$	$R^4 = CHO$ (46) + $R^6 = CHO$ (23)	285, 286
$R^1 = R^2 = R^3 = R^5 = OMe$	DMF, POCl$_3$	$R^4 = CHO$ (86)	234, 287
$R^1 = R^3 = R^5 = OMe, R^2 = OMe$	DMF, POCl$_3$	$R^4 = CHO$ (73)	274
$R^1 = R^2 = R^3 = Me, R^5 = OMe$	DMF, POCl$_3$	$R^4 = CHO$ (70)	217
$R^1 = OMe, R^2 = R^3 = R^5 = Me$	DMF, POCl$_3$	$R^4 = CHO$ (93)	234
$R^1 = R^2 = R^3 = OMe, R^5 = OEt$	—	$R^4 = CHO$ (—)	288
$R^1 = R^3 = R^5 = OMe, R^2 = OEt$	—	$R^4 = CHO$ (—)	288
C_{14}			
$R^1 = R^2 = R^3 = OMe, R^5 = CH_2CH(OAc)Me$	MFA, POCl$_3$	$R^4 = CHO$ (77)	289

TABLE I. BENZENES (*Continued*)

Substrate	Reagents	Product(s) and Yield(s) (%)	Refs.
C₁₅	MFA, POCl₃	(23)	290
C₁₇	DMF, POCl₃	(36) + (13)	291
C₁₇₋₂₁	DMF, POCl₃	**(I)** + **(II)**	291
	DMF, POCl₃	**I** (43) + **II** (9)	
	MFA, POCl₃	**I** (32) + **II** (21)	
	DMF, POCl₃	**I** (35) + **II** (12)	

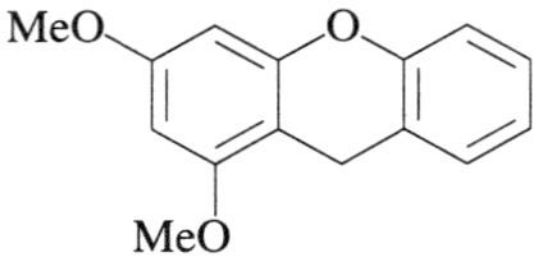

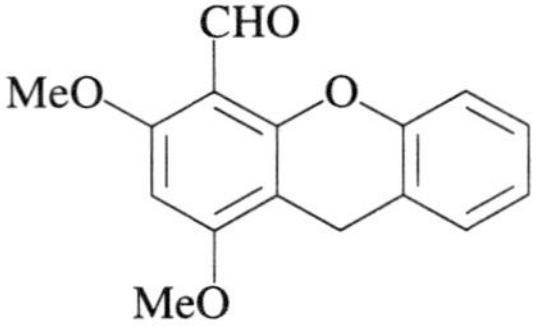

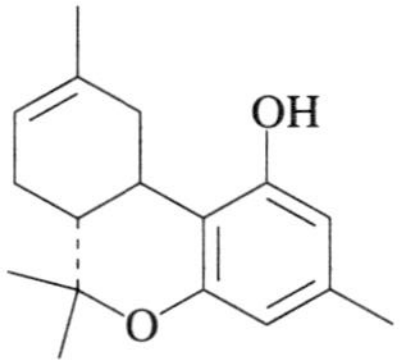

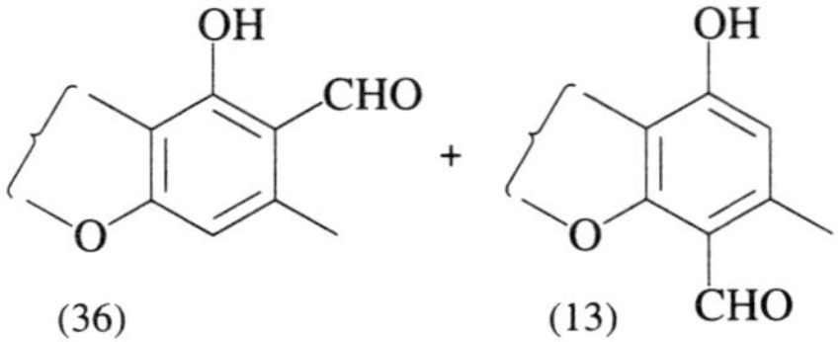

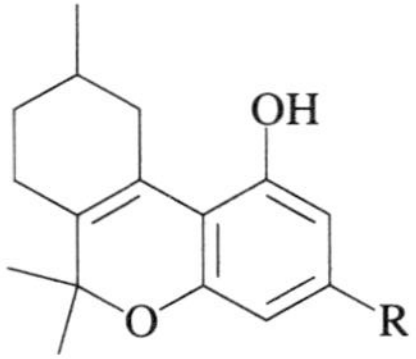

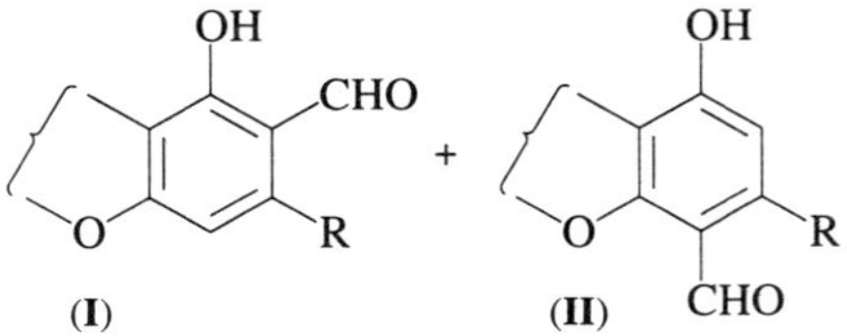

R
—
Me

n-C₅H₁₁

78

TABLE I. BENZENES (*Continued*)

	Substrate	Reagents	Product(s) and Yield(s) (%)	Refs.
C_{20}	$R^1 = R^3 = OH$, $R^5 = Me$,	DMF, POCl$_3$	$R^4 = CHO$ (75)	292
C_{21}	$R^1 = OMe$, $R^2 = CO_2Me$, $R^3 = n\text{-}C_5H_{11}$, $R^5 = OBn$	DMF, POCl$_3$	$R^4 = CHO$ (80)	293
C_{22}	$R^1 = Me$, $R^2 = CO_2Bn$, $R^3 = OBn$, $R^5 = OMe$	DMF, POCl$_3$	$R^6 = CHO$ (63) + $R^3 = OH$, $R^6 = CHO$ (3)	294
	$R^1 = R^3 = OMe$, $R^2 = R^5 = OBn$	DMF, POCl$_3$	$R^4 = CHO$ (61)	287
	$R^1 = R^3 = OBn$, $R^2 = R^5 = OMe$	DMF, POCl$_3$	$R^4 = CHO$ (27)	287
C_{27}		MFA, POCl$_3$	(88)	290
	$R^1 = OMe$, $R^2 = CO_2Bn$, $R^3 = n\text{-}C_5H_{11}$, $R^5 = OBn$	DMF, POCl$_3$	$R^4 = CHO$ (100)	293

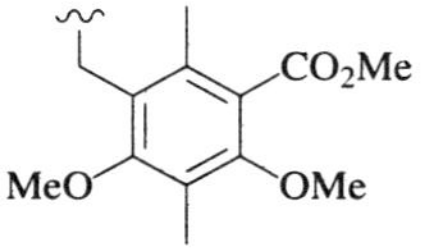

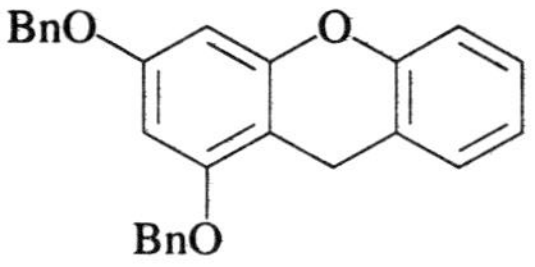

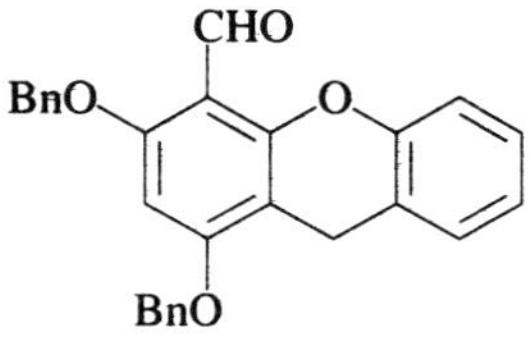

D3. R^1, R^2, R^4, R^5 Substituents

	Substrate	Reagents	Product(s) and Yield(s) (%)	Refs.
C_{10}	$R^1 = OMe$, $R^2 = R^4 = R^5 = Me$	DMF, POCl$_3$	$R^6 = CHO$ (66)	217

79

TABLE I. BENZENES (*Continued*)

Substrate	Reagents	Product(s) and Yield(s) (%)	Refs.
E. Pentasubstituted Benzenes			
C$_9$			
$R^1 = R^3 = R^5 = OH$, $R^4 = Me$, $R^2 = CO_2Me$	DMF, POCl$_3$	$R^6 = CHO$ (68)	295
C$_{11}$			
$R^1 + R^2 = OCH_2O$, $R^3 = OMe$, $R^4 + R^5 = OCH(Me)CH_2$	MFA, POCl$_3$	$R^6 = CHO$ (60)	296
$R^1 + R^2 = OCH_2O$, $R^3 = R^4 = OMe$, $R^5 = CH_2CHBrMe$	MFA, POCl$_3$	$R^6 = CHO$ (40)	296
$R^1 = OMe$, $R^2 = R^3 = R^4 = R^5 = Me$	DMF, POCl$_3$	$R^6 = CHO$ (68)	217

TABLE II. NAPHTHALENES

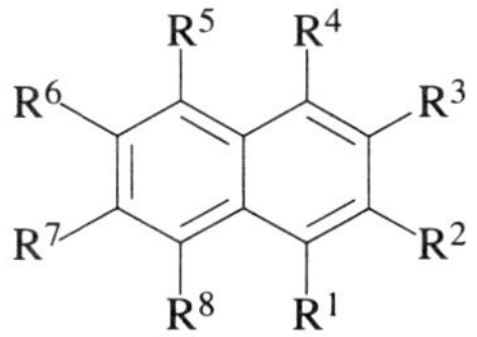

R^1 - R^8 = H except as indicated

Substrate	Reagents	Product(s) and Yield(s) (%)	Refs.
A. Naphthalene			
C_{10}	DMF, Tf_2O	R^1 = CHO (50)	47
B. Monosubstituted Naphthalenes			
B1. R^1 Substituents			
C_{10} R^1 = OH	Me_2NN=CHCHO, $POCl_3$ or $COCl_2$	R^4 = CH=CHN=NMe_2^+ Cl^- or $POCl_2^-$	40
C_{11} R^1 = OMe	DMF or MFA, $POCl_3$	R^4 = CHO (81)	297
	DMF, 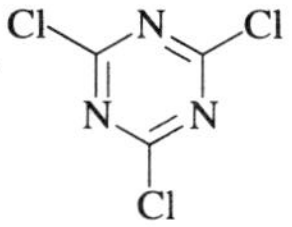 [pyrimidine]	R^4 = CHO (47)	36
	DMF, $POCl_3$	R^4 = CHO (81)	297
	DMF, $SOCl_2$	R^2 = CHO (37) 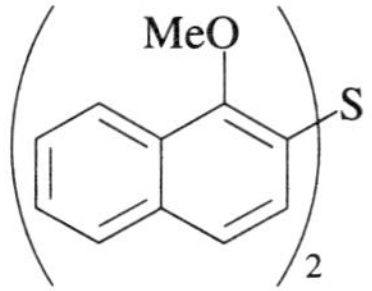 + [structure] (26)	42

TABLE II. NAPHTHALENES (*Continued*)

Substrate	Reagents	Product(s) and Yield(s) (%)	Refs.
C$_{12}$			
R^1 = NMe$_2$	DMF, (Cl$_2$PO)$_2$O	R^4 = CHO (96)	26
	MFA, POCl$_3$	R^4 = CHO (47)	36, 2
	DMF, POCl$_3$ (3:1)	R^4 = CHO (33) + R^2 = R^4 = CHO (30)	56
	DMF, POCl$_3$ (3:2)	R^4 = CHO (8) + R^2 = R^4 = CHO (76)	56
	MFA, POCl$_3$	R^4 = CHO (20)	56
	Et$_2$NCHO, POCl$_3$ (3:1)	R^4 = CHO (18) + R^2 = R^4 = CHO (19)	56
	Et$_2$NCHO, POCl$_3$ (3:2)	R^4 = CHO (40) + R^2 = R^4 = CHO (17)	56
	i-Pr$_2$NCHO, POCl$_3$ (3:1)	R^4 = CHO (59) + R^2 = R^4 = CHO (6)	56
	i-Pr$_2$NCHO, POCl$_3$ (3:2)	R^4 = CHO (46) + R^2 = R^4 = CHO (20)	56
R^1 = OEt	DMF, SOCl$_2$	R^4 = CHO (19) + (19)	42
C$_{14}$			
R^1 = NEt$_2$	DMF, POCl$_3$	R^4 = CHO (—)	191

B2. R^2 Substituents

Substrate	Reagents	Product(s) and Yield(s) (%)	Refs.
C$_{10}$			
R^2 = OH	MFA, POCl$_3$	R^1 = CHO (—)	189, 298
	H$_2$NCHO, AlCl$_3$	R^1 = CHO (—)	299
C$_{11}$			
R^2 = OMe	1. MFA, POCl$_3$ 2. NH$_2$OH	R^1 = CN (20)	197
	DMF, SOCl$_2$	R^1 = CHO (31) + (18)	42

TABLE II. NAPHTHALENES (*Continued*)

Substrate	Reagents	Product(s) and Yield(s) (%)	Refs.
C$_{12}$ R^2 = SMe	DMF, SO$_2$Cl$_2$ DMF, (Cl$_2$PO)$_2$O MFA, POCl$_3$	R^1 = CHO (—) R^1 = CHO (90) R^1 = CHO (34)	193 26 300
R^2 = OEt	MFA, POCl$_3$	R^1 = CHO (84)	179, 301, 302
	DMF, SOCl$_2$	R^1 = CHO (16) 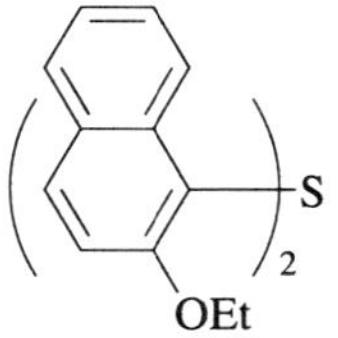 + (12)	42
C$_{13}$ R^2 = SEt	—	R^1 = CHO (—)	303
R^2 = OPr-*i*	DMF, SOCl$_2$	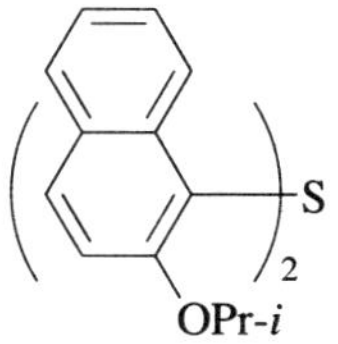 (10)	42
C$_{14}$ R^2 = OBu-*n*	DMF, SOCl$_2$	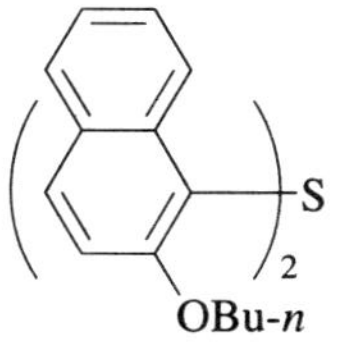 (11)	42
R^2 = NEt$_2$	DMF, POCl$_3$	R^6 = CHO (—)	191

TABLE II. NAPHTHALENES (Continued)

Substrate	Reagents	Product(s) and Yield(s) (%)	Refs.
C. Disubstituted Naphthalenes			
C1. R^1, R^2 Substituents			
C_{11} R^1 = OH, R^2 = Me	Me_2NN=CHCHO, $POCl_3$ or $COCl_2$	R^4 = CH=CHN=NMe_2^+ Cl^- or $POCl_2^-$	40
C2. R^1, R^3 Substituents			
C_{12} R^1 = Me, R^3 = OMe	DMF, $POCl_3$	R^4 = CHO (66)	304
C3. R^1, R^4 Substituents			
C_{12} R^1 = OMe, R^4 = Me	DMF or MFA, $POCl_3$	R^2 = CHO (22)	297
R^1 = R^4 = OMe	DMF, $POCl_3$	R^2 = CHO (100)	306, 305
C4. R^1, R^5 Substituents			
C_{10} R^1 = R^5 = OH	MFA, $POCl_3$	R^4 = CHO (—)	189
	1. MFA, $POCl_3$ 2. Et_3N, $PhNO_2$	(47)	59
C_{12} R^1 = R^5 = OMe	DMF, $POCl_3$	R^4 = CHO (93)	306, 307
	MFA, $POCl_3$	R^4 = CHO (81)	308

TABLE II. NAPHTHALENES (*Continued*)

Substrate	Reagents	Product(s) and Yield(s) (%)	Refs.
	C5. R^1, R^6 Substituents		
C_{12}			
$R^1 = R^6 = OMe$	—	$R^4 = CHO$ (—)	309
	DMF, $POCl_3$	$R^4 = CHO$ (61) + unknown isomer (7)	57
	MFA, $POCl_3$	$R^5 = CHO$ (54)	58
	C6. R^1, R^7 Substituents		
C_{12}			
$R^1 = R^7 = OMe$	DMF, $POCl_3$	$R^4 = CHO$ (76) + $R^8 = CHO$ (2)	310
	C7. R^1, R^8 Substituents		
C_{12}			
$R^1 + R^8 = (CH_2)_2$	MFA, $POCl_3$	$R^5 = CHO$ (85)	311
C_{14}			
$R^1 = R^8 = NMe_2$	DMF, $POCl_3$ (0.5 eq)	$R^4 = CHO$ (35)	312
	DMF, $POCl_3$ (1 eq)	$R^4 = R^5 = CHO$ (50) + $R^2 = R^5 = CHO$ (3)	312
	C8. R^2, R^3 Substituents		
C_{11}			
$R^2 = OH, R^3 = CO_2H$	MFA, $POCl_3$	$R^1 = CHO$ (—)	189
C_{12}			
$R^2 = R^3 = OMe$	DMF, $POCl_3$	$R^1 = CHO$ (23)	313
	C9. R^2, R^6 Substituents		
C_{11}			
$R^2 = OMe, R^6 = Br$	MeNHCHO, $POCl_3$	$R^1 = CHO$ (—)	314
C_{12}			
$R^2 = OMe, R^6 = Me$	MFA, $POCl_3$	$R^1 = CHO$ (86)	315
$R^2 = R^6 = OMe$	DMF, $POCl_3$	$R^1 = CHO$ (81) + $R^1 = R^5 = CHO$ (5)	297

TABLE II. NAPHTHALENES (*Continued*)

Substrate	Reagents	Product(s) and Yield(s) (%)	Refs.
C10. R^2, R^7 Substituents			
C_{10} $R^2 = R^7 = OH$	MFA, $POCl_3$	$R^1 = CHO$ (—)	189
C_{12} $R^2 = R^7 = OMe$	DMF, $POCl_3$	$R^1 = CHO$ (82)	316, 297
D. Trisubstituted Naphthalenes			
D1. R^1, R^2, R^6 Substituents			
C_{13} $R^1 = Me$, $R^2 = R^6 = OMe$	DMF, $POCl_3$	$R^5 = CHO$ (76)	297
D2. R^1, R^2, R^7 Substituents			
C_{13} $R^1 = Me$, $R^2 = R^7 = OMe$	DMF, $POCl_3$	$R^8 = CHO$ (83)	297
D3. R^1, R^4, R^5 Substituents			
C_{13} $R^1 = R^5 = OMe$, $R^4 = Me$	MFA, $POCl_3$	$R^8 = CHO$ (85)	308
D4. R^1, R^4, R^6 Substituents			
C_{13} $R^1 = Me$, $R^4 = R^6 = OMe$	DMF, $POCl_3$	$R^5 = CHO$ (70)	310
D5. R^1, R^4, R^8 Substituents			
C_{12} $R^1 = OH$, $R^4 = R^8 = OMe$	DMF, $POCl_3$	$R^1 = OCHO$ (62) + $R^2 = CHO$ (24)	306
C_{13} $R^1 = R^4 = R^8 = OMe$	DMF, $POCl_3$	$R^2 = CHO$ (13) + $R^5 = CHO$ (80)	306

TABLE II. NAPHTHALENES (*Continued*)

Substrate	Reagents	Product(s) and Yield(s) (%)	Refs.
	E. Tetrasubstituted Naphthalenes		
	E1. R^1, R^4, R^5, R^6 Substituents		
C_{15}			
$R^1 = R^6 = Me,$	MFA, POCl$_3$	$R^8 = CHO$ (91)	317
$R^4 + R^5 = CH(Me)CH_2O$			
	E2. R^1, R^4, R^5, R^8 Substituents		
C_{14}			
$R^1 = R^4 = R^5 = OMe, R^8 = Me$	DMF, POCl$_3$	$R^2 = CHO$ (99)	318
$R^1 = R^4 = R^5 = R^8 = OMe$	—	$R^2 = CHO$ (99)	319

TABLE III. OTHER POLYCYCLIC BENZENOID HYDROCARBONS

$C_6/C_6/C_6$

A.

Substrate	Reagents	Product(s) and Yield(s) (%)	Refs.
C_{14}			
1,5,10-Cl$_3$	MFA, POCl$_3$	1,5,10-Cl$_3$-9-CHO (—)	189
2,9-Cl$_2$	MFA, POCl$_3$	2,9-Cl$_2$-10-CHO (—)	189
—	DMF, POCl$_3$	9-CHO (63)	192
	MFA, POCl$_3$	9-CHO (92)	60, 179, 189, 301, 320
	DMF, PCl$_3$	9-CHO (49)	321
	DMF, PCl$_5$	9-CHO (50)	321
	DMF, SOCl$_2$	9-CHO (9) + (9-C$_{14}$H$_9$)$_2$S (19)	42
	DMF, SO$_2$Cl$_2$	9-CHO (—)	193
	DMF, Tf$_2$O	9-CHO (98)	47
C_{15}			
9-Me	MFA, POCl$_3$	9-Me-10-CHO	311
2-OMe	MFA, POCl$_3$	2-OMe-1-CHO (40) + 2-OMe-3-CHO (23)	42, 322, 323, 50
C_{16}			
1,2-OMe$_2$-10-Cl	MFA, POCl$_3$	1,2-(OMe)$_2$-10-Cl-9-CHO (—)	189
2,6-OMe$_2$-10-Cl	MFA, POCl$_3$	2,6-(OMe)$_2$-10-Cl-9-CHO)—)	189
2-OMe-9-Me	MFA, POCl$_3$	2-OMe-9-Me-1-CHO (48)	50

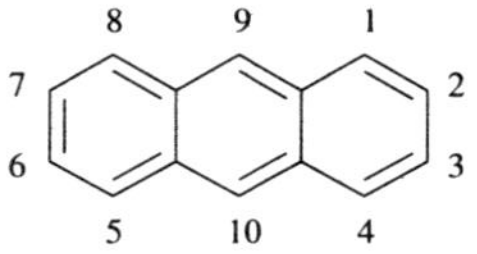

TABLE III. OTHER POLYCYCLIC BENZENOID HYDROCARBONS (*Continued*)

Substrate	Reagents	Product(s) and Yield(s) (%)	Refs.
C$_{17}$	MFA, POCl$_3$	(71)	324
C$_{21}$ 9-BzO	DMF, POCl$_3$	9-BzO-10-CHO (90)	325
2-OMe-9-Ph	MFA, POCl$_3$	2-OMe-9-Ph-1-CHO (50)	50
C$_{24}$	MFA, POCl$_3$	(98)	50
C$_{34}$	MFA, POCl$_3$	(24) + (45)	326

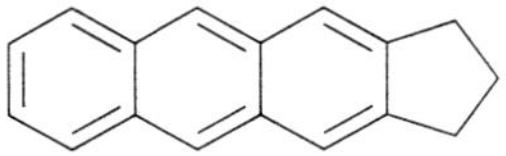

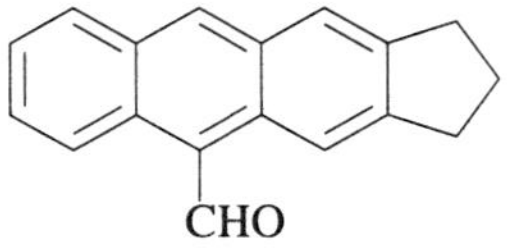

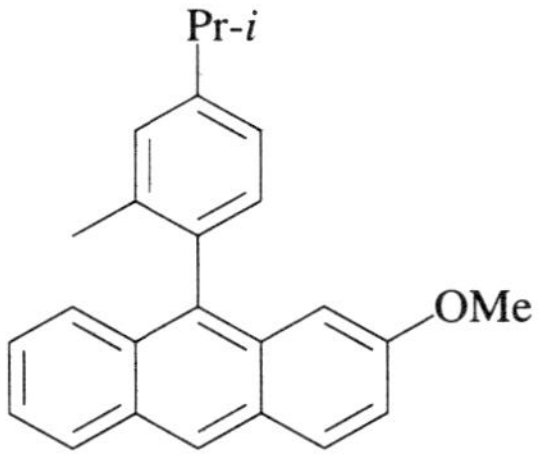

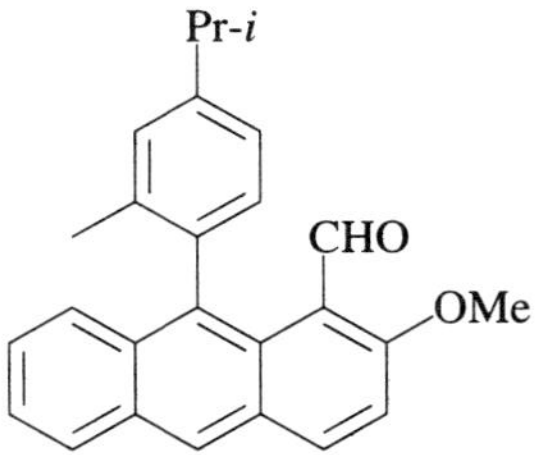

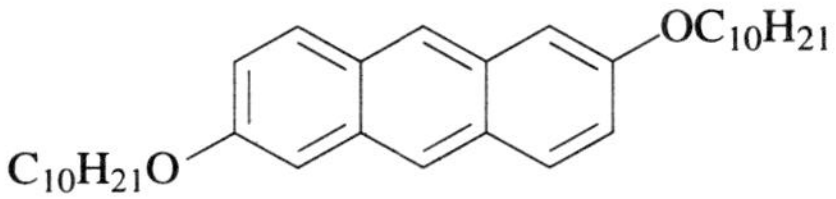

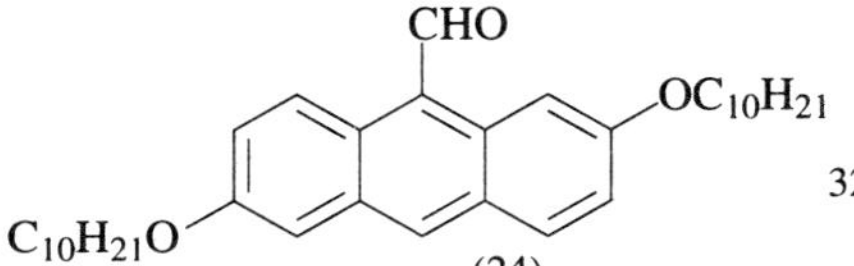

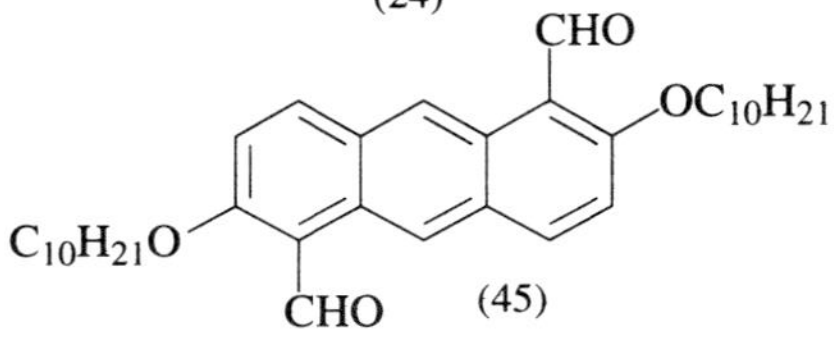

89

TABLE III. OTHER POLYCYCLIC BENZENOID HYDROCARBONS (Continued)

Substrate	Reagents	Product(s) and Yield(s) (%)	Refs.
B. 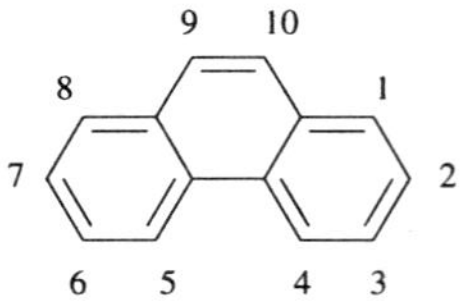			
C_{14} —	DMF, Tf_2O	3-CHO (25)	47
C_{15} 3-OMe	DMF, $POCl_3$	3-OMe-9-CHO (48)	61
C_{23}	DMF, $POCl_3$	(94) (100)	327

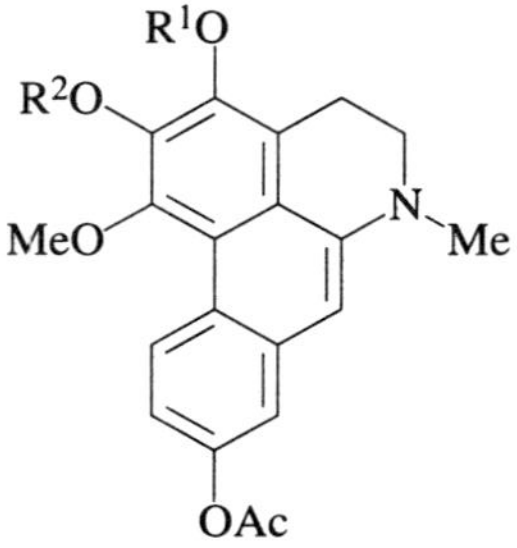

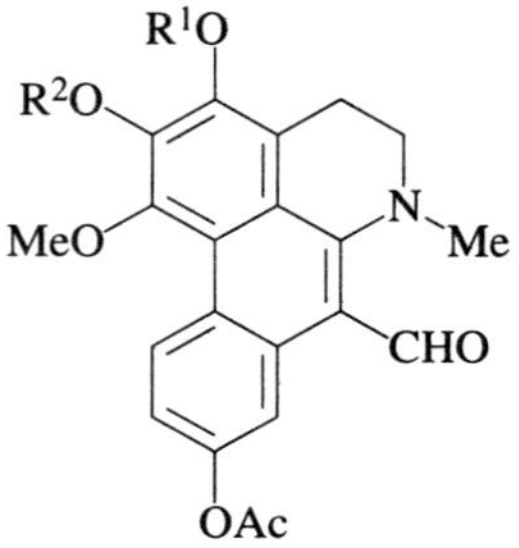

For the C_{23} substrate:

R^1	R^2
Me	COMe
COMe	Me

90

TABLE III. OTHER POLYCYCLIC BENZENOID HYDROCARBONS (*Continued*)

Substrate	Reagents	Product(s) and Yield(s) (%)	Refs.
	C₆/C₆/C₆/C₆		
C₁₆	MFA, POCl₃	(62)	328
	MFA, POCl₃	(64)	60
	MFA, POCl₃	(53)	329
C₁₇	MFA, POCl₃	(80)	330

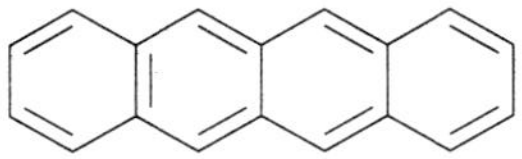

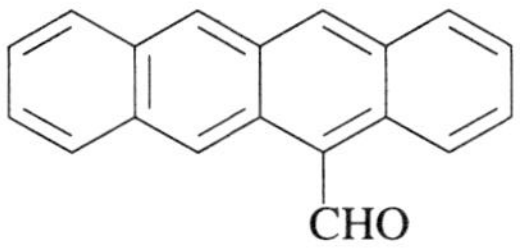

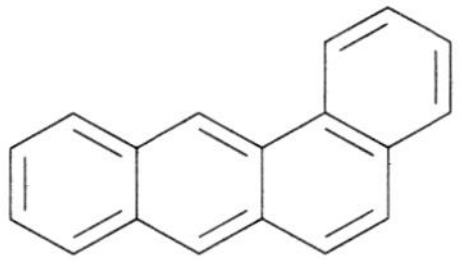

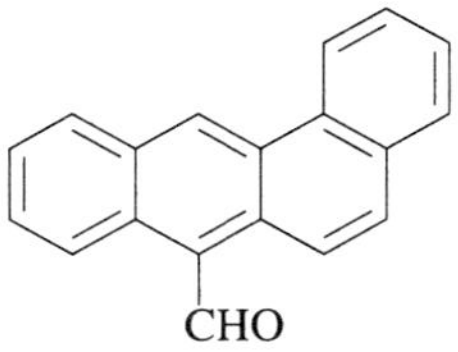

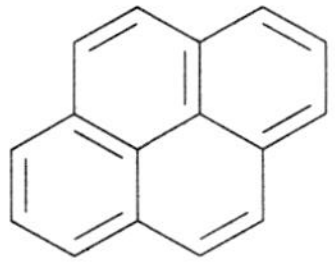

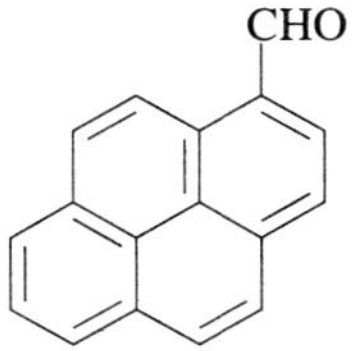

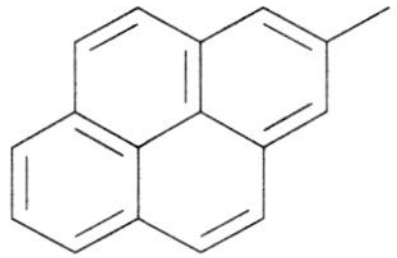

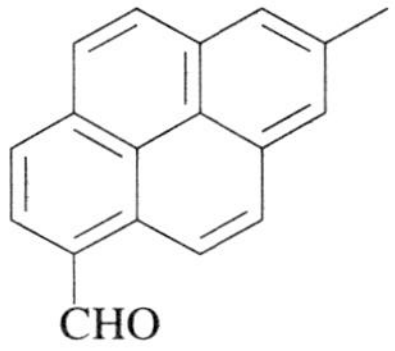

TABLE III. OTHER POLYCYCLIC BENZENOID HYDROCARBONS (Continued)

Substrate	Reagents	Product(s) and Yield(s) (%)	Refs.
C$_{20}$	C$_6$/C$_6$/C$_6$/C$_6$ MFA, POCl$_3$	(90)	331
C$_{20-21}$ 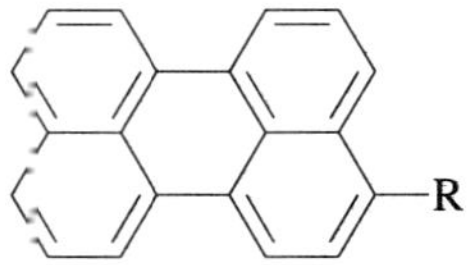 R: H, Me	MFA, POCl$_3$	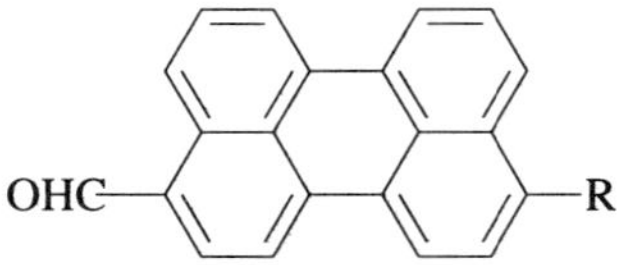 R (63) (55)	332
C$_{20-21}$ R: H, Me	C$_6$/C$_6$/C$_6$/C$_6$ MFA, POCl$_3$	CHO (53) (49)	333

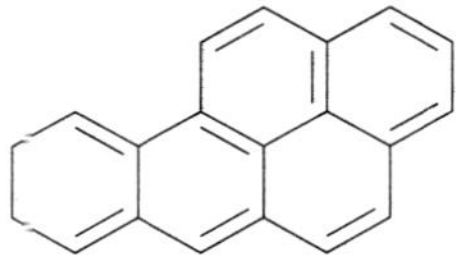

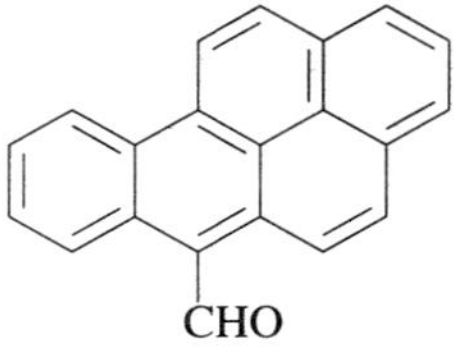

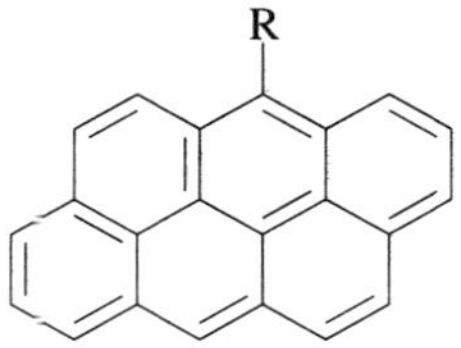

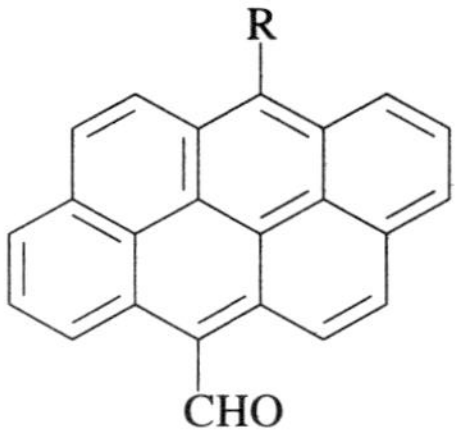

92

TABLE III. OTHER POLYCYCLIC BENZENOID HYDROCARBONS (*Continued*)

Substrate	Reagents	Product(s) and Yield(s) (%)	Refs.
C$_{24-25}$			
$\frac{R}{H}$ Me	MFA, POCl$_3$	CHO (75) (81)	334
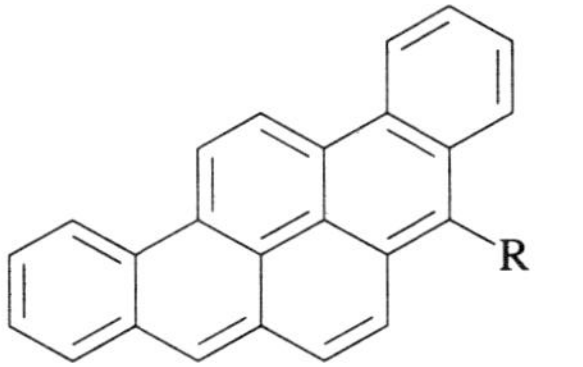 $\frac{R}{H}$ Me	MFA, POCl$_3$	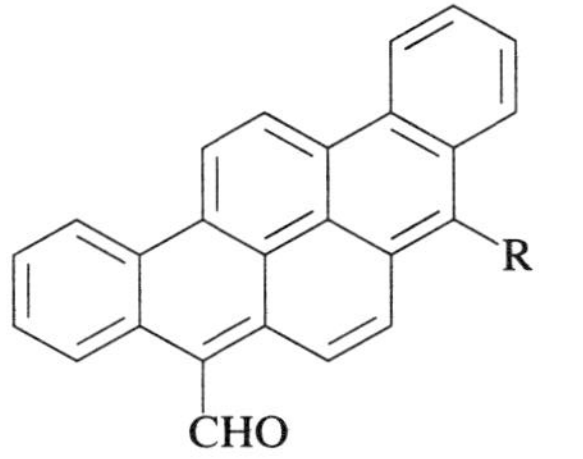 CHO (91) (46)	335

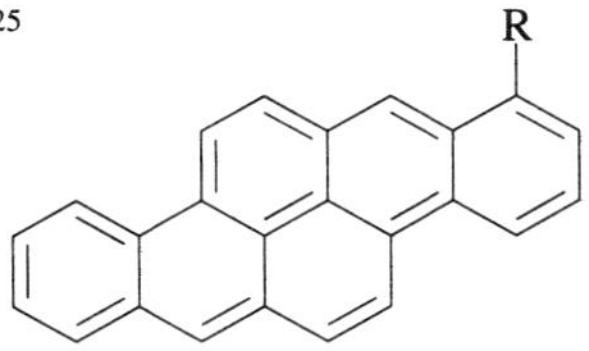

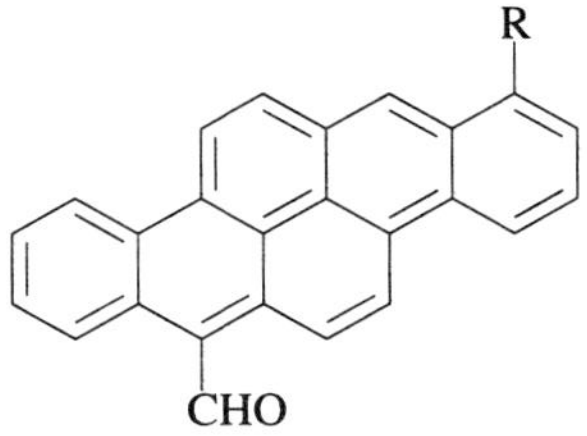

TABLE III. OTHER POLYCYCLIC BENZENOID HYDROCARBONS (*Continued*)

Substrate	Reagents	Product(s) and Yield(s) (%)	Refs.
C₂₄	MFA, POCl₃	(48) + (3)	336
C₂₅	MFA, POCl₃	(27)	336
C₂₆	MFA, POCl₃	(20)	337

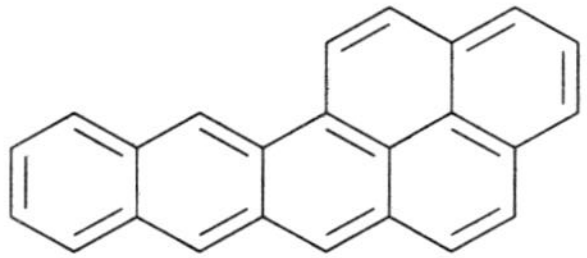

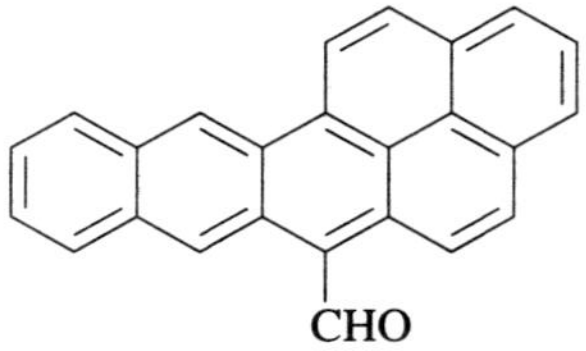

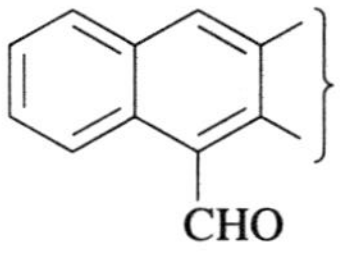

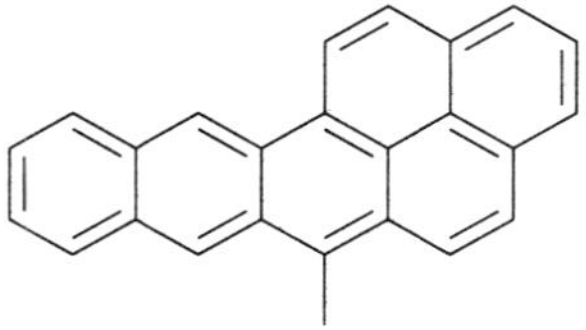

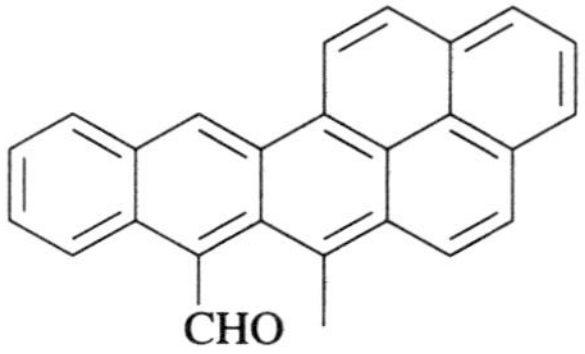

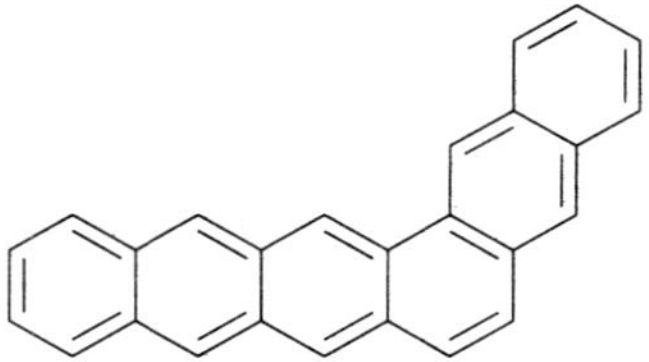

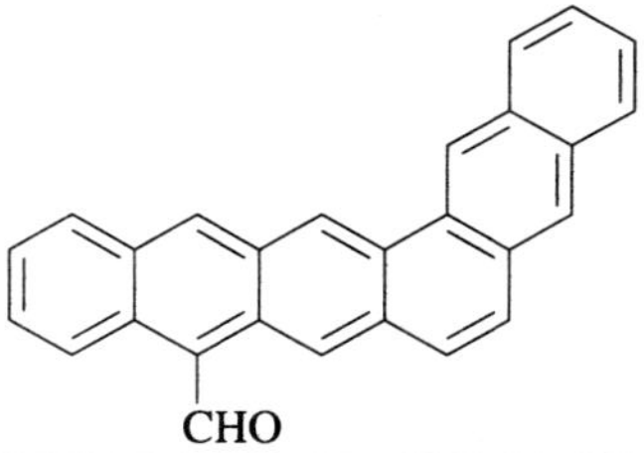

94

TABLE IV. CARBOCYCLIC ANIONS

Substrate	Reagents	Product(s) and Yield(s) (%)	Refs.
A. C$_5$			
C$_5$ (cyclopentadienyl anion)	$^+$NMe$_2$ ClO$_4^-$ NMe$_2$	(28)	64
C$_7$ SMe$_2^+$	DMF, POCl$_3$	OHC CHO SMe$_2^+$ (4)	338
C$_{7-15}$ R–C=O		R–C=O Me$_2$N	
R = Me	DMF, POCl$_3$, NaOMe	(93)	62, 339
R = OEt	DMF, POCl$_3$, NaOMe	(39)	62
R = (CH$_2$)$_5$OH	DMF, POCl$_3$	(54)	62
(1,3-dioxolane, Me, Et)	DMF, POCl$_3$	(58)	62, 339
(1,3-dioxolane, C$_5$H$_{11}$, Et)	DMF, POCl$_3$	(46)	62, 339

95

TABLE IV. CARBOCYCLIC ANIONS (*Continued*)

Substrate	Reagents	Product(s) and Yield(s) (%)	Refs.

C₈

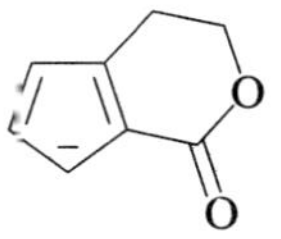

For C₈ row:

Substrate: (bicyclic lactone structure with fused cyclopentadiene ring)

Reagents: DMF, POCl₃, NaOMe

Product(s) and Yield(s) (%): (bicyclic lactone with =CH-NMe₂ substituent) (66)

Refs.: 62, 339

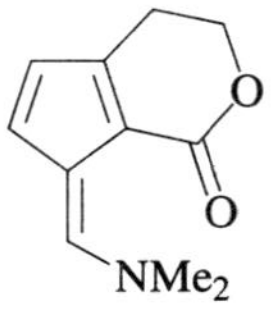

C₁₇₋₂₉

Substrate: (cyclopentadiene with R¹, R¹, Ph, Ph substituents)

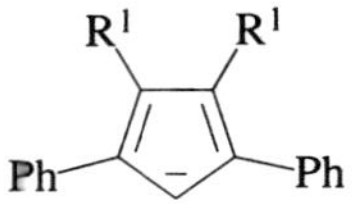

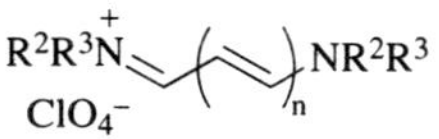

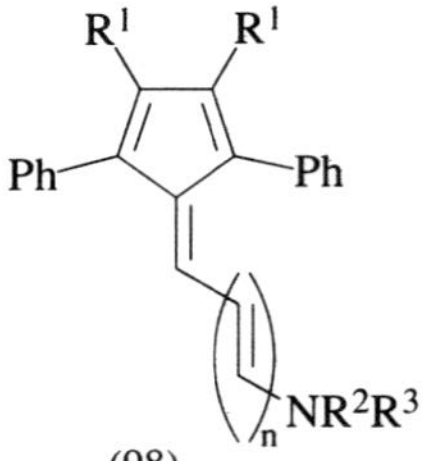

Reagents:

R²R³N⁺=CH-CH=(CH)ₙ-NR²R³ ClO₄⁻

R²	R³	n
Me	Me	0
Me	Me	1
Me	Me	0
Me	Me	2
(CH₂)₅		1
(CH₂)₅		2
Me	Ph	0
Me	Ph	1
Me	Ph	2

R¹
H
H
Ph

Product(s) and Yield(s) (%): (cyclopentadiene product with R¹, R³, Ph, Ph, and =CH-(CH)ₙ-NR²R³ substituent)

(98)
(98)
(76)
(97)
(81)
(85)
(57)
(70)
(90)

Refs.: 63

96

TABLE IV. CARBOCYCLIC ANIONS (*Continued*)

Substrate	Reagents	Product(s) and Yield(s) (%)	Refs.
C_{29} 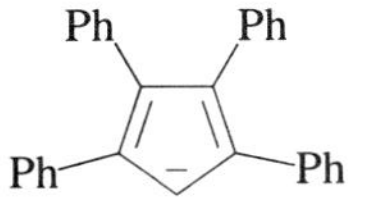	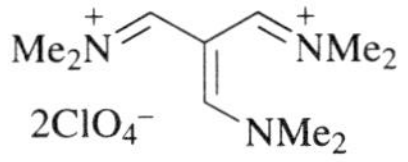 Me_2N⁺=⟍NMe_2 / NMe_2 2ClO_4⁻	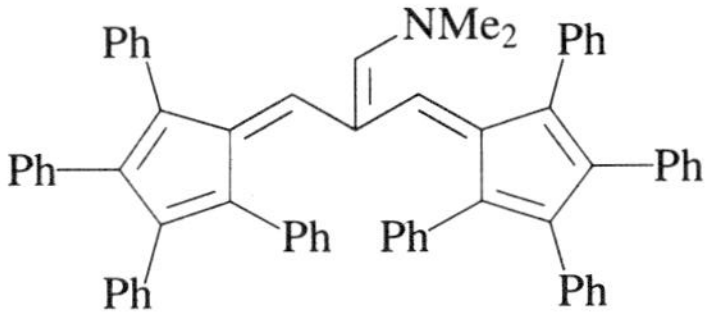 (89)	63
C_{21}	**C_5/C_6** Me_2N⁺=⟍NMe_2 / NMe_2 2ClO_4⁻	(98)	63
	R¹R²N⁺=⟍NR¹R² ClO_4⁻		63

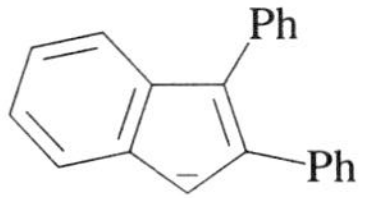

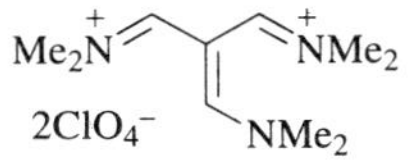

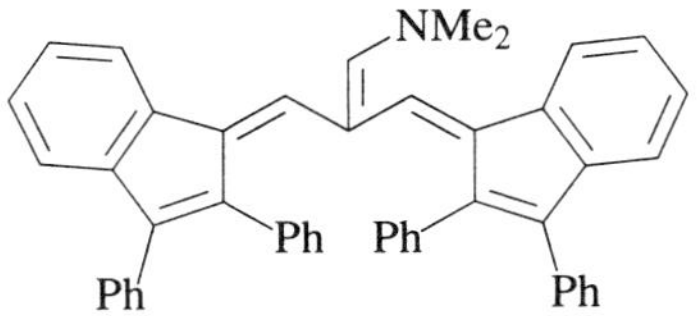

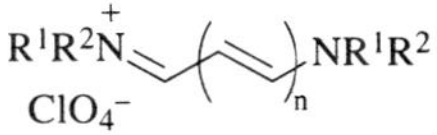

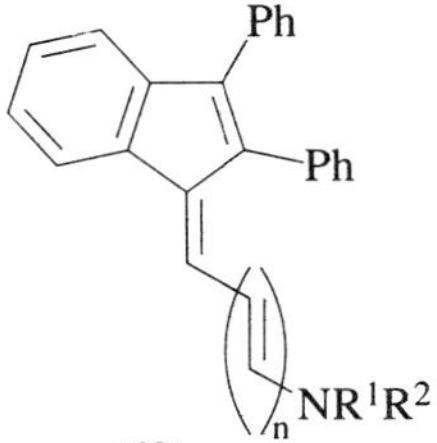

R²	R³	n	
Me	Me	0	(68)
Me	Me	1	(97)
Me	Me	2	(97)
(CH_2)_5		1	(80)
(CH_2)_5		2	(94)
Me	Ph	0	(53)
Me	Ph	1	(95)
Me	Ph	2	(91)

97

TABLE IV. CARBOCYCLIC ANIONS (*Continued*)

Substrate	Reagents	Product(s) and Yield(s) (%)	Refs.
C₁₃	**C₅/C₇**	(55)	64
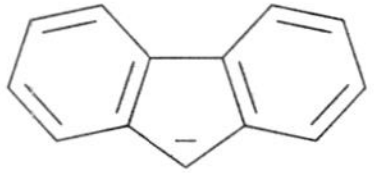 C₁₃	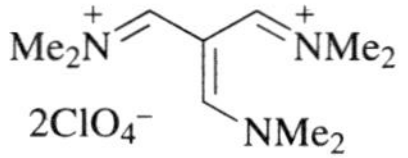 **C₅/C₆/C₆**	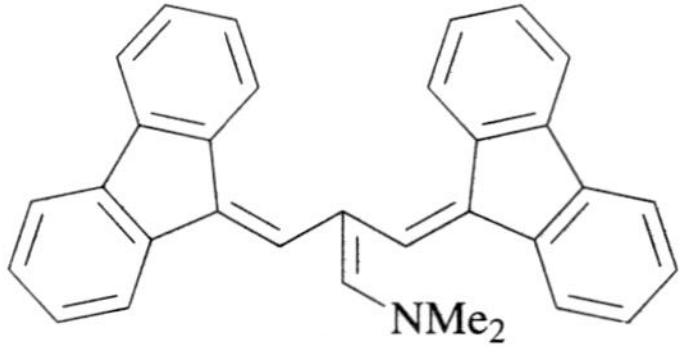 (87)	63
	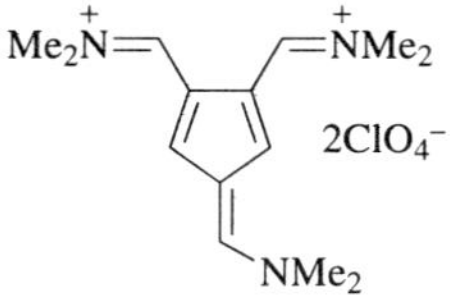	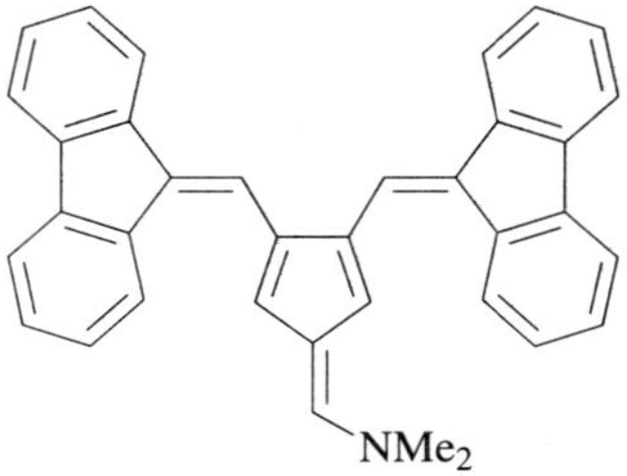 (23)	63

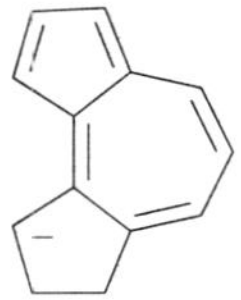

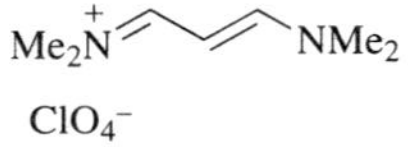

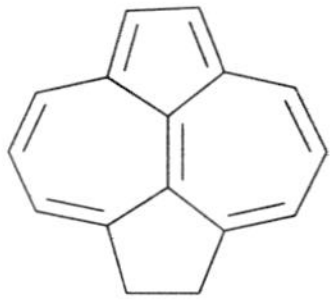

98

TABLE IV. CARBOCYCLIC ANIONS (Continued)

Substrate	Reagents	Product(s) and Yield(s) (%)	Refs.
C$_{13\text{-}14}$			

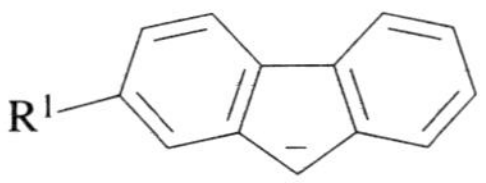

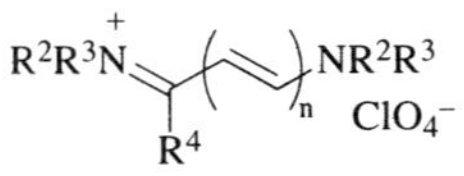

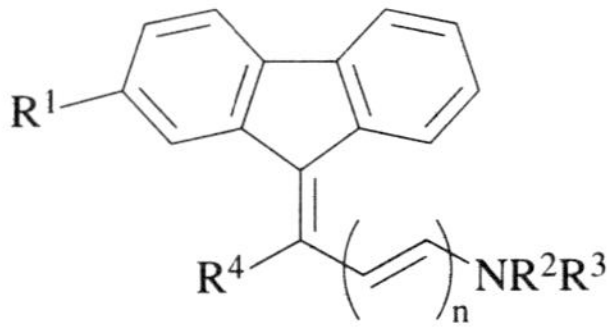

Substrate: fluorene with R^1

Reagents: $R^2R^3\overset{+}{N}\diagdown)_n\diagup^{NR^2R^3}\quad ClO_4^-$ with R^4

Product: fluorene-derived with R^1, R^4, $\diagdown)_n\diagup^{NR^2R^3}$

R^1	R^2	R^3	R^4	n		
H	Me	Me	H	1	(85)	63
	Me	Me	H	2	(90)	
	Me	Me	H	3	(90)	
	Me	Me	Me	2	(91)	
	(CH$_2$)$_5$		H	1	(80)	
	(CH$_2$)$_5$		H	2	(90)	
	Me	Ph	H	1	(87)	
	Me	Ph	H	2	(37)	
	Me	Me	Ph	2	(86)	
NO$_2$	Me	Me	H	0	(87)	
	Me	Me	H	1	(97)	
	Me	Me	H	2	(98)	
	Me	Ph	H	1	(87)	
	Me	Ph	H	2	(92)	
CN	Me	Me	H	0	(81)	
	Me	Me	H	1	(90)	
	Me	Ph	H	0	(68)	
	Me	Ph	H	1	(94)	
	Me	Ph	H	2	(91)	

99

TABLE IV. CARBOCYCLIC ANIONS (*Continued*)

Substrate	Reagents	Product(s) and Yield(s) (%)	Refs.
C$_{25}$ 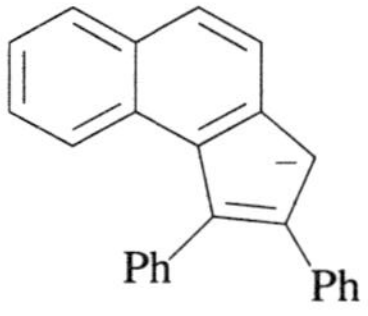	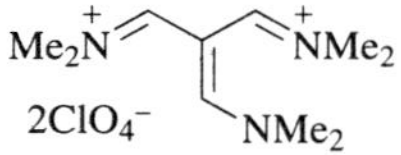 Me$_2$N$^+$=CH—...=NMe$_2$ ... NMe$_2$ 2ClO$_4^-$	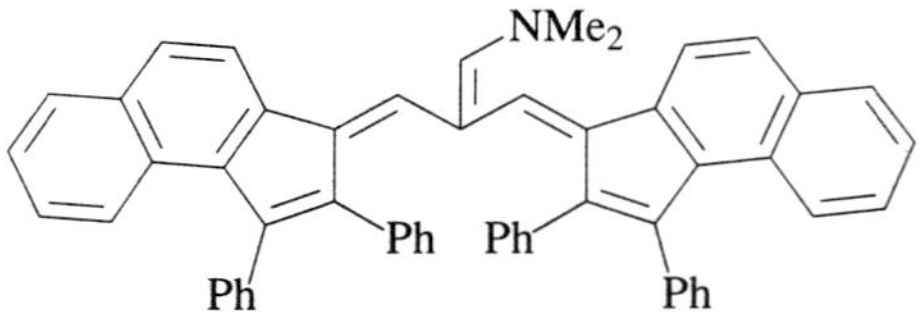 (81)	63
C$_{13}$	**C$_6$/C$_6$/C$_6$** Me$_2$N$^+$=...NMe$_2$ ClO$_4^-$	(91)	65, 340
	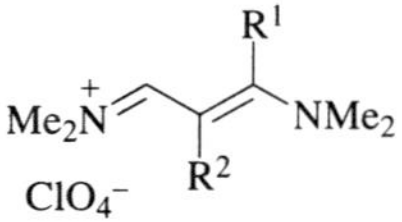 Me$_2$N$^+$=... R^1 ... R^2 ... NMe$_2$ ClO$_4^-$	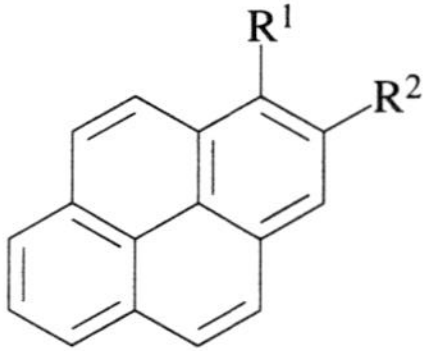	65

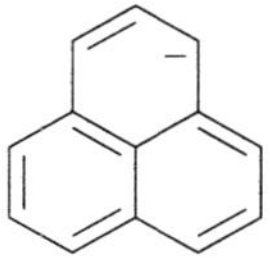

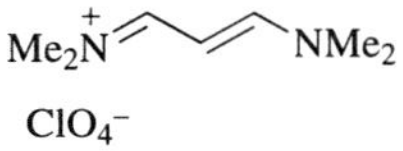

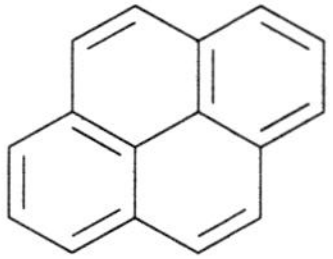

R^1	R^2	
H	Cl	(38)
H	NO$_2$	(30)
H	Me	(79)
Me	H	(88)
H	OMe	(63)
H	Et	(97)

TABLE IV. CARBOCYCLIC ANIONS (*Continued*)

Substrate	Reagents		Product(s) and Yield(s) (%)	Refs.
	R^1	R^2		
	Et	H	(87)	
	H	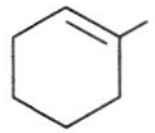 (3-methylcyclohexenyl)	(95)	
	H	Ph	(74)	
	Ph	H	(67)	
	H	4-ClC$_6$H$_4$	(87)	
	H	4-BrC$_6$H$_4$	(79)	
	H	4-MeOC$_6$H$_4$	(80)	
	H	1-C$_{10}$H$_7$	(83)	
	H	2-C$_{10}$H$_7$	(81)	
	Me$_2$N$^+$ =⟨ ⟩ NMe$_2$ ClO$_4^-$		(61)	65
	piperidino-cyclohexenyl =N$^+$ piperidine ClO$_4^-$		(83)	65

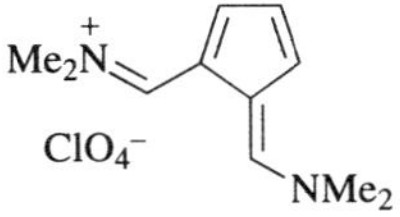

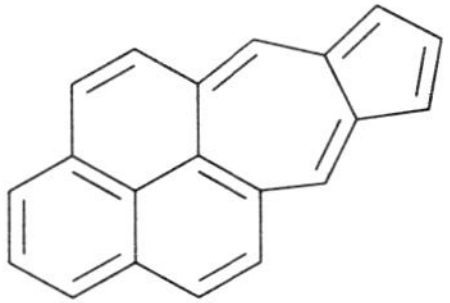

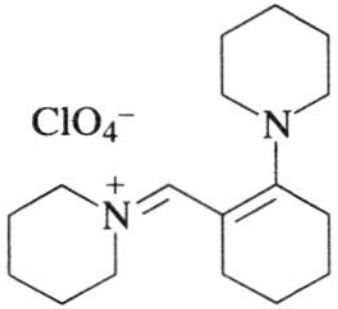

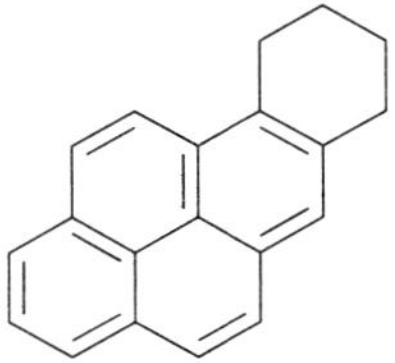

TABLE IV. CARBOCYCLIC ANIONS (Continued)

Substrate	Reagents	Product(s) and Yield(s) (%)	Refs.
	Me_2N^+=...NMe_2 ... NMe_2 ... Me_2N^+ ... $2ClO_4^-$	(68)	65
	$MeRN^+$—$\langle\ \rangle_n$—$NRMe$ ClO_4^-	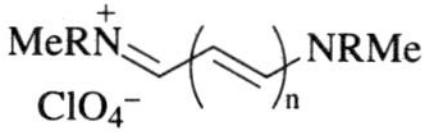 $\begin{array}{cc} R & n \\ \hline Me & 2 \\ Me & 3 \\ Ph & 1 \\ Ph & 2 \end{array}$	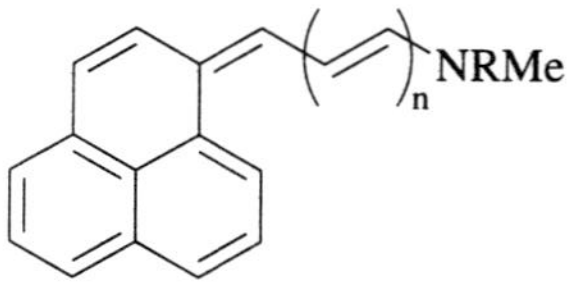 (72) (62) (95) (90)	65
C_{14} 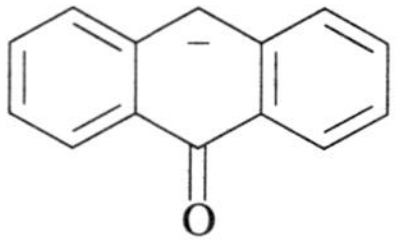	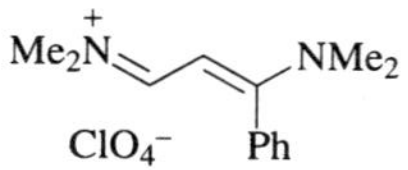 Me_2N^+ ... NMe_2 ... Ph ClO_4^-	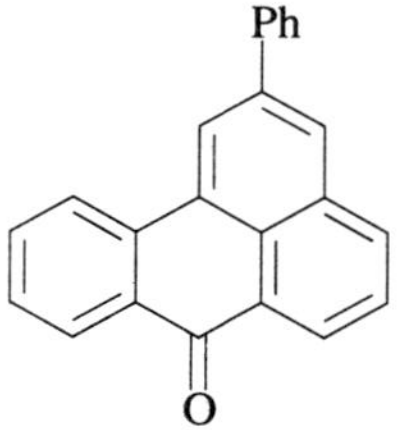 Ph ... O (87)	341

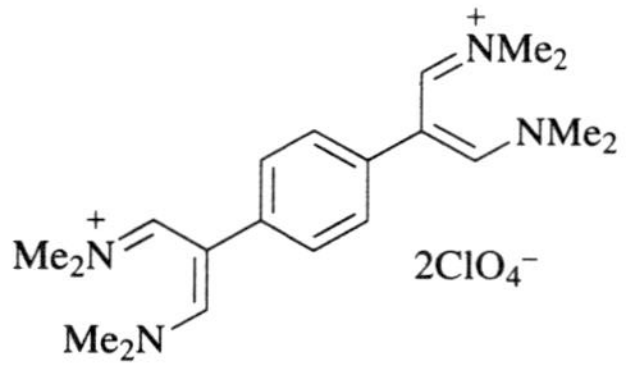

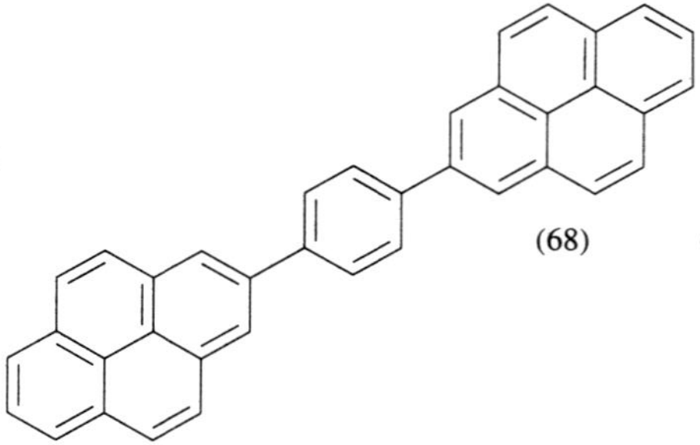

TABLE IV. CARBOCYCLIC ANIONS (Continued)

Substrate	Reagents	Product(s) and Yield(s) (%)	Refs.
C$_{19}$ Ph-pyrene	$Me_2\overset{+}{N}\diagup\diagup NMe_2$ ClO_4^-	Ph-pyrene (94)	65
C$_{17}$ fluorene (R^1, R^2, R^3, R^4)	C$_5$/C$_6$/C$_6$/C$_6$ $Me_2\overset{+}{N}\diagup(\diagup)_n NMe_2$ ClO_4^-	fluorene derivative with $=CH(CH=CH)_n NMe_2$ (R^1, R^2, R^3, R^4)	63

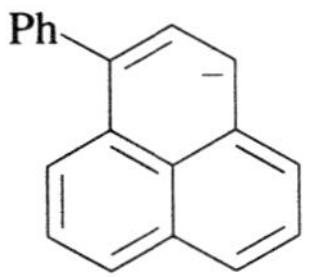

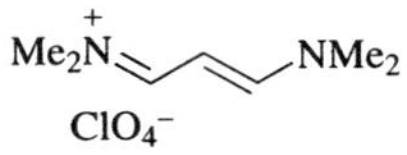

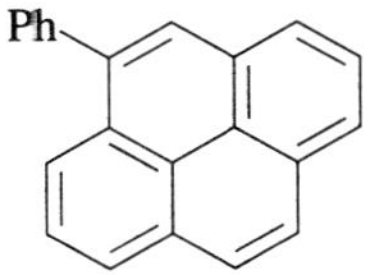

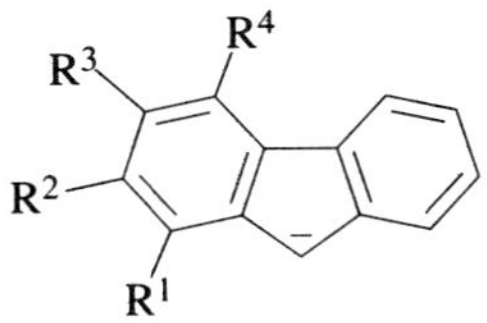

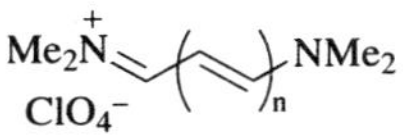

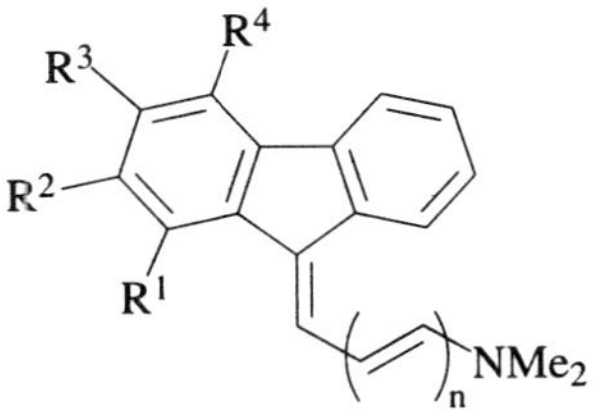

R^1	R^2	R^3	R^4	n	
CH=CHCH=CH	H		H	1	(95)
CH=CHCH=CH	H		H	2	(98)
H	CH=CHCH=CH		H	1	(70)
H	CH=CHCH=CH		H	2	(93)
H	H	CH=CHCH=CH		1	(96)
H	H	CH=CHCH=CH		2	(98)

TABLE IV. CARBOCYCLIC ANIONS (*Continued*)

Substrate	Reagents	Product(s) and Yield(s) (%)	Refs.
C$_{17}$	C$_6$/C$_6$/C$_6$/C$_6$ Me$_2$N$^+$=CH—CH=CH—NMe$_2$ ClO$_4^-$	(total 82)	65
C$_{19}$	C$_6$/C$_6$/C$_6$/C$_6$ Me$_2$N$^+$=CH—CH=CH—NMe$_2$ ClO$_4^-$	(total 33)	65

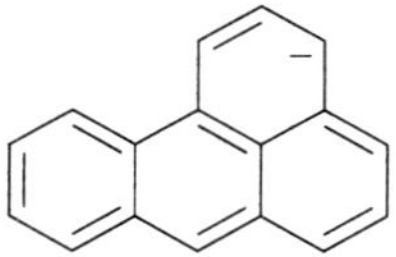

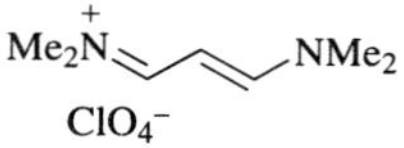

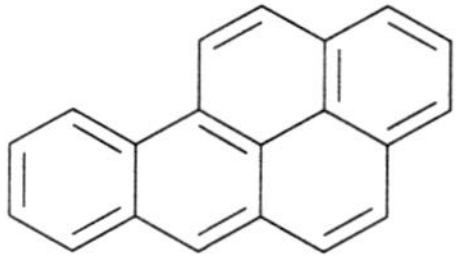

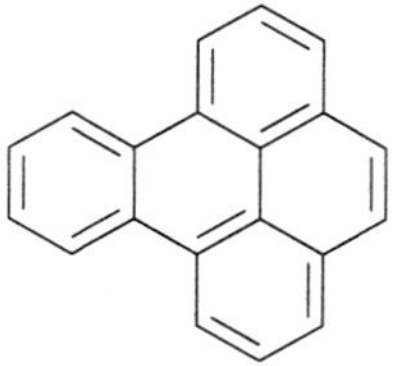

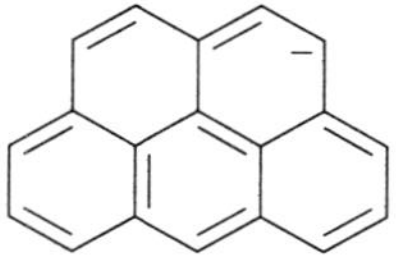

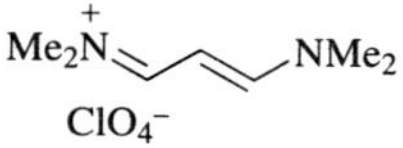

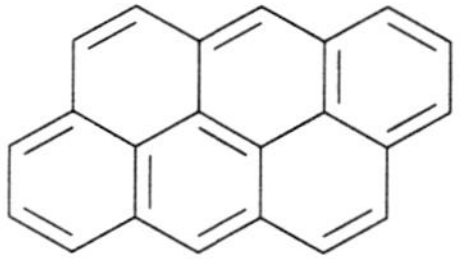

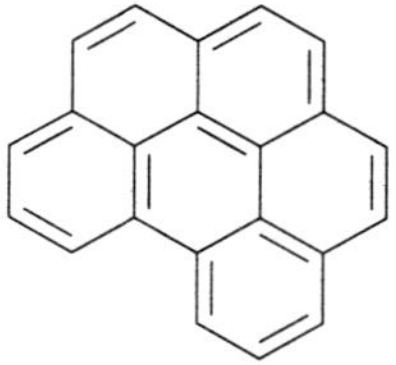

TABLE V. AZULENES

R^1 - R^8 = H except as indicated

Substrate	Reagents	Product(s) and Yield(s) (%)	Refs.
		A. Azulene	
C_{10}	DMF, POCl$_3$	R^1 = CHO (90-95)	342, 66, 343, 344
	DMF, POCl$_3$, 70°	R^1 = CHO (50) + R^1 = R^3 = CHO (43)	344
	DMF, POCl$_3$, 85°	R^3 = CHO (61)	66
	DMF, POCl$_3$	R^1 = CH=NMe$_2^+$ Cl$^-$ (77)	66
	—, PCl$_5$	R^1 = CHO (84)	345
	MFA, POCl$_3$	R^1 = CHO (85)	344
	DMA, POCl$_3$	R^1 = COMe (70)	66
	1. Me$_2$NCOBu-n, POCl$_3$	R^1 = C$_5$H$_{11}$-n (78)	67
	2. NaBH$_4$, BF$_3$		
	Me$_2$NCOC$_{11}$H$_{23}$-n, POCl$_3$	R^1 = COC$_{11}$H$_{23}$-n (82)	66
	Me$_2$NCOCH$_2$Cl	R^1 = COCH$_2$Cl (25)	346
	Et$_2$NCO(CH$_2$)$_2$CO$_2$Et, POCl$_3$	R^1 = CO(CH$_2$)$_2$CO$_2$Et (44)	67
	Et$_2$NCO(CH$_2$)$_3$CO$_2$Et, POCl$_3$	R^1 = CO(CH$_2$)$_3$CO$_2$Et (86)	67
	Me$_2$NCO(CH$_2$)$_6$CO$_2$Et, POCl$_3$	R^1 = CO(CH$_2$)$_6$CO$_2$Et (80)	66
	Me$_2$NCOPh, POCl$_3$	R^1 = COPh (22)	66
	1. PhN(Me)CH=CHCHO, POCl$_3$	R^1 = CH=CHCHO (95-97)	68
	2. ClO$_4^-$		
	3. OH$^-$		
	1. PhN(Me)(CH=CH)$_2$CHO, POCl$_3$	R^1 = (CH=CH)$_2$CHO (95-97)	68
	2. ClO$_4^-$		
	3. OH$^-$		

105

TABLE V. AZULENES (*Continued*)

Substrate	Reagents	Product(s) and Yield(s) (%)	Refs.
B. Monosubstituted Azulenes			
B1. R¹ Substituents			
C_{11} R^1 = Me	DMF, POCl$_3$	R^3 = CHO (95)	344, 345, 66
C_{14}			
R^1 = Bu-t	DMF, POCl$_3$	R^3 = CHO (80-100)	347
R^1 = (CH$_2$)$_4$OH	Et$_2$NCO(CH$_2$)$_2$CO$_2$Et, POCl$_3$	R^1 = (CH$_2$)$_4$Cl, R^3 = CO(CH$_2$)$_2$CO$_2$Et (67)	67
C_{15} R^1 = C$_5$H$_{11}$-n	1. Me$_2$NCOBu-n, POCl$_3$ 2. NaBH$_4$, BF$_3$	R^3 = C$_5$H$_{11}$-n (78)	67
R^1 = (CH$_2$)$_5$OH	Et$_2$NCO(CH$_2$)$_3$CO$_2$Et, POCl$_3$	R^1 = (CH$_2$)$_5$Cl, R^3 = CO(CH$_2$)$_3$CO$_2$Et (48)	67
C_{17} R^1 = Bn	DMF, POCl$_3$	R^3 = CHO (80-100)	347
C_{18} R^1 = CH(Me)Ph	DMF, POCl$_3$	R^3 = CHO (80-100)	347
C_{19} R^1 = C(Me)$_2$Ph	DMF, POCl$_3$	R^3 = CHO (80-100)	347
C_{23} R^1 = CHPh$_2$	DMF, POCl$_3$	R^3 = CHO (80-100)	347
C_{24} R^1 = C(Me)Ph$_2$	DMF, POCl$_3$	R^3 = CHO (80-100)	347
C_{29} R^1 = CPh$_3$	DMF, POCl$_3$	R^3 = CHO (80-100)	347
B2. R² Substituents			
C_{19}	DMF, POCl$_3$	(80)	348

TABLE V. AZULENES (Continued)

Substrate	Reagents	Product(s) and Yield(s) (%)	Refs.
B3. R^6 Substituents			
C_{18} $R^6 = C_8H_{17}\text{-}n$	DMF, POCl$_3$	$R^1 = CHO$ (24)	349
C. Disubstituted Azulenes			
C1. R^1, R^2 Substituents			
C_{21}	DMF, POCl$_3$	(59)	350
C2. R^1, R^5 Substituents			
C_{15} $R^1 = Et$, $R^5 = i\text{-}Pr$	DMF, POCl$_3$	$R^3 = CHO$ (—)	351
	DMA, POCl$_3$	$R^3 = COMe$ (—)	352
C3. R^4, R^7 Substituents			
C_{12} $R^4 = R^7 = Me$	—, POCl$_3$	$R^1 = CHO$ (91)	344
C_{13} $R^4 = Me$, $R^7 = Et$ + $R^4 = Et$, $R^7 = Me$	DMF, POCl$_3$	$R^1 = CHO$ (68) + $R^3 = CHO$ (13)	353
C_{14} $R^4 = Me$, $R^7 = i\text{-}Pr$	DMF, POCl$_3$	$R^1 = CHO$ (—)	345
$R^4 = Me$, $R^7 = i\text{-}Pr$ + $R^4 = i\text{-}Pr$, $R^7 = Me$	DMF, POCl$_3$	$R^1 = CHO$ (61) + $R^3 = CHO$ (15)	353

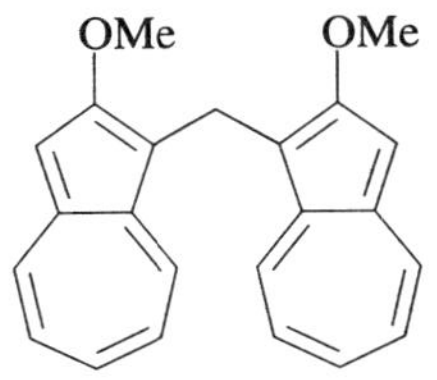

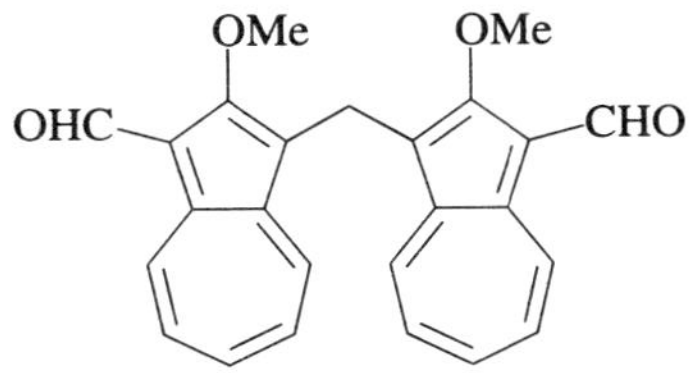

107

TABLE V. AZULENES (*Continued*)

Substrate	Reagents	Product(s) and Yield(s) (%)	Refs.
C4. R^4, R^8 Substituents			
C_{12}			
$R^4 = R^8 = Me$	DMF, POCl$_3$	$R^1 = CHO$ (94)	344
D. Trisubstituted Azulenes			
D1. R^1, R^4, R^7 Substituents			
C_{13}			
$R^1 = R^4 = R^7 = Me$	DMF, POCl$_3$	$R^3 = CHO$ (96)	344
C_{15}			
$R^1 = R^4 = Me, R^7 = i\text{-}Pr$	DMF, POCl$_3$	$R^3 = CHO$ (96)	344, 345
	DMF or MFA, POCl$_3$	$R^3 = CHO$ (—)	354
	MFA, POCl$_3$	$R^3 = CHO$ (87)	344
	MFA, POCl$_3$	$R^3 = CHO$ (29) + $R^5 = CHO$ (trace)	355
D2. R^2, R^4, R^6 Substituents			
C_{15}			
$R^2 = R^4 = Me, R^6 = i\text{-}Pr$	DMF, POCl$_3$	$R^1 = CHO$ (90)	344
D3. R^2, R^4, R^7 Substituents			
C_{15}			
$R^2 = R^4 = Me, R^7 = i\text{-}Pr$	DMF, POCl$_3$	$R^1 = CHO$ (98)	344
	MFA, POCl$_3$	$R^1 = CHO$ (88)	344
	DMF or MFA, POCl$_3$	$R^1 = CHO$ (—)	354
	DMF, POCl$_3$, 70°	$R^1 = R^3 = CHO$ (94)	344
D4. R^2, R^4, R^8 Substituents			
C_{15}			
$R^2 = i\text{-}Pr, R^4 = R^8 = Me$	DMF, POCl$_3$	$R^1 = CHO$ (98)	344
	MFA, POCl$_3$	$R^1 = CHO$ (84)	344
	DMF or MFA, POCl$_3$	$R^1 = CHO$ (—)	354

TABLE V. AZULENES (*Continued*)

Substrate	Reagents	Product(s) and Yield(s) (%)	Refs.
	D5. R^3, R^5, R^8 Substituents		
C_{15}			
$R^3 = R^8 = Me$, $R^5 = i\text{-}Pr$	1. PhN(Me)CH=CHCHO, POCl$_3$ 2. ClO$_4^-$ 3. OH$^-$	$R^1 = CH=CHCHO$ (—)	68
	1. PhN(Me)(CH=CH)$_2$CHO, POCl$_3$ 2. ClO$_4^-$ 3. OH$^-$	$R^1 = (CH=CH)_2CHO$ (—)	68
	D6. R^4, R^6, R^8 Substituents		
C_{15}			
$R^4 = R^6 = R^8 = Me$	DMF, POCl$_3$	$R^1 = CHO$ (99)	345, 66
	DMF, PCl$_5$	$R^1 = CHO$ (92)	66
	DMF, COCl$_2$	$R^1 = CHO$ (87)	345
	DMF, AlCl$_3$	$R^1 = CHO$ (50)	345
	DMF, POCl$_3$	$R^1 = CH=NMe_2^+$ Cl$^-$ (83)	356, 357
	PhN(Me)CH=CHCHO, POCl$_3$	$R^1 = CH=CHCH=N(Me)Ph^+$ Cl$^-$ (—)	357
	E. Tetrasubstituted Azulenes **E1. R^1, R^2, R^4, R^7 Substituents**		
C_{16}			
$R^1 = R^2 = R^4 = Me$, $R^7 = i\text{-}Pr$	DMF, POCl$_3$	$R^3 = CHO$ (92)	3 44
	E2. R^1, R^2, R^4, R^8 Substituents		
C_{16}			
$R^1 = R^4 = R^8 = Me$, $R^2 = i\text{-}Pr$	DMF, POCl$_3$	$R^3 = CHO$ (72)	344
	MFA, POCl$_3$	$R^3 = CHO$ (48)	344

TABLE V. AZULENES (*Continued*)

Substrate	Reagents	Product(s) and Yield(s) (%)	Refs.

E3. R^1, R^3, R^4, R^7 Substituents

C$_{16}$

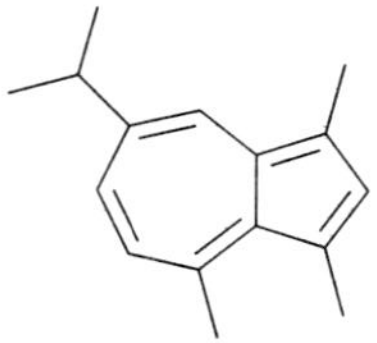

DMF, POCl$_3$

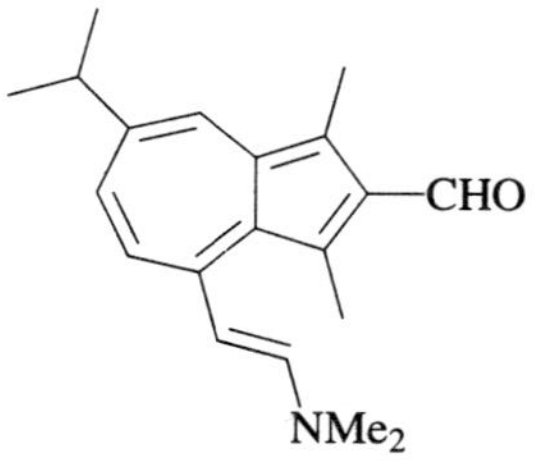

CHO   (76)

NMe$_2$

358

TABLE VI. OTHER POLYCYCLIC NONBENZENOID HYDROCARBONS

Substrate	Reagents	Product(s) and Yield(s) (%)	Refs.
C$_{12}$	**C$_7$/C$_7$** DMF, POCl$_3$	(38)	69
C$_{12}$	**C$_5$/C$_6$/C$_6$** DMF, POCl$_3$ DMF, Tf$_2$O	(11) 	359 47
C$_{14}$	**C$_5$/C$_6$/C$_7$** DMF, POCl$_3$	(91)	344

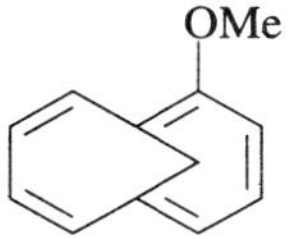

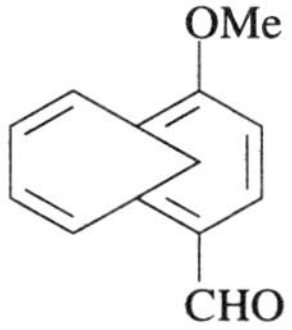

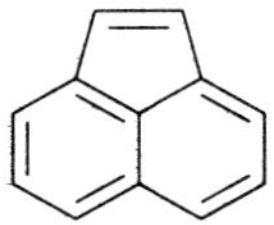

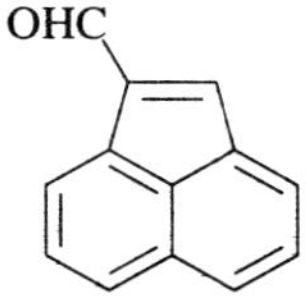

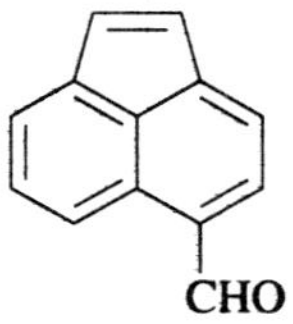

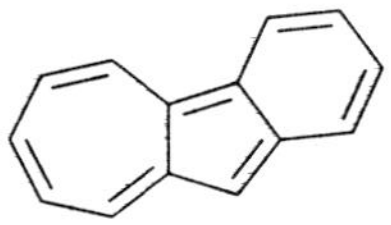

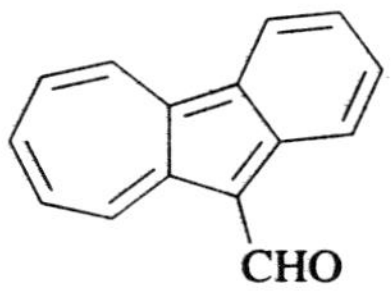

111

TABLE VI. OTHER POLYCYCLIC NONBENZENOID HYDROCARBONS (*Continued*)

Substrate	Reagents	Product(s) and Yield(s) (%)	Refs.
C_{24}	$C_3/C_3/C_5/C_8$		
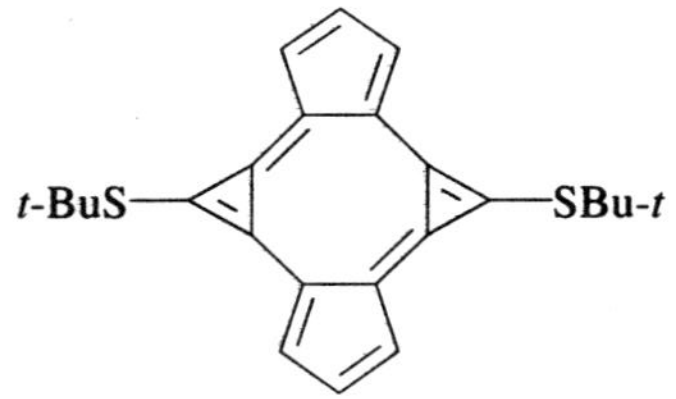	1. DMF, POCl$_3$ 2. NaClO$_4$ 3. NaHCO$_3$	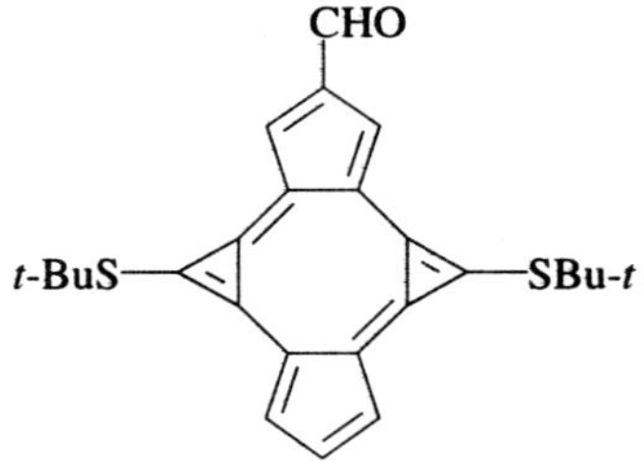 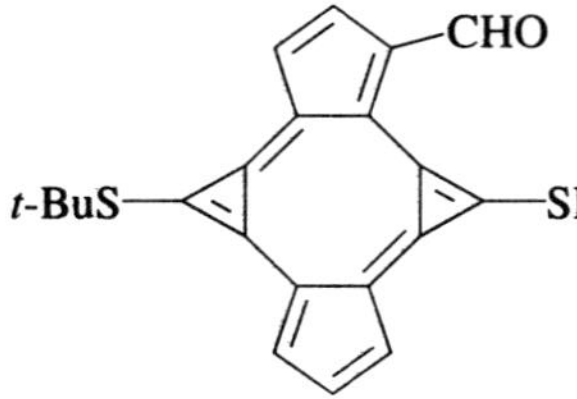 SBu-*t* (69) + *t*-BuS (13)	70
C_{20}	$C_5/C_5/C_6/C_6/C_7$		
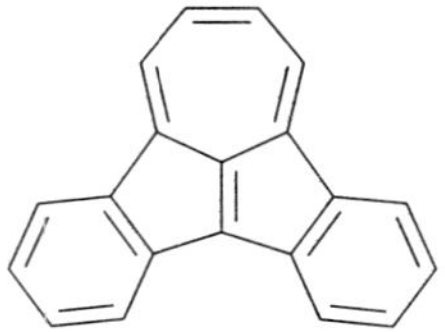	DMF, POCl$_3$	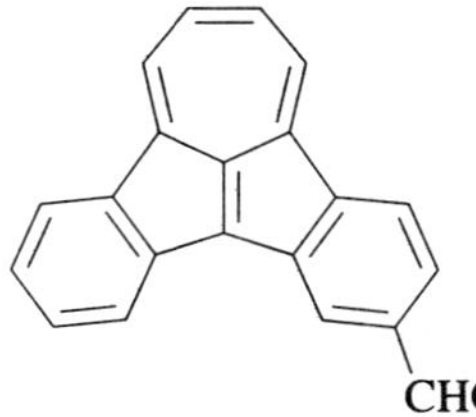 (84)	360

112

TABLE VII. CARBOCYCLIC ORGANOMETALLICS

Substrate	Reagents	Product(s) and Yield(s) (%)	Refs.
C_{17-18}		C_7	
R¹=Ph, R²=H; R¹=4-MeC₆H₄, R²=H; R¹=Ph, R²=Me	DMF, POCl₃	(70) (70) (60)	72, 361 72, 361 361
C_{11}		C_8	
	DMF, POCl₃	(60)	362
C_{16}		C_4/C_4	
	MFA, POCl₃	(50)	363

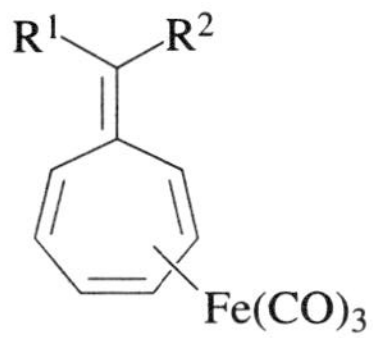

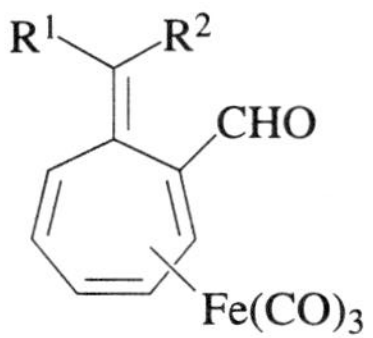

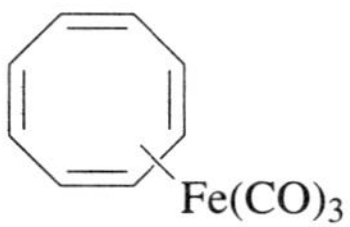

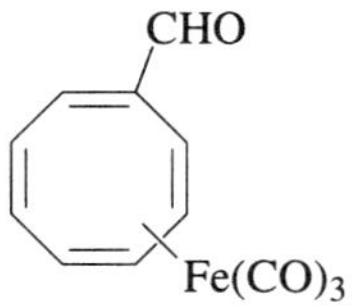

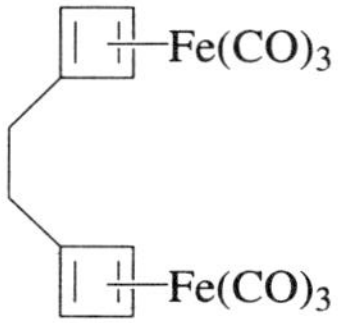

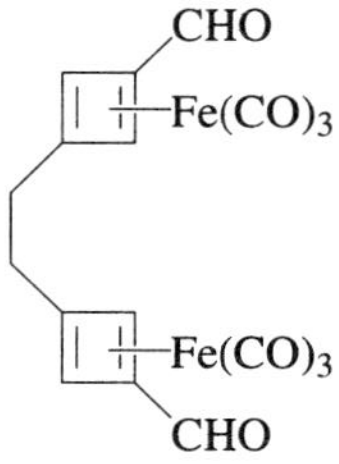

113

TABLE VII. CARBOCYCLIC ORGANOMETALLICS (*Continued*)

Substrate	Reagents	Product(s) and Yield(s) (%)	Refs.
C₄/C₅			
C₃₃ Co complex (Cp)(Ph, Ph, Ph, Ph-cyclobutadiene)	MFA, POCl₃	CHO-Cp–Co–(Ph, Ph, Ph, Ph-cyclobutadiene) (8)	73
C₅/C₅			
C₁₀ Fe(Cp)(Cp–R) R = Cl, Br	MFA, POCl₃	CHO–Fe–Cp–R (24) (42)	369
Ferrocene Fe(Cp)₂	MFA, POCl₃ (xs) MFA, POCl₃	CHO–Fe–Cp **I** + CHO–Fe–CHO **II** I / II : (0) (55) ; (77) (0)	71 368, 71, 365, 366, 367, 369

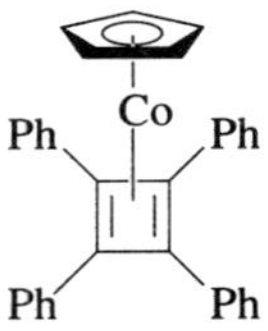

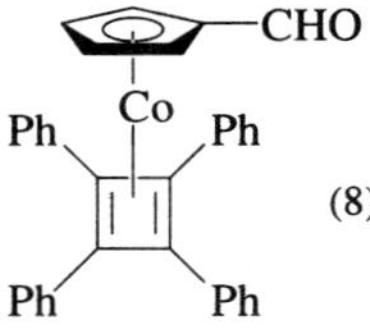

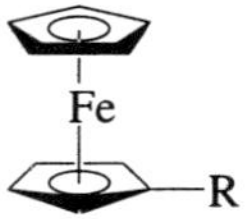

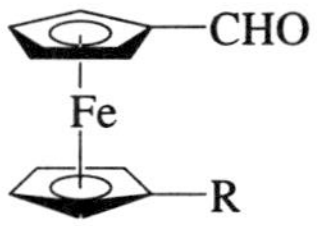

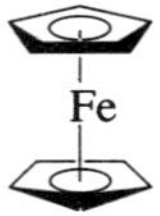

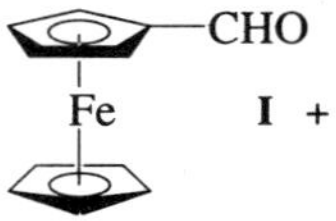

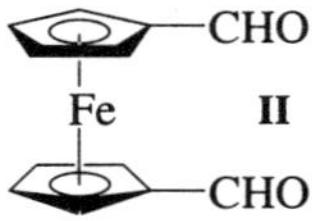

114

TABLE VII. CARBOCYCLIC ORGANOMETALLICS (*Continued*)

Substrate	Reagents	Product(s) and Yield(s) (%)	Refs.
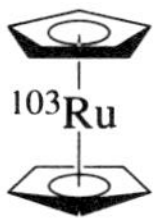 ^{103}Ru sandwich	—	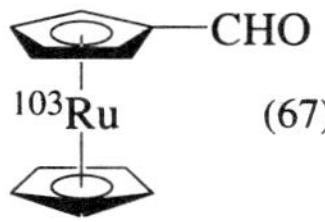 ^{103}Ru—CHO (67)	370
C$_{12-14}$ Fe sandwich, R,R	MFA, POCl$_3$	**I** + **II** + **III** (ferrocene aldehydes)	371
Fe sandwich, R	MFA, POCl$_3$	OHC—Fe—R	372

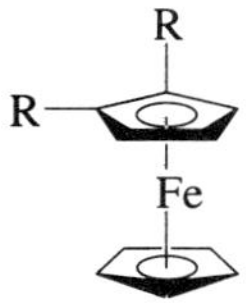

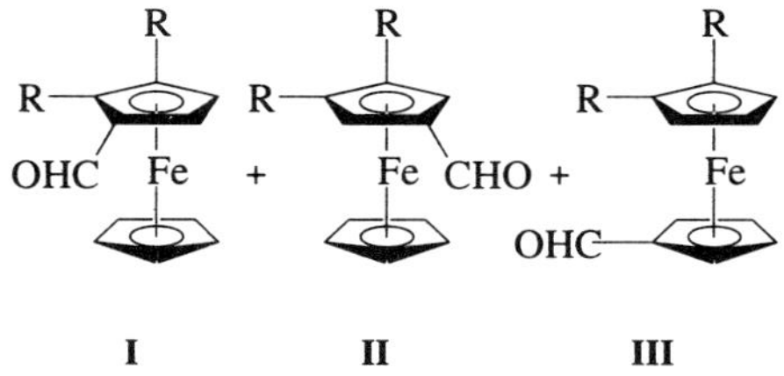

For the ferrocene substrate (Ref 371):

	I	**II**	**III**
R			
Me	(19)	(50)	(19)
(CH$_2$)$_4$	(20)	(45)	(11)

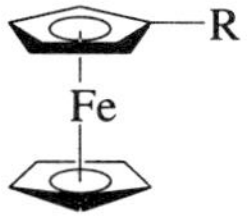

For the Fe substrate (Ref 372):

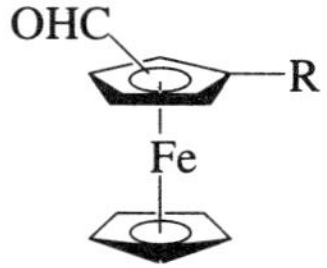

R	
Et	(58)
(CH$_2$)$_2$CO$_2$Me	(65)

TABLE VII. CARBOCYCLIC ORGANOMETALLICS (*Continued*)

Substrate	Reagents	Product(s) and Yield(s) (%)	Refs.
C$_{13-14}$ 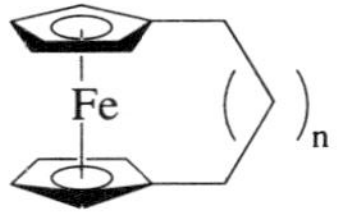		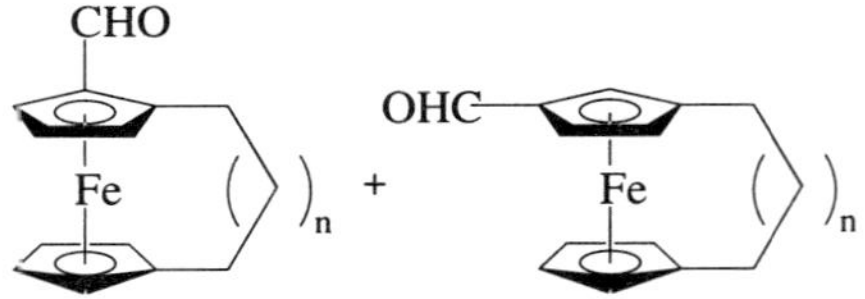	
(n = 1)	DMF, POCl$_3$	I (4) II (60)	373
	MFA, POCl$_3$	(0) (60)	372
	MeNHCHO, POCl$_3$	(—) (60)	373
(2)	DMF, POCl$_3$	(3) (64)	373
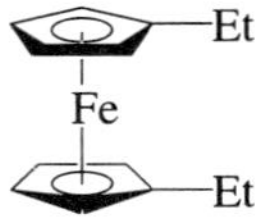	MFA, POCl$_3$	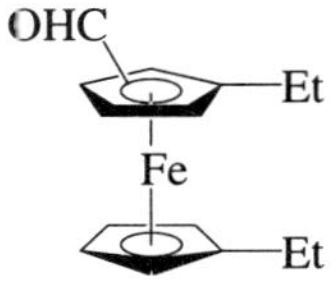 I + II	372
C$_{14}$	DMF, POCl$_3$	I (70) II (18)	373
	MFA, POCl$_3$	(53) (6)	374
	MeNHCHO, POCl$_3$, CH$_2$Cl$_2$	(70)	373
	MeNHCHO, POCl$_3$	(92) (8)	373

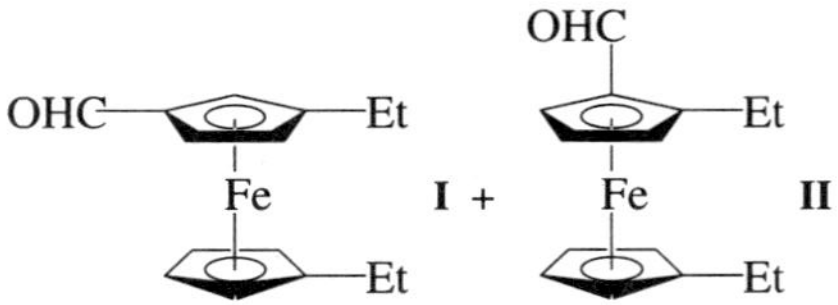

116

TABLE VII. CARBOCYCLIC ORGANOMETALLICS (*Continued*)

Substrate	Reagents	Product(s) and Yield(s) (%)	Refs.

C$_{19\text{-}22}$

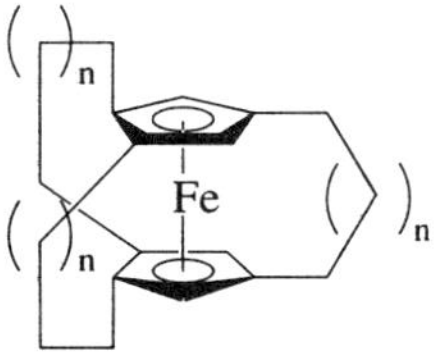

$$\begin{array}{c} n \\ \hline 1 \\ 2 \end{array}$$

DMF, POCl$_3$

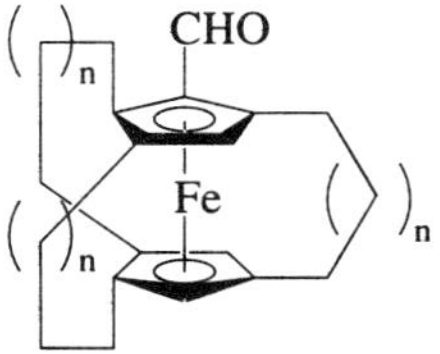

(91)
(96)

375
376

C$_5$/C$_5$/C$_5$/C$_5$

C$_{20}$

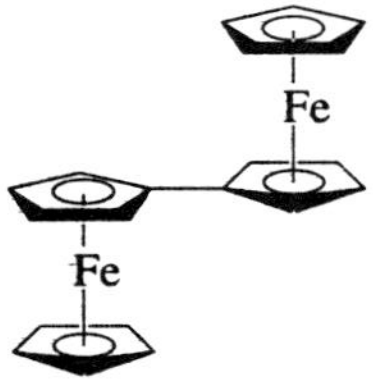

MFA, POCl$_3$

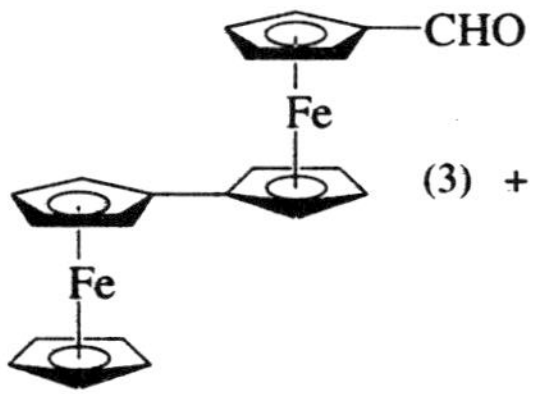

(3) +

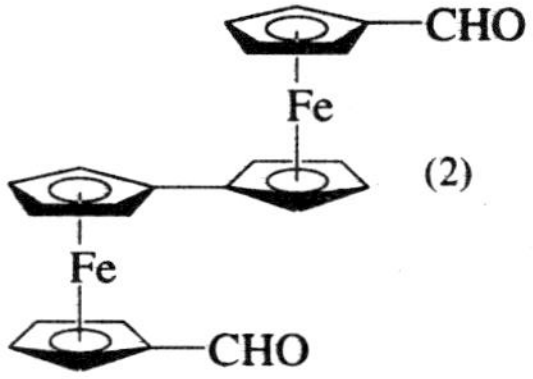

(2)

377

117

TABLE VIII. FURANS

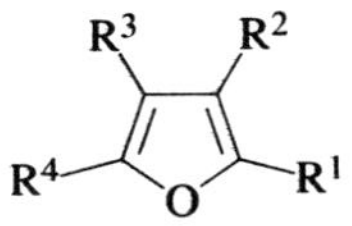

R¹ - R⁴ = H except as indicated

Substrate	Reagents	Product(s) and Yield(s) (%)	Refs.
A. Furan			
C₄	DMF, POCl₃	R¹ = CHO (64)	378
	DMF, (Cl₂PO)₂O	R¹ = CHO (71)	26
	Me₂N¹⁴CHO, POCl₃	R¹ = ¹⁴CHO (68)	76, 379, 380
	Me₂NCDO, POCl₃	R¹ = CDO (35)	76
B. Monosubstituted Furans			
B1. R¹ Substituents			
C₅ R¹ = Me	DMF, POCl₃	R⁴ = CHO (76)	378
	DMF, COCl₂	R⁴ = CHO (95)	381
	DMF, (Cl₂PO)₂O	R⁴ = CHO (77)	26
	1. DMF, POCl₃ 2. NH₄SH 3. BrCH₂CH=CH₂, NaH	R⁴ = CH₂SCH₂CH=CH₂ (50-60)	382
C₆ R¹ = Et	DMF, POCl₃	R⁴ = CHO (90)	384, 378, 383
C₇ R¹ = i-Pr	DMF, POCl₃	R⁴ = CHO (68)	76

118

TABLE VIII. FURANS (*Continued*)

Substrate	Reagents	Product(s) and Yield(s) (%)	Refs.
C$_{7-13}$ R^1 = (cyclopropyl-R) $\dfrac{R}{H \; Me \; Ph}$	DMF, POCl$_3$	R^4 = CHO (45) R^4 = CHO (63) R^4 = CHO (62)	385
C$_{8-17}$ R^1 = (pyrrolyl, N–H)	DMF, POCl$_3$	CHO (62) + CHO (14)	74
R^1 = (imidazolyl, N–Me)	DMF, POCl$_3$	(32)	386
R^1 = 4-ClC$_6$H$_4$ R^1 = 4-BrC$_6$H$_4$	DMF, POCl$_3$ DMF, POCl$_3$	R^4 = CHO (80) R^4 = CHO (76)	76 76
R^1 = (pyrrolyl-CO$_2$Me, N–H)	DMF, POCl$_3$	CO$_2$Me (88)	74
R^1 = 1-C$_{10}$H$_7$ R^1 = 2-C$_{10}$H$_7$	DMF, POCl$_3$ DMF, POCl$_3$	R^4 = CHO (82) R^4 = CHO (85)	76 76

119

TABLE VIII. FURANS (*Continued*)

Substrate	Reagents	Product(s) and Yield(s) (%)	Refs.
$E^1 =$ 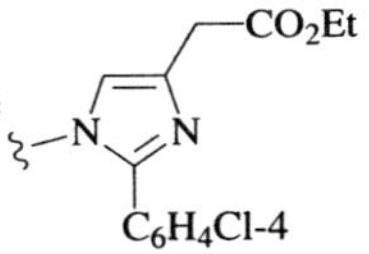 (imidazole with CO_2Et, C_6H_4Cl-4)	DMF, POCl$_3$	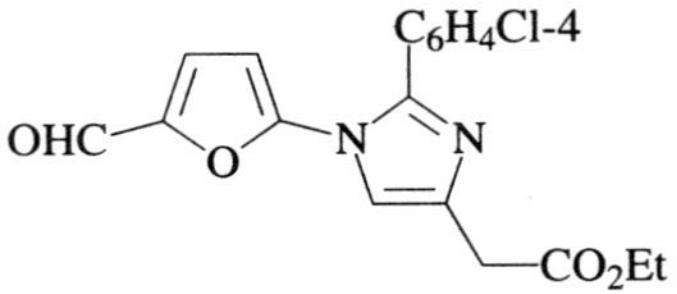 (furan-imidazole with C_6H_4Cl-4, CO_2Et, OHC) (49)	387
C$_{10}$ (diketone with furan and chlorofuran, OH)	DMF, POCl$_3$	(34) + (26)	388
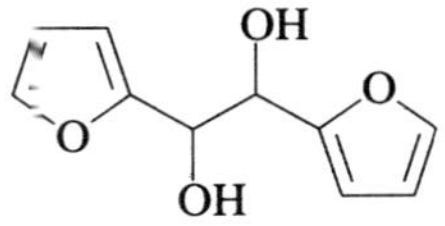 (diol with two furans, OH, OH)	DMF, POCl$_3$	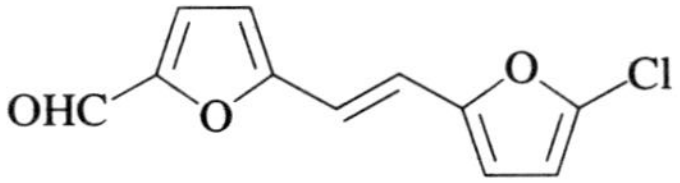 (Cl-furan-furan-CHO) (52)	388
C$_{11}$ $R^1 =$ (benzothiazole)	DMF, POCl$_3$	(benzothiazole-furan-OHC) (—)	389

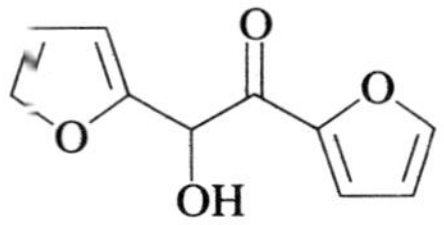

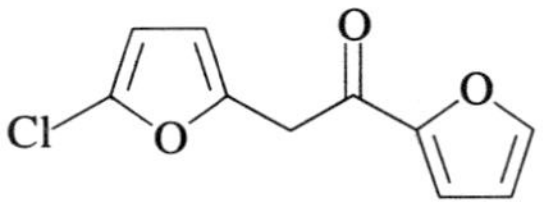

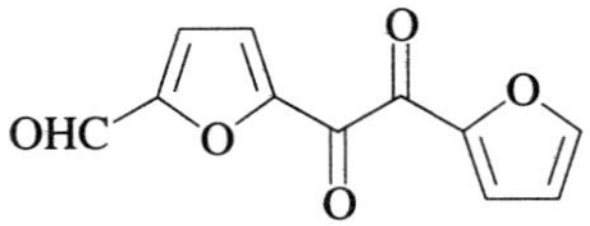

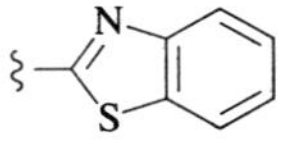

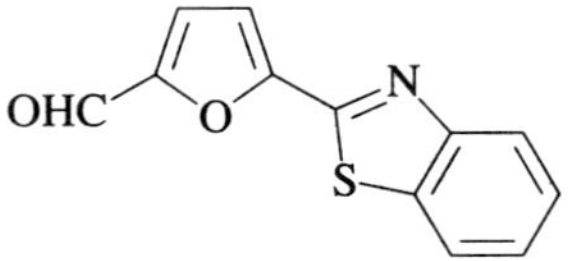

TABLE VIII. FURANS (*Continued*)

Substrate	Reagents	Product(s) and Yield(s) (%)	Refs.
C$_{12}$			
R^1 = CH$_2$OBn	1. DMF, POCl$_3$ 2. NH$_4$SH 3. LiAlH$_4$ 4. BrCH$_2$CH=CH$_2$, NaH	R^4 = CH$_2$SCH$_2$CH=CH$_2$ (50–60)	382
C$_{13}$			
R^1 = (CH$_2$)$_2$O$_2$CPh	DMF, POCl$_3$	R^4 = CHO (86)	390
R^1 = (CH$_2$)$_7$CO$_2$Me	—	R^4 = CHO (—)	391
C$_{14-15}$			
R^1 = CH=C(CO$_2$Me)Ar	DMF, POCl$_3$	R^4 = CHO	392
$\underline{\text{Ar}}$ 4-ClC$_6$H$_4$ 4-MeC$_6$H$_4$ 4-MeOC$_6$H$_4$		(70) (68) (91)	
C$_{16}$	DMF, POCl$_3$	R^4 = CHO (69)	393
C$_{30}$ R^1 = (CH$_2$)$_2$C$_{10}$H$_7$-1	DMF, POCl$_3$	 α anomer (64) β anomer (96)	394
B2. R^2 Substituents			
C$_5$			
R^2 = Me	DMF, POCl$_3$	R^1 = CHO (75) + R^4 = CHO (8)	76, 395
	1. DMF, POCl$_3$ 2. NH$_4$SH 3. LiAlH$_4$ 4. BrCH$_2$CH=CH$_2$, NaH	R^4 = CH$_2$SCH$_2$CH=CH$_2$ (50–60)	382

TABLE VIII. FURANS (*Continued*)

Substrate	Reagents	Product(s) and Yield(s) (%)	Refs.
C$_7$ R^2 = i-Pr	DMF, POCl$_3$	R^1 = CHO (61)	76
C$_8$	DMF, POCl$_3$	(56) + (12) + (8)	74
C$_{10}$	DMF, POCl$_3$	(56)	74

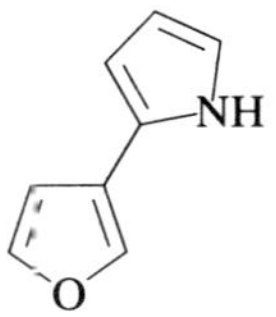

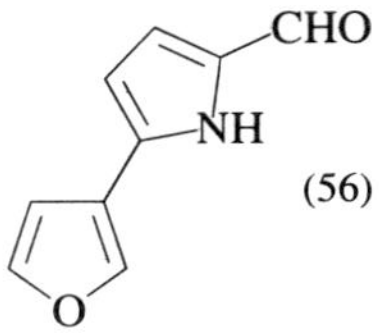

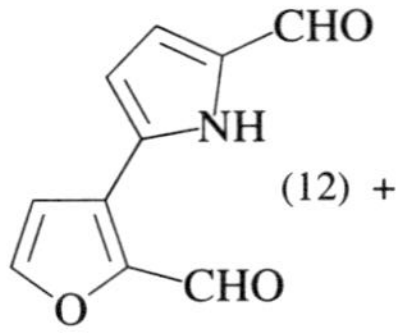

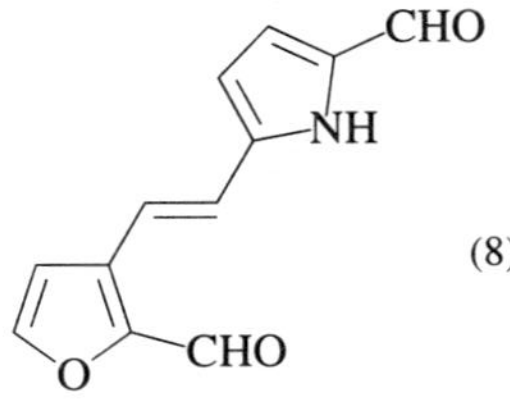

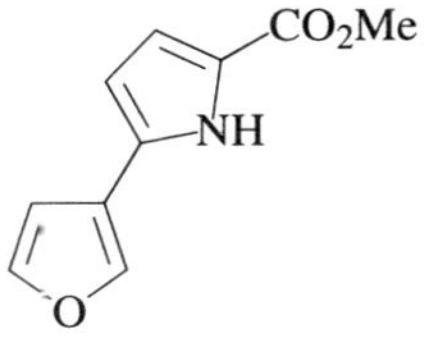

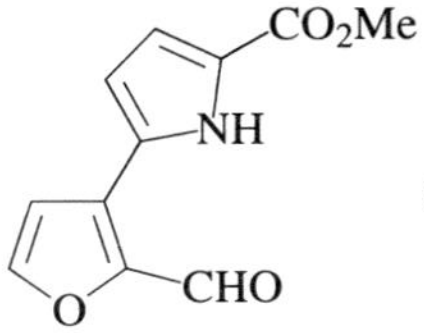

CHO, NH, O, CO$_2$Me

TABLE VIII. FURANS (Continued)

Substrate	Reagents	Product(s) and Yield(s) (%)	Refs.

C₂₄

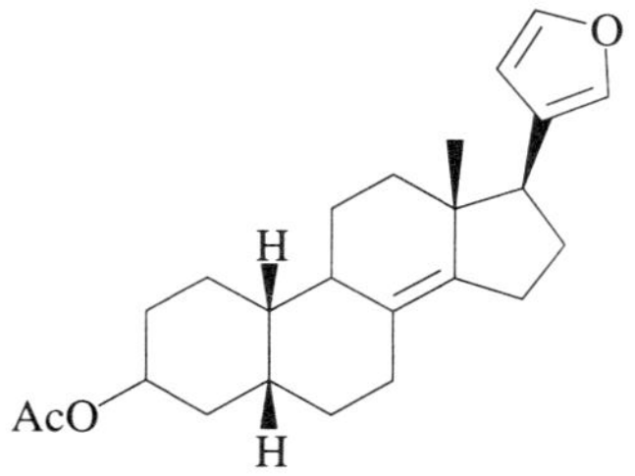

DMF, POCl₃

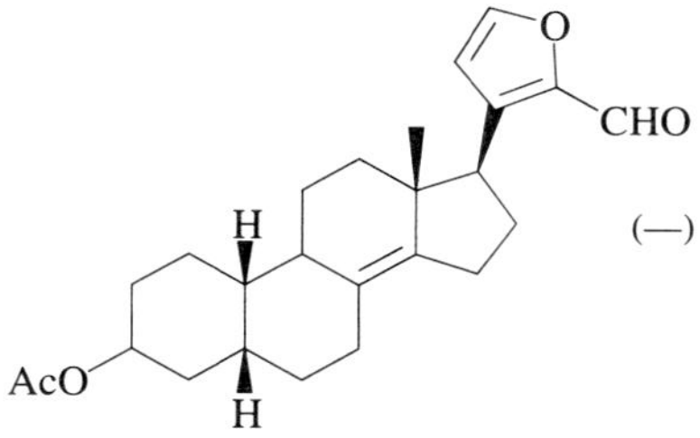

(—)   396

C. Disubstituted Furans
C1. R¹, R² Substituents

C₆

$R^1 = Me, R^2 = CN$	DMF, POCl₃	$R^4 = CHO$ (10-15)	397
$R^1 = R^2 = Me$	DMF, POCl₃	$R^4 = CHO$ (96)	398, 399, 400
	1. DMF, POCl₃ 2. NH₄SH 3. LiAlH₄ 4. BrCH₂CH=CH₂, NaH	$R^4 = CH_2SCH_2CH=CH_2$ (50-60)	382
	DMA, POCl₃	$R^4 = COMe$ (67)	400

C₇

| $R^1 = Me, R^2 = COMe$ | DMF, POCl₃ | $R^4 = CHO$ (20) | 397 |

C₈

| $R^1 = Me, R^2 = CO_2Et$ | DMF, POCl₃ | $R^4 = CHO$ (79) | 397 |

C₁₀

| $R^1 = n\text{-}Pr, R^2 = CO_2Et$ | DMF, POCl₃ | $R^4 = CHO$ (85) | 397 |
| $R^1 = R^2 = i\text{-}Pr$ | DMF, POCl₃ | $R^4 = CHO$ (68) | 76 |

C₁₂

| $R^1 = Me, R^2 = 4\text{-}MeOC_6H_4$ | DMF, POCl₃ | $R^4 = CHO$ (66) | 397 |

123

TABLE VIII. FURANS (*Continued*)

Substrate	Reagents	Product(s) and Yield(s) (%)	Refs.
C_{13}			
R^1 = Ph, R^2 = CO_2Et	DMF, $POCl_3$	R^4 = CHO (68)	397
R^1 = Me, R^2 = CH_2OBn	1. DMF, $POCl_3$ 2. NH_4SH 3. $LiAlH_4$ 4. $BrCH_2CH=CH_2$, NaH	R^4 = $CH_2SCH_2CH=CH_2$ (50-60)	382
C_{15}	DMF, $POCl_3$	CHO (86)	401, 402
C2. R^1, R^3 Substituents			
C_6			
R^1 = R^3 = Me	1. DMF, $POCl_3$ 2. NH_4SH 3. $LiAlH_4$ 4. $BrCH_2CH=CH_2$, NaH	R^4 = $CH_2SCH_2CH=CH_2$ (50-60)	382
C_{11}			
R^1 = Me, R^3 = Ph	DMF, $POCl_3$	R^4 = CHO (84)	403
C3. R^1, R^4 Substituents			
C_{16}			
R^1 = R^3 = Ph	DMF, $POCl_3$	R^4 = CHO (40)	387
	DMF, $POCl_3$	R^2 = CHO (40)	387

TABLE VIII. FURANS (*Continued*)

Substrate	Reagents	Product(s) and Yield(s) (%)	Refs.
C4. R², R³ Substituents			
C_8	DMF, POCl$_3$	(40)	404
C_{16} R² = R³ = Ph	DMF, POCl$_3$	R¹ = CHO (80)	405
C_{18} R² = R³ = 4-MeOC$_6$H$_4$	DMF, POCl$_3$	R¹ = CHO (—)	405
C_{20} R² = R³ = CH$_2$OBn	1. DMF, POCl$_3$ 2. NH$_4$SH 3. LiAlH$_4$ 4. BrCH$_2$CH=CH$_2$, NaH	R¹ = CH$_2$SCH$_2$CH=CH$_2$ (50-60)	382
D. Trisubstituted Furans **D1. R¹, R², R³ Substituents**			
C_7 R¹ = R² = R³ = Me	1. DMF, POCl$_3$ 2. NH$_4$SH 3. LiAlH$_4$ 4. BrCH$_2$CH=CH$_2$, NaH	R⁴ = CH$_2$SCH$_2$CH=CH$_2$ (50-60)	382
C_{15} R¹ = CH(Me)OBn, R² = R³ = OMe	DMF, POCl$_3$	R⁴ = CHO (13)	406

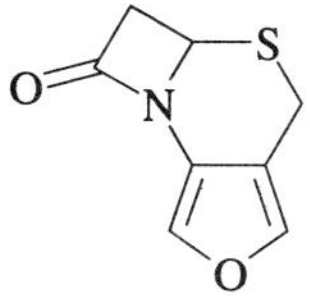

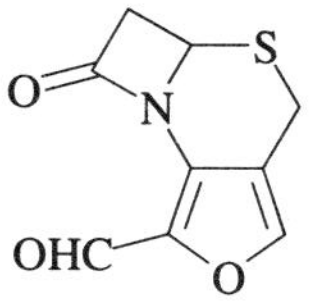

TABLE IX. THIOPHENES

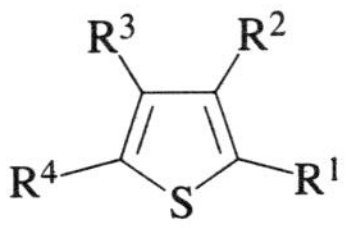

R^1 - R^4 = H except as indicated

Substrate	Reagents	Product(s) and Yield(s) (%)	Refs.
	A. Thiophene		
C_4	MFA, POCl$_3$	R^1 = CHO (76)	80, 79, 180, 407, 408
	DMF or MFA, POCl$_3$	R^1 = CHO (—)	409
	DMF, POCl$_3$	R^1 = CHO (72)	192, 409
	DMF, (Cl$_2$PO)$_2$O	R^1 = CHO (60)	26
	MFA, (Cl$_2$PO)$_2$O	R^1 = CHO (75)	26
	PhN(Et)CHO, POCl$_3$	R^1 = CHO (—)	80
	DMF, Ph$_3$PBr$_2$	R^1 = CHO (45)	35
	DMF, 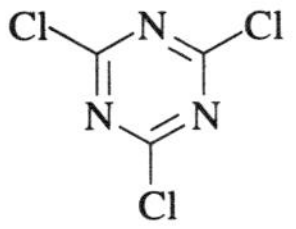 (triazine structure)	R^1 = CHO (5)	36
	Me$_2$NCDO, POCl$_3$	R^1 = CDO (30)	76
	1. MFA, POCl$_3$ 2. NH$_2$OH 3. Dehydrate	R^1 = CN (25)	197
	B. Monosubstituted Thiophenes **B1. R^1 Substituents**		
C_4 R^1 = Cl	MFA, POCl$_3$	R^4 = CHO (59)	407, 79, 80, 408

126

TABLE IX. THIOPHENES (Continued)

Substrate	Reagents	Product(s) and Yield(s) (%)	Refs.
R^1 = Br	DMF or MFA, $POCl_3$	R^4 = CHO (—)	409
	DMF, $POCl_3$	R^4 = CHO (43)	192, 409
	MFA, $POCl_3$	R^1 = Cl, R^4 = CHO (58)	79
	MFA, $POCl_3$	R^4 = CHO (44)	80
	MFA, $POBr_3$	R^4 = CHO (70)	80, 79
C₅ R^1 = Me	MFA, $POCl_3$	R^4 = CHO (80-85)	79, 80, 407
	DMF or MFA, $POCl_3$	R^4 = CHO (—)	409
	DMF, $POCl_3$	R^4 = CHO (66)	192, 409
C₅₋₉ R^1 = OR	MFA, $POCl_3$	R^4 = CHO (66)	
R — Me	MFA, $POCl_3$	(58)	410
	DMF, $(Cl_2PO)_2O$	(83)	26
	MFA, $(Cl_2PO)_2O$	(80)	26
Et	MFA, $POCl_3$	(55)	410
n-Pr	MFA, $POCl_3$	(59)	410
n-Bu	MFA, $POCl_3$	(40)	410
n-C_5H_{11}	MFA, $POCl_3$	(63)	410
C₆ R^1 = NHAc	MFA, $POCl_3$	R^4 = CHO (—)	411
	DMF, $POCl_3$	R^4 = CHO (47)	192
R^1 = Et	MFA, $POCl_3$	R^4 = CHO (75-80)	79
	MFA, $POCl_3$	R^4 = CHO (—)	408

TABLE IX. THIOPHENES (*Continued*)

Substrate	Reagents	Product(s) and Yield(s) (%)	Refs.
C$_{6-9}$ $\quad$ R^1 = NR$_2$ $\quad$ $\dfrac{\text{R}_2}{\text{Me}}$ (CH$_2$)$_4$ (CH$_2$)$_5$ [(CH$_2$)$_2$]O	DMF, POCl$_3$	R^4 = CHO (60) (45) (60) (80)	412
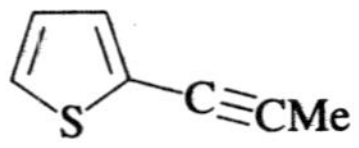 C$_7$ $\quad$ (thiophene–C≡CMe)	MFA, POCl$_3$	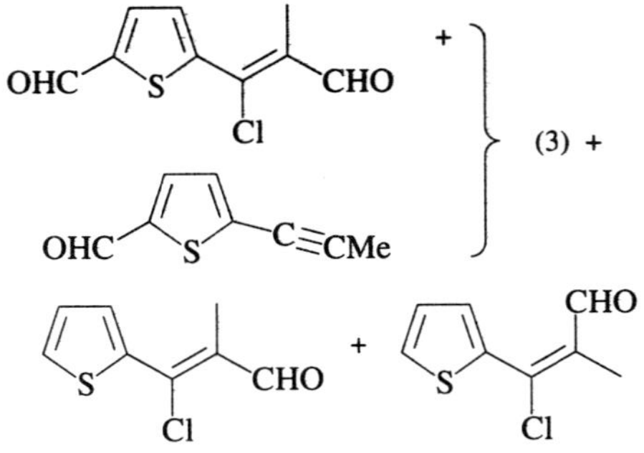 [isomeric chlorovinyl aldehyde products] (31) + (3) + (7)	413
C$_8$ $\quad$ R^1 = n-Pr	MFA, POCl$_3$	R^4 = CHO (80-85)	79, 414
R^1 = n-Bu	DMF or MFA, POCl$_3$	R^4 = CHO (—)	415
R^1 = i-Bu	MFA, POCl$_3$	R^4 = CHO (77)	408
R^1 = t-Bu	DMF, POCl$_3$	R^4 = CHO (76)	80, 192

TABLE IX. THIOPHENES (Continued)

Substrate	Reagents	Product(s) and Yield(s) (%)	Refs.
C_{8-26}			
$R^1 = Ar$ (Ar structure)			
Ar = 2-Thienyl	DMF, POCl$_3$	R^4 = CHO (48)	416
3-ClC$_6$H$_4$	MFA, POCl$_3$	R^4 = CHO (70)	417
4-ClC$_6$H$_4$	MFA, POCl$_3$	R^4 = CHO (81)	418
4-BrC$_6$H$_4$	DMF, POCl$_3$	R^4 = CHO (71)	419
	MFA, POCl$_3$	R^4 = CHO (80)	418
Ph	MFA, POCl$_3$	R^4 = CHO (87)	417, 418, 420
3-MeC$_6$H$_4$	DMF, POCl$_3$	R^4 = CHO (68)	419
4-MeC$_6$H$_4$	MFA, POCl$_3$	R^4 = CHO (78)	417
4-MeOC$_6$H$_4$	MFA, POCl$_3$	R^4 = CHO (83)	418
	MFA, POCl$_3$	R^4 = CHO (85)	418, 421
2,4-(Ph)$_2$C$_6$H$_3$	MFA, POCl$_3$	R^4 = CHO (—)	417
3,5-(4-MeC$_6$H$_4$)$_2$C$_6$H$_3$	MFA, POCl$_3$	R^4 = CHO (82)	417
(pyrrole with H–N, CO$_2$Me, thienyl substituent)	DMF, POCl$_3$	(pyrrole with CO$_2$Me, OHC, thienyl) (85)	74
(terthiophene: three fused/linked thiophene rings, S)	DMF, POCl$_3$	(terthiophene dialdehyde OHC···CHO) (75)	422, 416
C_9			
$R^1 = n$-C$_5$H$_{11}$	DMF or MFA, POCl$_3$	R^4 = CHO (—)	415
$R^1 = i$-C$_5$H$_{11}$	DMF or MFA, POCl$_3$	R^4 = CHO (—)	415

129

TABLE IX. THIOPHENES (Continued)

Substrate	Reagents	Product(s) and Yield(s) (%)	Refs.
C_{10} $\quad$ $R^1 = n\text{-}C_6H_{13}$	DMF or MFA, POCl$_3$	$R^4 = CHO$ (—)	415
C_{10-13} $\quad$ $R^1 = CH=CHAr$			
Ar			
$\quad$ 2-Thienyl	DMF, POCl$_3$	$R^4 = CHO$ (93)	423
$\quad$ 4-ClC$_6$H$_4$	DMF, POCl$_3$	$R^4 = CHO$ (—)	424
$\quad$ Ph	DMF, POCl$_3$	$R^4 = CHO$ (72–90)	424, 425
$\quad$ 4-MeOC$_6$H$_4$	DMF, POCl$_3$	$R^4 = CHO$ (61)	425
(bithienyl vinyl structure)	DMF, POCl$_3$	_(OHC-substituted structure)_ (58)	423
(bis(bithienyl) structure, ×2)	DMF, POCl$_3$	_(OHC-substituted structure, ×2)_ (42)	423
C_{11} $\quad$ $R^1 = Bn$	MFA, POCl$_3$	$R^4 = CHO$ (43)	408
$\quad$ $R^1 = CH_2C_6H_{11}$	DMF or MFA, POCl$_3$	$R^4 = CHO$ (—)	415
$\quad$ $R^1 = n\text{-}C_7H_{15}$	DMF or MFA, POCl$_3$	$R^4 = CHO$ (—)	415
C_{12} $\quad$ $R^1 = CH_2Bn$	DMF or MFA, POCl$_3$	$R^4 = CHO$ (—)	415
$\quad$ $R^1 = n\text{-}C_8H_{17}$	DMF or MFA, POCl$_3$	$R^4 = CHO$ (—)	415
(bis-thienyl piperazine structure)	DMF, POCl$_3$	_(OHC / CHO-substituted piperazine structure)_ (50)	412

TABLE IX. THIOPHENES (*Continued*)

Substrate	Reagents	Product(s) and Yield(s) (%)	Refs.
C_{13}			
$R^1 = (CH_2)_3C_6H_{11}$	DMF or MFA, POCl$_3$	$R^4 = $ CHO (—)	415
C_{14}			
$R^1 = N(Et)CH=C(CO_2Et)_2$	MFA, —	$R^4 = $ CHO (70)	426
$R^1 = n\text{-}C_{10}H_{21}$	DMF or MFA, POCl$_3$	$R^4 = $ CHO (—)	415
C_{15}			
$R^1 = CH_2(C_6H_4Bu\text{-}t\text{-}4)$	DMF or MFA, POCl$_3$	$R^4 = $ CHO (—)	415
$R^1 = n\text{-}C_{11}H_{23}$	MFA, POCl$_3$	$R^4 = $ CHO (57)	414
C_{16}			
$R^1 = (CH_2)_3CH(C_6H_{11})Et$	DMF or MFA, POCl$_3$	$R^4 = $ CHO (—)	415
$R^1 = n\text{-}C_{12}H_{25}$	MFA, POCl$_3$	$R^4 = $ CHO (37)	414
$C_{16\text{-}17}$			
$R^1 = C(Ar^1)=CH(Ar^2)$	DMF or MFA, POCl$_3$	$R^4 = $ CHO	424

Ar^1	Ar^2
Ph	4-ClC$_6$H$_4$
4-ClC$_6$H$_4$	Ph
Ph	Ph
4-MeOC$_6$H$_4$	4-ClC$_6$H$_4$

Substrate	Reagents	Product(s) and Yield(s) (%)	Refs.
		(—)	
		(—)	
		(—)	
		(—)	
C_{18}			
$R^1 = n\text{-}C_{14}H_{29}$	MFA, POCl$_3$	$R^4 = $ CHO (71)	414
$C_{21\text{-}23}$			

$R^1 = $

R			
H	DMF, POCl$_3$	$R^4 = $ CHO (60)	427
OMe	DMF, POCl$_3$	$R^4 = $ CHO (—)	427
CO$_2$H	DMF, POCl$_3$	$R^4 = $ CHO (50)	427
CO$_2$Me	DMF, POCl$_3$	$R^4 = $ CHO (92)	427

131

TABLE IX. THIOPHENES (*Continued*)

Substrate	Reagents	Product(s) and Yield(s) (%)	Refs.
C$_{22}$ R^1 = (CH$_2$)$_{13}$ [cyclopentyl]	DMF or MFA, POCl$_3$	R^4 = CHO (—)	415
B2. R^2 Substituents			
C$_4$ R^2 = Br	DMF, POCl$_3$	R^1 = CHO (70)	76
C$_5$ R^2 = Me	DMF, POCl$_3$	R^1 = CHO (33)	192, 76, 428
	DMF, POCl$_3$	R^1 = CHO (53) + R^4 = CHO (9)	429
	MFA, POCl$_3$	R^1 = CHO (83)	80
	MFA, POCl$_3$	R^1 = CHO (80-85)	79
	DMF, COCl$_2$	R^1 = CHO (99)	430
C$_6$ R^2 = OMe	DMF, POCl$_3$	R^1 = CHO (88)	78
R^2 = NHAc	DMF, POCl$_3$ moderate conditions	R^1 = CHO (73)	431
	DMF, POCl$_3$ forcing conditions	R^1 = CHO (4) + R^2 = N=CHNMe$_2$ (7) + R^1 = CHO, R^2 = N=CHNMe$_2$ (15)	431
[thiophene–CH$_2$CN]	DMF, POCl$_3$	[fused thieno-pyridine with CHO, Cl] (34)	431
C$_8$ R^2 = 2-Thienyl	—	R^4 = CHO (—)	432
R^2 = SBu-n	MFA, POCl$_3$	R^1 = CHO (72)	433
R^2 = SeBu-n	MFA, POCl$_3$	R^1 = CHO (70)	433

TABLE IX. THIOPHENES (*Continued*)

Substrate	Reagents	Product(s) and Yield(s) (%)	Refs.
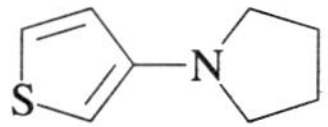 *(pyrrolidin-3-yl thiophene)*	DMF, POCl$_3$	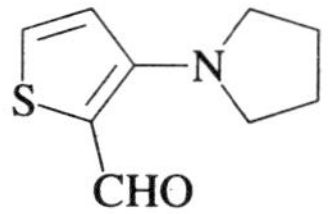 *(3-pyrrolidino-2-thiophenecarbaldehyde)* (25)	412
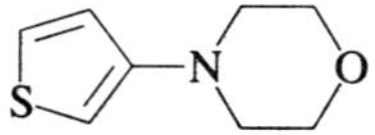 *(morpholin-3-yl thiophene)*	DMF, POCl$_3$	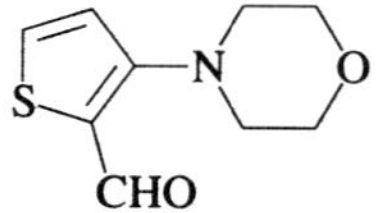 *(3-morpholino-2-thiophenecarbaldehyde)* (55)	412
C$_9$ *(bis-thienylmethane)*	DMF, POCl$_3$	*(dithienobenzene)* (33)	434
C$_{10}$ R^2 = Ph	DMF, POCl$_3$	R^1 = CHO + R^4 = CHO (80) 94:6	77
R^2 = (CH$_2$)$_3$CO$_2$Et	—	R^1 = CHO (—)	435
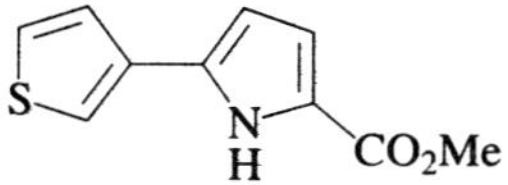 *(methyl 5-thienylpyrrole-2-carboxylate)*	DMF, POCl$_3$	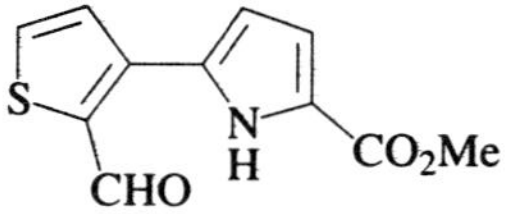 *(CHO + CO$_2$Me pyrrole product)* (4) 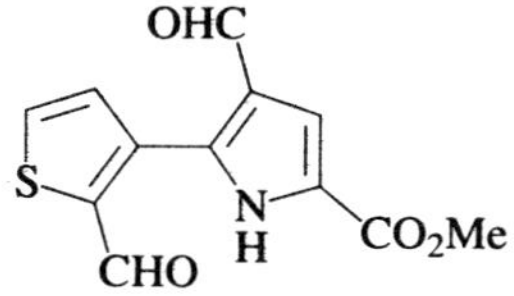 + *(85)*	74

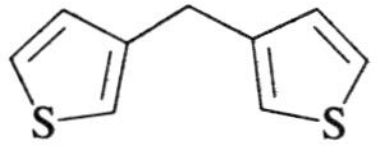

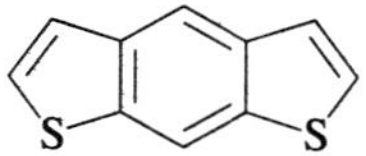

TABLE IX. THIOPHENES (Continued)

Substrate	Reagents	Product(s) and Yield(s) (%)	Refs.

C_{23}

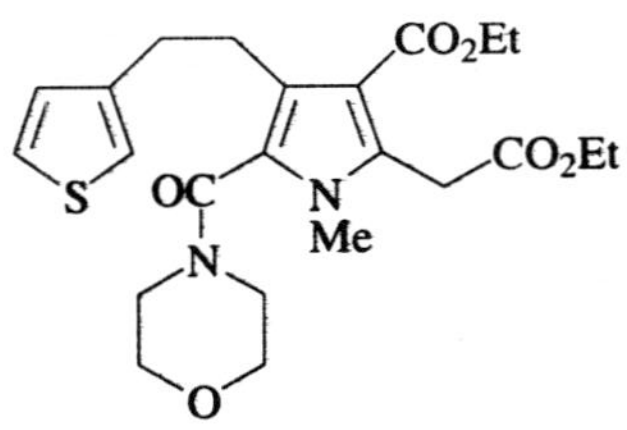

POCl₃

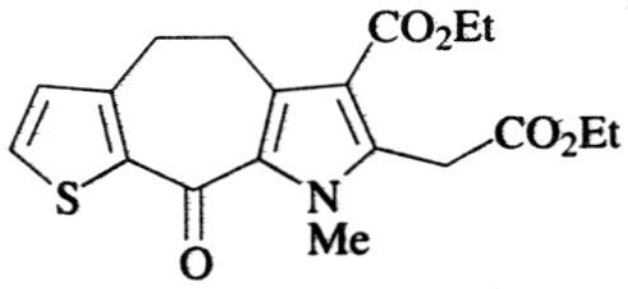

(56)

436

C. Disubstituted Thiophenes
C1. R¹, R² Substituents

C_6

$R^1 = OMe, R^2 = CO_2Me$	DMF, POCl₃	$R^1 = Cl, R^4 = CHO$ (71)	437
$R^1 = SMe, R^2 = CO_2Me$	DMF, POCl₃	$R^4 = CHO$ (69)	437

C_7

$R^1 = OMe, R^2 = CO_2Me$	DMF, POCl₃	$R^4 = CHO$ (60)	437
$R^1 = Et, R^2 = Me$	DMF, POCl₃	$R^4 = CHO$ (58)	428

C_8

$R^1 + R^2 = (CH_2)_4$	MFA, POCl₃	$R^4 = CHO$ (83)	438

C_{10}

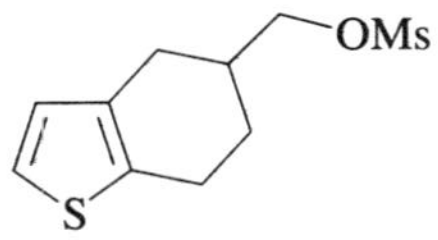

DMF, POCl₃

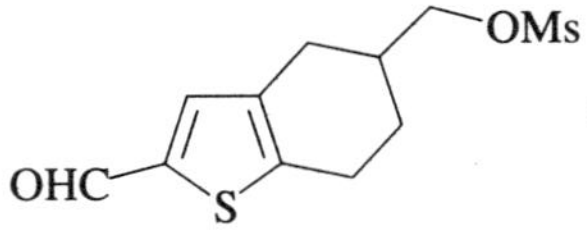

(65)

439

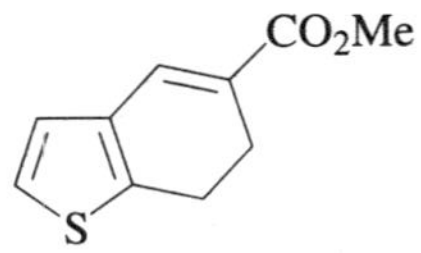

DMF, POCl₃

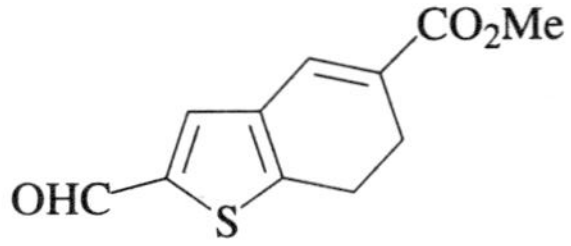

(88)

439

134

TABLE IX. THIOPHENES (*Continued*)

Substrate	Reagents	Product(s) and Yield(s) (%)	Refs.
C_{12}			
$R^1 = Me, R^2 = CH_2OPh$	1. DMF, $POCl_3$ 2. NH_4SH 3. $LiAlH_4$ 4. $BrCH_2CH=CH_2$	$R^4 = CH_2SCH_2CH=CH_2$ (50-70)	440
C2. R^1, R^3 Substituents			
C_6			
$R^1 = CO_2H, R^3 = OMe$	DMF, $POCl_3$	$R^4 = CHO$ (25)	437
C_{12}			
$R^1 = R^3 = t\text{-Bu}$	DMF, $POCl_3$	$R^4 = CHO$ (—)	441
C_{13}			
$R^1 = Me,$ $R^3 = (CH_2)_2(C_6H_3Br(Cl)\text{-}2,4)$	DMF, $POCl_3$	$R^4 = CHO$ (92)	442
C_{16}			
$R^1 = R^3 = Ph$	DMF, $POCl_3$ DMF or MFA, $POCl_3$	$R^4 = CHO$ (88) $R^4 = CHO$ (—)	443 417
C_{18}			
$R^1 = R^3 = 4\text{-MeC}_6H_4$	MFA, $POCl_3$	$R^4 = CHO$ (82)	417
C_{19}			
$R^1 = N(Me)Bn, R^3 = Bn$	DMF, $POCl_3$	$R^4 = CHO$ (74)	444
C3. R^1, R^4 Substituents			
C_6			
$R^1 = OMe, R^4 = CO_2H$	1. DMF, $POCl_3$ 2. MeOH	$R^1 = Cl, R^4 = CO_2Me$ (5)	437
$R^1 = R^4 = Me$	MFA, $POCl_3$	$R^2 = CHO$ (20)	80, 407

TABLE IX. THIOPHENES (*Continued*)

Substrate	Reagents	Product(s) and Yield(s) (%)	Refs.
C7 (MeO-methylthiophene structure)	MFA, POCl₃	(structure) (37)	83
R¹ = Me, R⁴ = NHAc	DMF, POCl₃	R³ = CHO (62)	78
R¹ = Et, R⁴ = OMe	DMF, POCl₃, 20°	R³ = CHO (61)	83
C7-8 (EtO-Me(Et)thiophene structure)	MFA, POCl₃, 50–60°	Ph(Me)N= ...Me(Et) (53)	83

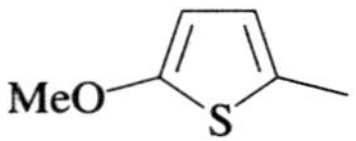

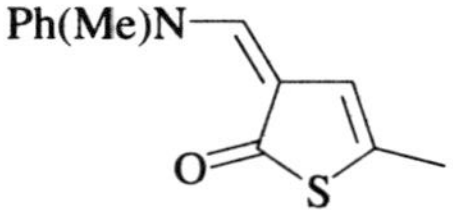

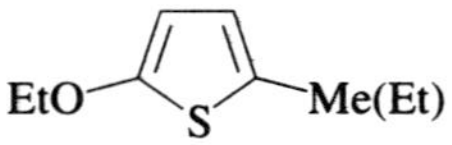

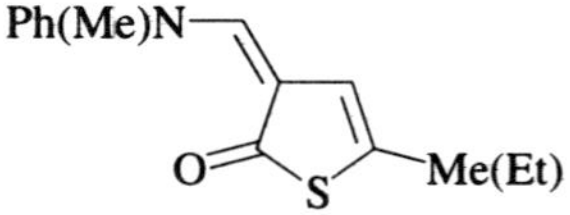

C4. R², R³ Substituents

Substrate	Reagents	Product(s) and Yield(s) (%)	Refs.
C6 R² = R³ = Me	DMF, POCl₃	R¹ = CHO (63)	76
R² = R³ = OMe	DMF, POCl₃	R¹ = CHO (69)	76
C7 R² = SCH₂CO₂Me, R³ = Br	MFA, POCl₃	R¹ = CHO (—)	445
C8-18 R² = R³ = NHCOR			
$\dfrac{\text{R}}{\text{CH}_2\text{Br}}$	DMF, POCl₃, Cl(CH₂)₂Cl, 83°	R¹ = CHO (—)	
Me	DMF, POCl₃, Cl(CH₂)₂Cl, 83°	R¹ = CHO (35)	
(imidazo-thiophene structure)	DMF, POCl₃ (xs), Cl(CH₂)₂Cl, 83°	(—)	446

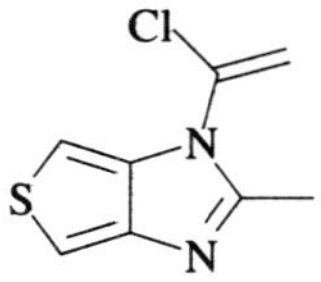

136

TABLE IX. THIOPHENES (Continued)

Substrate	Reagents	Product(s) and Yield(s) (%)	Refs.
	DMF, POCl$_3$ (xs), 0°	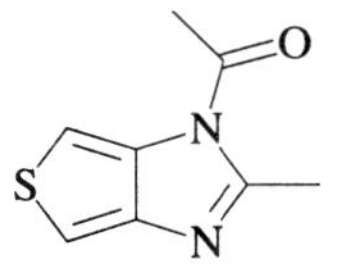 (33)	446
Et	DMF, POCl$_3$ (xs)	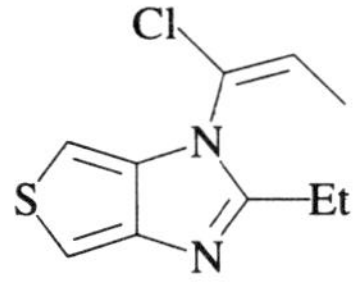 (39)	446
	DMF, POCl$_3$, Cl(CH$_2$)$_2$Cl, 83°	R^1 = CHO (60)	446
t-Bu	DMF, POCl$_3$, Cl(CH$_2$)$_2$Cl, 83°	R^1 = CHO (90)	446
Ph	DMF, POCl$_3$, Cl(CH$_2$)$_2$Cl, 83°	R^1 = CHO (80)	446
C$_{18}$ R^2 = R^3 = 4-MeOC$_6$H$_4$	DMF, POCl$_3$	R^1 = CHO (45)	76

D. Trisubstituted Thiophenes
D1. R^1, R^2, R^3 Substituents

Substrate	Reagents	Product(s) and Yield(s) (%)	Refs.
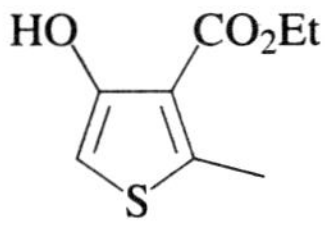 C$_8$	DMF, POCl$_3$, 100°	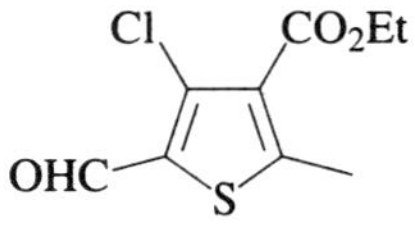 (53)	81, 82
	DMF, POCl$_3$, 30-50°	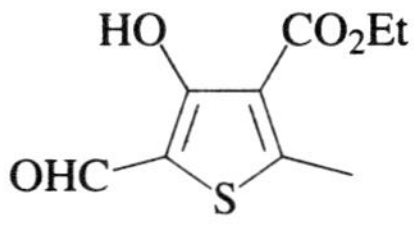 (78)	82

137

TABLE IX. THIOPHENES (Continued)

Substrate	Reagents	Product(s) and Yield(s) (%)	Refs.
C$_{20}$ (morpholino-3,4-diphenylthiophene)	Me$_2$NCOPh, POCl$_3$	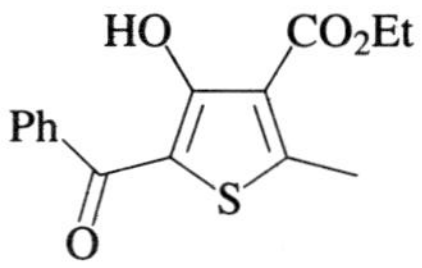 (41) HO, CO$_2$Et, Me, S, Ph, C=O	82
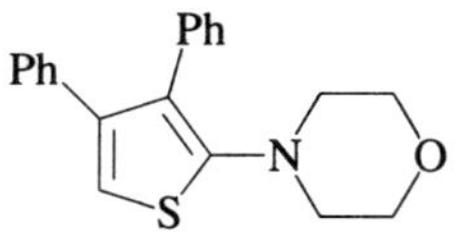	DMF, POCl$_3$	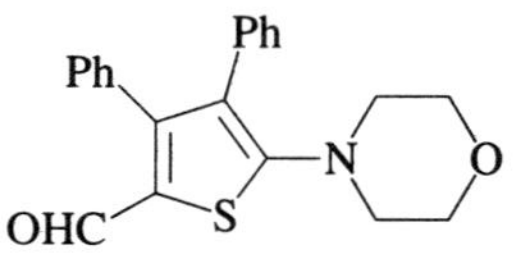 (65) morpholino, Ph, Ph, S, OHC	447
	1. PhN(Me)CH=CHCHO, POCl$_3$ 2. HClO$_4$	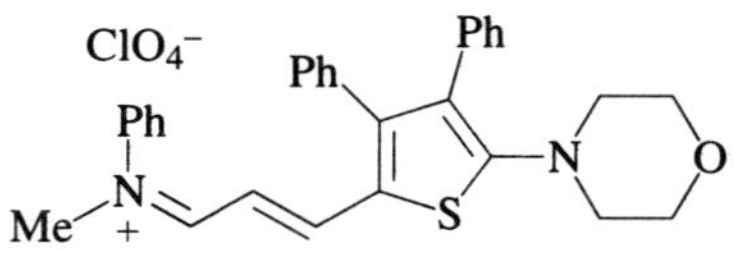 (7) morpholino, Ph, Ph, S, CH=CH–CH=N$^+$(Me)Ph, ClO$_4^-$	447

D2. R^1, R^2, R^4 Substituents

Substrate	Reagents	Product(s) and Yield(s) (%)	Refs.
C$_{8-16}$ 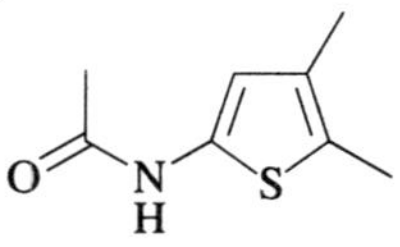 (2-acetamido-4,5-dimethylthiophene)	DMF, POCl$_3$	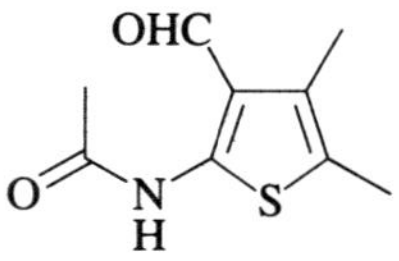 (73) OHC, Me, Me, S, NH–COMe	448, 78
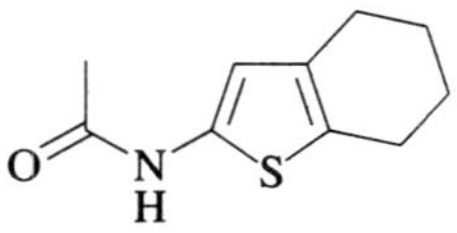 (2-acetamido-tetrahydrobenzothiophene)	DMF, POCl$_3$	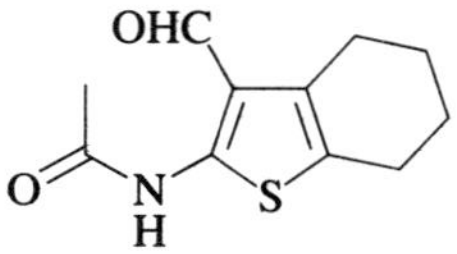 I (78) OHC, S, NH–COMe	448

138

TABLE IX. THIOPHENES (*Continued*)

Substrate	Reagents	Product(s) and Yield(s) (%)	Refs.
	DMF, POCl₃ (1:3)	**I** (79)	78
	DMF, POCl₃ (1:3), 1 h	**I** (8) + **II** (80)	78
	DMF, POCl₃ (1:3), 15 min	**I** (76) + **II** (12)	78
	DMF, POCl₃	(80)	448
	DMF, POCl₃	(80)	448
	DMF, POCl₃	(84)	448

TABLE X. SELENOPHENES

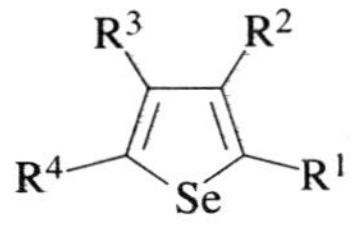

R¹ - R⁴ = H except as indicated

R^1 - R^4 = H except as indicated

Substrate	Reagents	Product(s) and Yield(s) (%)	Refs.
		A. Selenophene	
C₄	DMF, POCl₃	R^1 = CHO (70)	84
		B. Monosubstituted Selenophenes	
		B1. R^1 Substituents	
C₅ R^1 = Me	DMF, POCl₃	R^4 = CHO (88)	449
		B2. R^2 Substituents	
C₅ R^2 = Me	DMF, POCl₃	R^1 = CHO (72)	450
C₆ R^2 = NHAc	DMF, POCl₃	R^1 = CHO (3) + R^1 = CHO, R^2 = N=CHNMe₂ (24)	431
		$+$ (31)	
		C. Disubstituted Selenophenes	
		C1. R^1, R^3 Substituents	
C₁₆₋₁₈			
C2 R^1 = R^3 = Ar	MFA, POCl₃	R^4 = CHO	417
	Ar		
	Ph	(84)	
	4-MeC₆H₄	(81)	
	4-MeOC₆H₄	(69)	

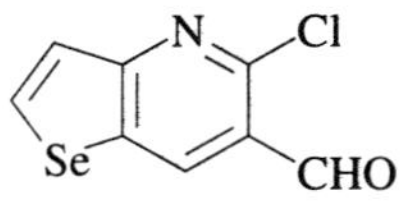

TABLE XI. PYRROLES

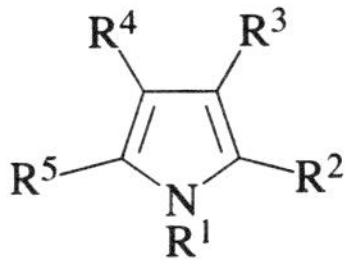

R^1 - R^4 = H except as indicated

Substrate	Reagents	Product(s) and Yield(s) (%)	Refs.
	A. Pyrrole		
C_4	DMF, POCl$_3$	R^2 = CHO (95)	181, 451-453
	DMF, Ph$_3$PBr$_2$	R^2 = CHO (40)	35
	DMF, Tf$_2$O	R^2 = CHO (75)	47
	DMF, 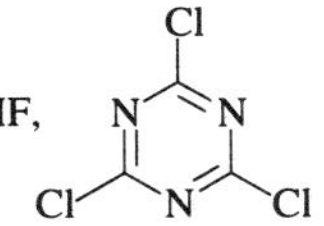 (structure)	R^2 = CHO (65)	36
	1. DMF, (COCl)$_2$ 2. MeCOCl, AlCl$_3$	R^2 = CHO, R^4 = COMe (80)	454
	1. DMF, (COCl)$_2$ 2. PhCOCl, AlCl$_3$	R^2 = CHO, R^4 = COPh (55)	454
	1. DMF, (COCl)$_2$ 2. Cl$_3$CCOCl, AlCl$_3$ 3. NaOMe, MeOH	R^2 = CHO, R^4 = CO$_2$Me (80)	454
	1. DMF, (COCl)$_2$ 2. EtOC(S)Cl, AlCl$_3$	R^2 = CHO, R^4 = COSEt (62)	454
	1. DMF, (COCl)$_2$ 2. MeOCHCl$_2$, AlCl$_3$	R^2 = R^4 = CHO (62)	454
	1. DMF, (COCl)$_2$ 2. NH$_2$OH 3. Dehydrate	R^2 = CN (64)	455

141

TABLE XI. PYRROLES (Continued)

Substrate	Reagents	Product(s) and Yield(s) (%)	Refs.
	1. DMF, POCl₃ 2. ClO₄⁻	R² = CH=NMe₂⁺ ClO₄⁻ (62)	456
	1. DMF, (COCl)₂ 2. NaH 3. RLi	R² = CHRNMe₂	91

$$R$$

t-Bu	(55)
s-Bu	(67-97)
n-Bu	(67-97)
Ph	(67-97)

Substrate	Reagents	Product(s) and Yield(s) (%)	Refs.
	DMA, COCl₂	R² = COMe (—)	103
	DMA, POCl₃	R² = COMe (49)	101
	Et₂NCOCH₂Cl, POCl₃	R² = COCH₂Cl (54)	87
	Me₂NCOCH₂Cl, POCl₃	R² = COCH₂Cl (75)	457
	R₂NCOCH₂CO₂Et, POCl₃		87

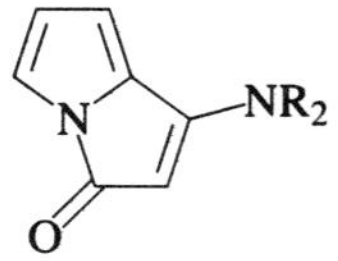

$$R$$

Me	(29)
Et	(31)

Substrate	Reagents	Product(s) and Yield(s) (%)	Refs.
	Me₂NCOCH₂NMe₂, POCl₃	R¹ = COCH₂CONMe₂ (—) + (6)	101

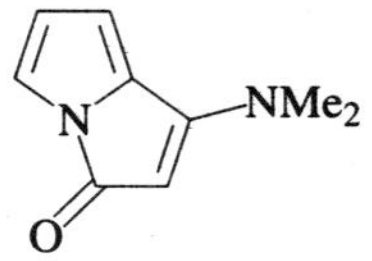

TABLE XI. PYRROLES (*Continued*)

Substrate	Reagents	Product(s) and Yield(s) (%)	Refs.
	$H_2NCO(CH_2)_2CO_2Me$, $POCl_3$	$R^2 = CO(CH_2)_2CO_2Me$ (73)	458
	$Et_2NCO(CH_2)_2CO_2Et$, $POCl_3$	$R^2 = CO(CH_2)_2CO_2Et$ (71)	87
	$POCl_3$, 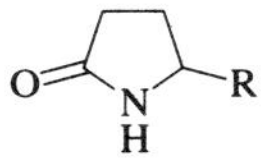	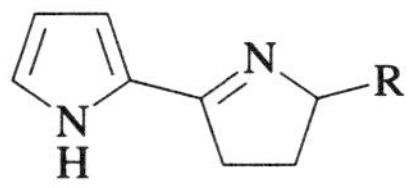	
	$\dfrac{R}{}$		
	H	(80)	459
	Me	(84)	460
	CO_2Me	(51)	460
	CO_2Et	(47)	88
	$Me_2NCH=C(Me)CHO$, $POCl_3$	$R^2 = CH=C(Me)CHO$ (51)	38
	1. $POCl_3$, 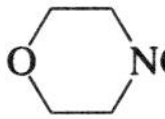 NCOAr	$R^2 = CH_2Ar$	461
	2. $LiBH_4$ or $Na(CN)BH_3$		
	$\dfrac{Ar}{}$		
	Ph	(90-92)	
	$4\text{-}MeC_6H_4$	(79-80)	
	$4\text{-}MeOC_6H_4$	(91-92)	
	$4\text{-}ClC_6H_4$	(80-82)	
	Me_2NCOPh, $POCl_3$	$R^2 = COPh$ (88)	86
	Et_2NCOPh, $POCl_3$	$R^2 = COPh$ (92)	86
	$Me_2NCO(C_6H_4NO_2\text{-}4)$, $POCl_3$	$R^2 = CO(C_6H_4NO_2\text{-}4)$ (91)	183

143

TABLE XI. PYRROLES (Continued)

Substrate	Reagents	Product(s) and Yield(s) (%)	Refs.
	POCl₃, O⟨ ⟩NCOAr	R² = COAr	
		Ar	
		Ph (88)	86, 183
		4-ClC₆H₄ (87)	183
		4-O₂NC₆H₄ (91)	183
		4-MeC₆H₄ (86)	183
		4-MeOC₆H₄ (88)	183
	1. Me₂NCOAr, POCl₃	R² = C(Ar)=NMe₂⁺ ClO₄⁻	462
	2. NaClO₄		
		Ar	
		Ph (66)	
		4-ClC₆H₄ (24)	
		4-O₂NC₆H₄ (24)	
		4-MeC₆H₄ (86)	
		4-MeOC₆H₄ (31)	

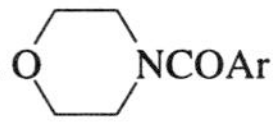

POCl₃, O⟨ ⟩ → (—) 89

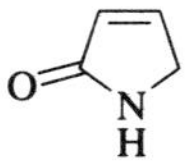

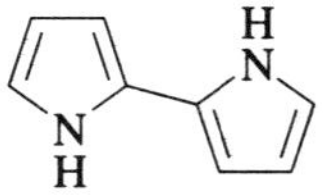

POCl₃, O⟨ ⟩ (with Me) → (79) 89

MeO, CO₂Et POCl₃, O⟨ ⟩ → (65) 89

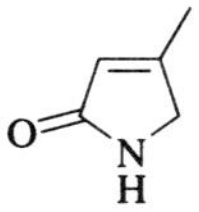

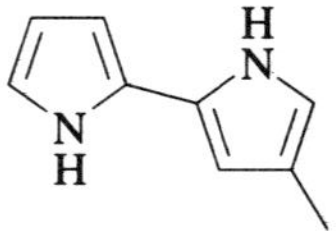

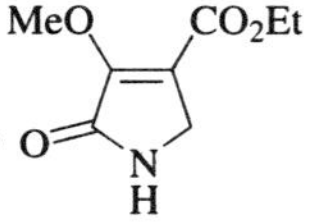

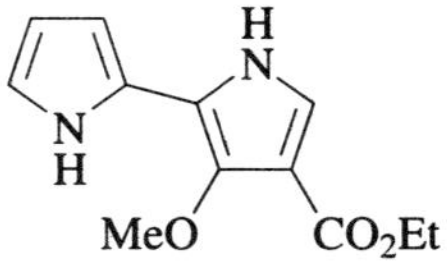

TABLE XI. PYRROLES (*Continued*)

Substrate	Reagents	Product(s) and Yield(s) (%)	Refs.
	POCl$_3$, 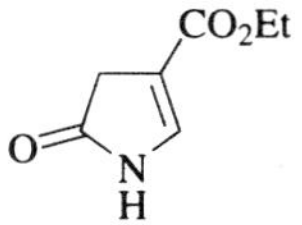	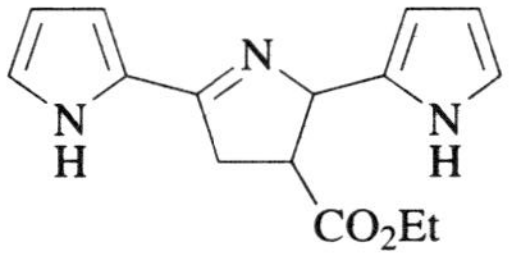 (36)	89
	POCl$_3$, 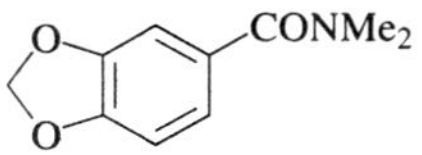	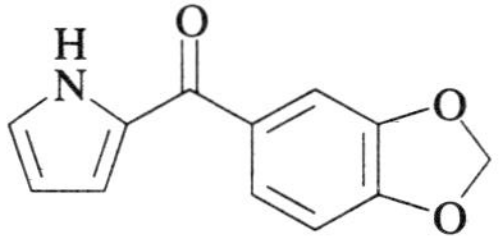 (80)	463

B. Monosubstituted Pyrroles
B1. R^1 Substituents

C$_5$ R^1 = Me

	DMF, POCl$_3$	R^2 = CHO (95)	453, 85, 452. 464, 465
	DMF, POCl$_3$	R^2 = CHO (84) + R^3 = CHO (6)	26
	DMF, (Cl$_2$PO)$_2$O	R^2 = CHO (88) + R^3 = CHO (5)	26
	(Me$_2$N=CHCl)$^+$ Cl$^-$	R^2 = CHO (88) + R^3 = CHO (5)	26
	1. DMF, (COCl)$_2$	R^2 = CN (67)	455
	2. NH$_2$OH		
	3. Dehydrate		
	DMA, COCl$_2$	R^2 = COMe (—)	103
	DMA, POCl$_3$	R^2 = COMe (40) + R^3 = COMe (16)	87
	Et$_2$NCOMe, POCl$_3$	R^2 = COMe (59) + R^3 = COMe (25)	87
	Me$_2$NCOCH$_2$Cl, POCl$_3$	R^2 = COCH$_2$Cl (40) + R^3 = COCH$_2$Cl (40)	457
	Ph(Me)NCOCH$_2$Cl, POCl$_3$	R^2 = COCH$_2$Cl (12) + R^3 = COCH$_2$Cl (3)	457

145

TABLE XI. PYRROLES (Continued)

Substrate	Reagents	Product(s) and Yield(s) (%)	Refs.
	POCl$_3$, 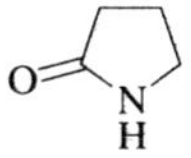	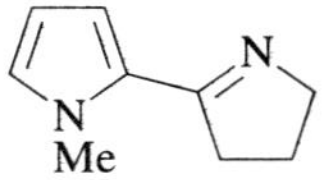 (34)	466
	POCl$_3$, 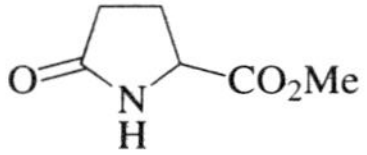	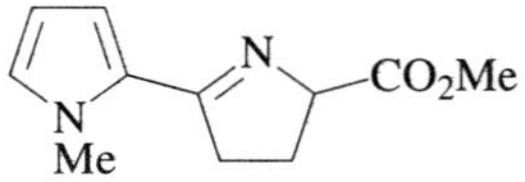 (69)	460, 466
	1. POCl$_3$, 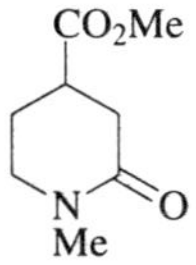 2. NaBH$_4$	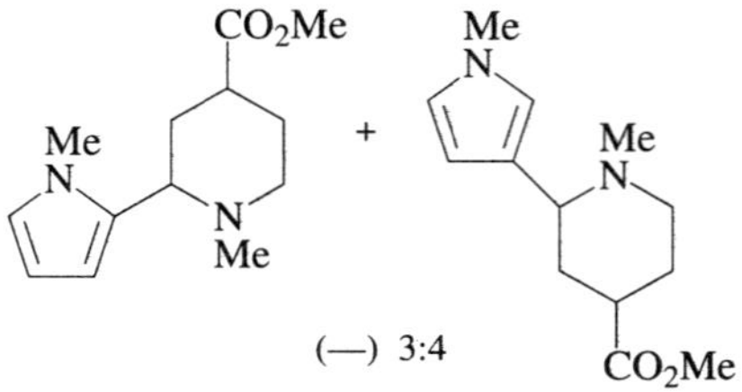 (—) 3:4	467
	Me$_2$NN=CHCHO, COCl$_2$	R^2 = CH=CHN=NMe$_2$$^+$ Cl$^-$ or PO$_2$Cl$_2$$^-$ (35)	40
	Me$_2$NCH=CHCHO, POCl$_3$	R^2 = CH=CHCHO (49)	38
	Me$_2$NCH=C(Me)CHO, POCl$_3$	R^2 = CH=C(Me)CHO (13)	38
	Me$_2$NCOPh, POCl$_3$	R^2 = COPh (86)	86
	Me$_2$NCO(C$_6$H$_4$Cl-4), POCl$_3$	R^2 = CO(C$_6$H$_4$Cl-4) (20)	468
	1. DMF, (COCl)$_2$ 2. ArCOCl, AlCl$_3$	R^2 = CHO, R^4 = COAr	469

Ar	
Ph	(40)
4-MeOC$_6$H$_4$	(52)
4-ClC$_6$H$_4$	(44)
4-FC$_6$H$_4$	(60)
4-O$_2$NC$_6$H$_4$	(63)

TABLE XI. PYRROLES (*Continued*)

	Substrate	Reagents	Product(s) and Yield(s) (%)	Refs.
C_6	$R^1 = COMe$	DMF, $POCl_3$	$R^2 = CHO$ (61)	85
	$R^1 = Et$	DMF, $POCl_3$	$R^2 = CHO$ (58) + $R^3 = CHO$ (27)	85
		Me_2NCOCH_2Cl, $POCl_3$	$R^2 = COCH_2Cl$ (30) + $R^3 = COCH_2Cl$ (35)	457
	$R^1 = NMe_2$	DMF, $POCl_3$	$R^2 = CHO$ (56)	470
C_7	$R^1 = CO_2Et$	DMF, $POCl_3$	$R^2 = CHO$ (54)	85
	$R^1 = i\text{-}Pr$	DMF, $POCl_3$	$R^2 = CHO$ (8) + $R^3 = CHO$ (71)	85
C_8	$R^1 = t\text{-}Bu$	DMF, $POCl_3$	$R^2 = CHO$ (5) + $R^3 = CHO$ (64)	85

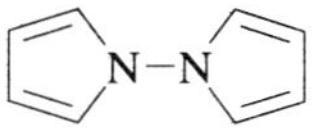

(56) 471

DMF, $POCl_3$

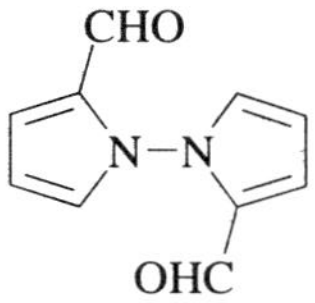

(56) 472

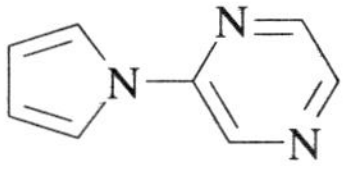

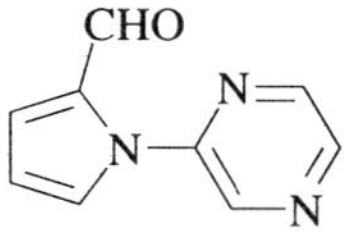

—

DMF, $POCl_3$

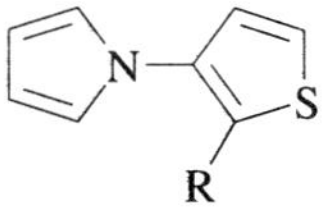

$C_{8\text{-}12}$

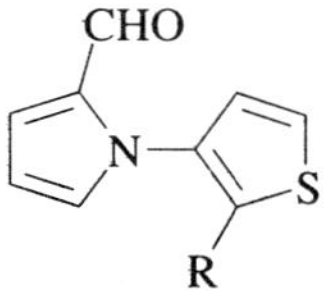

R	
H	(—) 473
CO_2H	(60) 474, 473
CO_2Me	(72) 474, 473

147

TABLE XI. PYRROLES (Continued)

Substrate	Reagents	Product(s) and Yield(s) (%)	Refs.
C₉			
(thiophene-pyrrole structure) R = CN, CO₂H	DMF, POCl₃	(CHO thiophene-pyrrole) (83) (49)	474, 475 474
(thiophene-pyrrole structure) R = CN, CO₂H	DMF, POCl₃	(CHO thiophene-pyrrole) (83) (49)	474
(chloropyridine-pyrrole structure)	—	(CHO chloropyridine-pyrrole) (83)	472
(pyridine-pyrrole structure)	DMF, POCl₃	(CHO pyridine-pyrrole) (96)	476

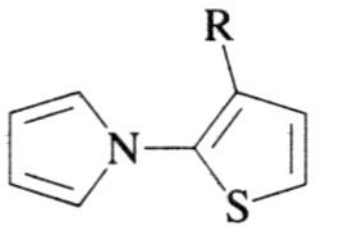

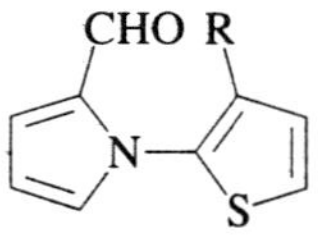

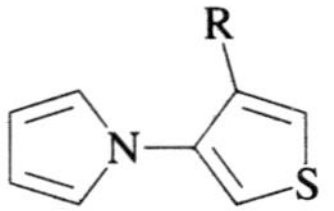

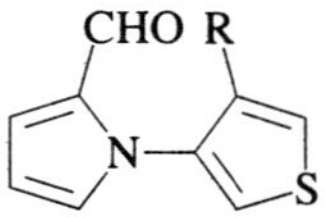

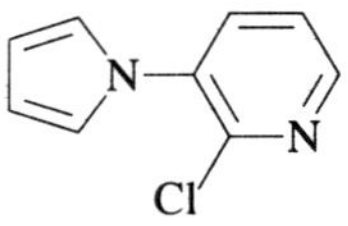

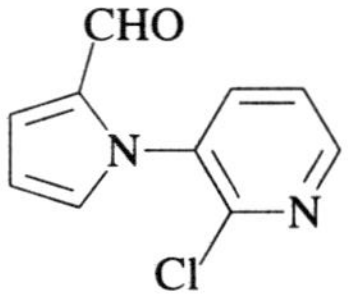

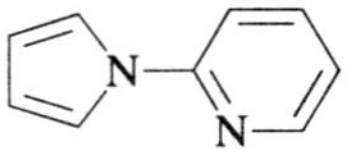

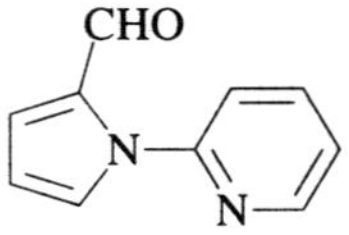

148

TABLE XI. PYRROLES (*Continued*)

Substrate	Reagents	Product(s) and Yield(s) (%)	Refs.
C₁₀			
$R^1 = 4\text{-}BrC_6H_4$	DMF, POCl₃	$R^2 = CHO\ (76) + R^3 = CHO\ (14)$	85
$R^1 = 2\text{-}O_2NC_6H_4$	DMF, POCl₃	$R^2 = CHO\ (100)$	477
$R^1 = 4\text{-}O_2NC_6H_4$	DMF, POCl₃	$R^2 = CHO\ (77) + R^3 = CHO\ (11)$	85
$R^1 = Ph$	DMF, POCl₃	$R^2 = CHO\ (84) + R^3 = CHO\ (8)$	85
	DMA, POCl₃	$R^2 = COMe\ (60) + R^3 = COMe\ (15)$	182
	Me₂NCOCH₂Cl, POCl₃	$R^2 = COCH_2Cl\ (33) + R^3 = COCH_2Cl\ (33)$	457
	Ph(Me)NCOCH₂Cl, POCl₃	$R^2 = COCH_2Cl\ (1) + R^3 = COCH_2Cl\ (9)$	457
$R^1 = SO_2Ph$	DMF, POCl₃	$R^2 = CHO\ (34)$	478
			479
C₁₁			
$R^1 = COPh$	DMF, POCl₃	$R^2 = CHO\ (74)$	85
$R^1 = (2\text{-}CH_2Cl)C_6H_4$	DMF, POCl₃, Cl(CH₂)₂Cl	$R^2 = CHO\ (70)$	480
	DMF, POCl₃	$R^2 = CHO\ (20) + R^3 = CHO\ (19)$	480
$R^1 = 4\text{-}MeC_6H_4$	DMF, POCl₃	$R^2 = CHO\ (84) + R^3 = CHO\ (8)$	85
$R^1 = 4\text{-}MeOC_6H_4$	DMF, POCl₃	$R^2 = CHO\ (81) + R^3 = CHO\ (12)$	85
	DMF, POCl₃	$R^2 = CHO\ (86) + R^3 = CHO\ (10)$	26, 85, 481
	DMF, (Cl₂PO)₂O	$R^2 = CHO\ (80) + R^3 = CHO\ (18)$	
	MFA, (Cl₂PO)₂O	$R^2 = CHO\ (75) + R^3 = CHO\ (22)$	
	Me₂NCOCH₂Cl, POCl₃	$R^2 = COCH_2Cl\ (27) + R^3 = COCH_2Cl\ (48)$	
	Ph(Me)NCOCH₂Cl, POCl₃	$R^2 = COCH_2Cl\ (1) + R^3 = COCH_2Cl\ (9)$	
$R^1 = CH_2(C_6H_4F\text{-}2)$	DMF, POCl₃	$R^2 = CHO\ (100)$	482, 483

149

TABLE XI. PYRROLES (Continued)

Substrate	Reagents	Product(s) and Yield(s) (%)	Refs.
C$_{12}$			
	DMF, POCl$_3$	(20) + (20)	484
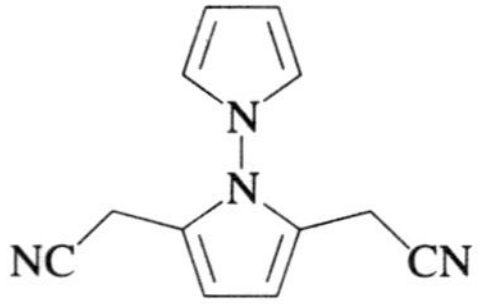	DMF, POCl$_3$	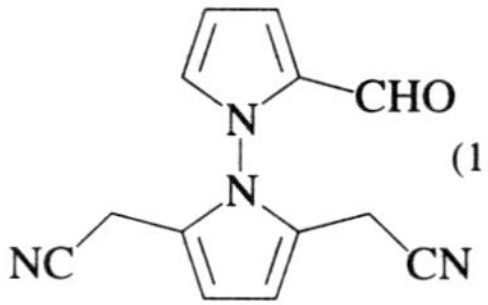 (11) 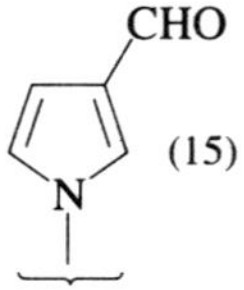 + (15)	485, 486
R^1 = C$_6$H$_4$CO$_2$Me-2	DMF, POCl$_3$	R^2 = CHO (76)	487, 480
R^1 = C$_6$H$_2$(Me)$_2$-2,6-NO$_2$-3	DMF, POCl$_3$	R^2 = CHO (75) + R^3 = CHO (16)	85
R^1 = C$_6$H$_2$(Me)$_2$-2,6-NO$_2$-4	DMF, POCl$_3$	R^2 = CHO (30) + R^3 = CHO (7)	85
R^1 = C$_6$H$_3$(Me)$_2$-2,6	DMF, POCl$_3$	R^2 = CHO (65) + R^3 = CHO (6)	85
R^1 = C$_6$H$_3$(OMe)$_2$-2,5	DMF, POCl$_3$	R^2 = CHO (70)	488
R^1 = CH$_2$C$_6$H$_4$SMe-2	DMF, POCl$_3$	R^2 = CHO (86)	489
	1. DMF, POCl$_3$ 2. NH$_2$OH 3. Dehydrate	R^2 = CN (80)	489

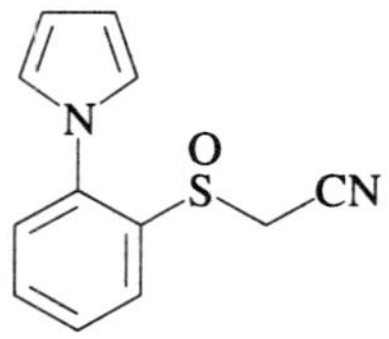

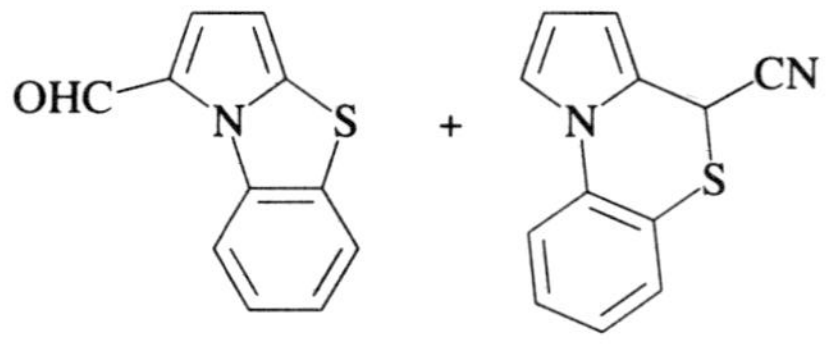

150

TABLE XI. PYRROLES (*Continued*)

Substrate	Reagents	Product(s) and Yield(s) (%)	Refs.
	DMF, POCl₃	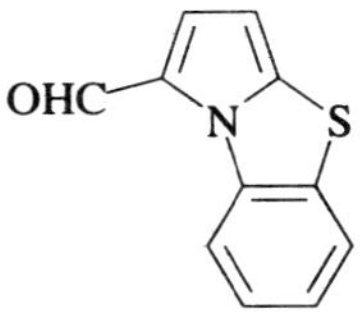 I (73) 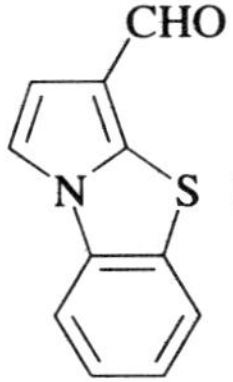 + II (9)	484
	DMF, (COCl)₂	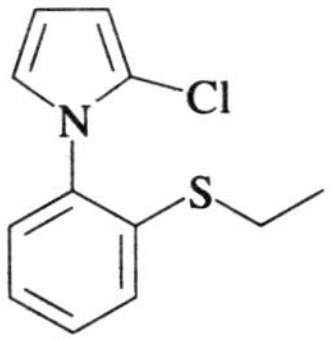 (32) + I (46) + II (10)	484
C₁₃ R¹ = (CH₂)₆CO₂Me	DMF, POCl₃	R² = CHO (70)	490
R¹ = (i-Pr)₃Si	DMF, (Cl₂PO)₂O	R² = CHO (66) + R³ = CHO (14)	26
	MFA, (Cl₂PO)₂O	R² = CHO (61-73) + R³ = CHO (3-7)	26
C₁₄ R¹ = 1-(8-nitronaphthyl)	DMF, POCl₃	R² = CHO (74)	491
	1. DMF, (COCl)₂ 2. NH₂OH 3. Dehydrate	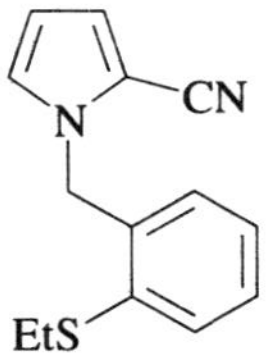 (70)	489

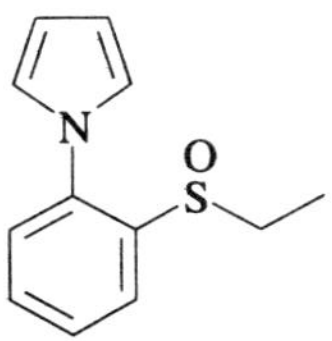

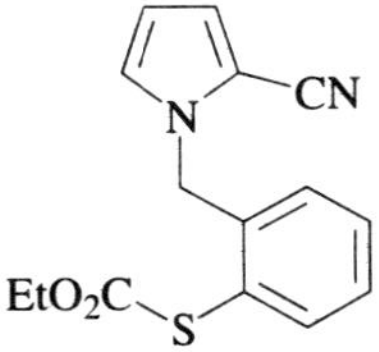

TABLE XI. PYRROLES (*Continued*)

Substrate	Reagents	Product(s) and Yield(s) (%)	Refs.
C_{17}	—	(—)	492
	1. DMF, (COCl)$_2$ 2. NH$_2$OH 3. Dehydrate	 $\begin{array}{c\|cc} X & I & II \\ \hline 4\text{-Cl} & (23) & (0) \\ 5\text{-Cl} & (55) & (0) \\ 4\text{-NO}_2 & (46) & (40) \\ 6\text{-NO}_2 & (40) & (42) \end{array}$	493
C_{18}	DMF, POCl$_3$	(28)	494

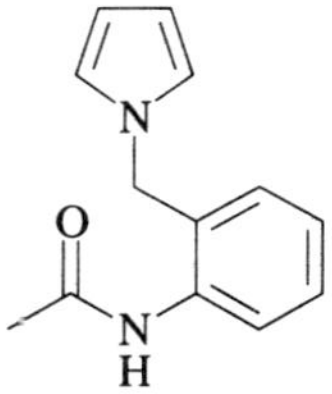

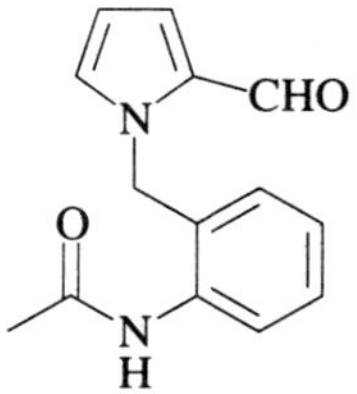

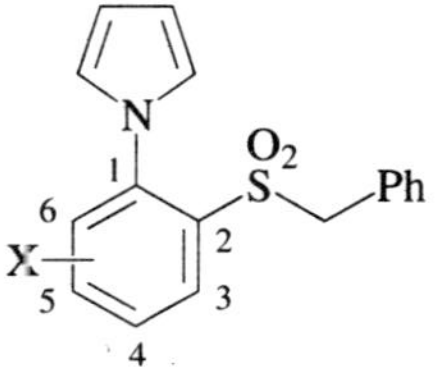

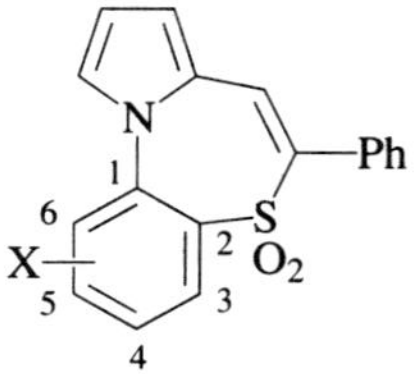

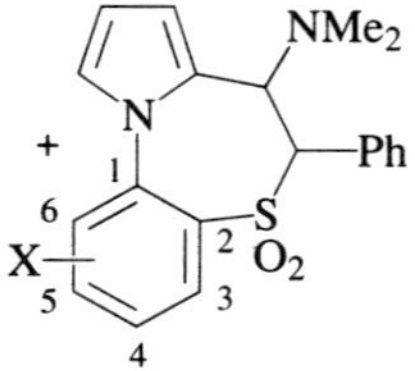

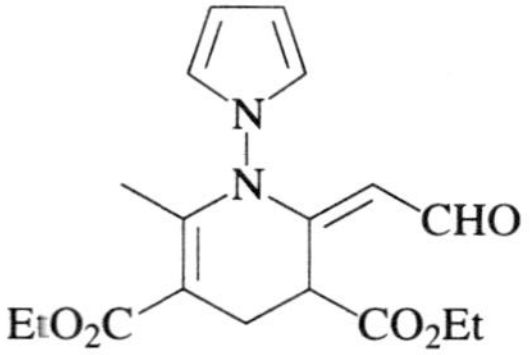

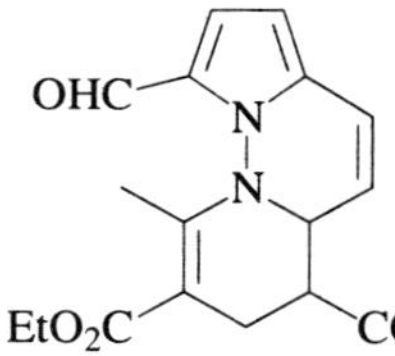

152

TABLE XI. PYRROLES (*Continued*)

Substrate	Reagents	Product(s) and Yield(s) (%)	Refs.

C$_{19}$

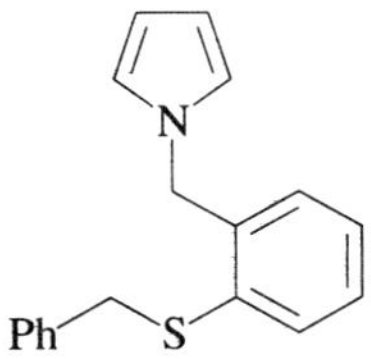

DMF, POCl$_3$ or (COCl)$_2$ — CHO (84) — 495

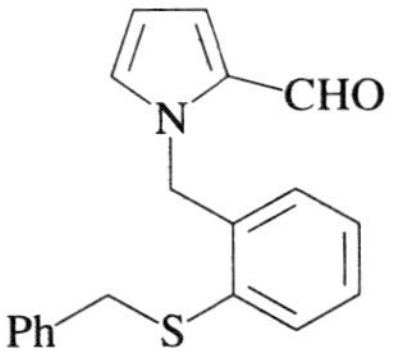

B2. R^2 Substituents

C$_4$

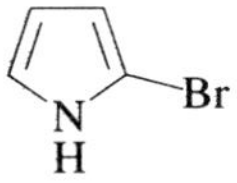

1. DMF, (COCl)$_2$
2. NaHCO$_3$

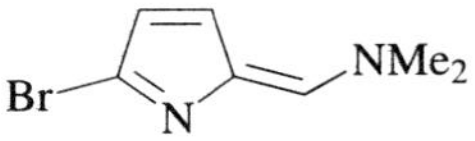

— NMe$_2$ (80) — 90

(*i*-Pr)$_2$NCHO, (COCl)$_2$ — $\overset{+}{N}$(Pr-*i*)$_2$ Cl$^-$ (95) — 90

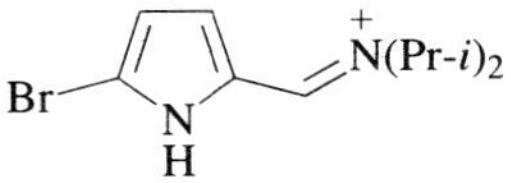

C$_5$ R^2 = Me

DMF, POCl$_3$ — R^5 = CHO (90) — 453
DMA, COCl$_2$ — R^5 = COMe (—) — 103
Me$_2$NCOCH$_2$Cl — R^5 = COCH$_2$Cl (45) — 457

POCl$_3$,

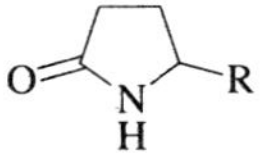

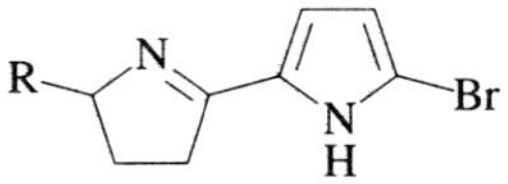

R
H
Me

(78)
(78)

460

TABLE XI. PYRROLES (*Continued*)

Substrate	Reagents	Product(s) and Yield(s) (%)	Refs.
C_6			
R^2 = COMe	DMF, POCl$_3$	R^5 = CHO, R^2 = CCl=CH$_2$ (31)	496
C_8			
R^2 = 2-furyl	DMF, POCl$_3$	R^5 = CHO (62) + R^5, $R^{5'}$ = CHO (14)	74, 75
R^2 = 3-furyl	DMF, POCl$_3$	R^5 = CHO (56) + R^5, $R^{2'}$ = CHO (12) 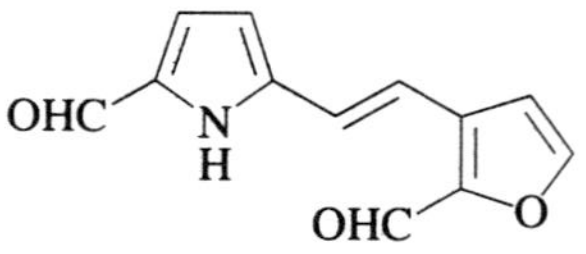 + (8)	74
R^2 = 2-thienyl	DMF, POCl$_3$	R^5 = CHO (85)	74, 75
R^2 = 3-thienyl	DMF, POCl$_3$	R^5 = CHO (81)	74
C_{10}			
R^2 = Ph	DMF, POCl$_3$	R^5 = CHO (71)	74
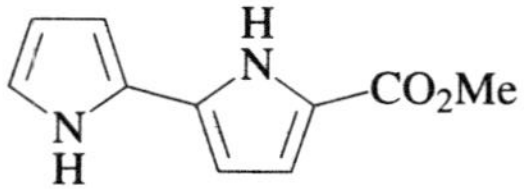	POCl$_3$, 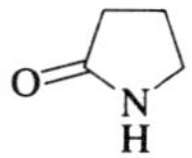 pyrrolidin-2-one	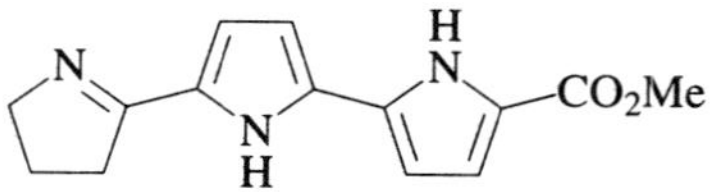 (25)	497
	POCl$_3$, 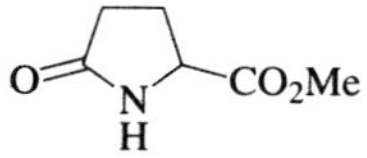 5-(CO$_2$Me)pyrrolidin-2-one	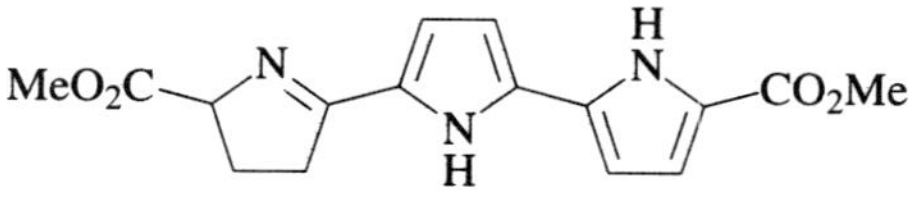 (76)	497
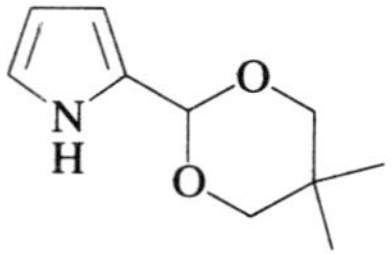	DMF, POCl$_3$	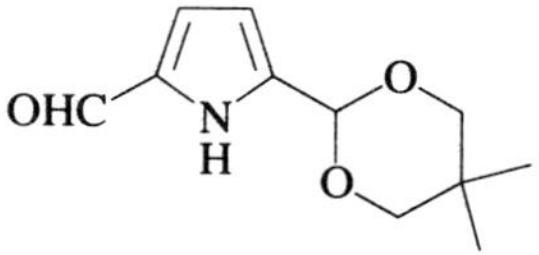 (56)	498

154

TABLE XI. PYRROLES (Continued)

Substrate	Reagents	Product(s) and Yield(s) (%)	Refs.
C$_{14}$ (pyrrole with OMe/SO$_2$Et-substituted aryl)	1. POCl$_3$, (N-Et pyrrolidinone) 2. NaBH$_4$	(55) (pyrrole with OMe/SO$_2$Et-aryl and N-Et pyrrolidine)	499
B3. R^3 Substituents			
C$_5$ (3-methylpyrrole)	POCl$_3$, (pyrrolidinone)	(40)	460
C$_8$ R^3 = CH$_2$CO$_2$Et	DMF, (COCl)$_2$, HMPT DMF, (COCl)$_2$	R^2 = CHO (60) + R^5 = CHO (40) R^2 = CHO (40) + R^5 = CHO (60)	94 94
	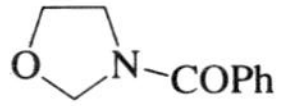 POCl$_3$, (N-COPh oxazolidinone)	R^2 = CHO (5) + R^5 = CHO (31)	500
C$_{10}$ (pyrrole with 1,3-dioxane)	DMF, POCl$_3$	(46) + (54)	501

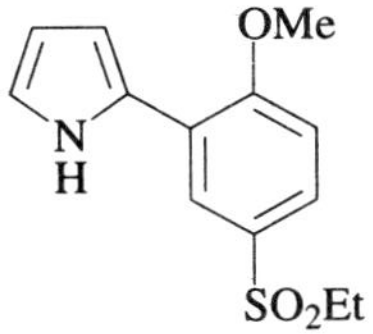

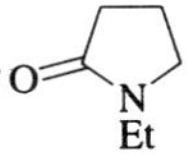

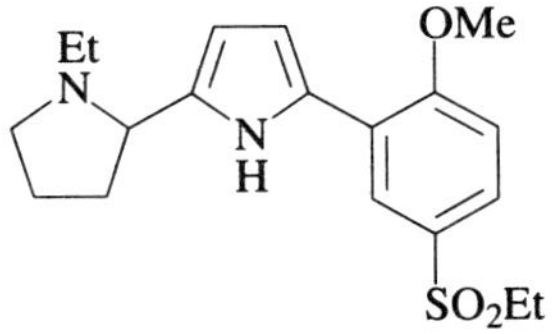

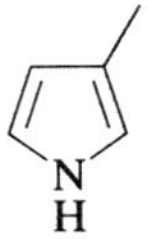

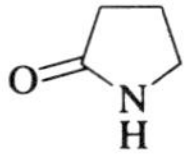

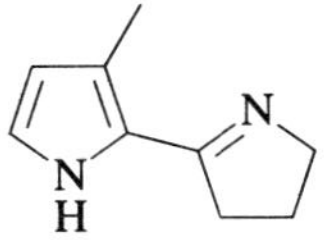

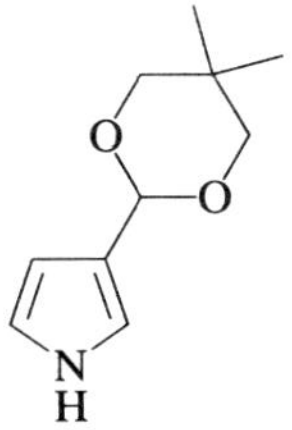

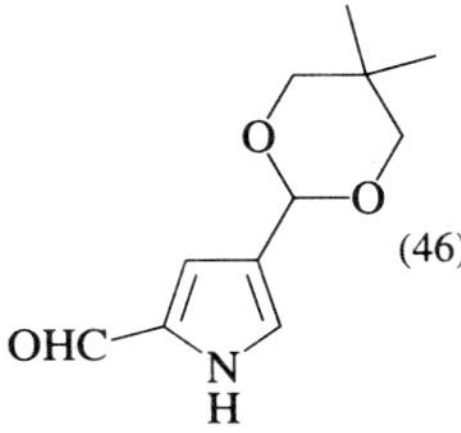

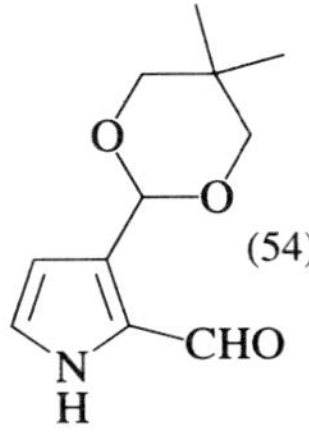

TABLE XI. PYRROLES (Continued)

Substrate	Reagents	Product(s) and Yield(s) (%)	Refs.

D. Disubstituted Pyrroles
C1. R¹, R² Substituents

C_6 $R^1 = R^2 = Me$

| | DMF, POCl₃ | R^5 = CHO (79) | 453 |
| | 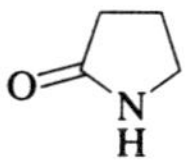 POCl₃, (pyrrolidinone)CONHMe | 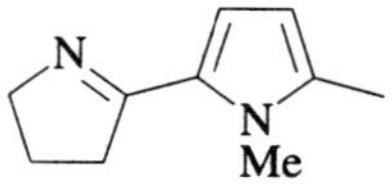 (66) | 466 |

C_8

| (pyrrole-CO₂Me, N-Me) | 1. SOCl₂, (tolyl)CONHMe 2. H⁺ | 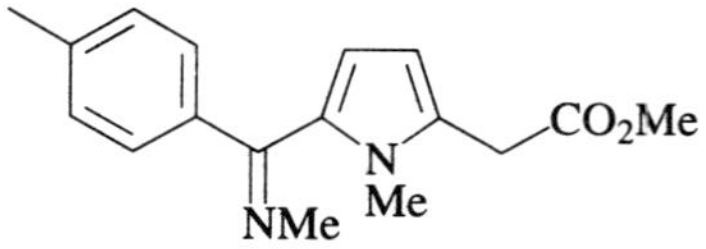 (30) | 502 |
| (dioxane-pyrrole, N-Me) | DMF, POCl₃ | (95) | 501 |

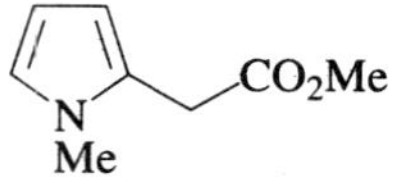

C_{11}

| (azepine-pyrrole CO₂Me) | POCl₃, (MeS-phenyl)CONMe₂ | | 503 |
| (pyrrolizine CO₂Pr-i) | Me₂NCOR, POCl₃ | | 503 |

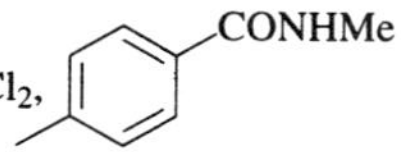

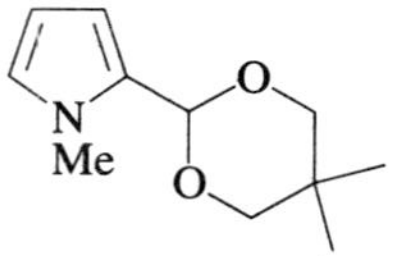

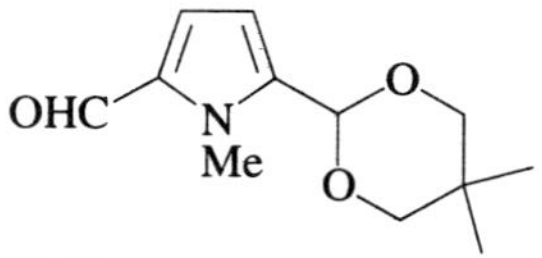

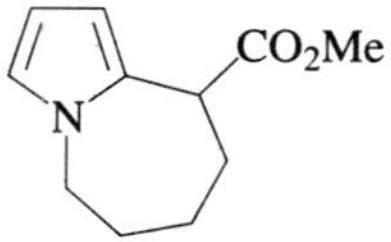

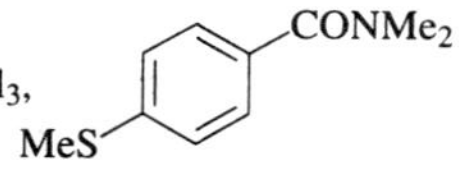

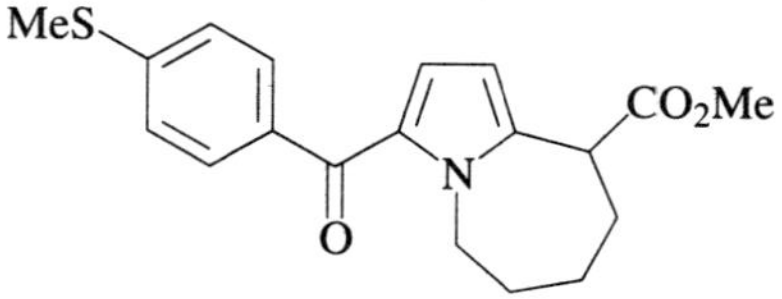

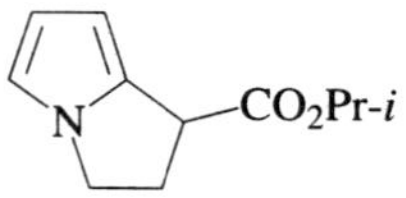

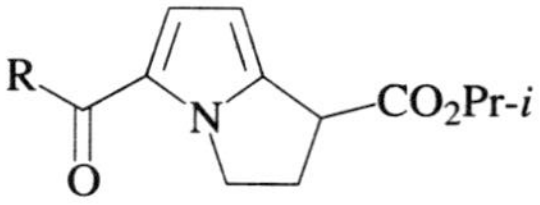

156

TABLE XI. PYRROLES (*Continued*)

Substrate	Reagents	Product(s) and Yield(s) (%)	Refs.
	R		
	c-C₄H₈	(88)	
	t-Bu	(25)	
	Ph	(85)	
	2-ClC₆H₄	(19)	
	3-ClC₆H₄	(70)	
	4-ClC₆H₄	(91)	
	3-FC₆H₄	(66)	
	4-FC₆H₄	(98)	
	2-MeC₆H₄	(37)	
	3-MeC₆H₄	(90)	
	4-MeC₆H₄	(66)	
	3-MeOC₆H₄	(66)	
	4-MeOC₆H₄	(100)	
	Bn	(70)	
	3-EtOC₆H₄	(45)	
	4-EtOC₆H₄	(52)	
	4-i-ProC₆H₄	(93)	
	2-C₁₀H₇	(88)	
	4-PhC₆H₄	(50)	
C₁₂	DMF, POCl₃	(40)	482

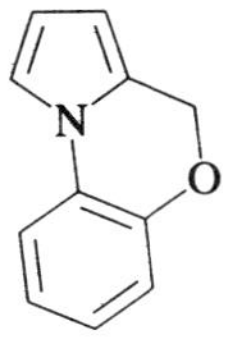

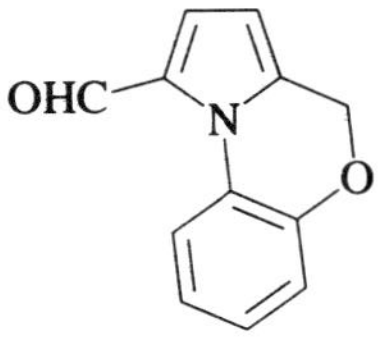

TABLE XI. PYRROLES (*Continued*)

Substrate	Reagents	Product(s) and Yield(s) (%)	Refs.
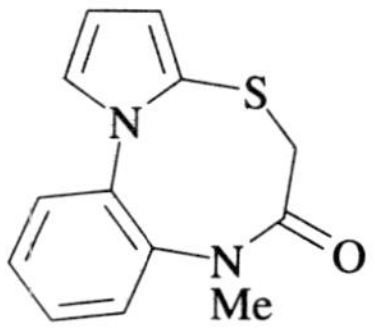 C$_{12-17}$	DMF, POCl$_3$	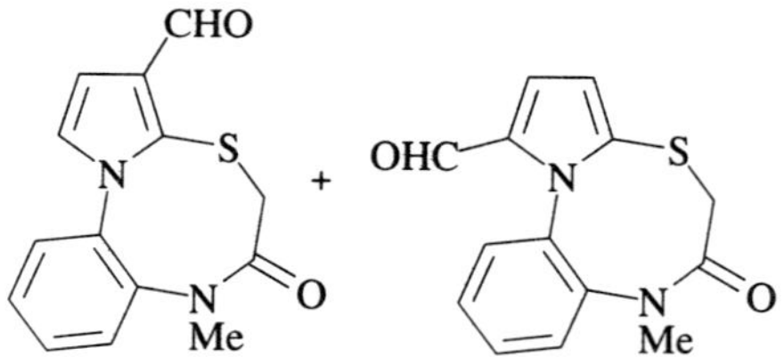 CHO + OHC (32)	504
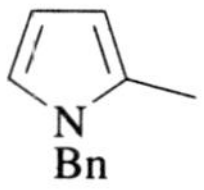	POCl$_3$, 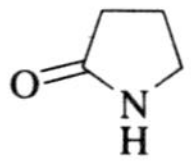	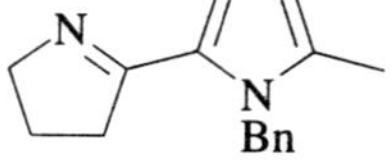 (21)	466
C$_{13}$	DMF, POCl$_3$	R = H (90) R = CO$_2$Et (16)	505
R^1 = Me, R^2 = CO(C$_6$H$_4$Me-4)	DMF, (COCl)$_2$	R^3 = CHO (trace) + R^5 = CHO (95)	506

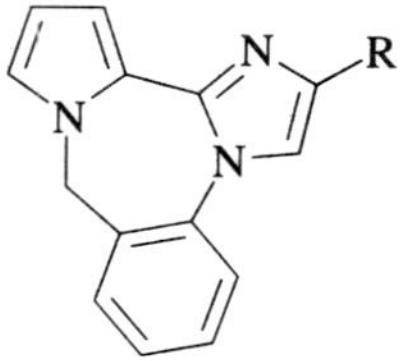

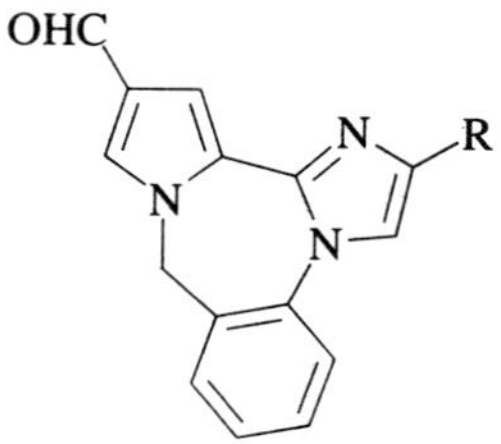

TABLE XI. PYRROLES (*Continued*)

Substrate	Reagents	Product(s) and Yield(s) (%)	Refs.
C2. R^1, R^3 Substituents			
C$_{11}$ 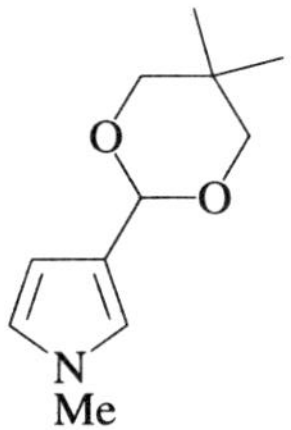 (pyrrole with 5,5-dimethyl-1,3-dioxane at 3-position, N-Me)	1. DMF, POCl$_3$ 2. Hydrolysis	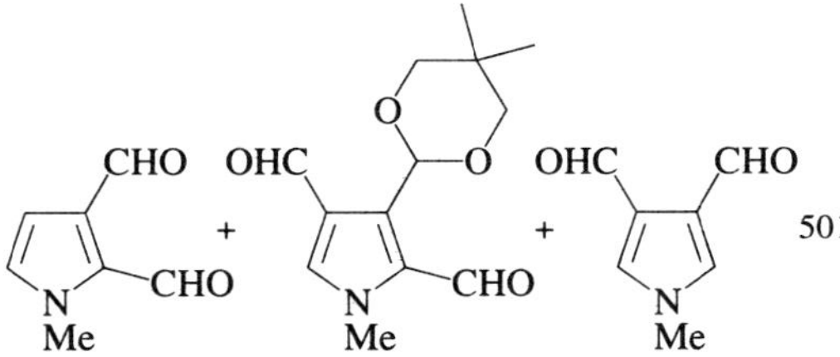 (three products: CHO/CHO N-Me pyrrole + dioxane-substituted diCHO N-Me pyrrole + OHC/CHO N-Me pyrrole) (total: 53)	501
C3. R^2, R^3 Substituents			
C$_8$ R^2 = Me, R^3 = CO$_2$Et	Me$_2$NCOPh, POCl$_3$	R^5 = COPh (98)	86
	POCl$_3$, 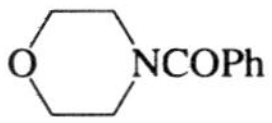 (morpholine-NCOPh)	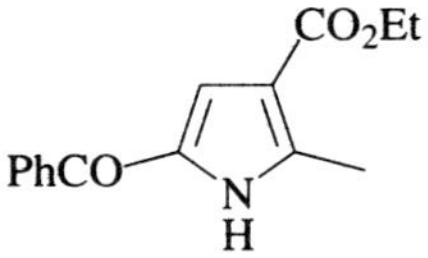 (90) PhCO-substituted pyrrole, CO$_2$Et, Me, NH	86
C$_{13}$ R^2 = CO$_2$Bn, R^3 = Me	DMF, POCl$_3$	R^4 = CHO (28) + R^5 = CHO (72)	507
C4. R^2, R^4 Substituents			
C$_6$ R^2 = R^4 = Me	PhNHCHO, POCl$_3$ DMF, POCl$_3$	R^5 = CHO (33) R^5 = CHO (75)	233 453, 508
	POCl$_3$, 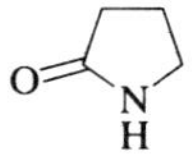 (2-pyrrolidinone)	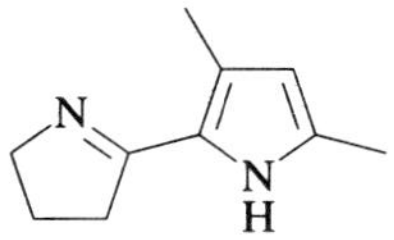 (66) dimethylpyrrole-dihydropyrrole	88

159

TABLE XI. PYRROLES (*Continued*)

Substrate	Reagents	Product(s) and Yield(s) (%)	Refs.
C$_7$			
R^2 = CO$_2$Me, R^4 = Me	DMF, POCl$_3$	R^5 = CHO (88)	509
C$_{11}$			
R^2 = Me, R^4 = 2-O$_2$N-3-ClC$_6$H$_3$	DMF, POCl$_3$	R^5 = CHO (—)	510
R^2 = Me, R^4 = 3-O$_2$N-4-ClC$_6$H$_3$	DMF, POCl$_3$	R^5 = CHO (—)	510
C$_{16}$			
R^2 = R^4 = Ph	MFA, POCl$_3$	R^5 = CHO (—)	511
R^2 = R^4 =	DMF, POCl$_3$	R^5 = CHO (77)	501
C$_{17}$ R^2 = Ph, R^4 = 4-MeOC$_6$H$_4$	MFA, POCl$_3$	R^5 = CHO (—)	511
C5. R^2, R^5 Substituents			
C$_6$			
R^2 = R^5 = Me	DMF, POCl$_3$	R^3 = CHO (62)	513, 512
	DMF, POCl$_3$	R^3 = CHO (18) + R^3 = R^4 = CHO (9)	514, 515
	DMA, POCl$_3$	R^3 = COMe (40)	516
	Et$_2$NCOCH$_2$CO$_2$Et, POCl$_3$	R^3 = CH(NEt$_2$)=CHCO$_2$Et (40)	87
	Me$_2$NN=CHCHO, POCl$_3$ or COCl$_2$	R^3 = CH=CHN=NMe$_2^+$ POCl$_2^-$ or Cl$^-$ (70)	40
C$_8$			
R^2 = Me, R^5 = CO$_2$Et	—	R^3 = CHO (—)	121
C$_{10}$			
R^2 = 2-furyl, R^5 = CO$_2$Me	DMF, POCl$_3$	R$^{2'}$ = CHO (88)	74, 75
R^2 = 3-furyl, R^5 = CO$_2$Me	DMF, POCl$_3$	R$^{2'}$ = CHO (56)	74
R^2 = 2-thienyl, R^5 = CO$_2$Me	DMF, POCl$_3$	R^3 = CHO (85) + R$^{2'}$ = CHO (4)	74, 75
R^2 = 3-thienyl, R^5 = CO$_2$Me	DMF, POCl$_3$	R^3 = CHO (85) + R$^{2'}$ = CHO (4)	74
C$_{12}$ R^2 = Ph, R^5 = CO$_2$Me	DMF, POCl$_3$	R^3 = CHO (62)	74

TABLE XI. PYRROLES (*Continued*)

C6. R³, R⁴ Substituents

Substrate	Reagents	Product(s) and Yield(s) (%)	Refs.
C$_6$			
R^3 = R^4 = Me	DMF, POCl$_3$	R^2 = CHO (79)	453, 518
C$_7$			
R^3 = Et, R^4 = CF$_3$	DMF, POCl$_3$	R^2 = CHO (86)	519
R^3 = Me, R^4 = COMe	DMF, POCl$_3$	R^2 = CHO (4) + R^2 = CHO, R^4 = C≡CH (34)	520
R^3 = Me, R^4 = Et	DMF, POCl$_3$	R^2 = CHO (51) + R^5 = CHO (33)	521
C$_8$			
R^3 = Me, R^4 = CO$_2$Et	DMF, POCl$_3$	R^2 = CHO (72)	516
	POCl$_3$,	(66)	460
R^3 = R^4 = Et	Me$_2$NCH=CHCHO, POCl$_3$	R^2 = CH=CHCHO (75)	522
C$_{8-12}$			
R^3 = Ar, R^4 = NO$_2$	DMF, POCl$_3$		523
Ar			
2-thienyl		R^2 = CHO (82)	
2,4-ClC$_6$H$_3$		R^2 = CHO (12)	
Ph		R^2 = CHO (76)	
4-MeOC$_6$H$_4$		R^2 = CHO (82)	
2,4-(MeO)$_2$C$_6$H$_3$		R^2 = CHO (90)	
3,5-(MeO)$_2$C$_6$H$_3$		R^2 = CHO (48) + R^3 = 2-CHO-3,5-(MeO)$_2$C$_6$H$_3$ (38)	
C$_{10}$			
R^3 = Et, R^4 = n-C$_3$F$_7$	DMF, POCl$_3$	R^2 = CHO (70)	519
C$_{16}$			
R^3 = R^4 = Ph	DMF, POCl$_3$	R^2 = CHO (83)	524

161

TABLE XI. PYRROLES (*Continued*)

Substrate	Reagents	Product(s) and Yield(s) (%)	Refs.

C7. R^3, R^5 Substituents

C$_{12-14}$

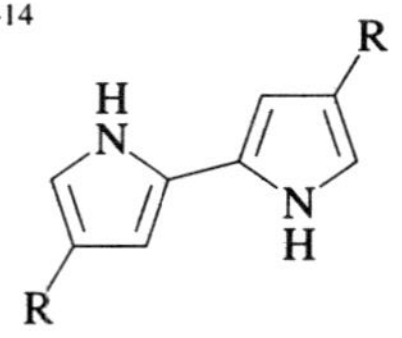

DMF, POCl$_3$

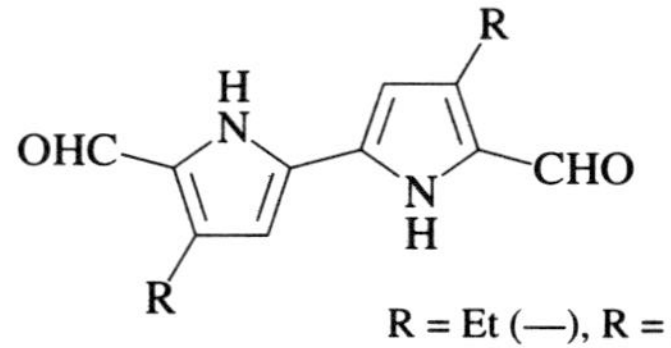

R = Et (—), R = n-Pr (86)

517

D. Trisubstituted Pyrroles
D1. R^1, R^2, R^3 Substituents

C$_{15}$

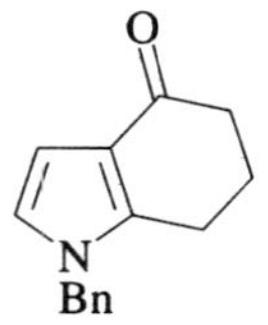

DMF, POCl$_3$

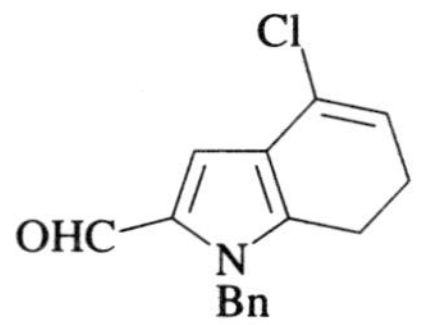

(50)

525

D2. R^1, R^2, R^4 Substituents

C$_7$ R^1 = R^2 = R^4 = Me

DMF, POCl$_3$

R^5 = CHO (18) + R^3 = R^5 = CHO (—)

514

C$_9$ R^1 = R^4 = Me, R^2 = CH$_2$CO$_2$Me

1. MeNHCOC$_6$H$_4$R-4, SOCl$_2$
2. H$^+$

R^5 = CH(=NMe)C$_6$H$_4$R-4

402

R	
H	(61)
Cl	(69)
NO$_2$	(65)
Me	(61)
MeO	(64)

C$_{10}$ R^1 = R^4 = Me, R^2 = CH$_2$CO$_2$Et

Me$_2$NCOC$_6$D$_4$-4

R^5 = CHO (—)

526

162

TABLE XI. PYRROLES (*Continued*)

Substrate	Reagents	Product(s) and Yield(s) (%)	Refs.
$R^1 = Me, R^2 = CH_2CO_2Et, R^4 = SMe$	$Me_2NCOAr, POCl_3$	$R^5 = COAr$	527
	Ar		
	Ph		
	4-ClC$_6$H$_4$		
	4-FC$_6$H$_4$		
	4-MeC$_6$H$_4$		
	4-EtOC$_6$H$_4$		
	4-n-PrC$_6$H$_4$		
	4-c-PrC$_6$H$_4$		

C$_{17}$

Substrate	Reagents	Product(s) and Yield(s) (%)	Refs.
$R^1 = Me, R^2 = R^4 = $	DMF, POCl$_3$	$R^5 = CHO$ (80)	501

D3. R^1, R^2, R^5 Substituents

C$_7$

Substrate	Reagents	Product(s) and Yield(s) (%)	Refs.
$R^1 = R^2 = R^5 = Me$	DMF, POCl$_3$	$R^3 = CHO$ (44) + $R^3 = R^4 = CHO$ (—)	514, 515

C$_{10-19}$

Substrate	Reagents	Product(s) and Yield(s) (%)	Refs.
$R^1 = Ar, R^2 = R^5 = Me$	DMF, POCl$_3$	$R^3 = CHO$	
Ar			
1-pyrrolyl		(67)	
2-pyridyl		(14)	
2-ClC$_6$H$_4$		(89)	
Ph		(73) + $R^3 = R^4 = CHO$ (13)	
2-NCC$_6$H$_4$		(84)	
4-Me-2-pyridyl		(68)	
2-Cl-6-MeC$_6$H$_3$		(79)	
2-MeC$_6$H$_4$		(90)	
2-MeOC$_6$H$_4$		(91)	
2-EtOC$_6$H$_4$		(93)	
4-Me$_2$NC$_6$H$_4$		(60)	
8-quinolyl		(62)	
2-BnOC$_6$H$_4$		(86)	

163

TABLE XI. PYRROLES (Continued)

Substrate	Reagents	Product(s) and Yield(s) (%)	Refs.
C_{13} R^1 = Bn, R^2 = R^5 = Me	DMF, POCl$_3$	R^3 = CHO (83)	531
C_{16} R^1 = 1-pyrrolyl, R^2 = R^5 = CH$_2$CO$_2$Et	DMF, POCl$_3$	R^3 = CHO (41)	485
C_{17} R^1 = R^2 = Ph, R^5 = Me	DMF, POCl$_3$	R^4 = CHO (98)	530
C_{20-21}	DMF, POCl$_3$	R = H (93), R = Me (67)	532

D4. R^1, R^3, R^4 Substituents

Substrate	Reagents	Product(s) and Yield(s) (%)	Refs.
C_7 R^1 = R^3 = R^4 = Me	DMF, POCl$_3$	R^2 = CHO (70)	453
C_{19} R^1 = i-Pr, R^3 = R^4 = 2-FC$_6$H$_4$	Me$_2$NCH=CHCHO, POCl$_3$	R^2 = CH=CHCHO (—)	533

D5. R^1, R^3, R^5 Substituents

Substrate	Reagents	Product(s) and Yield(s) (%)	Refs.
C_7 R^1 = R^3 = R^5 = Me	DMF, POCl$_3$	R^2 = CHO (65)	453

D6. R^2, R^3, R^4 Substituents

Substrate	Reagents	Product(s) and Yield(s) (%)	Refs.
C_7 R^2 = R^3 = R^4 = Me	DMF, POCl$_3$	R^2 = CHO (81)	453, 508, 534

TABLE XI. PYRROLES (*Continued*)

Substrate	Reagents	Product(s) and Yield(s) (%)	Refs.
C₈			
$R^2 = R^4 = Me$, $R^3 = COMe$	DMF, POCl₃, 80°	$R^5 = CHO$ (34) + $R^3 = C\equiv CH$, $R^5 = CHO$ (32)	535
	DMF, POCl₃, 0°	$R^3 = C(Cl)=CH_2$, $R^5 = CHO$ (70)	535
$R^2 = R^4 = Me$, $R^3 = Et$	DMF, POCl₃	$R^5 = CHO$ (76)	536, 508, 534
	Et₂NCOCH₂Cl, POCl₃	$R^5 = COCH_2Cl$ (34)	87
	Et₂NCOCH₂CO₂Et, POCl₃	$R^5 = C(NEt_2)=CHCO_2Et$ (36)	87
	POCl₃, [pyrrolidinone]	[pyrrole structure] (76)	87
$R^2 = R^3 = Me$, $R^4 = Et$	DMF, POCl₃	$R^5 = CHO$ (—)	534
C₉			
$R^2 = R^4 = Me$, $R^3 = CO_2Et$	DMF, POCl₃	$R^5 = CHO$ (93)	508, 516
	DMA, POCl₃	$R^5 = COMe$ (67)	516
	Et₂NCOCH₂Cl, POCl₃	$R^5 = COCH_2Cl$ (75)	87
	Et₂NCOCH₂CO₂Et, POCl₃	$R^5 = COCH_2CO_2Et$ (61)	87
	POCl₃, [piperidine amide, MeO₂C]	$R^5 = CO(CH_2)_2CO_2Me$ (40)	516
	R₂NCOAr, POCl₃	$R^5 = COAr$	

R	Ar		Refs.
Me	2-pyrrolyl	(86)	183, 537
Me	2-furyl	(81)	183, 537
Me	2-thienyl	(87)	183, 537
Me	3-pyridyl	(85)	183

165

TABLE XI. PYRROLES (*Continued*)

Substrate	Reagents		Product(s) and Yield(s) (%)	Refs.
	R	Ar		
	Me	Ph	(99)	86, 537, 538
	Me	3-ClC$_6$H$_4$	(88)	183, 537
	Me	4-ClC$_6$H$_4$	(90)	183, 537
	Me	3-O$_2$NC$_6$H$_4$	(90)	183, 537
	Me	4-O$_2$NC$_6$H$_4$	(96)	183, 537
	Me	4-MeC$_6$H$_4$	(82)	537
	Me	4-MeOC$_6$H$_4$	(86)	537
	Et	Ph	(95)	538, 86
	(CH$_2$)$_3$	Ph	(10)	538
	(CH$_2$)$_4$	Ph	(93)	538
	(CH$_2$)$_5$	Ph	(97)	538, 516
	[(CH$_2$)$_2$]$_2$O	Ph	(99)	86, 537-539
	[(CH$_2$)$_2$]$_2$O	4-ClC$_6$H$_4$	(87)	183, 537
	[(CH$_2$)$_2$]$_2$O	4-MeC$_6$H$_4$	(85)	183, 537
	[(CH$_2$)$_2$]$_2$O	4-MeOC$_6$H$_4$	(90)	183, 537
	[(CH$_2$)$_2$]$_2$NMe	Ph	(96)	538
POCl$_3$, PhCON⟨ ⟩NCOPh			R^5 = COPh (94)	538
1. O[(CH$_2$)$_2$]$_2$NCOAr, POCl$_3$			R^5 = CH$_2$Ar	461
2. LiBH$_4$ or Na(CN)BH$_3$				

Ar	
Ph	(90-93)
4-ClC$_6$H$_4$	(89-90)
4-MeC$_6$H$_4$	(87-88)
4-MeOC$_6$H$_4$	(91-93)

166

TABLE XI. PYRROLES (Continued)

Substrate	Reagents	Product(s) and Yield(s) (%)	Refs.
C_{11}			
$R^2 = R^4 = CO_2Et$, $R^3 = Me$	DMF, POCl$_3$	$R^5 = CHO$ (59)	516
C_{12}			
$R^2 = R^4 = Me$, $R^3 = Ph$	DMF, PhCOCl	$R^5 = CHO$ (88)	466
C_{14}			
$R^2 = CO_2Et$, $R^3 = Me$, $R^4 = Ar$	DMF, POCl$_3$	$R^5 = CHO$	510
Ar			
2-O$_2$N-3-ClC$_6$H$_3$		(—)	
3-O$_2$N-4-ClC$_6$H$_3$		(—)	
4-O$_2$NC$_6$H$_4$		(91)	
$R^2 = Me$, $R^3 = CO_2Et$, $R^4 = Ph$	MFA, POCl$_3$	$R^5 = CHO$ (—)	540
$R^2 = CO_2Bu\text{-}t$, $R^3 = (CH_2)_2CO_2Me$, $R^4 = Me$	DMF, POCl$_3$	$R^5 = CHO$ (75)	92, 541
C_{15}	DMF, PhCOCl	(13)	542
C_{16}	1. Decarboxylate 2. DMF, PhCOCl	COMe (74)	542
C_{17}	DMF, POCl$_3$	(100)	93, 516

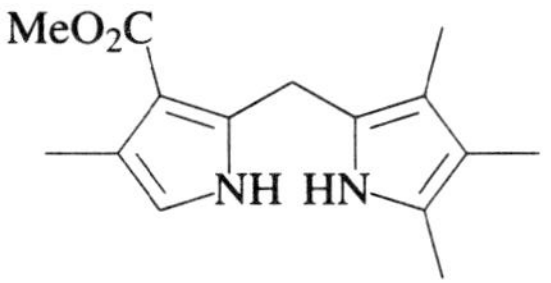

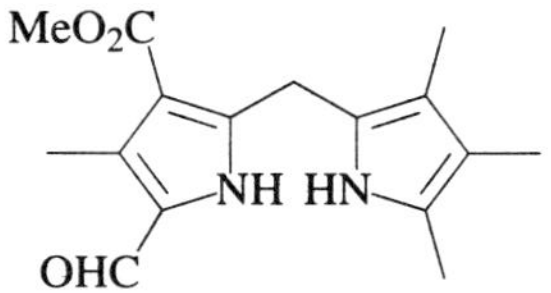

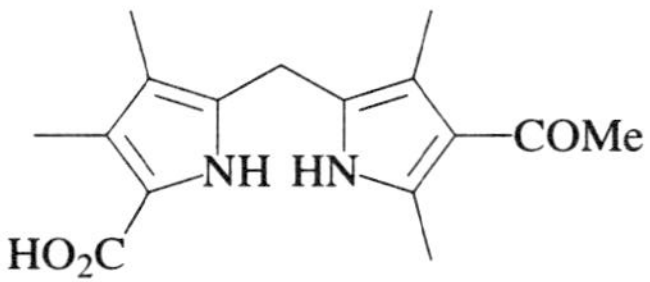

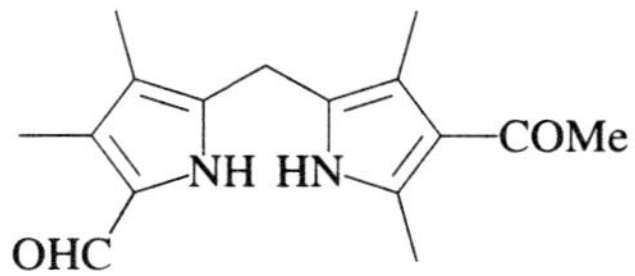

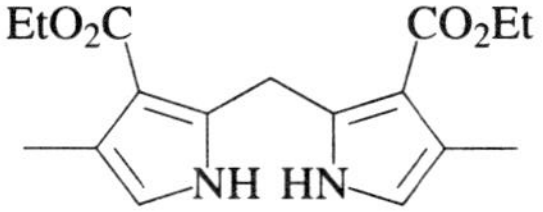

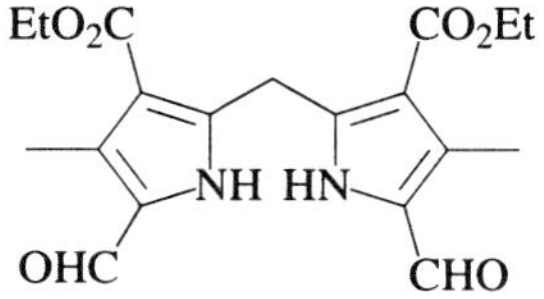

167

TABLE XI. PYRROLES (*Continued*)

Substrate	Reagents	Product(s) and Yield(s) (%)	Refs.
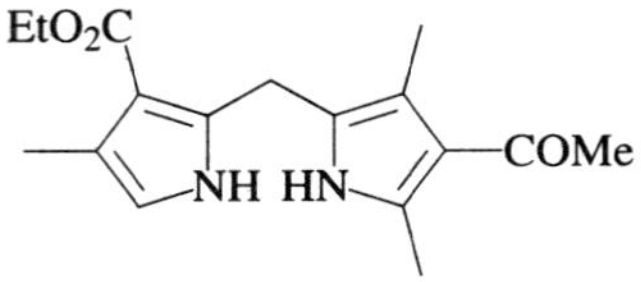 EtO$_2$C ... COMe, NH HN (dipyrromethane)	DMF, PhCOCl	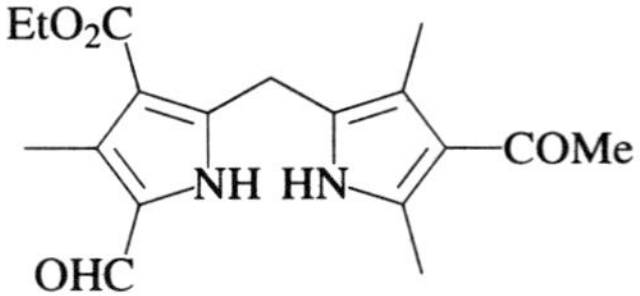 EtO$_2$C ... COMe, NH HN, OHC (27)	542
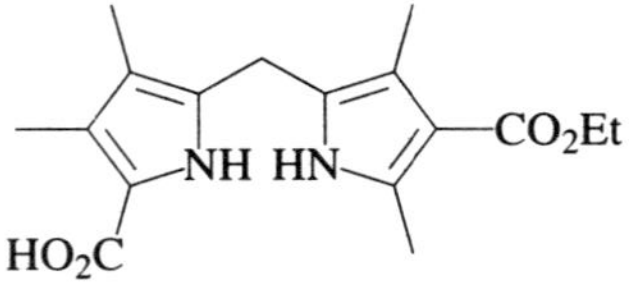 CO$_2$Et, NH HN, HO$_2$C	1. Decarboxylate 2. DMF, PhCOCl	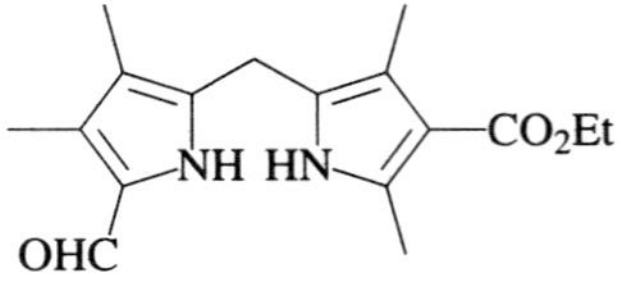 CO$_2$Et, NH HN, OHC (71)	542
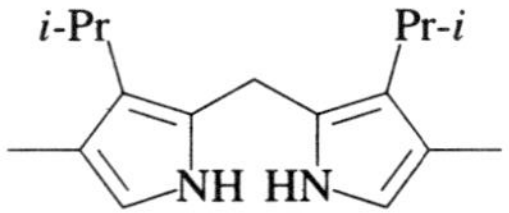 Pr-*i*, *i*-Pr, NH HN	DMF, POCl$_3$	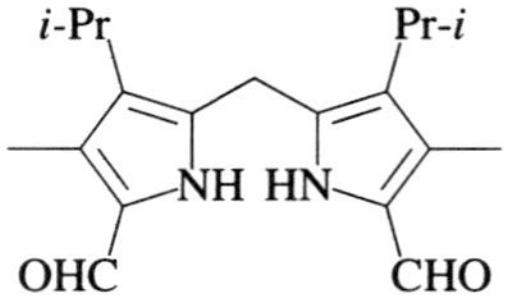 Pr-*i*, CHO, NH HN, OHC, *i*-Pr (66)	543
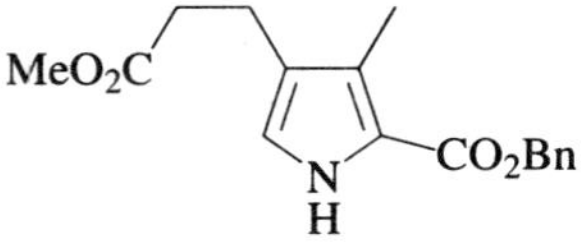 MeO$_2$C ... CO$_2$Bn, N H	DMF, POCl$_3$	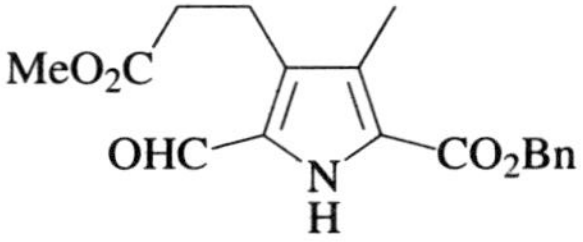 MeO$_2$C ... CO$_2$Bn, N H, OHC (89)	544
Et, CO$_2$Et, NH HN, HO$_2$C 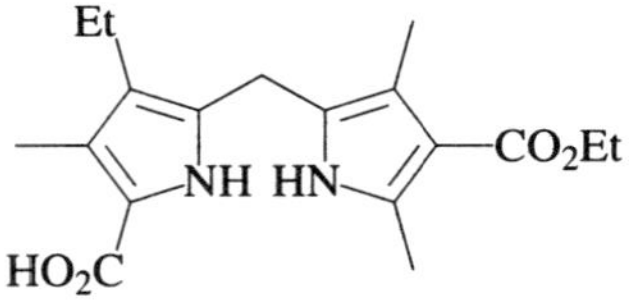	1. Decarboxylate 2. DMF, PhCOCl	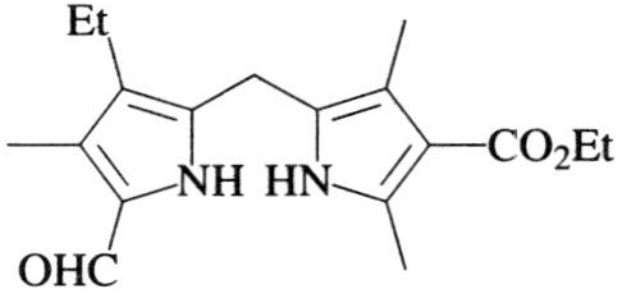 Et, CO$_2$Et, NH HN, OHC (56)	542

C$_{18}$

168

TABLE XI. PYRROLES (*Continued*)

Substrate	Reagents	Product(s) and Yield(s) (%)	Refs.
C$_{18-23}$ (EtO$_2$C... CO$_2$Et pyrrole structure)	DMF, PhCOCl	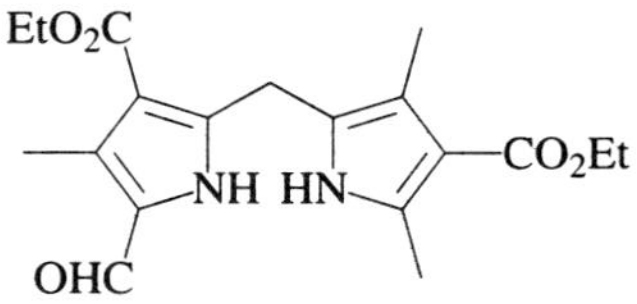 (EtO$_2$C... CO$_2$Et, OHC pyrrole structure) (22)	542
(bispyrrole, n = 6, 11)	DMF, POCl$_3$	(OHC... bispyrrole structure, n) (66), (91)	545, 546
C$_{19-20}$ (MeO$_2$C, COR, HO$_2$C pyrrole structure)	1. Decarboxylate 2. DMF, PhCOCl	(CO$_2$Me, COR, OHC pyrrole structure) R = Me (68), R = OEt (64)	542
C$_{19}$ (MeO$_2$C, CO$_2$Me, CO$_2$Bn pyrrole structure)	Me$_2$N^{13}CHO, —	(MeO$_2$C, CO$_2$Me, CO$_2$Bn, OH^{13}C pyrrole structure) (—)	548

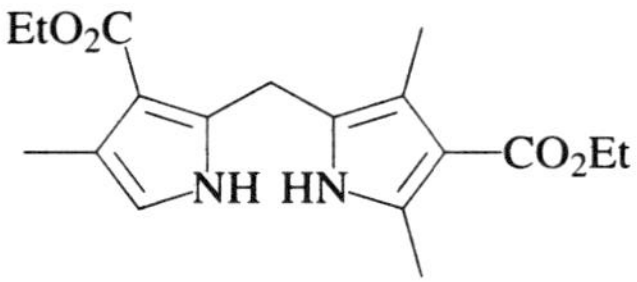

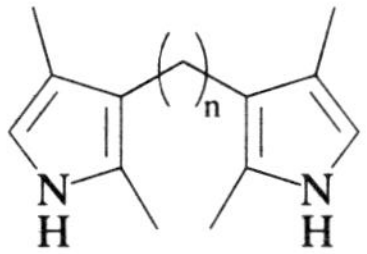

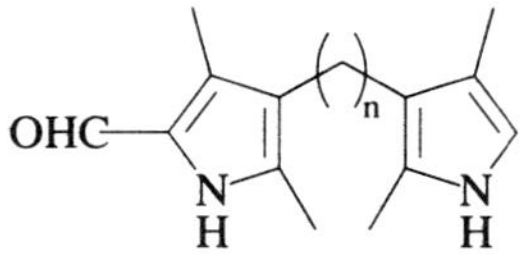

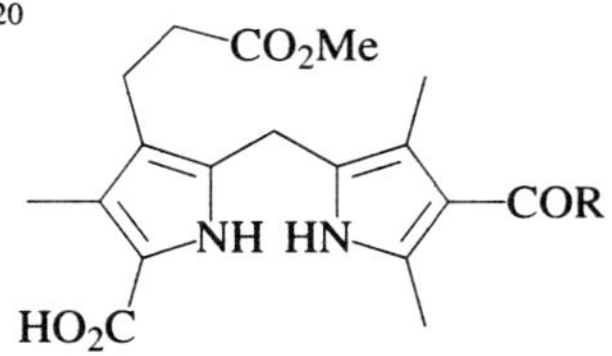

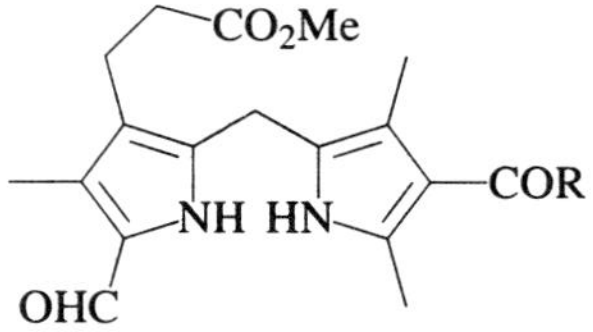

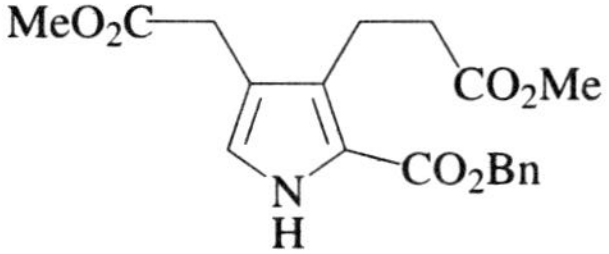

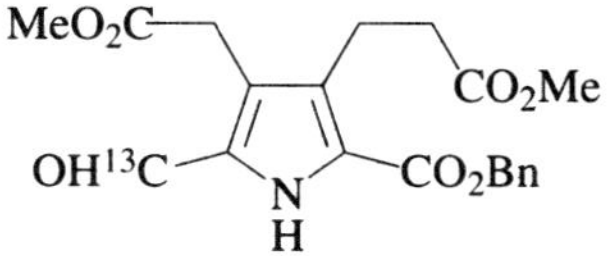

TABLE XI. PYRROLES (Continued)

Substrate	Reagents	Product(s) and Yield(s) (%)	Refs.
C_{20} 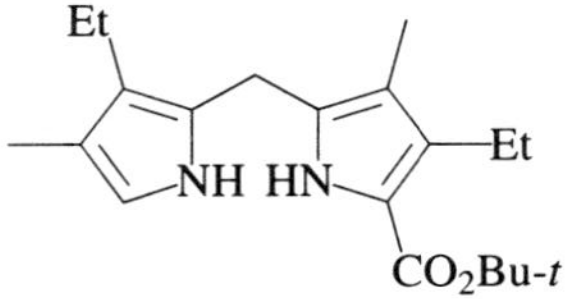	—	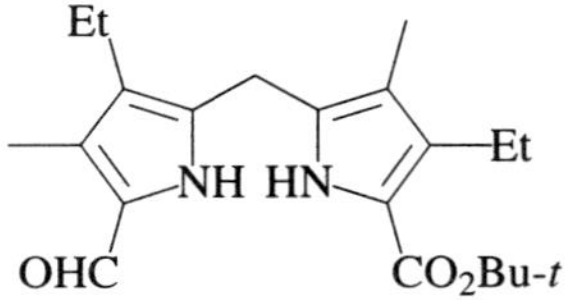 (—)	548
C_{22} 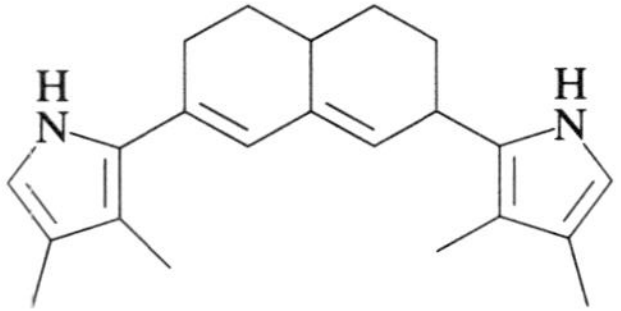	DMF, $POCl_3$	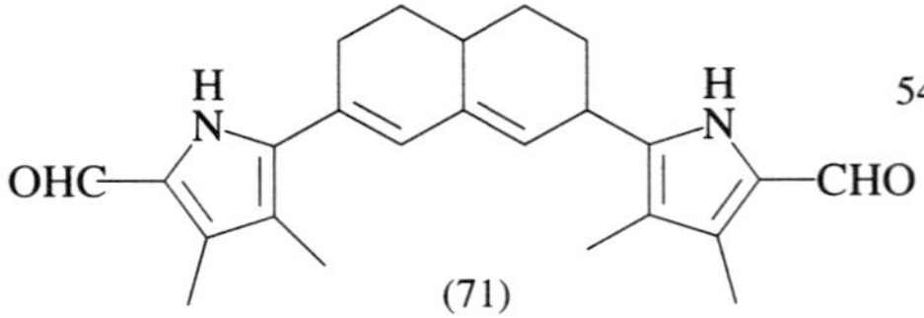 (71)	549
C_{23} 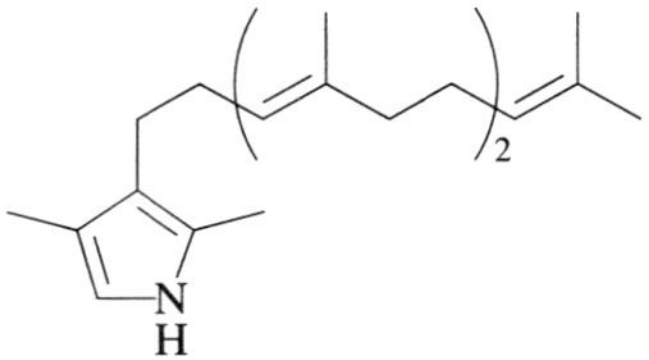	—	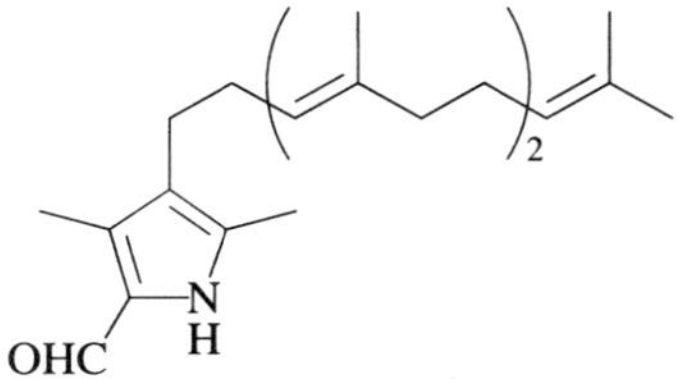 (74)	550
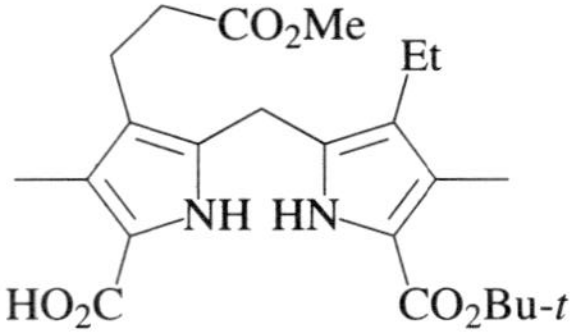	1. Decarboxylate 2. DMF, PhCOCl	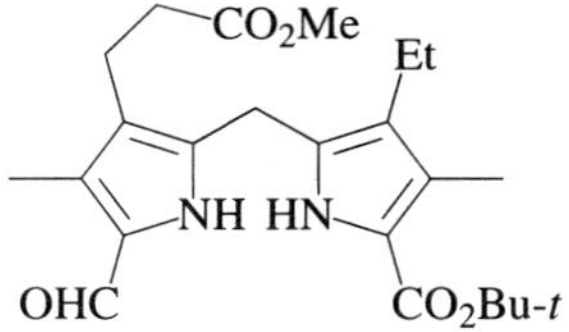 (66)	551

Structures (chemical diagrams):

- C_{20} substrate: pyrrole with Et, $CO_2Bu\text{-}t$, Et substituents; product bears OHC, $CO_2Bu\text{-}t$, Et groups.
- C_{22} substrate: bis-pyrrole fused system; product with OHC and CHO groups.
- C_{23} substrate: prenyl-substituted dimethylpyrrole; product with OHC.
- Lower substrate: pyrrole with CO_2Me, Et, $CO_2Bu\text{-}t$, HO_2C; product with CO_2Me, Et, $CO_2Bu\text{-}t$, OHC.

170

TABLE XI. PYRROLES (Continued)

Substrate	Reagents	Product(s) and Yield(s) (%)	Refs.
C_30-36 (structure, n = 4, 5, 7)	DMF, POCl$_3$	(structure, CHO/OHC)	552, 553
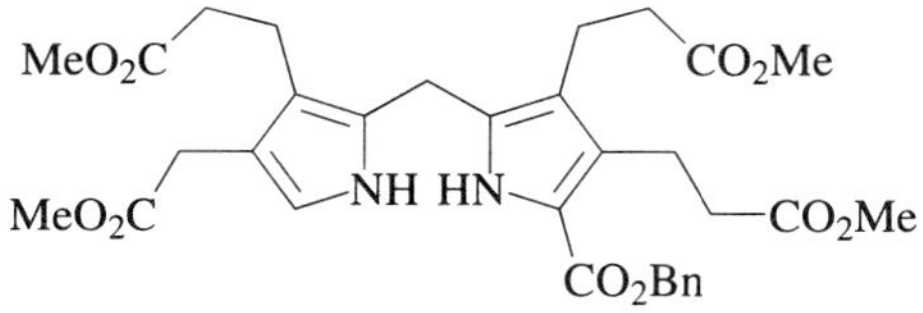 (structure, MeO$_2$C ... CO$_2$Me, CO$_2$Bn, NH HN)	—	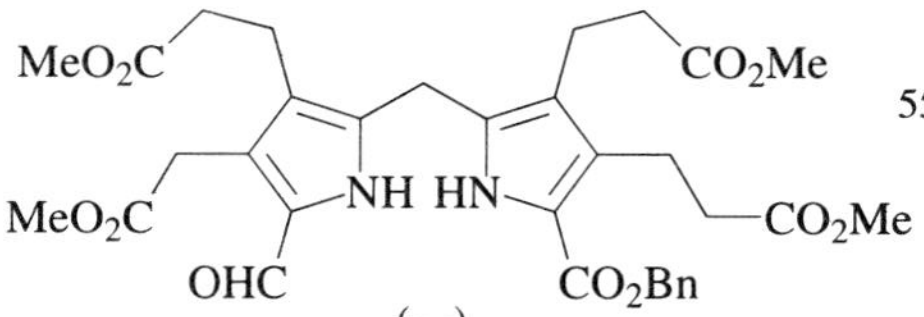 (—)	554

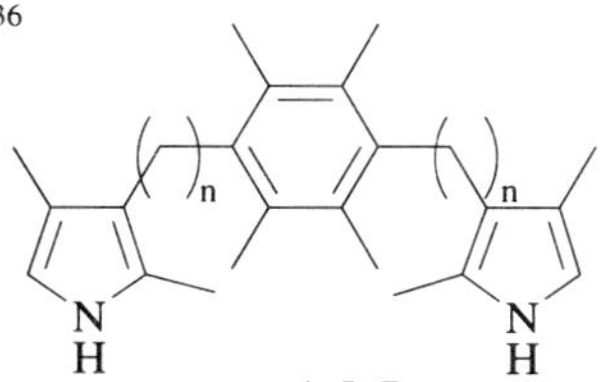

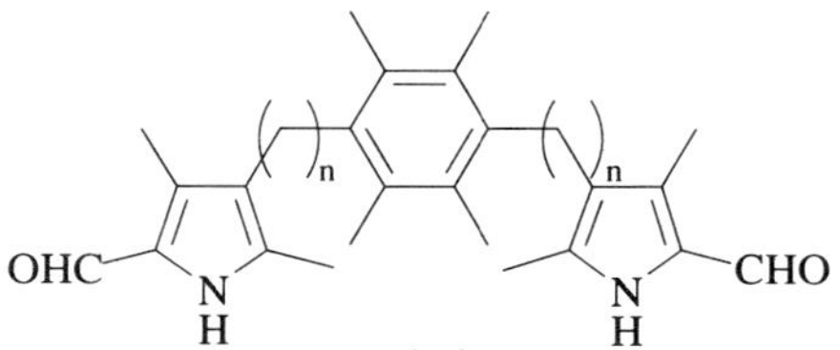

D6. R^2, R^3, R^5 Substituents

	Substrate	Reagents	Product(s) and Yield(s) (%)	Refs.
C$_7$	R^2 = R^3 = R^5 = Me	DMF, POCl$_3$	R^4 = CHO (99)	513
C$_9$	R^2 = CO$_2$Et, R^3 = R^5 = Me	DMF, POCl$_3$; PhNHCHO, POCl$_3$	R^4 = CHO (95); R^4 = CHO (—)	508, 555; 233
	R^3 = CO$_2$Et, R^2 = R^5 = Me	DMF, POCl$_3$; PhNHCHO, POCl$_3$	R^4 = CHO (93); R^4 = CHO (20)	516; 233
C$_{11}$	R^2 = R^3 = CO$_2$Et, R^5 = Me	DMF, POCl$_3$	R^4 = CHO (81)	556
C$_{14}$	R^2 = CO$_2$Et, R^3 = Ph, R^5 = Me	DMF, POCl$_3$	R^4 = CHO (97)	555
	R^2 = CO$_2$Bn, R^3 = R^5 = Me	DMF, POCl$_3$	R^4 = CHO (93)	557

TABLE XI. PYRROLES (*Continued*)

Substrate	Reagents	Product(s) and Yield(s) (%)	Refs.
E. Tetrasubstituted Pyrroles			
E1. R¹, R², R³, R⁴ Substituents			
C_{10} $R^1 = R^2 = R^4 = Me, R^3 = CO_2Et$	Me₂NCOPh, POCl₃	$R^5 = COPh$ (78)	86
C_{16} $R^1 = Bn, R^2 = R^4 = Me, R^3 = CO_2Et$	Me₂NCOPh, POCl₃	$R^5 = COPh$ (20)	86
E2. R¹, R², R³, R⁵ Substituents			
C_{13} $R^1 = Ph, R^2 = R^3 = R^5 = Me$	DMF, POCl₃	$R^4 = CHO$ (83)	530
C_{18} $R^1 = R^5 = Ph, R^2 = R^3 = Me$	DMF, POCl₃	$R^4 = CHO$ (77)	530
C_{23}	Me₂NCH=CHCHO, POCl₃	(—)	558

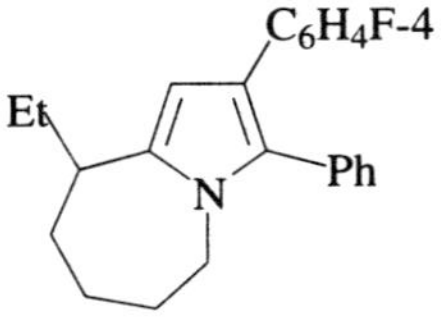

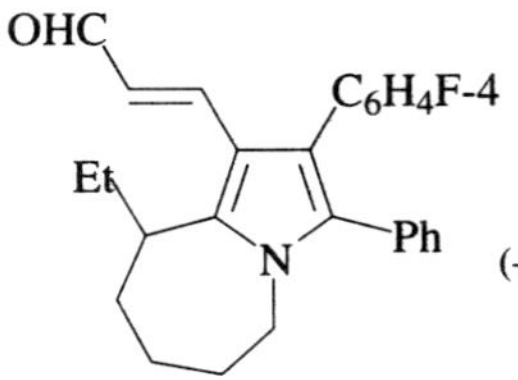

TABLE XII. BENZO[b]FURANS

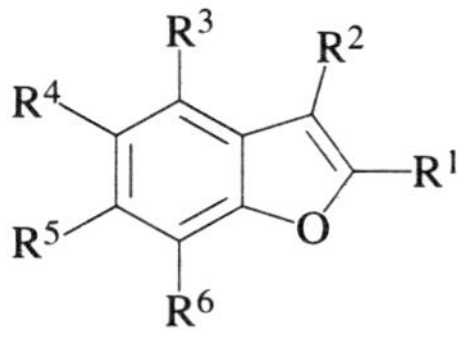

R¹ - R⁶ = H except as indicated.

Substrate	Reagents	Product(s) and Yield(s) (%)	Refs.
	A. Benzo[b]furan		
C_8	DMF, POCl$_3$	R^1 = CHO (62)	95, 156
	B. Monosubstituted Benzo[b]furans		
	B1. R^1 Substituents		
C_8 R^1 = Me	DMF, POCl$_3$	R^2 = CHO (78)	156
C_{10} R^1 = Et	DMF, POCl$_3$	R^2 = CHO (70)	156
	B2. R^4 Substituents		
C_9 R^4 = OMe	DMF, POCl$_3$	R^1 = CHO (38)	559
	C. Disubstituted Benzo[b]furans		
	C1. R^1, R^4 Substituents		
C_9 R^1 = Et, R^4 = Cl	DMF, POCl$_3$	R^2 = CHO (76)	298
C_{11} R^1 = Et, R^4 = Me	DMF, POCl$_3$	R^2 = CHO (84)	560
R^1 = Et, R^4 = OMe	DMF, POCl$_3$	R^2 = CHO (76)	298
C_{15} R^1 = Ph, R^4 = OMe	DMF, POCl$_3$	R^2 = CHO (75)	561

173

TABLE XII. BENZO[b]FURANS (*Continued*)

Substrate		Reagents	Product(s) and Yield(s) (%)	Refs.
		C2. R^2, R^3 Substituents		
C_{10}	R^2 = Me, R^3 = OMe	DMF, $POCl_3$	R^1 = CHO (80)	
		C3. R^2, R^4 Substituents		
C_{10}	R^2 = Me, R^4 = OMe	DMF, $POCl_3$	R^1 = CHO (50)	98
C_{11}	R^2 = NHAc, R^4 = OMe	DMF, $POCl_3$	R^1 = CHO (50)	562
C_{15}	R^2 = Ph, R^4 = OMe	DMF, $POCl_3$	R^1 = CHO (98)	561
		C4. R^2, R^5 Substituents		
C_{10}	R^2 = R^5 = Me	DMF, $POCl_3$	R^1 = CHO (75)	563
	R^2 = Me, R^5 = OMe	DMF, $POCl_3$	R^1 = CHO (90)	98
		C5. R^2, R^6 Substituents		
C_{10}	R^2 = Me, R^6 = OMe	DMF, $POCl_3$	R^1 = CHO (85)	98
		C6. R^3, R^5 Substituents		
C_9	R^3 = Me, R^5 = Cl	DMF, $POCl_3$	R^1 = CHO (97)	564
		D. Trisubstituted Benzo[b]furans		
		D1. R^1, R^2, R^4 Substituents		
C_{21}	R^1 = R^2 = Ph, R^4 = OMe	DMF, $POCl_3$	R^5 = CHO (78)	565
		D2. R^1, R^2, R^5 Substituents		
C_{21}	R^1 = R^2 = Ph, R^5 = OMe	DMF, $POCl_3$	R^4 = CHO (60)	565

TABLE XII. BENZO[b]FURANS (Continued)

Substrate	Reagents	Product(s) and Yield(s) (%)	Refs.
	D3. R^1, R^3, R^6 Substituents		
C_{11}			
$R^1 = R^6 = Me$, $R^3 = OMe$	DMF, POCl$_3$	$R^2 = CHO$ (79)	98
C_{12}			
$R^1 = Et$, $R^3 = Me$, $R^6 = OMe$	DMF, POCl$_3$	$R^5 = CHO$ (84)	98
	D4. R^2, R^3, R^5 Substituents		
C_{12}			
$R^2 = Et$, $R^3 = R^5 = Me$	DMA, POCl$_3$	$R^1 = COMe$ (26)	566
	D5. R^2, R^3, R^6 Substituents		
C_{11}			
$R^2 = R^3 = Me$, $R^6 = OMe$	DMF, POCl$_3$	$R^1 = CHO$ (67)	98
$R^2 = Me$, $R^3 = R^6 = OMe$	DMF, POCl$_3$	$R^1 = CHO$ (70)	98
	D6. R^2, R^5, R^6 Substituents		
C_{11}			
$R^2 = Me$, $R^5 = R^6 = OMe$	DMF, POCl$_3$	$R^1 = CHO$ (87)	98
	D7. R^3, R^4, R^5 Substituents		
C_{10}			
$R^3 = R^5 = Me$, $R^4 = Cl$	DMF, POCl$_3$	$R^1 = CHO$ (93)	564

TABLE XIII. BENZO[b]THIOPHENES

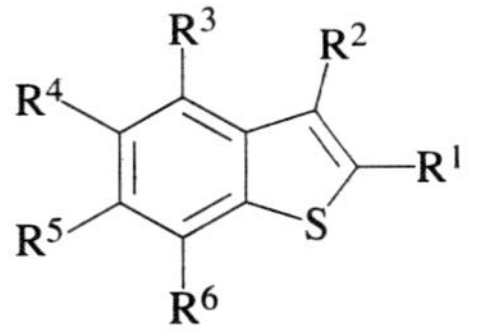

$R^1 - R^6 = H$ except as indicated.

Substrate	Reagents	Product(s) and Yield(s) (%)	Refs.
A. Benzo[b]thiophene			
C_8	MFA, POCl$_3$	R^2 = CHO (7)	96
B. Monosubstituted Benzo[b]thiophenes			
B1. R^1 Substituents			
C_{12} R^1 = N-morpholinyl	DMF, POCl$_3$	R^2 = CHO (80)	412
C_{16}	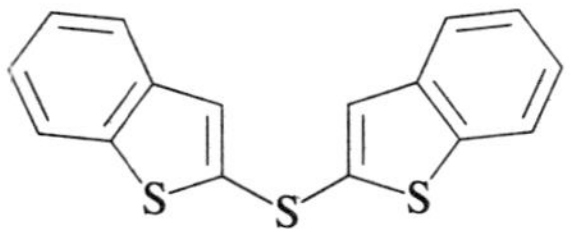 DMF, POCl$_3$	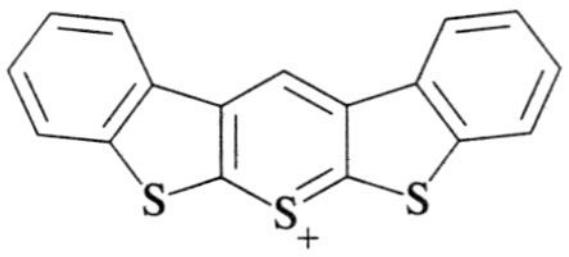 Cl$^-$ (37)	434
B2. R^2 Substituents			
C_8 R^2 = NH$_2$	DMF, POCl$_3$	R^1 = CHO, R^2 = N=CHNMe$_2$ (—)	567
C_9 R^2 = OMe	MFA, POCl$_3$, <45°	R^1 = CHO (40-45)	100
	MFA, POCl$_3$, 90°	R^1 = CHO, R^2 = Cl (90)	100
C_{10} R^2 = NHAc	DMF, POCl$_3$	R^1 = CHO (31)	568
	DMF, POCl$_3$	R^1 = CHO, R^2 = N=CHNMe$_2$ (35)	569

176

TABLE XIII. BENZO[b]THIOPHENES (Continued)

Substrate	Reagents	Product(s) and Yield(s) (%)	Refs.
C₁₁			
R² = N=CHNMe₂	DMF, POCl₃	R¹ = CHO (85)	567
B3. R³ Substituents			
C₉			
R³ = OMe	DMF, POCl₃	R⁶ = CHO (98)	99
	MFA, POCl₃	R⁶ = CHO (70)	100
C. Disubstituted Benzo[b]thiophenes			
C1. R², R³ Substituents			
C₁₀			
R² = Me, R³ = OMe	DMF, POCl₃	R¹ = CHO (35) + R⁶ = CHO (62)	99
C2. R², R⁵ Substituents			
C₁₀			
R² = OH, R⁵ = OEt	MFA, POCl₃	R¹ = CHO (—)	189
C3. R⁴, R⁵ Substituents			
C₁₀			
R⁴ + R⁵ = CH₂OCH₂	MFA, POCl₃	R¹ = CHO (66)	570
D. Trisubstituted Benzo[b]thiophenes			
D1. R², R⁴, R⁶ Substituents			
C₉			
C₂ R² = OH, R⁴ = Me, R⁶ = Cl	MFA, POCl₃	R¹ = CHO (—)	189

TABLE XIV. INDOLES

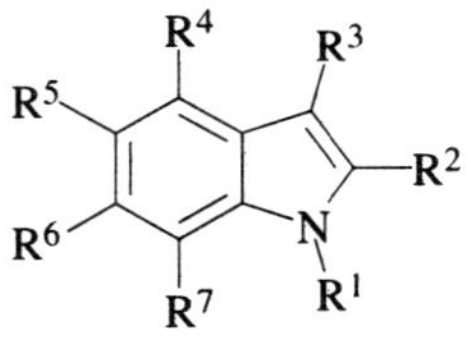

R^1 - R^7 = H except as indicated.

Substrate	Reagents	Product(s) and Yield(s) (%)	Refs.
	A. Indole		
C_8	DMF, POCl$_3$	R^3 = CHO (100)	572, 571, 97
	MFA, POCl$_3$	R^3 = CHO (53)	573
	DMF, (Cl$_2$PO)$_2$O	R^3 = CHO (97)	26
	DMF, PCl$_3$	R^3 = CHO (95)	321
	DMF, PCl$_5$	R^3 = CHO (97)	321
	DMF, SOCl$_2$	R^3 = CHO (65)	321
	DMF, BCl$_3$	R^3 = CHO (80)	574
	DMF, C$_3$N$_3$Cl$_3$[a]	R^3 = CHO (31)	36
	DMF, Ph$_3$PBr$_2$	R^3 = CHO (78)	55
	DMF, MeCOCl	R^3 = CHO (48)	575
	DMF, MeCOBr	R^3 = CHO (89)	575
	DMF, PhCOCl	R^3 = CHO (85)	575
	1. DMF, POCl$_3$ 2. NH$_2$OH 3. Dehydrate	R^3 = CN (54-59)	197
	1. N-Formylmorpholine, POCl$_3$ 2. NH$_2$OH 3. Dehydrate	R^3 = CN (31)	197
	1. DMF, POCl$_3$ 2. HBF$_4$	R^3 = CH=NMe$_2^+$ BF$_4^-$ (56)	576

TABLE XIV. INDOLES (*Continued*)

Substrate	Reagents	Product(s) and Yield(s) (%)	Refs.
	1. DMF, POCl$_3$ 2. base	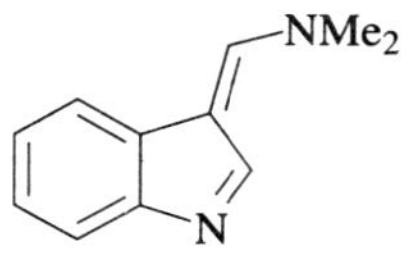 (61)	572
	1. DMF, POCl$_3$ 2. Na$_2$CO$_3$ 3. CNCH$_2$CO$_2$Me, DMF, MeOH	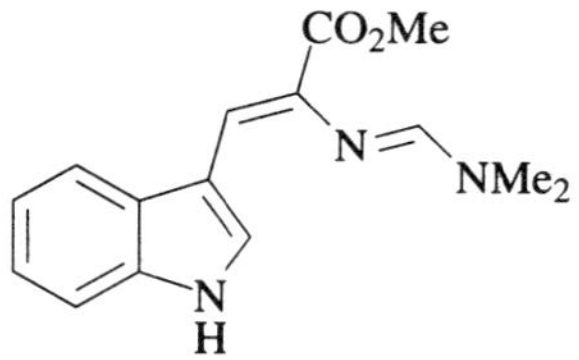 (80)	577
	1. DMF, POCl$_3$ 2. Na$_2$CO$_3$ 3. CNCH$_2$CO$_2$Me, DMF	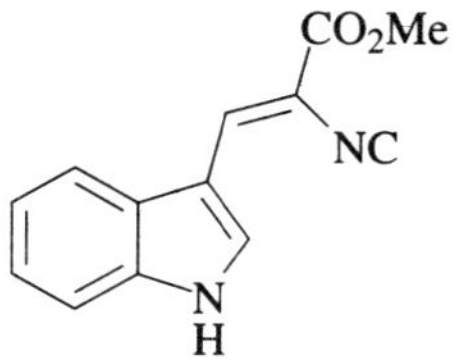 (60)	577
	DMA, COCl$_2$	R^1 = COMe + R^3 = COMe (—) 2:98	103
	DMA, POCl$_3$	R^3 = COMe (45)	578
	MeNHCOMe, POCl$_3$	R^3 = COMe (22)	101
	Me$_2$NCOCH$_2$Cl, POCl$_3$	R^3 = COCH$_2$Cl (37)	101
	Me$_2$NCOEt, POCl$_3$	R^3 = COEt (86)	101
	Me$_2$NCOPh, POCl$_3$	R^3 = COPh (51)	101, 578

179

TABLE XIV. INDOLES (Continued)

Substrate	Reagents	Product(s) and Yield(s) (%)	Refs.
	POCl₃,	(—)	102
	POCl₃,		
	R		
	H	(89)	579
	Me	(85)	102
	Et	(95)	579, 102
	Bu	(75)	579, 102
	Bn	(50)	579, 102
	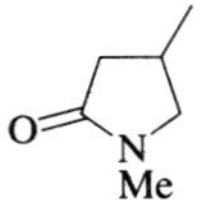 POCl₃,	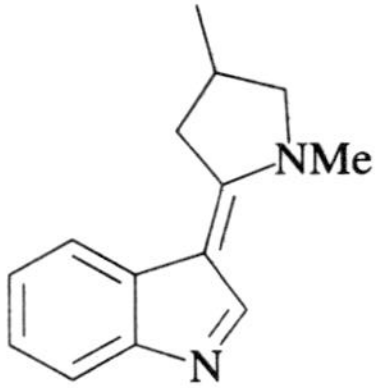 (53)	579, 102

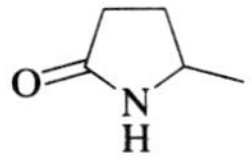

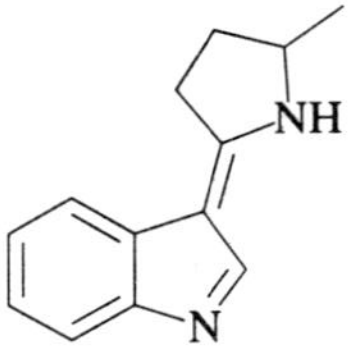

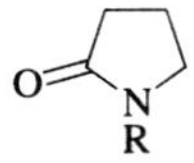

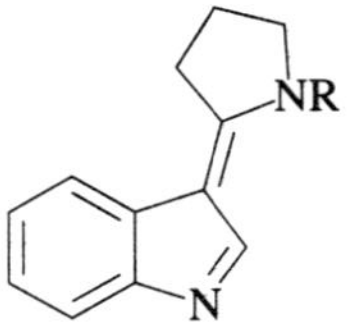

TABLE XIV. INDOLES (*Continued*)

Substrate	Reagents	Product(s) and Yield(s) (%)	Refs.
C₉ **R¹ = Me**	POCl₃,	(—)	102
	Me₂NN=CHCHO, COCl₂	R³ = CH=CHN=NMe₂⁺ Cl⁻ (90)	40

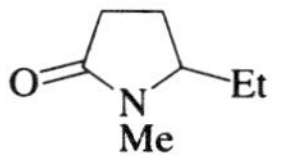

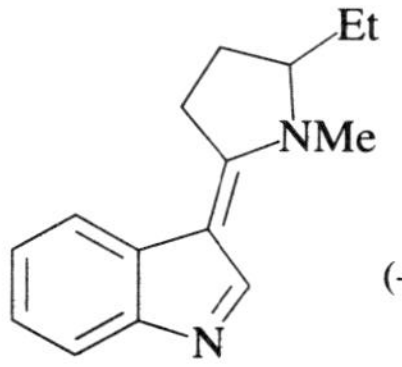

B. Monosubstituted Indoles
B1. R¹ Substituents

	DMF, POCl₃	R³ = CHO (87)	580
	DMF, (Cl₂PO)₂O	R³ = CHO (98)	26
	POCl₃, O=	(85)	581
C₁₀ **R¹ = OMe**	DMF, POCl₃	R³ = CHO (91)	582
R¹ = OAc	DMF, POCl₃	R¹ = OH, R³ = CHO (6) + R² = Cl, R³ = CHO (53)	583
R¹ = Et	DMA, POCl₃	R³ = COMe (76)	101

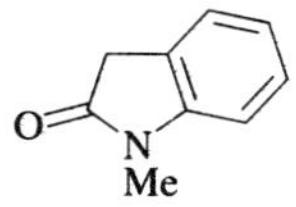

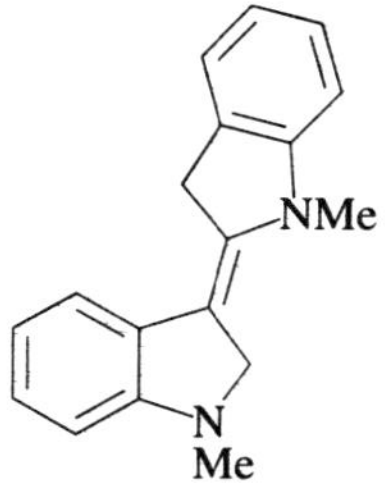

181

TABLE XIV. INDOLES (*Continued*)

Substrate	Reagents	Product(s) and Yield(s) (%)	Refs.
C$_{14}$			
R^1 = Ar			
Ar		R^3 = CHO	
2-O$_2$NC$_6$H$_4$	DMF, POCl$_3$	(92)	584
	DMF, PCl$_3$	(60)	321
	DMF, PCl$_5$	(60)	321
	DMF, SOCl$_2$	(24)	321
3-O$_2$NC$_6$H$_4$	DMF, POCl$_3$	(90)	584
	DMF, PCl$_3$	(60)	321
	DMF, PCl$_5$	(60)	321
	DMF, SOCl$_2$	(24)	321
4-O$_2$NC$_6$H$_4$	DMF, POCl$_3$	(98)	584, 585
	DMF, PCl$_3$	(60)	321
	DMF, PCl$_5$	(60)	321
	DMF, SOCl$_2$	(26)	321
R^1 = OTBDMS	—	**I** (39) + **II** (32)	586
C$_{15}$ R^1 = O$_2$CPh	1. DMF, POCl$_3$ (3 eq), 47 h 2. NaOH	**I** (98)	586
	1. DMF, POCl$_3$ (2 eq), 3 h 2. NaOH	**I** (14) + **II** (57)	586
C$_{21}$ R^1 = OTs	—	**I** (67)	586
C$_{22}$			
R^1 = tetraacetyl-β-D-arabinose	DMF, POCl$_3$	R^3 = CHO (86)	587
R^1 = tetraacetyl-β-D-glucose	DMF, POCl$_3$	R^3 = CHO (85)	588

182

TABLE XIV. INDOLES (*Continued*)

Substrate	Reagents	Product(s) and Yield(s) (%)	Refs.
	B2. R^2 Substituents		
C_9			
$R^2 = Me$	DMF, $POCl_3$	$R^3 = CHO$ (96)	589, 590
	DMF, $(Cl_2PO)_2O$	$R^3 = CHO$ (88)	26
	DMF, $PhCOCl$	$R^3 = CHO$ (88)	575
	DMA, $POCl_3$	$R^3 = COMe$ (98)	101
	DMA, $POCl_3$	$R^1 = COMe + R^3 = COMe$ (—) 2:98	103
	$Me_2NCOBu\text{-}t$, $POCl_3$	$R^3 = COBu\text{-}t$ (49)	101
	$Me_2NCOCH(Me)Et$, $POCl_3$	$R^3 = COCH(Me)Et$ (18)	101
	$Me_2NCOPr\text{-}i$, $POCl_3$	$R^3 = COPr\text{-}i$ (62)	101
	$Me_2NCOCH(Et)Pr\text{-}i$, $POCl_3$	$R^3 = COCH(Et)Pr\text{-}i$ (24)	101
	$POCl_3$, [N-Me pyrrolidinone]	[2-methyl-3-(1-methylpyrrolidin-2-ylidene)indole] (96)	579, 102
$R^2 = CONHNH_2$	DMF, $POCl_3$	$R^2 + R^3 = CONHNHCH_2$ (77)	591
C_{10}			
$R^2 = CH_2CN$	DMA, $POCl_3$	$R^3 = COMe$ (—)	592
$R^2 = Et$	DMF, $POCl_3$	$R^3 = CHO$ (70)	589
$R^2 = CH_2CO_2Me$	MFA, $POCl_3$	$R^3 = CHO$ (—)	593
$R^2 = NHCO_2Et$	DMF, $POCl_3$	$R^3 = CHO$ (88)	594
$R^2 = n\text{-}Pr$	DMF, $POCl_3$	$R^3 = CHO$ (75)	589
$R^2 = i\text{-}Pr$	DMF, $POCl_3$	$R^3 = CHO$ (88)	589
C_{12}			
$R^2 = CO_2Et$	DMF, $POCl_3$	$R^3 = CHO$ (—)	591
	MFA, $POCl_3$	$R^3 = CHO$ (100)	26

183

TABLE XIV. INDOLES (*Continued*)

Substrate	Reagents	Product(s) and Yield(s) (%)	Refs.
R^2 = 2-furyl	DMF, POCl$_3$, 30°	R^3 = CHO (85)	595
	DMF, POCl$_3$ (xs), 100°	R^3 = R^5 = CHO (63)	595
R^2 = t-Bu	DMF, POCl$_3$	R^3 = CHO (80)	590
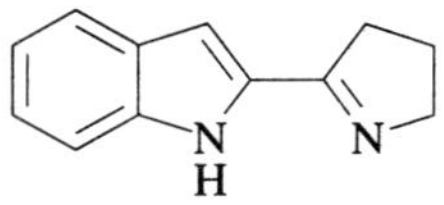 C$_{12\text{-}18}$	DMF, POCl$_3$	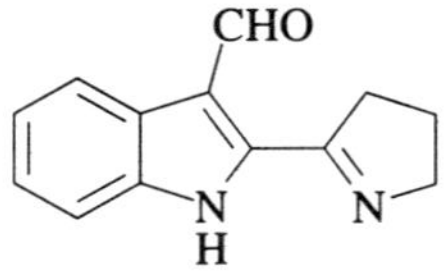 (30)	596
$\dfrac{R}{H}$	DMF, POCl$_3$, rt	R^3 = CHO (94)	235
	DMF, POCl$_3$, rt to reflux	2-methylcarbazole (80)	235
	DMF, POCl$_3$, heat	2-methylcarbazole or R^3 = CHO (high)	597
Me	DMF, POCl$_3$, rt	R^3 = CHO (96)	235
	DMF, POCl$_3$, rt to reflux	1,2-dimethylcarbazole (1)	235
Ph	DMF, POCl$_3$, rt	R^3 = CHO (98)	235
	DMF, POCl$_3$, rt to reflux	1-phenyl-2-methylcarbazole (3)	235
C$_{13}$ R^2 = C(Me)$_2$CH=CH$_2$	—	R^3 = CHO (—)	598
C$_{14}$ R^2 = C$_6$H$_{11}$	DMF, POCl$_3$	R^3 = CHO (79)	589
R^2 = CH$_2$(piperidyl-4)	DMF, POCl$_3$	R^3 = CHO (52)	599

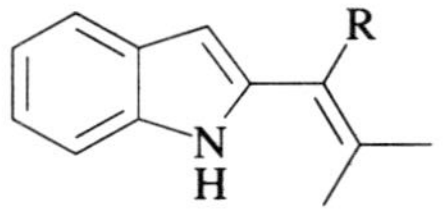

TABLE XIV. INDOLES (*Continued*)

Substrate	Reagents	Product(s) and Yield(s) (%)	Refs.
C_{14-15}			
R^2 = Ar			
Ar			
3-Cl-4-FC$_6$H$_3$	DMF, POCl$_3$	R^3 = CHO (68)	600
4-FC$_6$H$_4$	DMF, POCl$_3$	R^3 = CHO (76)	600
Ph	DMF, POCl$_3$	R^3 = CHO (97)	589
	MFA, POCl$_3$	R^3 = CHO (—)	601
	DMF, POCl$_3$	R^3 = CH=NMe$_2^+$ Cl$^-$ (—)	602
	DMF, PCl$_3$	R^3 = CHO (60)	321
	DMF, PCl$_5$	R^3 = CHO (60)	321
	DMF, SOCl$_2$	R^3 = CHO (20)	321
	DMF, PhCOCl	R^3 = CHO (96)	575
	DMF, POCl$_3$	R^3 = CHO (72)	600
C_{16}			
3,4-Me$_2$C$_6$H$_3$	DMF, POCl$_3$	(85)	599
R^2 = NHCO$_2$Bn	DMF, POCl$_3$	R^3 = CHO (66) + R^2 = N=CHNMe$_2$, R^3 = CHO (17)	594
	DMF, POCl$_3$	R^3 = CHO (67)	603
	Me$_2$NCOR, POCl$_3$	R^3 = COR	594
	R		
	Me	(41)	
	i-Pr	(46)	
	4-ClC$_6$H$_4$	(69)	
	Ph	(72)	
	4-MeOC$_6$H$_4$	(67)	
	Et$_2$NCOCH$_2$Cl, POCl$_3$	R^3 = COCH$_2$Cl (73)	594, 604

TABLE XIV. INDOLES (Continued)

Substrate	Reagents	Product(s) and Yield(s) (%)	Refs.
C₁₉	1. DMF, POCl₃ 2. NaBH₄	(89)	605
	DMF, POCl₃	(64) + H₂PO₄⁻ (27)	606
C₂₆	DMF, POCl₃	(11)	607

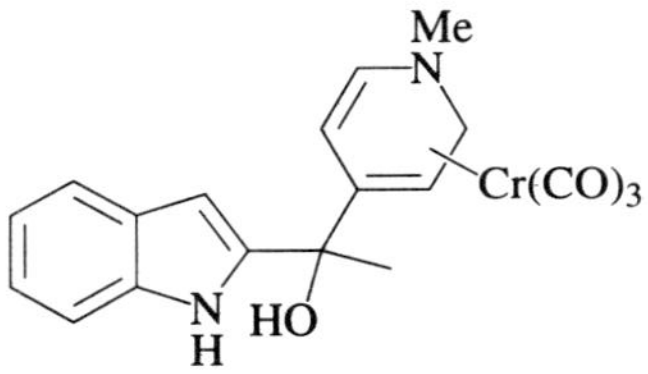

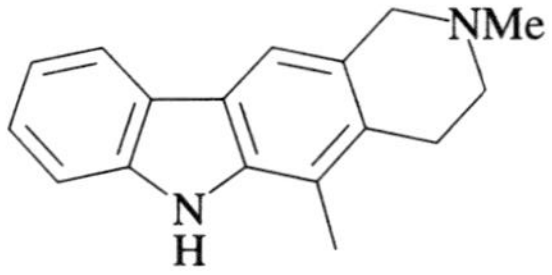

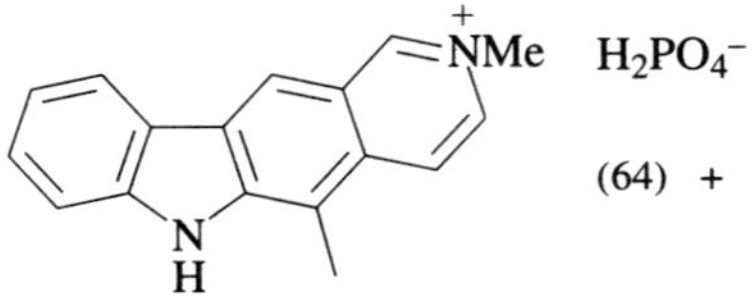

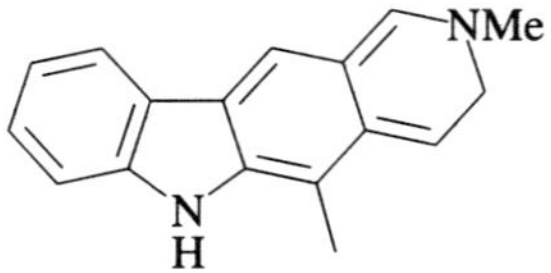

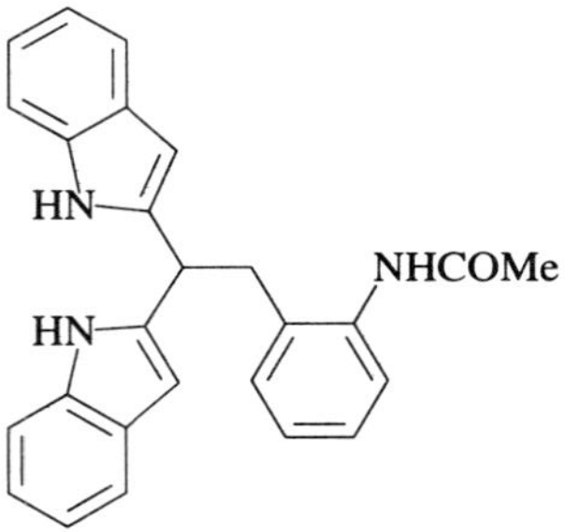

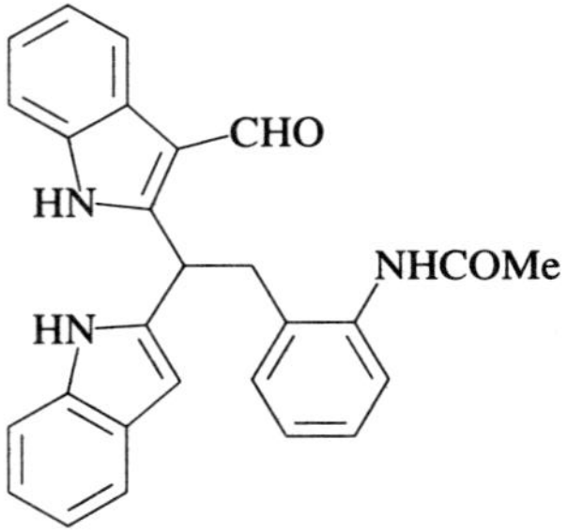

186

TABLE XIV. INDOLES (*Continued*)

Substrate	Reagents	Product(s) and Yield(s) (%)	Refs.
B3. R³ Substituents			
C_8 $R^3 = NH_2$	—	$R^2 = CHO$, $R^3 = N=CHNMe_2$ (—)	608
	Et_2NCHO, $POCl_3$	$R^2 = CHO$, $R^3 = N=CHNEt_2$ (48)	106
C_9 $R^3 = Me$	DMF, $POCl_3$	$R^1 = CHO$ (71) + $R^2 = CHO$ (22)	104, 105
	DMA, $POCl_3$	$R^1 = COMe + R^2 = COMe$ (—) 5:95	103
C_{11}	DMF, $POCl_3$		106
C_{12} $R^3 = t\text{-Bu}$	DMF, $POCl_3$	$R^1 = COMe$ (—)	103
C_{14} $R^3 = 4\text{-}FC_6H_4$	DMF, $POCl_3$	$R^2 = CHO$ (38)	108
B4. R⁴ Substituents			
C_8 $R^4 = Cl$	DMF, $POCl_3$	$R^3 = CHO$ (66)	609
	$POCl_3$, 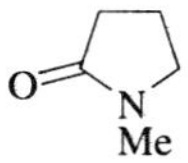	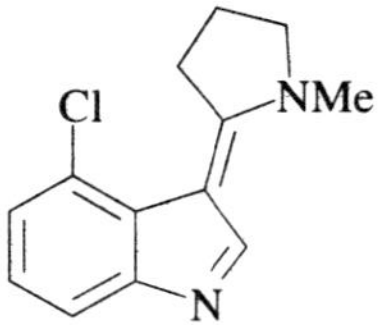 (49)	579, 102

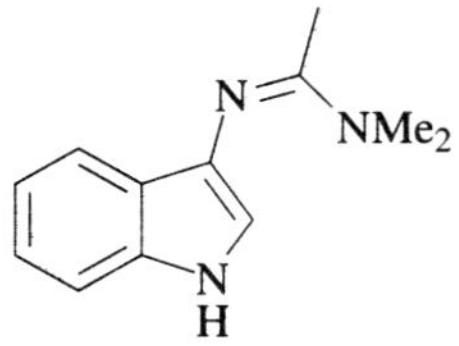

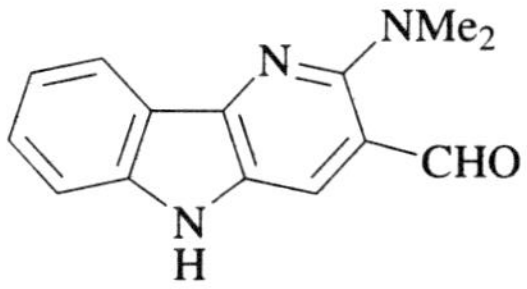

187

TABLE XIV. INDOLES (*Continued*)

Substrate	Reagents	Product(s) and Yield(s) (%)	Refs.
R⁴ = Br	DMF, POCl₃	R³ = CHO (67)	609
R⁴ = I	DMF, POCl₃	R³ = CHO (100)	609
C₉ R⁴ = Me	DMA, POCl₃	R¹ = COMe + R³ = COMe (—) 5:95	103
C₁₁ R⁴ = SCH₂CO₂Me	DMF, POCl₃	R³ = CHO (95)	610
C₁₃ R⁴ = 3-Pyridyl	DMF, POCl₃	R³ = CHO (92)	611
C₁₄	DMF, POCl₃		612
C₁₉	DMF, POCl₃	(96)	613

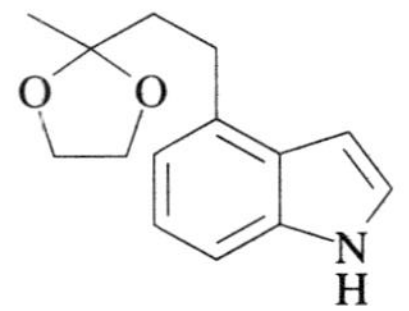

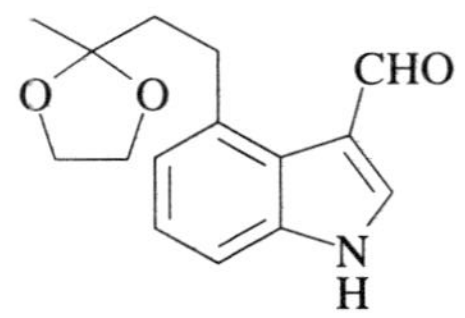

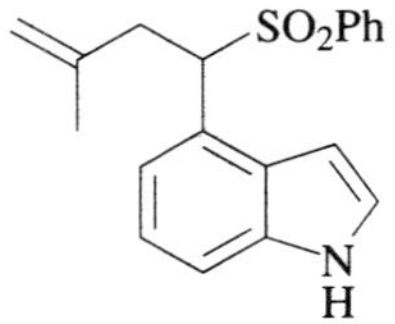

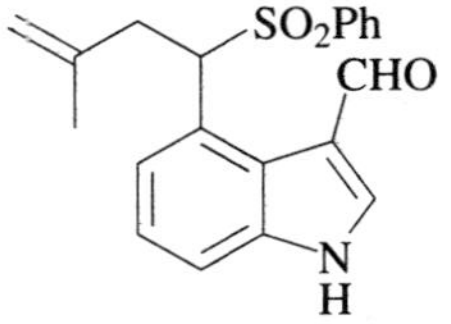

B5. R⁵ Substituents

Substrate	Reagents	Product(s) and Yield(s) (%)	Refs.
C₈ R⁵ = F	—	R³ = CHO (—)	614
R⁵ = Cl	DMF, POCl₃	R³ = CHO (93)	615
R⁵ = Br	DMF, POCl₃	R³ = CHO (95)	616
	MFA, POCl₃	R³ = CHO (13)	617
R⁵ = I	DMF, POCl₃	R³ = CHO (89)	618

TABLE XIV. INDOLES (*Continued*)

Substrate	Reagents	Product(s) and Yield(s) (%)	Refs.

C_{8-15}

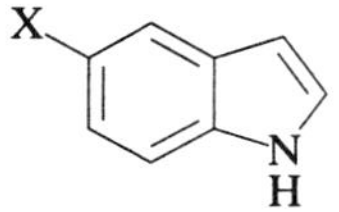

POCl₃,

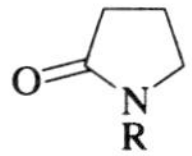

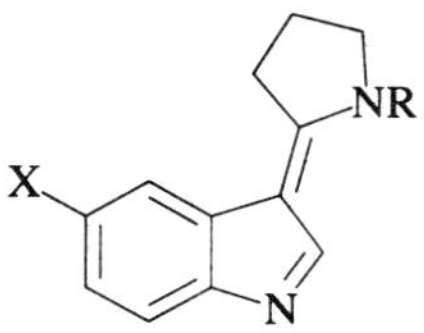

	R		
X			
F	Me	(69)	579, 102
Cl	Me	(76)	579, 102
	Bn	(79)	579, 102
Br	Me	(82)	579, 102
	Bn	(60)	579, 102
Me	Me	(93)	579, 102
MeO	Me	(88)	579
BnO	Me	(90)	579, 102

C_9	$R^5 = OMe$	DMF, POCl₃	$R^3 = CHO$ (94)	615, 97, 619-621
C_{10}	$R^5 = COCH_2Cl$	Et₂NCOCH₂Cl, POCl₃	$R^3 = COCH_2Cl$ (30)	622
	$R^5 = COMe$	Et₂NCOCH₂Cl, POCl₃	$R^3 = COCH_2Cl$ (—)	622
	$R^5 = Et$	—	$R^3 = CHO$ (—)	614
C_{12}	$R^5 = CH=CHCO_2Me$	DMF, POCl₃	$R^3 = CHO$ (72)	623
C_{15}	$R^5 = OBn$	DMF, POCl₃	$R^3 = CHO$ (86)	615, 621.
				624
		DMA, POCl₃	$R^3 = COMe$ (71)	101

189

TABLE XIV. INDOLES (*Continued*)

Substrate	Reagents	Product(s) and Yield(s) (%)	Refs.

C_{16-18}

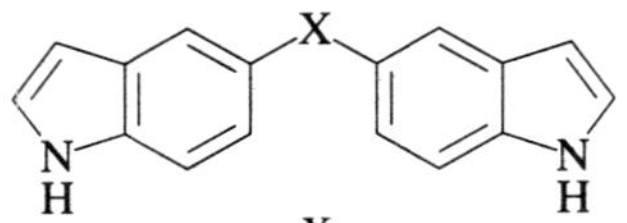

Substrate: (bis-indole linked by X) Reagents: DMF, POCl₃

Product (OHC / CHO bis-indole with X linker):

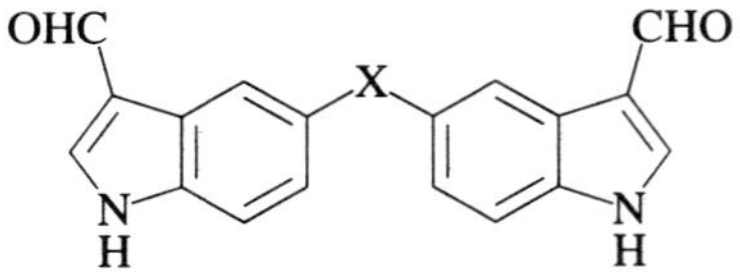

$\dfrac{X}{O}$	(75)	625
SO₂	(65)	626
(CH₂)₂	(85)	627

B6. R⁶ Substituents

C_8

R^6 = Cl	DMF, POCl₃	R^3 = CHO (53)	619
R^6 = Br	DMF, POCl₃	R^3 = CHO (94)	628
R^6 = I	DMF, POCl₃	R^3 = CHO (87)	629

C_9

| R^6 = Me | DMF, POCl₃ | R^3 = CHO (80) | 619 |
| R^6 = OMe | DMF, POCl₃ | R^3 = CHO (92) | 621, 615 |

C_{9-15}

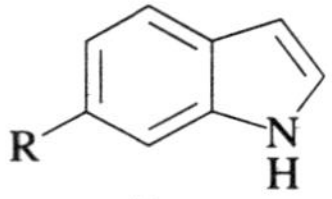

Substrate: indole with R

Reagents: POCl₃, (2-pyrrolidinone N-Me)

Product: (indole with NMe-pyrrolidine, R)

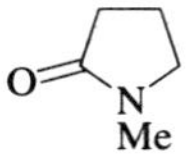

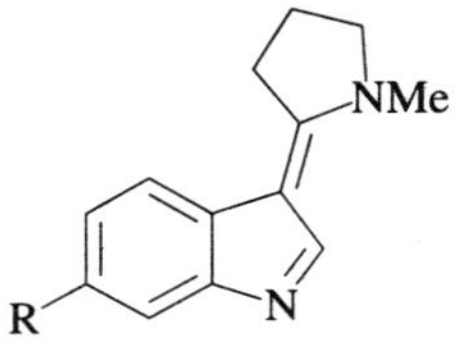

| $\dfrac{R}{Me}$ | (—) | 102 |
| BnO | (88) | 579, 102 |

TABLE XIV. INDOLES (*Continued*)

Substrate	Reagents	Product(s) and Yield(s) (%)	Refs.
C_{12} $R^6 = CH{=}CHCO_2Me$	DMF, POCl$_3$	$R^3 = CHO$ (96)	623
C_{15} $R^6 = OBn$	DMF, POCl$_3$	$R^3 = CHO$ (90)	621, 615
	DMF, PBr$_3$	$R^3 = CHO$ (82)	615
	DMF, PCl$_5$	$R^3 = CHO$ (27)	615

B7. R^7 Substituents

Substrate	Reagents	Product(s) and Yield(s) (%)	Refs.
C_9	POCl$_3$, R: Me / Bn	 (93) (97)	579
C_{11} $R^7 = OMe$	DMF, POCl$_3$	$R^3 = CHO$ (92)	630
$R^7 = Pr$	DMF, POCl$_3$	$R^3 = CHO$ (71)	631
C_{12} $R^7 = CH{=}CHCO_2Me$	DMF, POCl$_3$	$R^3 = CHO$ (96)	623
C_{13} $R^7 = C_5H_{11}$	DMF, POCl$_3$	$R^3 = CHO$ (56)	631

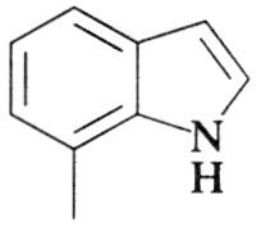

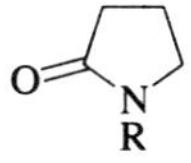

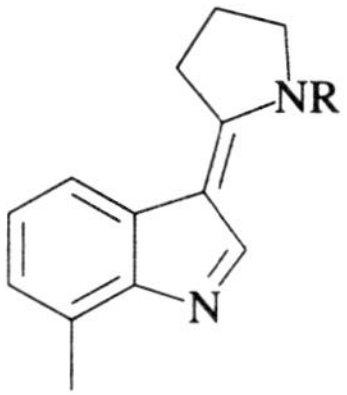

TABLE XIV. INDOLES (*Continued*)

Substrate	Reagents	Product(s) and Yield(s) (%)	Refs.
		C. Disubstituted Indoles	
		C1. R^1, R^2 Substituents	
C$_{10}$ R^1 = R^2 = Me	MFA, POCl$_3$	R^3 = CHO (—)	601
	DMF, (Cl$_2$PO)$_2$O	R^3 = CHO (95)	26
	1. DMF, POCl$_3$	R^3 = CHS (89)	632
	2. NaSH		
	PhNHCO(C$_6$H$_4$Cl-4), POCl$_3$	R^3 = CO(C$_6$H$_4$Cl-4) (—)	633
C$_{11}$	—		634
R^1 = OH, R^2 = CO$_2$Et	DMF, POCl$_3$	R^1 = H, R^3 = Cl (51)	635
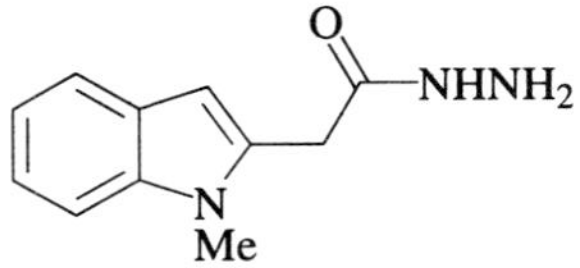	DMF, POCl$_3$	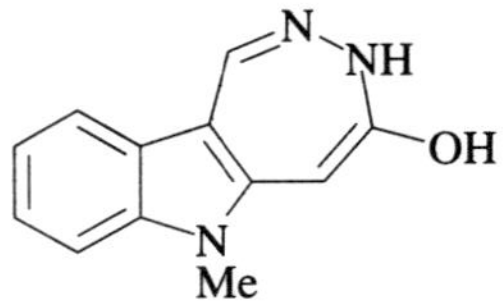 (90)	636
C$_{12}$	MFA, POCl$_3$	(91)	637, 638

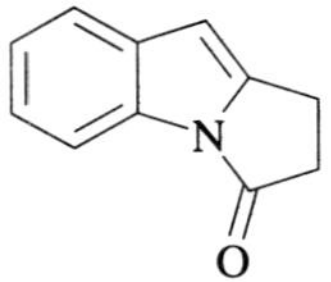

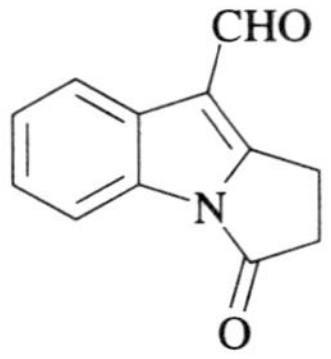

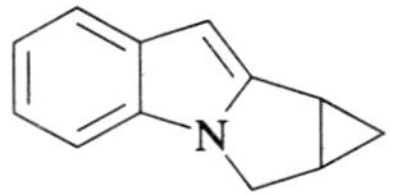

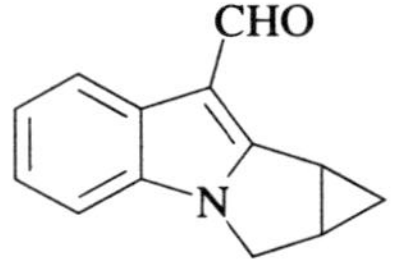

192

TABLE XIV. INDOLES (*Continued*)

Substrate	Reagents	Product(s) and Yield(s) (%)	Refs.
C$_{13}$			
R^1 = Me, R^2 = CH$_2$CO$_2$Et	DMF, POCl$_3$	R^3 = CHO (85)	636, 639
	DMA, POCl$_3$	R^3 = COMe (75)	639
	Me$_2$NCOPh, POCl$_3$	R^3 = COPh (85)	639
	Me$_2$NCO(C$_6$H$_4$Cl-4), POCl$_3$	R^3 = CO(C$_6$H$_4$Cl-4) (65)	639
C$_{15}$			
R^1 = Me, R^2 = Ph	MFA, POCl$_3$	R^3 = CHO (—)	601
	DMF, (Cl$_2$PO)$_2$O	R^3 = CHO (94)	26
	PhNHCO(C$_6$H$_4$Cl-4), POCl$_3$	R^3 = CO(C$_6$H$_4$Cl-4) (—)	633
C$_{15-16}$			
R^1 = Me, R^2 = C≡CR	—	R^3 = CHO	640

R	
Bu	(—)
	(—)

C$_{20}$			
R^1 = Ar, R^2 = Ph	DMF, POCl$_3$	R^3 = CHO	584
Ar			
2-O$_2$NC$_6$H$_4$		(68)	
3-O$_2$NC$_6$H$_4$		(95)	
4-O$_2$NC$_6$H$_4$		(98)	
C$_{23}$			
R^1 = Bn, R^2 = NHCO$_2$Bn	DMF, POCl$_3$	R^3 = CHO (66) + R^2 = N=CHNMe$_2$, R^3 = CHO (15)	594

C2. R^1, R^3, Substituents

C$_{10}$			
R^1 = R^3 = Me	DMF, (Cl$_2$PO)$_2$O	R^2 = CHO (87)	26

TABLE XIV. INDOLES (Continued)

Substrate	Reagents	Product(s) and Yield(s) (%)	Refs.
C_{12} $R^1 = CH_2CH=CH_2$, $R^3 = Me$	DMF, $POCl_3$	$R^2 = CHO$ (88)	107
C_{13} $R^1 = CH_2CH=CMe_2$, $R^3 = Me$	DMF, $POCl_3$	$R^2 = CHO$ (88)	107
C_{15} $R^1 = Me$, $R^3 = 4\text{-}FC_6H_4$	DMF, $POCl_3$	$R^2 = CHO$ (86)	108
C_{17} $R^1 = i\text{-}Pr$, $R^3 = 4\text{-}FC_6H_4$	DMF, $POCl_3$, 100°, 5 h	$R^2 = CHO$ (50-56)	108
	DMF, $POCl_3$, 80°, 18 h	$R^2 = CHO$ (39) + $R^5 = CHO$ (14) + $R^7 = CHO$ (7)	108
	$PhN(Me)CH=CHCHO$, $POCl_3$	$R^2 = CH=CHCHO$ (83)	641
	$POCl_3$, O	(80)	581

C3. R^1, R^6, Substituents

C_9 $R^1 = Me$, $R^6 = NO_2$	MFA, $POCl_3$	$R^3 = CHO$ (—)	601

C4. R^2, R^3, Substituents

C_{10} $R^2 = R^3 = Me$	Et_2NCHO, $POCl_3$	$R^1 = CHO$ (—)	642
	DMF, $POCl_3$	$R^1 = CHO$ (52)	104
	DMA, $POCl_3$	$R^1 = COMe + R^6 = COMe$ (—) 97:3	103

TABLE XIV. INDOLES (*Continued*)

Substrate	Reagents	Product(s) and Yield(s) (%)	Refs.
C_{12}			
$R^2 + R^3 = (CH_2)_4$	Et_2NCHO, $POCl_3$	$R^1 = CHO$ (—)	642
	DMF, $(Cl_2PO)_2O$	$R^1 = CHO$ (17) + (21)	26
	DMF, $POCl_3$	$R^1 = CHO$ (37) + $R^1 = CHO$, $R^7 = CHO$ (37) + (18)	643
	DMA, $POCl_3$	$R^1 = COMe + R^7 = COMe$ (—) 95:5	103

C5. R^2, R^5, Substituents

Substrate	Reagents	Product(s) and Yield(s) (%)	Refs.
C_9			
$R^2 = Me$, $R^5 = Br$	DMF, $POCl_3$	$R^3 = CHO$ (75)	590
C_{10}			
$R^2 = CON_3$, $R^5 = OMe$	DMF, $POCl_3$	$R^3 = CHO$ (80)	644
$R^2 = Me$, $R^5 = OMe$	DMF, $POCl_3$	$R^3 = CHO$ (76)	646, 645
C_{11}			
$R^2 = CO_2Et$, $R^5 = Br$	MFA, $POCl_3$	$R^3 = CHO$ (67)	617
$R^2 = CO_2Et$, $R^5 = OH$	DMF, $POCl_3$	$R^3 = CHO$ (80)	647
	DMF, $POCl_3$, 120°	$R^3 = CHO$ (77)	648
	DMF, $POCl_3$, 35°	$R^5 = OCHO$ (77)	648
	MFA, $POCl_3$	$R^3 = R^4 = CHO$ (3)	648

TABLE XIV. INDOLES (Continued)

Substrate	Reagents	Product(s) and Yield(s) (%)	Refs.
C_{12}			
$R^2 = CO_2Et$, $R^5 = OMe$	DMF, POCl$_3$	$R^3 = CHO$ (80)	647
$R^2 = CONMe_2$, $R^5 = OMe$	DMF, POCl$_3$	$R^3 = CHO$ (77)	649
C_{13}			
$R^2 = CO_2Et$, $R^5 = OEt$	DMF, POCl$_3$	$R^3 = CHO$ (85)	647
C_{14}			
$R^2 = 4\text{-}FC_6H_4$, $R^5 = F$	DMF, POCl$_3$	$R^3 = CHO$ (63)	600
$R^2 = CONEt_2$, $R^5 = OMe$	DMF, POCl$_3$	$R^3 = CHO$ (83)	649
C_{14-15}			
$\dfrac{X}{O}$ CH$_2$ NMe	DMF, POCl$_3$	(52) (96) (65)	649
C_{15}			
$R^2 = CONH(CH_2)_3NMe_2$, $R^5 = OMe$	DMF, POCl$_3$	$R^3 = CHO$ (66)	649
C_{16}			
$R^2 = NHCO_2Bn$, $R^5 = Cl$	DMF, POCl$_3$	$R^3 = CHO$ (41)	594
C_{17}			
$R^2 = NHCO_2Bn$, $R^5 = OMe$	DMF, POCl$_3$	$R^3 = CHO$ (45)	594

TABLE XIV. INDOLES (Continued)

Substrate	Reagents	Product(s) and Yield(s) (%)	Refs.
C_{18} [indole with 5-(p-NO_2-C$_6$H$_4$-S), 2-CO_2Et, N-H]	—	[3-CHO, 2-CO_2Et indole with 5-(p-NO_2-C$_6$H$_4$-S)] (—)	650
$R^2 = CO_2Et$, $R^5 = OBn$	DMF, POCl$_3$ MFA, POCl$_3$	R^3 = CHO (95) R^3 = CHO (96)	647 651
C_{19} [bis-indole linked by O; 2-CO_2Et, N-H]	DMF, POCl$_3$	[CHO-substituted bis-indole; 2-CO_2Et] (89)	625
C_{20} [2-Ph indole, 5-$PhSO_2$, N-H]	—	[3-CHO, 2-Ph indole, 5-$PhSO_2$] (—)	652

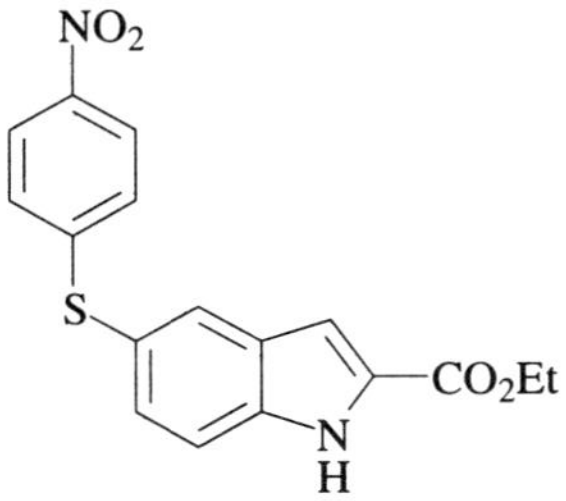

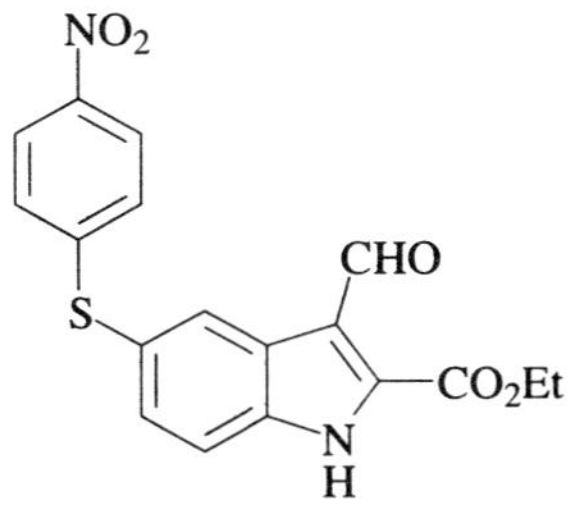

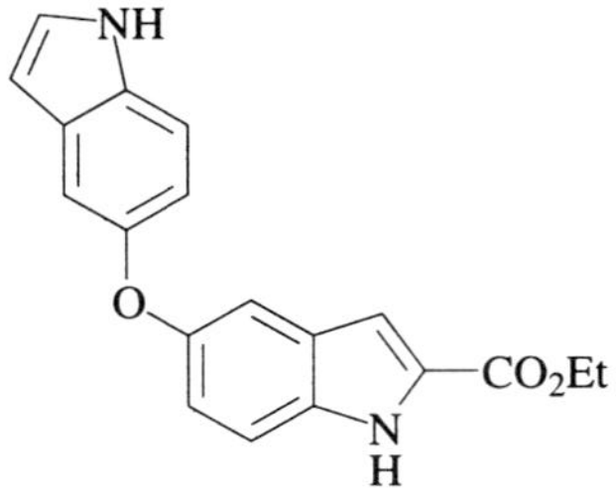

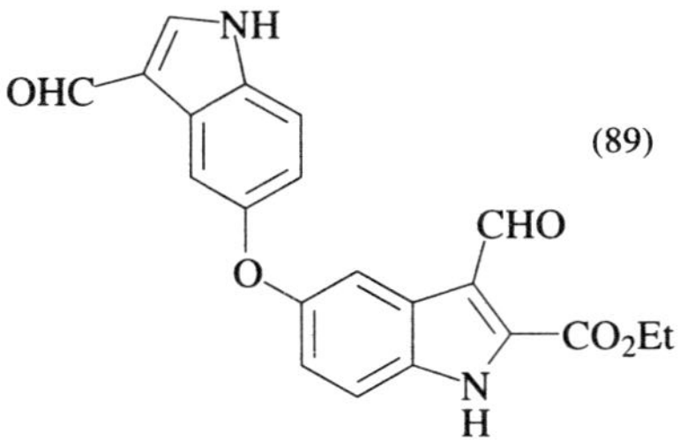

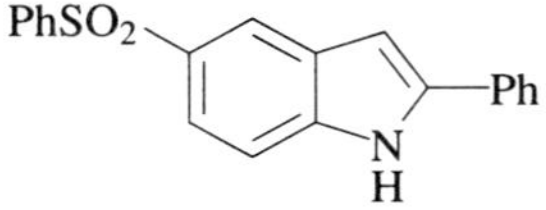

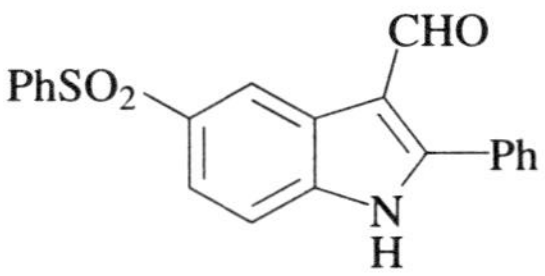

TABLE XIV. INDOLES (*Continued*)

Substrate	Reagents	Product(s) and Yield(s) (%)	Refs.
C$_{22}$	DMF, POCl$_3$	(22)	653
C$_{22-23}$	DMF, POCl$_3$		
$\dfrac{X}{SO_2}$ CH_2		(68) (85)	626 654

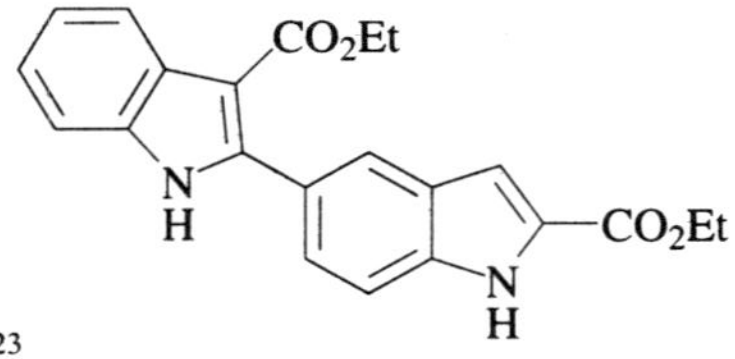

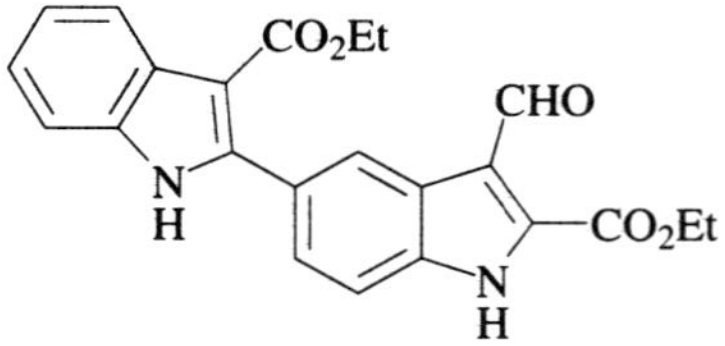

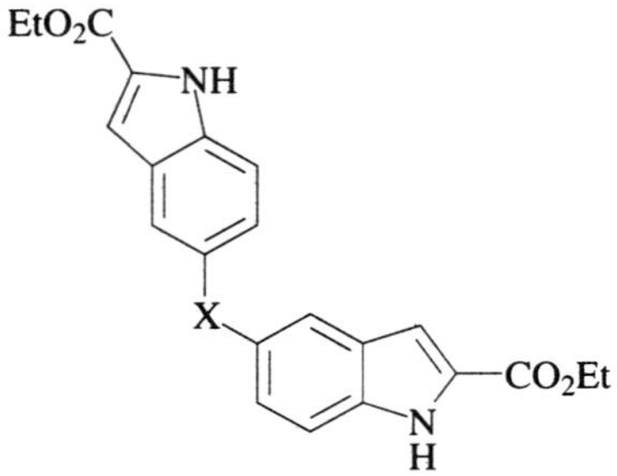

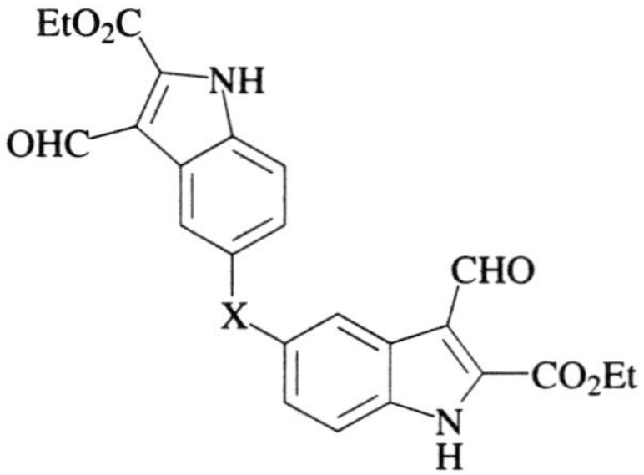

C6. R^2, R^6 Substituents

C$_9$ R^2 = Me, R^6 = Br	DMF, POCl$_3$	R^3 = CHO (88)	629
C$_{10}$ R^2 = Me, R^6 = OMe	DMF, POCl$_3$	R^3 = CHO (84)	655, 656
C$_{14}$ R^2 = 4-FC$_6$H$_4$, R^6 = F	DMF, POCl$_3$	R^3 = CHO (59)	600

TABLE XIV. INDOLES (*Continued*)

Substrate	Reagents	Product(s) and Yield(s) (%)	Refs.
C7. R³, R⁴ Substituents			
C_{19}	DMF, POCl₃	R^7 = CHO (65)	657
	MFA, POCl₃	R^7 = CHO (—)	657
	PhN(Me)CH=CHCHO, POCl₃	R^5 = CH=CHCHO (35) + R^7 = CH=CHCHO (6)	657
C8. R⁴, R⁵ Substituents			
C_{10} $R^4 + R^5 = O(CH_2)_2O$	DMF, POCl₃	R^3 = CHO (60)	658
C_{16} 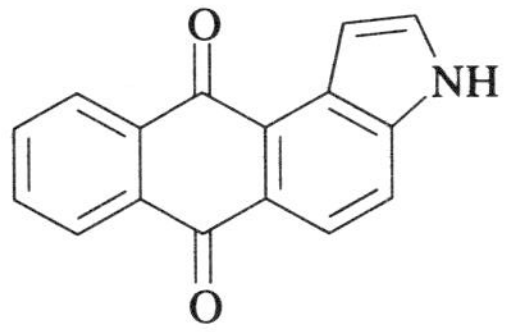	DMF, POCl₃	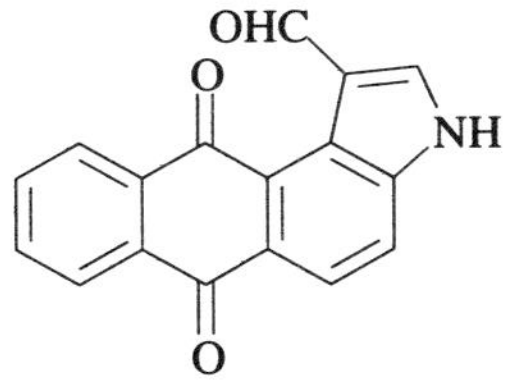 (71)	659
C9. R⁴, R⁶ Substituents			
C_{10} $R^4 = R^6 = OMe$	DMF, POCl₃	R^7 = CHO (56)	109

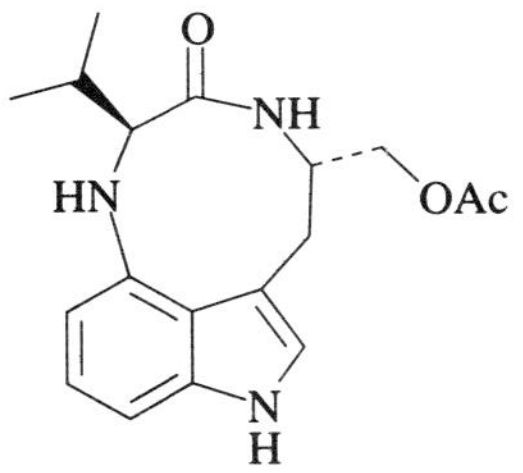

199

TABLE XIV. INDOLES (*Continued*)

Substrate	Reagents	Product(s) and Yield(s) (%)	Refs.
		C10. R^5, R^6 Substituents	
C_{16} 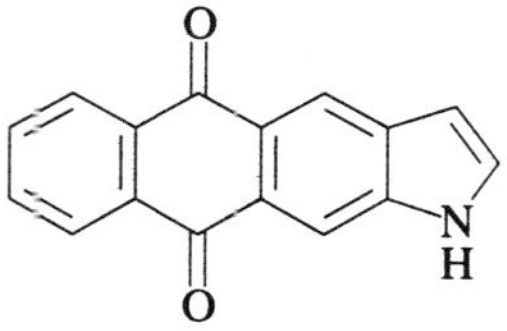	DMF, $POCl_3$	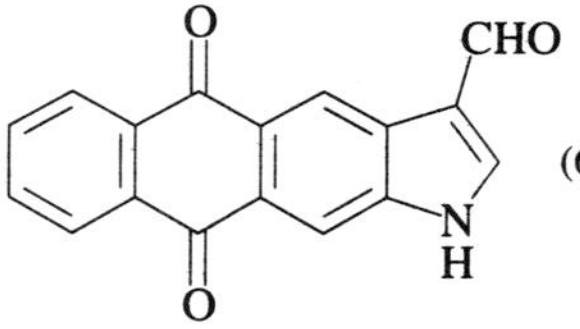 (68)	659
C_{24} 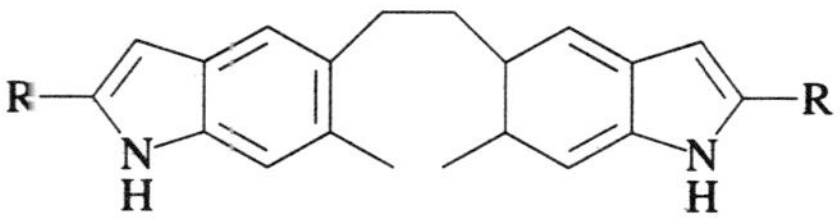 R = CO_2Et	DMF, $POCl_3$	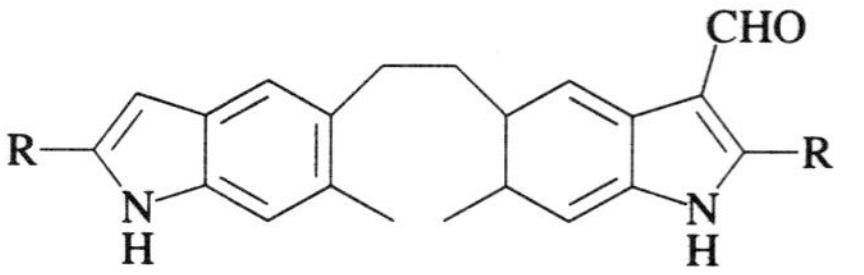 (75)	627
		C11. R^5, R^7 Substituents	
C_{15} R^5 = Br, R^7 = OTs	DMF, $POCl_3$	R^3 = CHO (—)	660
C_{17} 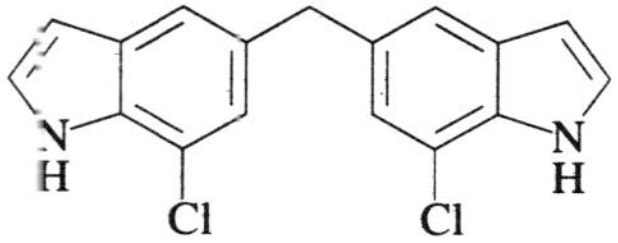	DMF, $POCl_3$	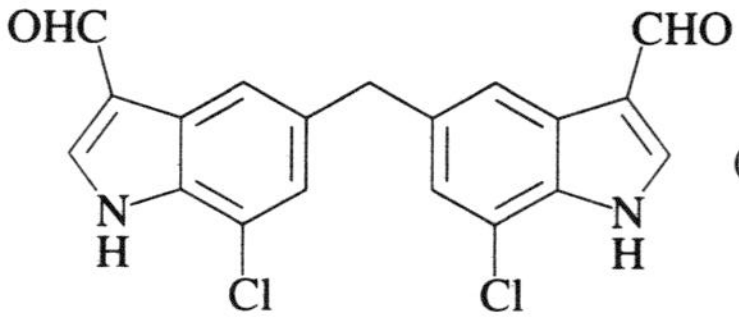 (95)	257
		C12. R^6, R^7 Substituents	
C_8 R^5 = Br, R^7 = I	DMF, $POCl_3$	R^3 = CHO (94)	629

TABLE XIV. INDOLES (*Continued*)

Substrate	Reagents	Product(s) and Yield(s) (%)	Refs.

D. Trisubstituted Indoles
D1. R¹, R², R³ Substituents

C₁₃

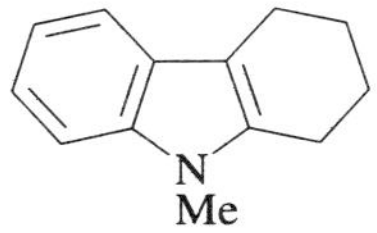

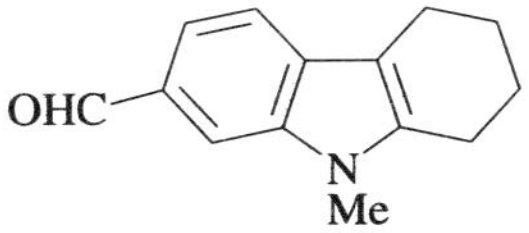

I   +   **II**

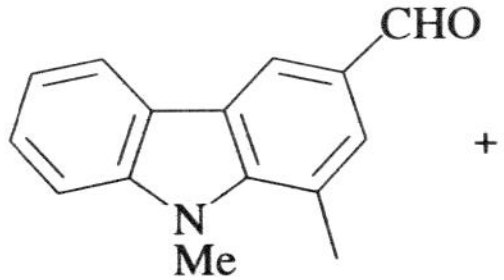

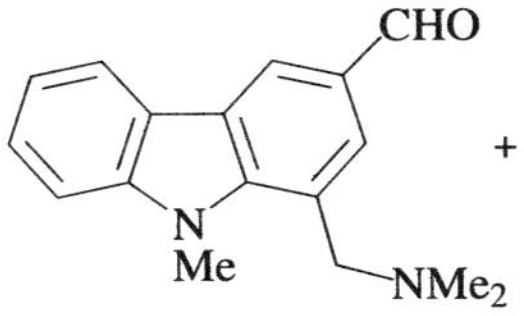

III   +   **IV**

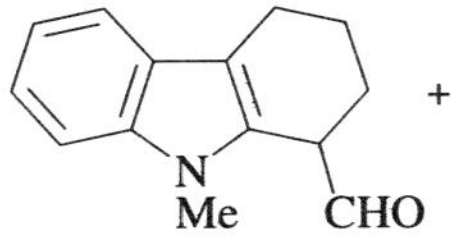

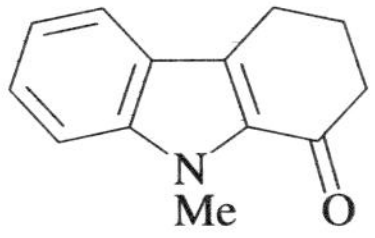

V

Reagents	I	II	III	IV	V	Refs.
Et₂NCHO, POCl₃						661, 642
DMF (1 eq), POCl₃ (xs) 100°	(42)	(27)			(2)	110
DMF (xs), POCl₃ (2 eq) 100°		(50)				110
DMF (xs), POCl₃ (2 eq) rt, 6 h				(89)		110
DMF, (Cl₂PO)₂O (1 eq), 80°	(17)		(11)			26
DMF, (Cl₂PO)₂O (1.8 eq), 80°	(21)		(25)			26
DMF, (Cl₂PO)₂O (xs), 0°					(96)	26

TABLE XIV. INDOLES (*Continued*)

Substrate	Reagents	Product(s) and Yield(s) (%)	Refs.

C₁₄

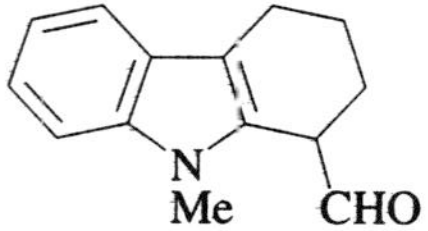

DMF (xs), POCl₃ (3 eq), rt, 3 h

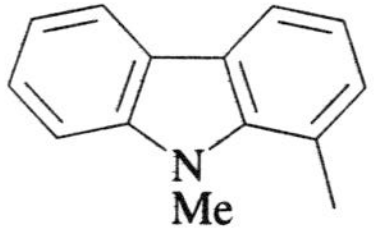

(73)

110

DMF (xs), POCl₃ (3 eq), rt, 20 min

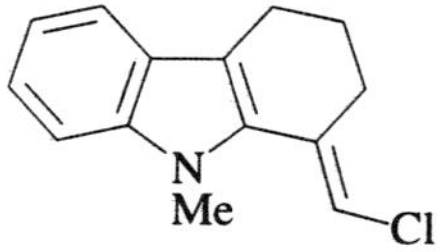

(85)

110

DMF (xs), POCl₃ (3 eq), 100°, 3 h

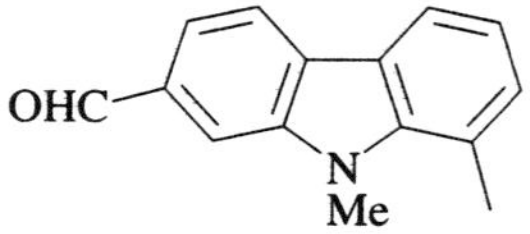

(50)

110

II   +

C₁₉

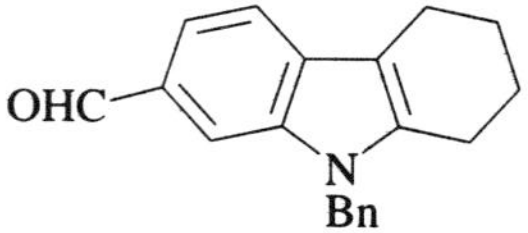

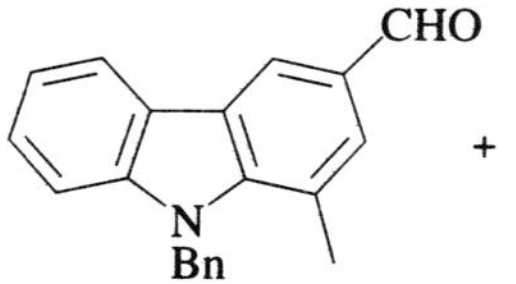

I   +

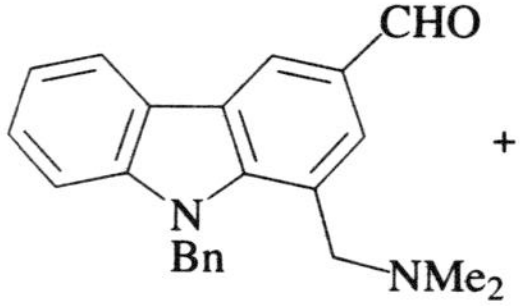

III   +

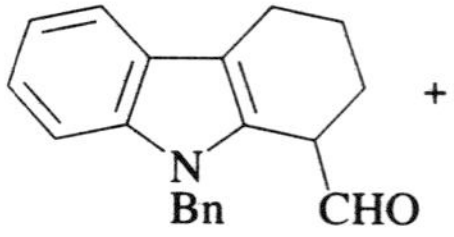

IV

TABLE XIV. INDOLES (Continued)

Substrate	Reagents	Product(s) and Yield(s) (%)						Refs.

Products V and VI:

V (carbazole, N-Bn, methyl); VI (OHC-carbazole, N-Bn, methyl)

		I	II	III	IV	V	VI	
(V structure, carbazole N-Bn, methyl)	DMF, POCl$_3$ (1 eq), 120°	(5)	(18)	(8)	(3)			662, 663
	DMF, POCl$_3$ (3 eq), 120°		(38)	(26)		(1)		662
	DMF, POCl$_3$ (3 eq), 0°				(90)			662, 663
	N-Formylpyrrolidine, POCl$_3$, 0°				(81)			664
	N-Formylpyrrolidine, POCl$_3$, 120°		(37)	(21)	(5)			664
	Et$_2$NCHO, POCl$_3$, 0°				(57)			664, 642
	Et$_2$NCHO, POCl$_3$, 120°		(26)	(6)	(25)		(15)	664
	MFA, POCl$_3$, rt				(4)		(6)	664
	MFA, POCl$_3$, 100°						(83)	664
	i-Pr$_2$NCHO, POCl$_3$, 120°						(77)	664

Substrate	Reagents	Product(s) and Yield(s) (%)	Refs.
C$_{20}$ (tetrahydrocarbazole, N-Bn, CHO)	DMF, POCl$_3$ (1.2 eq)	(60)	662
(methyl tetrahydrocarbazole, N-Bn)	DMF, POCl$_3$ (2.6 eq), 100°	(38) +	662

203

TABLE XIV. INDOLES (Continued)

Substrate	Reagents	Product(s) and Yield(s) (%)	Refs.

| | DMF, POCl$_3$ (2.6 eq), 0° | (15) + **I** (9)   **I** (88) | 662 |
| C_{21} 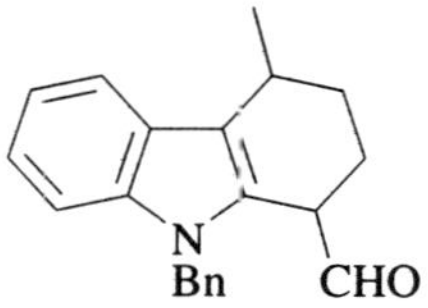 | DMF, POCl$_3$ (1.2 eq) | 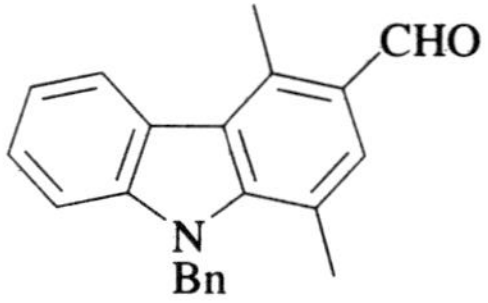 (44) | 662 |

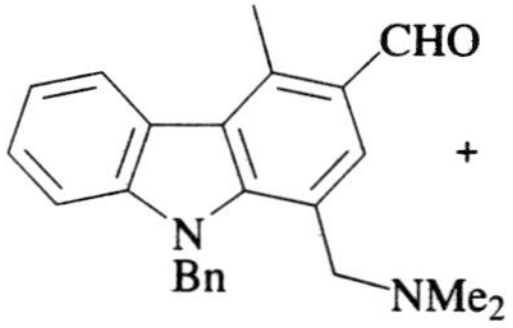

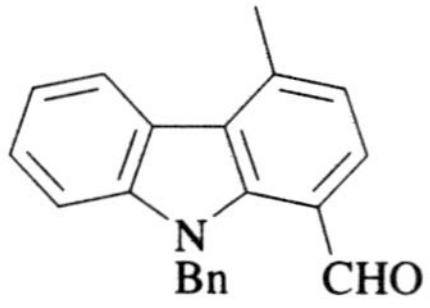

D2. R^1, R^2, R^5 Substituents

C_{11} R^1 = R^2 = Me, R^5 = OMe	DMF, POCl$_3$	R^3 = CHO (69)	645
C_{12} R^1 = R^2 = Me, R^5 = OAc	DMF, POCl$_3$	R^3 = CHO (68)	645, 665
C_{14}	—	(—)	666

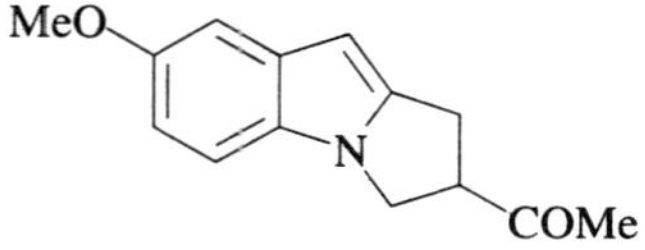

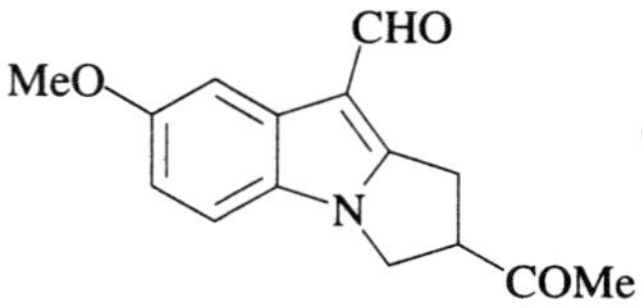

TABLE XIV. INDOLES (*Continued*)

Substrate	Reagents	Product(s) and Yield(s) (%)	Refs.
C_{15} $R^1 + R^2 = (CH_2)_3CHOAc$, $R^5 = OMe$	DMF, POCl$_3$	$R^3 = CHO$ (66)	667
C_{16} $R^1 = Ph$, $R^2 = Me$, $R^5 = OMe$	DMF, POCl$_3$	$R^3 = CHO$ (81)	645
C_{17} $R^1 = Ph$, $R^2 = Me$, $R^5 = OAc$	DMF, POCl$_3$	$R^3 = CHO$ (81)	645, 665
C_{17-18}	DMF, POCl$_3$	(70-90) (70-90)	668
C_{18} $R^1 + R^2 = (CH_2)_2CO$, $R^5 = OBn$	—	$R^3 = CHO$ (17)	669
$R^1 = Bn$, $R^2 = Me$, $R^5 = OAc$	DMF, POCl$_3$	$R^3 = CHO$ (47)	665
C_{26}	DMF, POCl$_3$	(58)	670

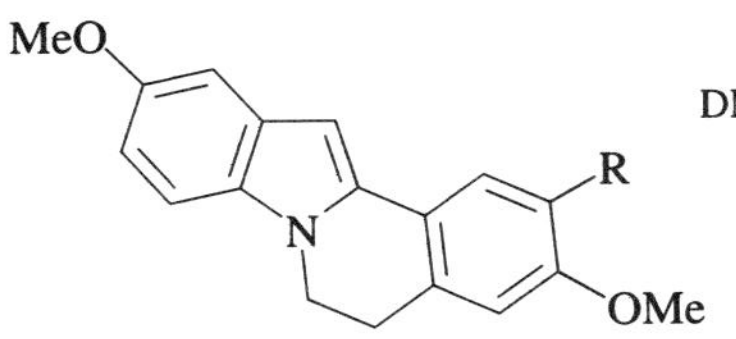

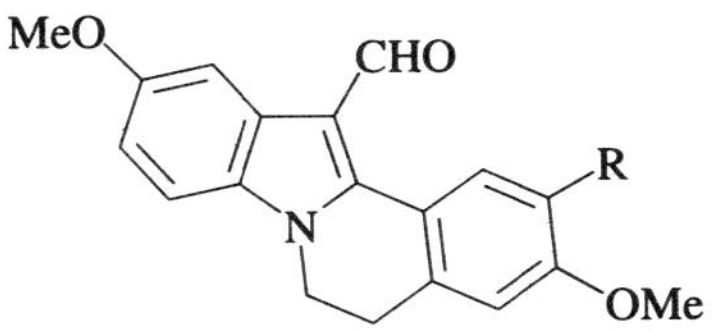

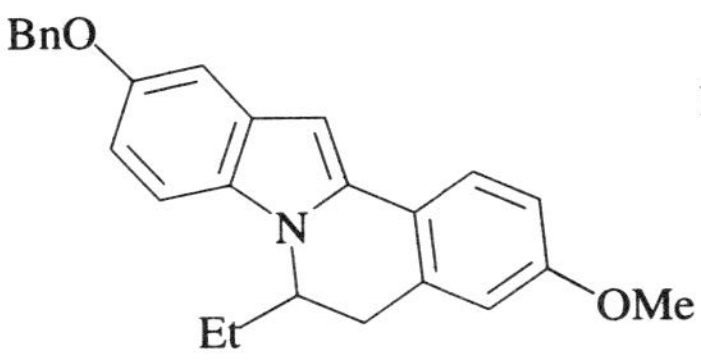

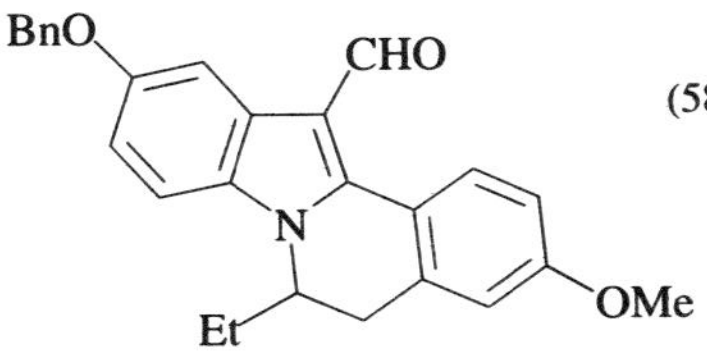

R / H / OMe

TABLE XIV. INDOLES (*Continued*)

Substrate	Reagents	Product(s) and Yield(s) (%)	Refs.

D3. R¹, R², R⁶ Substituents

C_{17-18}

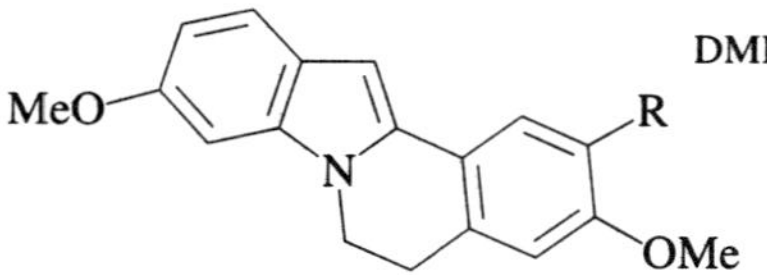

R
H
OMe

DMF, POCl₃

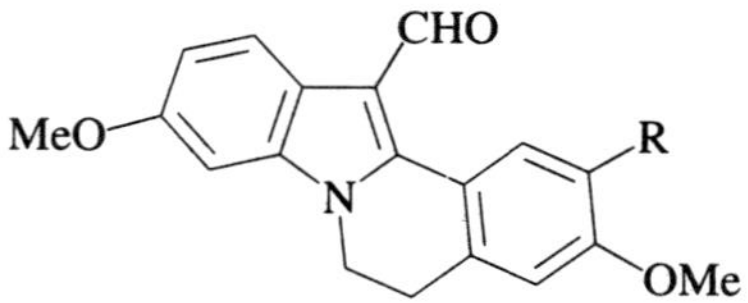

(70–90)
(70–90)

669

D4. R², R³, R⁵ Substituents

C_{12}

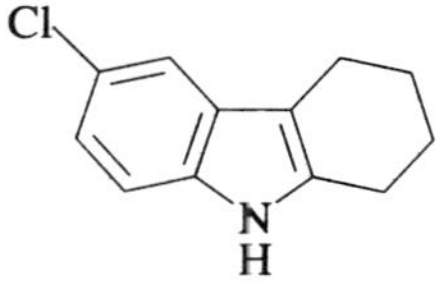

DMF, POCl₃

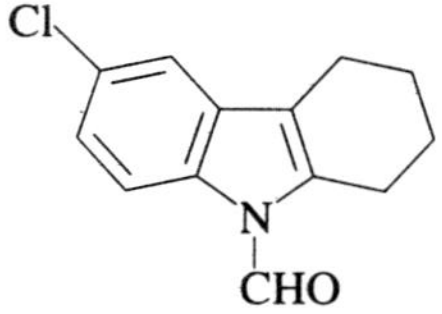

(1.5) +

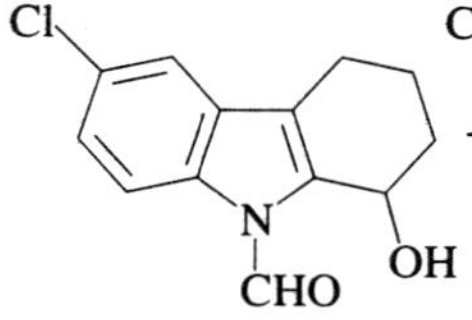

(15) + (45)

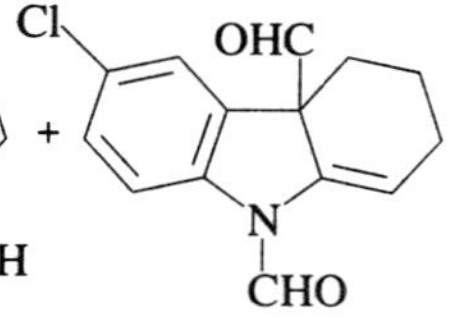

643

C_{13} R² = Me, R³ = CO₂Et, R⁵ = OMe   DMF, POCl₃

R⁶ = CHO (34)

645

206

TABLE XIV. INDOLES (*Continued*)

Substrate	Reagents	Product(s) and Yield(s) (%)	Refs.
C$_{14}$			
R^2 + R^3 = C(Me)$_2$(CH$_2$)$_3$, R^5 = Cl	DMF, POCl$_3$	R^1 = CHO (10) + R^4 = CHO (5) + R^6 = CHO (60)	643
C$_{21}$			
R^2 = R^3 = Ph, R^5 = Me	DMF, POCl$_3$	R^6 = CHO (90)	671
R^2 = R^3 = Ph, R^5 = OMe	—	R^6 = CHO (—)	672

D5. R^2, R^4, R^6 Substituents

Substrate	Reagents	Product(s) and Yield(s) (%)	Refs.
C$_{11\text{-}16}$			
R^2 = R, R^4 = R^6 = OMe			
R			
Me	DMF, POCl$_3$	R^3 = CHO (90) + R^7 = CHO (2)	109
4-BrC$_6$H$_4$	DMF, POCl$_3$	R^3 = CHO (33) + R^7 = CHO (35)	109
Ph	DMF, POCl$_3$	R^3 = CHO (38) + R^7 = CHO (36)	109

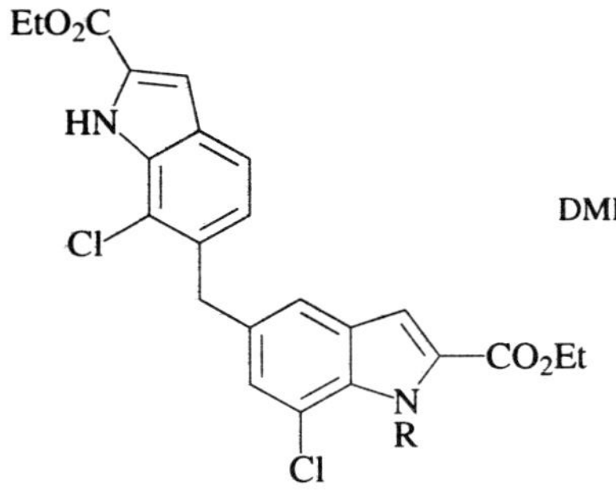	DMF, POCl$_3$	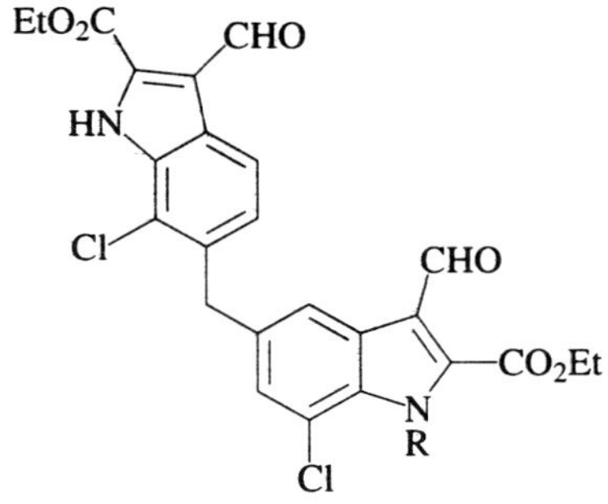	257

R
| H | (90) |
| Et | (40) |

TABLE XIV. INDOLES (*Continued*)

Substrate	Reagents	Product(s) and Yield(s) (%)	Refs.
D6. R², R⁵, R⁶ Substituents			
C_{11}			
$R^2 = CONHNH_2$, $R^5 = R^6 = OMe$	DMF, POCl$_3$	$R^2 + R^3 = CONHN=CH$ (80-85)	673
C_{12}			
$R^2 = CO_2Me$, $R^5 = R^6 = OMe$	DMF, POCl$_3$	$R^3 = CHO$ (52)	674
C_{13}			
$R^2 = CO_2Et$, $R^5 = R^6 = OMe$	DMF, POCl$_3$	$R^3 = CHO$ (85)	673
D7. R³, R⁴, R⁵ Substituents			
C_{11}			
$R^3 = Me$, $R^4 = R^5 = OMe$	DMF, POCl$_3$ Me$_2$NCO(C$_6$H$_4$Cl-4)	$R^2 = CHO$ (44) + $R^7 = CHO$ (46) $R^2 = CO(C_6H_4Cl$-4) (—) + $R^7 = CO(C_6H_4Cl$-4) (—)	109 675
D8. R⁴, R⁵, R⁶ Substituents			
C_8			
$R^4 = R^5 = R^6 = Br$	DMF, POCl$_3$	$R^3 = CHO$ (86)	676
D9. R⁵, R⁶, R⁷ Substituents			
C_8			
$R^5 = R^6 = R^7 = Br$	DMF, POCl$_3$	$R^3 = CHO$ (87)	676
E. Tetrasubstituted Indoles **E1. R¹, R², R³, R⁵ Substituents**			
C_{13}	DMF, POCl$_3$ (3 eq), 70°	X, Y = O (10) + X = Y = CHO (65) + X = OH, Y = CHO (12)	661

208

TABLE XIV. INDOLES (*Continued*)

Substrate	Reagents	Product(s) and Yield(s) (%)	Refs.
	DMF, POCl$_3$ (1.3 eq), 100°	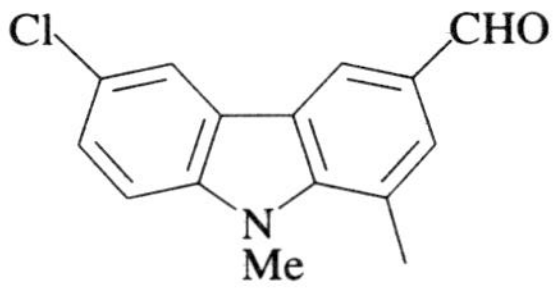 (2-8) 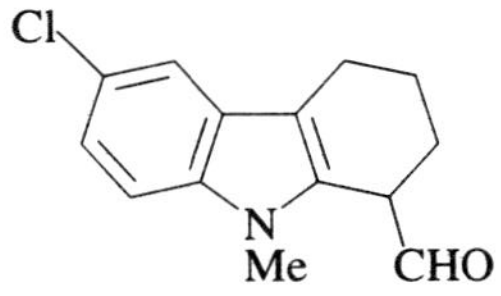 + (2-21) 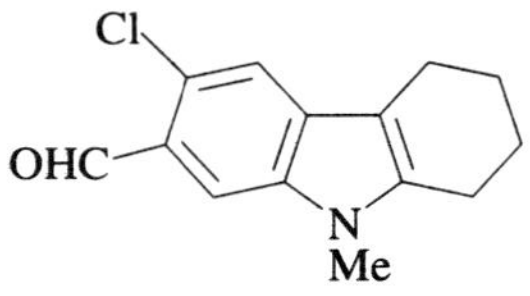 + (2)	661
C$_{14}$ R^1 = R^2 = Me, R^3 = CO$_2$Et, R^5 = OMe	DMF, POCl$_3$	R^6 = CHO (34)	645
C$_{19}$ R^1 = Ph, R^2 = Me, R^3 = CO$_2$Et, R^5 = OMe	DMF, POCl$_3$	R^6 = CHO (16)	645
C$_{20}$ R^1 = Bn, R^2 = Me, R^3 = CO$_2$Et, R^5 = OMe	DMF, POCl$_3$	R^6 = CHO (10)	645

TABLE XIV. INDOLES (*Continued*)

Substrate	Reagents	Product(s) and Yield(s) (%)	Refs.
E2. R^1, R^2, R^5, R^6 Substituents			
C_{13}			
$R^1 = R^2 = CO(CH_2)_2$, $R^5 = OMe$, $R^6 = Me$	DMF, $POCl_3$	$R^3 = CHO$ (96)	677
$R^1 = R^2 = (CH_2)_3$, $R^5 = OMe$, $R^6 = Me$	MFA, $POCl_3$	$R^3 = CHO$ (92)	678
$R^1 = Me$, $R^2 = CH_2OAc$, $R^5 = R^6 = OMe$	DMF, $POCl_3$	$R^3 = CHO$ (66)	679
C_{15}			
$R^1 = (CH_2)_2CN$, $R^2 = CO_2Me$, $R^5 = OMe$, $R^6 = Me$	DMF, $POCl_3$	$R^3 = CHO$ (90)	674
$R^1 + R^2 = (CH_2)_2CHOAc$, $R^5 = OMe$, $R^6 = Me$	DMF, $POCl_3$	$R^3 = CHO$ (98)	680
$R^1 + R^2 = (CH_2)_2CHOAc$, $R^5 = R^6 = OMe$	—	$R^3 = CHO$ (84)	681
C_{17}	DMF, $POCl_3$	(41)	682
C_{18-19}	DMF, $POCl_3$	(70-90) (70-90)	669

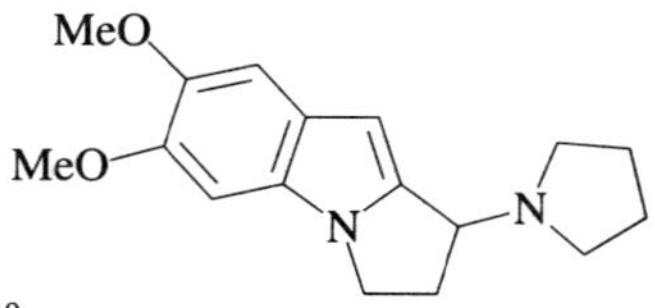

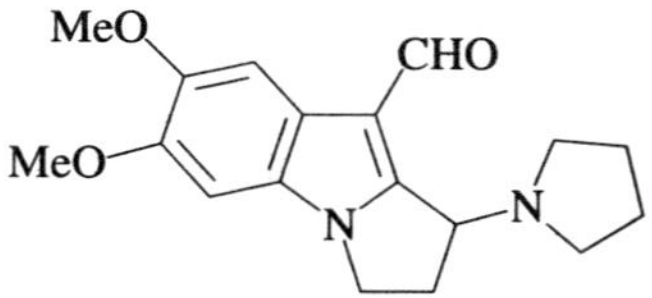

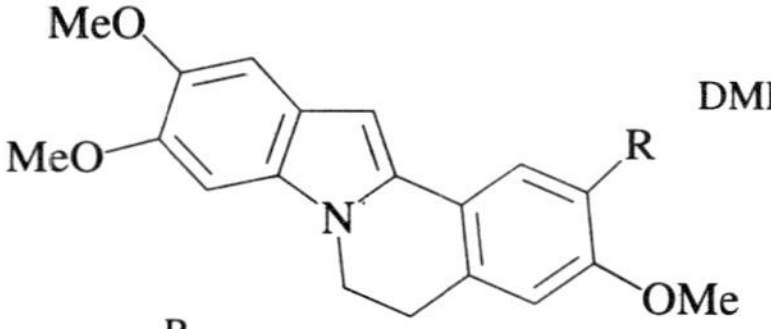

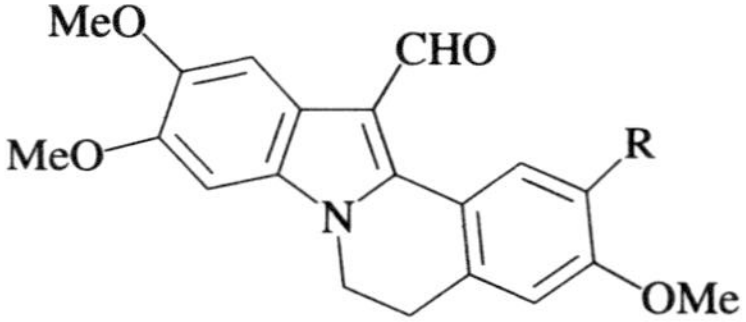

TABLE XIV. INDOLES (*Continued*)

Substrate	Reagents	Product(s) and Yield(s) (%)	Refs.
E3. R^2, R^3, R^4, R^6 Substituents			
C_{12} $R^2 = R^3 = R^4 = R^6 = OMe$	DMF, $POCl_3$	$R^7 = CHO$ (64)	109
C_{14} $R^2 = R^3 = CO_2Me$, $R^4 = R^6 = OMe$	DMF, $POCl_3$	$R^7 = CHO$ (70)	109
C_{20} $R^2 = R^3 = $ 2-pyridyl, $R^4 = R^6 = OMe$	DMF, $POCl_3$	$R^7 = CHO$ (62)	109
E4. R^2, R^3, R^5, R^6 Substituents			
C_{12} $R^2 = Me$, $R^3 = CO_2Et$, $R^5 = OH$, $R^6 = Br$	DMF, $POCl_3$	$R^4 = CHO$ (70)	155
C_{13} $R^2 = Me$, $R^3 = CO_2Et$, $R^5 = OH$, $R^6 = Me$	DMF, $POCl_3$	$R^4 = CHO$ (40)	155
E5. R^4, R^5, R^6, R^7 Substituents			
C_8 $R^4 = R^5 = R^6 = R^7 = F$	DMF, $POCl_3$	$R^3 = CHO$ (89)	683
F. Pentasubstituted Indoles			
F1. R^1, R^2, R^4, R^5, R^6 Substituents			
C_{20} 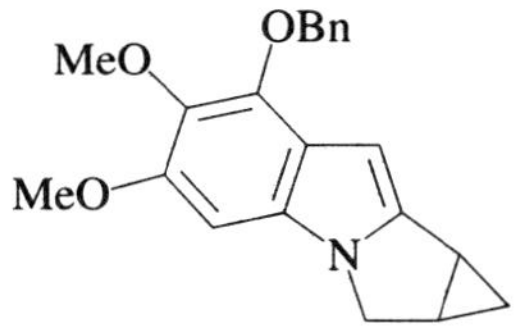	DMF, $POCl_3$	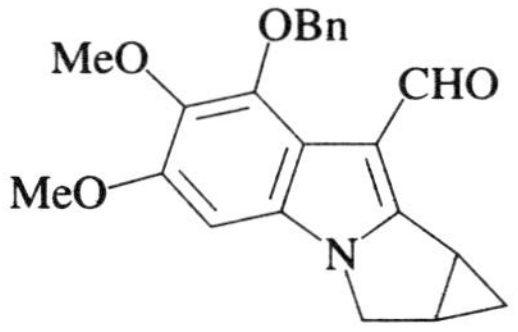 (63)	684, 685

211

TABLE XIV. INDOLES (*Continued*)

Substrate	Reagents	Product(s) and Yield(s) (%)	Refs.
	F2. R^1, R^2, R^4, R^6, R^7 Substituents		
C$_{19}$ 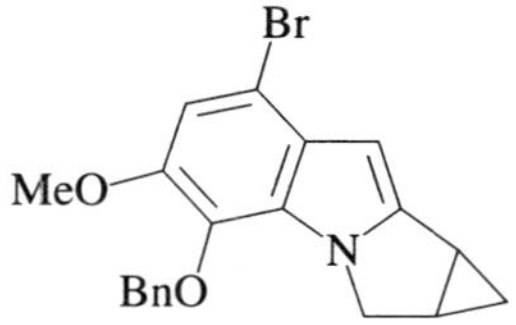	DMF, POCl$_3$	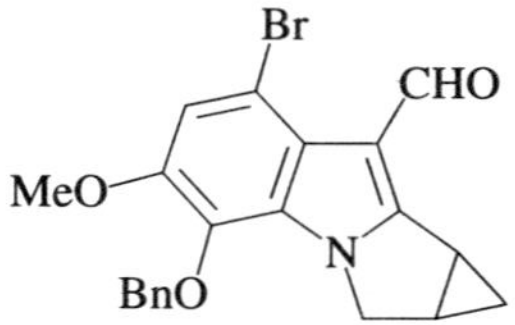 (73)	638, 673
	G. Hexasubstituted Indoles		
	G1. R^1, R^2, R^4, R^5, R^6, R^7 Substituents		
C$_{15}$ 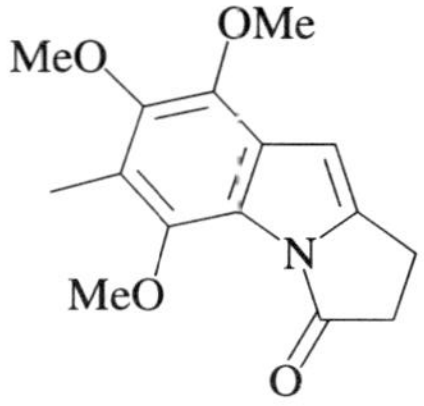	DMF, POCl$_3$	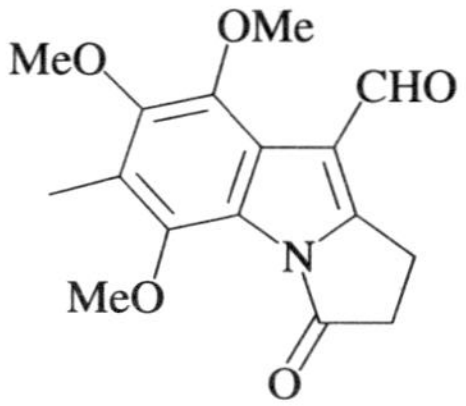 (80)	686

TABLE XV. CARBAZOLES

R^1 - R^9 = H except as indicated.

Substrate	Reagents	Product(s) and Yield(s) (%)	Refs.
	A. Carbazole		
C_{12}	Me_2NCOPh, $POCl_3$	R^9 = COPh (80)	111
	B. Monosubstituted Carbazoles		
	B1. R^1 Substituents		
C_{12} R^1 = Me	MFA, $POCl_3$	R^3 = CHO (33) + R^6 = CHO (36)	112
	B2. R^2 Substituents		
C_{12} R^2 = OH	DMF, $POCl_3$	R^1 = R^3 = CHO (30)	687
	B3. R^9 Substituents		
C_{13-19} R^9 = Me	MFA, $POCl_3$	R^3 = CHO (81)	112, 113
	DMF, $POCl_3$	R^3 = CHO (60)	688, 689
	MFA, $POCl_3$	R^3 = CHO (87)	690
	MFA, $POCl_3$	R^3 = CHO (—) + R^3 = R^6 = CHO (31)	691
	MFA, $POCl_3$	R^3 = CHO (81)	113
	MFA, $POCl_3$	R^3 = CHO (85)	113
	MFA, $POCl_3$	R^3 = CHO (86)	664
	DMF, $POCl_3$	R^3 = CHO (63)	692
	Et_2NCHO, $POCl_3$	R^3 = CHO (25-32)	664
	i-Pr_2NCHO, $POCl_3$	R^3 = CHO (25-32)	664
	N-Formylpyrrolidine, $POCl_3$	R^3 = CHO (25-32)	664

213

TABLE XV. CARBAZOLES (*Continued*)

Substrate	Reagents	Product(s) and Yield(s) (%)	Refs.
	C. Disubstituted Carbazoles		
	C1. R^1, R^4 Substituents		
C_{14} $R^1 = R^4 = Me$	MFA, POCl$_3$	$R^3 = CHO$ (44)	112, 693–695
	MFA, POCl$_3$ (xs)	$R^3 = CHO$ (42)	695
	C2. R^1, R^6 Substituents		
C_{13} $R^1 = OH$, $R^6 = Me$	MFA, POCl$_3$	$R^2 = CHO$ (12)	114
	C3. R^1, R^7 Substituents		
C_{13} $R^1 = OH$, $R^7 = Me$	MFA, POCl$_3$	$R^2 = CHO$ (10)	114
	C4. R^1, R^8 Substituents		
C_{13} $R^1 = OH$, $R^8 = Me$	MFA, POCl$_3$	$R^2 = CHO$ (18)	114
	C5. R^1, R^9 Substituents		
C_{14} $R^1 = R^9 = Me$	DMF, POCl$_3$	$R^3 = CHO$ (74)	110
C_{20} $R^1 = Me$, $R^9 = Bn$	DMF, POCl$_3$	$R^3 = CHO$ (42) + $R^6 = CHO$ (28)	662, 663
	C6. R^2, R^6 Substituents		
C_{13} $R^2 = OH$, $R^6 = Me$	MFA, POCl$_3$	$R^3 = CHO$ (18)	114
	C7. R^2, R^7 Substituents		
C_{13} $R^2 = OH$, $R^7 = Me$	MFA, POCl$_3$	$R^3 = CHO$ (10) + $R^9 = CHO$ (53)	114

214

TABLE XV. CARBAZOLES (*Continued*)

Substrate	Reagents	Product(s) and Yield(s) (%)	Refs.
C8. R^2, R^8 Substituents			
C_{13}			
$R^2 = OH$, $R^8 = Me$	MFA, POCl₃	$R^3 = CHO$ (12)	114
C9. R^3, R^9 Substituents			
C_{14-18}			
R^3 = Me, R^9 = Me	MFA, POCl₃	$R^6 = CHO$ (—)	113
R^3 = Me, R^9 = Et	MFA, POCl₃	$R^6 = CHO$ (60)	690
R^3 = Me, R^9 = n-Bu	MFA, POCl₃	$R^6 = CHO$ (85)	113
R^3 = Me, R^9 = i-C₅H₁₁	MFA, POCl₃	$R^6 = CHO$ (—)	113
D. Trisubstituted Carbazoles			
D1. R^1, R^4, R^6 Substituents			
C_{15}			
$R^1 = R^4 = R^6 = Me$	MFA, POCl₃	$R^3 = CHO$ (29)	695
$R^1 = R^4 = Me$, $R^6 = OMe$	MFA, POCl₃	$R^3 = CHO$ (60)	693
C_{17}			
$R^1 = R^4 = Me$, $R^6 = NMe_2$	—	$R^6 = CHO$ (—)	696
	DMF, POCl₃, 0°	$R^9 = CHO$ (29)	692
	DMF, POCl₃, 100°	$R^3 = CHO$ (9) + $R^5 = CHO$ (3)	692
D2. R^1, R^4, R^7 Substituents			
C_{15}			
$R^1 = R^4 = R^7 = Me$	MFA, POCl₃	$R^3 = CHO$ (71)	693
$R^1 = R^4 = Me$, $R^7 = OMe$	MFA, POCl₃	$R^6 = CHO$ (13) + $R^8 = CHO$ (3)	693
D3. R^1, R^4, R^9 Substituents			
C_{15}			
$R^1 = R^4 = R^9 = Me$	MFA, POCl₃	$R^3 = CHO$ (19) + $R^6 = CHO$ (4)	693
C_{21}			
$R^1 = R^4 = Me$, $R^9 = Bn$	DMF, POCl₃	$R^3 = CHO$ (66) + $R^6 = CHO$ (14)	662, 692

TABLE XV. CARBAZOLES (Continued)

Substrate	Reagents	Product(s) and Yield(s) (%)	Refs.
E. Tetrasubstituted Carbazoles			
E1. R^1, R^4, R^6, R^7 Substituents			
C_{15} $R^1 = R^4 = Me$, $R^6 + R^7 = OCH_2O$	MFA, POCl$_3$	$R^3 = CHO$ (50)	112
E2. R^1, R^4, R^6, R^9 Substituents			
C_{16} $R^1 = R^4 = R^9 = Me$, $R^6 = OMe$	MFA, POCl$_3$	$R^3 = CHO$ (68)	693
C_{23} $R^1 = R^4 = Me$, $R^6 = NMe_2$, $R^9 = Bn$	DMF, POCl$_3$	$R^3 = R^5 = CHO$ (3) + $R^5 = CHO$ (7)	692
F. Pentasubstituted Carbazoles			
F1. R^1, R^2, R^3, R^5, R^8 Substituents			
C_{17}	MFA, POCl$_3$	(100)	697

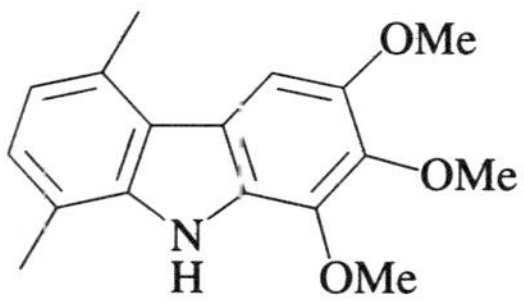

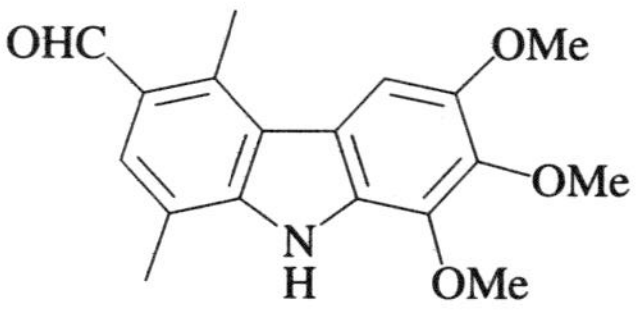

TABLE XVI. OTHER HETEROCYCLES WITH ONE FULLY CONJUGATED RING

Substrate	Reagents	Product(s) and Yield(s) (%)	Refs.
	C_2N_3 **1,2,3-Triazoles**		
C_{6-7} (pyrazolo-triazole, R substituent)	DMF, $COCl_2$	pyrrole-2,5-dicarbaldehyde (R, OHC, CHO) R = H (35); R = Me (30)	698
	C_2N_2O **1,2,3-Oxadiazolium-5-olates (Sydnones)**		
C_{3-12} (sydnone, R)		4-CHO sydnone (R)	
R = Me	MFA, $POCl_3$	(77)	699
4-ClC$_6$H$_4$	DMF, $POCl_3$	(34)	127
4-BrC$_6$H$_4$	MFA, $POCl_3$	(20)	700
Ph	DMF, $POCl_3$	(40)	127
	DMF, $POCl_3$	(46)	127
C$_6$H$_{11}$	MFA, $POCl_3$	(55)	701, 700
2-MeC$_6$H$_4$	DMF, $POCl_3$	(69)	702
4-MeC$_6$H$_4$	DMF, $POCl_3$	(30)	127
4-MeOC$_6$H$_4$	DMF, $POCl_3$	(61)	127
	DMF, $POCl_3$	(76)	127
Bn	MFA, $POCl_3$	(81)	699, 700
4-EtOC$_6$H$_4$	MFA, $POCl_3$	(15)	701
	MFA, $POCl_3$	(52)	700

217

TABLE XVI. OTHER HETEROCYCLES WITH ONE FULLY CONJUGATED RING (*Continued*)

Substrate	Reagents	Product(s) and Yield(s) (%)	Refs.
		C_2N_2S	
		1,2,3-Thiadiazolium-5-olates	
C_9			
4-MeOC$_6$H$_4$ (isothiazolium-olate structure)	DMF, POCl$_3$	4-MeOC$_6$H$_4$ (CHO-substituted structure) (92)	703
		C_3N_2	
		Pyrazoles	
R^1 - R^4 = H except as indicated		R^1 - R^4 = as in substrate except as indicated	
C_{5-24}			
R^1 = R^2 = Me	DMF, POCl$_3$	R^3 = CHO (58)	115
R^1 = R^4 = Me	DMF, POCl$_3$	R^3 = CHO (57)	115
R^1 = 3-O$_2$NC$_5$H$_4$	DMF, POCl$_3$	R^3 = CHO (9)	704
R^1 = 4-ClC$_6$H$_4$, R^4 = NH$_2$	DMF, POCl$_3$	R^2 = CHO, R^4 = N=CHNMe$_2$ (70)	116
	N-Formylpiperidine, POCl$_3$	R^2 = CHO, R^4 = N=CH(N-piperidyl) (76)	116
R^1 = 4-O$_2$NC$_5$H$_4$, R^4 = NH$_2$	DMF, POCl$_3$	R^2 = CHO, R^4 = N=CHNMe$_2$ (92)	116
R^1 = Ph, R^4 = NH$_2$	DMF, POCl$_3$	R^2 = CHO, R^4 = N=CHNMe$_2$ (84)	116
	N-Formylpiperidine, POCl$_3$	R^2 = CHO, R^4 = N=CH(N-piperidyl) (96)	116
R^1 = 3-CF$_3$C$_6$H$_4$	—	R^3 = CHO (—)	705
R^1 = Me, R^2 = NH$_2$, R^4 = 4-FC$_6$H$_4$	DMF, POCl$_3$	R^3 = CHO (35)	706
R^1 = Me, R^2 = NH$_2$, R^4 = 4-ClC$_6$H$_4$	DMF, POCl$_3$	R^3 = CHO (60)	706
R^1 = 4-ClC$_6$H$_4$, R^3 = Me, R^4 = NH$_2$	DMF, POCl$_3$	R^2 = CHO, R^4 = N=CHNMe$_2$ (71)	116
R^1 = 4-O$_2$NC$_5$H$_4$, R^3 = Me, R^4 = NH$_2$	DMF, POCl$_3$	R^2 = CHO, R^4 = N=CHNMe$_2$ (80)	116
R^1 = Ph, R^2 = Me, R^4 = NH$_2$	DMF, POCl$_3$	R^3 = CHO (64)	707
R^1 = Ph, R^3 = Me, R^4 = NH$_2$	DMF, POCl$_3$	R^2 = CHO, R^4 = N=CHNMe$_2$ (70)	116

218

TABLE XVI. OTHER HETEROCYCLES WITH ONE FULLY CONJUGATED RING (*Continued*)

Substrate	Reagents	Product(s) and Yield(s) (%)	Refs.
$R^1 = R^2 = Me$, $R^4 = $ 2-oxypyridyl	—	$R^3 = CHO$ (—)	708
$R^1 = $ 4-ClC$_6$H$_4$, $R^3 = Ph$, $R^4 = NH_2$	DMF, POCl$_3$	$R^2 = CHO$, $R^4 = N{=}CHNMe_2$ (94)	116
$R^1 = $ 4-O$_2$NC$_6$H$_4$, $R^3 = Ph$, $R^4 = NH_2$	DMF, POCl$_3$	$R^2 = CHO$, $R^4 = N{=}CHNMe_2$ (99)	116
$R^1 = R^3 = Ph$, $R^4 = NH_2$	DMF, POCl$_3$	$R^2 = CHO$, $R^4 = N{=}CHNMe_2$ (83)	116
$R^1 = $ 4-MeC$_6$H$_4$, $R^3 = Ph$, $R^4 = NH_2$	DMF, POCl$_3$	$R^2 = CHO$, $R^4 = N{=}CHNMe_2$ (91)	116
$R^1 = $ 4-MeOC$_6$H$_4$, $R^3 = Ph$, $R^4 = NH_2$	DMF, POCl$_3$	$R^2 = CHO$, $R^4 = N{=}CHNMe_2$ (74)	116
$R^1 = R^2 = Ph$, $R^4 = $ 4-(1-phenylpyrazolyl)	DMF, POCl$_3$	$R^3 = CHO$ (83)	704

Pyrazol-5-ones

C$_4$

Substrate	Reagents	Product(s) and Yield(s) (%)	Refs.
	1. DMF, POCl$_3$ 2. NH$_4$Cl	(80)	709
	1. DMF, POCl$_3$ 2. ArNH$_2$		709

Ar	
2-ClC$_6$H$_4$	(80)
4-O$_2$NC$_6$H$_4$	(83)
Ph	(85)
2-MeOC$_6$H$_4$	(91)
2,5-(MeO)$_2$C$_6$H$_3$	(87)

TABLE XVI. OTHER HETEROCYCLES WITH ONE FULLY CONJUGATED RING (*Continued*)

Substrate		Reagents	Product(s) and Yield(s) (%)	Refs.
C$_{4-14}$ 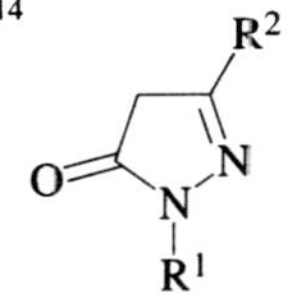				

R^1	R^2			
Me	H	DMF, POCl$_3$	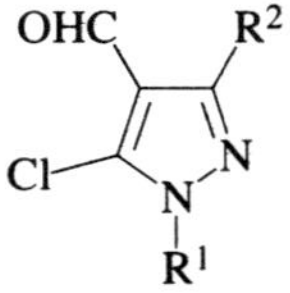 **I** (72)	710
Me	Me	DMF, POCl$_3$	**I** (91)	118, 121, 710, 711
		DMF, POCl$_3$, 3 h	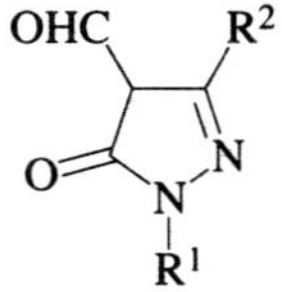 **II** (27)	711
		DMF, POCl$_3$, prolonged standing	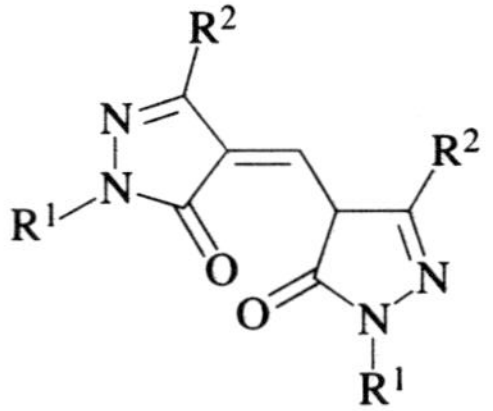 **III** (—)	711

TABLE XVI. OTHER HETEROCYCLES WITH ONE FULLY CONJUGATED RING (*Continued*)

Substrate		Reagents	Product(s) and Yield(s) (%)	Refs.
R^1	R^2			
H	*i*-Pr	DMF, POCl₃	I (48)	712
Me	*i*-Pr	DMF, POCl₃	I (87)	710
H	*i*-Bu	DMF, POCl₃	II (68)	713
H	*t*-Bu	DMF, POCl₃	I (30)	712
Ph	Cl	DMF, POCl₃	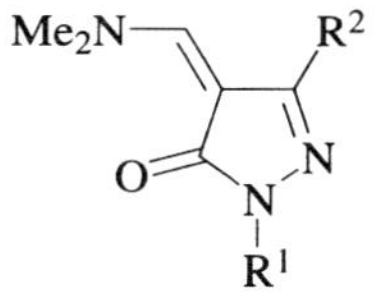 IV (76)	711
Ph	H	DMF, POCl₃	IV (89)	711
H	Ph	DMF, POCl₃	II (99)	713
Ph	OH	DMF, POCl₃	IV (82)	711
Ph	NH₂	DMF, POCl₃	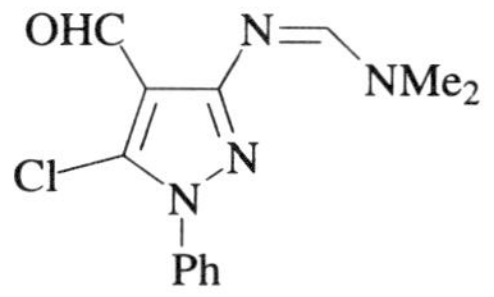 (60)	714
2,4-F₂C₆H₃	Me	1. — 2. Et₂NSF₃	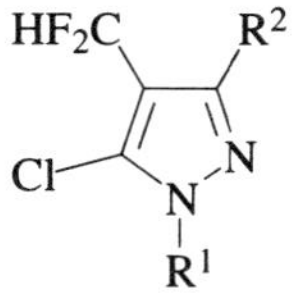 (—)	715
2-pyridyl	Me	DMF, POCl₃	I (77)	121

221

TABLE XVI. OTHER HETEROCYCLES WITH ONE FULLY CONJUGATED RING (*Continued*)

Substrate R¹	R²	Reagents	Product(s) and Yield(s) (%)	Refs.
4-O₂NC₆H₄	Me	DMF, POCl₃	**I** (30)	121
		DMF, POCl₃	**I** (—)	716
Me	4-ClC₆H₄	DMF, POCl₃	**I** (77)	710
Me	Ph	DMF, POCl₃	**I** (80)	710
		DMF, POCl₃	**I** (60)	711
Ph	Me	DMF, POCl₃	**I** (94)	121, 716
		DMF, POCl₃	**II** (86)	713
		DMF, POCl₃	**II** (—) + **III** (—)	717
		DMF, POCl₃	**I** (6) + **IV** (22-48)	718
		DMF, POCl₃	**I** (—) + **IV** (51)	719
		DMF, POCl₃	**IV** (78)	119, 720, 721
4-MeC₆H₄	Me	DMF, POCl₃	**IV** (—)	721
Me	4-MeOC₆H₄	DMF, POCl₃	**I** (82)	710
Ph	NMe₂	DMF, POCl₃	**IV** (72)	119
Ph	CO₂Et	DMF, POCl₃	**IV** (76)	119, 721
Ph	*i*-Bu	DMF, POCl₃	**II** (90)	713
Ph	*t*-Bu	DMF, POCl₃	**I** (79)	184
Ph	3-pyridyl	DMF, POCl₃	**II** (43)	722, 723
Ph	Ph	DMF, POCl₃	**I** (60)	724, 725
		DMF, POCl₃	**II** (72)	713
Ph	NH(C₆H₄Cl-4)	DMF, POCl₃	**IV** (84)	726
Ph	4-MeOC₆H₄	DMF, POCl₃	**IV** (—)	721

222

TABLE XVI. OTHER HETEROCYCLES WITH ONE FULLY CONJUGATED RING (*Continued*)

Substrate	Reagents	Product(s) and Yield(s) (%)	Refs.
C_{6-16}			
	DMF, $POCl_3$	(40)	117
	DMF, $POCl_3$	(52)	727
	DMF, $POCl_3$	(80)	729, 728
	MFA, $POCl_3$	(53)	728
	DMF, $POCl_3$	(—)	730

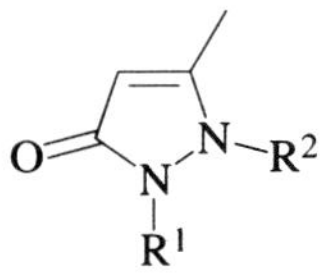

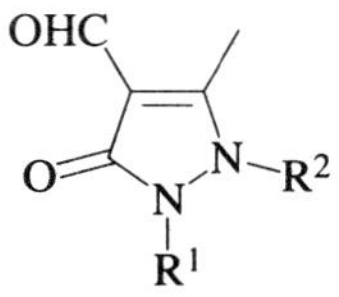

R^1	R^2
Me	Me
Me	Ph
Ph	Me
Ph	Ph

| C_9 | DMF, $POCl_3$ | (42) | 731 |

| C_{10-15} | DMF, $POCl_3$ | | 732 |

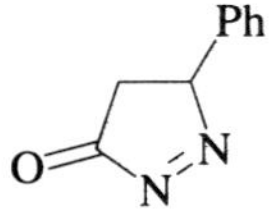

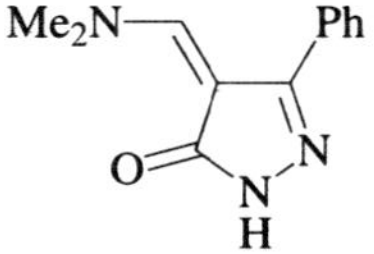

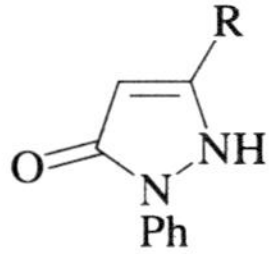

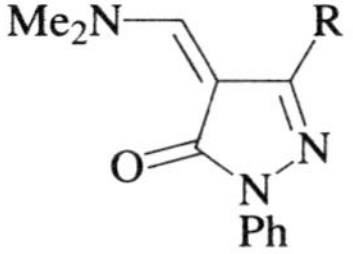

R	
Me	(70)
n-Pr	(60)
Ph	(80)

223

TABLE XVI. OTHER HETEROCYCLES WITH ONE FULLY CONJUGATED RING (*Continued*)

Substrate	Reagents	Product(s) and Yield(s) (%)	Refs.
		Pyrazole-3,5-diones	
C_{15}			
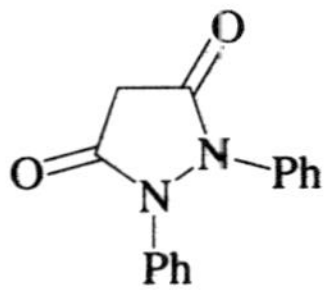	DMF, $POCl_3$	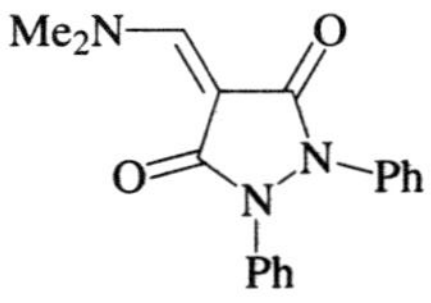 (—)	731
		Imidazoles	
C_{11-19}			
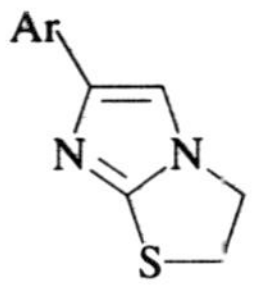		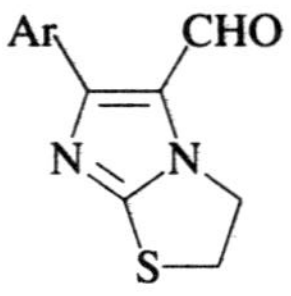 I	
Ar			
2,4-$Cl_2C_6H_3$	DMF, $POCl_3$	**I** (64)	120
4-ClC_6H_4	DMF, $POCl_3$	**I** (60-80)	733
4-MeC_6H_4	DMF, $POCl_3$	**I** (60-80)	733
4-PhC_6H_4	DMF, $POCl_3$	**I** (60-80)	733
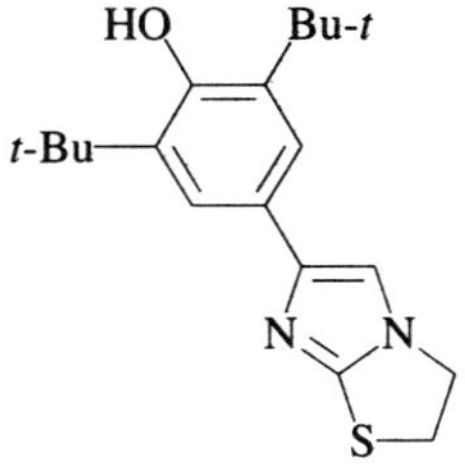	DMF, $POCl_3$	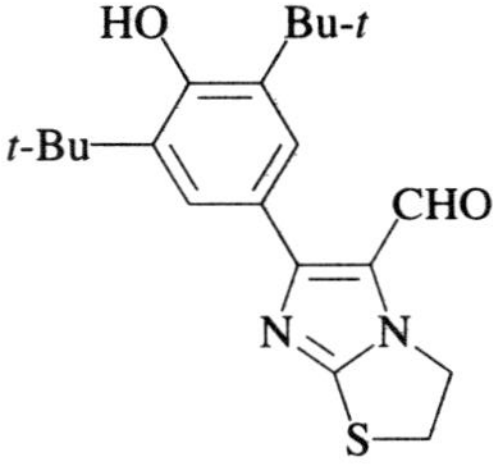 (62)	733

TABLE XVI. OTHER HETEROCYCLES WITH ONE FULLY CONJUGATED RING (*Continued*)

Substrate	Reagents	Product(s) and Yield(s) (%)	Refs.
C_{18}			
(phenol with 2,6-di-*t*-Bu, imidazole–SMe)	1. DMF, POCl$_3$ 2. NH$_2$OH 3. Dehydrate	(imidazo-thiazoline, Ar, CN) (15)	734
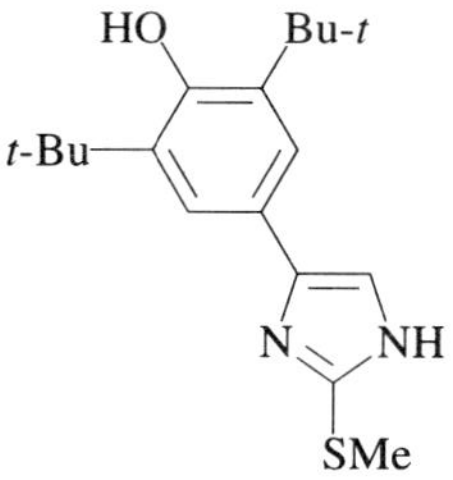	1. DMF, POCl$_3$ 2. NH$_2$OH 3. SOCl$_2$	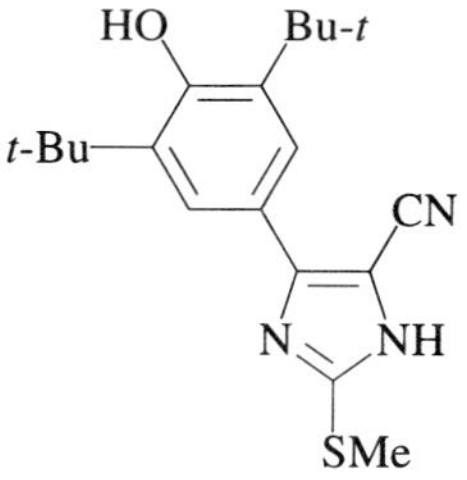 (imidazole with CN, NH, SMe, di-*t*-Bu phenol) (—)	735
Imidazol-(5H)-4-ones			
C_{10-15}			
(imidazolinone, C_6H_{11}, R)	DMF, POCl$_3$	OHC product; R = Me (58) R = Ph (90)	121, 736 121
Thiohydantoins			
C_9			
(thiohydantoin, Ph)	DMF, POCl$_3$	(Cl, CHO, thiohydantoin, Ph) (40)	737

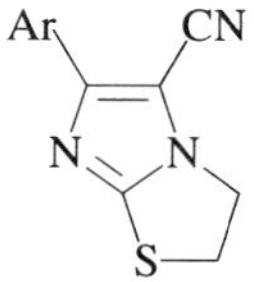

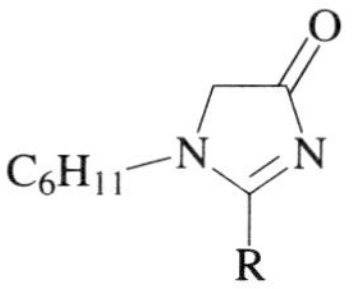

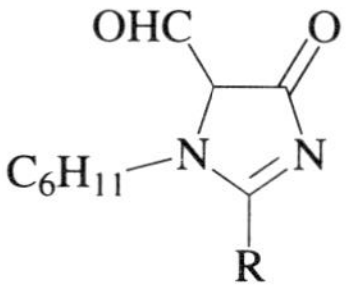

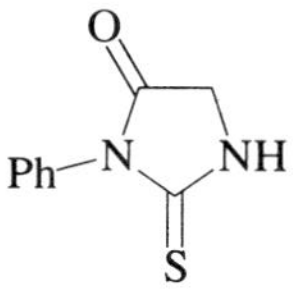

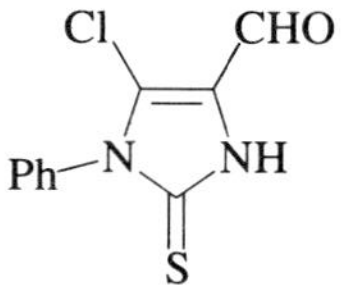

TABLE XVI. OTHER HETEROCYCLES WITH ONE FULLY CONJUGATED RING (*Continued*)

Substrate	Reagents	Product(s) and Yield(s) (%)	Refs.
		C_3NO **1,2-Oxazoles**	
C_4 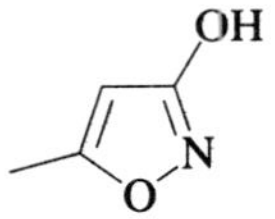	DMF, $COCl_2$	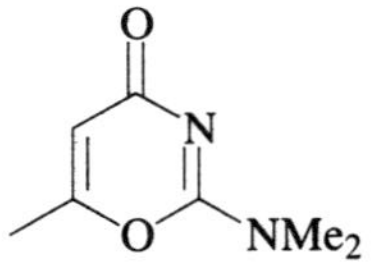 (45)	738
C_9 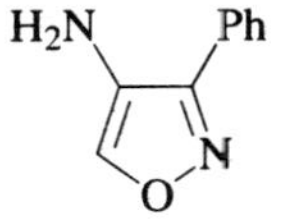	DMF, $POCl_3$	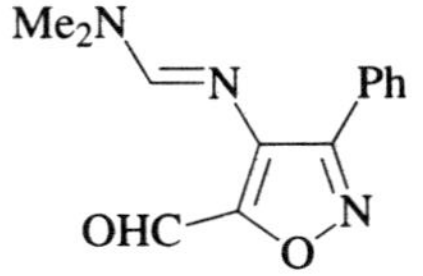 (82)	739
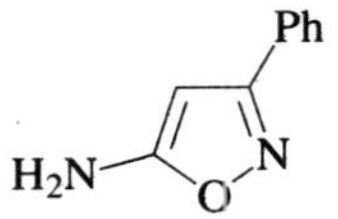	DMF, $POCl_3$	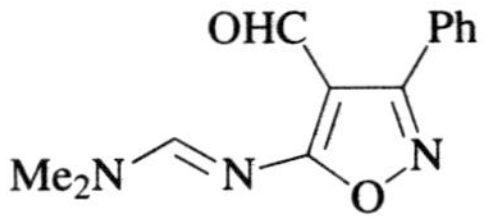 (82)	740
		1,2-Oxazol-(4*H*)-5-ones	
C_{4-10} 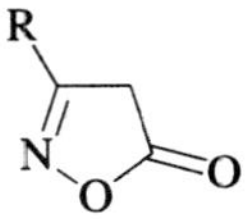		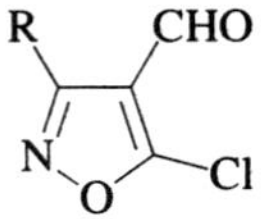 **I** 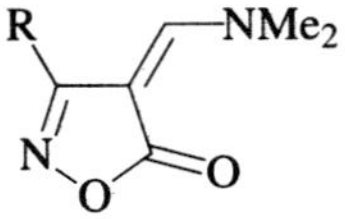 **II** 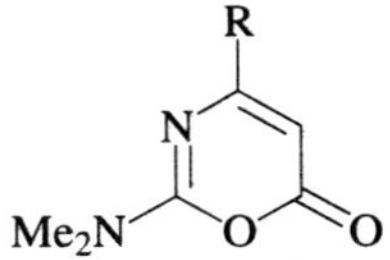 **III** 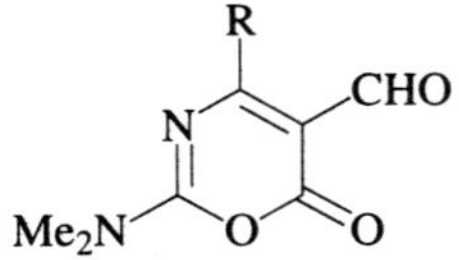 **IV**	

TABLE XVI. OTHER HETEROCYCLES WITH ONE FULLY CONJUGATED RING (*Continued*)

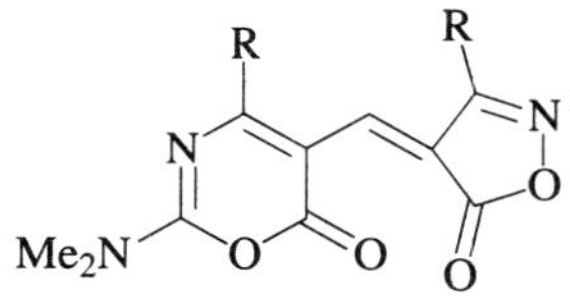

V, 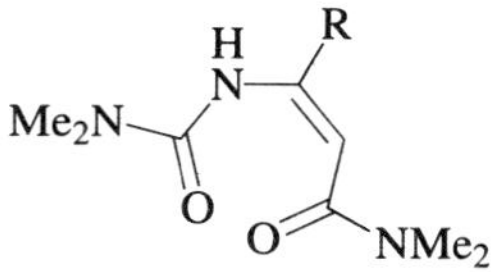 VI, 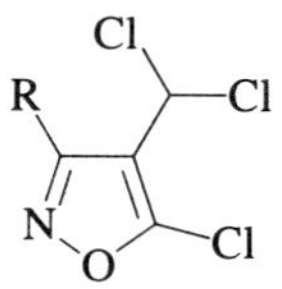 VII, 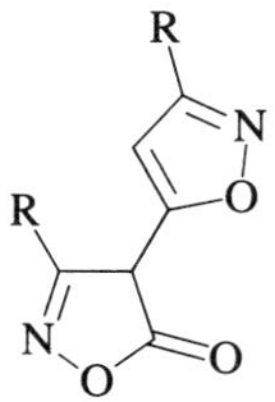 VIII (structures)

Substrate	Reagents	Product(s) and Yield(s) (%)	Refs.
R			
Me	DMF, POCl$_3$ (1 eq)	**III** (53) + **IV** (5) + **VI** (9)	122
	DMF, POCl$_3$ (2 eq)	**III** (16) + **IV** (42) + **VI** (6)	122
n-Pr	DMF, POCl$_3$ (1 eq)	**III** (62) + **IV** (6) + **VI** (6)	122
	DMF, POCl$_3$ (2 eq)	**IV** (66) + **V** (11)	122
t-Bu	DMF, POCl$_3$ (1 eq)	**I** (4) + **II** (70)	122
	DMF, POCl$_3$ (2 eq)	**I** (64) + **IV** (17)	122
Ph	DMF, POCl$_3$ (1 eq)	**I** (4) + **II** (50) + **IV** (15)	122
	DMF, POCl$_3$ (1 eq)	**II** (35)	741
	DMF, POCl$_3$ (1 eq)	**VI** (59) + **VIII** (30)	742
	DMF, POCl$_3$ (2 eq)	**I** (15) + **IV** (80)	122
	DMF, POCl$_3$ (12 eq)	**IV** (71)	741
	DMF (8 eq), POCl$_3$ (17 eq)	**I** (46)	741
	DMF, POCl$_3$ (xs)	**VIII** (78)	742

227

TABLE XVI. OTHER HETEROCYCLES WITH ONE FULLY CONJUGATED RING (*Continued*)

Substrate	Reagents	Product(s) and Yield(s) (%)	Refs.
R			
4-O$_2$NC$_6$H$_4$	DMF, POCl$_3$ (1 eq)	**II** (67)	122
	DMF, POCl$_3$ (2 eq)	**I** (11) + **II** (22) + **IV** (54)	122
4-MeC$_6$H$_4$	DMF, POCl$_3$ (1 eq)	**VII** (58) + **VIII** (29)	742
	DMF, POCl$_3$ (xs)	**VIII** (79)	742
4-MeOC$_6$H$_4$	DMF, POCl$_3$ (1 eq)	**VII** (60) + **VIII** (28)	742
	DMF, POCl$_3$ (xs)	**VIII** (78)	742

C$_{10-17}$

	DMF, POCl$_3$		743

R^1	R^2	R^3	
Me	CO$_2$Me	Me	E (82) + Z (8)
n-Pr	CO$_2$Et	Me	E (80) + Z (14)
n-Pr	(CH$_2$)$_3$		(67)
Ph	H	Me	(57)
Ph	(CH$_2$)$_3$		(74)
Ph	(CH$_2$)$_4$		(75)
Ph	CO$_2$Et	Me	E (51) + Z (28)
Ph	Ph	H	E (72)

C$_{12}$

DMF, POCl$_3$ (—) 741

TABLE XVI. OTHER HETEROCYCLES WITH ONE FULLY CONJUGATED RING (*Continued*)

Substrate	Reagents	Product(s) and Yield(s) (%)	Refs.
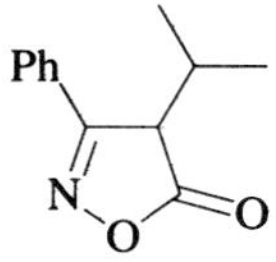	Et₂NCHO, POCl₃	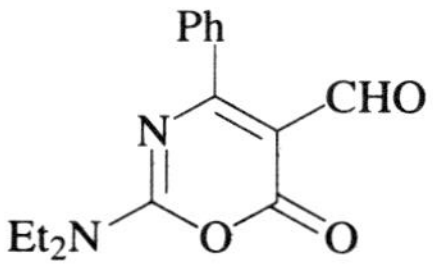 (77)	741
	DMF, POCl₃	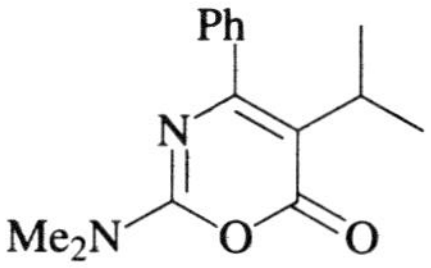 (61)	743
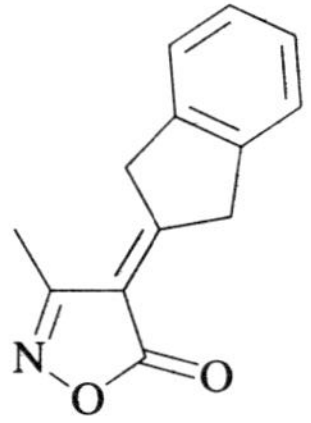	DMF, POCl₃	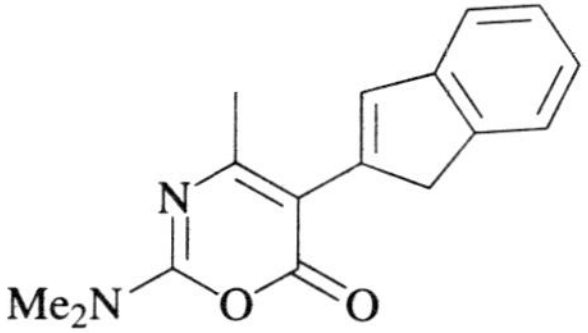 (47) 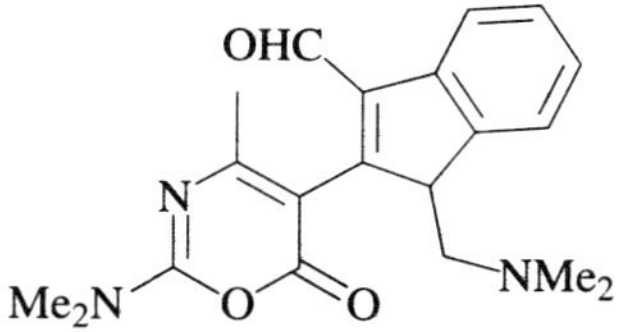 + (15)	743

C₁₄

TABLE XVI. OTHER HETEROCYCLES WITH ONE FULLY CONJUGATED RING (*Continued*)

1,2-Oxazol-(2H)-5-ones

Substrate (C$_{6-20}$): 4-R^2, 3-R^1, 1,2-oxazol-5(2H)-one

Reagents: R$_2$NCHO, POCl$_3$ (n eq)

Products: **I** = R^1, R^2, R$_2$N-substituted oxazinone; **II** = R^1, R^2, NMe$_2$ open-chain product

R^1	R^2	R	n	Product(s) and Yield(s) (%)	Refs.
H	CO$_2$Et	Me	1	**I** (69)	122
H	CO$_2$Et	Me	2	**I** (8) + **II** (76)	122
CO$_2$Et	Me	Me	1	**I** (70)	122
(CH$_2$)$_4$		Me	1	**I** (85)	122, 744
(CH$_2$)$_4$		Et	1	**I** (85)	122
Me	CH$_2$CO$_2$Me	Me	1	**I** (62)	122
Ph	H	Me	1	**I** (69)	122
H	Ph	Me	1	**I** (69)	122
H	Ph	Et	1	**I** (93)	122
H	Ph	R$_2$N = morpholino	1	**I** (60)	122
H	Ph	Me	2	**I** (73) + **II** (11)	122
H	Ph	Et	1	**I** (93)	122
H	Ph	R$_2$N = morpholino	1	**I** (60)	122
H	4-ClC$_6$H$_4$	Me	1	**I** (62)	122
H	4-ClC$_6$H$_4$	Me	2	**I** (48) + **II** (49)	122
H	4-MeC$_6$H$_4$	Me	1	**I** (87)	122
H	4-MeC$_6$H$_4$	Me	2	**I** (28) + **II** (57)	122
H	4-MeC$_6$H$_4$	R$_2$N = morpholino	1	**I** (45)	122
Me	Ph	Me	1	**I** (73)	122, 744
Ph	Me	Me	1	**I** (87)	122, 744
Me	Bn	Me	1	**I** (80)	122, 744
Ph	Et	Me	1	**I** (72)	122, 744

230

TABLE XVI. OTHER HETEROCYCLES WITH ONE FULLY CONJUGATED RING (*Continued*)

Substrate			Reagents	Product(s) and Yield(s) (%)	Refs.

R^1	R^2	R	n		
Ph	CO_2Et	Me	1	**I** (74)	122
Me	CH_2Bn	Me	1	**I** (87)	122
Ph	Bn	Me	1	**I** (62)	122
Ph	$n\text{-}C_8H_{17}$	Me	1	**I** (90)	122
Me	$n\text{-}C_{16}H_{33}$	Me	1	**I** (77)	122

1,3-Oxazoles

C_9

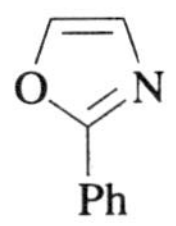

DMF, $POCl_3$ — (58) — 123

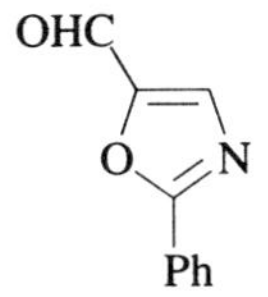

C_{10}

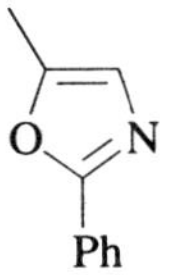

DMF, $POCl_3$ — (—) — 745

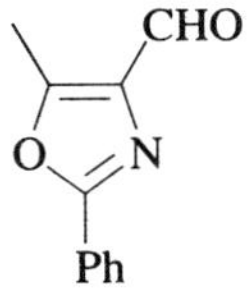

C_{17}

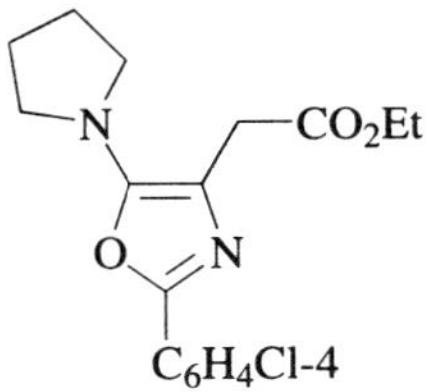

DMF, $POCl_3$ — (71) — 124

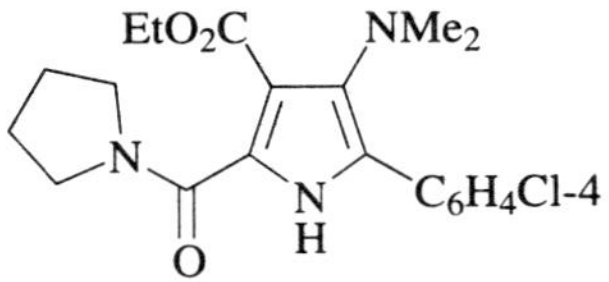

TABLE XVI. OTHER HETEROCYCLES WITH ONE FULLY CONJUGATED RING (*Continued*)

Substrate	Reagents	Product(s) and Yield(s) (%)	Refs.

C_{17-21}

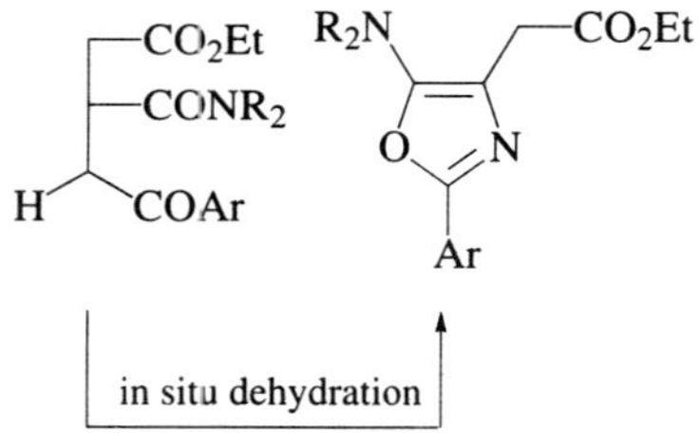

DMF, POCl₃ → in situ dehydration

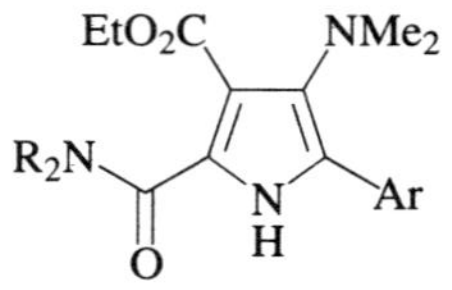

Ar		
4-ClC₆H₄	(80)	
4-ClC₆H₄	(82)	
4-FC₆H₄	(57)	
4-ClC₆H₄	(95)	
Ph	(95)	

R₂	
(CH₂)₄	
(CH₂)₂C(CH₂)₂	
(CH₂)₂C(CH₂)₂	
(CH₂)₂N(Me)(CH₂)₂	
N[(CH₂)₂CO₂Me]₂	

124

1,3-Oxazol-(4H)-5-ones

C_9

DMF, POCl₃ → in situ dehydration

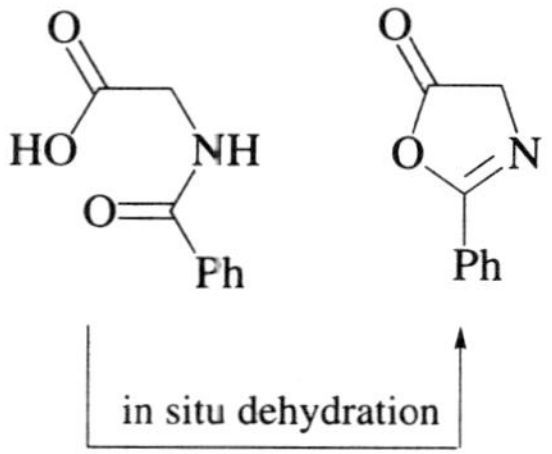

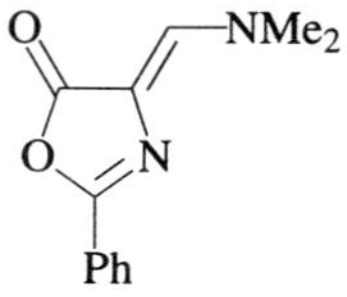

(93)   746

TABLE XVI. OTHER HETEROCYCLES WITH ONE FULLY CONJUGATED RING (*Continued*)

Substrate	Reagents	Product(s) and Yield(s) (%)	Refs.
C₃NS			
1,3-Thiazoles			
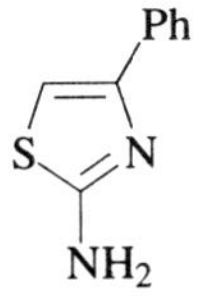 C₉	DMF, POCl₃	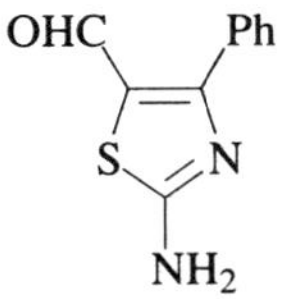 (12)	125
1,3-Thiazol-(3H)-2-ones			
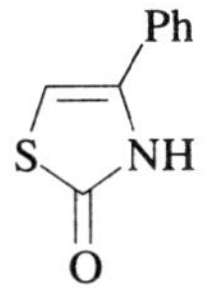 C₉	—	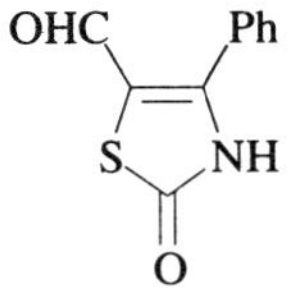 (40)	747
1,3-Thiazol-(5H)-4-ones			
C₃	DMF, POCl₃	(—)	748
	—	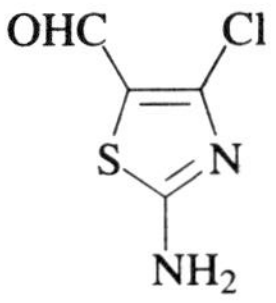 (—)	749

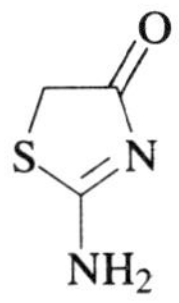

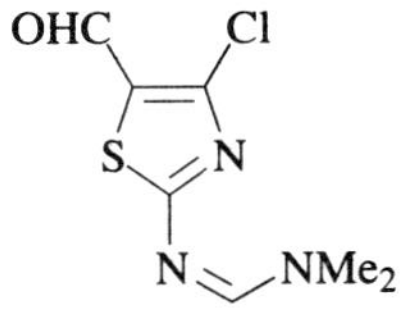

233

TABLE XVI. OTHER HETEROCYCLES WITH ONE FULLY CONJUGATED RING (*Continued*)

Substrate	Reagents	Product(s) and Yield(s) (%)	Refs.
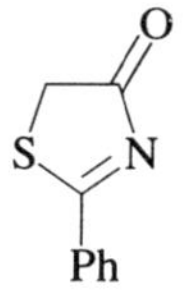 C₉	DMF, POCl₃	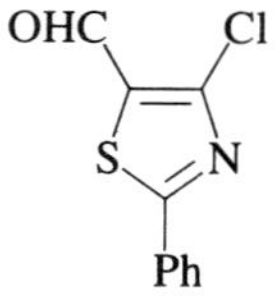 (63)	121
1,3-Thiazol-(3,5H)-4-ones			
	DMF, POCl₃	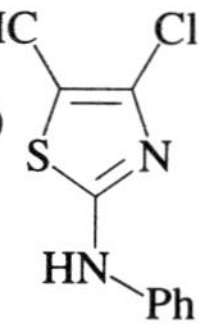 (55)	750
1,3-Thiazol-(4H)-5-ones			
C₃	DMF, POCl₃, 50° DMF, POCl₃, 70°	 R = NH₂ (—) R = N=CHNMe₂ (—)	751 751

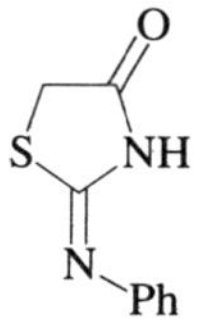

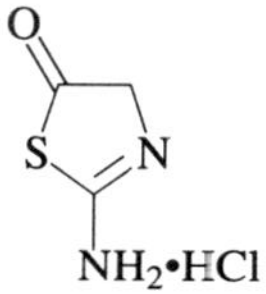

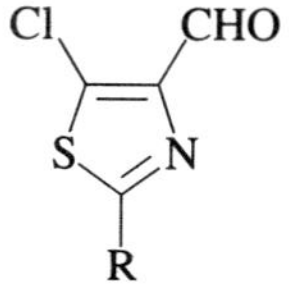

TABLE XVI. OTHER HETEROCYCLES WITH ONE FULLY CONJUGATED RING (*Continued*)

Substrate	Reagents	Product(s) and Yield(s) (%)	Refs.
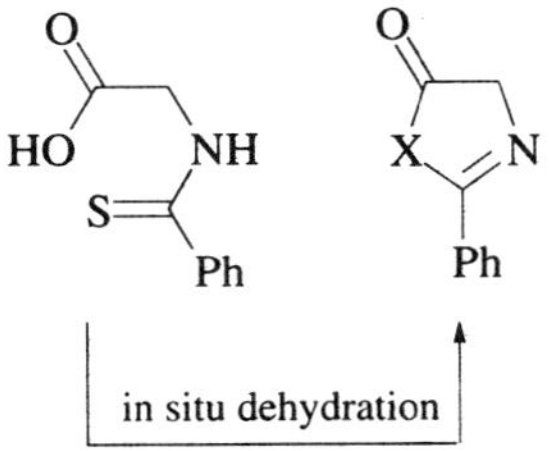 C_9	DMF, POCl$_3$	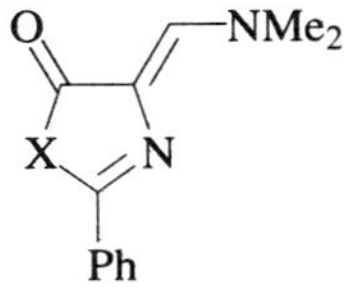 $X = O\ (30) + X = S\ (35)$	746
C_3 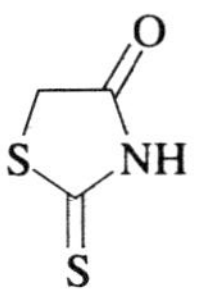	—	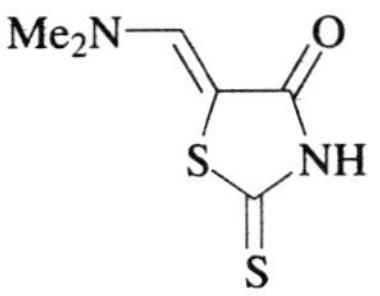 (59)	751

1,3-Thiazol-(3,5H)-2-thion-4-ones

C_4N
Pyrrol-(3H)-2-ones

Substrate	Reagents	Product(s) and Yield(s) (%)	Refs.
$C_{4\text{-}11}$ 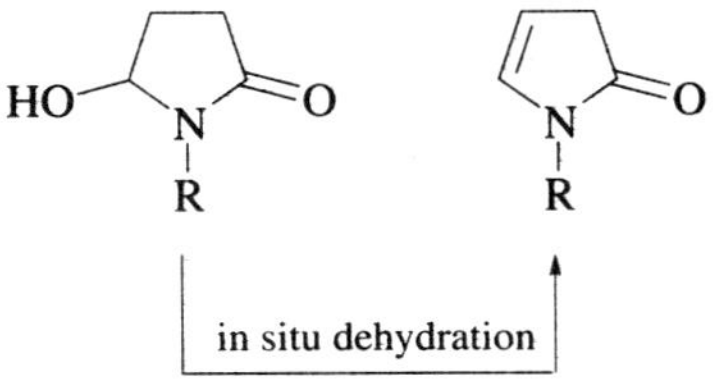	in situ dehydration	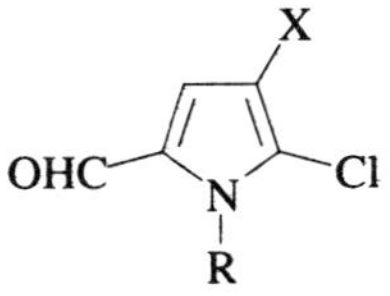 **I**, X = H **II**, X = CHO	752

in situ dehydration

X = O, S

TABLE XVI. OTHER HETEROCYCLES WITH ONE FULLY CONJUGATED RING (*Continued*)

Substrate	Reagents	Product(s) and Yield(s) (%)	Refs.
R			
H	DMF, POCl$_3$ (3:6), 25°	I (2)	
	DMF, POCl$_3$ (5:10), 50°	II (2)	
Me	DMF, POCl$_3$ (3:6), 25°	I (3) + II (tr)	
	DMF, POCl$_3$ (5:10), 50°	I (5) + II (62)	
n-C$_5$H$_{11}$	DMF, POCl$_3$ (3:6), 25°	I (35) + II (tr)	
	DMF, POCl$_3$ (5:10), 50°	I (4) + II (56)	
Bn	DMF, POCl$_3$ (3:6), 25°	I (57) + II (tr)	753
	DMF, POCl$_3$ (5:10), 50°	II (87)	
	DMF, POCl$_3$ (3:3), 25°	III (46) + IV (9)	
	DMF, POCl$_3$ (5:5), 25°	III (25) + IV (9)	
C$_8$	DMF, POCl$_3$	(72)	753
	DMF, POBr$_3$	(75)	754

Product structures:

III — OHC—[pyrrole, N–H]—Cl

IV — pyrrole (N–H) bearing CHO, Cl, and OHC substituents

Product (72): EtO$_2$C- / -OH / CHO substituted pyrrole (N–H), methyl

Product (75): EtO$_2$C- substituted pyrrolinone (N–H, =O) with =NMe$_2$ methylene, methyl

Substrates:

(pyrrolinone with N–CO$_2$Bu-*t*, =O)

(EtO$_2$C-, methyl-substituted pyrrolinone, N–H, =O)

TABLE XVI. OTHER HETEROCYCLES WITH ONE FULLY CONJUGATED RING (*Continued*)

Substrate	Reagents	Product(s) and Yield(s) (%)	Refs.
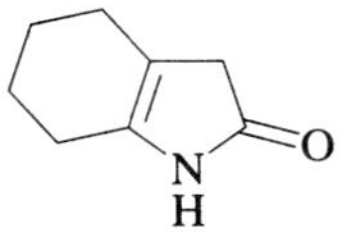 C$_{11}$	DMF, POCl$_3$	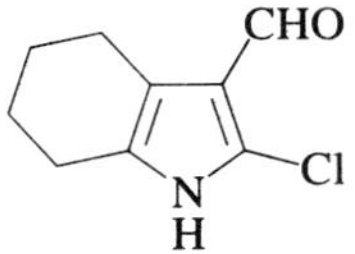 (50–60)	755
C$_{12}$	DMF, POCl$_3$	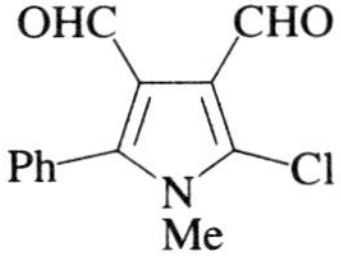 (79)	710
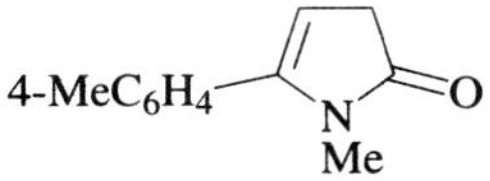	DMF, POCl$_3$	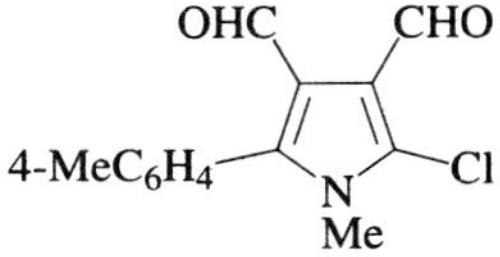 (67)	710
C$_{14}$ 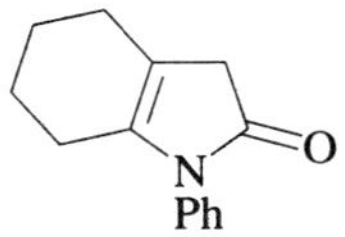	DMF, POCl$_3$	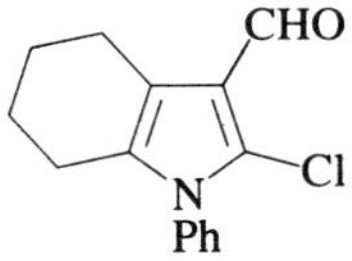 (50–60)	755

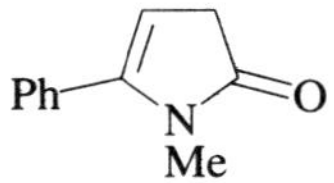

Pyrrol-(5H)-2-ones

Substrate	Reagents	Product(s) and Yield(s) (%)	Refs.
C$_6$ 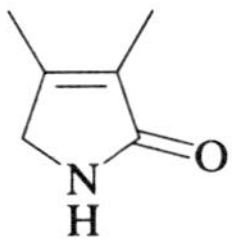	DMA, POBr$_3$	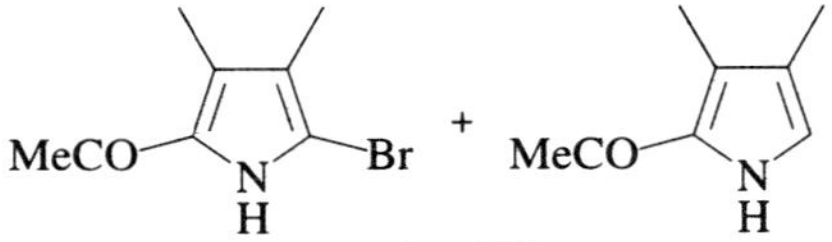 (total 17)	754

TABLE XVI. OTHER HETEROCYCLES WITH ONE FULLY CONJUGATED RING (*Continued*)

Substrate	Reagents	Product(s) and Yield(s) (%)	Refs.
C₇ (3-ethyl-4-methyl-pyrrol-2-one)	DMA, POCl₃	[pyrrole: MeCO, Cl, ethyl, methyl, N–H] (50)	754
C₉ (CO₂Me-substituted pyrrolone)	1. DMF, POBr₃ 2. ClO₄⁻	[Me₂N, Br, N–H pyrrolium] ClO₄⁻ (50)	756
	1. DMF, POBr₃ 2. ClO₄⁻ 3. base	[Me₂N, Br, CO₂Me-propyl, N] (75)	756
	1. DMF, POBr₃ 2. ClO₄⁻ 3. base	[MeO₂C, Me₂N, Br, N] (80)	756

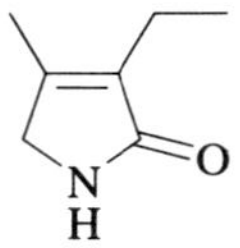

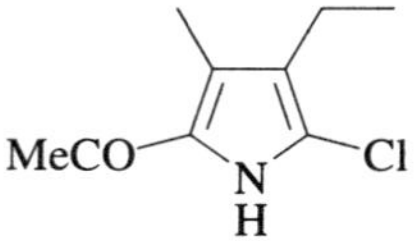

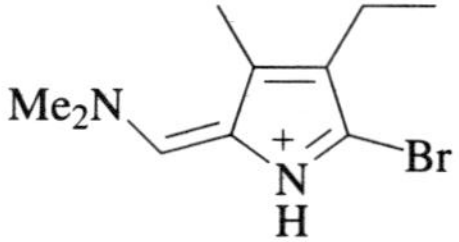

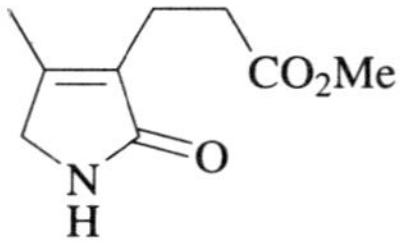

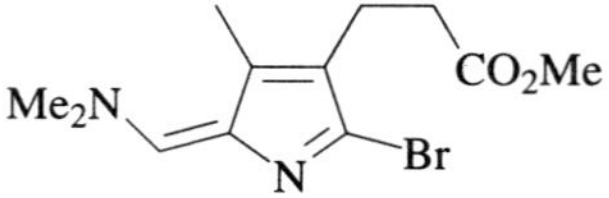

Pyrrol-(2H)-3-ones

Substrate	Reagents	Product(s) and Yield(s) (%)	Refs.
C₈ (EtO₂C-substituted pyrrolinone)	DMF, POCl₃	[EtO₂C, OH, CHO, methyl, N–H] (3) + [EtO₂C, Cl, CHO, methyl, N–H] (10)	753

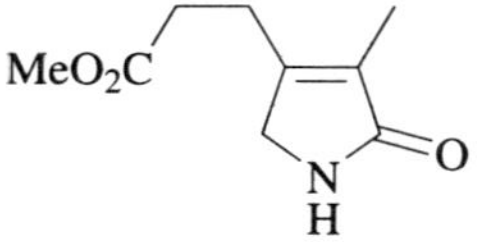

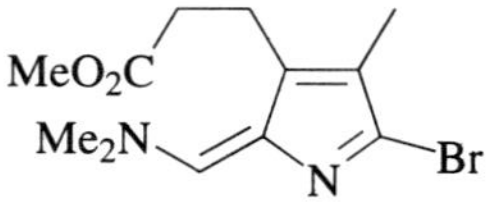

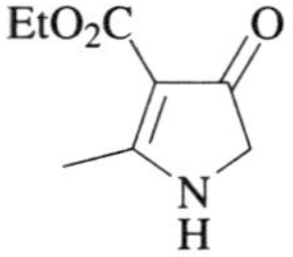

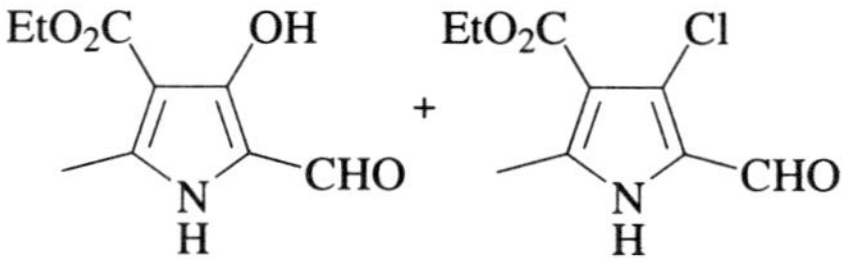

TABLE XVI. OTHER HETEROCYCLES WITH ONE FULLY CONJUGATED RING (*Continued*)

Substrate	Reagents	Product(s) and Yield(s) (%)	Refs.
Thiophen-(5H)-2-imines			
C_5	DMF, $POCl_3$	(83)	757
Thiophen-(5H)-4-ones			
C_7	—	(—)	758, 759
C_3N_2S **1,2,6-Thiadiazoles**			
C_{15}	DMF, $POCl_3$	(74)	760
C_4N_2 **Pyridazin-3-ones**			
C_{5-12}	DMF, $POCl_3$		761

239

TABLE XVI. OTHER HETEROCYCLES WITH ONE FULLY CONJUGATED RING (*Continued*)

Substrate	Reagents	Product(s) and Yield(s) (%)	Refs.

C_{8-14}

R	
H	(59)
Me	(88)
Ph	(88)
Bn	(71)

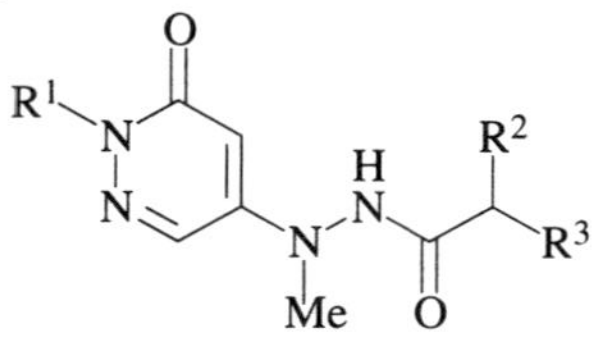

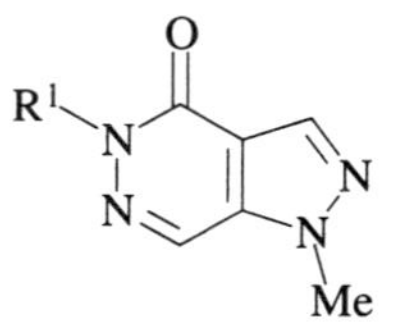

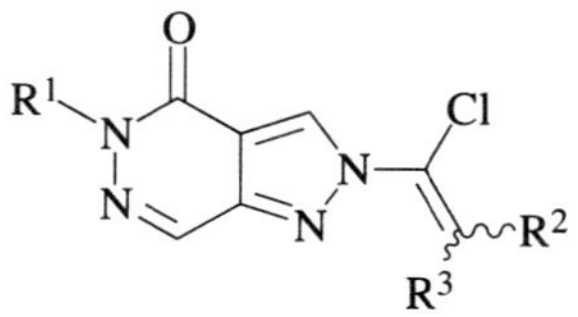

DMF, POCl$_3$

I +

II

R^1	R^2	R^3	
Me	H	H	**I** (48) + **II** (18)
Me	H	Me	**I** (48) + (*E*)-**II** (15) + (*Z*)-**II** (4)
Me	Me	Me	**I** (51) + **II** (22)
Ph	H	H	**I** (63) + **II** (15)
Bn	H	H	**I** (58) + **II** (24)

761

TABLE XVI. OTHER HETEROCYCLES WITH ONE FULLY CONJUGATED RING (*Continued*)

Substrate	Reagents	Product(s) and Yield(s) (%)	Refs.
Pyrimidines			
C$_{16-18}$	PhN(Me)CH=CHCHO, POCl$_3$	(70) (56)	641
Pyrimid-(1*H*)-6-ones			
C$_4$ 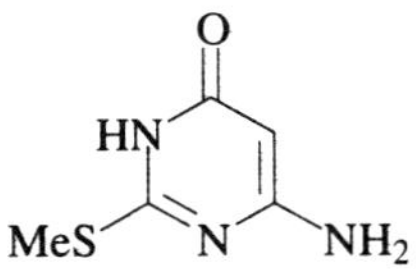	DMF, POCl$_3$	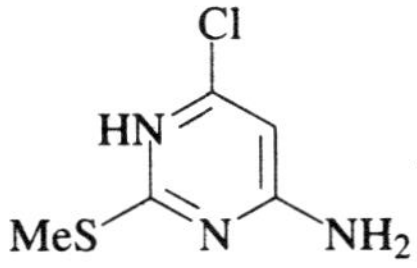 (26) 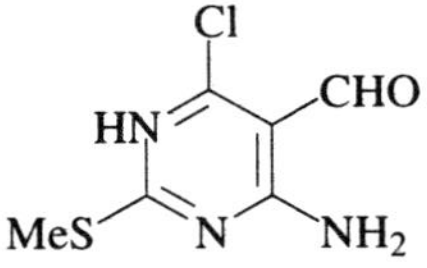 + (11)	762
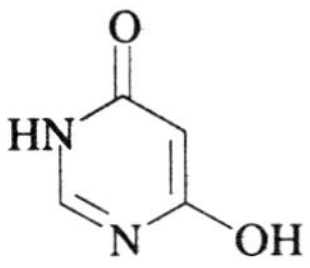	DMF, COCl$_2$	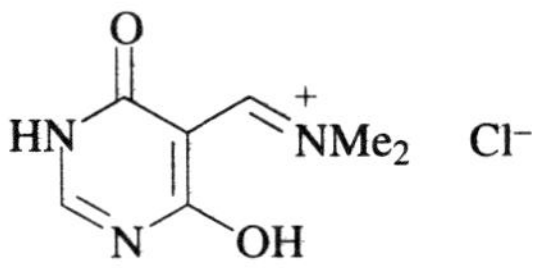 (93)	763
	1. DMF, COCl$_2$ 2. H$_2$O	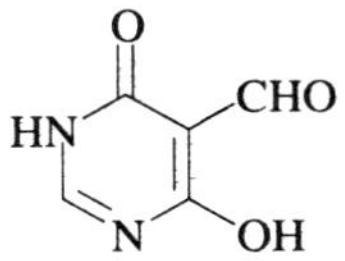 (54)	763

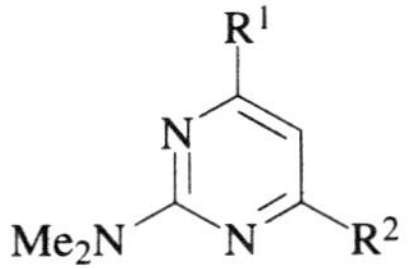

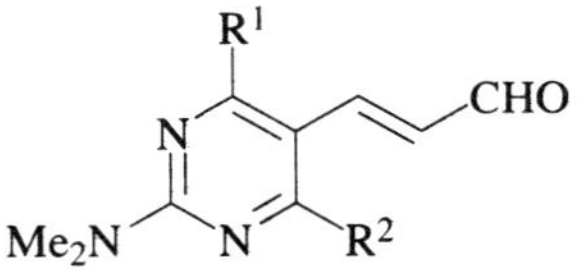

TABLE XVI. OTHER HETEROCYCLES WITH ONE FULLY CONJUGATED RING (*Continued*)

Substrate	Reagents	Product(s) and Yield(s) (%)	Refs.
		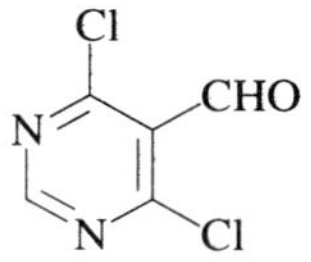 (48)	762
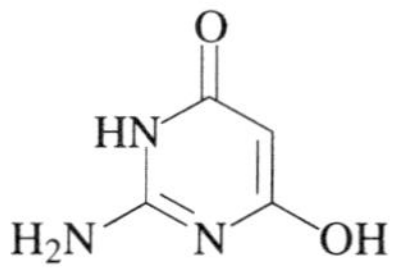	DMF, POCl$_3$	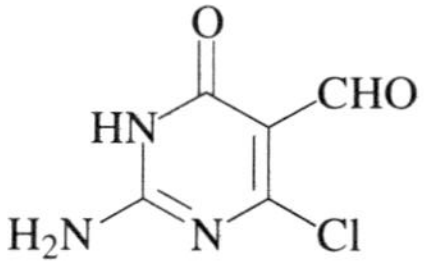 (28)	764
	DMF, POCl$_3$	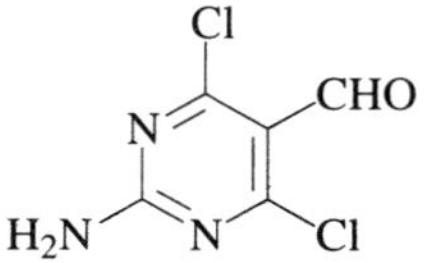 (28)	129, 762, 765
	DMF, POCl$_3$	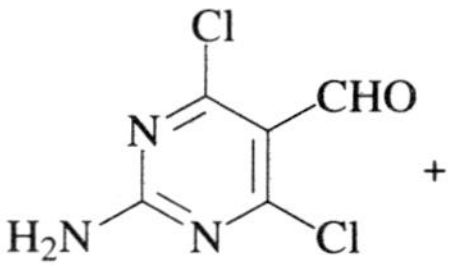 (51) 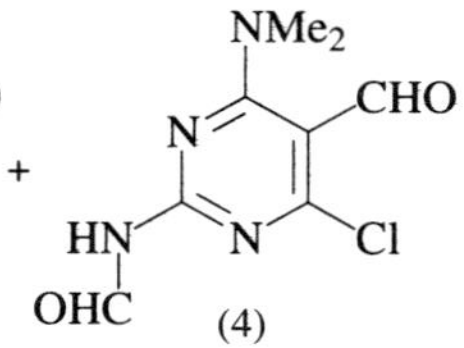 + (4)	766
	DMF, POCl$_3$	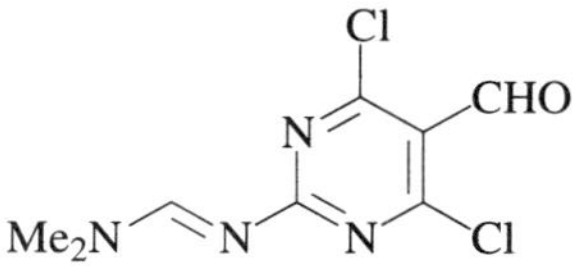 (65)	765, 129

TABLE XVI. OTHER HETEROCYCLES WITH ONE FULLY CONJUGATED RING (*Continued*)

Substrate	Reagents	Product(s) and Yield(s) (%)	Refs.
C$_{5-10}$ 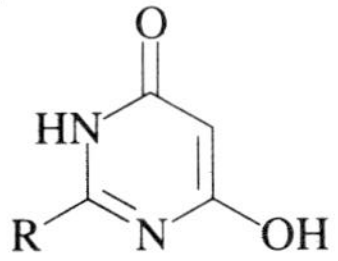	DMF, COCl$_2$	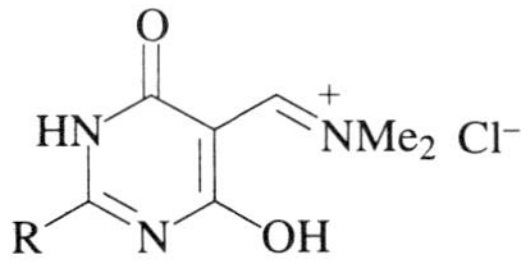 R = Me (98) R = Ph (93)	763
C$_{17-20}$ R: Me, Et, n-Pr, n-Bu	1. DMF, COCl$_2$ 2. H$_2$O	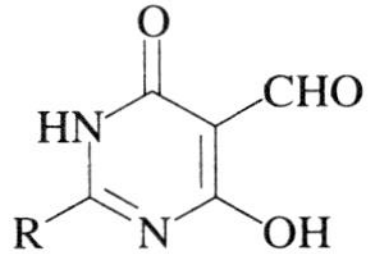 R = Me (78) R = Ph (60)	763
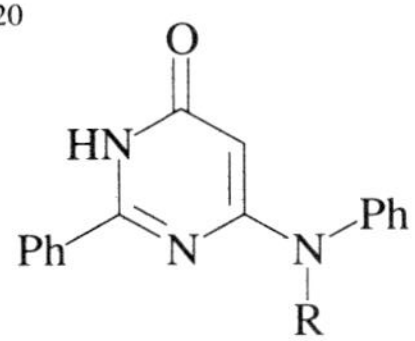	DMF, POCl$_3$	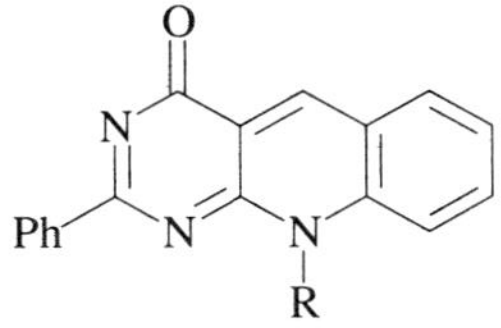 (96) (98) (93) (99)	130
Pyrimi-(1,3H)-2,6-diones			
C$_4$ 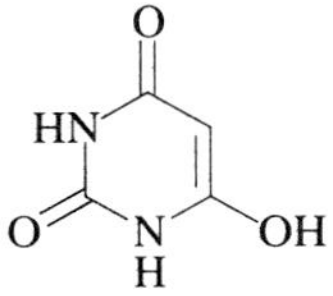	DMF, POCl$_3$	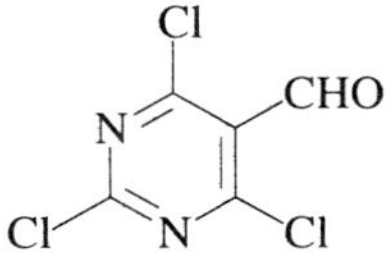 (—)	767, 768

TABLE XVI. OTHER HETEROCYCLES WITH ONE FULLY CONJUGATED RING (*Continued*)

Substrate	Reagents	Product(s) and Yield(s) (%)	Refs.
	DMF, POCl₃	(82)	769
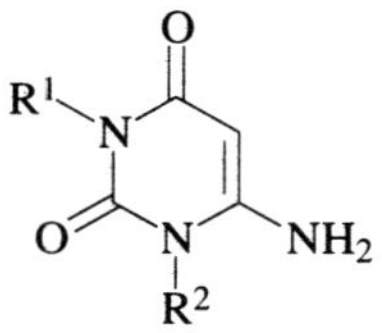 C₅₋₁₁	DMF, POCl₃	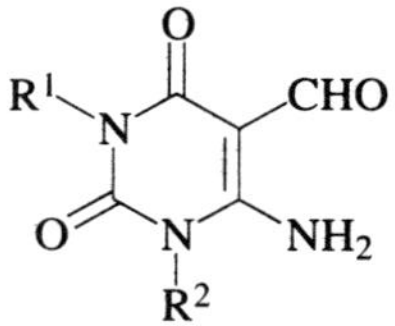 (24) **I** 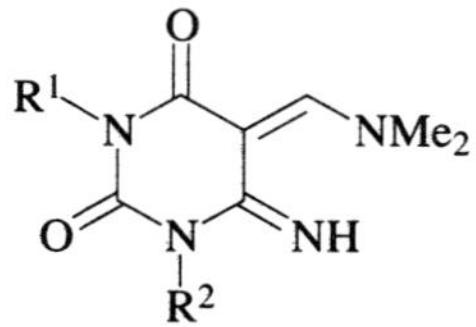 **II** 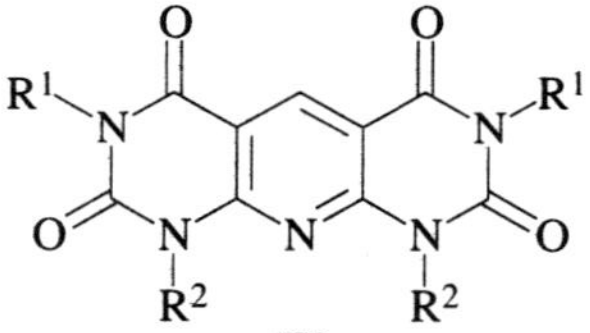 **III**	762

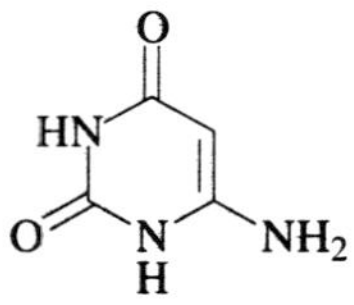

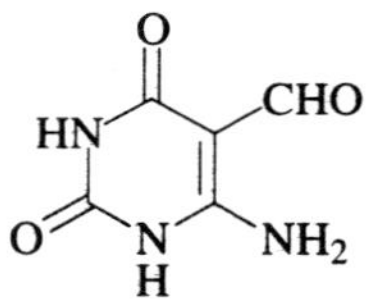

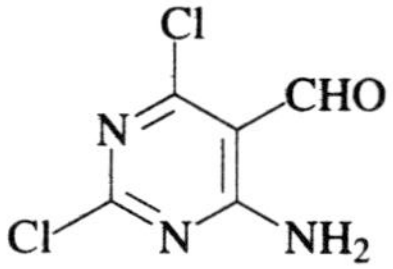

244

TABLE XVI. OTHER HETEROCYCLES WITH ONE FULLY CONJUGATED RING (Continued)

Substrate		Reagents	Product(s) and Yield(s) (%)	Refs.
R^1	R^2			
Me	H	DMF, POCl$_3$	I (82)	769
H	Me	DMF, POCl$_3$	I (81)	769
Me	Me	DMF, POCl$_3$	I (85) + III (6)	769
		DMF, —	II (95)	770
		DMF, POCl$_3$	II (95)	771
		DMF, COCl$_2$	II (95)	771
		DMF, SOCl$_2$	II (60)	771
		DMF, PhCOCl	II (80)	771
		DMF, Me$_2$SO$_4$	III (20–26)	771
Me	Ph	DMF, POCl$_3$	I (86) + III (7)	769

I

II

C$_{5\text{-}16}$

Substrate		Reagents	Product(s) and Yield(s) (%)	Refs.
R^1	R^2			
Me	H	DMF, POCl$_3$	I (68)	772
Me	Me	DMF, POCl$_3$ (xs)	I (30)	773
Me	Me	DMF, POCl$_3$	II (50)	773
H	n-Bu	DMF, POCl$_3$ (xs)	I (12)	773
H	Ph	DMF, POCl$_3$ (xs)	I (88)	773
H	C$_6$H$_{11}$	DMF, POCl$_3$ (xs)	I (45)	773
H	C$_6$H$_{11}$	DMF, POCl$_3$	II (45)	773
CH(Me)Ph	H	DMF, POCl$_3$	I (—) (R) and (S)	774
Ph	Ph	DMF, POCl$_3$ (xs)	I (45)	773
Ph	Ph	DMF, POCl$_3$	II (40)	773
C$_6$H$_{11}$	C$_6$H$_{11}$	DMF, POCl$_3$ (xs)	I (56)	773
C$_6$H$_{11}$	C$_6$H$_{11}$	DMF, POCl$_3$	II (73)	773

245

TABLE XVI. OTHER HETEROCYCLES WITH ONE FULLY CONJUGATED RING (*Continued*)

Substrate	Reagents	Product(s) and Yield(s) (%)	Refs.
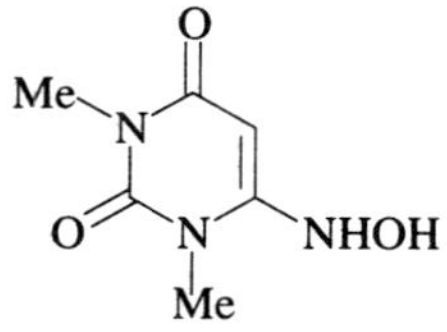 C₆ (structure)	DMF, POCl₃	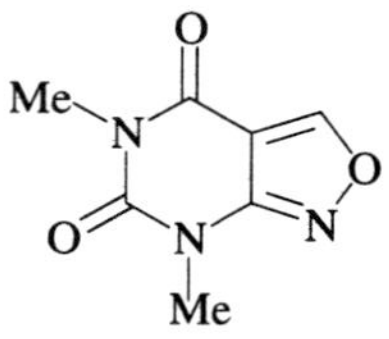 (59) (structure)	775
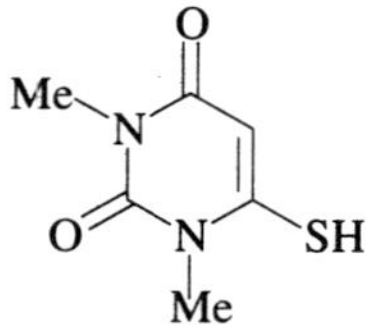 (structure)	DMF, POCl₃	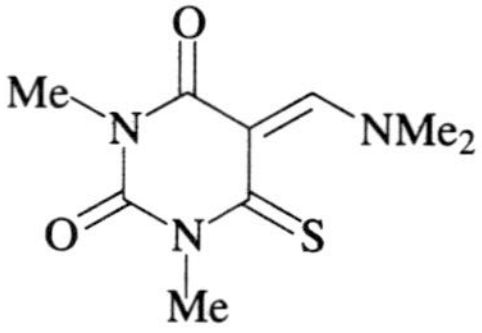 (40) (structure)	776
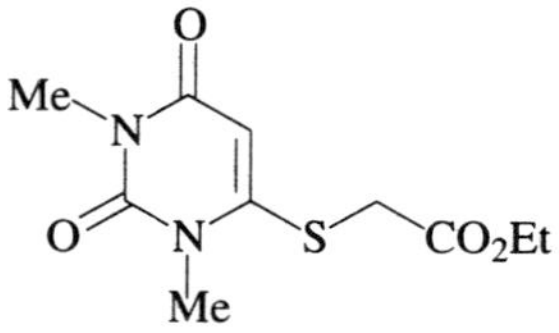 C₁₀ (structure)	DMF, POCl₃	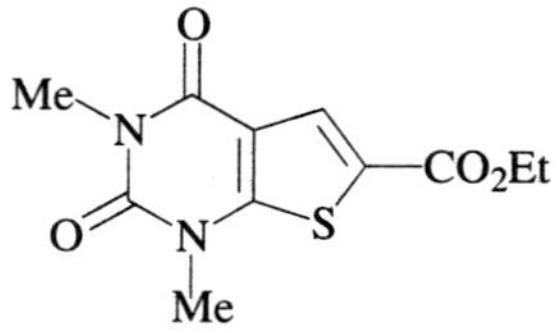 (67) (structure)	429
C₆₋₁₂ (structure)		(structure)	

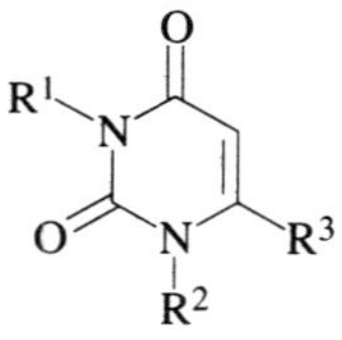

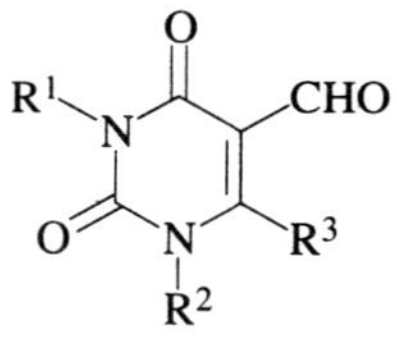

TABLE XVI. OTHER HETEROCYCLES WITH ONE FULLY CONJUGATED RING (*Continued*)

Substrate	Reagents	Product(s) and Yield(s) (%)	Refs.

Substrate (pyrimidine with R[1], R[2], R[3]):

R[1]	R[2]	R[3]			
Me	Me	H	DMF, POCl$_3$	(93)	777
Me	Me	Me	DMF, POCl$_3$ (xs)	(44)	773
Me	C$_6$H$_{11}$	H	DMF, POCl$_3$ (xs)	(60)	773
Me	Ph	Me	DMF, POCl$_3$ (xs)	(70)	773
C$_6$H$_{11}$	Me	Me	DMF, POCl$_3$ (xs)	(65)	773

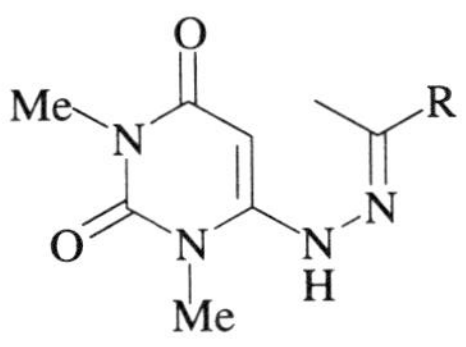

R			
Me	Me$_2$NNO, POCl$_3$	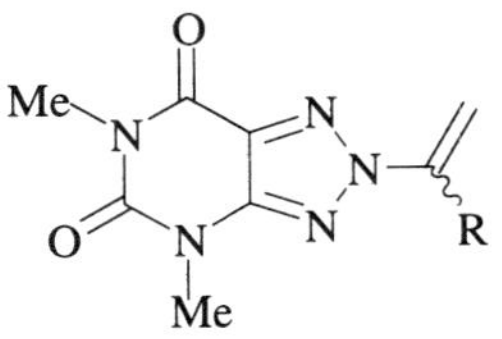 (57)	778
4-ClC$_6$H$_4$		(31)	
Ph		(65)	
4-MeC$_6$H$_4$		(52)	

Product: (57)

C$_8$

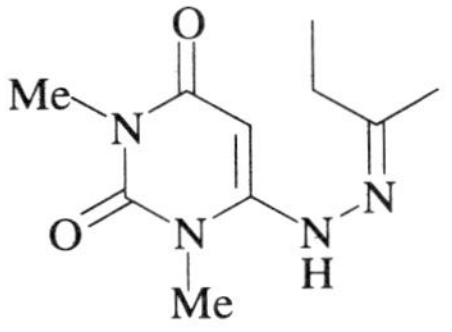

	Me$_2$NNO, POCl$_3$	(44)	778

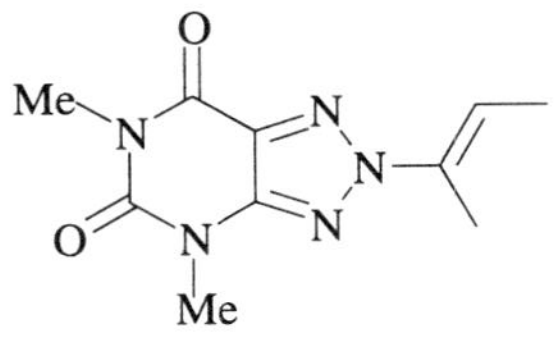

247

TABLE XVI. OTHER HETEROCYCLES WITH ONE FULLY CONJUGATED RING (*Continued*)

Substrate	Reagents	Product(s) and Yield(s) (%)	Refs.

C_{10-17}

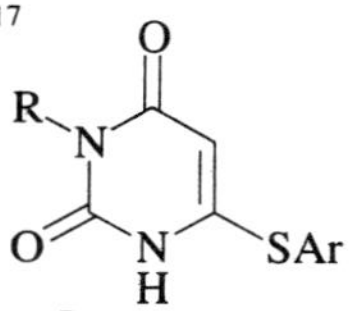

DMF, POCl$_3$

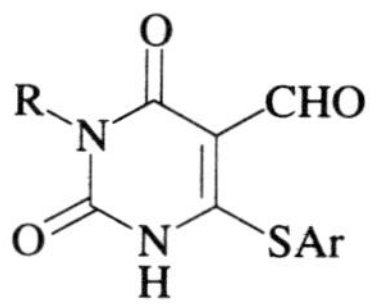

R	Ar	
H	4-ClC$_6$H$_4$	(96)
H	Ph	(86)
H	4-MeC$_6$H$_4$	(94)
Me	4-ClC$_6$H$_4$	(94)
Me	Ph	(93)
Me	2-MeC$_6$H$_4$	(75)
Me	3-MeC$_6$H$_4$	(70)
Me	4-MeC$_6$H$_4$	(92)
Ph	4-ClC$_6$H$_4$	(99)
Ph	Ph	(67)
Ph	4-MeC$_6$H$_4$	(76)

779

C_{11-12}

DMF, POCl$_3$

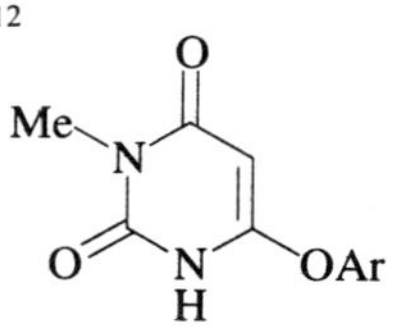

Ar	
3-ClC$_6$H$_4$	(79)
4-ClC$_6$H$_4$	(90)
Ph	(84)
3-MeC$_6$H$_4$	(89)
4-MeC$_6$H$_4$	(81)

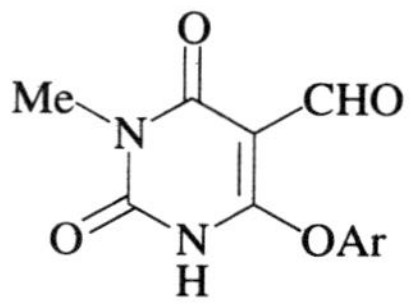

780

248

TABLE XVI. OTHER HETEROCYCLES WITH ONE FULLY CONJUGATED RING (*Continued*)

Substrate	Reagents	Product(s) and Yield(s) (%)	Refs.
C$_{11-13}$	DMF, POCl$_3$		781

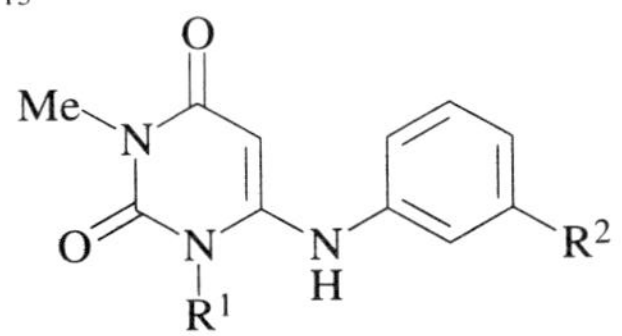

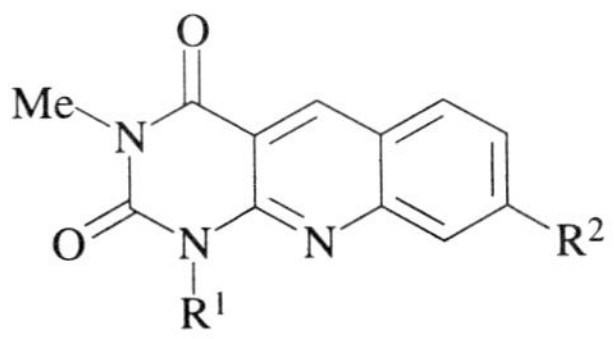

R^1	R^2		Yield
H	OH		(96)
H	Me		(88)
H	OMe		(90)
Me	OH		(78)
Me	OMe		(66)

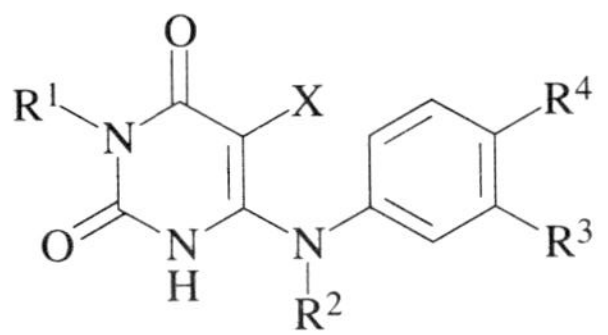

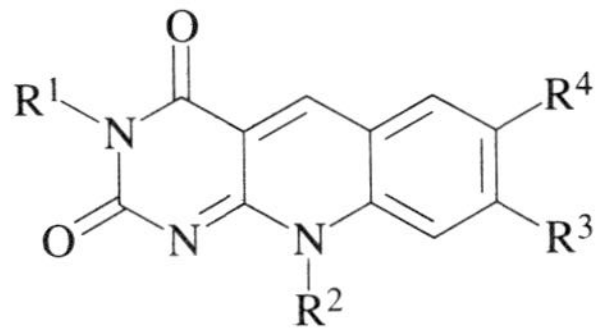

C$_{11-18}$

R^1	R^2	R^3	R^4	X	Reagents	Yield	Refs.
H	Me	H	H	H	DMF, POCl$_3$	(95)	772
H	H	Me	Me	H	DMF, POCl$_3$	(85)	782
H	Et	H	H	H	DMF, POCl$_3$	(87)	772
H	Me	Me	Me	H	DMF, POCl$_3$	(80)	782
H	n-Pr	H	H	H	DMF, POCl$_3$	(90)	772
H	n-Bu	H	H	H	DMF, POCl$_3$	(92)	772
Me	Me	H	H	H	DMF, POCl$_3$	(96)	772
Me	Et	H	H	H	DMF, POCl$_3$	(88)	772

249

TABLE XVI. OTHER HETEROCYCLES WITH ONE FULLY CONJUGATED RING (Continued)

Substrate					Reagents	Product(s) and Yield(s) (%)	Refs.
R^1	R^2	R^3	R^4	X			
Me	Et	H	H	NO_2	DMF, POCl₃, 130°	(63)	783
Me	Me	Me	Me	H	DMF, POCl₃	(89)	784
Me	n-Pr	H	H	H	DMF, POCl₃	(86)	772
Me	n-Pr	H	H	NO_2	DMF, POBr₃, 130°	(81)	783
Me	n-Bu	H	H	H	DMF, POCl₃	(89)	772
Me	n-Bu	H	H	NO_2	DMF, POBr₃, 130°	(72)	783
H	ribityl	Me	Me	H	1. Ac₂O 2. DMF, POCl₃	R^2 = tetraacetylribityl (91)	785

C_{12-17}

R¹	R²	R³	X	Reagents		Refs.
H	Et	H	H	Me₂NNO, POCl₃	(70)	786
Me	Me	Br	H	Me₂NNO, POCl₃	(60)	786
Me	Me	H	H	Me₂NNO, POCl₃	(68)	786
Me	Et	H	H	Me₂NNO, POCl₃	(65)	786
Me	Et	Br	H	Me₂NNO, POCl₃	(67)	786
Me	Et	Me	H	Me₂NNO, POCl₃	(70)	786
Me	Me	H	NO_2	DMF, POCl₃	(73)	787
Me	Me	H	NO_2	DMF, POBr₃, 80°	(87)	783
Me	Et	H	NO_2	DMF, POCl₃	(71)	787
Me	Et	H	NO_2	DMF, POBr₃, 80°	(73)	783
Me	n-Pr	H	NO_2	DMF, POCl₃	(72)	787
Me	n-Pr	H	NO_2	DMF, POBr₃, 80°	(55)	783

250

TABLE XVI. OTHER HETEROCYCLES WITH ONE FULLY CONJUGATED RING (*Continued*)

Substrate				Reagents	Product(s) and Yield(s) (%)	Refs.
R^1	R^2	R^3	X			
C$_{13}$						
Me	n-Bu	H	NO_2	DMF, $POCl_3$	(65)	787
Me	n-Bu	H	NO_2	DMF, $POBr_3$, 80°	(68)	783
Me	Et	Me	NO_2	DMF, $POCl_3$	(82)	787
Me	Ph	H	NO_2	DMF, $POCl_3$	(88)	787

C$_{13}$

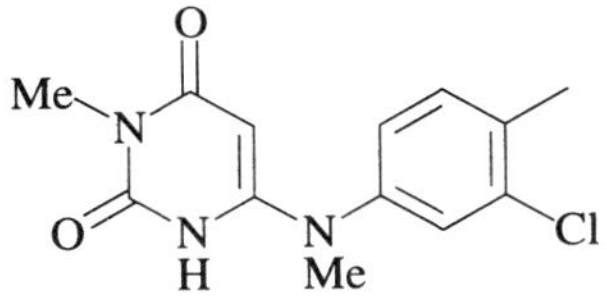

Substrate	Reagents	Product(s) and Yield(s) (%)	Refs.
(structure)	DMF, $POCl_3$	(48) + (29)	784
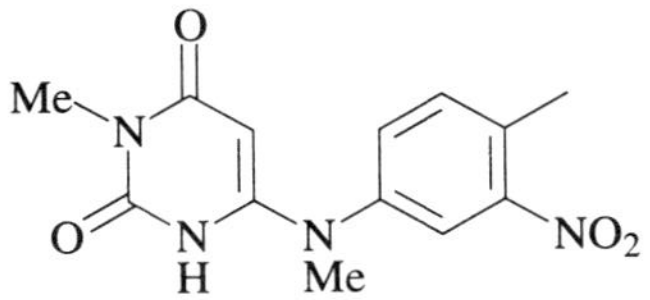 (structure)	DMF, $POCl_3$	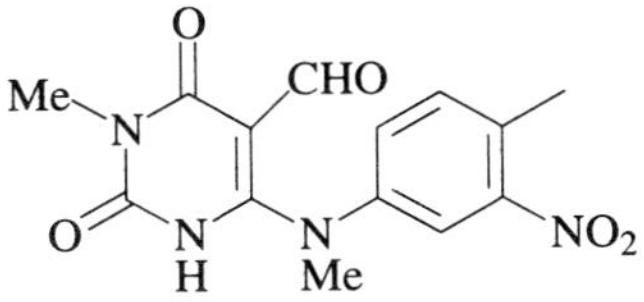 (83)	784

C$_{13-14}$

Substrate	Reagents	Product(s) and Yield(s) (%)	Refs.
(structure)	DMF, $POCl_3$	R = H (82); R = Me (56)	781

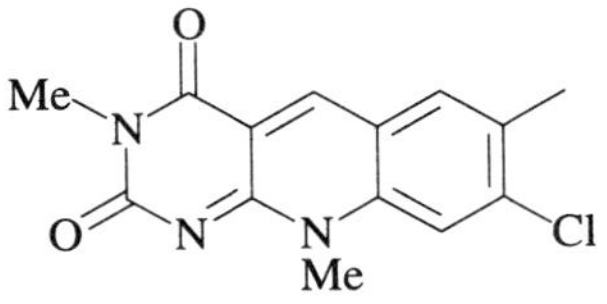

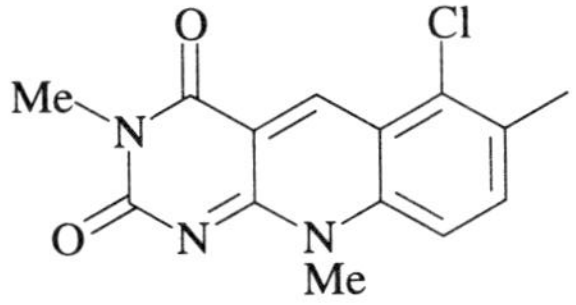

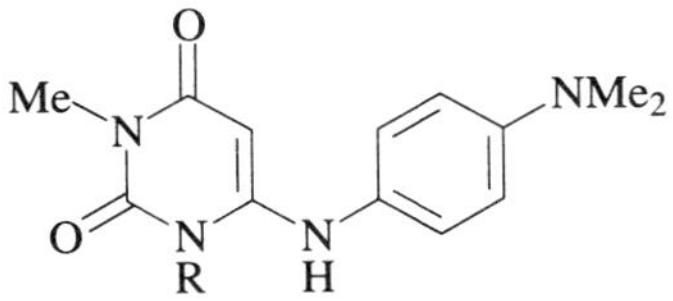

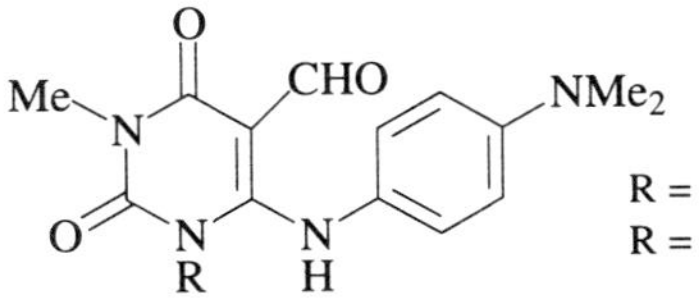

TABLE XVI. OTHER HETEROCYCLES WITH ONE FULLY CONJUGATED RING (*Continued*)

Substrate	Reagents	Product(s) and Yield(s) (%)	Refs.
	DMF, POCl$_3$	I, X = H II, X = CHO	781
C$_{14-20}$ n = 4, 6, 8, 10	DMF, POCl$_3$	I (74) + II (11) I (77) + II (14) (—)	788
C$_{20-40}$		I +	789

252

TABLE XVI. OTHER HETEROCYCLES WITH ONE FULLY CONJUGATED RING (*Continued*)

Substrate	Reagents	Product(s) and Yield(s) (%)	Refs.

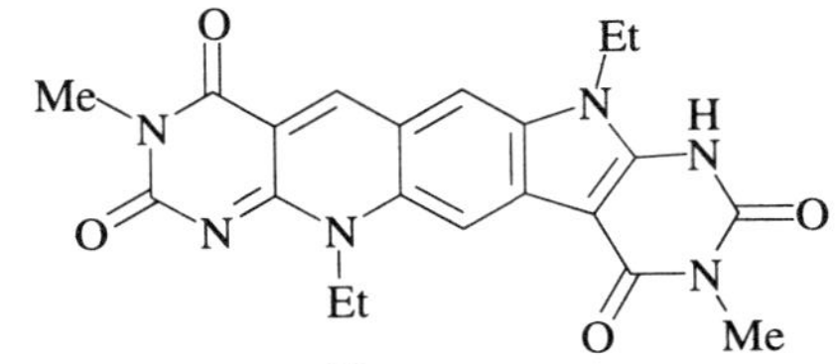

II

R			
Et	DMF, POCl$_3$ (7:1)	I (10)	
Et	DMF, POCl$_3$ (1:12)	I (14) + II (7)	
n-C$_8$H$_{17}$	DMF, POCl$_3$ (7:1)	I (20)	
n-C$_{12}$H$_{25}$	DMF, POCl$_3$ (7:1)	I (28)	

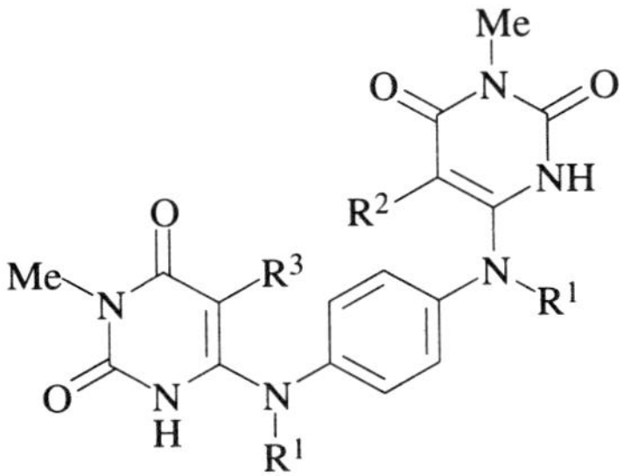

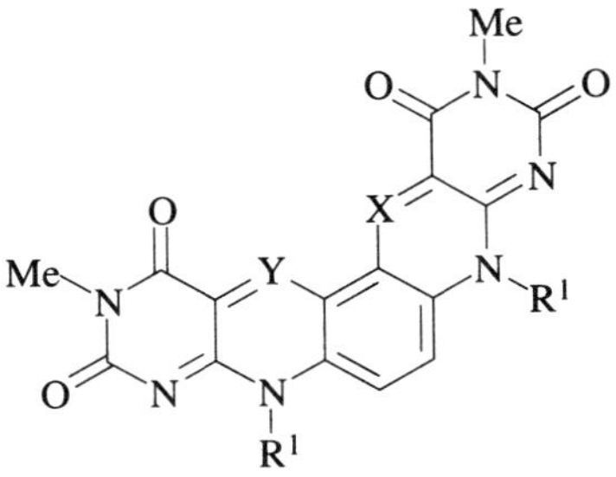

I, X = Y = CH
II, X = Y = N
III, X = CH, Y = N

R^1	R^2	R^3		
Et	H	H	DMF, POCl$_3$	I (34)	790, 791
Et	NO$_2$	NO$_2$	DMF, POCl$_3$ (1:1)	II (7) + III (2)	789
Et	NO$_2$	H	DMF, POCl$_3$ (5:1)	II (31) + III (3)	789
n-C$_8$H$_{17}$	NO$_2$	H	DMF, POCl$_3$ (5:1)	II (6) + III (2)	789
n-C$_8$H$_{17}$	H	H	DMF, POCl$_3$	I (36)	790, 791
n-C$_{12}$H$_{25}$	NO$_2$	H	DMF, POCl$_3$ (5:1)	II (4) + III (1)	789
n-C$_{12}$H$_{25}$	H	H	DMF, POCl$_3$	I (35)	790, 791

253

TABLE XVI. OTHER HETEROCYCLES WITH ONE FULLY CONJUGATED RING (*Continued*)

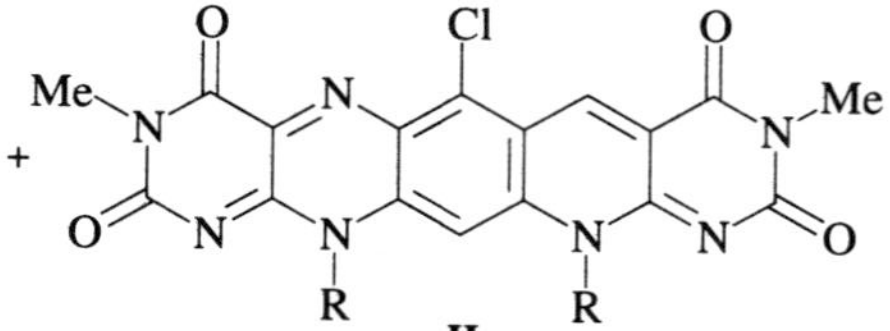

Substrate	Reagents	Product(s) and Yield(s) (%)	Refs.

C_{31-50}

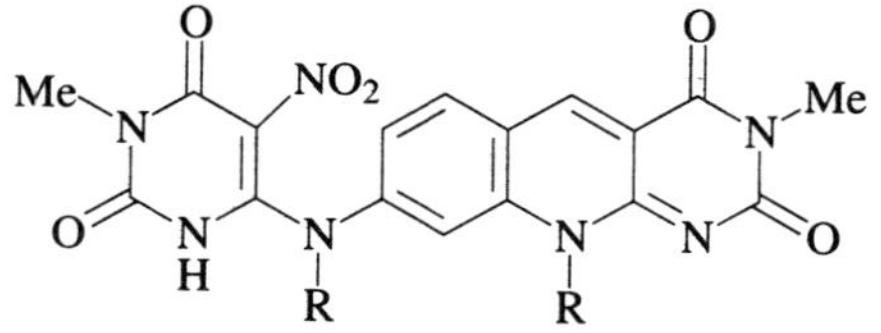

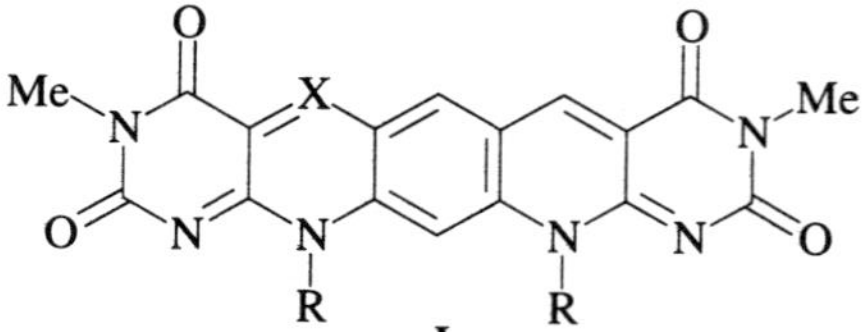

792

R	
n-C_8H_{17}	DMF, POCl$_3$ (5:1)
n-$C_{12}H_{25}$	DMF, POCl$_3$ (5:1)
	DMF, POCl$_3$ (1:1)
n-$C_{18}H_{37}$	DMF, POCl$_3$ (5:1)

I, X = N (7) + **I**, X = CH (2-3)
I, X = N (12) + **I**, X = CH (2-3)
I, X = CH (3) + **II** (15)
I, X = N (12) + **I**, X = CH (2-3)

C_4NS
1,3-Thiazin-5-ones

C_{10-12}

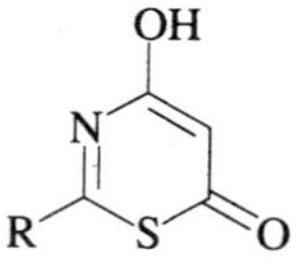

DMF, POCl$_3$

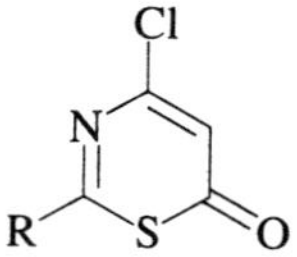

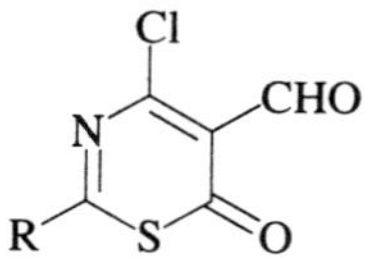

793

TABLE XVI. OTHER HETEROCYCLES WITH ONE FULLY CONJUGATED RING (*Continued*)

Substrate	Reagents	Product(s) and Yield(s) (%)		Refs.

R

		I	**II**	
4-BrC$_6$H$_4$		(56)	(32)	
4-O$_2$NC$_6$H$_4$		(62)	(16)	
4-MeOC$_6$H$_4$		(64)	(5)	
3,4-(MeO)$_2$C$_6$H$_4$		(21)	(59)	
4-Me$_2$NC$_6$H$_4$		(10)	(73)	

C$_{16}$

C$_4$S$_2$ DMF, POCl$_3$ (32)   443

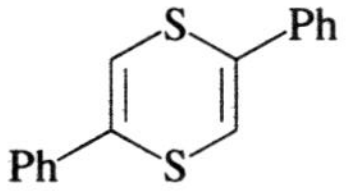

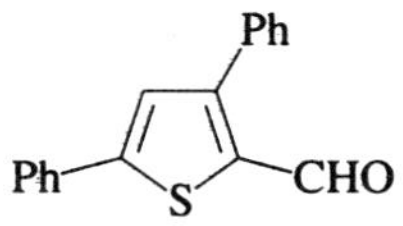

C$_{14}$

C$_5$B   MFA, POCl$_3$ (65)   794

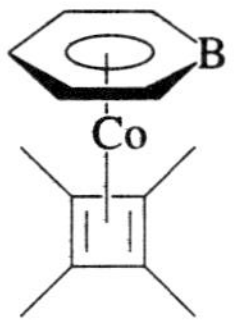

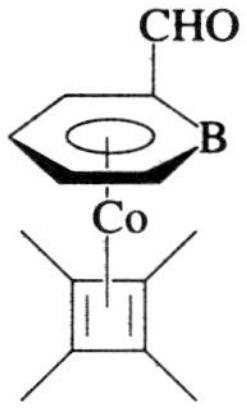

TABLE XVI. OTHER HETEROCYCLES WITH ONE FULLY CONJUGATED RING (*Continued*)

Substrate	Reagents	Product(s) and Yield(s) (%)	Refs.

Pyridines

C_{14}

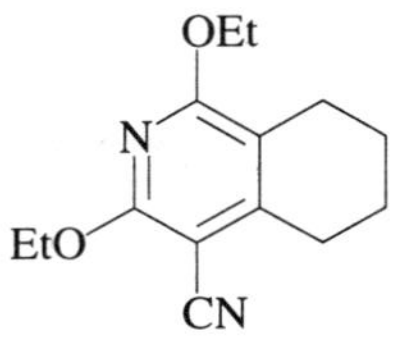

DMF, POCl$_3$

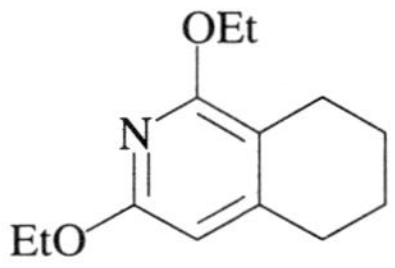

(0-80)

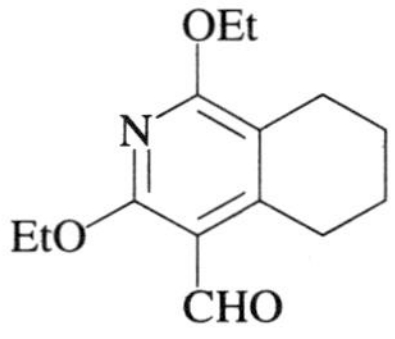

(5-82)

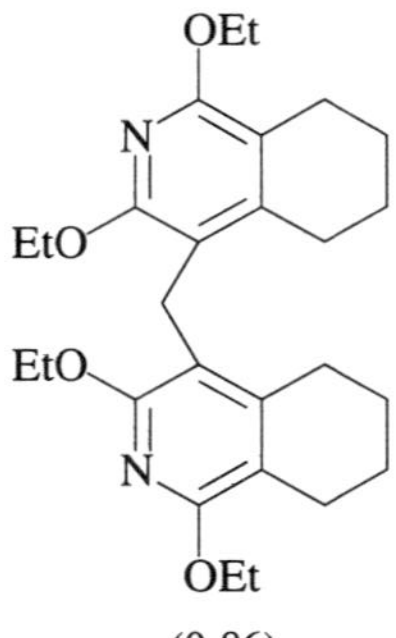

(0-86)

795

Pyrid-2-ones

C_6

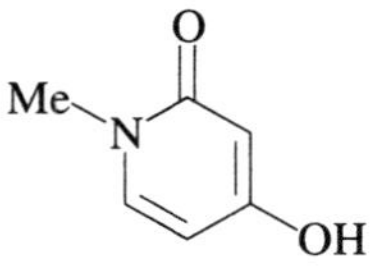

DMF, POCl$_3$

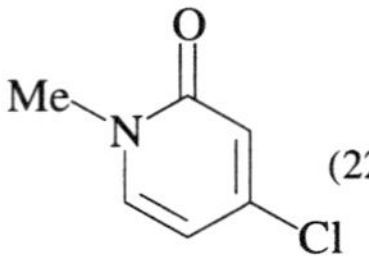

(22) +

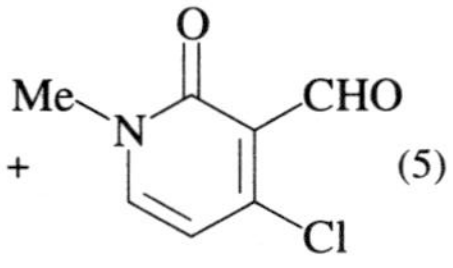

(5)

128

C_{7-13}

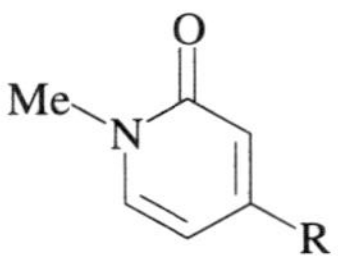

DMF, POCl$_3$

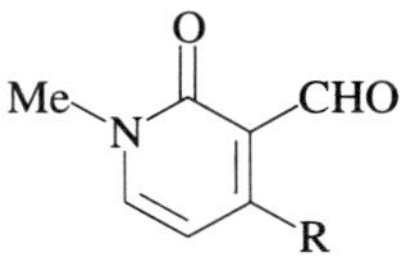

TABLE XVI. OTHER HETEROCYCLES WITH ONE FULLY CONJUGATED RING (*Continued*)

Substrate	Reagents	Product(s) and Yield(s) (%)	Refs.

C_{11-14}

R^1	R^2	R^3	R^4
H	H	H	H
Me	Cl	H	H
Me	H	OMe	H
Me	Me	H	Me

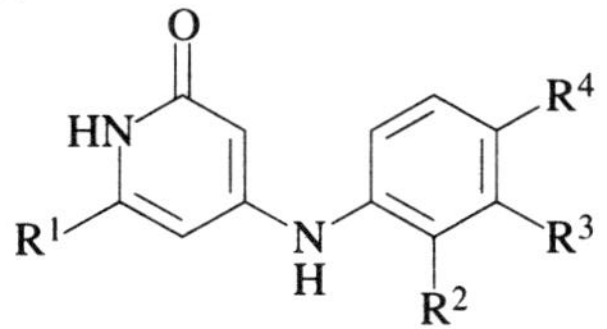

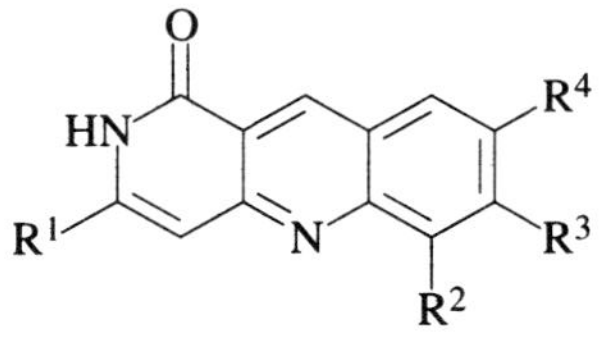

DMF, POCl$_3$, 90°

R
OMe
OEt
OBn

(73) 128, 796
(73) 128
(67) 128

(80)
(69)
(66)
(83)

797

C_{12-13}

R
Bn
2,5-Me$_2$C$_6$H$_3$

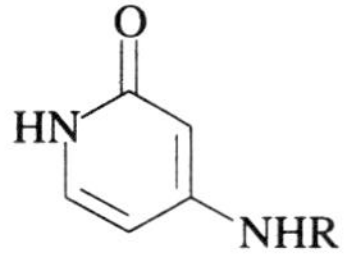

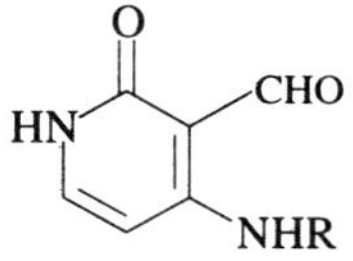

DMF, POCl$_3$

(29)
(31)

797

TABLE XVI. OTHER HETEROCYCLES WITH ONE FULLY CONJUGATED RING (*Continued*)

Substrate	Reagents	Product(s) and Yield(s) (%)	Refs.
C$_{16}$ 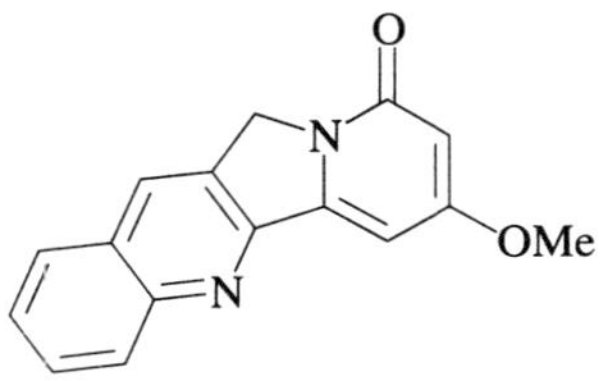	DMF, POCl$_3$	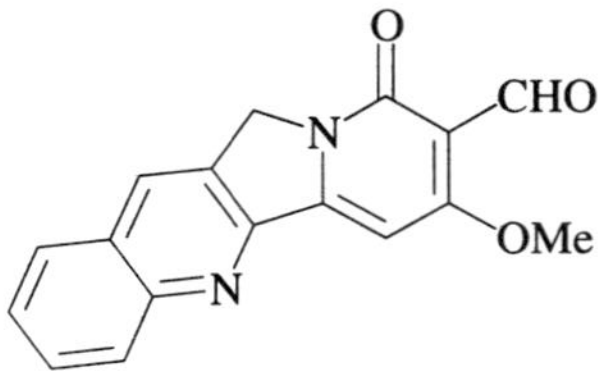 (82)	798
C$_{17}$ 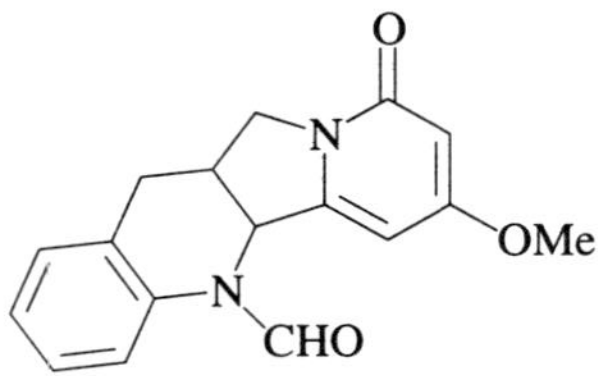	DMF, POCl$_3$	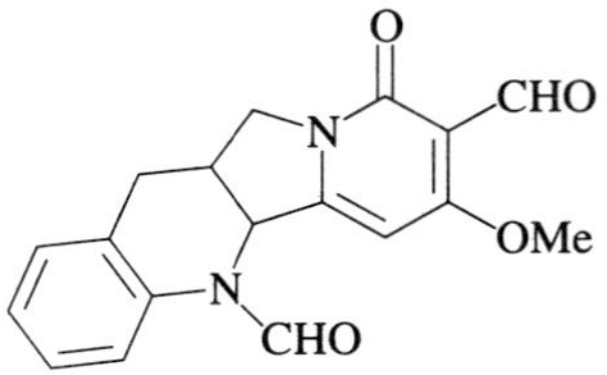 (75)	796, 798
C$_{18}$ 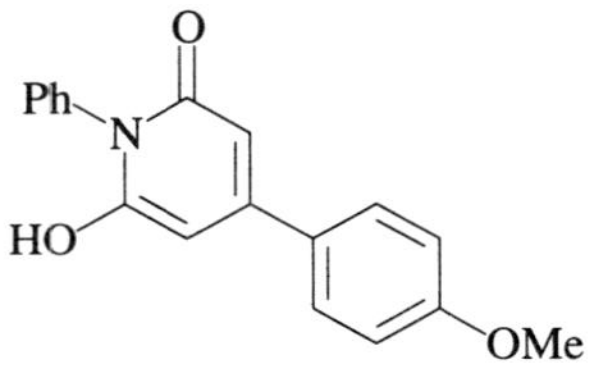	DMF, POCl$_3$	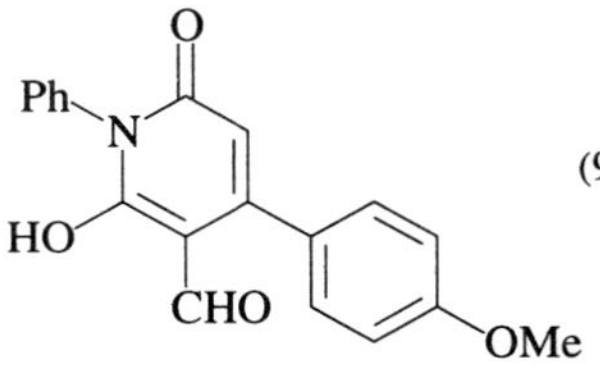 (99)	799

258

TABLE XVII. OTHER HETEROCYCLES WITH TWO FULLY CONJUGATED RINGS

Substrate	Reagents	Product(s) and Yield(s) (%)	Refs.
	C_2N_2S/C_3S_2		
$C_{10\text{-}14}$ 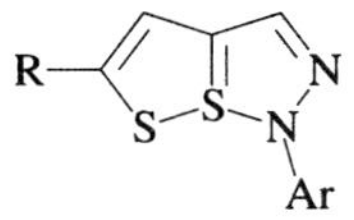	DMF, $POCl_3$	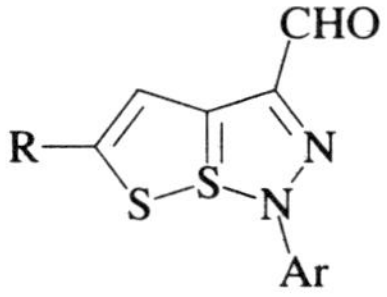	800
R: H, Ar: Ph		(15)	
R: *t*-Bu, Ar: 4-$O_2NC_6H_4$		(18)	
R: *t*-Bu, Ar: Ph		(57)	
	C_3N_2/C_3NS		
$C_{5\text{-}12}$ 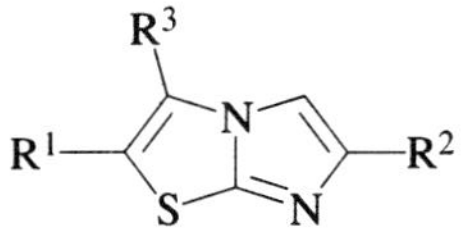		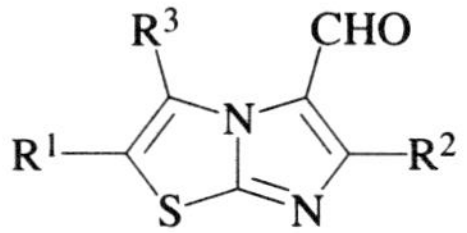	
R^1: H, R^2: Br, R^3: H	DMF, $POBr_3$	(73)	800
R^1: H, R^2: Cl, R^3: H	DMF, $POCl_3$	(57)	802, 801
R^1: Me, R^2: Cl, R^3: H	DMF, $POCl_3$	(60)	803
R^1: H, R^2: Me, R^3: H	DMF, $POCl_3$	(60)	137
R^1: Me, R^2: Me, R^3: H	DMF, $POCl_3$	(65)	805, 804
R^1: H, R^2: 4-ClC_6H_4, R^3: H	DMF, $POCl_3$	(60)	803
R^1: H, R^2: Ph, R^3: H	DMF, $POCl_3$	(81)	137, 805
R^1: H, R^2: Ph, R^3: Me	DMF, $POCl_3$	(65)	209

TABLE XVII. OTHER HETEROCYCLES WITH TWO FULLY CONJUGATED RINGS (*Continued*)

Substrate	Reagents	Product(s) and Yield(s) (%)	Refs.
		C$_3$N$_2$/C$_4$S, Se	
C$_7$ 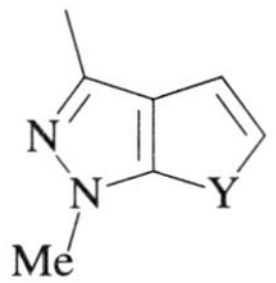	DMF, POCl$_3$	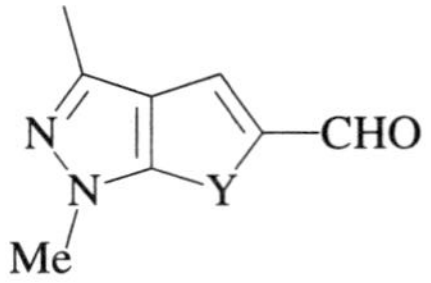 Y = S (86) Y = Se (88)	806
		C$_3$S$_2$/C$_3$S$_2$	
C$_{5-18}$ 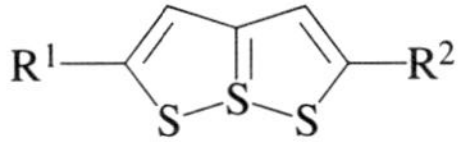		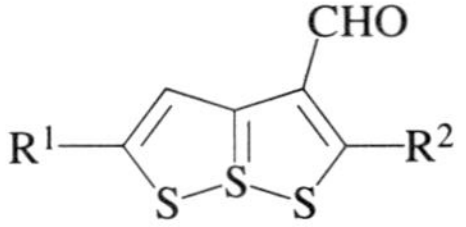	
	Me$_2$NCHS, POCl$_3$	(21)	138
	Me$_2$NCHS, POCl$_3$	(21)	138
	Me$_2$NCHS, POCl$_3$	(77)	138
	Me$_2$NCHS, POCl$_3$	(77)	138
	DMF, POCl$_3$	(50)	807

R^1	R^2
H	H
D	D
t-Bu	H
t-Bu	D
Ph	4-MeOC$_6$H$_4$

		C$_3$NS/C$_4$N	
C$_{5-12}$ 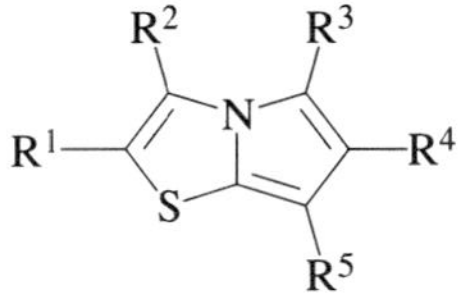		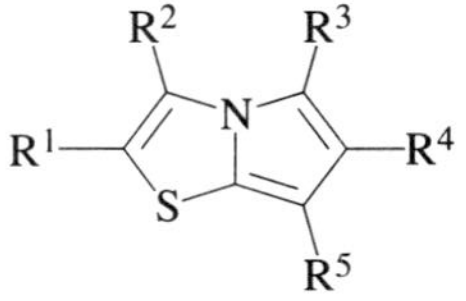	

TABLE XVII. OTHER HETEROCYCLES WITH TWO FULLY CONJUGATED RINGS (*Continued*)

R^1	R^2	R^3	R^4	R^5	Reagents	Product(s) and Yield(s) (%) $R^1 - R^5$ as in substrate except as indicated	Refs.
H	H	H	Me	H	DMF, POCl$_3$	R^3 = CHO (88)	136, 143
					DMF, POCl$_3$ (2.7 eq)	R^3 = CHO (95) + R^5 = CHO (5)	136
					DMF, POCl$_3$ (6 eq)	R^3 = CHO (70) + R^5 = CHO (30)	136
					1. DMF, POCl$_3$ 2. H$_2$S or NaSH, H$_2$O	R^3 = CHS (89)	632, 808
					1. Me$_2$NCDO, POCl$_3$ 2. H$_2$S, H$_2$O	R^3 = CDS (89)	808
Me	H	H	Me	H	DMF, POCl$_3$	R^3 = CHO (43)	143
					1. DMF, POCl$_3$ 2. H$_2$S, H$_2$O	R^3 = CHS (89)	808
H	Me	H	Me	H	DMF, POCl$_3$	R^3 = CHO (80)	809
					1. DMF, POCl$_3$ 2. H$_2$S, H$_2$O	R^3 = CHS (49) + R^5 = CHS (5)	808
H	H	Me	Me	H	DMF, POCl$_3$	R^3 = CHO (54)	136, 143
					1. DMF, POCl$_3$ 2. H$_2$S, H$_2$O	R^3 = CHS (65)	808
H	H	H	Me	Me	DMF, POCl$_3$	R^3 = CHO (84)	136, 143
					1. DMF, POCl$_3$ 2. H$_2$S, H$_2$O	R^3 = CHS (71)	808
Me	Me	H	Me	H	1. DMF, POCl$_3$ 2. H$_2$S, H$_2$O	R^3 = CHS (77)	808
					1. Me$_2$NCDO, POCl$_3$ 2. H$_2$S, H$_2$O	R^3 = CDS (83)	808
H	Me	Me	Me	H	1. DMF, POCl$_3$ 2. H$_2$S, H$_2$O	R^5 = CHS (92)	808
Me	Me	Me	Me	H	1. DMF, POCl$_3$ 2. H$_2$S, H$_2$O	R^5 = CHS (75)	808

TABLE XVII. OTHER HETEROCYCLES WITH TWO FULLY CONJUGATED RINGS (Continued)

	Substrate					Reagents	Product(s) and Yield(s) (%)	Refs.
R^1	R^2	R^3	R^4	R^5				

C_{14}

R^1	R^2	R^3	R^4	R^5	Reagents	Product	Ref.
H	Me	H	t-Bu	H	1. DMF, $POCl_3$ 2. H_2S, H_2O	R^3 = CHS (12) + R^5 = CHS (17)	808
H	Me	H	4-BrC_6H_4	H	DMF, $POCl_3$	R^3 = CHO (78)	809
H	4-ClC_6H_4	H	Me	H	DMF, $POCl_3$	R^3 = CHO (78) + R^5 = CHO (12)	809

C_4N/C_4N

C_{14}

DMF, $POCl_3$ — CHO (90) — 810

C_4N/C_4O

$C_{9\text{-}16}$

DMF, $POCl_3$ — 135

R^1	R^2	R^3	R^4	R^5		Product(s)
H	H	H	H	CO_2Et		R^2 = CHO (58)
H	H	Me	H	CO_2Et		R^2 = CHO (71)
Ph	H	COMe	H	H		R^4 = CHO (71)
Ph	H	H	H	CO_2Et	8 h	R^3 = CHO (44)
					60 h	R^3 = CHO (43)
Ph	H	Me	H	CO_2Et		R^5 = CHO (67)
4-MeC_6H_4	H	COMe	H	H		R^4 = CHO (72)
4-MeC_6H_4	H	H	H	CO_2Et		R^3 = CHO (54)
4-MeC_6H_4	H	Me	H	CO_2Et		R^5 = CHO (62)

R^1 - R^5 as in substrate except as indicated

TABLE XVII. OTHER HETEROCYCLES WITH TWO FULLY CONJUGATED RINGS (*Continued*)

Substrate	Reagents	Product(s) and Yield(s) (%)	Refs.

C₄N/C₄S

C_{6-9}

Substrate structure with R^2, R^3, R^4, N–H, S, R^1

R^1	R^2	R^3	R^4
H	H	H	H
H	H	H	CO_2Et

Reagents: DMF, POCl₃

Product(s):
R^1 - R^4 as in substrate except as indicated
R^1 = CHO (1) + R^3 = CHO (3) + R^4 = CHO (51)
R^1 = CHO (78) + R^3 = CHO (3)

Refs.: 134

C₄S/C₄S

C_{6-8}

Substrate structure with R^2, R^3, S, S, R^1, R^4

R^1	R^2	R^3	R^4
H	H	H	H
CO_2Me	H	H	H

Reagents: DMF, POCl₃

Product(s):
R^1 - R^4 as in substrate except as indicated
R^3 = CHO (39) + R^4 = CHO (17)
R^3 = CHO (20) + R^4 = CHO (20)

Refs.: 131

C_{18-22}

Substrate structure with R^1, R^2, S, S, R^3, R^4

R^1	R^2	R^3	R^4
i-PrS	*i*-PrS	*i*-PrS	*i*-PrS
i-PrS	*i*-PrS	*t*-BuS	*t*-BuS
t-BuS	*t*-BuS	*t*-BuS	*t*-BuS

Reagents:
DMF, POCl₃
DMF, POCl₃ (xs)
DMF, POCl₃

Product(s):
R^1 - R^4 as in substrate except as indicated
R^1 = CHO (33) + R^1 = R^2 = CHO (14)
R^1 = R^2 = CHO (51)
R^1 = CHO (40)

Refs.:
811, 812
811
812

263

TABLE XVII. OTHER HETEROCYCLES WITH TWO FULLY CONJUGATED RINGS (*Continued*)

Substrate	Reagents	Product(s) and Yield(s) (%)	Refs.
		C₄S/C₄Se	
C_6	MFA, POCl₃	(42) + (28)	133
C_6		**C₄Se/C₄Se**	
	DMF, POCl₃	(77)	132
	DMF, POCl₃	(37) + (31)	132
	DMF, POCl₃	(96)	132
C_6	—	**C₂N₃/C₄N₂** (85)	813
C_7	DMF, POCl₃	**C₂N₃/C₅N** (8)	146

TABLE XVII. OTHER HETEROCYCLES WITH TWO FULLY CONJUGATED RINGS (*Continued*)

Substrate	Reagents	Product(s) and Yield(s) (%)	Refs.
		C_3N_2/C_5N	
C_{6-7}	DMF, $POCl_3$	**I** + **II**	144
R: H		**I** (20) + **II** (8)	
R: Me		**II** (68)	
C_{13-14}	DMF, $POCl_3$		
R^1 = H, R^2 = Ph		(—)	814
R^1 = Me, R^2 = 2-$O_2NC_6H_4$		(80)	815
		C_4N/C_4N_2	
C_6 (from)	1. DMF, $POCl_3$ 2. $NaHCO_3$	(70)	816, 817
	DMF, $POCl_3$	Cl^-	816

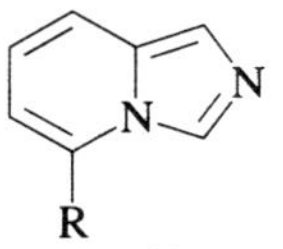

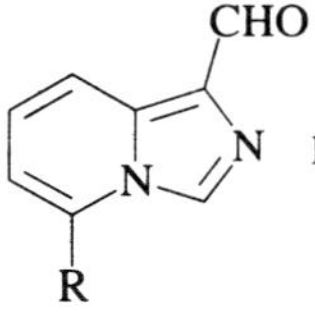

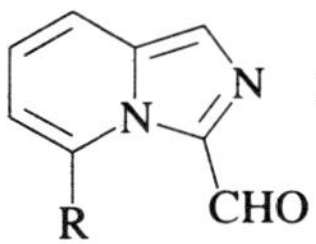

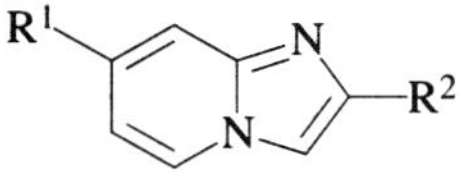

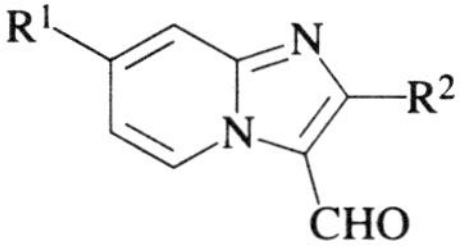

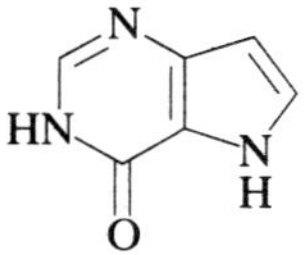

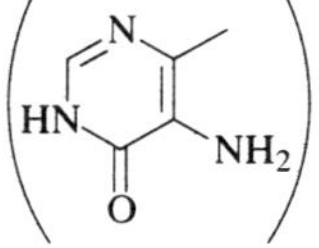

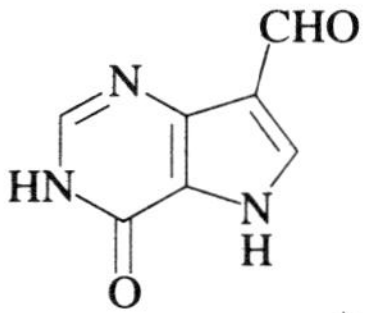

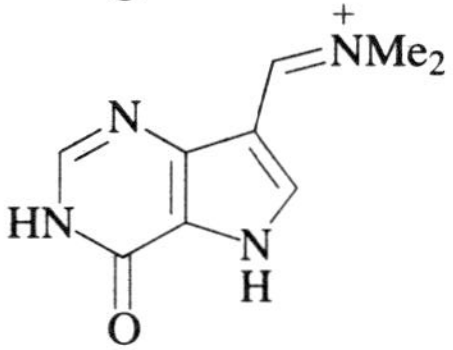

TABLE XVII. OTHER HETEROCYCLES WITH TWO FULLY CONJUGATED RINGS (*Continued*)

C$_{8-14}$

Substrate				Reagents	Product(s) and Yield(s) (%)	Refs.
R^1	R^2	R^3	R^4		R^1 - R^4 as in substrate except as indicated	
Me	Me	H	H	DMF, POCl$_3$	R^4 = CH=NMe$_2^+$ Cl$^-$ (90)	818, 819
				1. DMF, POCl$_3$ 2. NaOH	R^4 = CHO (95)	818
H	Et	H	H	DMF, POCl$_3$	R^4 = CH=NMe$_2^+$ Cl$^-$ (74)	818
				1. DMF, POCl$_3$ 2. NaOH	R^4 = CHO (85)	818
Me	Et	H	H	DMF, POCl$_3$	R^4 = CH=NMe$_2^+$ Cl$^-$ (95)	818
				1. DMF, POCl$_3$ 2. NaOH	R^4 = CHO (92)	818
H	n-Pr	H	H	DMF, POCl$_3$	R^4 = CH=NMe$_2^+$ Cl$^-$ (94)	818
				1. DMF, POCl$_3$ 2. NaOH	R^4 = CHO (95)	818
Me	Me	Me	H	DMF, POCl$_3$	R^4 = CH=NMe$_2^+$ Cl$^-$ (90)	818
				1. DMF, POCl$_3$ 2. NaOH	R^4 = CHO (73)	818
H	CH$_2$CH=CH$_2$	H	H	DMF, POCl$_3$	R^4 = CH=NMe$_2^+$ Cl$^-$ (87)	818
				1. DMF, POCl$_3$ 2. NaOH	R^4 = CHO (89)	818
Me	CH$_2$CH=CH$_2$	H	H	DMF, POCl$_3$	R^4 = CH=NMe$_2^+$ Cl$^-$ (83)	818
				1. DMF, POCl$_3$ 2. NaOH	R^4 = CHO (98)	818

266

TABLE XVII. OTHER HETEROCYCLES WITH TWO FULLY CONJUGATED RINGS (*Continued*)

Substrate				Reagents	Product(s) and Yield(s) (%)	Refs.
R¹	R²	R³	R⁴		R^1 - R^4 as in substrate except as indicated	
Et	Et	H	H	DMF, POCl₃	R^4 = CH=NMe₂⁺ Cl⁻ (33)	818
				1. DMF, POCl₃ 2. NaOH	R^4 = CHO (76)	818
Me	CH₂CH=CH₂	Me	H	DMF, POCl₃	R^4 = CH=NMe₂⁺ Cl⁻ (78)	818
				1. DMF, POCl₃ 2. NaOH	R^4 = CHO (87)	818
Me	n-Bu	H	H	DMF, POCl₃	R^4 = CH=NMe₂⁺ Cl⁻ (87)	818
				1. DMF, POCl₃ 2. NaOH	R^4 = CHO (90)	818
Me	Ph	H	H	1. DMF, POCl₃ 2. NaOH	R^4 = CHO (63)	818
Me	Ph	Me	H	1. DMF, POCl₃ 2. NaOH	R^4 = CHO (85)	818

C₉

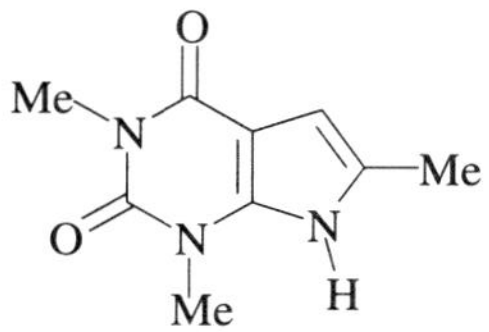

DMF, POCl₃

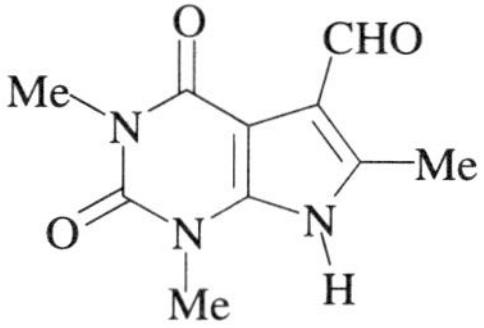

(51)

818

C₉₋₁₄

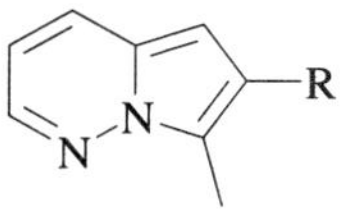

R
Me
Ph

DMF, POCl₃

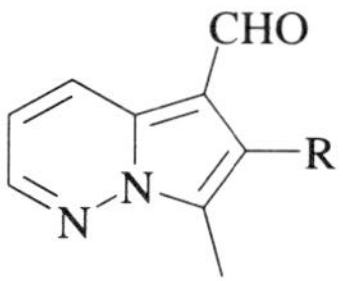

(72)
(88)

820

TABLE XVII. OTHER HETEROCYCLES WITH TWO FULLY CONJUGATED RINGS (*Continued*)

Substrate	Reagents	Product(s) and Yield(s) (%)	Refs.
C$_{9-15}$ 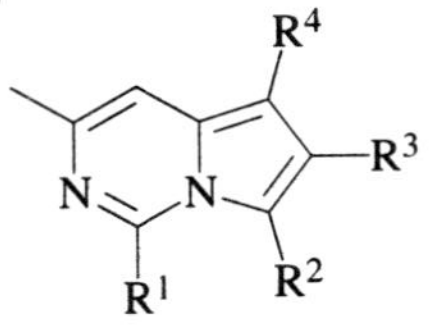 *(pyrrolopyridine, R^1–R^4 substituents)*			145

R^1	R^2	R^3	R^4	Reagents	Product(s) and Yield(s) (%)
					R^1 - R^4 as in substrate except as indicated
Me	H	H	H	DMF, POCl$_3$	R^2 = CHO (62)
				1. DMF, POCl$_3$ 2. NaSH	R^2 = CHS (58)
H	Me	Me	H	DMF, POCl$_3$	R^4 = CHO (68)
Me	H	Me	H	DMF, POCl$_3$	R^2 = CHO (20) + R^3 = CHO (2)
Me	Me	Me	H	DMF, POCl$_3$	R^4 = CHO (5)
Me	H	Ph	H	DMF, POCl$_3$	R^4 = CHO (5)

C$_4$S/C$_4$N$_2$

Substrate	Reagents	Product(s) and Yield(s) (%)	Refs.
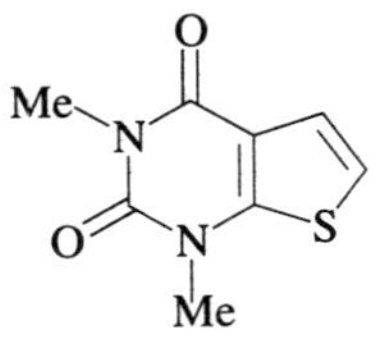 C$_8$ *(thienopyrimidinedione structure)*	DMF, POCl$_3$	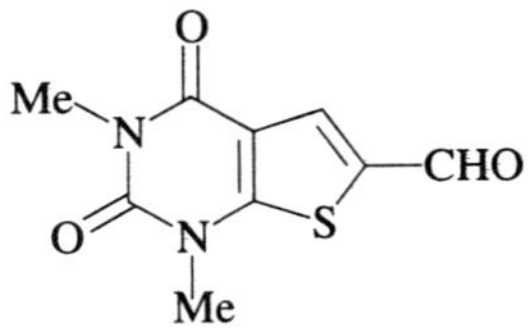 CHO (92)	821

TABLE XVII. OTHER HETEROCYCLES WITH TWO FULLY CONJUGATED RINGS (*Continued*)

Substrate		Reagents	Product(s) and Yield(s) (%)	Refs.
		C₄N/C₅N (No bridgehead nitrogen)		

C₈₋₁₅

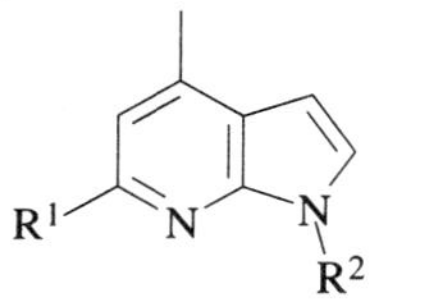

R¹	R²			
H	H	DMF, POCl₃	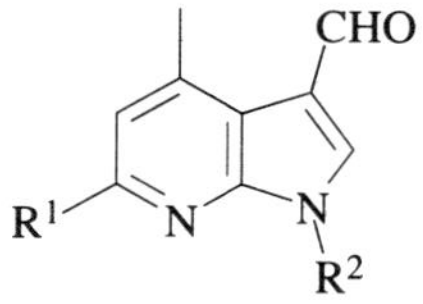 **I** (25)	139
Cl	H	DMF, POCl₃	**I** (25)	139
H	Me	DMF, POCl₃	**I** (17)	139
H	*n*-Bu	DMF, POCl₃	**I** (48)	139
H	Ph	1. DMF, POCl₃	(76)	822
		2. NaBH₄		
H	Bn	DMF, POCl₃	**I** (38)	139
H	4-MeOC₆H₄	DMF, POCl₃	**I** (71)	139

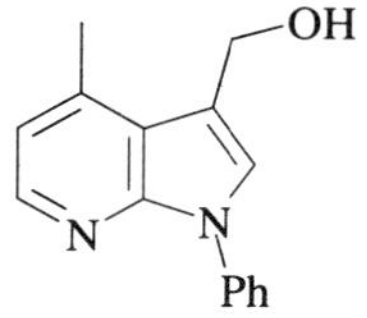

TABLE XVII. OTHER HETEROCYCLES WITH TWO FULLY CONJUGATED RINGS (*Continued*)

Substrate	Reagents	Product(s) and Yield(s) (%)	Refs.

C_{8-23}

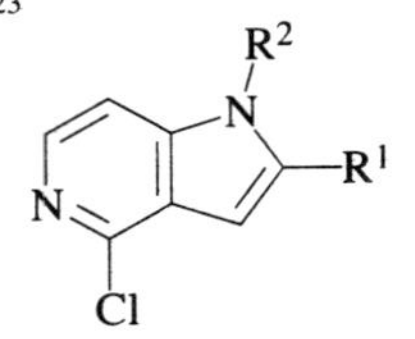

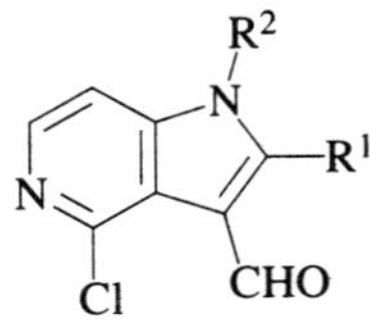

R^1	R^2		
H	Me	(94)	823
CH(Me)Ph	Me	(16)	824
CH(Me)C$_6$H$_4$Me-4	Me	(39)	824
CH(Me)C$_6$H$_4$Me-4	Bn	(32)	824

DMF, POCl$_3$

C$_4$N/C$_5$N, Bridgehead nitrogen

C_{7-19}

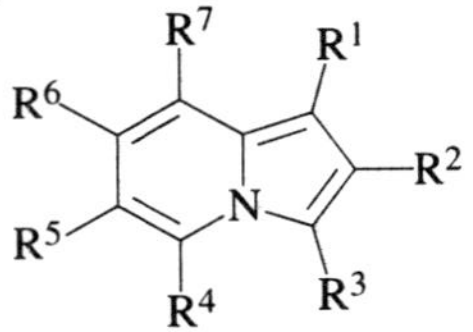

R^1 - R^7 = H except as indicated

R^2 = Me

	Reagents	R^1 - R^7 as in substrate except as indicated	Refs.
	DMF, POCl$_3$	R^1 = CHO (tr) + R^3 = CHO (44)	144
	1. DMF, POCl$_3$ 2. NaSH	R^3 = CHS (77)	141, 142
	MFA, POCl$_3$	R^3 = CHO (8)	825
	DMF, POCl$_3$, 70°	R^1 = R^3 = CHO (96)	141, 142
	1. DMF, POCl$_3$ 2. NaSH	R^3 = CHS (86)	141, 142
			632

TABLE XVII. OTHER HETEROCYCLES WITH TWO FULLY CONJUGATED RINGS (*Continued*)

Substrate	Reagents	Product(s) and Yield(s) (%)	Refs.
R^1 - R^7 = H except as indicated		R^1 - R^7 as in substrate except as indicated	
	1. DMF, $POCl_3$	R^3 = CHSe (3)	143
	2. NaSeH		
$R^1 = R^2$ = Me	DMF, $POCl_3$	R^3 = CHO (84)	141, 142
	1. DMF, $POCl_3$	R^3 = CHS (81)	141, 142
	2. NaSH		
	1. DMA, $POCl_3$	R^3 = C(S)Me (66)	632
	2. NaSH		
	1. DMF, $POCl_3$	R^3 = CHSe (43)	143
	2. NaSeH		
$R^2 = R^3$ = Me	DMF, $POCl_3$	R^1 = CHO (62)	141, 142
	MFA, $POCl_3$	R^1 = CHO (36)	825
	1. DMF, $POCl_3$	R^1 = CHS (76)	141, 142
	2. NaSH		
	1. DMF, $POCl_3$	R^1 = CHSe (12)	143
	2. NaSeH		
$R^2 = R^5$ = Me	1. DMF, $POCl_3$	R^3 = CHS (86)	141, 142
	2. NaSH		
$R^2 = R^6$ = Me	1. DMF, $POCl_3$	R^3 = CHS (83)	141, 142
	2. NaSH		
	1. DMF, $POCl_3$	R^3 = CHSe (46)	143
	2. NaSeH		
$R^2 = R^7$ = Me	1. DMF, $POCl_3$	R^3 = CHS (82)	141, 142
	2. NaSH		
	1. DMF, $POCl_3$	R^3 = CHSe (46)	143
	2. NaSeH		
$R^2 = R^3$ = Me, R^5 = Et	DMF, $POCl_3$	R^1 = CHO (77)	826
R^2 = *t*-Bu	1. DMF, $POCl_3$	R^1 = CHS (5) + R^3 = CHS (90)	141, 142
	2. NaSH		

TABLE XVII. OTHER HETEROCYCLES WITH TWO FULLY CONJUGATED RINGS (*Continued*)

Substrate	Reagents	Product(s) and Yield(s) (%)	Refs.
R^1 - R^7 = H except as indicated		R^1 - R^7 as in substrate except as indicated	
R^1 = Me, R^2 = t-Bu	1. DMF, POCl$_3$ 2. NaSeH	R^3 = CHSe (40)	143
R^2 = t-Bu, R^6 = Me	1. DMF, POCl$_3$ 2. NaSeH	R^3 = CHSe (37)	143
R^2 = R^6 = Ph, R^4 = Me	DMF, POCl$_3$	R^1 = R^3 = CHO (50)	827
C$_{15}$	1. DMA, POCl$_3$ 2. HClO$_4$	ClO$_4^-$ (—)	828
C$_4$N/C$_6$			
C$_8$	—	(48)	829
C$_{8-15}$	DMF, POCl$_3$		

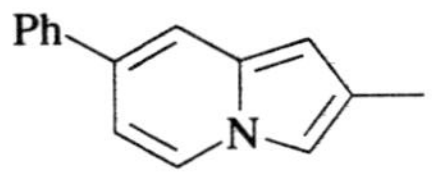

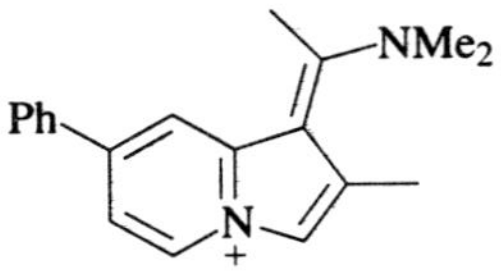

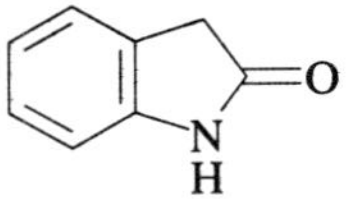

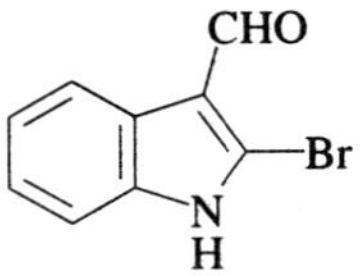

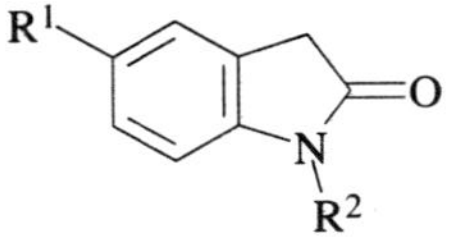

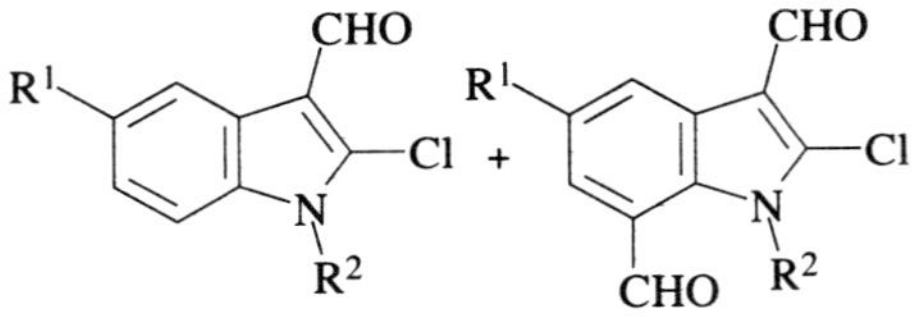

TABLE XVII. OTHER HETEROCYCLES WITH TWO FULLY CONJUGATED RINGS (*Continued*)

Substrate	Reagents	Product(s) and Yield(s) (%)	Refs.

R^1	R^2			
H	H		**I** (50-80)	830, 753, 755
H	Me		**I** (50-80) + **II** (10-15)	830
OMe	H		**I** (77)	753
H	Ph		**I** (75)	753
H	4-ClC$_6$H$_4$CO		**I** (50-80) + **II** (10-15)	830
H	COPh		**I** (50-80)	830
			I (50-80)	830

C$_{14}$

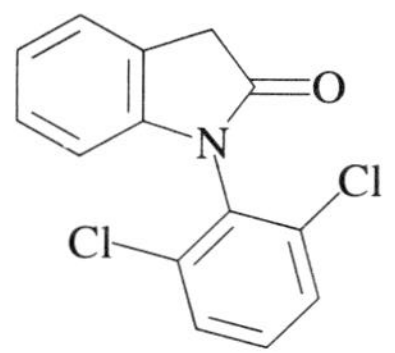	DMF, POCl$_3$	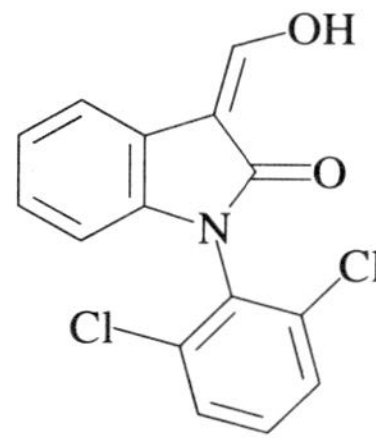 (—)	831

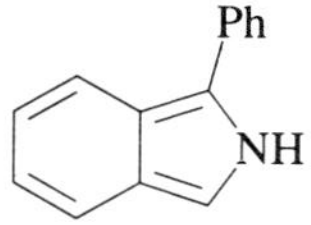	DMF, POCl$_3$	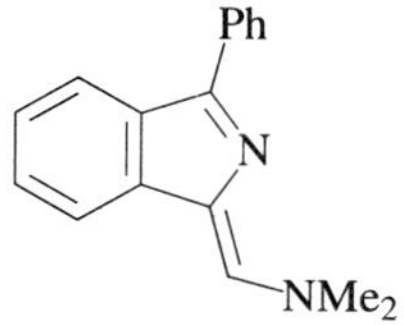 (68)	185

	1. DMF, POCl$_3$ 2. H$_2$O, OH$^-$	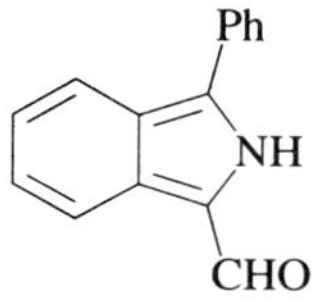 (33)	185

TABLE XVII. OTHER HETEROCYCLES WITH TWO FULLY CONJUGATED RINGS (*Continued*)

Substrate	Reagents	Product(s) and Yield(s) (%)	Refs.

C$_{16}$

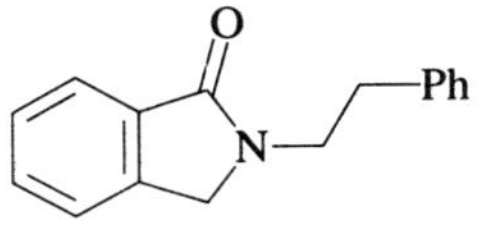

1. DMF, POCl$_3$
2. ClO$_4^-$

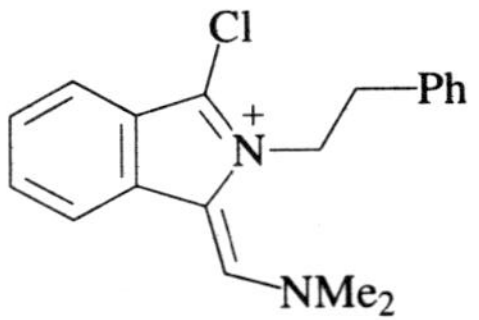

ClO$_4^-$ (80)

756

C$_4$O/C$_5$N

C$_9$

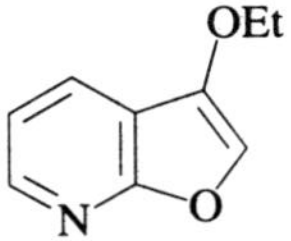

DMF, POCl$_3$

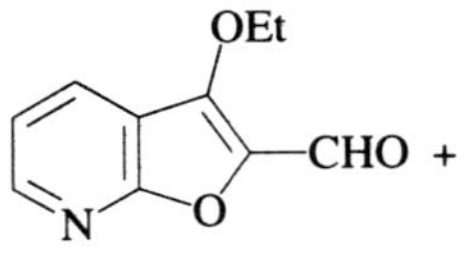

(22)

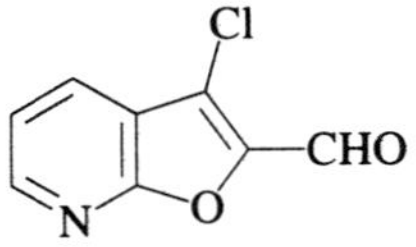

Cl

CHO 140

(43)

DMF, POCl$_3$

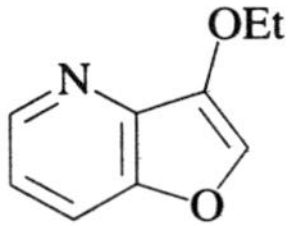

(30)

Cl

CHO 140

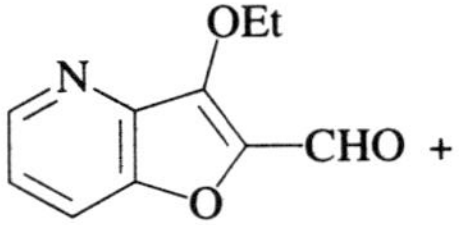

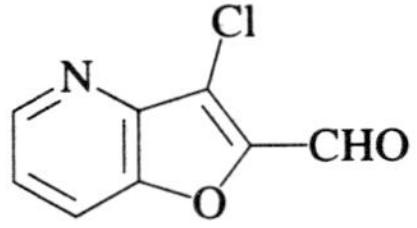

(35)

C$_4$O/C$_6$

C$_8$

DMF, POCl$_3$

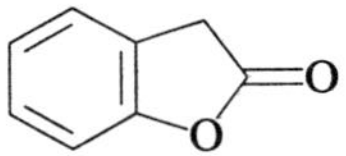

(46)

(10)

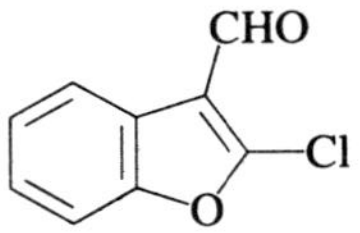

832

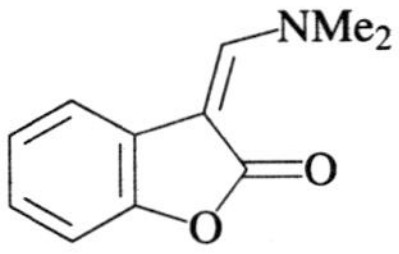

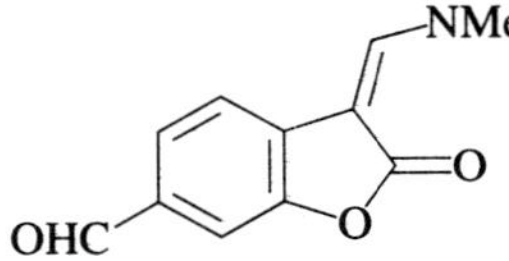

(24)

274

TABLE XVII. OTHER HETEROCYCLES WITH TWO FULLY CONJUGATED RINGS (*Continued*)

Substrate	Reagents	Product(s) and Yield(s) (%)	Refs.
	C₄N/C₆N		
C₉	DMF, POCl₃	(73)	147
	DMA, POCl₃	(43)	833
	C₄N₂/C₅N		
C₈	DMF, POCl₃	(50)	151
C₉	—	(—)	834

275

TABLE XVII. OTHER HETEROCYCLES WITH TWO FULLY CONJUGATED RINGS (*Continued*)

Substrate	Reagents	Product(s) and Yield(s) (%)	Refs.

C$_{8-15}$

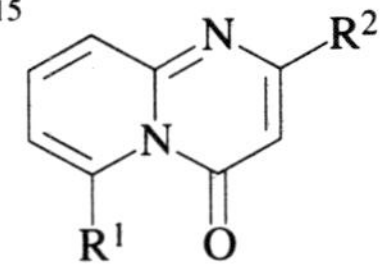

DMF, POCl$_3$

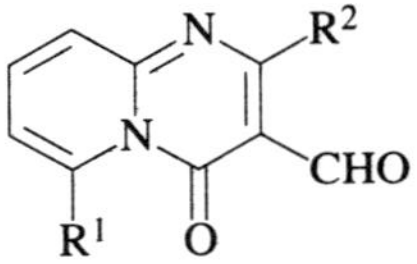

R^1	R^2	
H	Cl	(94)
Me	Cl	(93)
H	OMe	(79-95)
H	NHBu	(90)
H	N-piperidyl	(89)
Me	NHBu	(89)
Me	N-piperidyl	(93)

835

C$_{10-16}$

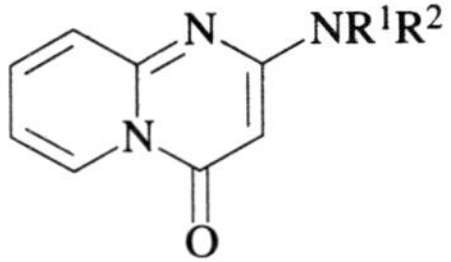

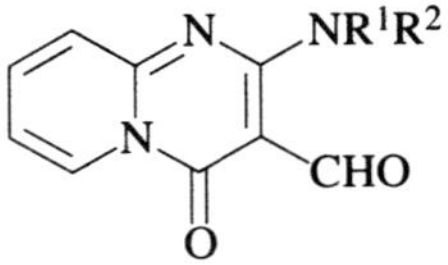

I +   **II** +

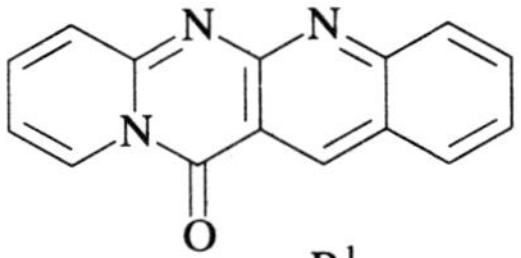

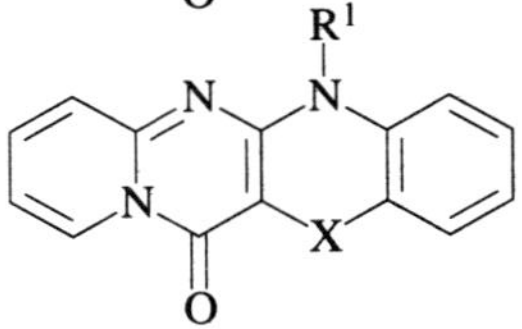

III, X = CH$_2$
IV, X = CO

152

276

TABLE XVII. OTHER HETEROCYCLES WITH TWO FULLY CONJUGATED RINGS (Continued)

Substrate	Reagents	Product(s) and Yield(s) (%)	Refs.
C$_{8-14}$ R^1 / R^2 (structure with O_2S, N–H, R)			
Me Me	DMF, POCl$_3$	**I** (92)	
Et Et	DMF, POCl$_3$	**I** (87)	
(CH$_2$)$_4$	DMF, POCl$_3$	**I** (83)	
H Ph	DMF, POCl$_3$, 45°	**I** (77) + **II** (8)	
	DMF, POCl$_3$, 90°	**I** (11) + **II** (57)	
Me Ph	DMF, POCl$_3$	**III** (37) + **IV** (41)	
Et Ph	DMF, POCl$_3$	**III** (23) + **IV** (22)	
C$_4$NS/C$_6$	—	R = H (—) R = Ph (—)	836
C$_{11}$ (MeO, Cl, N, OMe isoquinoline structure)	DMF, POCl$_3$	**C$_5$N/C$_6$** (structure, CHO, MeO, Cl, N, OMe) (40)	837
C$_{12}$ 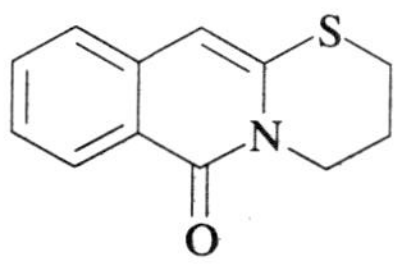 (structure with S, N, =O)	DMF, POCl$_3$	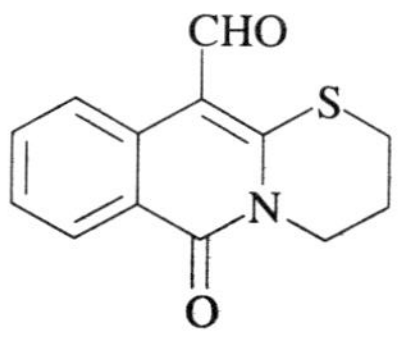 (structure with CHO, S, N, =O) (71)	838

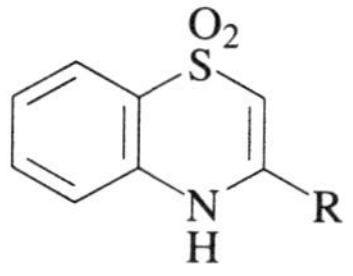

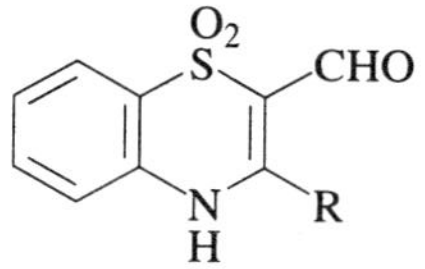

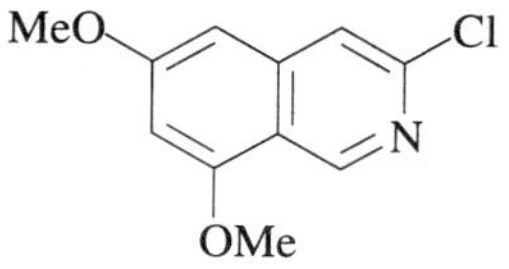

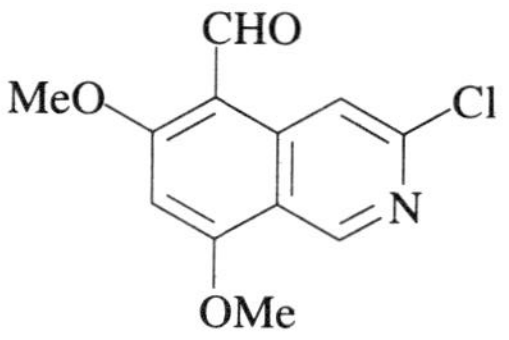

277

TABLE XVII. OTHER HETEROCYCLES WITH TWO FULLY CONJUGATED RINGS (*Continued*)

Substrate	Reagents	Product(s) and Yield(s) (%)	Refs.
		C_5O/C_6	

C₉

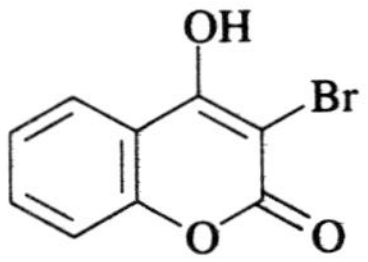

DMF, POCl₃

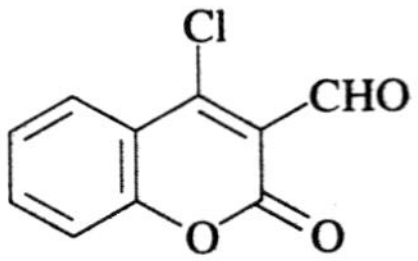

(54) +

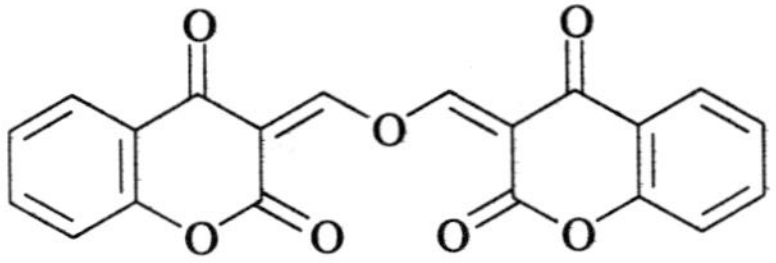

(40)

839

C₉₋₁₄

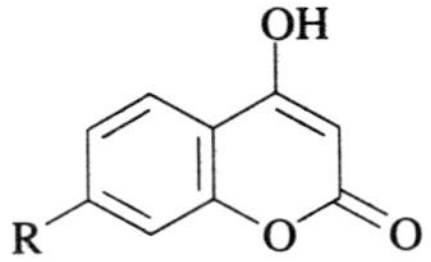

DMF, POCl₃

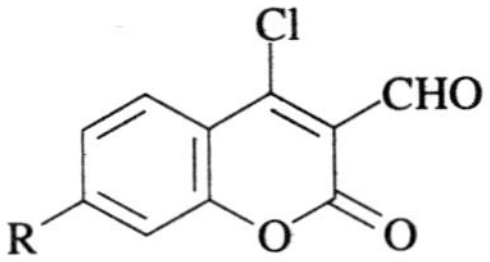

R		
H	(96)	840, 841
OMe	(70)	841
NMe₂	(75–80)	841
N-piperidyl	(75)	841

TABLE XVII. OTHER HETEROCYCLES WITH TWO FULLY CONJUGATED RINGS (*Continued*)

Substrate	Reagents	Product(s) and Yield(s) (%)	Refs.

C$_{9-17}$

R^1 - R^6 = H except as indicated

R^1 - R^6 as in substrate except as indicated

R^3 = R^6 = NO$_2$, R^5 = Cl	DMF, POCl$_3$	R^6 = Cl (—) + R^6 = CHO (—)	48
R^3 = R^6 = NO$_2$, R^5 = OH	DMF, POCl$_3$	R^6 = Cl, R^6 = CHO (—) + R^5 = R^6 = Cl (—)	48
R^4 = OH, R^5 = Me	MFA, POCl$_3$	R^3 = CHO (9)	842
R^2 = R^4 = OH, R^5 = Me	MFA, POCl$_3$	R^1 = CHO (54)	842
R^2 = R^5 = Me, R^4 = OH	MFA, POCl$_3$	R^1 = CHO (7) + R^3 = CHO (—)	842
R^2 = R^4 = OMe	DMF, POCl$_3$	R^1 = CHO (76)	149
R^2 = R^4 = OMe, R^5 = Me	MFA, POCl$_3$	R^1 = CHO (—) + R^3 = CHO (32)	842
R^2 = R^3 = R^4 = OMe, R^5 = Me	MFA, POCl$_3$	R^1 = CHO (—)	842
R^5 = NH(C$_5$H$_{11}$-n)	DMF, POCl$_3$	R^6 = CHO (88)	843
R^5 = NHBn	DMF, POCl$_3$	R^6 = CHO (80)	843, 844
R^5 = NHCH$_2$Bn	DMF, POCl$_3$	R^6 = CHO (81)	843
R^5 = NH(C$_6$H$_3$Me$_2$-2,4)	DMF, POCl$_3$	R^6 = CHO (67)	843

C$_{14-16}$

	DMF, POCl$_3$		

279

TABLE XVII. OTHER HETEROCYCLES WITH TWO FULLY CONJUGATED RINGS (Continued)

Substrate	Reagents	Product(s) and Yield(s) (%)	Refs.

C$_{11}$

R^1	R^2	R^3		
H	H	Cl	(69)	843
H	H	H	(89)	150, 843
Me	H	H	(98)	843
OMe	H	H	(96)	843
H	Me	H	(86)	150
H	H	Me	(78)	843
H	H	OMe	(86)	843
H	H	OEt	(88)	843

C$_6$/C$_5$N$_2$

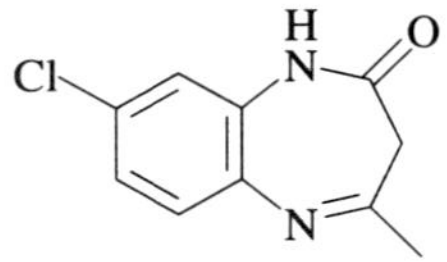

	DMF, POCl$_3$	(—)	845

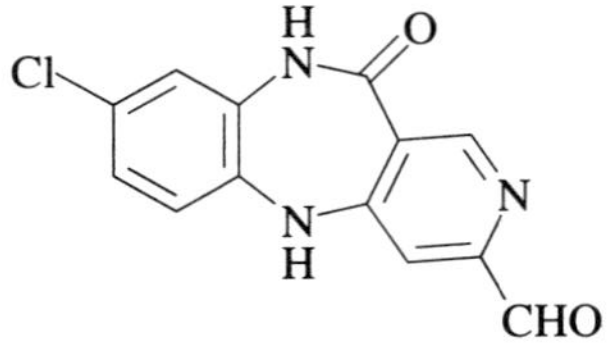

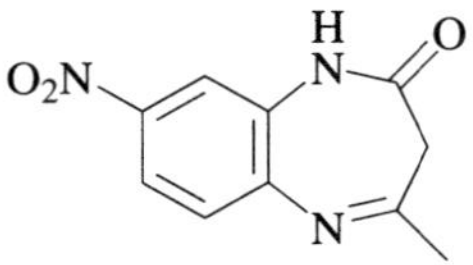	DMF, POCl$_3$	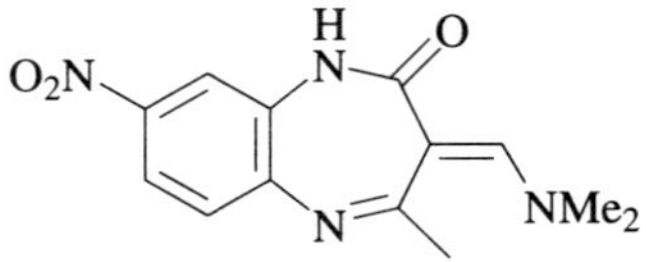 (—)	845

C$_{15-16}$

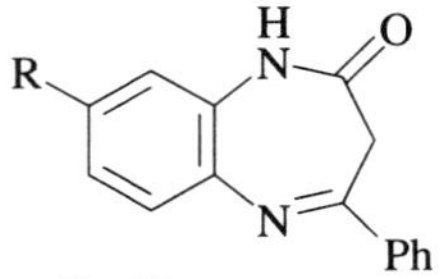

	DMF, POCl$_3$		
R = Cl		(20–80)	846
R = H		(52)	847
R = OMe		(30)	846

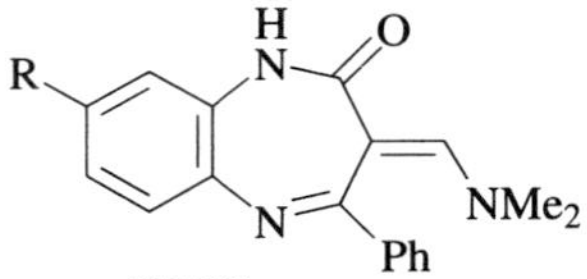

280

TABLE XVIII. OTHER HETEROCYCLES WITH THREE OR MORE FULLY CONJUGATED RINGS

Substrate	Reagents	Product(s) and Yield(s) (%)	Refs.
$C_3N_2/C_3N_2/C_4N_2$			
C_{16-21} 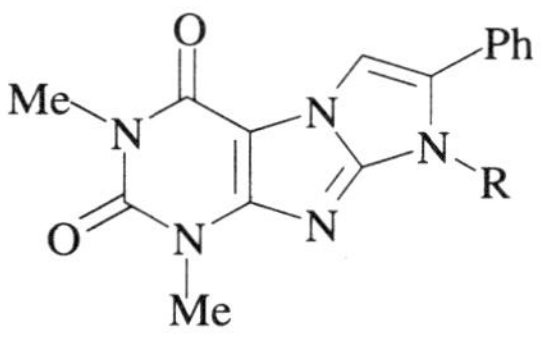 R = Me, Bu, Ph	DMF, POCl$_3$	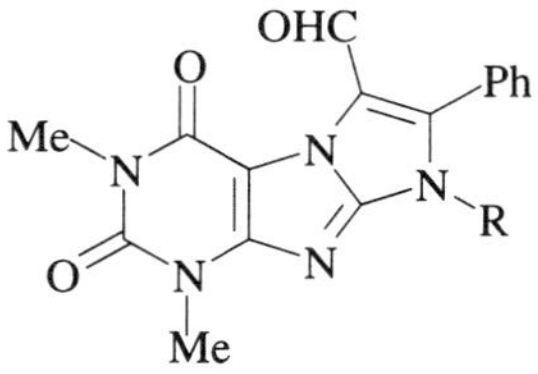 (77-95)	848
$C_3N_2/C_4N_2/C_5N$			
C_{13-34} 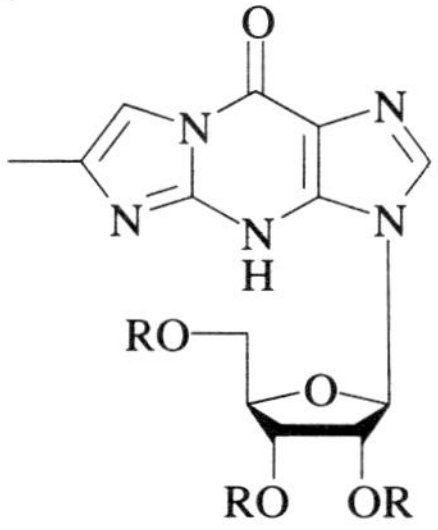 R = H	1. Ac$_2$O, Py 2. DMF, POCl$_3$	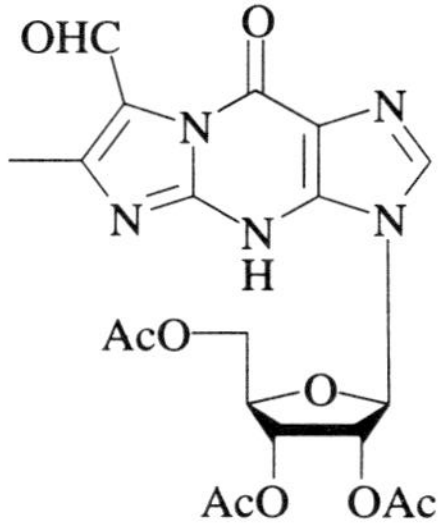 **I** (92)	170
R = COMe	DMF, POCl$_3$, –25°	**I** (92)	171

281

Substrate	Reagents	Product(s) and Yield(s) (%)	Refs.
R = Bn	DMF, POCl₃, -30°	(63)	171
	DMF, POCl₃, rt	(—)	171
C₁₅	DMF, POCl₃	(65)	171
C₁₆	DMF, POCl₃	(100)	849, 850

282

TABLE XVIII. OTHER HETEROCYCLES WITH THREE OR MORE FULLY CONJUGATED RINGS (*Continued*)

Substrate	Reagents	Product(s) and Yield(s) (%)	Refs.
C3N2/C3N2/C6			
C_{11-16} 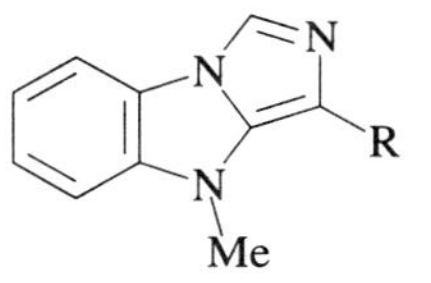 R = Me, Ph	DMF, POCl$_3$	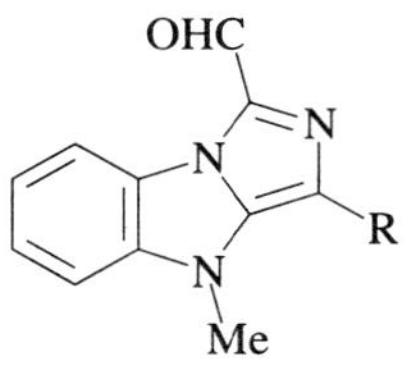 OHC ... (72)	164
C_{21} 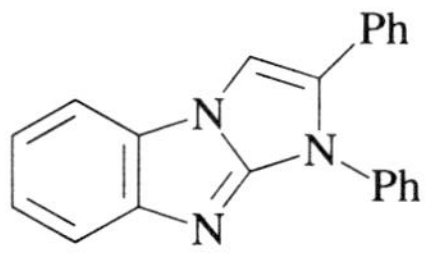 Ph, N–Ph	DMF, POCl$_3$	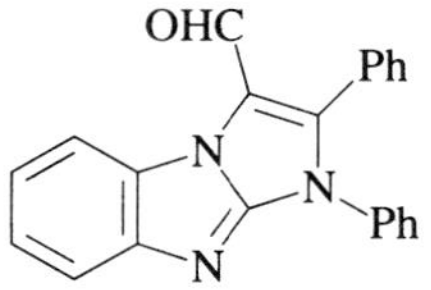 OHC Ph N–Ph (88)	165
C3N2/C3NS/C6			
C_{15-16} 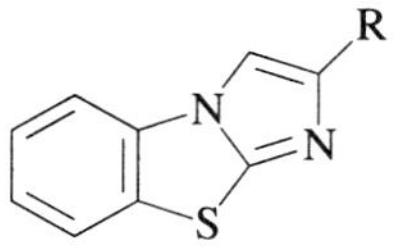	DMF, POCl$_3$	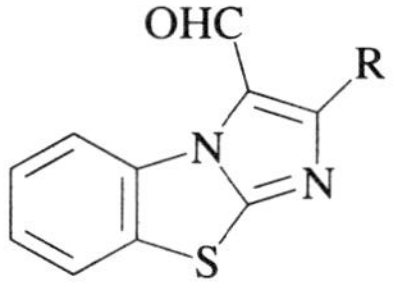 OHC R	
		R	
		Cl (—)	801
		4-ClC$_6$H$_4$ (91)	166
		4-BrC$_6$H$_4$ (62)	166
		Ph (89)	166
		4-MeC$_6$H$_4$ (92)	166
		4-MeOC$_6$H$_4$ (91)	166

TABLE XVIII. OTHER HETEROCYCLES WITH THREE OR MORE FULLY CONJUGATED RINGS (*Continued*)

Substrate	Reagents	Product(s) and Yield(s) (%)	Refs.
	C₃N₂/C₄N/C₅N		
C₉	DMF, POCl₃	(37) + (25)	167
	C₃NS/C₄N/C₆		
C₁₀	DMF, POCl₃	(62)	484
	C₃NS/C₄S/C₆		
C₁₀	—	(—)	163

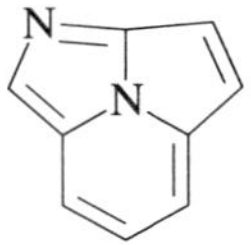

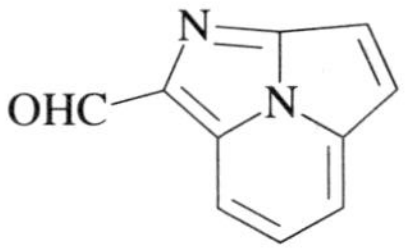

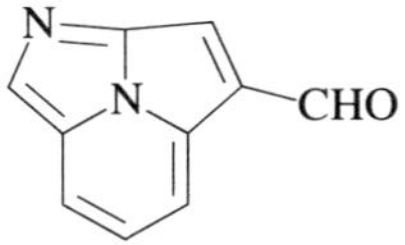

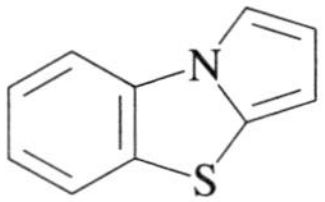

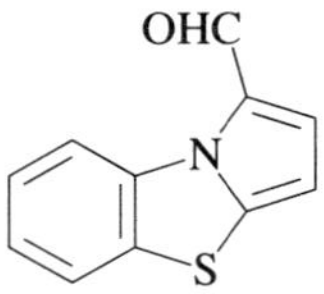

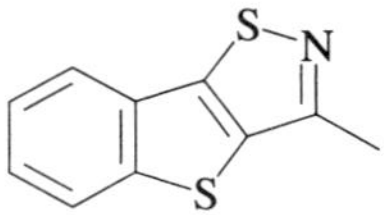

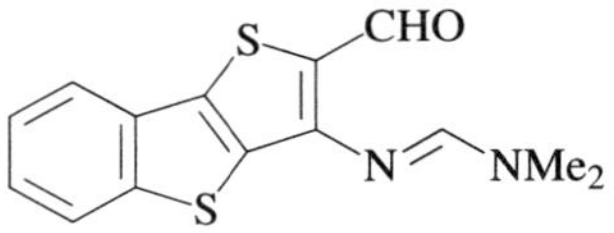

Substrate	Reagents	Product(s) and Yield(s) (%)	Refs.
C₄N/C₄N/C₄N₂			
C₁₁₋₁₆ R = Me R = Ph	DMF, POCl₃	 **I** + **II** **I** (20) + **II** (27) **I** (28) + **II** (28)	145
C₄N/C₄N/C₆			
C₁₀ R¹ = R² = R³ = H	DMF, POCl₃, 1:1 DMF, POCl₃, 3:1 DMA, POCl₃	R¹ - R³ as in substrate except as indicated R¹ = CHO (2) + R³ = CHO (27) R¹ = R³ = CHO (4) + R² = R³ = CHO (81) R¹ = COMe (9) + R² = COMe (20) + R³ = COMe (34) + R² = R³ = COMe (6)	851 851 852
	DMF, POCl₃	 (74)	853

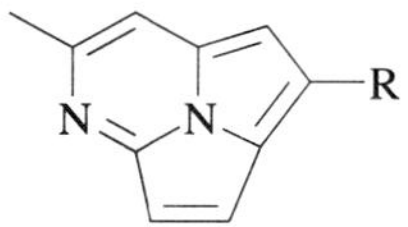

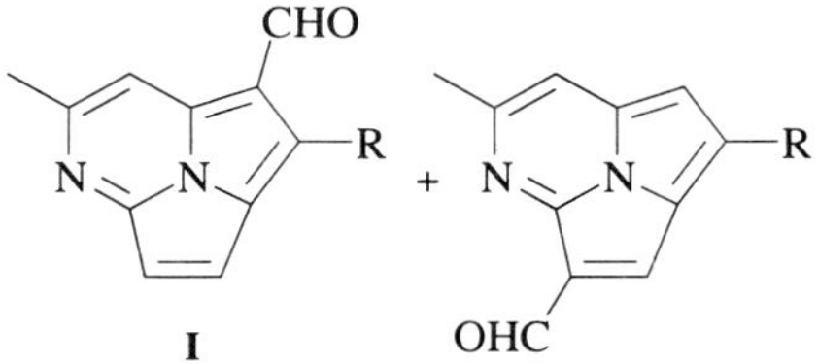

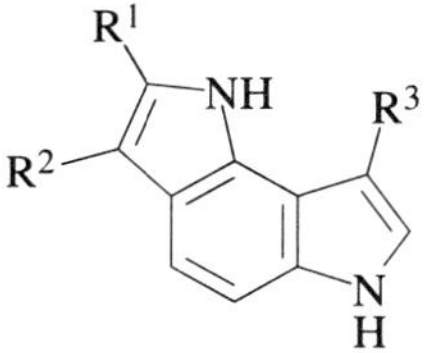

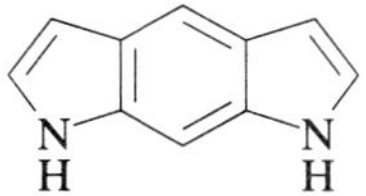

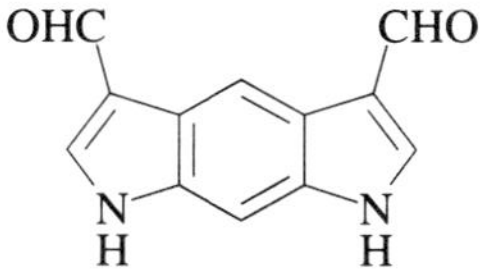

285

TABLE XVIII. OTHER HETEROCYCLES WITH THREE OR MORE FULLY CONJUGATED RINGS (*Continued*)

Substrate	Reagents	Product(s) and Yield(s) (%)	Refs.
		C₄N/C₄O/C₆	

C_{10-13}

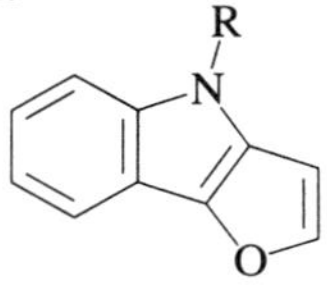

DMF, POCl₃

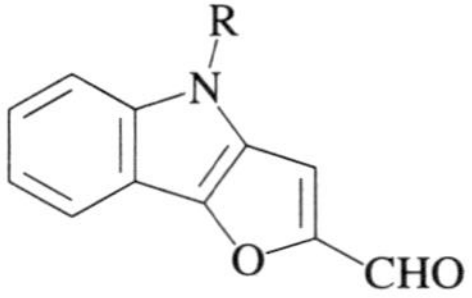

R	
H	(81)
Me	(82)
Et	(85)
n-Pr	(72)

162

| | | **C₄N/C₄S/C₄N₂** | |

C_9

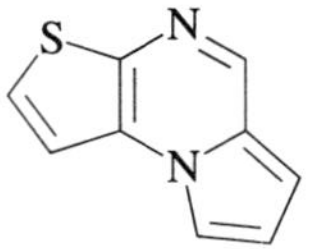

DMF, POCl₃

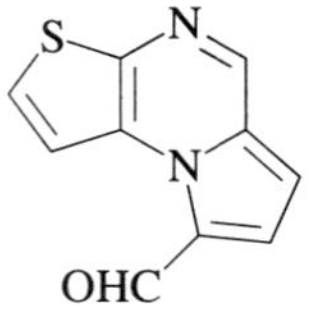

(82)

168

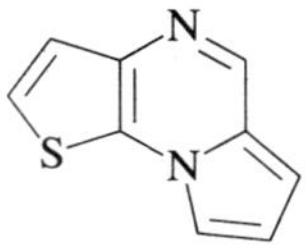

DMF, POCl₃

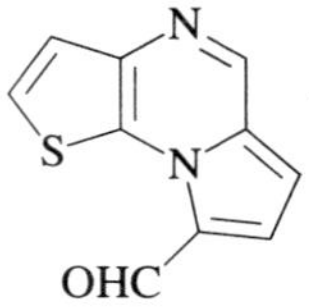

(28)   (9)

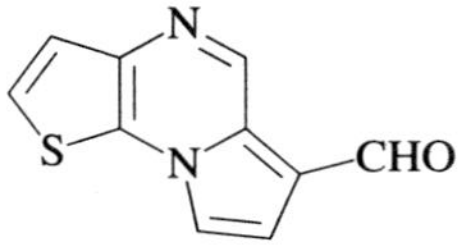

169

TABLE XVIII. OTHER HETEROCYCLES WITH THREE OR MORE FULLY CONJUGATED RINGS (*Continued*)

Substrate	Reagents	Product(s) and Yield(s) (%)	Refs.
		$C_4S/C_4S/C_6$ and $C_4S/C_4Se/C_6$	

C_{10-11}

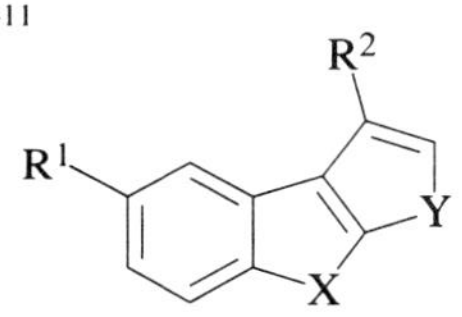

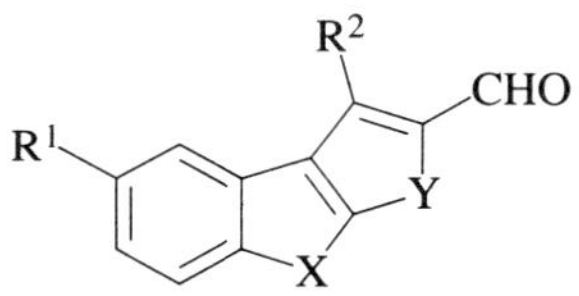

R^1	R^2	X	Y			
H	H	S	S	MFA, POCl$_3$	(83)	161
H	Br	S	S	MFA, POCl$_3$	(95)	854
Me	H	S	S	MFA, POCl$_3$	(91)	854
H	H	S	Se	—	(↑)	708
H	H	Se	S	—	(↑)	708
H	H	Se	Se	—	(↑)	708

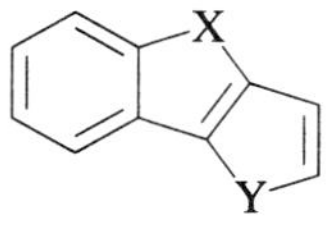

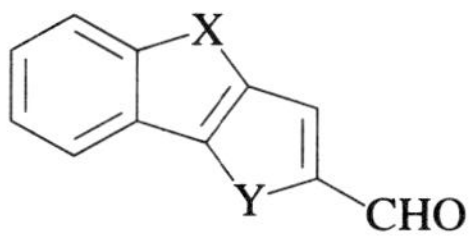

X	Y			
S	S	MFA, POCl$_3$	(84)	161
S	Se	—	(↑)	708
Se	S	—	(↑)	708
Se	Se	—	(↑)	708

TABLE XVIII. OTHER HETEROCYCLES WITH THREE OR MORE FULLY CONJUGATED RINGS (*Continued*)

Substrate	Reagents	Product(s) and Yield(s) (%)	Refs.
	C₃N₂/C₄N₂/C₄N₂		
C₁₇	DMF, POCl₃	(82)	855
	C₃N₂/C₅N/C₆		
C₁₃	DMF, POCl₃	(84)	856
	C₄N/C₄N₂/C₅N		
C₁₁	DMF, POCl₃	(47)	857

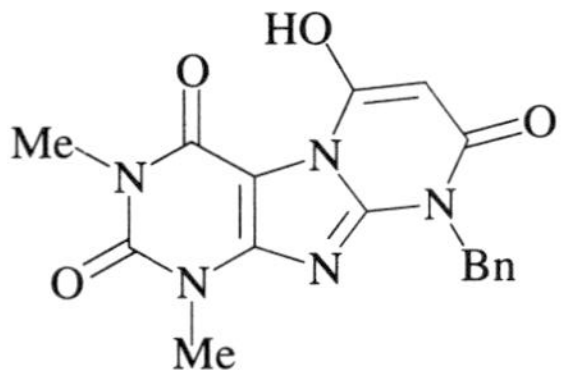

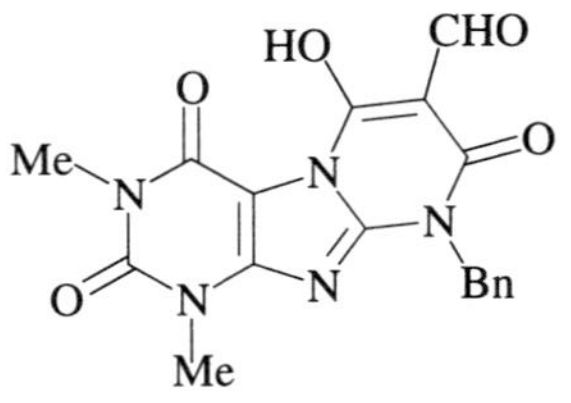

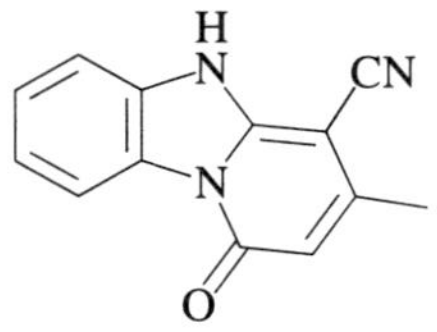

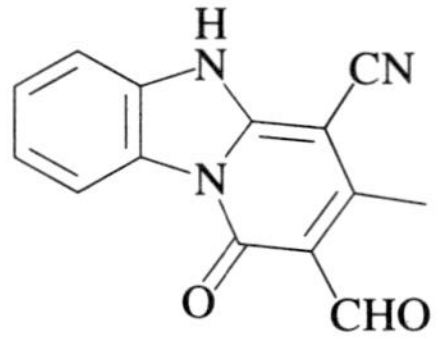

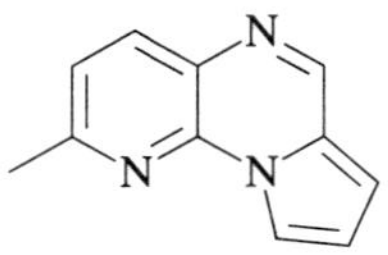

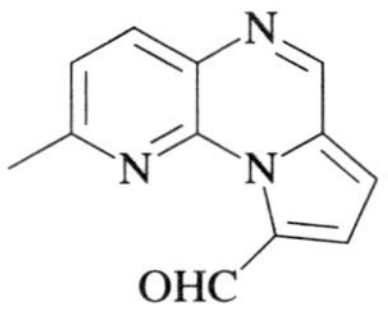

TABLE XVIII. OTHER HETEROCYCLES WITH THREE OR MORE FULLY CONJUGATED RINGS (*Continued*)

Substrate	Reagents	Product(s) and Yield(s) (%)	Refs.
	C₄S/C₅/C₇		
C₁₅	DMF, POCl₃	(100)	858
	C₄N/C₅N/C₆		
C₁₁	DMF, POCl₃	(41)	859
	from DMF, POCl₃	(97)	860
	C₄N/C₆/C₆		
C₁₂	DMF, POCl₃	(94)	153

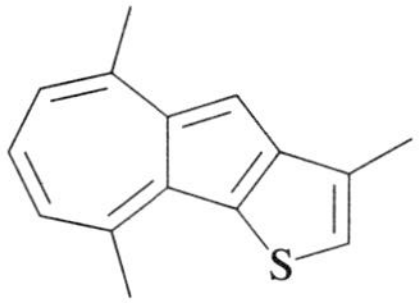

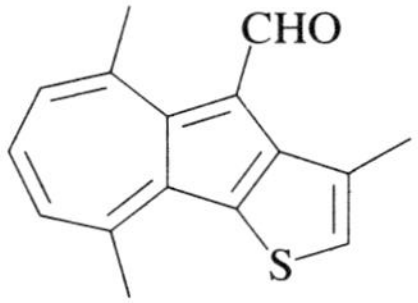

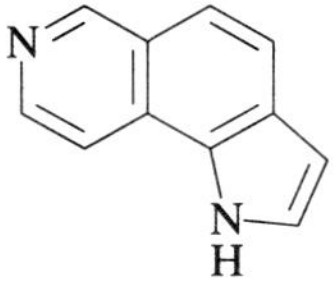

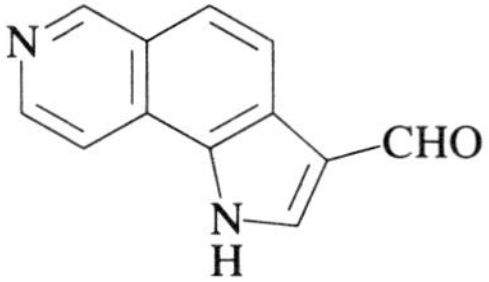

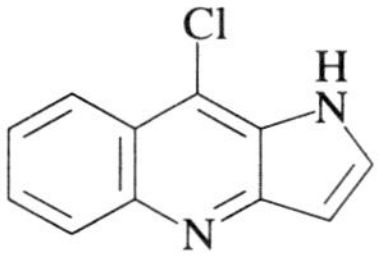

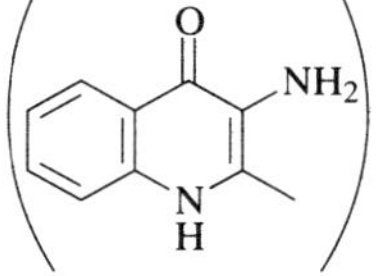

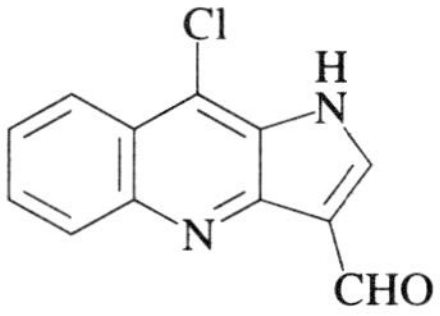

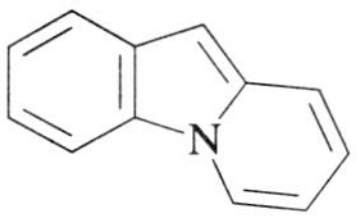

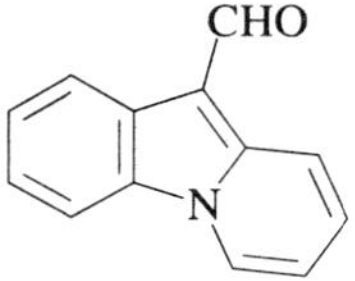

289

TABLE XVIII. OTHER HETEROCYCLES WITH THREE OR MORE FULLY CONJUGATED RINGS (*Continued*)

Substrate	Reagents	Product(s) and Yield(s) (%)	Refs.

C_{12-18}

R^1	R^2	R^3	R^4		
H	H	H	H	DMF, POCl$_3$	
CO$_2$Et	H	H	H	—	
CO$_2$Et	Me	H	H	DMF, POCl$_3$	
Me	CO$_2$Et	H	OH	DMF, POCl$_3$	
Ph	H	H	H	DMF, POCl$_3$	

R^1 - R^4 as in substrate except as indicated

Product	Refs.
R^2 = CHO (50)	154
R^2 = CHO (6-8) + R^3 = CHO (78-80)	861
R^2 = CHO (69)	154
R^4 = CHO (82)	154
R^3 = CHO (80)	155
R^2 = CHO (75)	154

Reagents	R	Refs.
DMF, POCl$_3$	CHO (72)	154
Me$_2$NCOMe, POCl$_3$	COMe (—)	862
Et$_2$NCOMe, POCl$_3$	COMe (—)	862
Me$_2$NCOCH$_2$Cl, POCl$_3$	COCH$_2$Cl (—)	862
Et$_2$NCOCH$_2$Cl, POCl$_3$	COCH$_2$Cl (—)	862
Me$_2$NCOPh, POCl$_3$	COPh (—)	862
Et$_2$NCOPh, POCl$_3$	COPh (—)	862

TABLE XVIII. OTHER HETEROCYCLES WITH THREE OR MORE FULLY CONJUGATED RINGS (*Continued*)

Substrate	Reagents	Product(s) and Yield(s) (%)	Refs.
C_{13} R CO₂Et Ph	DMF, POCl₃ DMF, POCl₃	R¹ CHO (65) CHO (74)	154 154
C₄O/C₆/C₆			
C_{13-14}	DMF, POCl₃	(—)	157
R = Me R = Et	DMF, POCl₃ DMF, POCl₃	**I** (—) **I** (—)	157 157

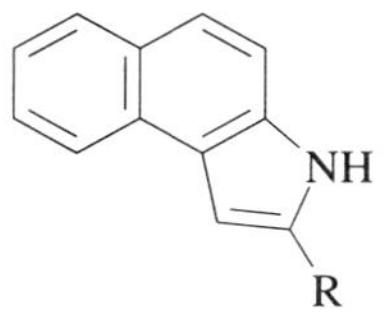

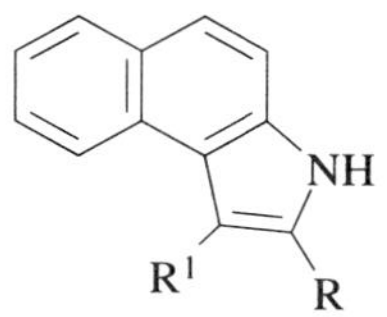

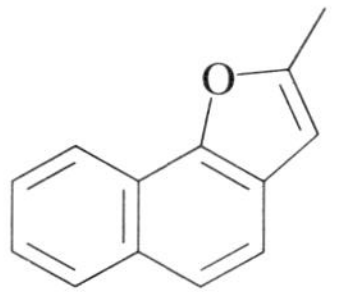

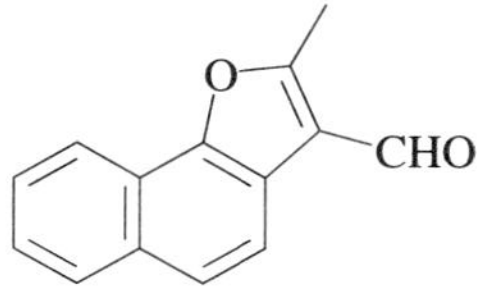

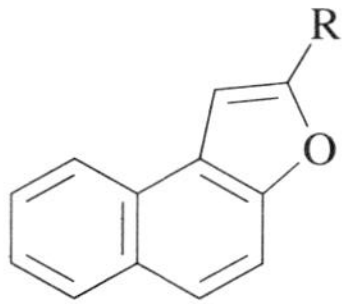

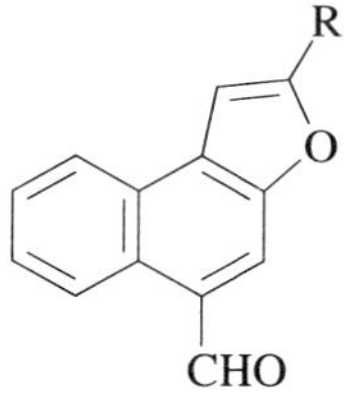

TABLE XVIII. OTHER HETEROCYCLES WITH THREE OR MORE FULLY CONJUGATED RINGS (*Continued*)

Substrate	Reagents	Product(s) and Yield(s) (%)	Refs.
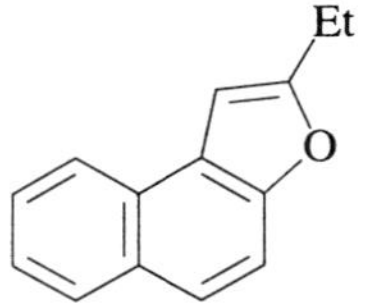 C$_{12}$ (Et-substituted naphthofuran)	DMF, POCl$_3$	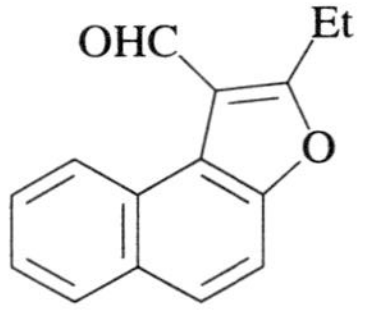 (OHC, Et-substituted naphthofuran) (50)	156
(naphthothiophene)	DMF, POCl$_3$	(CHO-substituted naphthothiophene) (69)	863
C$_4$S/C$_6$/C$_6$			
C$_{17}$ (Ph-substituted)	DMF, POCl$_3$	(CHO, Ph-substituted) (92)	864
C$_5$/C$_5$O/C$_6$			
C$_4$N/C$_4$N$_2$/C$_7$	DMF, POCl$_3$	(24) + (CHO product)	865

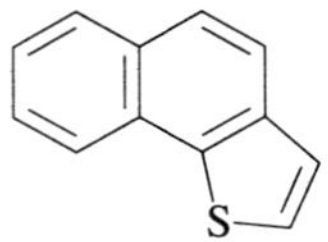

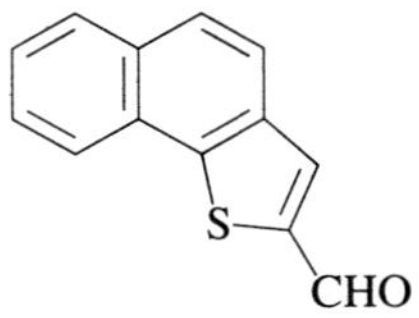

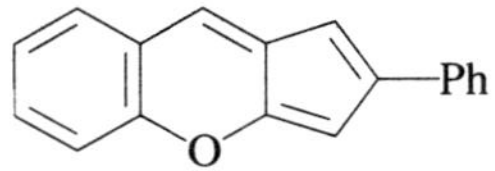

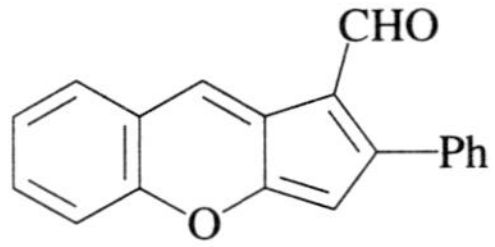

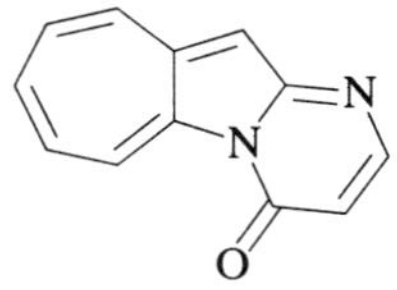

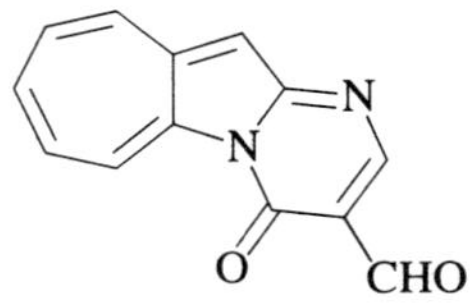

TABLE XVIII. OTHER HETEROCYCLES WITH THREE OR MORE FULLY CONJUGATED RINGS (*Continued*)

Substrate	Reagents	Product(s) and Yield(s) (%)	Refs.
	DMF, POCl₃	(18)	865
	DMF, POCl₃	(44)	866

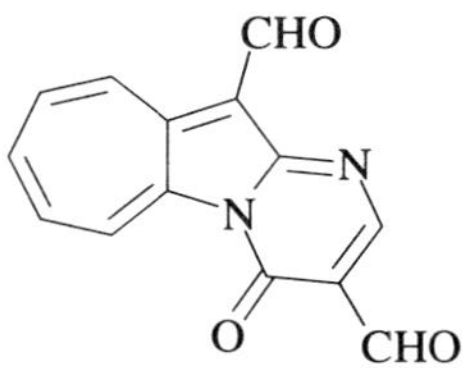

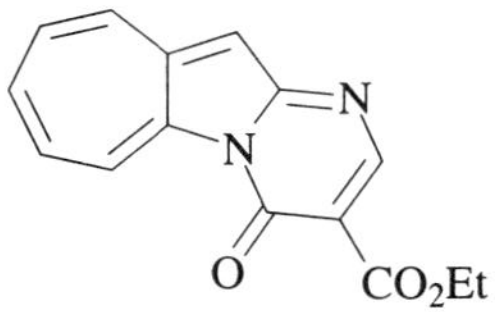

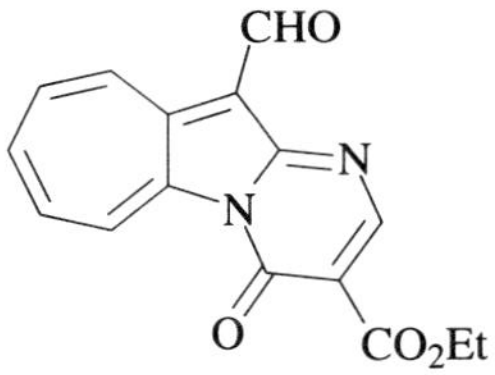

C₄N₂/C₆/C₆, C₄NO/C₆/C₆, and C₄NS/C₆/C₆

C$_{13-14}$

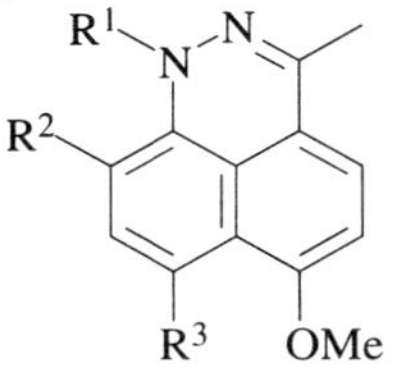

R¹	R²	R³
H	H	H
Me	H	H

DMF, POCl₃

DMA, POCl₃
DMF, POCl₃
1. DMF, POCl₃
2. HClO₄

R¹ - R³ as in substrate except as indicated

R¹ = CHO (5) + R² = CHO (33) +
   R³ = CHO (3) + R¹ = R³ = CHO (30)
R² = COMe (75)
R³ = CHO (87)
R³ = CH=NMe₂⁺ ClO₄⁻ (66)

293

C$_{13-19}$

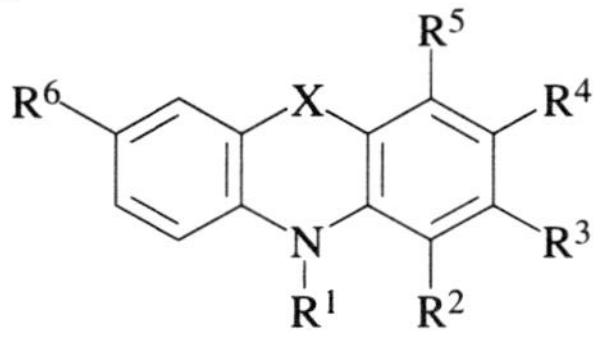

Substrate structure (labels: R^5, R^4, R^3, R^6, X, N, R^1, R^2)

R^1	R^2	R^3	R^4	R^5	R^6	X	Reagents	Product(s) and Yield(s) (%)	Refs.
								X, R^1 - R^3 as in substrate except as indicated	
Me	H	H	H	H	H	O	DMF, POCl$_3$	R^4 = CHO (—)	867
Me	H	H	H	H	H	S	DMF, POCl$_3$	R^4 = CHO (—), R^2 = H or CHO (—)	202
Me	H	H	Cl	H	H	S	MFA, POCl$_3$	R^4 = CHO (83)	869, 868
Me	H	H	H	H	H	S	DMF, POCl$_3$	R^6 = CHO (—)	202
Me	H	H	H	H	H	NMe	DMF, POCl$_3$	R^4 = CHO (63)	872, 871, 870
Me	Me	H	H	H	H	S	DMF, POCl$_3$	R^6 = CHO (67)	873
Me	H	Me	H	H	H	S	DMF, POCl$_3$	R^4 = CHO (82)	873
Me	H	H	Me	H	H	S	DMF, POCl$_3$	R^4 = CHO (55)	873
Me	H	H	H	Me	H	S	MFA, POCl$_3$	R^4 = CHO (48)	869
Et	H	H	H	H	H	S	MFA, POCl$_3$	R^4 = CHO (84)	874, 869
Et	OMe	H	H	H	H	S	DMF, POCl$_3$	R^4 = CHO (87)	875
Me	Me	H	H	Me	H	S	DMF, POCl$_3$	R^6 = CHO (67)	873
Me	Me	H	H	H	Me	S	DMF, POCl$_3$	R^4 = CHO (85)	873
Et	H	H	H	Me	H	S	MFA, POCl$_3$	R^4 = CHO (72)	869
Ph	H	H	H	H	H	S	MFA, POCl$_3$	R^4 = CHO (63)	869
Ph	H	H	H	Me	H	S	MFA, POCl$_3$	R^4 = CHO (49)	869

294

TABLE XVIII. OTHER HETEROCYCLES WITH THREE OR MORE FULLY CONJUGATED RINGS (*Continued*)

Substrate	Reagents	Product(s) and Yield(s) (%)	Refs.
		$C_5N/C_5S/C_6$	
C_{19} 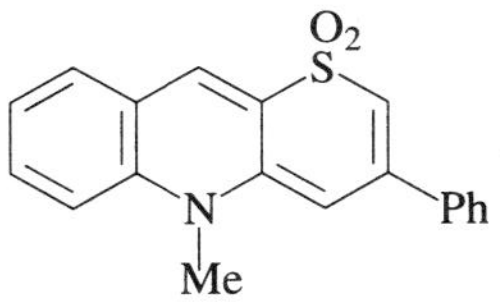	1. DMF, $POCl_3$ 2. HBF_4 3. OH^-	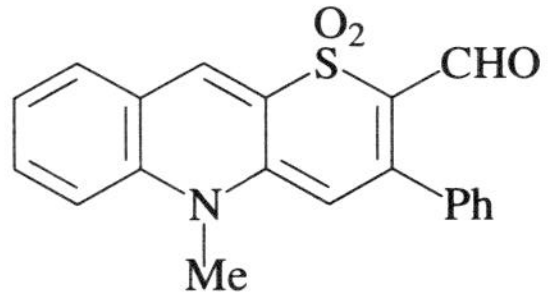 (17)	173
		$C_5N/C_6/C_6$	
C_{17} 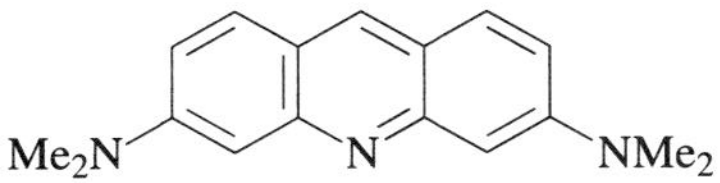	DMF, $POCl_3$	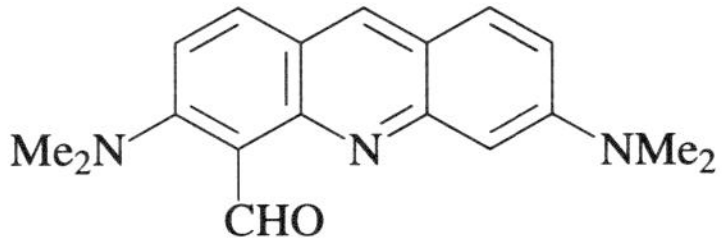 (28)	172
	DMF, $POCl_3$ (xs)	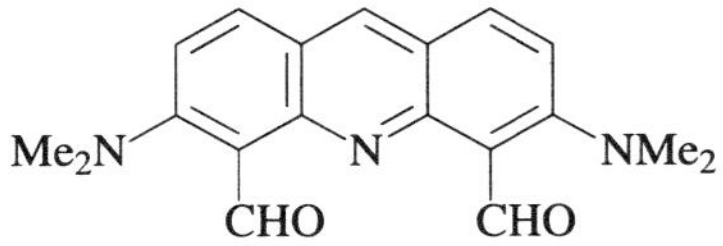 (62)	172
		$C_4N/C_4N/C_6/C_6$	
C_{14} 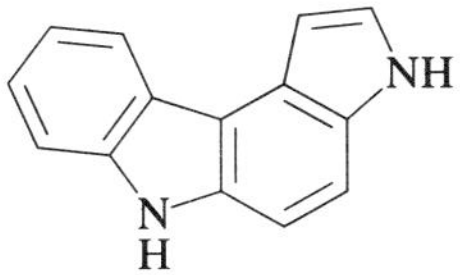		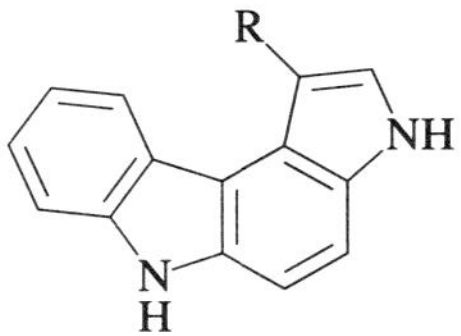	
	DMA, $POCl_3$	R = COMe (10)	159
	Me_2NCOCH_2Cl, $POCl_3$	R = $COCH_2Cl$ (40)	159

Substrate	Reagents	Product(s) and Yield(s) (%)	Refs.
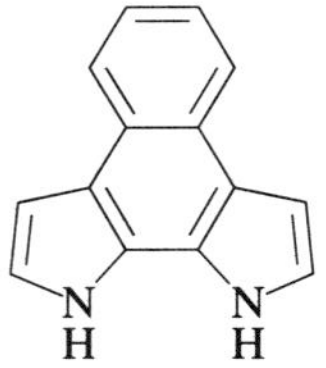	DMF, POCl$_3$, 1:3	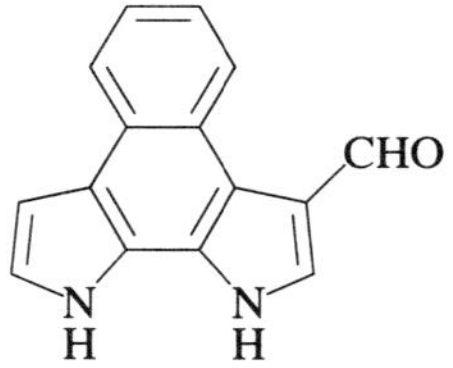 (50)	160
	DMF, POCl$_3$, 1:15	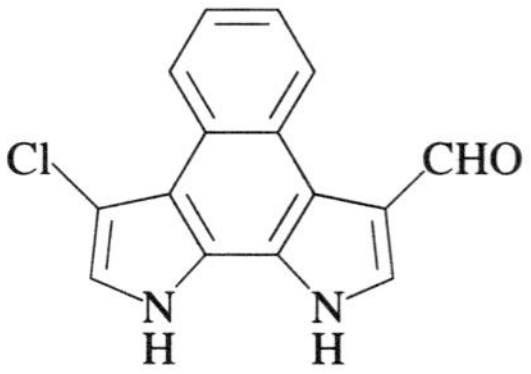 (64)	160
C$_4$N/C$_4$O/C$_6$/C$_6$			
C$_{14}$			
	DMF, POCl$_3$	(88)	876
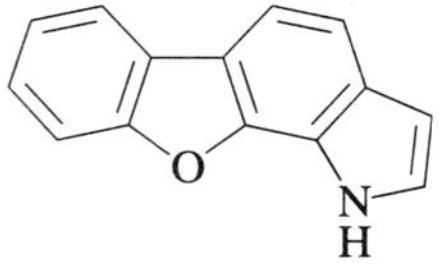	DMF, POCl$_3$	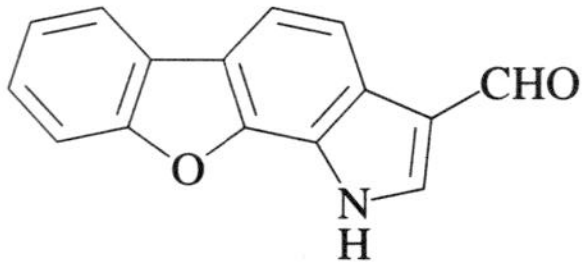 (100)	876

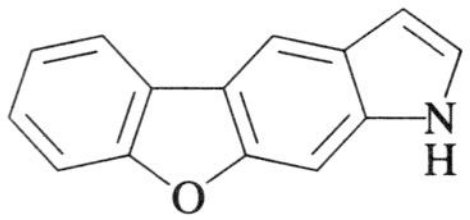

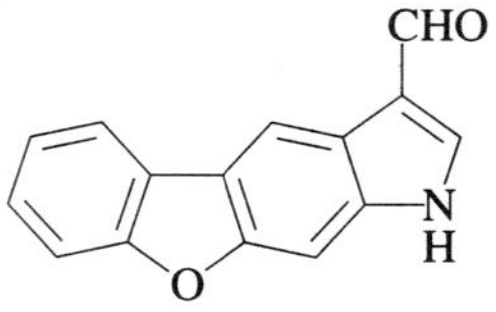

TABLE XVIII. OTHER HETEROCYCLES WITH THREE OR MORE FULLY CONJUGATED RINGS (*Continued*)

Substrate	Reagents	Product(s) and Yield(s) (%)	Refs.
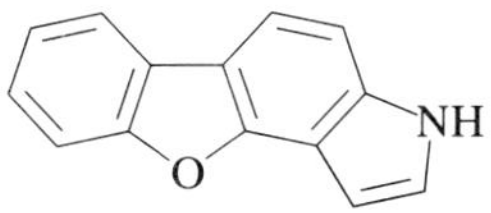	DMF, POCl$_3$	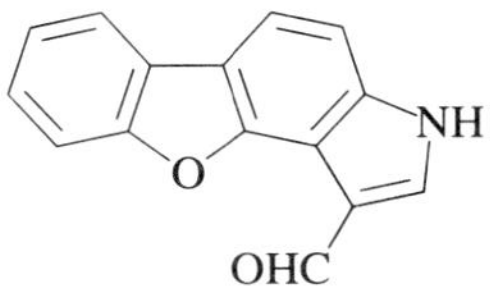 (100)	876
	—	(—)	877
C$_{14}$	C$_4$N/C$_4$S/C$_6$/C$_6$		
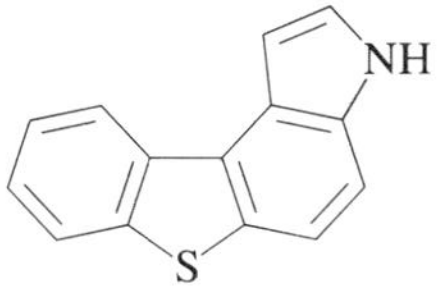	—	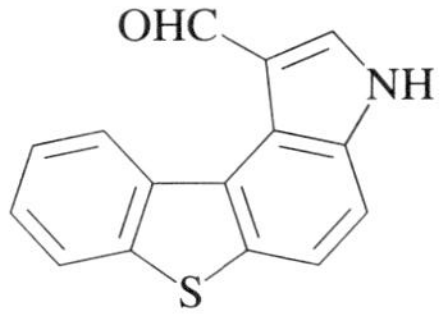 (—)	878
	C$_4$N/C$_4$NS/C$_6$/C$_6$		
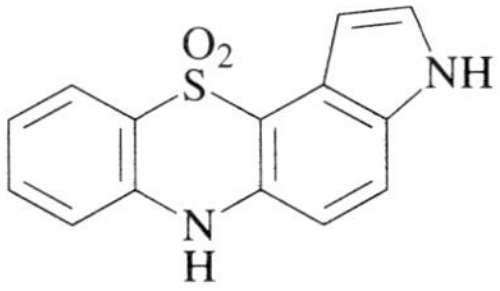	MFA, POCl$_3$	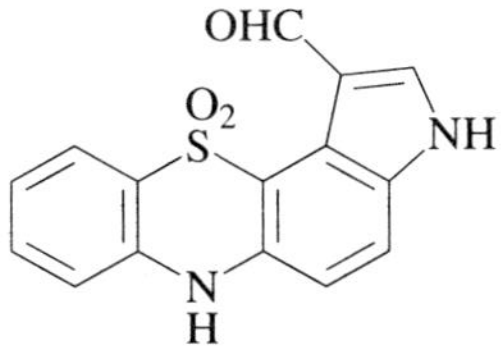 (59)	879
C$_{14}$			

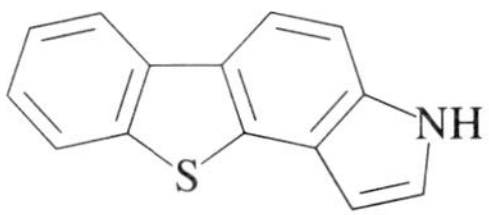

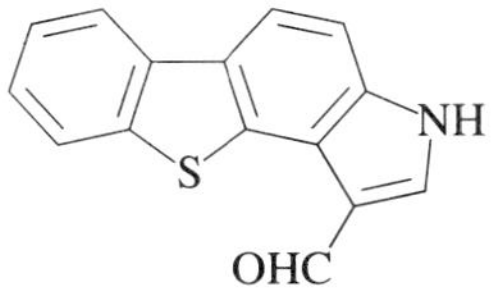

TABLE XVIII. OTHER HETEROCYCLES WITH THREE OR MORE FULLY CONJUGATED RINGS (*Continued*)

Substrate	Reagents	Product(s) and Yield(s) (%)	Refs.
C₁₆			

$C_4N/C_6/C_6/C_6$

—

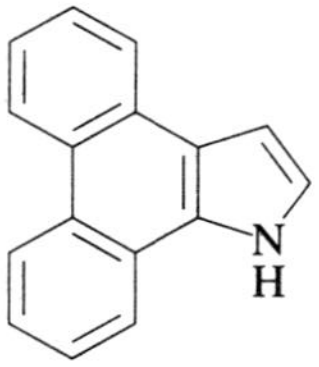

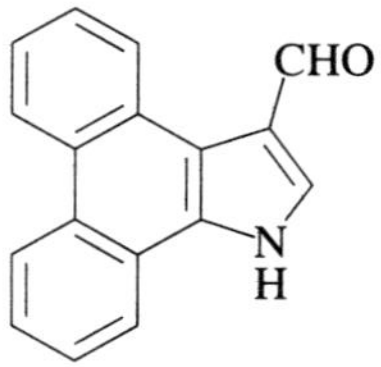

(85)

158

$C_4O/C_6/C_6/C_6/C_6$

C₂₄

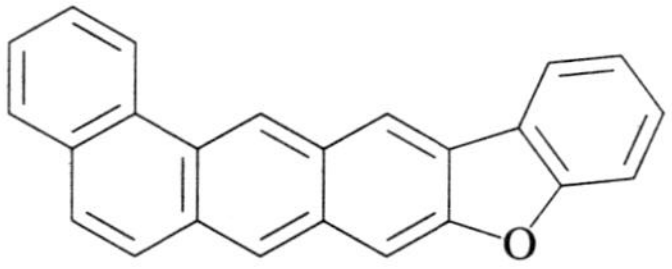

MFA, POCl₃

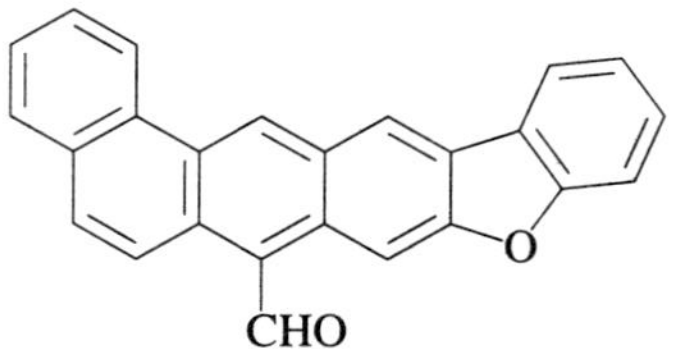

(2)

880

TABLE XIX. PORPHYRINS

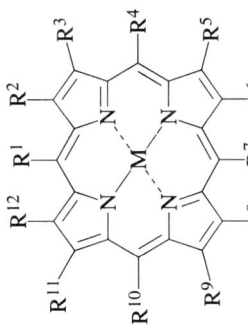

R¹ - R¹² = H except as indicated.
M = H₂ in substrate except as indicated,
and is unchanged in the product unless
indicated otherwise.

Substrate	Reagents	Product(s) and Yield(s) (%)	Refs.
C_{20}			
M = Cu	DMF, POCl₃	$R^4 = CHO$ (—)	881
C_{28}			
$R^1 = R^4 = R^7 = R^{10} = CO_2Me$	DMF, POCl₃	$R^3 = CHO$ (70)	882
C_{32}			
$R^2 = R^5 = R^8 = R^{11} = Me,$ $R^3 = R^6 = R^9 = R^{12} = Et, M = Ni$	DMF, POCl₃	$R^1 = CHO$ (95)	174, 175
$R^2 = R^5 = R^8 = R^{11} = Me,$ $R^3 = R^6 = R^9 = R^{12} = Et, M = Cu$	DMF, POCl₃	$R^1 = CHO, M = H_2$ (24) + $R^1 = R^7 = CHO$ (43) + $R^1 = R^4 = R^7 = CHO$ (5)	883
$R^3 = R^5 = R^9 = R^{11} = Me,$ $R^2 = R^6 = R^8 = R^{12} = Et, M = Ni$	DMF, POCl₃	$R^1 = R^4 = CHO$ (16) + $R^1 = R^7 = CHO$ (13) + $R^1 = R^4 = R^7 = CHO$ (16) + $R^1 = R^4 = R^7 = R^{10} = CHO$ (<1)	884, 885
$R^3 = R^5 = R^9 = R^{11} = Me,$ $R^2 = R^6 = R^8 = R^{12} = Et, M = Cu$	1. DMF, POCl₃ 2. NaBH₄	$R^1 = CH_2NMe_2$ (26) + $R^4 = CH_2NMe_2$ (65)	886
	DMF, POCl₃	$R^1 = CHO + R^4 = CHO$ (95)	886
	1. DMF, POCl₃ 2. NaBH₄	$R^1 = CH_2NMe_2 + R^4 = CH_2NMe_2$ (80-90), 1:2	886

299

TABLE XIX. PORPHYRINS (Continued)

Substrate	Reagents	Product(s) and Yield(s) (%)	Refs.
$R^2 + R^3 = R^5 + R^6 =$ $R^{11} + R^{12} = (CH_2)_4$, M = Ni	$Me_2NCH=CHCHO$, $POCl_3$	$R^4 = CH=CHO$ (65)	176
$R^2 = R^5 = R^9 = R^{11} = Me$, $R^6 = R^8 = (CH_2)_2CO_2Me$, M = Fe	DMF, $POCl_3$	$R^4 = CHO$ (13) + $R^4 = R^{12} = CHO$ (17)	887
$R^2 = R^5 = R^9 = R^{11} = Me$, $R^6 = R^8 = (CH_2)_2CO_2Me$, M = Cu	$i\text{-}Pr_2NCHO$, $POCl_3$	$R^1 = CHO$ (13) + $R^3 = CHO$ (28) + $R^4 = CHO$ (5) + $R^{12} = CHO$ (28)	888
	DMF, $POCl_3$	$R^1 = CHO, M = H_2 + R^3 = CHO, M = H_2$ (16) + $R^4 = CHO, M = H_2$ (2) + $R^{12} = CHO, M = H_2$ (0.5) + $R^3 = R^{12} = CHO, M = H_2$ (6) + $R^4 = R^{12} = CHO, M = H_2$ (1)	889
$R^2 = R^5 = R^9 = R^{11} = Me$, $R^6 = R^8 = (CH_2)_2CO_2Me$, M = Ni	1. $Me_2NCH=CHCHO$, $POCl_3$ 2. H_2SO_4 (conc.)	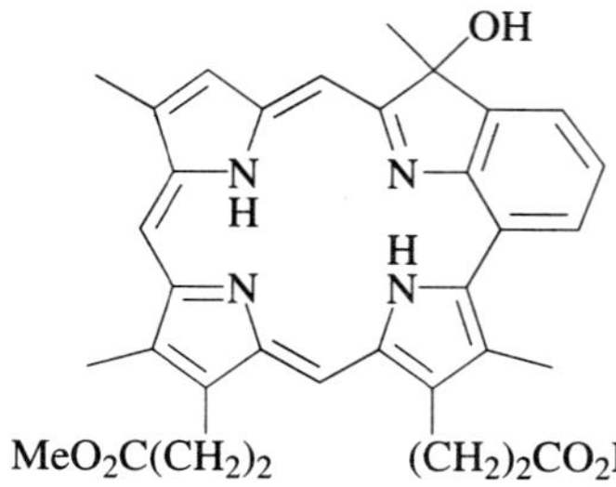 (10) 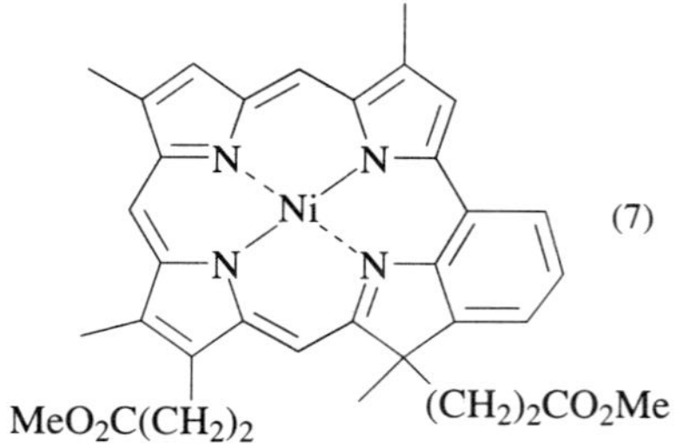 +	176

(10) +

(7)

300

TABLE XIX. PORPHYRINS (*Continued*)

Substrate	Reagents	Product(s) and Yield(s) (%)	Refs.
C$_{33}$			
R^1 = R^3 = R^5 = R^9 = R^{11} = Me, R^2 = R^6 = R^8 = R^{12} = Et, M = Ni	1. DMF, POCl$_3$ 2. MeNH$_2$	R^4 = CH=NMe + R^7 = CH=NMe (—), 9:1	890
R^1 = R^2 = R^6 = R^8 = R^{12} = Me, R^3 = R^5 = R^9 = R^{11} = Et, M = Ni	1. DMF, POCl$_3$ 2. MeNH$_2$	R^4 = CH=NMe (46) + R^7 = CH=NMe (46)	890
R^1 + R^{12} = (CH$_2$)$_2$, R^2 = R^6 = R^8 = Me, R^3 = R^5 = R^9 = R^{11} = Et, M = Ni	1. DMF, POCl$_3$ 2. MeNH$_2$	R^4 = CH=NMe (14) + R^7 = CH=NMe (57) + R^{10} = CH=NMe (tr)	890
R^2 + R^3 = (CH$_2$)$_5$, R^5 + R^6 = R^{11} + R^{12} = (CH$_2$)$_4$, M = Ni	Me$_2$NCH=CHCHO, POCl$_3$	R^4 = CH=CHCHO (72)	176
	Me$_2$NCH=CHCHO, POCl$_3$. (xs)	R^1 = R^4 = CH=CHCHO (44)	176

C$_{33-35}$

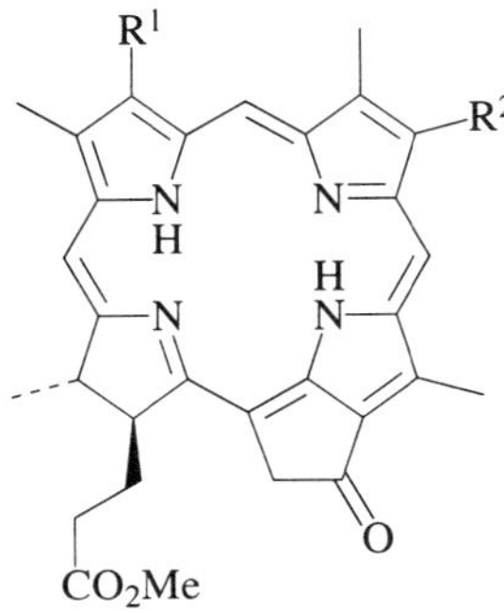

R^1 = CHOHMe or Et
R^2 = Et, *n*-Pr, *i*-Bu

| | DMF, POCl$_3$ | (—) | 891 |

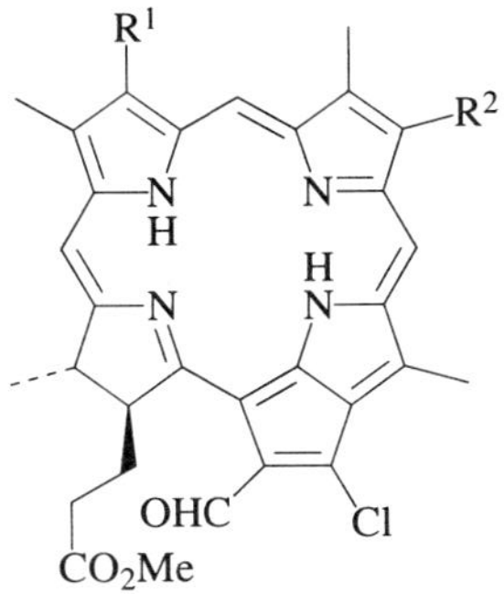

TABLE XIX. PORPHYRINS (*Continued*)

Substrate	Reagents	Product(s) and Yield(s) (%)	Refs.
C_{34}			
$R^2 = R^5 = R^6 = R^{11} = Me$, $R^3 = (CH_2)_2OH$, $R^6 = R^8 = (CH_2)_2CO_2Me$, $M = Cu$	$i\text{-}Bu_2NCHO$, $POCl_3$	$R^3 = (CH_2)_2Cl$, $R^{12} = CHO$ (44) + $R^3 = (CH_2)_2Cl$ (10) + $meso$-CHO (20)	177
$R^2 = R^5 = R^9 = R^{11} = Me$, $R^3 = R^{12} = Et$, $R^6 = R^8 = n\text{-}Pr$, $M = Cu$	DMF, $POCl_3$	$R^7 = CHO$ (75)	892
$C_{34\text{-}37}$			

Substrate	Reagents	Product(s) and Yield(s) (%)	Refs.
I, $R^1 = Et$, $R^2 + R^3 = COCH_2$, $R^4 = H$, $M = Ni$	$Me_2NCH=CHCHO$, $POCl_3$	**I**, $R^1 = Et$, $R^2 + R^3 = COCH_2$, $R^4 = CH=CHCHO$ (83)	176
I, $R^1 = CH=CH_2$, $R^2 + R^3 = COCH_2$, $R^4 = H$, $M = Ni$	$Me_2NCH=CHCHO$, $POCl_3$	**I**, $R^1 = CH=CH_2$, $R^2 + R^3 = COCH_2$, $R^4 = CH=CHCHO$ (63)	176
I, $R^1 = CH=CH_2$, $R^2 = R^4 = H$, $R^3 = CH_2CO_2Me$, $M = Fe$	DMF, $POCl_3$	**I**, $R^1 = CH=CHCHO$, $R^2 = R^4 = H$, $R^3 = CH_2CO_2Me$ (40)	887
I, $R^1 = CH=CH_2$, $R^2 = CO_2Me$, $R^3 = CH_2CO_2Me$, $R^4 = H$, $M = Cu$	DMF, $POCl_3$	**I**, $R^1 = CH=CHCHO$, $R^2 = CO_2Me$, $R^3 = CH_2COMe$, $R^4 = H$ (35)	893

TABLE XIX. PORPHYRINS (*Continued*)

Substrate	Reagents	Product(s) and Yield(s) (%)	Refs.
I, $R^1 = (CH_2)_2OH$, $R^2 = CO_2Me$, $R^3 = CH_2CO_2Me$, $R^4 = H$, $M = Cu$	DMF, $POCl_3$	**I**, $R^1 = (CH_2)_2Cl$, $R^2 = CO_2Me$, $R^3 = CH_2CO_2Me$, $R^4 = CHO$ (65)	894
I, $R^1 = CH=CH_2$, $R^2 = CO_2Me$, $R^3 = CH_2CO_2Me$, $R^4 = H$, $M = Fe$	DMF, $POCl_3$	**I**, $R^1 = CH=CHCHO$, $R^2 = CO_2Me$, $R^3 = CH_2CO_2Me$, $R^4 = H$ (7) + **I**, $R^1 = CH=CHCHO$, $R^2 = CO_2Me$, $R^3 = CH_2CO_2Me$, $R^4 = CHO$ (28)	887
I, $R^1 = Et$, $R^2 = CO_2Me$, $R^3 = CH_2CO_2Me$, $R^4 = H$, $M = Ni$	$Me_2NCH=CHCHO$, $POCl_3$	**I**, $R^1 = Et$, $R^2 = CO_2Me$, $R^3 = CH_2CO_2Me$, $R^4 = CH=CHCHO$ (89)	176
C$_{36}$			
$R^2 = R^3 = R^5 = R^6 = R^8 = R^9 = R^{11} = R^{12} = Et$, $M = Ni$	$Me_2NCH=CHCHO$, $POCl_3$	$R^4 = CH=CHCHO$ (85)	176
$R^2 = R^3 = R^5 = R^6 = R^8 = R^9 = R^{11} = R^{12} = Et$, $M = Cu$	$Me_2NCH=CHCHO$, $POCl_3$ (xs) DMF, $POCl_3$	$R^1 = R^4 = CH=CHCHO$ (55) $R^1 = CHO$ (20) + $R^1 = R^4 = CHO$ (11) + $R^1 = R^7 = CHO$ (10) + $R^1 = R^4 = R^7 = CHO$ (9) + $R^1 = R^4 = R^7 = R^{10} = CHO$ (1)	176 885
	DMF, $POCl_3$ $Me_2NCH=CHCHO$, $POCl_3$	$R^1 = CHO$ (72) $R^1 = CH=CHCHO$ (57)	892 176
$R^2 = R^3 = R^5 = R^6 = R^8 = R^9 = R^{11} = R^{12} = Et$, $M = Cu$	DMF, $POCl_3$	$R^1 = R^7 = CHO + R^1 = R^4 = CHO$ (23) + $R^1 = R^4 = R^7 = CHO$ (9) + $R^1 = R^4 = R^7 = R^{10} = CHO$ (<1)	884

303

TABLE XIX. PORPHYRINS (Continued)

Substrate	Reagents	Product(s) and Yield(s) (%)	Refs.
$R^2 = R^5 = R^9 = R^{11} = Me$, $R^3 = R^{12} = Et$, $R^6 = R^8 = (CH_2)_2CO_2Me$, $M = Cu, Ni, Fe, Co$	DMF, POCl$_3$	various meso-CHO	895
$R^2 = R^5 = R^9 = R^{11} = Me$, $R^3 = R^{12} = CH=CH_2$, $R^6 = R^8 = (CH_2)_2CO_2Me$, $M = Fe$	DMF, POCl$_3$	$R^{12} = CH=CHCHO$ (5) + $R^3 = R^{12} = CH=CHCHO$ (15)	887

I

Substrate	Reagents	Product(s) and Yield(s) (%)	Refs.
I, $R^1 = R^2 = H$, $M = Cu$	DMF, POCl$_3$	I, $R^1 = H$, $R^2 = CHO$ (—) + $R^1 = R^2 = CHO$ (—)	896
I, $R^1 = R^2 = H$, $M = Ni$	Me$_2$NCH=CHCHO, POCl$_3$	I, $R^1 = H$, $R^2 = CH=CHCHO$ (81)	176
$R^1 = CH=CH_2$, $R^2 = R^3 = R^5 = R^6 =$ $R^8 = R^9 = R^{11} = R^{12} = Et$, $M = Co$	DMF, POCl$_3$, 20°	$R^1 = CH=CHCHO$, $R^7 = Cl$ (4) + $R^1 = CH=CHCHO$ (85)	897
$R^1 = CH=CH_2$, $R^2 = R^3 = R^5 = R^6 =$ $R^8 = R^9 = R^{11} = R^{12} = Et$, $M = Ni$	DMF, POCl$_3$, 50° —	$R^1 = R^7 = CH=CHCHO$ (20) $R^1 = CH=CHCHO$ (—)	897 898

C$_{38}$

TABLE XIX. PORPHYRINS (*Continued*)

Substrate	Reagents	Product(s) and Yield(s) (%)	Refs.
C$_{39}$ (structure)	Me$_2$NCH=CHCHO, POCl$_3$	(structure) (90)	176
C$_{40}$ R^2 = R^5 = R^8 = R^{11} = Me, R^3 = R^6 = R^9 = R^{12} = (CH$_2$)$_2$CO$_2$Me, M = Ni	DMF, POCl$_3$	R^1 = CHO (83)	174
C$_{44}$ R^3 = R^5 = R^9 = R^{11} = Me, R^4 = R^{10} = 4-O$_2$NC$_6$H$_4$, R^2 = R^6 = R^8 = R^{12} = Et, M = Ni	DMF, POCl$_3$	R^1 = CHO (44)	899
R^2 = R^3 = R^5 = R^6 = R^8 = R^9 = R^{11} = R^{12} = Pr, M = Ni	1. DMF, POCl$_3$ 2. MeNH$_2$ 3. H$_2$SO$_4$	R^4 = CH=NMe, M = H$_2$ (83)	900
C$_{44-58}$ R^1 = R^4 = R^7 = R^{10} = Ph, M = Cu	DMF, POCl$_3$	R^3 = CHO (99)	901
R^1 = R^4 = R^7 = R^{10} = Ph, M = Co	DMF, POCl$_3$	R^4 = CH=NMe$_2$$^+$ $^-$OPOCl$_2$ (97)	901
	1. DMF, POCl$_3$ 2. H$_2$SO$_4$ (conc) 3. NaOAc	R^3 = CHO, M = H$_2$ (65)	901

305

TABLE XIX. PORPHYRINS (*Continued*)

Substrate	Reagents	Product(s) and Yield(s) (%)	Refs.
$R^1 = R^4 = R^7 = R^{10} = 4\text{-MeC}_6\text{H}_4$,	DMF, POCl$_3$	R^3 = CHO (50) + dialdehyde (2)	902
$\quad$ M = H$_2$			
$\quad$ M = Cu	DMF, POCl$_3$	R^3 = CHO (58)	902
$\quad$ M = Co	1. DMF, POCl$_3$	R^3 = CHO, M = Co(Py)$_2$ (71)	902
	2. Py		
$\quad$ M = Ni	DMF, POCl$_3$	R^3 = CHO (81)	902
$\quad$ M = Pd	DMF, POCl$_3$	R^3 = CHO (87)	902
$\quad$ M = Pt	DMF, POCl$_3$	R^3 = CHO (84)	902
$\quad$ M = FeCl$_2$	DMF, POCl$_3$	R^3 = CHO (16)	902
$\quad$ M = Al(OH)$_2$	DMF, POCl$_3$	R^3 = CHO (59)	902
$\quad$ M = Cr(OH)$_2$	1. DMF, POCl$_3$	R^3 = CHO, M = Cr(Py)$_2$ (63)	902
	2. Py		
$\quad$ M = CoClPy	DMF, POCl$_3$	R^3 = CHO (65)	902
$\quad$ M = MnOMe	DMF, POCl$_3$	R^3 = CHO, M = H$_2$ (14)	902
C$_{46}$			
$R^3 = R^5 = R^9 = R^{11} = \text{Me}$,	DMF, POCl$_3$	R^1 = CHO (90)	899
$\quad R^4 = R^{10} = 4\text{-MeC}_6\text{H}_4$,			
$\quad R^2 = R^6 = R^8 = R^{12} = \text{Et}, \text{M} = \text{Ni}$			
$R^3 = R^5 = R^9 = R^{11} = \text{Me}$,	DMF, POCl$_3$	R^1 = CHO (90)	899
$\quad R^4 = R^{10} = \text{Ts}$,			
$\quad R^2 = R^6 = R^8 = R^{12} = \text{Et}, \text{M} = \text{Ni}$			
C$_{48}$			
$R^3 = R^5 = R^9 = R^{11} = \text{Me}$,	DMF, POCl$_3$	R^1 = CHO (60)	899
$\quad R^4 = R^{10} = 4\text{-Me}_2\text{NC}_6\text{H}_4$,			
$\quad R^2 = R^6 = R^8 = R^{12} = \text{Et}, \text{M} = \text{Ni}$			

TABLE XIX. PORPHYRINS (Continued)

Substrate	Reagents	Product(s) and Yield(s) (%)	Refs.
C_{56}			
$R^2 = R^3 = R^5 = R^6 = R^8 = R^9 = R^{11} = R^{12} = Et,$ $R^4 = R^{10} = 1\text{-}(2\text{-naphtholyl}),\ M = Ni$ + $R^2 = R^3 = R^5 = R^6 = R^8 = R^9 = R^{11} = R^{12} = Et,$ $R^7 = R^{10} = 1\text{-}(2\text{-naphtholyl}),\ M = Ni$	DMF, POCl$_3$	$R^1 = CHO$ (65)	903
C_{64}			
$R^1 = R^4 = R^7 = R^{10} =$ $C_6H_4NHCOBu\text{-}t,\ M = Cu$	1. DMF, POCl$_3$ 2. NaBH$_4$	$R^1 = CH_2OH$ (—)	904
	DMF, POCl$_3$	$R^1 = CHO$ (66)	905
C_{73}	DMF, POCl$_3$	$R^2 = CHO$ (17) + $R^1 = R^{1'} = CHO$ (2) + $R^2 = R^{2'} = CHO$ (33) + $R^2 = R^{1'} = CHO$ (16)	906

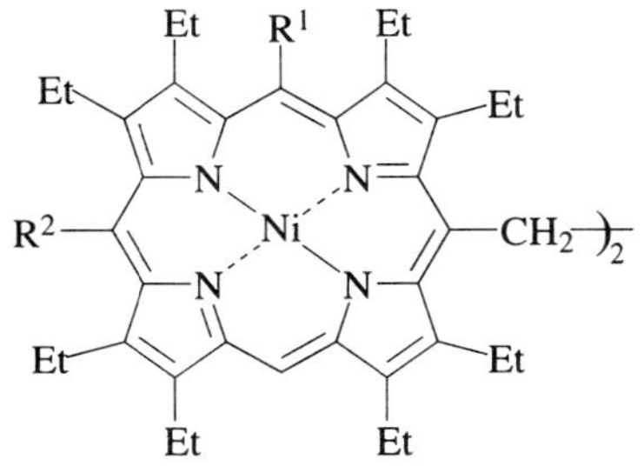

REFERENCES

[1] Fischer, O.; Müller, A.; Vilsmeier, A. *J. Prakt. Chem.* **1925**, *109*, 69.

[2] Vilsmeier, A.; Haack, A. *Ber. Dtsch. Chem. Ges.* **1927**, *60B*, 119.

[3] Simpson, A. J. Ph.D. Thesis, University of Northumbria at Newcastle, 1994.

[4] Vilsmeier, A. *Chem.-Ztg.* **1951**, *75*, 133.

[5] Bayer, O. in *Methoden Der Organischen Chemie (Houben-Weyl)*, Thieme, Stuttgart, 1954.

[6] Bredereck, H.; Gompper, R.; van Schuh, H. G.; Theilig, G. *Angew. Chem.* **1959**, *71*, 753.

[7] Eilingsfeld, H.; Seefelder, M.; Weidinger, H. *Angew. Chem.* **1960**, *72*, 836.

[8] Minkin, V. I.; Dorofeenko, G. N. *Uspekhi Khim.* **1960**, *29*, 1301 (*Chem. Abstr.* **1961**, *55*, 12265h).

[9] Oda, R.; Yamamoto, K. *Kagaku (Kyôto)* **1960**, *15*, 384 (*Chem. Abstr.* **1960**, *54*, 24325c).

[10] de Maheas, M.-R. *Bull. Soc. Chim. Fr.* **1962**, 1989.

[11] Hafner, K.; Häfner, K. H.; König, C.; Kreuder, M.; Ploss, G.; Schulz, G.; Strum, E.; Vöpel, K. H. *Angew. Chem., Int. Ed. Engl.* **1963**, *2*, 123.

[12] Gore, P. H. in *Friedel Crafts and Related Reactions*, Olah, G. A., Ed., Vol. III, Wiley, New York, 1964.

[13] Hazebroucq, G. *Ann. Pharm. Fr.* **1966**, *24*, 793.

[14] Jutz, C. *Chem. Lab. Betrieb.* **1968**, *19*, 289 and 337.

[15] Ulrich, H. in *The Chemistry of Imidoyl Halides, Plenum Press, New York, 1968*.

[16] Kuehne, M. E. in *Enamines: Synthesis, Structure and Reactions*, Cook, A. G., Ed., Dekker, New York, 1969.

[17] Seshadri, S. *J. Sci. Ind. Res.* **1973**, *32*, 128.

[18] Jutz, C. in *Advances in Organic Chemistry: Methods and Results*, Taylor, E. C., Ed., Boehme, H.; Viehe, H. G., Eds., Vol. 9, Interscience, New York, **1976**, p. 225.

[19] Meth-Cohn, O.; Tarnowski, B. *Adv. Heterocycl. Chem.* **1982**, *31*, 207.

[20] Simchen, G. in *Methoden Der Organischen Chemie (Houben-Weyl)*, Thieme, Stuttgart, 1983.

[21] Marson, C. M. *Tetrahedron* **1992**, *48*, 3659.

[22] Meth-Cohn, O.; Stanforth, S. P. in *Comprehensive Organic Synthesis*, Trost, B. M.; Fleming, I. Eds.; Vol. 2, Heathcock, C. H., Vol. Ed., Pergamon, Oxford, **1991**, p. 777.

[23] Meth-Cohn, O. *Heterocycles* **1993**, *35*, 539.

[24] Kantlehner, W. in *Advances in Organic Chemistry: Methods and Results*, Taylor, E. C., Ed., Vol. 9, Boehme, H.; Viehe, H. G., Vol. Eds., Interscience, New York, **1976**, pp. 5 and 65.

[25] Liebscher, J.; Hartmann, H. *Synthesis* **1979**, 241.

[26] Downie, I. M.; Earle, M. J.; Heaney, H.; Shuhaibar, K. F. *Tetrahedron* **1993**, *49*, 4015.

[27] Martin, G. J.; Poignnant, S. *J. Chem. Soc., Perkin Trans. 2* **1972**, 1964.

[28] Martin, G. J.; Poignnant, S. *J. Chem. Soc., Perkin Trans. 2* **1974**, 642.

[29] Fritz, H.; Oehl, R. *Justus Liebigs Ann. Chem.* **1971**, *749*, 159.

[30] Rivett, D. E; Rosevear, J.; Wilshire, J. F. K. *Aust. J. Chem.* **1979**, *32*, 1601.

[31] Zollinger, H.; Bosshard, H. H. *Helv. Chim. Acta* **1959**, *42*, 1659.

[32] Campaigne, E.; Archer, W. L. *Org. Synth.* **1953**, *33*, 27.

[33] Grundmann, C.; Dean, J. M. *Angew. Chem., Int. Ed. Engl.* **1965**, *4*, 955.

[34] Ueda, C.; Mitsui, A.; Maeda, S.; Wataba, H. *Jpn. Kokai* 52125138, 1977 (*Chem. Abstr.* **1978**, *88*, 89367n).

[35] Bestmann, H. J.; Lienert, J.; Mott, L. *Justus Liebigs Ann. Chem.* **1968**, *718*, 24.

[36] Oda, R.; Yamamoto, K. *Nippon Kagaku Zasshi* **1962**, *83*, 1292 (*Chem. Abstr.* **1963**, *59*, 11399g).

[37] Shah, R. C.; Deshpande, R. K.; Chaubal, J. S. *J. Chem. Soc.* **1932**, 642.

[38] Ullrich, F.-W.; Breitmaier, E. *Synthesis* **1983**, 641.

[39] Jutz, C. *Chem. Ber.* **1958**, *91*, 850.

[40] Ipach, I.; Lerche, H.; Mayring, L.; Severin, T. *Chem. Ber.* **1979**, *112*, 2565.

[41] Tsujimoto, M.; Akahori, H.; Hasegawa, K.; Asano, M. *PCT Int. Appl.WO 8303840, 1983* (Chem. Abstr. **1984**, *100*, 69872m).

[42] Wolter, G.; Kosler, W. *Z. Chem.* **1970**, *10*, 401.

[43] Hirota, T.; Fujita, H.; Sasaki, K.; Namba, T.; Hayakawa, S. *Heterocycles* **1986**, *24*, 771.

[44] Hirota, T.; Tashima, Y.; Sasaki, K.; Namba, T.; Hayakawa, S. *Heterocycles* **1987**, *26*, 2717.

[45] Hirota, T.; Koyama, T.; Namba, T.; Yamato, M.; Matsumura, T. *Chem. Pharm. Bull.* **1978**, *26*, 245.

[46] Wiegrebe, W.; Sasse, D.; Reinhart, H.; Faber, L. *Z. Naturforsch., Teil B* **1970**, *B25*, 1408.

[47] Garcia Martinez, A.; Martinez Alvarez, R.; Osio Barcina, J.; De la Moya Cerero, S.; Teso Villar, E.; Garcia Fraile, A.; Hanack, M.; Subramanian, L. R. *J. Chem. Soc., Chem. Commun.* **1990**, 1571.

[48] Jacob III, P.; Anderson III, G.; Meshul, C. K.; Shulgin, A. T.; Castagnoli, N., Jr. *J. Med. Chem.* **1977**, *20*, 1235.

[49] Mielicki, J.; Szwarc, J. *Pol. PL 97750*, 1978 (*Chem. Abstr.* **1979**, *90*, 170161d).

[50] Iwata, M.; Emoto, S. *Bull. Chem. Soc. Jpn.* **1974**, *47*, 1687.

[51] Mergler, M.; Tanner, R.; Gosteli, J.; Grogg, P. *Tetrahedron Lett.* **1988**, *29*, 4005.

[52] Mangoni, L. *Ann. Chim. (Rome)* **1958**, *48*, 930 (*Chem. Abstr.* **1959**, *53*, 8050i).

[53] Neumann, W. P.; Hillgärtner, H.; Baines, K. M.; Dicke, R.; Vorspohl, K.; Kobs, U.; Nussbeutel, U. *Tetrahedron* **1989**, *45*, 951.

[54] Shawcross, A. P.; Stanforth, S. P. *Tetrahedron* **1989**, *45*, 7063.

[55] Shawcross, A. P.; Stanforth, S. P. *Tetrahedron* **1988**, *44*, 1461.

[56] Buyukliev, R. T.; Pojarlieff, I. G. *Dokl. Bolg. Akad. Nauk* **1987**, *40*, 71 (*Chem. Abstr.* **1988**, *109*, 149021j).

[57] Buu-Hoï, N. P.; Lavit, D. *Bull. Soc. Chim. Fr.* **1955**, 1419.

[58] Hoan, N. *C. R. Hebd. Seances Acad. Sci.* **1954**, *238*, 1136 (*Chem. Abstr.* **1955**, *49*, 3914b).

[59] Barton, D. H. R.; Dawes, C. C.; Franceschi, G.; Foglio, M.; Ley, S. V.; Magnus, P. D.; Mitchell, W. L.; Temperelli, A. *J. Chem. Soc., Perkin Trans. 1* **1980**, 643.

[60] Fieser, L. F.; Hartwell, J. L. *J. Am. Chem. Soc.* **1938**, *60*, 2555.

[61] Bronovitskaya, V. P.; Kizil'shtein, A. A.; Berlin, A. Ya. *Zh. Org. Khim.* **1968**, *4*, 340 (*Chem. Abstr.* **1968**, *68*, 114284e).

[62] Fujisawa, T.; Sakai, K. *Tetrahedron Lett.* **1976**, 3331.

[63] Jutz, C.; Amschler, H. *Chem. Ber.* **1964**, *97*, 3331.

[64] Jutz, C.; Schweiger, E. *Synthesis* **1974**, 193.

[65] Jutz, C.; Kirchlechner, R.; Seidel, H.-J. *Chem. Ber.* **1969**, *102*, 2301.

[66] Hafner, K.; Bernard, C. *Justus Liebigs Ann. Chem.* **1959**, *625*, 108.

[67] Anderson, A. G.; Breazeale, R. D. *J. Org. Chem.* **1969**, *34*, 2375.

[68] Jutz, C. *Angew. Chem.* **1958**, *70*, 270.

[69] Neidlein, R.; Hartz, G. *Helv. Chim. Acta* **1985**, *68*, 255.

[70] Yoshida, Z. *Jpn. Kokai Tokkyo Koho*, 60188362, 1985 (*Chem. Abstr.* **1986**, *104*, 88187r).

[71] Graham, P. J. *US Patent*, 2849469, 1958 (*Chem. Abstr.* **1959**, *53*, 4298a).

[72] Johnson, B. F. G.; Lewis, J.; McArdle, P.; Randall, G. L. P. *J. Chem. Soc., Dalton Trans.* **1972**, 456.

[73] Rausch, M. D.; Genetti, R. A. *J. Org. Chem.* **1970**, *35*, 3888.

[74] Bouka-Poba, J-P.; Farnier, M.; Guilard, R. *Can. J. Chem.* **1981**, *59*, 2962.

[75] Bouka-Poba, J-P.; Farnier, M.; Guilard, R. *Tetrahedron Lett.* **1979**, 1717.

[76] Chadwick, D. J.; Chambers, J.; Hargraves, H. E.; Meakins, G. D.; Snowden, R. L. *J. Chem. Soc., Perkin Trans. 1* **1973**, 2327.

[77] Gjos, N.; Gronowitz, S. *Acta Chem. Scand.* **1970**, *24*, 99.

[78] Meth-Cohn, O.; Narine, B.; Tarnowski, B. *J. Chem. Soc., Perkin Trans. 1* **1981**, 1531.

[79] King, W. J.; Nord, F. F. *J. Org. Chem.* **1948**, *13*, 635.

[80] Weston, A. W.; Michaels, Jr., R. J. *J. Am. Chem. Soc.* **1950**, *72*, 1422.

[81] Clarke, K.; Fox, W. R.; Scrowston, R. M. *J. Chem. Soc., Perkin Trans. 1* **1980**, 1029.

[82] Shvedov, V. I.; Vasil'eva, V. K.; Grinev, A. N. *Khim. Geterotsikl. Soedin.* **1972**, 427 (*Chem. Abstr.* **1972**, *77*, 88356u).

[83] Gol'dfarb, Y. A.; Kalik, M. A. *Khim. Geterotsikl. Soedin* **1971**, *7*, 171 (*Chem. Abstr.* **1971**, *75*, 48798s).

[84] Yur'ev, Y. K.; Mezentsova, N. N. *Zh. Obshch. Khim.* **1957**, *27*, 179 (*Chem. Abstr.* **1957**, *51*, 12878b).

[85] Jones, R. A.; Candy, C. F.; Wright, P. H. *J. Chem. Soc. C* **1970**, 2563.

[86] White, J.; McGillivray, G. *J. Chem. Soc., Perkin Trans. 2* **1984**, 1179.

[87] Ermili, A.; Castro, A. J.; Westfall, P. A. *J. Org. Chem.* **1965**, *30*, 339.

[88] Atkinson, J. H.; Grigg, R.; Johnson, A. W. *J. Chem. Soc.* **1964**, 893.

[89] Bordner, J.; Rapoport, H. *J. Org. Chem.* **1965**, *30*, 3824.

[90] Bray, B. L.; Hess, P.; Muchowski, J. M.; Scheller, M. E. *Helv. Chim. Acta* **1988**, *71*, 2053.

[91] Muchowski, J. M.; Scheller, M. E. *Synth. Commun.* **1987**, *17*, 863.

[92] Smith, K. M.; Pandey, R. K. *J. Chem. Soc., Perkin Trans. 1* **1987**, 1229.

[93] Byun, Y. -S.; Lightner, D. A. *J. Heterocycl. Chem.* **1991**, *28*, 1683.

[94] Faber, K.; Anderson, H. J.; Loader, C. E.; Daley, A. S. *Can. J. Chem.* **1984**, *62*, 1046.

[95] Suu, V. T.; Buu-Hoï, N. P.; Xuong, N. D. *Bull. Soc. Chim. Fr.* **1962**, 1875.

[96] Ghaisas, V. V. *J. Org. Chem.* **1957**, *22*, 703.

[97] Klohr, S. E.; Cassady, J. M. *Synth. Commun.* **1988**, *18*, 671.

[98] Royer, R.; Demerseman, P.; Rossignol, J.-F.; Cheutin, A. *Bull. Soc. Chim. Fr.* **1971**, 2072.

[99] Clarke, K.; Scrowston, R. M.; Sutton, T. M. *J. Chem. Soc., Perkin Trans. 1* **1973**, 623.

[100] Ricci, A.; Balucani, D.; Buu-Hoï, N. P. *J. Chem. Soc. C* **1967**, 779.

[101] Anthony, W. C. *J. Org. Chem.* **1960**, *25*, 2049.

[102] Anthony, W. C.; Szmuszkovicz, J. *US Patent 3361759*, 1968 (*Chem. Abstr.* **1968**, *69*, 27242s).

[103] Cipiciani, A.; Clementi, S.; Linda, P.; Marino, G.; Savelli, G. *J. Chem. Soc., Perkin Trans. 2* **1977**, 1284.

[104] Chatterjee, A.; Biswas, K. M. *J. Org. Chem.* **1973**, *38*, 4002.

[105] Whittle, C. W.; Castle, R. N. *J. Pharm. Sci.* **1963**, *52*, 645.

[106] Simakov, S. V.; Velezheva, V. S.; Kozik, T. A.; Suvorov, N. N. *Khim. Geterotsikl. Soedin.* **1985**, 76 (*Chem. Abstr.* **1985**, *103*, 53907c).

[107] Bhuyan, P. J.; Boruah, R. C.; Sandhu, J. S. *Tetrahedron Lett.* **1989**, *30*, 1421.

[108] Walkup, R. E.; Linder, J. *Tetrahedron Lett.* **1985**, *26*, 2155.

[109] Black, D. St. C.; Kumar, N.; Wong, L. C. H. *Synthesis* **1986**, 474.

[110] Narasimhan, N. S.; Dhauale, D. D. *Indian J. Chem., Sect. B* **1984**, *23B*, 199.

[111] Raison, C. G. *J. Chem. Soc.* **1949**, 3319.

[112] Dalton, L. K.; Demerac, S.; Teitei, T. *Aust. J. Chem.* **1969**, *22*, 185.

[113] Buu-Hoï, N. P.; Hoán, N. *J. Am. Chem. Soc.* **1951**, *73*, 98.

[114] Patel, B. P. J. *J. Indian Chem. Soc.* **1985**, *62*, 534.

[115] Mal'tseva, S. P.; Borodulina, Z. A.; Smepanov, B. I. *Zh. Org. Khim.* **1973**, *9*, 815 (*Chem. Abstr.* **1973**, *79*, 18631x).

[116] Simay, A.; Takacs, K.; Horvath, K.; Dvortsak, P. *Acta Chim. Acad. Sci. Hung.* **1980**, *105*, 127 (*Chem. Abstr.* **1981**, *95*, 42979t).

[117] Rockley, J. E.; Summers, L. A. *Aust. J. Chem.* **1981**, *34*, 1117.

[118] Kishida, M.; Hamaguchi, H.; Akita, T. *Jpn. Kokai Tokkyo Koho JP 63267762*, 1988 (*Chem. Abstr.* **1989**, *111*, 57728h).

[119] Koshelev, Yu. N.; Kvitko, I. Ya.; Efros, L. S. *Zh. Org. Khim.* **1972**, *8*, 1750 (*Chem. Abstr.* **1972**, *77*, 139876v).

[120] Andreani, A.; Rambaldi, M.; Andreani, F.; Bossa, R.; Galatulas, I. *Eur. J. Med. Chem.* **1988**, *23*, 385.

[121] Becher, J.; Olesen, P. H.; Knudsen, N. A.; Tofflund, H. *Sulfur Lett.* **1986**, *4*, 175.

[122] Beccalli, E. M.; Marchesini, A. *J. Org. Chem.* **1987**, *52*, 3426.

[123] Belen'kii, L. I.; Ceskis, M. *Khim. Geterotsikl. Soedin.* **1984**, 881 (*Chem. Abstr.* **1984**, *101*, 191754k).

[124] Kubota, I.; Moriya, T.; Matsumoto, K. *Chem. Pharm. Bull.* **1990**, *38*, 570.

[125] Chatterjee, A.; Padhi, K. *Indian J. Chem., Sect. B* **1979**, *18B*, 82.

[126] Nohara, A. *Tetrahedron Lett.* **1974**, 1187.

[127] Yeh, M.-Y.; Tien, H.-J.; Huang, L.-Y.; Chen, M.-H. *J. Chin. Chem. Soc. (Taipei)* **1983**, *30*, 29 (*Chem. Abstr.* **1983**, *99*, 105178m).

[128] Sugasawa, T.; Sasakura, K.; Toyoda, T. *Chem. Pharm. Bull.* **1974**, *22*, 763.

[129] Seela, F.; Steker, H. *Heterocycles* **1985**, *23*, 2521.

[130] Yoneda, F.; Mori, K.; Ono, M.; Kadokawa, Y.; Nagao, E.; Yamaguchi, H. *Chem. Pharm. Bull.* **1980**, *28*, 3514.

[131] Wynberg, H.; Feijen, J. *Rec. Trav. Chim. Pays Bas* **1970**, *89*, 77.

[132] Konar, A. *Chem. Scr.* **1983**, *22*, 177.

[133] Gronowitz, S.; Konar, A.; Litvinov, V. P. *Izv. Akad. Nauk SSSR, Ser. Khim.* **1981**, 1363 (*Chem. Abstr.* **1981**, *95*, 132712e).

[134] Soth, S.; Farnier, M.; Fournari, P. *Bull. Soc. Chim. Fr.* **1975**, 2511.

[135] Kralovicova, E.; Krutosikova, A.; Kovac, J.; Dandarova, M. *Collect. Czech. Chem. Commun.* **1986**, *51*, 106.

[136] McKenzie, S.; Molloy, B. B.; Reid, D. H. *J. Chem. Soc. C* **1966**, 1908.

[137] Andreani, A.; Bonazzi, D.; Rambaldi, M.; Greci, L. *Boll. Chim. Farm.* **1979**, *118*, 694 (*Chem. Abstr.* **1980**, *93*, 150175v).

[138] Duguay, G.; Reid, D. H.; Wade, K. O.; Webster, R. G. *J. Chem. Soc. C* **1971**, 2829.

[139] Yakhontov, L. N.; Liberman, S. S.; Krasnokutskaya, D. M.; Uritskaya, M. Y.; Azimov, V. A.; Sokolova, M. S.; Gerchikov, L. N. *Khim.-Farm. Zh.* **1974**, *8*, 5 (*Chem. Abstr.* **1975**, *82*, 57581n).

[140] Shiotani, S.; Morita, H.; Ishida, T.; In, Y. *J. Heterocycl. Chem.* **1988**, *25*, 1205.

[141] McKenzie, S.; Reid, D. H. *J. Chem. Soc. C* **1970**, 145.

[142] McKenzie, S.; Reid, D. H. *J. Chem. Soc., Chem. Commun.* **1966**, 401.

[143] Reid, D. H.; Webster, R. G.; McKenzie, S. *J. Chem. Soc., Perkin Trans. 1* **1979**, 2334.

[144] Fuentes, O.; Paudler, W. W. *J. Heterocycl. Chem.* **1975**, *12*, 379.

[145] Buchan, R.; Fraser, M.; Shand, C. *J. Org. Chem.* **1976**, *41*, 351.

[146] Davies, L. S.; Jones, G. *Tetrahedron Lett.* **1969**, 1549.

[147] Flitsch, W.; Gurke, A.; Mueter, B. *Chem. Ber.* **1975**, *108*, 2969.

[148] Mozhaeva, T. Ya.; Samsonova, O. L.; Savel'ev, V. L. *Khim. Geterotsikl. Soedin.* **1988**, 1287 (*Chem. Abstr.* **1989**, *110*, 212552).

[149] Ishii, H.; Kobayashi, J.-I.; Ishikawa, T. *Chem. Pharm. Bull.* **1983**, *31*, 3330.

[150] Heber, D. *Arch. Pharm.* **1987**, *320*, 595.

[151] Eldredge, C. H.; Mee, J. D. *Fr. Demande FR* 2027006, 1970 (*Chem. Abstr.* **1971**, *75*, 50420f).

[152] Roma, G.; Di Braccio, M.; Balbi, A.; Mazzei, M.; Ermili, A. *J. Heterocycl. Chem.* **1987**, *24*, 329.

[153] Soldatenkov, A. T.; Bagdadi, M. V.; Radzhan, P. K.; Brindkha, O. S.; Edogiaverie, S. L.; Fomichev, A. A.; Prostakov, N. S. *Zh. Org. Khim.* **1983**, *19*, 1326 (*Chem. Abstr.* **1983**, *99*, 175563g).

[154] Hosmane, R. S.; Hiremath, S. P.; Schneller, S. W. *J. Heterocycl. Chem.* **1974**, *11*, 29.

[155] Hiremath, S. P.; Badami, P. S.; Purohit, M. G. *Indian J. Chem., Sect. B* **1983**, *22B*, 437.

[156] Bisagni, M.; Buu-Hoï, N. P.; Roger, R. *J. Chem. Soc.* **1955**, 3688.

[157] Chaterjea, J. N.; Lal, S.; Jha, U.; Rycroft, D. S.; Carnduff, J. *J. Chem. Res. (S)* **1979**, 356.

[158] Mirzametova, R. M.; Buyanov, V. N.; Suvorov, N. N. *Deposited Doc.*, VINITI, 1868 **1978** (*Chem. Abstr.* **1980**, *92*, 6355r).

[159] Sikharulidze, M. I.; Khoshtariya, T. E.; Kurkovskaya, L. N.; Suvorov, N. N. *Khim. Geterotsikl. Soedin.* **1981**, 497 (*Chem. Abstr.* **1981**, *95*, 97628h).

[160] Trapaidze, M. V.; Samsoniya, Sh. A.; Kuprashvili, N. A.; Mamaladze, L. M.; Suvorov, N. N. *Khim. Geterotsikl. Soedin.* **1988**, 603 (*Chem. Abstr.* **1989**, *110*, 114709h).

[161] Chapman, N. B.; Hughes, C. G.; Scrowston, R. M. *J. Chem. Soc. C* **1971**, 463.

[162] Tanaka, A.; Yakushijin, K.; Yoshina, S. *J. Heterocycl. Chem.*, **1977**, *14*, 975.

[163] Clarke, K.; Fox, W. R.; Scrowston, R. M. *J. Chem. Res. (S)* **1980**, 33.

[164] Aryuzina, V. M.; Shchukina, M. N. *Khim.-Farm. Zh.* **1972**, *6*, 22 (*Chem. Abstr.* **1972**, *77*, 48403a).

[165] Priimenko, B. A.; Klyuer, N. A.; Sapozhnikov, Yu. M.; Garmash, S. N.; Kochergin, P. M. *Khim. Geterotsikl. Soedin.* **1981**, 1251 (*Chem. Abstr.* **1981**, *95*, 220012g).

[166] El-Shorbaji, A.-N.; Sakai, S.-i.; El-Gendy, M. A.; Omar, N.; Farag, H. H. *Chem. Pharm. Bull.* **1989**, *37*, 2971.

[167] Fuentes, O.; Paudler, W. W. *J. Org. Chem.* **1975**, *40*, 3065.

[168] Effi, Y.; Lancelot, J. -C.; Rault, S.; Robba, M. *J. Heterocycl. Chem.* **1986**, *23*, 17.

[169] Rioult, J. P.; Cugnon de Sevricourt, M.; Rault, S.; Robba, M. *J. Heterocycl. Chem.* **1984**, *21*, 1449.

[170] Itaya, T. *Chem. Pharm. Bull.* **1987**, *35*, 4372.

[171] Itaya, T.; Morisue, M.; Takeda, M.; Kumazawa, Y. *Chem. Pharm. Bull.* **1990**, *38*, 2656.

[172] Sakovich, G. S.; Konyukhov, V. N.; Degtyarev, V. F.; Pushkareva, Z. V. *Khim. Geterotsikl. Soedin.* **1972**, 213 (*Chem. Abstr.* **1972**, *76*, 153546b).

[173] Pagani, G. A. *J. Chem. Soc., Perkin Trans. 2* **1974**, 1392.

[174] Morgan, A. R.; Rampersaud, A.; Garbo, G. M.; Keck, R. W.; Selman, S. H. *J. Med. Chem.* **1989**, *32*, 904.

[175] Johnson, A. W.; Oldfield, D. *J. Chem. Soc. C* **1966**, 794.

[176] Graça, M.; Vicente, H.; Smith, K. M. *J. Org. Chem.* **1991**, *56*, 4407.

[177] Smith, K. M.; Fujinari, E. M.; Langry, K. C.; Parish, D. W.; Tabba, H. D. *J. Am. Chem. Soc.* **1983**, *105*, 6638.

[178] Laird, T. in *Comprehensive Organic Chemistry*, Barton, D. H. R.; Ollis, W. D. Eds., Vol. 1, Stottart, J. F. Vol. Ed., Pergamon, Oxford, **1979**, p. 1105.

[179] Fieser, A. L.; Hartwell, J. L.; Jones, J. E.; Wood, J. H.; Bost, R. W. *Org. Synth.* **1940**, *20*, 11.

[180] Weston, A. W.; Michaels, Jr., R. J. *Org. Synth.* **1951**, *31*, 108.

[181] Silverstein, R. M.; Ryskiewicz, E. E.; Willard, C. *Org. Synth.* **1956**, *36*, 74.

[182] Croce, P. D.; Ferraccioli, R.; Ritieni, A. *Synthesis* **1990**, 212.

[183] White, J.; McGillivray, G. *J. Org. Chem.* **1977**, *42*, 4284.

[184] Becher, J.; Jørgensen, P. L.; Frydendahl, H.; Fält-Hansen, B. *Synthesis* **1991**, 609.

[185] Bonnett, R.; Hursthouse, M. B.; North, S. A.; Trotter, J. *J. Chem. Soc., Perkin Trans. 2* **1984**, 833.

[186] Buu-Hoï, N. P.; Xuong, N. D.; Sy, M.; Lejeune, G.; Tien, N. B. *Bull. Soc. Chim. Fr.* **1955**, 1594.

[187] Townsend, C. A.; Salituro, G. M. *J. Chem. Soc., Chem. Commun.* **1984**, 1631.

[188] Cheung, G. K.; Downie, I. M.; Earle, M. J.; Heaney, H.; Matough, M. F. S.; Shuhaibar, K. F.; Thomas, D. *Synlett* **1992**, 77.

[189] I. G. Farbenind. A.-G. Fr. 648069, 1929 (*Chem. Abstr.* **1929**, *23*, 2446).

[190] Dashunin, V. M.; Ivannikova, M. P.; Pedorenko, A. V.; Tovbina, M. S. *Tr. Vses. Nauch.-Issled. Inst. Sin. Natur. Dushistykh Veshchestv* **1971**, 85 (*Chem. Abstr.* **1973**, *78*, 83965c).

[191] E. I. du Pont de Nemours & Co., Brit. 607920, 1948 (*Chem. Abstr.* **1949**, *43*, 2232i).

[192] Campaigne, E.; Archer, W. L. *J. Am. Chem. Soc.* **1953**, *75*, 989.

[193] Kojtscheff, T.; Wolf, F.; Wolter, G. *Z. Chem.* **1966**, *6*, 148.

[194] Stepanov, B.; Migachev, G. I. *Zh. Obshch. Khim.* **1968**, *38*, 194 (*Chem. Abstr.* **1968**, *69*, 35642).

[195] Nordmann, H. G.; Kröhnke, F. *Angew. Chem., Int. Ed. Engl.* **1969**, *8*, 756.

[196] Sumitomo Chemical Co., Ltd., *Jpn. Kokai Tokkyo Koho* JP 56133220, 1981 (*Chem. Abstr.* **1982**, *96*, 68596m).

[197] Liebscher, J.; Bechstein, U. *Z. Chem.* **1983**, *23*, 214.

[198] Degutis, J.; Degutiene, A. *Zh. Org. Khim.* **1978**, *14*, 2060 (*Chem. Abstr.* **1979**, *90*, 87371b).

[199] Katz, H. E.; Schilling, M. L. *J. Am. Chem. Soc.* **1989**, *111*, 7554.

[200] Degen, H. G.; Grychtol, K. Ger. Offen. DE 2901845, 1980 (*Chem. Abstr.* **1981**, *94*, 4952a).

[201] Barton, J. W.; Buhaenko, M.; Moyle, B.; Ratcliffe, N. M. *J. Chem. Soc., Chem. Commun.* **1988**, 488.

[202] Farcasan, V.; Oprean, I.; Bodea, C. *Rev. Roum. Chim.* **1970**, *15*, 1433 (*Chem. Abstr.* **1971**, *74*, 87903w).

[203] Jostes, F. Ger. 547108, 1929 (*Chem. Abstr.* **1932**, *26*, 3515).

[204] Florvall, L.; Ask, A.-L.; Ross, S. B.; Ögren, S.-O.; Holm, A.-C. *Acta Pharm. Suec.* **1983**, *20*, 255 (*Chem. Abstr.* **1984**, *100*, 85319s).

[205] Freyer, W.; Teuchner, K.; Daehne, S. *J. Prakt. Chem.* **1981**, *323*, 324.

[206] Abraham, U.; Willand, C. S.; Kohler, W.; Robello, D. R.; Williams, D. J.; Handley, L. *J. Am. Chem. Soc.* **1990**, *112*, 7083.

[207] Kawamura, S. Japan. JP 49004531, 1974 (*Chem. Abstr.* **1975**, *82*, 45053d).

[208] Renfrew, E. E.; Genta, G. R. L. U.S. 4008262, 1977 (*Chem. Abstr.* **1977**, *86*, 157048v).

[209] Seus, E. J. *J. Org. Chem.* **1965**, *30*, 2818.

[210] Dix, J. P.; Voegtle, F. *Chem. Ber.* **1980**, *113*, 457.

[211] Kazankov, M. V.; Ginodman, L. G.; Mustafina, M. Ya. *Zh. Org. Khim.* **1983**, *19*, 153 (*Chem. Abstr.* **1983**, *99*, 38148m).

[212] Kutulya, L.A.; Shevchenko, A. E.; Pchelinova, L. D. *Khim. Geterotsikl. Soedin.* 1982, 807 (*Chem. Abstr.* **1982**, *97*, 144194s).

[213] Hashimoto, K.; Omura, T.; Suzuki, Y.; Hayashi, Y. U.S. 4833123, 1989 (*Chem. Abstr.* **1989**, *111*, 235256n).

[214] Akata, M.; Sugafuji, J.; Nakamura, M. Jpn. Kokai Tokkyo Koho JP 63142062, 1988 (*Chem. Abstr.* **1988**, *109*, 212382v).

[215] Okwute, S. K.; Okogun, J. I.; Okorie, D. A. *Tetrahedron* **1984**, *40*, 2541.

[216] Behringer, H.; Duesberg, P. *Chem. Ber.* **1963**, *96*, 377.

[217] Buu-Hoï, N.; Lejeune, G.; Sy, M. *C. R. Hebd. Seances Acad. Sci.* **1955**, *240*, 2241.

[218] Sommers, A. H.; Michaels, Jr., R. J.; Weston, A. W. *J. Am. Chem. Soc.* **1952**, *74*, 5546.

[219] Szilagyi, J.; Halmos, J.; Szabo, S.; Szileczky, F.; Mezei, J.; Szabolcsi, T.; Kortvelyessi, G.; Nyitrai, J.; Miskolczi, I.; Szatmari, K. Hung. Teljes HU 55741, 1991 (*Chem. Abstr.* **1991**, *115*, 207658j).

[220] El'tsov, A. V. *Zh. Obshch. Khim.* **1962**, *32*, 1525 (*Chem. Abstr.* **1963**, *58*, 4539h).

[221] Silhankova, A.; Ferles, M.; Maly, J. *Collect. Czech. Chem. Commun.* **1978**, *43*, 1484.

[222] Katsumi, I.; Kondo, H.; Yamashita, K.; Hidaka, T.; Hosoe, K.; Yamashita, T.; Watanabe, K. *Chem. Pharm. Bull.* **1986**, *34*, 121.

[223] Fatope, M. O.; Abraham, D. J. *J. Med. Chem.* **1987**, *30*, 1973.

[224] Rosenmund, P.; Gektidis, S.; Brill, H.; Kalbe, R. *Tetrahedron Lett.* **1989**, *30*, 61.

[225] Florvall, L.; Kumar, Y.; Ask, A. L.; Fagervall, I.; Renyi, l.; Ross, S. B. *J. Med. Chem.* **1986**, *29*, 1406.

[226] Sallmann, A.; Pfister, R. Ger. Offen. DE 2041273, 1971 (*Chem. Abstr.* **1971**, *74*, 99912m).

[227] Adamczyk, M.; Chen, Y. Y.; Fishpaugh, J. R. *Org. Prep. Proc. Int.* **1991**, *23*, 365.

[228] Gomm, W.; Beecken, H. Ger. Offen. DE 2308706, 1974 (*Chem. Abstr.* **1975**, *82*, 45057h).

[229] Coates, C. A., Jr.; Weaver, M. A. U.S. 4161601, 1979 (*Chem. Abstr.* **1979**, *91*, 142088n).

[230] Frishberg, M. D. U. S. 4180663, 1979 (*Chem. Abstr.* **1980**, *92*, 148519w).

[231] Bite, P.; Diszler, E. *Magy. Kem. Foly.* **1971**, *77*, 417 (*Chem. Abstr.* **1971**, *75*, 118434w).

[232] Tsien, R. Y.; Grynkiewicz, G. Eur. Pat. Appl. EP 178775, 1986 (*Chem. Abstr.* **1986**, *105*, 116551w).

[233] Nenitzescu, C. D.; Isacescu, D. *Bull. Soc. Chim. Romania*, **1930**, *11*, 135 (*Chem. Abstr.* **1930**, *24*, 2442).

[234] Godfrey, I. M.; Sargent, M. V.; Elix, J. A. *J. Chem. Soc., Perkin Trans. 1* **1974**, 1353.

[235] Bergan, J.; Pelcman, B. *Tetrahedron* **1988**,*44*, 5215.

[236] Hirota, T.; Fujita, H.; Sasaki, K.; Namba, T.; Hayakawa, S. *J. Heterocycl. Chem.* **1986**, *23*, 1347.

[237] Narasimham, N. S.; Mukhopadhyay, T.; Kusurkav, S. S. *Indian J. Chem., Sect. B* **1981**, *20B*, 546.

[238] Sumitomo Chemical Co., Ltd., Jpn. Kokai Tokkyo Koho JP 58125749, 1983 (*Chem. Abstr.* **1984**, *100*, 35848t).

[239] Noack, H.; Roemhild, H. J.; Guenther, E.; Hartmann, H. Ger. (East) DD 224592, 1985 (*Chem. Abstr.* **1986**, *105*, 21413z).

[240] Hirota, T.; Fujita, H.; Sasaki, K.; Namba, T. *J. Heterocycl. Chem.* **1986**, *23*, 1715.

[241] Renfrew, E. E.; Pons, H. W. U.S. US 3909198, 1975 (*Chem. Abstr.* **1976**, *84*, 19184u).

[242] CIBA Ltd., Brit. GB 1147125, 1969 (*Chem. Abstr.* **1969**, *71*, 82646t).

[243] Niegel, H.; Mayer, R.; Roland, J.; Ernst, A.; Wirth, H. Ger. (East) DD 152802, 1981 (*Chem. Abstr.* **1982**, *97*, 25164x).

[244] Hirota, T.; Fujita, H.; Sasaki, K.; Namba, T.; Hayakawa, S. *Heterocycles* **1986**, *24*, 771.

[245] Nishikuri, M.; Yamazaki, M.; Kojima, K.; Seino, J.; Kenmochi, H. Ger. Offen. DE 3020473, 1980 (*Chem. Abstr.* **1981**, *94*, 14102v).

[246] Arnold, Z.; Holy, A. *Collect. Czech. Chem. Commun.* **1963**, *28*, 869.

[247] Eszenyi, T.; Timar, T. *Synth. Commun.* **1990**, *20*, 3219.

[248] Yamada, Y.; Akahori, H.; Ito, N.; Nishizawa, I.; Yamaguchi, T. Jpn. Kokai Tokkyo Koho JP 02273651, 1990 (*Chem. Abstr.* **1991**, *114*, 163731a).

[249] Yamada, Y.; Akahori, H.; Ito, N.; Nishizawa, I.; Yamaguchi, T. Jpn. Kokai Tokkyo Koho JP 02275845, 1990 (*Chem. Abstr.* **1991**, *114*, 163733a).

[250] Brown, E.; Loriot, M.; Robin, J. -P. *Tetrahedron Lett.* **1982**, *23*, 949.

[251] Imaki, S.; Takuma, Y.; Oiishi, M. Jpn. Kokai Tokkyo Koho JP 62077344, 1987 (*Chem. Abstr.* **1987**, *107*, 197790q).

[252] Bick, I. R. C.; Russel, R. A. *Aust. J. Chem.* **1969**, *22*, 1563.

[253] Fleischhacker, W.; Richter, B.; Urban, E. *Monatsh. Chem.* **1989**, *120*, 765.

[254] Brownell, W. B.; Weston, A. W. *J. Am. Chem. Soc.* **1951**, *73*, 4971.

[255] Coppola, G. M.; Schuster, H. F. *J. Heterocycl. Chem.* **1989**, *26*, 957.

[256] Singh, I. P.; Shukla, V. K.; Dwivedi, A. K.; Khanna, N. M. *Indian J. Chem., Sect. B* **1989**, *28B*, 692.

[257] Zeghough, D.; Samsoniya, Sh. A.; Kadzhrishvili, D. O.; Mamaladze, L. M. *Khim. Geterotsikl. Soedin.* **1989**, 1363 (*Chem. Abstr.* **1990**, *113*, 40384m).

[258] Dallacker, F.; Leuther, P.; Wabinski-Westerop, K. *Chem.-Ztg.* **1989**, *113*, 97.

[259] Sasaki, I.; Yamashita, K. *Agric. Biol. Chem.* **1979**, *43*, 137 (*Chem. Abstr.* **1979**, *90*, 186833k).

[260] Velusamy, T. P.; Rao, G. S. K. *Indian J. Chem., Sect. B* **1981**, *20B*, 351.

[261] Ohishi, Y.; Mukai, T.; Nagahara, M.; Yajima, M.; Kajikawa, N. *Chem. Pharm. Bull.* **1989**, *37*, 2398.

[262] Dallacker, F.; Eschelbach, F.-E. *Justus Liebigs Ann. Chem.* **1965**, *689*, 171.

[263] Nichols, D. E.; Hoffman, A. J.; Oberlender, R. A.; Riggs, R. M. *J. Med. Chem.* **1986**, *29*, 302.

[264] De Paulis, T.; Kumar, Y.; Johansson, L.; Raemsby, S.; Hall, H.; Saellemark, M.; Aengeby-Moeller, K.; Oegren, S. O. *J. Med. Chem.* **1986**, *29*, 61.

[265] Martani, A.; Buu-Hoï, N. P. *Bull. Soc. Chim. Fr.* **1969**, 187.

[266] Guillaumet, G.; Hretani, M.; Coudert, G.; Averbeck, D.; Averbeck, S. *Eur. J. Med. Chem.* **1990**, *25*, 45.

[267] Guillaumet, G.; Hretani, M.; Coudert, G. *Tetrahedron Lett.* **1988**, *29*, 2665.

[268] Chaterjea, J. N.; Banerji, K. D.; Prasad, N. *Chem. Ber.* **1963**, *96*, 2356.

[269] Akesson, C. E.; Rahwan, R. G.; Brumbaugh, R. J.; Witiak, D. T. *Res. Commun. Chem. Pathol. Pharmacol.* **1980**, *27*, 265 (*Chem. Abstr.* **1980**, *93*, 36840h).

[270] Schroff, L. G.; Zsom, R. L. J.; Van der Weerdt, A. J. A.; Schrier, P. I.; Geerts, J. P.; Nibbering, N. M. M.; Verhoeven, J. W.; De Boer, T. J. *Recl. Trav. Chim. Pays-Bas* **1976**, *95*, 89.

[271] Pyne, S. G. *J. Chem. Soc. Chem. Commun.* 1986, 1686.

[272] Reddy, P. A.; Rao, G. S. K. *Indian J. Chem., Sect. B* **1980**, *19B*, 578.

[273] Mukherjee, A.; Akhtar, M. S.; Sharma, V. L.; Seth, M.; Bhaduri, A. P.; Agnihotri, A.; Mehrotra, P. K.; Kamboj, V. P. *J. Med. Chem.* **1989**, *32*, 2297.

[274] Sargent, M. V.; *J. Chem. Soc., Perkin Trans. 1* **1982**, 403.

[275] Inigo, A. C.; Sanz, V.; Seoane, E. *An. Quim., Ser. C* **1985**, *81*, 116 (*Chem. Abstr.* **1986**, *105*, 133578c).

[276] Albericio, F.; Kneib-Cordonier, N.; Biancalana, S.; Gera, L.; Masada, R.; Hudson, D.; Barang, G. *J. Org. Chem.* **1990**, *55*, 3730.

[277] Afzal, S. M.; Whalley, W. B. *Pak. J. Sci. Res.* **1984**, *33*, 56 (*Chem. Abstr.* **1985**, *102*, 78674s).

[278] Eichinger, K.; Hartmann, F. Austrian AT 385272, 1988 (*Chem. Abstr.* **1988**, *109*, 94747t).

[279] Reisch, J.; Kamal, G. M.; Gunaherath, B. *J. Chem. Soc., Perkin Trans 1* **1989**, 1047.

[280] Danishefsky, S.; Phillips, G.; Ciufolini, M. *Carbohydr. Res.* **1987**, *171*, 317.

[281] Benington, F.; Morin, R. D.; Clark, Jr., L. C. *J. Org. Chem.* **1955**, *20*, 102.

[282] Syper, L.; Kloc, K.; Mlochowski, J. *J. Prakt. Chem.* **1985**, *327*, 808.

[283] Alonso, R.; Brossi, A. *Tetrahedron Lett.* **1988**, *29*, 735.

[284] Ohtaka, H.; Hori, M.; Iemura, R.; Yumioka, H. *Chem. Pharm. Bull.* **1989**, *37*, 3122.

[285] Shirasaka, T.; Takuma, Y.; Imaki, N. *Synth. Commun.* **1990**, *20*, 1223.

[286] Morita, Y.; Imaki, S.; Takayanagi, H.; Takuma, Y.; Shirasaka, T. Jpn. Kokai Tokkyo Koho JP 6256484, 1987 (*Chem. Abstr.* **1987**, *107*, 96706m).

287 Iinuma, M.; Matoba, Y.; Tanaka, T.; Mizuno, M. *Chem. Pharm. Bull.* **1986**, *34*, 1656.

288 Paul, E. G.; Wang, P. S. -C. *J. Org. Chem.* **1979**, *44*, 2307.

289 Govindachari, T. R.; Patankar, S. J.; Viswanathan, N. *Indian J. Chem.* **1971**, *9*, 507.

290 Pillai, R. K. M.; Naiksatam, P.; Johnson, F.; Rajagopalan, R.; Watts, P. C.; Cricchio, R.; Borras, S. *J. Org. Chem.* **1986**, *51*, 717.

291 Eiden, F.; Gerstlauer, C. *Arch. Pharm.* **1982**, *315*, 551.

292 Gunzinger, J.; Tabacchi, R. *Helv. Chim. Acta* **1985**, *68*, 1940.

293 Comber, M. F.; Sargent, M. V.; Skelton, B. W.; White, A. H. *J. Chem. Soc., Perkin Trans. 1* **1989**, 441.

294 Elix, J. A.; Yu, J.; Tønsberg, T. *Aust. J. Chem.* **1991**, *44*, 157.

295 Carvalho, C. E.; Russo, A. V.; Sargent, M. V. *Aust. J. Chem.* **1985**, *38*, 777.

296 Devakumar, C.; Mukerjee, S. K. *Indian J. Chem., Sect. B* **1986**, *25B*, 1150.

297 Buu-Hoï, N. P.; Lavit, D. *J. Chem. Soc.* **1954**, 2776.

298 Kalischer, G.; Scheyer, H.; Keller, K. Ger. 514415, 1930 (*Chem. Abstr.* **1931**, *25*, 1536).

299 Kalischer, G.; Keller, K. Ger. 519806, 1928 (*Chem. Abstr.* **1931**, *25*, 3012).

300 Buu-Hoï, N. P.; Lavit, D. *J. Chem. Soc.* **1953**, 485.

301 Wood, J. H.; Bost, R. W. *J. Am. Chem. Soc.* **1937**, *59*, 1721.

302 Kulka, K. *Am. Perfumer Aromat.* **1957**, *69*, 42 (*Chem. Abstr.* **1957**, *51*, 9521b).

303 Eckstein, M.; Zajaczkowska, J. *Diss. Pharm. Pharmacol.* **1971**, *23*, 141 (*Chem. Abstr.* **1971**, *75*, 35835b).

304 Chaterjea, J. N.; Lal, S.; Jha, U.; Carnduff, J. *Indian J. Chem., Sect. B* **1981**, *20B*, 264.

305 Buu-Hoï, N. P.; Lavit, D. *J. Chem. Soc.* **1956**, 1743.

306 Tonoue, Y.; Terada, A.; Matsumoto, Y. *Bull. Soc. Chem. Jpn.* **1989**, *62*, 2736.

307 Doetz, K. H.; Popall, M. *Chem. Ber.* **1988**, *121*, 655.

308 Buu-Hoï, N. P.; Lavit, D. *J. Org. Chem.* **1955**, *20, 1191.*

309 Basu, B.; Mukherjee, D. *J. Chem. Soc., Chem. Commun.* **1984**, 105.

310 Buu-Hoï, N. P.; Lavit, D. *J. Org. Chem.* **1956**, *21, 1257.*

311 Fieser, L. F.; Jones, J. E. *J. Am. Chem. Soc.* **1942**, *64*, 1666.

312 Vistorobskii, N. V.; Pozharskii, A. F. *Zh. Org. Khim.* **1989**, *25*, 2154 (*Chem. Abstr.* **1990**, *112*, 234930m).

313 Buu-Hoï, N. P.; Lavit, D. *J. Org. Chem.* **1956**, *21*, 21.

314 Hoan, N. *Bull. Soc. Chim. Fr.* **1953**, 309.

315 Buu-Hoï, N. P.; Lavit, D.; Collard, J. *J. Org. Chem.* **1958**, *23, 542.*

316 Morrison, G. A.; Laundon, B. *J. Chem. Soc. C* **1971**, 1694.

317 Best, W. M.; Wege, D. *Aust. J. Chem.* **1986**, *39*, 647.

318 Terada, A.; Tanoue, Y.; Hatada, A.; Sakamoto, H. *Bull. Chem. Soc. Jpn.* **1987**, *60*, 205.

319 Terada, A.; Tagami, Y. Jpn. Kokai Tokkyo Koho JP 63112531, 1988 (*Chem. Abstr.* **1989**, *110*, 192452h).

320 Kalischer, G.; Scleyer, H.; Keller, K. Ger. 519444, 1931 (*Chem. Abstr.* **1931**, *25*, 2734).

321 Khan, M. A.; Kosar, F. S. *Bol. Soc. Quim. Peru* **1986**, *52*, 128 (*Chem. Abstr.* **1988**, *108*, 186414q).

322 Ferrari, J. L.; Hunsberger, I. M.; Gutowsky, H. S. *J. Am. Chem. Soc.* **1965**, *87*, 1247.

323 Ferrari, J. L.; Hunsberger, I. M. *J. Org. Chem.* **1960**, *25*, 485.

324 Buu-Hoï, N. P.; Lavit, D. *Recl. Trav. Chim. Pays-Bas* **1958**, *77*, 467.

325 Nédélec, L.; Rigaudy J. *Bull. Soc. Chim. Fr.* **1960**, 1204.

326 Fages, F.; Desvergne, J. P.; Bouas-Laurent, H. *Bull. Soc. Chim. Fr.* **1985**, 959.

327 Castedo, L.; Granja, J. A.; Rodriguez de Lera, A.; Villaverde, M. C. *J. Heterocycl. Chem.* **1988**, *25*, 1561.

328 Buu-Hoï, N. P.; Lavit, D. *Recl. Trav. Chim. Pays-Bas* **1957**, *76*, 674.

329 Vollman, H.; Becker, H.; Corell, M.; Streeck, H.; Langbein, G. *Justus Liebigs Ann. Chem.* **1937**, *531*, 1.

330 Treibs, W.; Möbius, G. *Justus Liebigs Ann. Chem.* **1958**, *619*, 122.

331 Fieser, L. F.; Hershberg, E. B. *J. Am. Chem. Soc.* **1938**, *60*, 2542.

332 Buu-Hoï, N. P.; Long, C. T. *Recl. Trav. Chim. Pays-Bas* **1956**, *75*, 1121.

[333] Buu-Hoï, N. P.; Lavit, D. *Recl. Trav. Chim. Pays-Bas* **1957**, *76*, 200.

[334] Buu-Hoï, N. P.; Lavit, D. *Recl. Trav. Chim. Pays-Bas* **1957**, *76*, 321.

[335] Buu-Hoï, N. P.; Lavit, D. *Recl. Trav. Chim. Pays-Bas* **1957**, *76*, 1194.

[336] Buu-Hoï, N. P.; Lavit, D.; Jacquignon, P.; Chalvet, O. *Recl. Trav. Chim. Pays-Bas* **1958**, *77*, 462.

[337] Buu-Hoï, N. P.; Lavit, D. *Recl. Trav. Chim. Pays-Bas* **1957**, *76*, 419.

[338] Yoshida, Z.; Yoneda, S.; Hazama, M. *J. Org. Chem.* **1972**, *37*, 1364.

[339] Fujisawa, T.; Sakai, K.; Kobori, T. Japan. Kokai JP 52106840, 1977 (*Chem. Abstr.* **1978**, *88*, 74118h).

[340] Jutz, C.; Kirchlechner, R. *Angew. Chem. Int. Ed. Engl.* **1966**, *5*, 516.

[341] Jutz, C.; Wagner, R. M. *Angew. Chem. Int. Ed. Engl.* **1972**, *11*, 315.

[342] Hafner, K.; Bernard, C. *Angew. Chem.* **1957**, *69*, 533.

[343] Hafner, K. *Angew. Chem.* **1958**, *70*, 419.

[344] Treibs, W.; Neupert, H.-J.; Hiebsch, J. *Chem. Ber.* **1959**, *92*, 141.

[345] Hafner, K.; Bernhard, C. Ger. 1059447, 1959 (*Chem. Abstr.* **1961**, *55*, 7379g).

[346] Brunken, J.; Poppe, E.-J. *Chem. Ber.* **1960**, *93*, 2572.

[347] Franke, H. E.; Muehlstaedt, M. *J. Prakt. Chem.* **1967**, *35*, 249.

[348] Kuroda, S.; Maeda, S.; Hirooka, S.; Ogiso, M.; Yamazaki, K.; Shimao, I.; Yasunami, M. *Tetrahedron Lett.* **1989**, *30*, 1557.

[349] Tachibana, Y.; Obara, K.; Masuyama, Y. Jpn. Kokai Tokkyo Koho JP 62198636, 1987 (*Chem. Abstr.* **1988**, *109*, 230392c).

[350] Asao, T.; Ito, S.; Morita, N. *Tetrahedron Lett.* **1989**, *29*, 2839.

[351] Yasunami, M.; Takase, K.; Shinji, A.; Mimura, T.; Torri, K.; Okutsu, M.; Kobayashi, T.; Meguro, T. Eur. Pat. Appl. EP 182491, 1986 (*Chem. Abstr.* **1986**, *105*, 214094u).

[352] Yasunami, M.; Takase, K.; Suzuki, K.; Hiwatari, O.; Nakamura, T.; Okutsu, M. Jpn. Kokai Tokkyo Koho JP 01102041, 1989 (*Chem. Abstr.* **1989**, *111*, 232165w).

[353] Mukherjee, D.; Dunn, L. C.; Houk, K. N. *J. Am. Chem. Soc.* **1979**, *101*, 251.

[354] Treibs, W.; Hiebsch, A.; Heupert, H.-J. *Naturwissenschaften* **1957**, *44*, 352.

[355] Ukita, T.; Miyazaki, M.; Hashi, M. *Chem. Pharm. Bull.* **1958**, *6*, 223.

[356] Hafner, K.; Schneider, J. *Justus Liebigs Ann. Chem.* **1959**, *624*, 37.

[357] Hafner, K. *Angew. Chem.* **1959**, *71*, 378.

[358] Satoh, K.; Yamaguchi, M.; Ogura, I. *Nippon Kagaku Kaishi* **1982**, 1199 (*Chem. Abstr.* **1983**, *98*, 107574r).

[359] Buu-Hoï, N. P.; Lavit, D. *Recl. Trav. Chim. Pays-Bas* **1958**, *77*, 724.

[360] Toda, T.; Minabe, M.; Yoshida, M.; Tobita, K. *J. Org. Chem.* **1990**, *55*, 1297.

[361] Johnson, B. F. G.; Lewis, J.; McArdle, P.; Randall, G. L. P. *J. Chem. Soc., Dalton Trans.* **1972**, 2076.

[362] Johnson, B. F. G.; Lewis, J.; Randall, G. L. P. *J. Chem. Soc. A* **1971**, 422.

[363] Adams, C. M.; Holt, E. M. *Organometallics* **1990**, *9*, 980.

[364] Frederic, J.; Toma, Š. *Collect. Czech. Chem. Commun.* **1987**, *52*, 174.

[365] Schlögl, K. *Monatsh. Chem.* **1957**, *88*, 601.

[366] Broadhead, G. D.; Osgerby, J. M.; Pauson, P. L. *Chem. Ind. (London)* **1957**, 209.

[367] Rosenblum, M. *Chem. Ind. (London)* **1957**, 72.

[368] Broadhead, G. D.; Osgerby, J. M.; Pauson, P. L. *J. Chem. Soc.* **1958**, 650.

[369] Graham, P. J.; Lindsey, R. V.; Parshall, G. W.; Peterson, M. L.; Whitman, G. M. *J. Am. Chem. Soc.* **1957**, *79*, 3416.

[370] Wenzel, M.; Asindraza, P.; Schachschneider, G. *J. Labelled Compd. Radiopharm.* **1983**, *20*, 1061 (*Chem. Abstr.* **1984**, *100*, 103616t).

[371] Krajnik, P.; Schögl, K.; Widhalm, M. *Monatsh. Chem.* **1990**, *121*, 413.

[372] Schögl, K.; Peterlik, M.; Seller, H. *Monatsh. Chem.* **1962**, *93*, 1309.

[373] Sutherland, R. G.; Sutton, J. R.; Horspool, W. M. *J. Organomet. Chem.* **1976**, *122*, 393.

[374] Tainturier, G.; Tirouflet, J. *Bull. Soc. Chim. Fr.* **1966**, 600.

[375] Watanabe, J.; Hisatome, M.; Yamakawa, K. *Tetrahedron Lett.* **1987**, *28*, 1427.

[376] Hisatome, M.; Watanabe, J.; Kawajiri, Y.; Yamakawa, K.; Iitaka, Y. *Organometallics* **1990**, *9*, 497.

[377] Rausch, M. D.; Gund, T. M. *J. Organomet. Chem.* **1970**, *24*, 463.

[378] Tragnelis, V. J.; Mishel, Jr., J. J.; Sowa, J. R. *J. Org. Chem.* **1957**, *22*, 1269.

[379] Ozdowska, Z. J.*Labelled Compd. Radiopharm.* **1978**, *14*, 361 (*Chem. Abstr.* **1978**, *89*, 197288s).

[380] Clementi, S.; Fringuelli, F.; Linda, P.; Marino, G.; Savelli, G.; Taticchi, A. *J. Chem. Soc., Perkin Trans. 2* **1973**, 2097.

[381] Sumitomo Chemical Co., Ltd., Jpn. Kokai Tokkyo Koho JP 57091982, 1982 (*Chem. Abstr.* **1982**, *97*, 162801s).

[382] Klein, L. L. *J. Org. Chem.* **1985**, *50*, 1770.

[383] Kuo, Y.-H.; Shih, K.-S. *Chem. Pharm. Bull.* **1991**, *39*, 181.

[384] Martin, S. F.; Pacofsky, G. J.; Gist, R. P.; Lee, W. C. *J. Am. Chem. Soc.* **1989**, *111*, 7634.

[385] Karakhanov, R. A.; Zefirova, V. A.; Snoval'nikova, T. I. *Khim. Geterotsikl. Soedin.* **1972**, 1591 (*Chem. Abstr.* **1973**, *78*, 71792f).

[386] Stoyanov, V. M.; El'chaninov, M. M.; Simonov, A. M.; Pozharskii, A. F. *Khim. Geterotsikl. Soedin.* **1989**, 1396 (*Chem. Abstr.* **1990**, *113*, 23761q).

[387] Moriya, T.; Seki, M.; Takabe, S.; Matsumoto, K.; Takashima, K.; Mori, T.; Odawara, A.; Takeyama, S. *J. Med. Chem.* **1988**, *31*, 1197.

[388] Pennanen, S. *Acta Chem. Scand.* **1973**, *27*, 3133.

[389] Farcasan, V.; Paiu, F.; Iusan, C. *Stud. Univ. Babes-Bolyai, [Ser.] Chem.* **1977**, *22*, 15 (*Chem. Abstr.* **1977**, *87*, 53005h).

[390] Kraus, G. A.; Hagen, M. D. *J. Org. Chem.* **1985**, *50*, 3252.

[391] Eliasson, R.; Nedenskov, P. Ger. Offen. DE 2313506, 1973 (*Chem. Abstr.* **1974**, *80*, 14581d).

[392] Karminski-Zamola, G.; Bajic, M.; Tkalcic, M.; Fiser-Jakic, L. *Heterocycles* **1986**, *24*, 733.

[393] Duval, O.; Gomes, L. M. *Bull. Soc. Chim. Fr.* **1987**, 131.

[394] Maeba, I.; Hara, O.; Suzuki, M.; Furukawa, H. *J. Org. Chem.* **1987**,*52*, 2368.

[395] Kutney, J. P.; Hanssen, H. W.; Nair, G. V. *Tetrahedron* **1971**, *27*, 3323.

[396] Schubert, G.; Wunderwald, M.; Ponsold, K.; Eibisch, H. Ger. (East) DD 236736, 1986 (*Chem. Abstr.* **1987**, *106*, 18941z).

[397] Marquet, J. P.; Bisagni, E.; Andre-Louisfert, J. *Bull. Soc. Chim. Fr.* **1973**, 2323.

[398] Martin, S. F.; Gluchowski, C.; Campbell, C. L.; Chapman, R. C. *Tetrahedron* **1988**, *44*, 3171.

[399] Ziegler, F. E.; Thottathil, J. K. *Tetrahedron Lett.* **1981**, *22*, 4883.

[400] Vegh, D.; Zalupsky, P.; Kovac, J. *Synth. Commun.* **1990**, *20*, 1113.

[401] Ali, A.; Guile, S. D.; Saxton, J. E.; Thornton-Pett, M. *Tetrahedron* **1991**, *47*, 6407.

[402] Ali, A.; Saxton, J. E. *Tetrahedron Lett.* **1989**, *30*, 3197.

[403] Ito, K.; Yakushijin, K.; Yoshina, T.; Akira, Y. K. *J. Heterocycl. Chem.* **1978**, *15*, 301.

[404] Beeby, P. J.; Edwards, J. A. *Tetrahedron Lett.* **1976**, 3261.

[405] Yoshina, S. Japan. Kokai JP 48091057, 1973 (*Chem. Abstr.* **1974**, *80*, 145999u).

[406] Klein, L. L.; Shanklin, M. S. *J. Org. Chem.* **1988**, *53*, 5202.

[407] Buu-Hoï, N. P.; Lavit, D. *J. Chem. Soc.* **1950**, 2130.

[408] Buu-Hoï, N. P.; Hoán, N.; Lavit, D. *J. Chem. Soc.* **1952**, 4590.

[409] Emerson, W. S.; Patrick, T. M., Jr. U.S. 2581009, 1952 (*Chem. Abstr.* **1952**, *46*, 9610g).

[410] Profft. E. *Justus Liebigs Ann. Chem.* **1959**, *622*, 196.

[411] Cymerman-Craig, J.; Willis, D. *Chem. Ind. (London)* **1953**, 797.

[412] Hartmann, H.; Scheithauer, S. *J. Prakt. Chem.* **1969**, *311*, 827.

[413] Carpita, A.; Lezzi, A.; Rossi, R.; Marchetti, F.; Merlino, S. *Tetrahedron* **1985**, *41*, 621.

[414] Buu-Hoï, N. P.; Xuong, N. D.; Royer, R.; Lavit, D. *J. Chem. Soc.* **1953**, 547.

[415] Buu-Hoï, N. P.; Lavit, D.; Xuong, N. D. *J. Chem. Soc.* **1955**, 1581.

[416] Nakayama, J.; Murabayashi, S.; Hoshino, M. *Heterocycles* **1986**, *24*, 2639.

[417] Demerseman, P.; Buu-Hoï, N. P.; Royer, R. *J. Chem. Soc.* **1954**, 4193.

[418] Polyakov, V. K.; Zaplyuisvechka, Z. P.; Tsukerman, S. V. *Khim. Geterotsikl. Soedin.* **1974**, 136 (*Chem. Abstr.* **1974**, *80*, 108280g).

[419] Beaton, C. M.; Chapman, N. B.; Clarke, K.; Willis, J. M. *J. Chem. Soc., Perkin Trans. 1* **1976**, 2355.

[420] Gjoes, N.; Gronowitz, S. *Acta Chem. Scand.* **1972**, *26*, 1851.

[421] Frimm, R.; Fisera, L.; Kovac, J. *Collect. Czech. Chem. Commun.* **1973**, *38*, 1809.

[422] Nakayama, J.; Nakamura, Y.; Tajiri, T.; Hoshino, M. *Heterocycles* **1986**, *24*, 637.

[423] Nakayama, J.; Fujimori, T. *Heterocycles* **1991**, *32*, 991.

[424] Ham, N. H.; Buu-Hoï, N. P.; Xuong, N. D. *J. Chem. Soc.* **1954**, 1690.

[425] Karminski-Zamola, G.; Bajic, M. *Heterocycles* **1985**, *23*, 1497.

[426] Giral, L.; Bompart, J.; Puygrenier, M. Eur. Pat. Appl. EP 161235, 1985 (*Chem. Abstr.* **1986**, *104*, 88515w).

[427] Trakhtenberg, P. L.; Lipkin, A. E. *Khim. Geterotsikl. Soedin.* **1974**, 61 (*Chem. Abstr.* **1974**, *80*, 108337f).

[428] Lamy, J.; Lavit, D.; Buu-Hoï, N. P. *J. Chem. Soc.* **1958**, 4202.

[429] Gronowitz, S.; Moses, P.; Hörnfeldt, A.-B.; Häkansson, R. *Ark. Kemi* **1961**, *17*, 165 (*Chem. Abstr.* **1962**, *57*, 8528h)

[430] Kato, S.; Ishizaki, M. Jpn. Kokai Tokkyo Koho JP 61293979, 1986 (*Chem. Abstr.* **1987**, *106*, 213755u).

[431] Paulmier, C.; Outurquin, F. *J. Chem. Res. (S)* **1977**, 318.

[432] Popolitova, E. A.; Lipkin, A. E. *Khim.-Farm. Zh.* **1976**, *10*, 75 (*Chem. Abstr.* **1976**, *85*, 94168r).

[433] Litvinov, V. P.; Gol'dfarb, Ya. L.; Bogdanov, V. S.; Konyaeva, I. P.; Sukiasyan, A. N. *J. Prakt. Chem.* **1973**, *315*, 850.

[434] Ahmed, M.; Ashby, J.; Meth-Cohn, O. *J. Chem. Soc., Chem. Commun.* **1970**, 1094.

[435] Terad, A.; Wachi, K.; Amamiya, Y. Jpn. Kokai Tokkyo Koho JP 01050879, 1989 (*Chem. Abstr.* **1989**, *111*, 115021k).

[436] Goudie, A. C.; Rosenberg, H. E.; Ward, R. W. *J. Heterocycl. Chem.* **1983**, *20*, 1027.

[437] Hallberg, A.; Gronowitz, S. *Chem. Scr.* **1980**, *16*, 38.

[438] Buu-Hoï, N. P.; Khenissi, M. *Bull. Soc. Chim. Fr.* **1958**, 359.

[439] Amemiya, Y.; Terada, A.; Wachi, K.; Miyazawa, H.; Hatakeyama, N.; Matsuda, K.; Oshima, T. *J. Med. Chem.* **1989**, *32*, 1265.

[440] Klein, L. L. *J. Am. Chem. Soc.* **1985**, *107*, 2573.

[441] Belenky, L. I.; Loktionov, A. A.; Krayushkin, M. M. *Stud. Org. Chem. (Chem. Heterocycl. Compd).*, Kovác, J.; Zálupsky, P. Eds. **1987**, *35*, 221.

[442] Gronowitz, S.; Stenhammer, K.; Svensson, L. *Heterocycles* **1981**, *15*, 947.

[443] Parham, W. E.; Traynells, V. J. *J. Am. Chem. Soc.* **1954**, *76*, 4960.

[444] Gillon, D. W.; Forest, I. J.; Meakins, G. D.; Tirel, M. D.; Wallis, J. D. *J. Chem. Soc., Perkin Trans. 1* **1983**, 341.

[445] Mazaki, Y.; Kobayashi, K. *Tetrahedron Lett.* **1989**, *30*, 3315.

[446] Outurquin, F.; Paulmier, C. *Bull. Soc. Chim. Fr.* **1983**, 153.

[447] Hartmann, M. *J. Prakt. Chem.* **1967**, *36*, 50.

[448] Shvedov, V. I.; Kharizomenova, I. A.; Grinev, A. N. *Khim. Geterotsikl. Soedin.* **1974**, 58 (*Chem. Abstr.* **1974**, *80*, 95865z).

[449] Yur'ev, Yu. K.; Mezentsova, N. N.; Vas'kovskii, V. E. *Zh. Obshch. Khim.* **1957**, *27*, 3155 (*Chem. Abstr.* **1958**, *52*, 9065g).

[450] Yur'ev, Yu. K.; Sadovaya, N. K.; Gal'bershtam, M. A. *Zh. Obshch. Khim.* **1958**, *28*, 620 (*Chem. Abstr.* **1958**, *52*, 17234b).

[451] Silverstein, R. M.; Ryskiewicz, E. E.; Chaikin, S. W. *J. Am. Chem. Soc.* **1954**, *76*, 4485.

[452] Silverstein, R. M.; Ryskiewicz, E. E.; Willard, C.; Koehler, R. C. *J. Org. Chem.* **1955**, *20*, 668.

[453] De Groot, J. A.; Gorter-La Roy, G. M.; Van Koeveringe, J. A.; Lugtenburg, J. *Org. Prep. Proc. Int.* **1981**, *13*, 97.

[454] Anderson, H. J.; Loader, C. E.; Foster, A. *Can. J. Chem.* **1980**, *58*, 2527.

[455] Barnett, G. H.; Anderson, H. J.; Loader, C. E. *Can. J. Chem.* **1980**, *58*, 409.

[456] Kira, M. A.; Bruckner-Wilhelms, A. *Acta Chim. (Budapest)* **1968**, *56*, 47 (*Chem. Abstr.* **1968**, *69*, 86888e).

[457] Croce, P. D.; La Rosa, C.; Ritieni, A. *Synthesis* **1989**, 783.

[458] Julia, M.; Pascal, Y. R. *Chim. Ther.* **1970**, *5*, 279 (*Chem. Abstr.* **1971**, *74*, 34488).

[459] Rapoport, H.; Castagnoli, Jr., N. *J. Am. Chem. Soc.* **1962**, *84*, 2178.

[460] Rapoport, H.; Bordner, J. *J. Org. Chem.* **1964**, *29*, 2727.

[461] McGillivray, G.; Smal, E. *J. Chem. Soc., Perkin Trans. 1* **1983**, 633.

[462] Jones, R. A.; Candy, C. F. *J. Chem. Soc. B* **1971**, 1405.

463 Dolby, L. J.; Nelson, S. J.; Senkovich, D. *J. Org. Chem.* **1972**, *37*, 3691.

464 Ryskiewicz, E. R.; Silverstein, R. M. *J. Am. Chem. Soc.* **1954**, *76*, 5802.

465 Clementi, S.; Linda, P.; Marino, G. *J. Chem. Soc., Chem. Commun.* **1972**, 427.

466 Brown, D.; Griffiths, D.; Rider, M. E.; Smith, R. C. *J. Chem. Soc., Perkin Trans. 1* **1986**, 455.

467 Bosch, J.; Mauleon, D.; Granados, R. *An. Quim., Ser. C* **1982**, *78*, 236 (*Chem. Abstr.* **1982**, *97*, 162867t).

468 Carson, J. R.; Davis, N. M. *J. Org. Chem.* **1981**, *46*, 839.

469 Massa, S.; Artico, M.; Corelli, F.; Mai, A.; Di Santo, R.; Cortes, S.; Marongiu, M. E.; Pani, A.; La Colla, P. *J. Med. Chem.* **1990**, *33*, 2845.

470 Hinz, W.; Jones, R. A.; Anderson, T. *Synthesis* **1986**, 620.

471 Flitsch, W.; Schulten, W. *Synthesis* **1977**, 414.

472 Lancelot, J. -C.; Ladurée, D.; Robba, M. *Chem. Pharm. Bull.* **1985**, *33*, 3122.

473 Effi, Y.; Cugnon de Sevricourt, M.; Rault, S.; Robba, M. *Heterocycles* **1981**, *16*, 1519.

474 Rault, S.; Effi, Y.; Cugnon de Sevricourt, M.; Lancelot, J.-C.; Robba, M. *J. Heterocycl. Chem.* **1983**, *20*, 17.

475 Rault, S.; Effi, Y.; Lancelot, J. C.; Robba, M. *Heterocycles* **1986**, *24*, 575.

476 Stanforth, S. P. unpublished results.

477 Cheeseman, G. W. H.; Rafiq, M. *J. Chem. Soc. C* **1971**, 2732.

478 Anderson, H. J.; Loader, C. E.; Xu, R. X.; Le, N.; Gogan, N. J.; McDonald, R.; Edwards, L. G. *Can. J. Chem.* **1985**, *63*, 896.

479 Effland, R. C.; Davis, L.; Kapples, K. J.; Olsen, G. E. *J. Heterocycl. Chem.* **1990**, *27*, 1015.

480 Cheeseman, G. W. H.; Eccleshall, S. A.; Thornton, T. *J. Heterocycl. Chem.* **1985**, *22*, 809.

481 Anderson, H. J.; Griffiths, S. J. *Can. J. Chem.* **1967**, *45*, 2227.

482 Effland, R. C.; Davis, L. *J. Heterocycl. Chem.* **1985**, *22*, 1071.

483 Effland, R. C.; Davis, L.; Schaub, W. U.S. 4045448, 1977 (*Chem. Abstr.* **1977**, *87*, 201597v).

484 Bates, D. K.; Sell, B. A.; Picard, J. A. *Tetrahedron Lett.* **1987**, *28*, 3535.

485 Flitsch, W.; Lerner, H.; Zimmermann, H. *Chem. Ber.* **1977**, *110*, 2765.

486 Flitsch, W.; Lerner, H. *Tetrahedron Lett.* **1974**, 1677.

487 Massa, S.; Corelli, F.; Stefancich, G. *J. Heterocycl. Chem.* **1981**, *18*, 829.

488 Giuliano, R.; Porretta, G. C.; Scalzo, M.; Vomero, S.; Artico, M.; Dolfini, E.; Morasca, L. *Farmaco, Ed. Sci.* **1972**, *27*, 908 (*Chem. Abstr.* **1973**, *78*, 40526).

489 Nacci, V.; Garofalo, A.; Fiorini, I. *J. Heterocycl. Chem.* **1986**, *23*, 769.

490 Pailer, M.; Schlaeger, I. *Monatsh. Chem.* **1978**, *109*, 313.

491 Cheeseman, G. W. H.; Eccleshall, S. A. *J. Heterocycl. Chem.* **1986**, *23*, 65.

492 De Martino, G.; Massa, S.; Scalzo, M.; Giuliano, R.; Artico, M. *J. Chem. Soc. D* **1971**, 1549.

493 Nacci, V.; Filacchioni, G.; Porretta, G. C.; Stefancich, G.; Guaitani, A. *Farm. Ed. Sci.* **1973**, *28*, 494 (*Chem. Abstr.* **1973**, *79*, 78762a).

494 Flitsch, W.; Lewinski, U.; Temme, R.; Wibbeling, B. *Justus Liebigs Ann. Chem.* **1990**, 623.

495 Nacci, V.; Garofalo, A.; Fiorini, I. *J. Heterocycl. Chem.* **1985**, *22*, 259.

496 Flitsch, W.; Koszinowski, J.; Witthake, P. *Chem. Ber.* **1979**, *112*, 2465.

497 Rapoport, H.; Castagnoli, N., Jr.; Holden, K. G. *J. Org. Chem.* **1964**, *29*, 883.

498 Loader, C. E.; Anderson, H. J. *Synthesis* **1978**, 295.

499 Van Wijngaarden, I.; Kruse, C. G.; Van Hes, R.; Van der Heyden, J. A. M.; Tulp, M. Th. M. *J. Med. Chem.* **1987**, *30*, 2099.

500 Demopoulos, V. J.; Rekka, E.; Retsas, S. *Pharmazie* **1990**, *45*, 403 (*Chem. Abstr.* **1990**, *113*, 231135r).

501 Loader, C. E.; Barnett, G. H.; Anderson, H. J. *Can. J. Chem.* **1982**, *60*, 383.

502 Mills, J. E.; Cosgrove, R. M.; Shah, R. D.; Maryanoff, C. A.; Paragamian, V. *J. Org. Chem.* **1985**, *49*, 546.

503 Muchowski, J. M.; Unger, S. H.; Ackrell, J.; Cheung, P.; Cooper, G. F.; Cook, J.; Gallegra, P.; Halpern, O.; Koehler, R.; Kluge, A. F.; Van Horn, A. R.; Antonio, Y.; Carpio, H.; Franco, F.; Galeazzi, E.; Garcia, I.; Greenhouse, R.; Guzmán, A.; Iriarte, J.; Leon, A.; Peña, A.; Peréz, V.; Valdéz, D.; Ackerman, N.; Ballaron, S. A.; Murthy, D. V. K.; Rovito, J. R.; Tomolonis, A. J.; Young, J. M.; Rooks, W. H. *J. Med. Chem.* **1985**, *28*, 1037.

[504] Cheeseman, G. W. H.; Varvounis, G. *J. Heterocycl. Chem.* **1987**, *24*, 1157.

[505] Duceppe, J.-S.; Gauthier, J. *J. Heterocycl. Chem.* **1985**, *22*, 305.

[506] Sagami Chemical Research Center, Jpn. Kokai Tokkyo Koho JP 56049353, 1981 (*Chem. Abstr.* **1981**, *95*, 150424s).

[507] Smith, K. M.; Bobe, F. W.; Minetian, O. M.; Hope, H.; Yanuk, M. D. *J. Org. Chem.* **1985**, *50*, 790.

[508] Ghighi, E.; Drusiani, A. *Chem. Zentr.* **1957**, *128*, 3250.

[509] Pasquier, C.; Gossauer, A.; Keller, W.; Kratky, C. *Helv. Chim. Acta* **1987** *70*, 2098.

[510] Suminori, U.; Kariyone, K.; Kunihiko, T. Jpn. 6720708, 1964 (*Chem. Abstr.* **1968**, *69*, 27239w).

[511] Rogers, M. A. T. *J. Chem. Soc.* **1943**, 596.

[512] Gardner, T. S.; Wenis, E.; Lee, J. *J. Org. Chem.* **1958**, *23*, 823.

[513] Hinman, R. L.; Theodoropulos, S. *J. Org. Chem.* **1963**, *28*, 3052.

[514] Ghighi, E.; Drusiani, A. *Chem. Zentr.* **1958**, *129*, 7737.

[515] Ghighi, E.; Drusiani, A. *Atti. Accad. Sci. Ist. Bologna* **1957**, *4*, 14 (*Chem. Abstr.* **1958**, *52*, 11818a).

[516] Kleinspehn, C. G.; Briod, A. E. *J. Org. Chem.* **1961**, *26*, 1652.

[517] Vogel, E.; Balci, M.; Pramod, K.; Kock, P.; Lex, J.; Ermer, O. *Angew. Chem. Int. Ed. Engl.* **1987**, *26*, 928.

[518] Paine, J. B., III; Dolphin, D. *J. Org. Chem.* **1988**, *53*, 2787.

[519] Aoyagi, K.; Toi, H.; Aoyama, Y.; Ogoshi, H. *Chem. Lett.* **1988**, 1891.

[520] Mironov, A. F.; Akimenko, L. V.; Rumyantseva, V. D.; Evstigneeva, R. P. *Khim. Geterotsikl. Soedin.* **1975**, 423 (*Chem. Abstr.* **1975**, *83*, 28041b).

[521] Schnierle, F.; Reinhard, H.; Dieter, N.; Lippacher, E.; Von Dobeneck, H. *Justus Liebigs Ann. Chem.* **1968**, *715*, 90.

[522] Gosmann, M.; Franck, B. *Angew. Chem. Int. Ed. Engl.* **1986**, *25*, 1100.

[523] Ono, N.; Muratani, E.; Ogawa, T. *J. Heterocycl. Chem.* **1991**, *28*, 2053.

[524] Dannhardt, G.; Steindl, L. *Arch. Pharm.* **1986**, *319*, 749.

[525] Remers, W. A.; Weiss, M. J. *J. Am. Chem. Soc.* **1965**, *87*, 5262.

[526] Sun, T. H.; Abbot, F. S.; Burton, R.; Orr, J. *J. Labelled Compd. Radiopharm.* **1982**, *19*, 1043.

[527] Muchowski, J. M.; Galeazzi, E.; Greenhouse, R.; Guzmán, A.; Peréz, V.; Ackerman, N.; Ballaron, S. A.; Rovito, J. R.; Tomolonis, A. J.; Young, J. M.; Rooks, W. H. *J. Med. Chem.* **1989**, *32*, 1202.

[528] Tilford, C. H.; Hudak, W. J.; Lewis, R. E. *J. Med. Chem.* **1971**, *14*, 328.

[529] Vorkapic-Furac, J.; Mintas, M.; Burgemeister, T.; Mannschreck, A. *J. Chem. Soc., Perkin Trans. 2* **1989**, 713.

[530] Rips, R.; Buu-Hoï, N. P. *J. Org. Chem.* **1959**, *24*, 372.

[531] Fritz, H.; Schenk, S. *Justus Liebigs Ann. Chem.* **1975**, 255.

[532] Perche, J. C.; Saint-Ruf, G. *Bull. Soc. Chim. Fr.* **1974**, 1117.

[533] Huebsch, W.; Angerbauer, R.; Fey, P.; Bischoff, H.; Petzinna, D.; Schmidt, D.; Thomas, G. Eur. Pat. Appl. EP 334147, 1989 (*Chem. Abstr.* **1990**, *112*, 178656n).

[534] Clezy, P. S.; Fookes, C. J. R.; Prashar, J. K. *Aust. J. Chem.* **1989**, *42*, 775.

[535] Kozhich, D. T.; Akimenko, L. V.; Mironov, A. F.; Evstigneeva, R. P. *Zh. Org. Khim.* **1977**, *13*, 2604.

[536] Wijesekera, T. P.; Paine, J. B.; Dolphin, D. *J. Org. Chem.* **1985**, *50*, 3832.

[537] White, J. *J. Chem. Soc., Perkin Trans. 2* **1984**, 1607.

[538] White, J.; McGillivray, G. *J. Chem. Soc., Perkin Trans. 2* **1982**, 259.

[539] White, J.; McGillivray, G. *J. Chem. Soc., Perkin Trans. 2* **1979**, 943.

[540] Cook, A. H.; Majer, J. R. *J. Chem. Soc.* **1944**, 482.

[541] Smith, K. M.; Miura, M.; Morris, I. K. *J. Org. Chem.* **1986**, *51*, 4660.

[542] Clezy, P. S.; Liepa, A. J. *Aust. J. Chem.* **1970**, *23*, 2461.

[543] Sessler, J. L.; Hugdahl, J.; Johnson, M. R. *J. Org. Chem.* **1986**, *51*, 2838.

[544] Sambrotta, L.; Rezzano, I.; Buldain, G.; Frydman, B. *Tetrahedron* **1989**, *45*, 6645.

[545] Kozyrev, A. N.; Mironov, A. F.; Davila, J.; Harriman, A. *J. Heterocycl. Chem.* **1988**, *25*, 885.

[546] Wijesekera, T. P.; Paine, J. B, III.; Dolphin, D. *J. Org. Chem.* **1988**, *53*, 1345.

[547] Buldain, G.; Valasinas, A. *J. Labelled Compd. Radiopharm.* **1982**, *19*, 1.

[548] Smith, K. M.; Kishore, D. *J. Chem. Soc., Chem. Commun.* **1982**, 888.

[549] LeGoff, E.; Weaver, O. G. *J. Org. Chem.* **1987**, *52*, 710.

[550] Zhestkov, V. P.; Mironov, A. F.; Ustynyuk, L. A.; Evstigneeva, R. P. *Bioorg. Khim.* **1975**, *1*, 1673 (*Chem. Abstr.* **1976**, *84*, 121581h).

[551] Jackson, A. H.; Kenner, G. W.; Smith, K. M. *J. Chem. Soc. C* **1971**, 502.

[552] Wijesekera, T. P.; David, S.; Paine, J. B., III; James, B. R.; Dolphin, D. *Can. J. Chem.* **1988**, *66*, 2063.

[553] David, S.; Dolphin, D.; James, B. R.; Paine, J. B., III; Wijesekera, T. P.; Einstein, F. W. B.; Jones, T. *Can. J. Chem.* **1986**, *64*, 208.

[554] Battersby, A. R.; Fookes, C. J. R.; Gustafson-Potter, K. E.; Matcham, G. W. J.; McDonald, E. *J. Chem. Soc., Chem. Commun.* **1979**, 1155.

[555] Chu, E. J.-C.; Chu, T. C. *J. Org. Chem.* **1954**, *19*, 266.

[556] Nicolaus, R. A.; Mangoni, L. *Ann. Chim. (Rome)* **1956**, *46*, 865 (*Chem. Abstr.* **1957**, *51*, 6600i).

[557] Smith, K. M.; Bisset, G. M. F. *J. Org. Chem.* **1979**, *44*, 2077.

[558] Philipps, T.; Angerbauer, R.; Fey, P.; Huebsch, W.; Bischoff, H.; Petzinna, D.; Schmidt, D.; Thomas, G. Eur. Pat. Appl. EP 352575, 1990 (*Chem. Abstr.* **1990**, *113*, 58927a).

[559] Trofimov, F. A.; Mukhanova, T. I.; Grinev, A. N. *Khim. Geterotsikl. Soedin.* **1973**, 597 (*Chem. Abstr.* **1973**, *79*, 78492r).

[560] Neidlein, R.; Bernhard, E. *Arch. Pharm.* **1978**, *311*, 714.

[561] Grinev, A. N.; Zotova, S. A. *Khim. Geterotsikl. Soedin.* **1976**, 1023 (*Chem. Abstr.* **1976**, *85*, 159795q).

[562] Shevchenko, L. I.; Trofimov, F. A. *Khim. Geterotsikl. Soedin.* **1987**, 179.

[563] Royer, R.; Rene, L. *Bull. Soc. Chim. Fr.* **1970**, 1029.

[564] Royer, R.; Rene, L. *Bull. Soc. Chim. Fr.* **1970**, 3601.

[565] Hishmat, O. H.; Abd el Rahman, A. H. *Aust. J. Chem.* **1974**, *27*, 2499.

[566] Deorha, D. S.; Gupta, P. *Chem. Ber.* **1964**, *97*, 616.

[567] Carrington, D. E. L.; Clarke, K.; Scrowston, R. M. *J. Chem. Soc., Perkin Trans. 1* **1972**, 3006.

[568] Bushkov, A. Ya.; Lantsova, O. I.; Bren, V. A. *Izv. Sev-Kavk. Nauchn. Tsentra Vyssh. Shk., Estestv. Nauki* **1985**, 57 (*Chem. Abstr.* **1986**, *105*, 208818s).

[569] Bushkov, A. Ya.; Lantsova, O. I.; Bren, V. A.; Minkin, V. I. *Khim. Geterotsikl. Soedin.* **1985**, 565 (*Chem. Abstr.* **1985**, *103*, 37451u).

[570] Dallacker, F.; Kaiser, E.; Uddrich, P. *Justus Liebigs Ann. Chem.* **1965**, *689*, 179.

[571] Tyson, F. T.; Shaw, J. T. *J. Am. Chem. Soc.* **1952**, *74*, 2273.

[572] Smith, G. F. *J. Chem. Soc.* **1954**, 3842.

[573] Shabica, A. C.; Howe, E. E.; Ziegler, J. B.; Tishler, M. *J. Am. Chem. Soc.* **1946**, *68*, 1156.

[574] Gertsev, V. V. *Zh. Vses. Khim. O-va.* **1982**, *27*, 341 (*Chem. Abstr.* **1982**, *97*, 162750z).

[575] Horning, D. E.; Muchowski, J. M. *Can. J. Chem.* **1970**, *48*, 193.

[576] Genkina, N. K.; Gendina, N. G.; Suvorov, N. N. *Zh. Org. Khim.* **1985**, *21*, 415 (*Chem. Abstr.* **1985**, *103*, 36886c).

[577] Moriya, T.; Yoneda, N. *Chem. Pharm. Bull.* **1982**, *30*, 158.

[578] Murakami, Y.; Tani, M.; Suzuki, M.; Sudoh, K.; Uesato, M.; Tanaka, K.; Yokoyama, Y. *Chem. Pharm. Bull.* **1985**, *33*, 4707.

[579] Youngdale, G. A.; Anger, D. G.; Anthony, W. C.; Da Vanzo, J. P; Greig, M. E.; Heinzelman, R. V.; Keasling, H. H.; Szmuszkovicz, J. *J. Med. Chem.* **1964**, *7*, 415.

[580] Reynolds, G. A.; Van Allan, J. A. *J. Org. Chem.* **1971**, *36*, 600.

[581] Bergman, J.; Eklund, N. *Tetrahedron* **1980**, *36*, 1445.

[582] Somei, M.; Ohnishi, H.; Shoken, Y. *Chem. Pharm. Bull.* **1986**, *34*, 677.

[583] Acheson, R. M.; Hunt, P. G.; Littlewood, D. M.; Murrer, B. A.; Rosenberg, H. E. *J. Chem. Soc., Perkin Trans. 1* **1978**, 1117.

[584] Khan, M. A.; Rocha, E. K. *Chem. Pharm. Bull.* **1979**, *27*, 528.

[585] Canas-Rodriquez, A.; Leeming, P. R. Brit. GB 1220628, 1971 (*Chem. Abstr.* **1971**, *75*, 5690h).

[586] Kawaski, T.; Kodama, A.; Nishida, T.; Shimizu, K.; Somei, M. *Heterocycles* **1991**, *32*, 221.

[587] Preobrazhenskaya, M. N.; Tolkachev, V. N.; Kudryasheva, V. A.; Yartseva, I. V. *Zh. Org. Khim.* **1975**, *11*, 199 (*Chem. Abstr.* **1975**, *82*, 140433m).

[588] Preobrazhenskaya, M. N.; Tolkachev, V. N.; Geling, O. N.; Kostyuchenko, N. P. *Zh. Org. Khim.* **1974**, *10*, 1764 (*Chem. Abstr.* **1974**, *81*, 136835a).

[589] Moody, C. J.; Ward, J. G. *J. Chem. Soc., Perkin Trans. 1* **1984**, 2895.

[590] Kamenov, L.; Yudin, L. G.; Budylin, V. A.; Kost, A. N. *Khim. Geterotsikl. Soedin.* **1970**, 923 (*Chem. Abstr.* **1971**, *74*, 12932w).

[591] Nogrady, T.; Morris, L. *Can. J. Chem.* **1969**, *47*, 1999.

[592] Akimoto, H.; Kawai, A.; Nomura, H.; Nagao, M.; Kawashi, T.; Sugimura, T. *Chem. Lett.* **1977**, 1061.

[593] Maiti, B. C.; Thomson, R. H.; Mahendran, M. *J. Chem. Res. (S)* **1978**, 126.

[594] Sato, Y.; Tanaka, T.; Nagasaki, T. *Yakugaku Zasshi* **1970**, 90, 618 (*Chem. Abstr.* **1970**, *73*, 35318u).

[595] Holla, B. S.; Ambekar, S. Y. *Indian J. Chem., Sect. B* **1976**, *14b*, 579.

[596] Street, J. D.; Harris, M.; Bishop, D. I.; Heatley, F.; Beddoes, R. L.; Mills, O. S.; Joule, J. A. *J. Chem. Soc., Perkin Trans. 1* **1987**, 1599.

[597] Bergman, J.; Pelcman, B. *Tetrahedron Lett.* **1985**, *26*, 6389.

[598] Tomita, K.; Terada, A.; Tachikawa, R. *Heterocycles* **1976**, *4*, 733.

[599] Scopes, D. I. C.; Allen, M. S.; Hignett, G. J.; Wilson, N. D. V.; Harris, M.; Joule, J. A. *J. Chem. Soc., Perkin Trans. 1* **1977**, 2376.

[600] Joshi, K. C.; Pathak, V. N.; Chaturvedi, R. K. *Pharmazie* **1986**, *41*, 634 (*Chem. Abstr.* **1987**, *107*, 58777c).

[601] Wolff, P. Ger. 614325, 1935 (*Chem. Abstr.* **1935**, *29*, 5861).

[602] Hoyle, V. A., Jr. US 3855209, 1974 (*Chem. Abstr.* **1975**, *82*, 113186b).

[603] Sato, H.; Tanaka, T.; Takagi, H. Jpn. 7012136, 1970 (*Chem. Abstr.* **1970**, *73*, 25293t).

[604] Sato, H.; Tanaka, T.; Takagi, H. Jpn. 7012135, 1970 (*Chem. Abstr.* **1970**, *73*, 25292s).

[605] Kutney, J. P.; Noda, M.; Lewis, N. G.; Monteiro, B.; Mostowicz, D.; Worth, B. R. *Heterocycles* **1981**, *16*, 1469.

[606] Kutney, J. P.; Noda, M.; Lewis, N. G.; Monteiro, B.; Mostowicz, D.; Worth, B. R. *Can J. Chem.* **1982**, *60*, 2426.

[607] Ishii, H.; Murakami, K.; Sakurada, E.; Hosoya, K.; Murakami, Y. *J. Chem. Soc., Perkin Trans. 1* **1988**, 2377.

[608] Suvorov, N. N.; Chernov, V. A.; Velezheva, V. S.; Ershova, A. Yu.; Simakov, S. V.; Sevodin, V. P. *Khim.-Farm. Zh.* **1981**, *15*, 27 (*Chem. Abstr.* **1982**, *96*, 85492s).

[609] Somei, M.; Kizu, K.; Kunimoto, M.; Yamada, F. *Chem. Pharm. Bull.* **1985**, *33*, 3696.

[610] Matsumoto, M.; Watanabe, N. *Heterocycles* **1987**, *26*, 1743.

[611] Somei, M.; Amari, H.; Makita, Y. *Chem. Pharm. Bull.* **1986**, *34*, 3971.

[612] Muratake, H.; Natsume, M. *Heterocycles* **1985**, *23*, 1111.

[613] Hatanaka, N.; Ozaki, O.; Matsumoto, M. *Tetrahedron Lett.* **1986**, *27*, 3169.

[614] Stepanov, B. I.; Stepanova, G. P.; Avramenko, G. V.; Stekhova, S. A.; Koz'min, Yu. P. *Khim. Tekhnol. Krasheniya, Sint. Krasitelei Polim. Mater.*, Ed. Mel'nikov, B. N., Ivanov. Khim.-Tekhnol. Inst.: Ivanovo, Ussr, **1977**, 130 (*Chem. Abstr.* **1979**, *91*, 212549q).

[615] Young, E. H. P. *J. Chem. Soc.* **1958**, 3493.

[616] Still, I. W. J.; Strautmanis, J. R. *Tetrahedron Lett.* **1989**, *30*, 1041.

[617] Cavallini, G.; Ravenna, V. *Farm. Ed. Sci.* **1958**, *13*, 113 (*Chem. Abstr.* **1958**, *52*, 20126i).

[618] Somei, M.; Saida, Y.; Funamoto, T.; Ohta, T. *Chem. Pharm. Bull.* **1987**, *35*, 3146.

[619] Moriya, T.; Hagio, K.; Yoneda, N. *Chem. Pharm. Bull.* **1980**, *28*, 1891.

[620] Hamabuchi, S.; Hamada, H.; Hironaka, A.; Somei, M. *Heterocycles* **1991**, *32*, 443.

[621] Young, E. H. P. Brit. 807620, 1959 (*Chem. Abstr.* **1959**, *53*, 12302b).

[622] Avramenko, V. G.; Mosina, G. S.; Suvorov, N. N. *Tr. Mosk. Khim.-Tekhnol. Inst.* **1970**, 129 (*Chem. Abstr.* **1971**, *75*, 88424y).

[623] Somei, M.; Saida, Y.; Komura, N. *Chem. Pharm. Bull.* **1986**, *34*, 4116.

[624] Ash, A. S. F.; Wragg, W. R. *J. Chem. Soc.* **1958**, 3887.

[625] Samsoniya, Sh. A.; Tabidze, D. M.; Kereselidze, Dzh. A.; Suvorov, N. N. *Khim. Geterotsikl. Soedin.* **1983**, 55 (*Chem. Abstr.* **1983**, *98*, 160542e).

[626] Samsoniya, Sh. A.; Chikvaidze, I. Sh.; Kereselidze, Dzh. A.; Suvorov, N. N. *Khim. Geterotsikl. Soedin.* **1983**, 1653 (*Chem. Abstr.* **1983**, *98*, 125033t).

[627] Cheshmaritashvili, M. G.; Samsoniya, Sh. A.; Kurkovskaya, L. N.; Suvorov, N. N. *Khim. Geterotsikl. Soedin.* **1984**, 73 (*Chem. Abstr.* **1984**, *100*, 191694v).

[628] Schmidt, U.; Wild, J. *Justus Liebigs Ann. Chem.* **1985**, 1882.

[629] Ohta, T.; Yamato, Y.; Tahira, H.; Somei, M. *Heterocycles* **1987**, *26*, 2817.

[630] Yamada, F.; Saida, Y.; Somei, M. *Heterocycles* **1986**, *24*, 2619.

[631] Glennon, R. A.; Chaurasia, C.; Titeler, M. *J. Med. Chem.* **1990**, *33*, 2777.

[632] Mckenzie, S.; Reid, D. H. *J. Chem. Soc., Chem. Commun.* **1966**, 401.

[633] Wolff, P.; Werner, W. Ger. 614326, 1935 (*Chem. Abstr.* **1935**, *29*, 5861).

[634] Flitsch, W.; Russkamp, P. *Heterocycles* **1984**, *22*, 541.

[635] Nagayoshi, T.; Saeki, S.; Hamana, M. *Chem. Pharm. Bull.* **1984**, *32*, 3678.

[636] Monge, A.; Martcnez, M. T.; Palop, J. A.; Fernández-Alvarez, E. *J. Heterocycl. Chem.* **1984**, *21*, 381.

[637] Jones, G. B.; Moody, C. J. *J. Chem. Soc., Perkin Trans. 1* **1989**, 2449.

[638] Jones, G. B.; Moody, C. J. *J. Chem. Soc. Chem. Commun.* **1988**, 166.

[639] Monge, A.; Palop, J. A.; Goɢi, T.; Martcnez, A.; Fernández-Alvarez, E. *J. Heterocycl. Chem.* **1985**, *22*, 1145.

[640] Prikhod'ko, T. A.; Vasilevskii, S. F.; Shvartsberg, M. S. *Izv. Akad. Nauk Sssr, Ser. Khim.* **1984**, 2602 (*Chem. Abstr.* **1985**, *102*, 113337h).

[641] Lee, G. T.; Amedio, J. C., Jr.; Underwood, R.; Prasad, K.; Repic, O. *J. Org. Chem.* **1992**, *57*, 3250.

[642] Kucherova, N. F.; Evdakov, V. P.; Kocketkov, N. K. *Zh. Obshch. Khim.* **1957**, *27*, 1049 (*Chem. Abstr.* **1958**, *52*, 3763c).

[643] Murakami, Y.; Ishii, H. *Chem. Pharm. Bull.* **1981**, *29*, 699.

[644] Pandit, R. S.; Seshadri, S. *Indian J. Chem.* **1974**, 943.

[645] Martin, G. J.; Poignant, S.; Filleux-Blanchard, M. L.; Quemeneur, M. T. *Tetrahedron Lett.* **1970**, 5061.

[646] Gillard, J. W.; Bélanger, P. *J. Med. Chem.* **1987**, *30*, 2051.

[647] Monge, A.; Parrado, P.; Font, M.; Fernández-Alvarez, E. *J. Med. Chem.* **1987**, *30*, 1029.

[648] Julia, M.; Lallemand, J. Y. *Bull. Soc. Chim. Fr.* **1973**, 2046.

[649] Betrabet, A. M.; Mani, K. V. S.; Seshadri, S.; Nimbkar, A. Y.; Rao, M. R. R. *Indian J. Chem.* **1970**, *8*, 704.

[650] Megrelishvili, N. Sh.; Chikvaidze, I. Sh.; Samsoniya, Sh. A.; Suvorov, N. N. *Soobshch. Akad. Nauk Gruz. SSR* **1991**, *141*, 121 (*Chem. Abstr.* **1991**, *115*, 49313m).

[651] Koo, J.; Avakian, S.; Martin, G. J. *J. Org. Chem.* **1959**, *24*, 179.

[652] Beretta, P.; Valbusa, L. Ger. Offen. DE 2429230, 1975 (*Chem. Abstr.* **1975**, *83*, 12227f).

[653] Chikvaidze, I. Sh.; Mumladze, E. A.; Samsoniya, Sh. A.; Suvorov, N. N. *Izv. Akad. Nauk Gruz. SSR, Ser. Khim.* **1985**, *11*, 109 (*Chem. Abstr.* **1986**, *104*, 168309x).

[654] Samsoniya, Sh. A.; Chikvaidze, I. Sh.; Suvorov, N. N. *Soobshch. Akad. Nauk Gruz. SSR* **1980**, *99*, 613 (*Chem. Abstr.* **1981**, *94*, 174791q).

[655] Cardwell, K.; Hewitt, B.; Magnus, P. *Tetrahedron Lett.* **1987**, *28*, 3303.

[656] Cardwell, K.; Hewitt, B.; Ladlow, M.; Magnus, P. *J. Am. Chem. Soc.* **1988**, *110*, 2242.

[657] Okuno, S.; Irie, K.; Nishino, H.; Iwashima, A.; Koshimizu, K. *Agric. Biol. Chem.* **1990**, *54*, 1885 (*Chem. Abstr.* **1990**, *113*, 206465r).

[658] Partsvaniya, D. A.; Akhvlediani, R. N.; Zhigachev, V. E.; Gordeev, E. N.; Kuleshova, L. N.; Suvorov, N. N. *Khim. Geterotsikl. Soedin.* **1987**, 919 (*Chem. Abstr.* **1988**, *108*, 150385s).

[659] Vorob'eva, S. L.; Buyanov, V. N.; Levina, I. I.; Suvorov, N. N. *Khim. Geterotsikl. Soedin.* **1989**, 783 (*Chem. Abstr.* **1990**, *112*, 178542x).

[660] Murakami, Y.; Takahashi, H.; Nakazawa, Y.; Koshimizu, M.; Watanabe, T.; Yokoyama, Y. *Tetrahedron Lett.* **1989**, *30*, 2019.

[661] Murakami, Y.; Ishii, H. *Chem. Pharm. Bull.* **1981**, *29*, 711.

[662] Yokoyama, Y.; Okuyama, N.; Iwadate, S.; Momoi, T.; Murakami, Y. *J. Chem. Soc., Perkin Trans. 1* **1990**, 1319.

[663] Murakami, Y.; Yokoyama, Y.; Okuyama, N. *Tetrahedron Lett.* **1983**, *24*, 2189.

[664] Murakami, Y.; Yokoama, Y.; Miura, T.; Nozawa, S.; Takeda, E.; Suzuki, H. *Heterocycles* **1988**, *27*, 2341.

[665] Grinev, A. N.; Shvedov, V. I.; Chizhov, A. K.; Vlasova, T. F. *Khim. Geterotsikl. Soedin.* **1975**, 1250 (*Chem. Abstr.* **1976**, *84*, 30797s).

[666] Yamada, Y.; Hirata, T.; Matsui, M. *Tetrahedron Lett.* **1969**, 101.

[667] Casner, M. L.; Remers, W. A.; Bradner, W. T. *J. Med. Chem.* **1985**, *28*, 921.

[668] Ambros, R.; Von Angerer, S.; Wiegrebe, W. *Arch. Pharm.* **1988**, *321*, 743.

[669] Allen, Jr., G. R.; Weiss, M. *J. Org. Chem.* **1965**, *30*, 2904.

[670] Ambros, R.; Schneider, M. R.; Von Angerer, S. *J. Med. Chem.* **1990**, *33*, 153.

[671] Abd El-Rahman, A. El R.; Khodeir, M. N.; Sarhan, A. A.; Kandeel, E. El D. M.; Abdov, S. B. *Pol. J. Chem.* **1988**, *62*, 879.

[672] Abd El-Rahman, A. H.; Kandeel, E. El D. M.; Amer, F. A. K.; El Desoky, El S. I. *Pol. J. Chem.* **1988**, *62*, 489 (*Chem. Abstr.* **1989**, *111*, 232498g).

[673] Monge, A.; Font, M.; Parrado, P.; Fernández-Alvarez, E. *An. Quim., Ser. C* **1988**, *84*, 270 (*Chem. Abstr.* **1989**, *111*, 23463c).

[674] Remers, W. A.; Roth, R. H.; Weiss, M. J. *J. Am. Chem. Soc.* **1964**, *86*, 4612.

[675] Black, D. St. C.; Craig, D. C.; Kumar, N. *J. Chem. Soc., Chem. Commun.* **1989**, 425.

[676] Ohta, T.; Somei, M. *Heterocycles* **1989**, *29*, 1663.

[677] Flitsch, W.; Russkamp, P.; Langer, W. *Justus Liebigs Ann. Chem.* **1985**, 1413.

[678] Allan, Jr., G. R.; Poletto, J. F.; Weiss, M. J. *J. Org. Chem.* **1965**, *30*, 2897.

[679] Iyengar, B. S.; Remers, W. A.; Catino, J. J. *J. Med. Chem.* **1989**, *32*, 1866.

[680] Hodges, J. C.; Remers, W. A.; Bradner, W. T. *J. Med. Chem.* **1981**, *24*, 1184.

[681] Fost, D. L.; Ekwuribe, N. N.; Remers, W. A. *Tetrahedron Lett.* **1973**, 131.

[682] Allan, Jr., G. R.; Weiss, M. J. *J. Heterocycl. Chem.* **1970**, *7*, 193.

[683] Filler, R.; Woods, S. M.; White, W. L. *Can J. Chem.* **1989**, *67*, 1837.

[684] Jones, G. B.; Moody, C. J. *J. Chem. Soc., Perkin Trans. 1* **1989**, 2455.

[685] Jones, G. B.; Moody, C. J. *J. Chem. Soc., Chem. Commun.* **1989**, 186.

[686] Flitsch, W.; Langer, W. *Justus Liebigs Ann. Chem.* **1988**, 391.

[687] Bhagwanth, M. R. R.; Rao, A. V. R.; Venkataraman, K. *Indian J. Chem.* **1969**, *7*, 1065.

[688] Langendoen, A.; Koomen, G. J.; Pandit, U. K. *Heterocycles* **1987**, *26*, 91.

[689] Langendoen, A.; Plug, J. P. M.; Koomen, G. J.; Pandit, U. K. *Tetrahedron* **1989**, *45*, 1759.

[690] Buu-Hoï, N. P. *J. Org. Chem.* **1951**, *16*, 1327.

[691] Garuti, L.; Roberti, M.; Leoni, A.; Brigidi, P. *Pharmazie* **1990**, *45*, 863 (*Chem. Abstr.* **1991**, *114*, 185314u).

[692] Gansser, C.; Viel, C.; Viossat, B.; Rodier, N.; Merienne, C. *Farm. Ed. Sci.* **1988**, *43*, 301 (*Chem. Abstr.* **1989**, *110*, 175520n).

[693] Dalton, L. K.; Demerac, S.; Elmes, B. L.; Loder, J. W.; Swan, J. M.; Teitei, T. *Aust. J. Chem.* **1967**, *20*, 2715.

[694] Cranwell, P. A.; Saxton, J. E. *Chem. Ind. (London)* **1962**, 45.

[695] Cranwell, P. A.; Saxton, J. E. *J. Chem. Soc.* **1962**, 3482.

[696] Viossat, P. B.; Rodier, N.; Gansser, C.; Viel, C. *Acta Cryst.* **1988**, *C44*, 581.

[697] Hall, R. J.; Jackson, A. H.; Oliveira-Campus, A. M. F.; Queiroz, M. -J. R. P.; Shannon, P. V. R. *Heterocycles* **1990**, *31*, 401.

[698] Dulcere, J.-P.; Tawil, M.; Santelli, M.; *J. Org. Chem.* **1990**, *55*, 571.

[699] Neumann, B. M.; Henning, H. G.; Gloyna, D.; Bandlow, M. *J. Prakt. Chem.* **1976**, *318*, 823.

[700] Dallazker, F.; Kern, J. *Chem. Ber.* **1966**, *99*, 3830.

[701] Thoman, C. J.; Denys, S. J.; Voaden, J.; Hunsberger, I. M. *J. Org. Chem.* **1964**, *29*, 2044.

[702] Yeh, M.-Y.; Tien, H.-J.; Tung, C.-H.; Hwang, C.-C.; Wu, T. S. *J. Chin. Chem. Soc. (Taipai)* **1988**, *35*, 459 (*Chem. Abstr.* **1989**, *110*, 172341t).

[703] Masuda, K.; Adachi, J.; Nomura, K. *J. Chem. Soc., Perkin Trans. 1* **1979**, 2349.

[704] Finar, I. L.; Lord, G. H. *J. Chem. Soc.* **1959**, 1819.

[705] Anderson, P. L.; Paolella, N. A. U.S. 4220792, 1980 (*Chem. Abstr.* **1980**, *93*, 239403r).

[706] Bernard, M. K.; Wrzeciono, U. *Pharmazie* **1988**, *43*, 723 (*Chem. Abstr.* **1989**, *110*, 212680m).

[707] Haeufel, J.; Breitmaier, E. *Angew. Chem. Int. Ed. Engl.* **1974**, *13*, 604.

[708] Haga, T.; Toki, T.; Koyanagi, T.; Toru, Okada, H.; Imai, O.; Morita, M. Jpn. Kokai Tokkyo Koho Jp 02096568, 1990 (*Chem. Abstr.* **1990**, *113*, 132173v).

[709] Thiruvikraman, S. V.; Seshadri, S. *Indian J. Chem. Sect. B* **1984**, *23b*, 768.

[710] Becher, J.; Jorgensen, P. L.; Pluta, K.; Krake, N. J.; Fält-Hansen, B. *J. Org. Chem.* **1992**, *57*, 2127.

[711] Porai-Koshits, B. A.; Kvito, I. Ya.; Shutkova, E. A. *Pharm. Chem. J. (New York)* **1970**, 138 (*Chem. Abstr.* **1970**, *73*, 3844w).

[712] Malhotra, N.; Fält-Hansen, B.; Becher, J. *J. Heterocycl. Chem.* **1991**, *28*, 1837.

[713] Wallace, D. J.; Straley, J. M. *J. Org. Chem.* **1961**, *26*, 3825.

[714] Barnela, S. B.; Pandit, R. S.; Seshadri, S. *Indian J. Chem., Sect. B* **1976**, *14B*, 668.

[715] Enomoto, M.; Nagano, E.; Sato, R.; Sakaki, M. Eur. Pat. Appl. EP 422639, 1991 (*Chem. Abstr.* **1991**, *115*, 114522v).

[716] Chandramohan, M. R.; Sardessai, M. S.; Shah, S. R.; Seshadri, S. *Indian J. Chem.* **1969**, *7*, 1006.

[717] Dymek, W.; Janik, B.; Zimon, R. *Acta Polon. Pharm.* **1963**, *20*, 9 (*Chem. Abstr.* **1964**, *61*, 8293h).

[718] Wrzeciono, U.; Szostak-Rzepiak, B. *Pharmazie* **1975**, *30*, 582 (*Chem. Abstr.* **1976**, *84*, 4848q).

[719] Barnela, S. B.; Pandit, R. S.; Seshadri, S. *Indian J. Chem., Sect. B* **1976**, *14B*, 665.

[720] Kira, M. A.; Bruckner-Wilhelms, A. *Acta Chim. Acad. Sci. Hung.* **1968**, *56*, 47 (*Chem. Abstr.* **1968**, *69*, 86888e).

[721] Farbenfabriken Bayer AG, Brit. 887509, 1962 (*Chem. Abstr.* **1962**, *57*, 12502).

[722] Abdel-Megeed, M. F. *Indian J. Chem., Sect. B* **1987**, *26B*, 827.

[723] Abdel-Megeed, M. F. *Rev. Roum. Chim.* **1987**, *32*, 591 (*Chem. Abstr.* **1988**, *108*, 186636).

[724] El-Shekeil, A.; Babaqi, A.; Hassan, M.; Shiba, S. *Heterocycles* **1988**, *27*, 2577.

[725] Babaqi, A.; El-Shekeil, A.; Hassan, M.; Shiba, S. *Heterocycles* **1988**, *27*, 2119.

[726] Purnaprajna, V.; Seshadri, S. *Indian J. Chem., Sect. B* **1976**, *14B*, 971.

[727] Ledrut, J.; Combes, G. *Bull. Soc. Chim. Fr.* **1957**, 877.

[728] Ito, I. *Yakugaku Zasshi* **1956**, *76*, 167 (*Chem. Abstr.* **1956**, *50*, 13939f).

[729] Kokkinos, K.; Markopoulos, C. *J. Prakt. Chem.* **1980**, *322*, 543.

[730] Takahashi, T.; Kanematsu, K. *Chem. Pharm. Bull.* **1958**, *6*, 374.

[731] Ledrut, J. H. T.; Combes, G. *Chim. Ther.* **1968**, *3*, 248 (*Chem. Abstr.* **1969**, *70*, 37708h).

[732] Mach-Phouc-Sinh; Buu-Hoï, N. P. *Chim. Ther.* **1967**, *2*, 106 (*Chem. Abstr.* **1967**, *67*, 100064a).

[733] Andreani, A.; Bonazzi, D.; Rambaldi, M. *Arch. Pharm.* **1982**, *315*, 451.

[734] Isomura, Y.; Ito, N.; Sakamoto, S.; Homma, H.; Abe, T.; Kubo, K. *Chem. Pharm. Bull.* **1983**, *31*, 3179.

[735] Isomura, Y.; Abe, T. Jpn. Kokai Tokkyo Koho JP 62142162, 1987 (*Chem. Abstr.* **1988**, *108*, 21889k).

[736] Iddon, B. in *Stud. Org. Chem. (Chem. Heterocycl. Compd.)*, Kovác, J.; Zálupsky, P., Eds., *33*, **1987**, 24.

[737] Bahar, M. H.; Sabata, B. K. *Indian J. Chem., Sect. B* **1981**, *20B*, 328.

[738] Tomita, K.; Murakami, T. Jpn. Kokai Tokkyo Koho JP 54020504, 1979 (*Chem. Abstr.* **1979**, *91*, 157755b).

[739] Sakamoto, T.; Yamanaka, H.; Shiozawa, A.; Tanaka, W.; Miyazaki, H. *Chem. Pharm. Bull.* **1980**, *28*, 1832.

[740] Yamanaka, H.; Sakamoto, T.; Shiozawa, A. *Heterocycles* **1977**, *7*, 51.

[741] Anderson, D. J. *J. Org. Chem.* **1986**, *51*, 945.

[742] Kallury, R. K. M. R.; Devi, P. S. U. *Tetrahedron Lett.* **1977**, 3655.

[743] Beccalli, E. M.; Marchesini, A.; Pilati, T. *Tetrahedron* **1989**, *45*, 7485.

[744] Beccalli, E. M.; Marchesini, A.; Molinari, H. *Tetrahedron Lett.* **1986**, *27*, 627.

[745] Makkay, C.; Literati-Kiraly, I. *Stud. Univ. Babes-Bolyai, [Ser.] Chem.* **1978**, *23*, 52 (*Chem. Abstr.* **1979**, *90*, 203923j).

[746] Kvitko, I. Ya.; Smirnova, V. A.; El'tsov, A. V. *Khim. Geterotsikl. Soedin.* **1980**, 36 (*Chem. Abstr.* **1980**, *92*, 215325v).

[747] Isomura, Y.; Sakamoto, S.; Yoshida, M.; Abe, T. Jpn. Kokai Tokkyo Koho JP 63112572, 1988 (*Chem. Abstr.* **1988**, *109*, 110423s).

[748] Egli, R. Ger. Offen. DE 3015121, 1980 (*Chem. Abstr.* **1981**, *94*, 158317j).

[749] Egli, R. Braz. Pedido PI BR 8101464, 1981 (*Chem. Abstr.* **1982**, *96*, 105794x).

[750] Pawar, R. A.; Rajput, A. P. *Indian J. Chem., Sect. B* **1989**, *28B*, 866.

[751] Egli, R. US 4395544, 1983 (*Chem. Abstr.* **1983**, *99*, 159949g).

[752] Guzmán, A.; Romero, M.; Muchowski, J. M. *Can. J. Chem.* **1990**, *68*, 791.

[753] Schulte, K. E.; Reisch, J.; Stoess, U. *Angew. Chem. Int. Ed. Engl.* **1965**, *4*, 1081.

[754] Messerschmitt, T.; Von Specht, U.; Von Dobeneck, H. *Justus Liebigs Ann. Chem.* **1971**, *751*, 50.

[755] Schulte, K. E.; Reisch, J.; Stoess, U. *Arch. Pharm.* **1972**, *305*, 523.

[756] Von Dobeneck, H.; Messerschmitt, T. *Justus Liebigs Ann. Chem.* **1971**, *751*, 32.

[757] Ito, N.; Aiga, H.; Nishihara, M.; Yano, N.; Nishida, T.; Nagayoshi, T. Jpn. Kokai Tokkyo Koho JP 60208976, 1985 (*Chem. Abstr.* **1986**, *104*, 148727d).

[758] Egli, R.; Henzi, B. Brit. UK Pat. Appl. GB 2163768, 1986 (*Chem. Abstr.* **1986**, *105*, 135467q).

[759] Egli, R.; Henzi, B. Ger. Offen. DE 3529831, 1986 (*Chem. Abstr.* **1987**, *106*, 6402m).

[760] Pagani, G. A. *J. Chem. Soc., Perkin Trans. 1* **1974**, 2050.

[761] Kaji, K.; Nagashima, H.; Hirose, Y.; Oda, H. *Chem. Pharm. Bull.* **1985**, *33*, 982.

[762] Klötzer, W.; Herberz, M. *Monatsh. Chem.* **1965**, *96*, 1567.

[763] Bredereck, H.; Simchen, G.; Wagner, H.; Santos, A. A. *Justus Liebigs Ann. Chem.* **1972**, *766*, 73.

[764] Aoyagi, T.; Yanada, R.; Bessho, K.; Yoneda, F.; Armarego, W. L. F. *J. Heterocycl. Chem.* **1991**, *28*, 1537.

[765] Seela, F.; Steker, H. *Helv. Chim. Acta* **1986**, *69*, 1602.

[766] Bell, L.; McGuire, H. M.; Freeman, G. A. *J. Heterocycl. Chem.* **1983**, *20*, 41.

[767] Dehnert, J. Ger. Offen. DE 3603797, 1987 (*Chem. Abstr.* **1988**, *108*, 7511z).

[768] Schlager, L. H. Ger. Offen. DE 2310334, 1973 (*Chem. Abstr.* **1973**, *79*, 146549p).

[769] Cherdantseva, N. M.; Nesterov, V. M.; Safonova, T. S. *Khim. Geterotsikl. Soedin.* **1983**, 834 (*Chem. Abstr.* **1983**, *99*, 139895h).

[770] Kitamura, N.; Ohnishi, A. Eur. Pat. Appl. EP 163599, 1985 (*Chem. Abstr.* **1986**, *104*, 186439u).

[771] Bredereck, H.; Effenberger, F.; Simchen, G. *Chem. Ber.* **1964**, *97*, 1403.

[772] Yoneda, F.; Sakuma, Y. *J. Chem. Soc., Chem. Commun.* **1976**, 203.

[773] Senda, S.; Hirota, K.; Yang, G. -N.; Shirahashi, M. *Yakugaku Zasshi* **1971**, *91*, 1372 (*Chem. Abstr.* **1972**, *76*, 126915q).

[774] Tanaka, K.; Kimura, T.; Okada, T.; Chen, X.; Yoneda, F. *Chem. Pharm. Bull.* **1987**, *35*, 1397.

[775] Nishigaki, S.; Kanamori, Y.; Senga, K. *Chem. Pharm. Bull.* **1978**, *26*, 2497.

[776] Hirota, K.; Asao, T.; Fujioka, T.; Senda, S. *Nippon Kagaku Kaishi*, **1981**, 721 (*Chem. Abstr.* **1981**, *95*, 150597a).

[777] Botta, M.; De Angelis, F.; Corelli, F.; Menichincheri, M.; Nicoletti, R.; Marongiu, M. E.; Pani, A.; La Colla, P. *Arch. Pharm.* **1991**, *324*, 203.

[778] Senga, K.; Kanamori, Y.; Nishigaki, S.; Yoneda, F. *Chem. Pharm. Bull.* **1976**, *24*, 1917.

[779] Yoneda, F.; Tsukuda, K.; Kawazoe, M.; Sone, A.; Koshiro, A. *J. Heterocycl. Chem.* **1981**, *18*, 1329.

[780] Yoneda, F.; Hirayama, R.; Yamashita, M. *Chem. Lett.* **1980**, 1157.

[781] Grauert, R. W. *Arch. Pharm.* **1982**, *315*, 949.

[782] Link, P. A. J.; Van der Plas, H. C.; Müller, F. *J. Heterocycl. Chem.* **1985**, *22*, 841.

[783] Sakuma, Y.; Matsushita, Y.; Yoneda, F.; Nitta, Y. *Heterocycles* **1978**, *9*, 1767.

[784] Grauert, R. W. *Justus Liebigs Ann. Chem.* **1979**, 1802.

[785] Janda, M.; Hemmerich, P. *Angew. Chem. Int. Ed. Engl.* **1976**, *15*, 443.

[786] Yoneda, F.; Shinozuka, K.; Sakuma, Y.; Senga, K. *Heterocycles* **1977**, *6*, 1179.

[787] Yoneda, F.; Shinozuka, K.; Hiromatsu, K.; Matsushita, R.; Sakuma, Y.; Hamana, M. *Chem. Pharm. Bull.* **1980**, *28*, 3576.

[788] Yoneda, F.; Tanaka, K.; Yamato, H.; Moriyama, K.; Nagamatsu, T. *J. Am. Chem. Soc.* **1989**, *111*, 9199.

[789] Yoneda, F.; Koga, M.; Tanaka, K.; Yano, Y. *J. Heterocycl. Chem.* **1989**, *26*, 1221.

[790] Yoneda, F.; Koga, M.; Yano, Y. *J. Chem. Soc., Perkin Trans. 1* **1988**, 1813.

[791] Yoneda, F.; Koga, M.; Ibuka, T.; Yano, Y. *Chem. Pharm. Bull.* **1986**, *34*, 2653.

[792] Yoneda, F.; Koga, M. *J. Heterocycl. Chem.* **1988**, *25*, 549.

[793] Mironova, G. A.; Kirillova, E. N.; Kuklin, V. N.; Smorygo, N. A.; Ivin, B. A. *Khim. Geterotsikl. Soedin.* **1984**, 1328 (*Chem. Abstr.* **1985**, *102*, 149205c).

[794] Herberich, G. E.; Naithani, A. K. *J. Organomet. Chem.* **1983**, *241*, 1.

[795] Kasturi, T. R.; Jois, H. R. Y.; Mathew, L. *Synthesis* **1984**, 743.

[796] Sugasawa, T.; Toyoda, T.; Sasakura, K. *Tetrahedron Lett.* **1972**, 5109.

[797] Rivalle, C.; Bisagni, E. *J. Heterocycl. Chem.* **1980**, *17*, 245.

[798] Sugasawa, T.; Toyoda, T.; Sasakura, K. *Chem. Pharm. Bull.* **1974**, *22*, 771.

[799] Panditrao, P. R.; Deval, S. D.; Gupte, S. M.; Samant, S. D.; Deodhar, K. D. *Indian J. Chem., Sect. B* **1981**, *20B*, 929.

[800] Christie, R. M.; Reid, D. H. *J. Chem. Soc., Perkin Trans. 1* **1977**, 848.

[801] Costakis, E.; Hamogeorgakis, M.; Tsatsas, G. *Chem. Chron.* **1978**, *7*, 171.

[802] Paolini, J. P.; Tendvay, L. J. *J. Med. Chem.* **1969**, *12*, 1031.

[803] Andreani, A.; Rambaldi, M.; Carloni, P.; Greci, L.; Stipa, P. *J. Heterocycl. Chem.* **1989**, *26*, 525.

[804] Andreani, A.; Mascellani, G.; Rambaldi, M.; Rugarli, P. Eur. Pat. Appl. EP 164635, 1985 (*Chem. Abstr.* **1986**, *104*, 224897b).

[805] Andreani, A.; Mascellani, G.; Rambaldi, M.; Rugarli, P. *Eur. J. Med. Chem.* **1987**, *22*, 19.

[806] Koshelev, Yu. N.; Reznichenko, A. V.; Efros, L. S.; Kvitko, I. Ya. *Zh. Org. Khim.* **1973**, *9*, 2201 (*Chem. Abstr.* **1974**, *80*, 27165w).

[807] Bignebat, J.; Quiniou, H. *C. R. Hebd. Acad. Sci.* **1969**, *269C*, 1129.

[808] Mackie, R. K.; McKenzie, S.; Reid, D. H.; Webster, R. G. *J. Chem. Soc., Perkin Trans. 1* **1973**, 657.

[809] Brindley, J. C.; Gillan, D. G.; Meakins, G. D. *J. Chem. Soc., Perkin Trans. 1* **1986**, 1255.

[810] Mukai, T.; Kumagai, T.; Tanaka, S. Jpn. Kokai Tokkyo Koho JP 62207275, 1987 (*Chem. Abstr.* **1988**, *108*, 186728v).

[811] Tsubouchi, A.; Matsumura, N.; Inoue, H.; Hamasaki, N.; Yoneda, S.; Yanagi, K. *J. Chem. Soc., Chem. Commun.* **1989**, 223.

[812] Tsubouchi, A.; Matsumura, N.; Inoue, H.; Yanagi, K. *J. Chem. Soc., Perkin Trans. 1* **1991**, 909.

[813] Lippmann, E.; Strauch, P.; Tenor, E.; Thomas, E. Ger. (East) DD 264439, 1989 (*Chem. Abstr.* **1989**, *111*, 115203w).

[814] Pentimalli, L.; Bozzini, S. *Boll. Sci. Fac. Chim. Ind. Bologna,* **1965**, *23*, 181 (*Chem. Abstr.* **1965**, *63*, 14848).

[815] Ollis, W. D.; Stanforth, S. P.; Ramsden, C. A. *J. Chem. Soc., Perkin Trans. 1* **1989**, 961.

[816] Sizova, O. S.; Britikova, N. E.; Novitskii, K. Yu.; Shcherbakova, L. I.; Pershin, G. N. *Khim.-Farm. Zh.* **1980**, *14*, 63 (*Chem. Abstr.* **1980**, *93*, 239357d).

[817] Britikova, N. E.; Novitskii, K. Yu. *Khim. Geterotsikl. Soedin.* **1977**, 1672 (*Chem. Abstr.* **1978**, *88*, 105263h).

[818] Senda, S.; Hirota, K. *Chem. Pharm. Bull.* **1974**, *22*, 2921.

[819] Kikugawa, K.; Kawashima, T. *Chem. Pharm. Bull.* **1971**, *19*, 2629.

[820] Fraser, M. *J. Org. Chem.* **1972**, *37*, 3027.

[821] Hirota, K.; Shirahashi, M.; Senda, S.; Yogo, M. *J. Heterocycl. Chem.* **1990**, *27*, 717.

[822] Uritskaya, M. Ya.; Vasil'eva, V. V.; Liberman, S. S.; Yakhontov, L. N. *Khim.-Farm. Zh.* **1973**, *7*, 8 (*Chem. Abstr.* **1974**, *80*, 47880d).

[823] Robaut, C.; Rivalle, C.; Rautureau, M.; Lhoste, J. M.; Bisagni, E.; Chermann, J. C. *Tetrahedron* **1985**, *41*, 1945.

[824] Chi Hung Nguyen; Bisagni, E.; Lhoste, J. M. *Can. J. Chem.* **1986**, *64*, 545.

[825] Rossiter, E. D.; Saxton, J. E. *J. Chem. Soc. C* **1953**, 3654.

[826] Turchinovich, G. Yu. *Tr. Kievsk. Politekhn. Inst.* **1963**, *43*, 81 (*Chem. Abstr.* **1965**, *62*, 10559).

[827] Yufit, D. S.; Struchkov, Yu. T.; Prostakov, N. S.; Kuznetsov, V. I. *Khim. Geterotsikl. Soedin.* **1986**, 1048 (*Chem. Abstr.* **1987**, *106*, 176106k).

[828] Gibson, W. K.; Leaver, D. *J. Chem. Soc. C* **1966**, 324.

[829] Somei, M.; Hirai, Y.; Fujii, S.; Fujita, A. Jpn. Kokai Tokkyo Koho JP 02085251, 1990 (*Chem. Abstr.* **1990**, *113*, 110930n).

[830] Marchetti, L.; Andreani, A. *Ann. Chim. (Rome)* **1973**, *63*, 681 (*Chem. Abstr.* **1975**, *82*, 72723d).

[831] Predvoditeleva, G. S.; Kartseva, T. V.; Oleshko, O. N.; Shvedov, V. I.; Syubaev, R. D.; Sharts, G. Ya.; Alekseeva, L. M.; Anisimova, O. S.; Chistyakov, V. V.; Sheinker, L. M. *Khim.-Farm. Zh.* **1987**, *21*, 441 (*Chem. Abstr.* **1987**, *107*, 126578r).

[832] Coppola, G. M. *J. Heterocycl. Chem.* **1981**, *18*, 845.

[833] Toda, T.; Ryu, S.; Hagiwara, Y.; Nozoe, T. *Bull. Chem. Soc. Jpn.* **1975**, *48*, 82.

[834] Burova, O. A.; Bystryakova, I. D.; Smirnova, N. M.; Safonova, T. S. *Khim. Geterotsikl. Soedin.* **1990**, 662 (*Chem. Abstr.* **1991**, *114*, 6438t).

[835] Horváth, Á.; Hermecz, I. *J. Heterocycl. Chem.* **1986**, *23*, 1295.

[836] Pagani, G.; Bradamante, S. *Tetrahedron Lett.* **1968**, 1041.

[837] Koyama, T.; Hirota, T.; Shinohara, Y.; Yamato, M.; Ohmori, S. *Chem. Pharm. Bull.* **1975**, *23*, 497.

[838] Kubo, K.; Ito, N.; Isomura, Y.; Sozu, I.; Homma, H.; Murakami, M. *Yakugaku Zasshi* **1979**, *99*, 993 (*Chem. Abstr.* **1980**, *92*, 146698e).

[839] M-ur-Rahman; Khan, K. -Z.; Siddiqi, Z. S.; Zaman, A. *Indian J. Chem., Sect. B* **1990**, *28B*, 941.

[840] Andrieux, J.; Battioni, J. -P.; Giraud, M.; Mohlo, D. *Bull. Soc. Chim. Fr.* **1973**, 2093.

[841] Weissenfels, M.; Hantschmann, A.; Steinführer, T.; Birkner, E. *Z. Chem.* **1989**, *29*, 166.

[842] Naik, R. M.; Thakor, V. M. *J. Org. Chem.* **1957**, *22*, 1630.

[843] Tabakovic, K.; Tabakovic, I.; Ajdini, N.; Leci, O. *Synthesis* **1987**, 308.

[844] Heber, D. *Arch. Pharm.* **1987**, *320*, 577.

[845] Proshkina, V. N.; Solomko, Z. F.; Bozhanova, N. Ya. *Khim. Geterotsikl. Soedin.* **1988**, 1288 (*Chem. Abstr.* **1989**, *111*, 39322s).

[846] Solomko, Z. F.; Proshkina, V. N.; Avramenko, V. I.; Plastun, I. A.; Bozhanova, N. Ya. *Khim. Geterotsikl. Soedin.* **1984**, 1262 (*Chem. Abstr.* **1985**, *102*, 6436t).

[847] Solomko, Z. F.; Proshkina, V. N.; Bozhanova, N. Ya.; Loban, S. V.; Babichenko, L. N. *Khim. Geterotsikl. Soedin.* **1984**, 223 (*Chem. Abstr.* **1984**, *100*, 209762c).

[848] Sheinkman, A. K.; Strokin, Yu. V.; Priimenko, B. A.; Klyuev, N. A. *Vopr. Khim. Tekhnol.* **1983**, *73*, 64 (*Chem. Abstr.* **1985**, *102*, 6375x).

[849] Itaya, T.; Mizutani, A. *Tetrahedron Lett.* **1985**, *26*, 347.

[850] Itaya, T.; Mizutani, A.; Takeda, M.; Shioyama, C. *Chem. Pharm. Bull.* **1989**, *37*, 284.

[851] Samsoniya, Sh. A.; Lomtatidze, Z. Sh.; Dolidze, S. V.; Suvorov, N. N. *Khim.-Farm. Zh.* **1984**, *18*, 1452 (*Chem. Abstr.* **1985**, *102*, 220770r).

[852] Samsoniya, Sh. A.; Targamadze, N. L.; Kozik, T. A.; Suvorov, N. N. *Khim. Geterotsikl. Soedin.* **1982**, 206 (*Chem. Abstr.* **1982**, *96*, 181174k).

[853] Samsoniya, Sh. A.; Kadzhrishvili, D. O.; Gordeev, E. N.; Suvorov, N. N. *Khim. Geterotsikl. Soedin.* **1984**, 1219 (*Chem. Abstr.* **1985**, *102*, 6243c).

[854] Chapman, N. B.; Hughes, C. G.; Scrowston, R. M. *J. Chem. Soc. C* **1971**, 1308.

[855] Blythin, D. J.; Kaminski, J. J.; Domalski, M. S.; Spitler, J.; Solomon, D. M.; Conn, D. J.; Wong, S. C.; Verbiar, L. L.; Bober, L. A.; Chiu, P. J. S.; Watnick, A. A.; Siegel, M. I.; Hilbert, J. M.; McPhail, A. T. *J. Med. Chem.* **1986**, *29*, 1099.

[856] Rida, S. M.; Soliman, F. S. G.; Badaway, El S. A. M.; Kappe, T. *J. Heterocycl. Chem.* **1988**, *25*, 1725.

[857] Lancelot, J. C.; Laduree, D.; Robba, M. *Chem. Pharm. Bull.* **1985**, *33*, 4242.

[858] Hayashi, S.; Kurokawa, S.; Matsuura, T. *Bull. Chem. Soc. Jpn.* **1969**, *42*, 1404.

859 Ponasenkova, T. F.; Akhvlediani, R. N.; Kurkovskaya, L. N.; Dikopolova, V. V.; Surorov, N. N. *Khim. Geterotsikl. Soedin.* **1984**, 490 (*Chem. Abstr.* **1984**, *101*, 72640k).

860 Glushkov, R. G.; Davydova, N. K.; Marchenko, N. B. *Khim. -Farm. Zh.* **1987**, *21*, 985 (*Chem. Abstr.* **1988**, *108*, 167347d).

861 Voronina, E. I.; Ostapchuk, G. M.; Ivanova, T. M.; Babushkina, T. A.; Shagalov, L. B.; Surorov, N. N. *Khim. Geterotsikl. Soedin.* **1984**, 343 (*Chem. Abstr.* **1984**, *101*, 6978a).

862 Kaminsky, D.; Klutchko, S.; Von Strandtmann, M. U.S. 3937828, 1976 (*Chem. Abstr.* **1976**, *84*, 164807d).

863 Clarke, K. *J. Chem. Soc., Perkin Trans. 1* **1973**, 2956.

864 Porshnev, Yu. N.; Churkina, V. A.; Cherkashin, M. I. *Khim. Geterotsikl. Soedin.* **1976**, 751 (*Chem. Abstr.* **1976**, *85*, 123723a).

865 Abe, N. *Bull. Chem. Soc. Jpn.* **1987**, *60*, 1053.

866 Mezheritskii, V. V.; Elisevich, D. M.; Dorofeenko, G. N. *Zh. Org. Khim.* **1981**, *17*, 2444 (*Chem. Abstr.* **1982**, *96*, 104167h).

867 Brack, A. Brit. GB 1202808, 1970 (*Chem. Abstr.* **1970**, *73*, 131985e).

868 Buu-Hoï, N. P.; Hoán, N. *J. Chem. Soc.* **1951**, 1834.

869 Cauquil, G.; Casadevall, A. *C. R. Hebd. Seances Acad. Sci.* **1955**, *240*, 1784.

870 Khristiansen, M. G.; Il'chenko, A. Ya.; Rozum, Yu. S.*Dopov. Akad. Nauk Ukr. RSR, Ser. B* **1970**, *32*, 829 (*Chem. Abstr.* **1971**, *74*, 99989s).

871 Pokhodenko, V. D.; Koshechko, V. G.; Inozemtsev, A. N. *J. Chem. Soc., Chem. Commun.* **1985**, 72.

872 Koshechko, V. G.; Inozemtsev, A. N. *Khim. Geterotsikl. Soedin.* **1984**, 835 (*Chem. Abstr.* **1984**, *101*, 170330d).

873 Cauquil, G.; Casadevall, E.; Greze, R. *Bull. Soc. Chim. Fr.* **1964**, 590.

874 Sengupta, D.; Anand, N. *Indian J. Chem., Sect. B* **1986**, *25B*, 72.

875 Buu-Hoï, N. P.; Lobert, B.; Saint-Ruf, G. *J. Chem. Soc. C* **1969**, 2137.

876 Khoshtariya, T. E.; Kakhabrishvili, M. L.; Sikharulidze, M. I.; Kurkovskaya, L. N.; Suvorov, N. N. *Khim. Geterotsikl. Soedin.* **1985**, 355 (*Chem. Abstr.* **1985**, *103*, 53975g).

877 Khoshtariya, T. E.; Kintsurashvili, L. A.; Kurkovskaya, L. N.; Suvorov, N. N. *Soobshch. Akad. Nauk Gruz. SSR* **1984**, *116*, 341 (*Chem. Abstr.* **1985**, *103*, 71219j).

878 Kintsurashvili, L. A.; Khoshtariya, T. E.; Kurkovskaya, L. N.; Suvorov, N. N. *Soobshch. Akad. Nauk Gruz. SSR* **1982**, *107*, 341 (*Chem. Abstr.* **1983**, *98*, 125917j).

879 Khoshtariya, T. E.; Palavandishvili, G. A.; Sikharulidze, M. I.; Kurkovskaya, L. N.; Suvorov, N. N. *Khim. Geterotsikl. Soedin.* **1984**, 1335 (*Chem. Abstr.* **1985**, *102*, 45860w).

880 Buu-Hoï, N. P.; Lavit, D. *J. Chem. Soc.* **1959**, 38.

881 Schlözer, R.; Fuhrhop, J.-H. *Angew. Chem. Int. Ed. Engl.* **1975**, *14*, 388.

882 Niedercorn, F.; Ledon, H.; Tkatchenko, I. *New J. Chem.* **1988**, *12*, 897.

883 Ponomarev, G. V.; Kirillova, G. V.; Maravin, G. B.; Babushkina, T. A.; Suboch, V. P. *Khim. Geterotsikl. Soedin.* **1979**, 767 (*Chem. Abstr.* **1979**, *91*, 193275d).

884 Smith, K. M.; Bisset, G. M. F.; Case, J. J.; Tabba, H. D. *Tetrahedron Lett.* **1980**, *21*, 3747.

885 Smith, K. M.; Graham, M. F.; Tabba, H. D. *J. Chem. Soc., Perkin Trans. 1* **1982**, 581.

886 Ponomarev, G. V.; Kirillova, G. V.; Lazukova, L. B.; Babushkina, T. A. *Khim. Geterotsikl. Soedin.* **1982**, 1507 (*Chem. Abstr.* **1983**, *98*, 89034h).

887 Nichol, A. W. *J. Chem. Soc. C* **1970**, 903.

888 Smith, K. M.; Langry, K. C. *J. Chem. Soc., Perkin Trans. 1* **1983**, 439.

889 Brockmann, H., Jr., Bleisener, K.-M.; Inhoffen, H. H. *Justus Liebigs Ann. Chem.* **1968**, *718*, 148.

890 Ponomarev, G. V.; Shul'ga, A. M. *Khim. Geterotsikl. Soedin.* **1987**, 922 (*Chem. Abstr.* **1988**, *109*, 6281h).

891 Smith, K. M.; Goff, D. A.; Simpson, D. J. *J. Am. Chem. Soc.* **1985**, *107*, 4946.

892 Inhoffen, H. H.; Fuhrhop, J.-H.; Voigt, H.; Brockmann, Jr., H. *Justus Liebigs Ann. Chem.* **1966**, *695*, 133.

893 Wray, V.; Juergens, U.; Brockmann, Jr., H. *Tetrahedron* **1979**, *35*, 2275.

[894] Smith, K. M.; Bisset, G. M. F.; Bushell, M. J. *J. Org. Chem.* **1980**, *45*, 2218.

[895] Kirillova, G. V.; Babushkina, T. A.; Suboch, V. P., Ponomarev, G. V. *Khim. Geterotsikl. Soedin.* **1978**, 1215 (*Chem. Abstr.* **1979**, *90*, 87418x).

[896] Bushell, M. J.; Evans, B.; Kenner, G. W.; Smith, K. M. *Heterocycles* **1977**, *7*, 67.

[897] Ponomarev, G. V.; Kirillova, G. V.; Maravin, G. B.; Babushkina, T. A.; Suboch, V. P. *Khim. Geterotsikl. Soedin.* **1979**, 776 (*Chem. Abstr.* **1980**, *92*, 94359y).

[898] Arnold, D.; Gaete-Holmes, R.; Johnson, A. W. *J. Chem. Soc., Chem. Commun.* **1978**, 73.

[899] Gunter, M. J.; Robinson, B. C. *Aust. J. Chem.* **1990**, *43*, 1839.

[900] Shul'ga, A. M.; Ponomarev, G. V. *Khim. Geterotsikl. Soedin.* **1984**, 922 (*Chem. Abstr.* **1985**, *102*, 78627d).

[901] Ponomarev, G. V.; Maravin, G. B. *Khim. Geterotsikl. Soedin.* **1982**, 59 (*Chem. Abstr.* **1982**, *97*, 38727h).

[902] Buchler, J. W.; Dreher, C.; Herget, G. *Justus Liebigs Ann. Chem.* **1988**, 43.

[903] Ogoshi, H.; Saita, K.; Sakurai, K. -i.; Watanabe, T.; Toi, H.; Aoyama, Y.; Okamoto, Y. *Tetrahedron Lett.* **1986**, *27*, 6365.

[904] Tsuchida, H. Jpn. Kokai Tokkyo Koho JP 59179128, 1984 (*Chem. Abstr.* **1985**, *102*, 148998b).

[905] Eshima, K.; Hasegawa, E.; Matsushita, Y.-i.; Nishide, H.; Tsuchida, E. *Nippon Kagaku Kaishi* **1988**, 1836 (*Chem. Abstr.* **1989**, *110*, 127478g).

[906] Arnold, D.; Johnson, A. W.; Winter, M. *J. Chem. Soc., Perkin Trans. 1* **1977**, 1643.

CHAPTER 2

[6 + 4] CYCLOADDITION REACTIONS

James H. Rigby

Department of Chemistry, Wayne State University, Detroit, Michigan

CONTENTS

Organic Reactions, Vol. 49, Edited by Leo A. Paquette et al.
ISBN 0-471-15655-8   © 1997 Organic Reactions, Inc. Published by John Wiley & Sons, Inc.

ACKNOWLEDGMENTS

I wish to express my sincere gratitude to Ms. Melanie Brown for her expert technical assistance during the preparation of this chapter.

INTRODUCTION

Higher-order cycloaddition reactions possess many of the attributes that have made the Diels–Alder reaction so useful in synthesis, including high stereoselectivity, rapid increase in molecular complexity, and the ability to accommodate substantial functionalization in both reaction partners. The limiting feature of many higher-order processes, however, is a lack of periselectivity that translates directly into relatively low chemical yields of the desired cycloadducts.

$[6\pi + 4\pi]$ Cycloaddition reactions are typical higher-order transformations in that they exhibit many of the attractive features delineated above, but afford only modest yields of adducts in many instances. The engagement of the 6π and 4π components in these reactions often results in multiple, competitive pericyclic events that yield numerous cycloaddition products. The obvious synthetic potential offered by this class of reactions has prompted recent developments, such as metal mediation, that have successfully addressed the periselectivity issue and, as a result, have considerably broadened the synthetic scope and utility of the reaction.[1]

The thermally allowed [6 + 4] cycloaddition of 2,4,6-cycloheptatrien-1-one (tropone) (1) has been well studied and offers substantial opportunities for assembling functionally rich and stereochemically homogeneous bicyclic systems (Eq. 1). Typically, the triene partner is heated at 80–140° in the presence of ex-

cess diene (trienophile) to afford a bicyclo[4.4.1]undecene ring system that is sufficiently functionalized to permit subsequent manipulations. The scope of this
transformation for preparative purposes is somewhat restricted, however, since
only a limited set of diene partners will effectively participate. For example,
electron-rich dienes constitute the majority of reactants that afford meaningful
yields of bicyclo[4.4.1]undecane products upon reaction with **1**.

A variety of substituted fulvene species also participate as effective 6π partners
in a closely related set of [6 + 4] cycloaddition reactions (Eq. 2). The range of

$$(Eq.\ 2)$$

useful 4π partners is reasonably broad and rapid access to functionalized polycycles of considerable synthetic interest can be achieved with these transformations.

More recently, transition metal promoted versions of the [6 + 4] cycloaddition
have been developed.[2,3] Group VI metals (Cr and Mo) have been identified as capable promoters for this transformation, with chromium(0) emerging as the metal
of choice for most applications (Eq. 3). The ring-forming event in this case can

$$(Eq.\ 3)$$

be achieved through either thermal or photochemical activation, and the metal-
promoted process has proven to accommodate a much wider range of triene and
trienophilic participants than the thermal metal-free reaction. Unlike the cycloaddition reactions of tropone, the metal-mediated reactions are relatively insensitive to the electronic nature of the diene partner, and high chemical yields of
stereochemically homogeneous products are typical for these reactions. It is noteworthy that, in certain instances, reactions employing sub-stoichiometric quantities of metal have been reported.[4]

This chapter covers the thermal, metal-free [6 + 4] cycloaddition chemistry
of tropone and related trienes, fulvenes, and metal-promoted reactions through
mid-1995. Aspects of both the metal-mediated and metal-free versions of the
[6 + 4] cycloaddition reaction have been reviewed.[1-3]

MECHANISM AND STEREOCHEMISTRY

Tropone–Diene Cycloadditions

2,4,6-Cycloheptatrien-1-one (tropone) (**1**) reacts with cyclopentadiene at 80°
to form a single 1:1 adduct **3**, displaying *exo* stereochemistry (Eq. 4).[5,6] No evi-

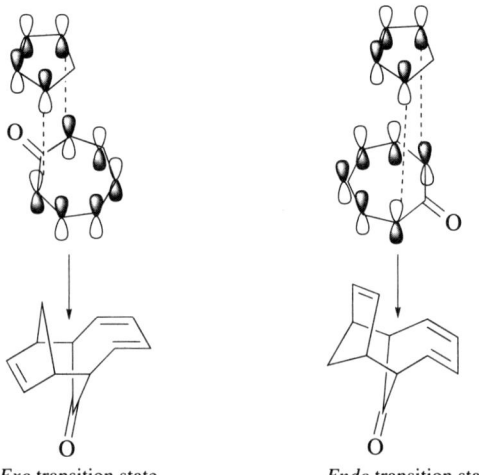

(Eq. 4)

3 (100%)   **4** (0%)

dence for the alternative *endo* product **4** is observed. It is noteworthy that the *exo* stereochemical preference in the [6 + 4] cycloaddition had been predicted, employing Woodward–Hoffmann orbital symmetry selection rules, a year before it was observed experimentally.[7] Subsequently, it has been found that virtually all metal-free [6 + 4] cycloadditions of cyclic trienes afford *exo* products.

The stereochemical preferences as well as the periselectivity of the [6 + 4] cycloaddition can be rationalized by consideration of the HOMO and LUMO interactions of the diene and triene participants, respectively.[8] Orbital combinations for the *exo* and *endo* transition states are presented in Figure 1.[9] An unfavorable repulsive secondary orbital interaction develops during an *endo* approach of the diene to the 6π partner, which is avoided in the *exo* transition state. Studies of this reaction at high pressure suggest that favorable secondary orbital interactions in the *exo* transition state may be involved (activation volume, -7.5 cm^3 mol^{-1}).[10] However, other work appears to be more consistent with only a minimal attractive secondary orbital interaction in these processes.[11–13]

Exo transition state   *Endo* transition state

Figure 1. Frontier molecular orbitals for the cycloaddition of cyclopentadiene to tropone.

From the earliest investigations on the [6 + 4] tropone–diene cycloaddition, it was noted that the periselectivity of the reaction varied as a function of reaction temperature.[14–16] For example, heating tropone (**1**) and 2,5-dimethyl-3,4-

diphenylcyclopentadienone at 60° affords a high yield of the *exo*-[6 + 4] adduct **5**. In contrast, the same combination heated at 100° provides primarily the *endo* [4 + 2] adduct **6** (Eq. 5).[15] Furthermore, heating [6 + 4] adducts at elevated temperatures can often effect cycloreversion to the component reactants.[5]

(Eq. 5)

These and related observations are attributed to the operation of a concerted, kinetic *exo* [6 + 4] pathway and a thermodynamic *endo*-[4 + 2] pathway in these reactions. Additionally, other kinetic studies on the [6 + 4] tropone–diene cycloaddition ($\Delta H^{\ddagger}$ = 15.3 kcal mol^{-1}; $\Delta S^{\ddagger}$ = −35 eu at 100°) conclude that the higher-order reaction channel possesses a late transition state, consistent with a concerted process, and mechanistically resembles the Diels–Alder reaction.[10–13]

An alternative reaction pathway, outlined in Eq. 6, has been proposed to explain the observed stereo- and periselectivity profiles of the tropone–diene cycloaddition, but direct experimental support for this mechanism is currently lacking.[17]

(Eq. 6)

Fulvene–Diene Cycloadditions

Fulvenes, like their troponoid counterparts, engage dienes in multiple peri-cyclic reactions, and Eq. 7 reveals typical examples of two of these reaction chan-

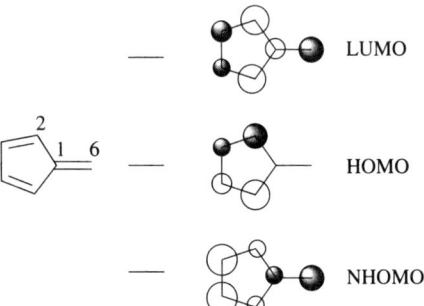

(Eq. 7)

nels.[18,19] Fulvenes can participate as either 6π or 2π reactants in these transfor-mations, and the factors governing which reactivity is expressed in a particular case have been elucidated, employing frontier molecular orbital considerations. The relevant fulvene orbitals are displayed in Figure 2.[20,21]

Figure 2. The frontier molecular orbitals of fulvene.

The controlling orbitals in the reaction of a fulvene with an electron-deficient diene are the fulvene HOMO and the diene LUMO. The large coefficients at C-2 and C-3, as well as the node through C-1 and C-6, dictate that the fulvene will participate as a 2π partner in this situation (see Eq. 7). On the other hand, LUMO-controlled reactions with electron-rich diene partners should react at C-6 and C-2, affording [6 + 4] adducts. Occasionally, products derived from reaction at C-1 and C-6 under these circumstances are also observed. Furthermore, strongly electron-donating substituents located at the C-6 position of fulvene ele-vate the next highest occupied molecular orbital (NHOMO) sufficiently to permit the [6 + 4] mode of cycloaddition to prevail with electron-deficient 4π systems. Examples of this type of reaction are presented subsequently.

Metal-Promoted Cycloadditions

In contrast to thermal, metal-free [6 + 4] cycloadditions, reactions of metal-complexed trienes are known to furnish exclusively *endo* products, rendering the two pathways stereocomplementary. A typical example of this transformation is depicted in Eq. 8, in which photoactivated cycloaddition initially affords the

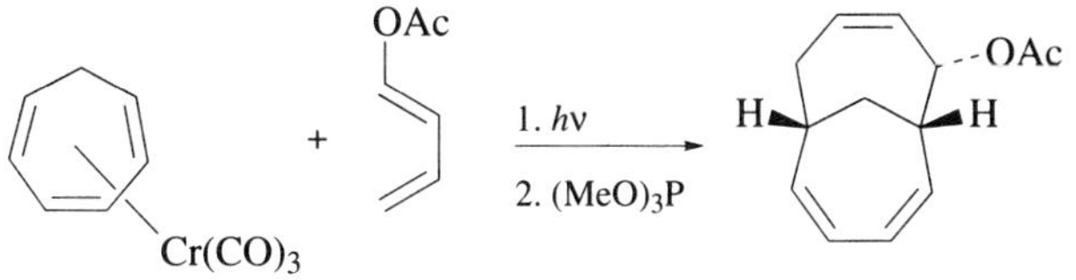

$$\text{(Eq. 8)}$$

cycloadduct–Cr(CO)$_3$ complex, and subsequent treatment with trimethyl phosphite provides the metal-free organic product.[22]

Two mechanistic pathways are proposed for the metal-promoted [6 + 4] cycloaddition that differ primarily in the way in which the initial coordinatively unsaturated intermediates are generated.[23–25]

Scheme 1 depicts the carbon monoxide extrusion based mechanism in which the coordinatively unsaturated species **8** is produced by light-induced dissociation of one CO ligand from complex **2**.[25] The resultant 16-electron intermediate then coordinates with the diene, and bond reorganization affords the Cr(II) complex **9**. At this juncture, recapture of the previously dissociated CO ligand produces the observed [6 + 4] cycloadduct–chromium tricarbonyl complex **10**. Support for this pathway comes from matrix-isolation studies in which species related to **8** were observed to lead to bicyclo[4.4.1]undecane products.[25]

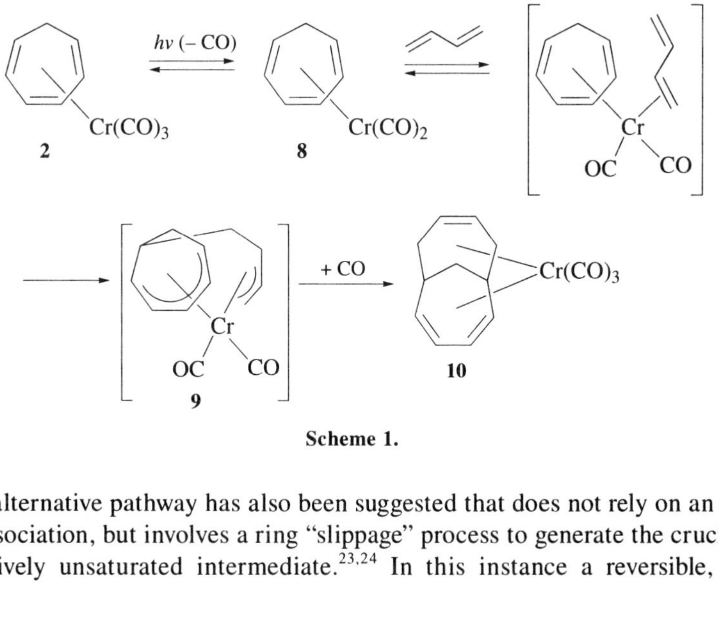

Scheme 1.

An alternative pathway has also been suggested that does not rely on an initial CO dissociation, but involves a ring "slippage" process to generate the crucial coordinatively unsaturated intermediate.[23,24] In this instance a reversible, light-induced hapticity change ($\eta^6 \rightarrow \eta^4$) occurs to afford 16-electron complex **11**, which then coordinates to the diene. Next, a bond-reorganization event similar to that proposed in Scheme 1 occurs to produce a bis(allyl) intermediate which collapses to the observed cycloadduct complex. This pathway is presented in Scheme 2. The absence of any detectable CO evolution has been cited as evidence in support of this pathway.[24] In addition, the observation that purging the reaction mixture with an inert gas during photolysis results in enhanced yields of cycloadducts is also more consistent with the mechanism presented in Scheme 2.[23] Diminished yields of product would be anticipated under these conditions if the pathway in Scheme 1 were operational in these reactions.[25]

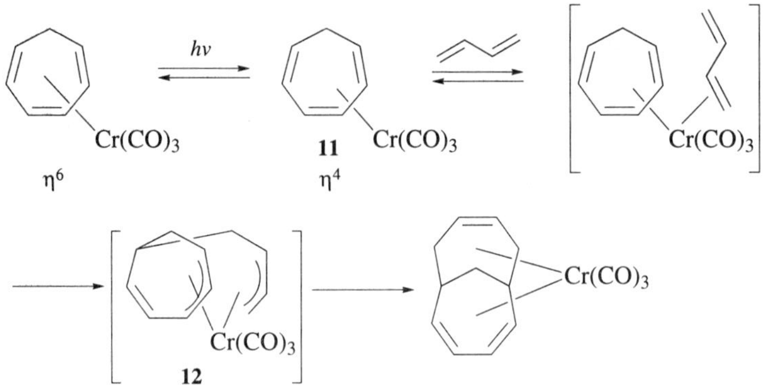

Scheme 2.

Both mechanistic pathways are consistent with the *endo* nature of the resultant cycloadducts because neither species **9** nor **12** is geometrically capable of accommodating an *exo*-oriented diene component.

SCOPE AND LIMITATIONS

Tropone–Diene Cycloadditions

The thermally allowed cycloaddition of tropone and related cyclic trienes with appropriate 4π reaction partners can offer the opportunity for rapid access to functionalized bicyclic products that are often difficult or impossible to make in other ways. In contrast, most other cyclic triene substrates such as 1,3,5-cycloheptatriene[15] and azepine[26] are frequently poor 6π partners in thermal, metal-free [6 + 4] cycloaddition reactions and offer little synthetic advantage.

The thermally induced cycloaddition between tropone (**1**) and (*E*)-1-acetoxy-1,3-butadiene illustrates many of the salient features of the metal-free [6π + 4π] process.[16,27] Typically, chemical yields are in the range of 60% (occasionally as high as 80%), and the bicyclo[4.4.1]undecatrienone products **13** are isolated in diastereomerically homogeneous form.

(Eq. 9)

(59%) **13**

The *exo* isomer is formed to the complete exclusion of the corresponding *endo* species in virtually every known example. Extended reaction times or higher reaction temperatures tend to enhance the yield of other pericyclic products at the expense of higher-order adducts, and an alternative $[4\pi + 2\pi]$ pathway often prevails under harsher conditions. A good illustration of this phenomenon is the reaction of tropone with (E)-1-trimethylsilyloxy-1,3-butadiene at various temperatures (Eq. 10).[16] In benzene at reflux, the electron-rich diene affords an excellent

14 (80%)

(Eq. 10)

yield of the [6 + 4] cycloadduct, but higher reaction temperature leads to the [4 + 2] adduct nearly exclusively.

Furthermore, only minor structural changes on either reaction partner can have a profound impact on the periselectivity of these transformations. For example, (Z)-1-acetoxy-1,3-butadiene affords only a small quantity of [6 + 4] cycloadduct accompanied by a much larger amount of a mixture of isomeric products arising from a $[4\pi + 2\pi]$ cycloaddition between tropone, participating as the 4π component, and the diene (Eq. 11a).[16] Other important factors that

(12%)   (50%)

(Eq. 11a)

affect these reactions include the electronic nature of the diene and the steric environment at the bond-forming centers in the triene. For instance, electron-deficient dienes are normally poor participants in these reactions, and substituents at the 2 position of the tropone partner are known to suppress the higher-order cycloaddition pathway (Eq. 11b).[27,28]

(Eq. 11b)

Electronically biased substituents located at sites on the tropone nucleus remote from those participating in bond formation can strongly influence the regiochemical course of the cycloaddition event, although chemical yields tend to be modest in most cases (Eq. 12).[29] It is noteworthy that the regioselectivities ex-

(Eq. 12)

hibited by tropones bearing electron-withdrawing groups in [6 + 4] cycloaddition qualitatively parallel those observed in the Diels–Alder [4 + 2] reaction, whereas 3- and 4-methoxytropone exhibit both low regioselectivity and poor chemical yields, again stressing the crucial role played by the electronic nature of the reactants.

Tethering the diene and triene components together has been an effective ploy for circumventing some of the difficulties encountered with the presence of substituents at the 2-position of tropone.[27,30] For example, employing a three-carbon spacer permits rapid entry into the ABC tricyclic system of the ingenane diterpenes (Eq. 13). As in the intermolecular version, *exo* stereoselectivity prevails in these transformations.

15   **16** (80%)

(Eq. 13)

1,3-Dipoles have also been examined as 4π partners in [6 + 4] cycloadditions with tropone; however, only small quantities of higher-order adducts have been isolated in these reactions.[8]

(4%)   (54%)

(Eq. 14)

Fulvene–Diene Cycloadditions

Fulvenes, like the cyclic trienes considered previously, are capable of undergoing multiple, competitive pericyclic reactions with dienes and other 4π reactants. To a large extent, fulvenes participate in these transformations as either a 6π or 2π component, and the factors governing which of these reactivities is expressed in a particular situation have been defined by employing frontier molecular orbital theory (see Fig. 2).[20]

Electron-deficient dienes normally engage fulvenes in a [$4\pi + 2\pi$] cycloaddition at one of the endocyclic double bonds,[18] while electron-rich dienes react as 4π components in a [$6\pi + 4\pi$] process.[19] An illustration of each of these modes of reaction is given for 6-phenylfulvene (**17**) in Eq. 15. Electron-rich fulvenes

(Eq. 15)

(65%)

such as **18** are also known to react with electron-deficient diene partners primarily in the [$6\pi + 4\pi$] mode (Eq. 16).[31]

(Eq. 16)

Steric hindrance at the fulvene C-6 position can also influence cycloaddition regiochemistry, as illustrated in Eq. 17.[32] For example, 6-dimethylaminofulvene

(Eq. 17)

(**18**) reacts with the electron-deficient heterodiene 3,6-diphenyltetrazine in the expected [6 + 4] fashion to afford a diazaazulene product after spontaneous loss of the elements of nitrogen and dimethylamine, whereas 6,6-diphenylfulvene yields compound **19** via an endocyclic [4 + 2] cycloaddition.[33]

The influence of intramolecularity on the periselectivity of fulvene–diene cycloaddition is quite intriguing. Reaction of the quinodimethane precursor **20** with 6,6-dimethylfulvene affords only the product of a [4 + 2] cycloaddition at an endocyclic double bond.[34] When these same reactants are tethered, as in compound **21**, only the [6 + 4] adduct is formed as a mixture of cyclopentadiene isomers. The dramatic change in periselectivity in this case has been ascribed to conformational restrictions imposed on the reactant by the carbon chain connecting the fulvene moiety and the benzocyclobutane unit.[35]

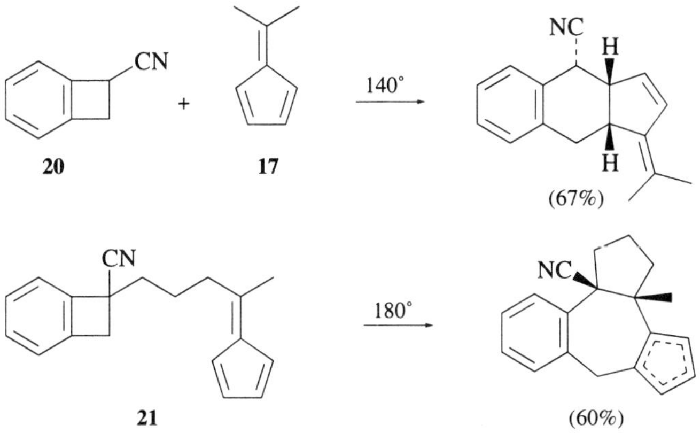

(Eq. 18)

Electron-donor substituents on the diene moiety exhibit a profound influence on the regiochemical course of the intramolecular cycloaddition in the fulvene series, as depicted in Eq. 19. With X = H in fulvene **22a**, three cycloadducts are produced at elevated temperature; however, when X = NEt$_2$, only the [6 + 4] adduct **23** is isolated, even at moderate reaction temperatures.[36]

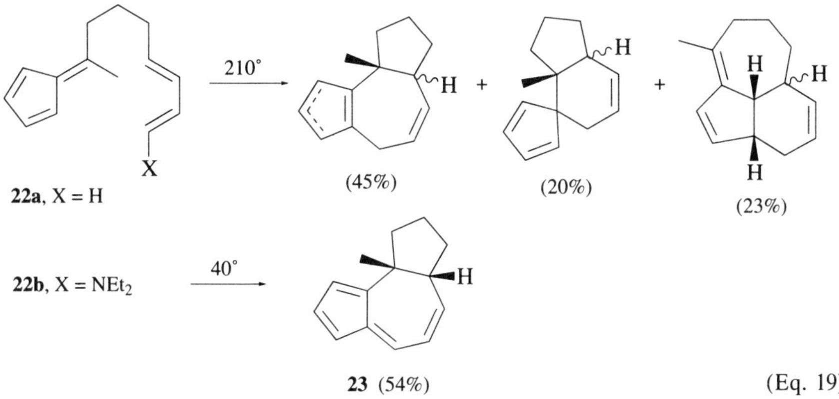

(Eq. 19)

Cycloadditions involving 1,3-dipoles have received some attention in the fulvene series, and the reactivity patterns observed in the fulvene–diene transformations are repeated. For example, diazomethane, a 1,3-dipole with well-established nucleophilic character, adds exclusively in [6 + 4] fashion to 6,6-dimethylfulvene (**17**), as expected based on FMO considerations.[21,37] Tropone also can be a serviceable diene partner in combination with 6,6-dimethylfulvene.[38]

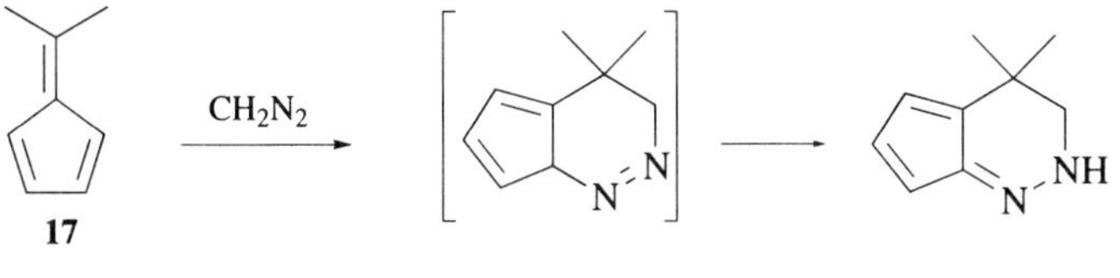

(Eq. 20)

Metal-Promoted Cycloadditions

Transition metal promoted [$6\pi + 4\pi$] cycloadditions of cyclic trienes offer numerous advantages over the thermal, metal-free versions described above. The inevitable problem of regioselectivity that is a prominent feature of the latter set of transformations does not play a significant role in the course of the metal-promoted process.

The requisite triene–chromium(0) complexes, such as **24**, employed in these reactions are quite stable and can be prepared in a number of ways; the most versatile method employs tris(acetonitrile)tricarbonylchromium(0)[39] as the "Cr(CO)₃" source (Eq. 21). Group VI metals (Cr, Mo, W) have been examined as

$$\text{(Eq. 21)}$$

24 (86%)

promoters for the [6 + 4] reaction, with chromium(0) emerging as the metal of choice. It is noteworthy that tungsten does not appear to promote cycloaddition.[23]

Equation 22 depicts many of the salient features of the photoinitiated, metal-promoted [6 + 4] cycloaddition. In concurrence with the thermal, metal-free version, diastereoselection is virtually complete; however, the metal-mediated process affords *endo* products exclusively. It is particularly significant that the efficiency of these transformations is relatively insensitive to the electronic nature of the reactants. Both electron-rich and electron-poor dienes afford high yields of cycloadducts. Indeed, little rate difference has been noted in competitive studies between electron-rich and electron-poor dienes in their reaction with complex **24**. In most cases, the initially formed cycloadduct–metal complex is demetalated with trimethyl phosphite prior to isolation, and all yields reported in this review are for isolated, metal-free cycloadducts. Trimethylphosphine or air/diethyl ether (see *Caution*, p. 351) have also been effective for demetalating cycloadduct–metal complexes.

$$\text{(Eq. 22)}$$

Considerable stereochemical information can be generated during these cycloadditions. Readily available 7-*exo*- or 7-*endo*-substituted complexes[40,41] afford adducts possessing as many as five contiguous stereogenic centers.[23] In stark contrast to the thermal, metal-free reactions which, in most instances, cannot tolerate substituents at triene bond-forming centers, the metal-promoted version provides adducts from substituted triene complexes in good to moderate yields and with high levels of regiocontrol (Eq. 24).[23]

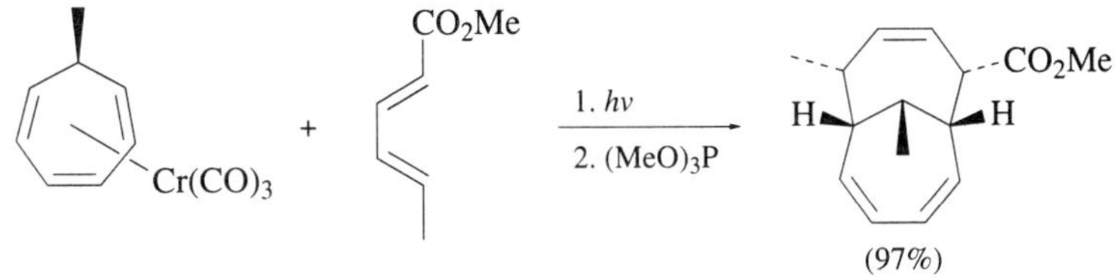

(Eq. 23)

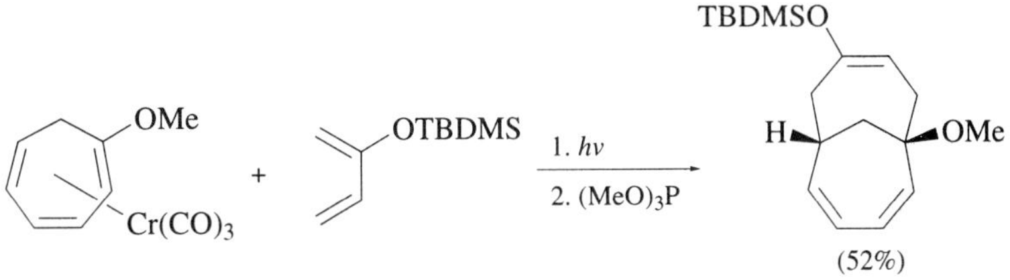

(Eq. 24)

Substituents at either the 2 or 3 positions of the triene ligand tend not to have much impact on the regiochemical course of cycloaddition (Eq. 25), but chemical yields remain quite high, and good to excellent levels of asymmetric induction can be achieved during cycloaddition by incorporating a chiral auxiliary onto either the diene or triene moiety.

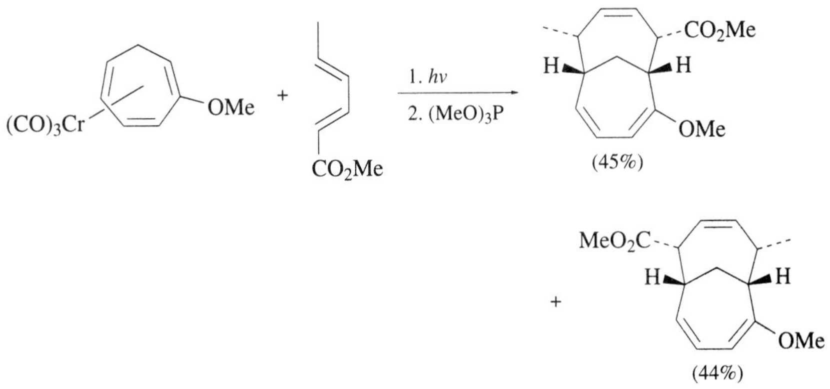

(Eq. 25)

Metal-promoted cycloadditions can also be effected with heterocyclic triene complexes (Eq. 26); this example also illustrates an alternative method for

$$X = NCO_2Me, SO_2$$

(Eq. 26)

$$X = NCO_2Me \ (75\%)$$
$$X = SO_2 \ (78\%)$$

demetalation by passing air through an ether solution of the complex.[42] The resultant cycloadducts are amenable to conversion into 10-membered carbocycles by heteroatom extrusion, as depicted in Eq. 27. In this case, the cycloaddition is best

(Eq. 27)

achieved using a uranium glass filter (350 nm), while the extrusion proceeds only with quartz-filtered light.

It is noteworthy that the metal-promoted [6 + 4] cycloaddition can be effected thermally as well as photochemically. In a typical example, heating a mixture of complex 24 with methyl sorbate at 140° in di-*n*-butyl ether affords a cycloadduct that is identical in all ways to the product obtained from the corresponding photochemical reaction (Eq. 22).[23] Efforts to carry out catalytic [6 + 4] reactions have met with limited success, but the related [6 + 2] reaction works well with substoichiometric quantities of metal.[43]

COMPARISON WITH OTHER METHODS

The majority of methods for the construction of the bicyclo[4.4.1]undecane ring system have focused on the synthesis of 1,6-methano[10]annulenes[44] and the potent tumor-promoting diterpene, ingenol.[45] Most of these reports involve multistep procedures for the assembly of the bicyclic architecture; however, the [6π + 4π] cycloaddition method has the distinct advantage of affording the target ring system in only one operation.

Ingenol

1,6-methano[10]annulene

Typical of the approaches to the 1,6-methano[10]annulene system is the model study directed toward the novel marine natural product spiniferin-1.[46] Birch reduction followed by enol ether hydrolysis and reduction converts 6-methoxytetralin into compound **25**. Functional group manipulation and cyclo-propanation afford **26**, which upon enolization triggers tautomerization to the cy-cloheptatriene target **27** (Eq. 28).

(Eq. 28)

Recently, a variation on this theme has been developed for preparing 1,6-methano[10]annulenes substituted at the C-11 position.[47] In this approach, diacid **28** is transformed into propellane **29** in three steps. Ring contraction, oxidation, and isomerization to the annulene complete the synthesis.

Various approaches specifically focused on the bicyclo[4.4.1]undecane sub-structure of the ingenane diterpenes have also been reported. Several of these feature a cycloheptannulation step that is achieved via sequential bis-alkylation of a cyclohexanone or cycloheptanone precursor.

In one case, the ABC ring system of ingenol was assembled starting from 6-methoxytetralone and involving a six-step bis-alkylation sequence leading to dione **30**. Several additional steps, including an intriguing ring contraction–ring expansion, afforded the tricycle **31**.[48]

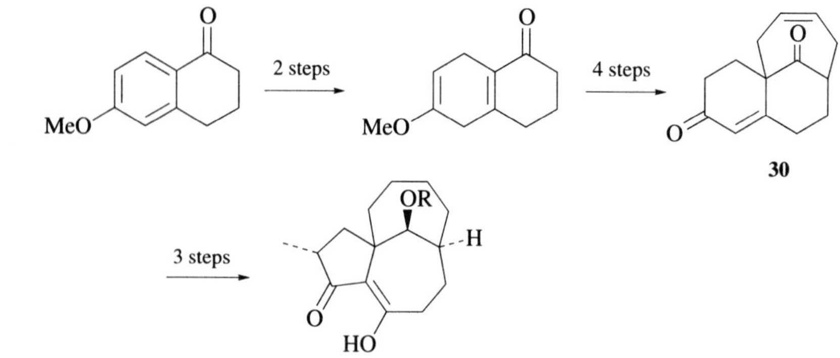

30

31

A somewhat related sequence for constructing a modestly functionalized bicyclo[4.4.1]undecanone species commencing from a cycloheptanone starting material is depicted in Eq. 29.[49]

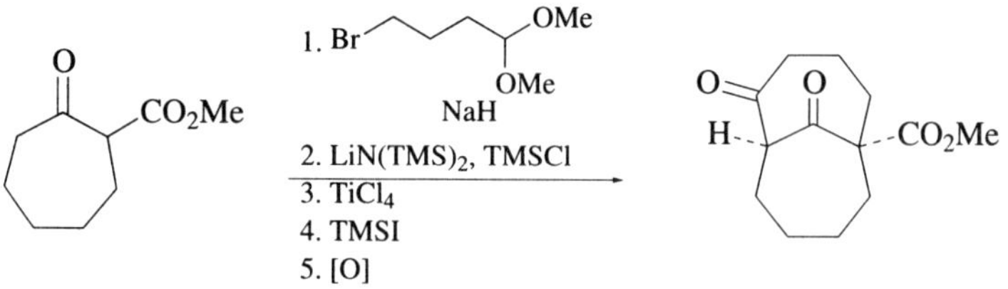

1. Br~~~~~OMe / OMe
   NaH
2. LiN(TMS)$_2$, TMSCl
3. TiCl$_4$
4. TMSI
5. [O]

(Eq. 29)

Ring contraction via Claisen rearrangement has also been effectively employed for the synthesis of the bicyclo[4.4.1]undecane system. In this instance, the resultant product possesses the strained *trans* interbridgehead stereochemistry characteristic of the naturally occurring ingenane diterpenes.[50]

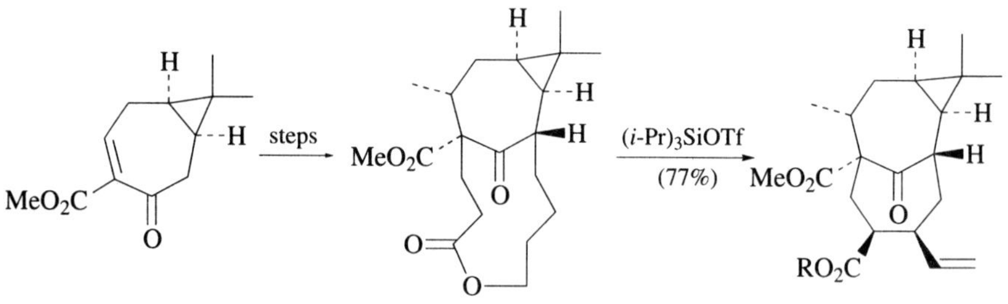

A [2 + 2] photocycloaddition has been used as the key step in a second entry into the ingenane bicyclo[4.4.1]undecane ring system that was elegantly designed to produce the strained "in, out" isomer directly.[51]

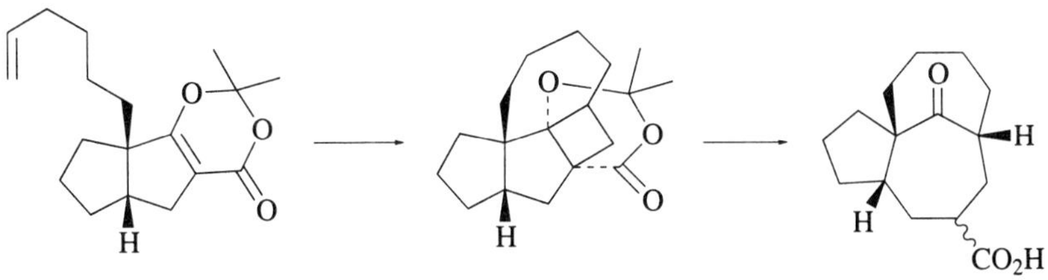

The fulvene–diene [6 + 4] cycloaddition yields a bicyclo[5.3.0]decane ring system. In light of the prominence of the hydroazulene substructure in numerous biologically active sesquiterpenes, as well as in the tumor-promoting diterpenes phorbol and ingenol, a large number of alternative entries into this important ring system have emerged and several overviews of this area have appeared.[52-55]

Most traditional methods for assembly of the hydroazulene system can be categorized into one of four basic strategies (with illustrative references): annulation of a cyclopentane ring onto a preexisting seven-membered ring (Eq. 30);[56] cycloheptannulation onto a five-membered ring (Eq. 31);[57] transannular cyclization of a cyclodecane (Eq. 32);[58] and rearrangement of bicyclo[4.3.1]decane or bicyclo[4.4.0]decane precursors (Eq. 33).[52]

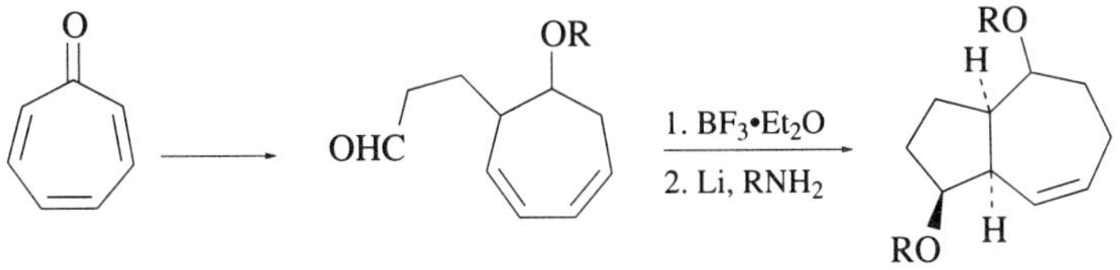

(Eq. 30)

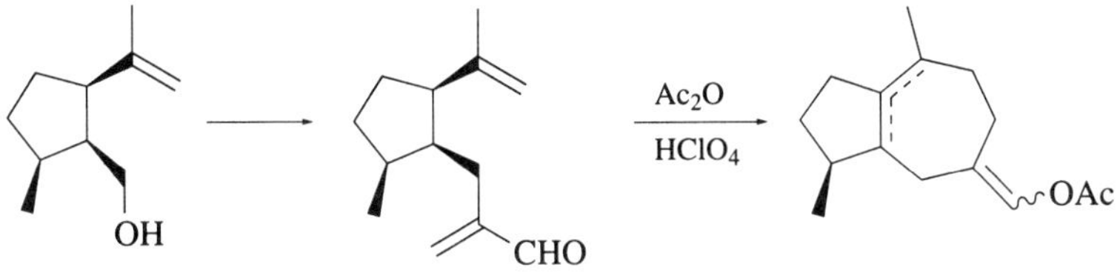

(Eq. 31)

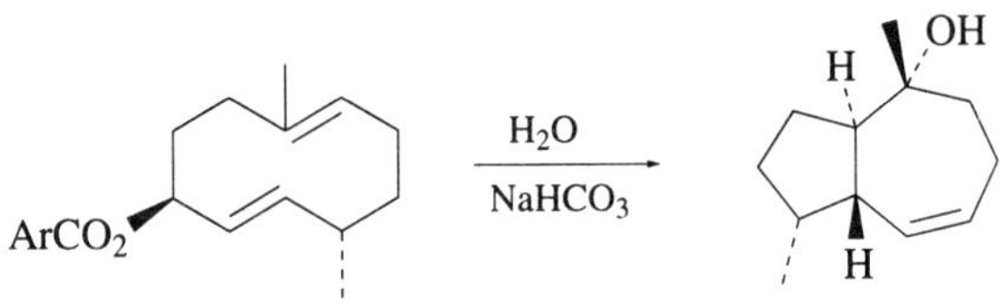

(Eq. 32)

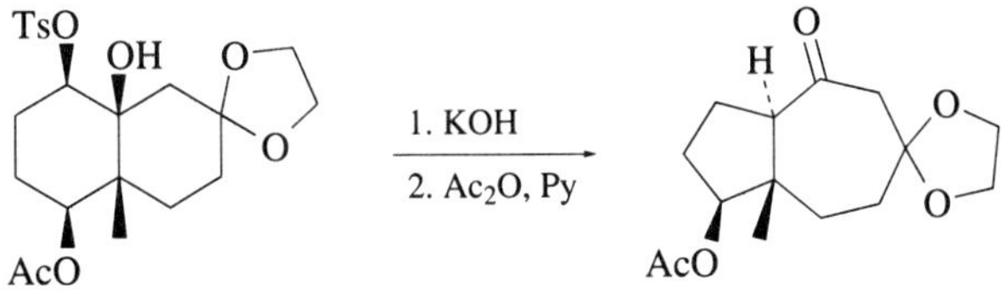

(Eq. 33)

Several more recent contributions have appeared that feature novel methods for executing the key ring-forming process. An anionic, intermolecular ring-opening procedure affords highly functionalized hydroazulene products from oxabicyclo[3.2.1]octane systems.[59]

(80%)

A novel and very promising entry into hydroazulenes exploits a Rh(I)-mediated, intramolecular [5 + 2] cycloaddition of an alkyne to a vinylcyclo-propane moiety.[60]

(83%)

Although the choice of which approach to employ for assembling a bicy-clo[5.3.0]decane ring system ultimately depends on the specific substitution pattern required, the fulvene–diene [6 + 4] cycloaddition offers the advantages of convergency, ease of substrate synthesis, and well-defined reaction characteristics that are attractive in many situations.

EXPERIMENTAL CONDITIONS

For most thermal, metal-free [6 + 4] cycloadditions, a freshly distilled, non-polar solvent of appropriate boiling point is preferred. Benzene, toluene, and xylene are most frequently employed. Since forcing conditions can often promote other pericyclic events (i.e., [4 + 2] cycloadditions) at the expense of the high-order pathway, it is recommended that lower-boiling solvents be screened first for efficacy with each substrate. While the optimum conditions for thermal, metal-free [6 + 4] cycloadditions vary greatly as a function of substrate, transition-metal-mediated photocycloadditions have well-defined conditions for achieving optimal results. Freshly distilled hexanes is the solvent of choice for most applications. Occasionally, a small amount of ether cosolvent may be needed to solubilize a particular complex. Maximum yields of adducts are obtained when the reaction mixture is purged with an inert gas (Ar or N_2) prior to and during photolysis. Pyrex or uranium glass-filtered light affords optimum yields in all cases examined. Employing quartz-filtered light gives inferior yields, owing to the instability of the cycloadduct complexes to the reaction conditions.

With electron-rich diene partners, maximum product yields are obtained when the reaction mixture is stirred under a blanket of carbon monoxide for several hours after completion of the irradiation step.

Demetalation of most cycloadduct–metal complexes is accomplished with trimethyl phosphite; however, in some cases better results are obtained by bubbling air through an ether solution of the crude reaction product. *CAUTION. Lower-alkyl ethers readily form explosive peroxides under these conditions. An ether so treated should never be distilled; peroxides can be removed from it by passing it through a column of basic activated alumina[60a] or by treating it with indicating molecular sieves.[60b]*

EXPERIMENTAL PROCEDURES

7β-Acetoxy-(1Hα,6Hα)-bicyclo[4.4.1]undeca-2,4,8-trien-11-one   (Thermal, Metal-Free Tropone-Diene Cycloaddition).[61]   A solution of 2,4,6-cycloheptatrien-1-one (tropone) (10 g, 94 mmol) and (E)-1-acetoxy-1,3-butadiene (15.8 g, 140 mmol) in xylene (300 mL) was heated at reflux for five days. Removal of solvent in vacuo and purification of the crude product by flash-column chromatography (4/1, hexanes/ethyl acetate) afforded 12.1 g (59%) of the cycloadduct as a pale yellow oil: IR (neat) 3031, 2937, 1735, 1707, 1437 cm^{-1}; ^{1}H NMR (CDCl$_3$) δ 2.07 (s, 3 H), 2.56 (m, 2 H), 3.54 (br q, J = 6.3 Hz, 1 H), 3.67 (br t, J = 6.3 Hz, 1 H), 5.67 (m, 4 H), 5.78 (m, 1 H), 6.10 (m, 2 H). High-resolution mass spectrum: Calcd for C$_{13}$H$_{14}$O$_3$: 218.0943. Found: 218.0940.

1,2,3,8,9,11a-Hexahydro-3a,8-methano-3aH-cyclopentacyclodecen-12-one (Intramolecular, Thermal Tropone–Diene Cycloaddition).[27]   2-(Hepta-4,6-dienyl)cyclohepta-2,4,6-trien-1-one (200 mg, 1 mmol) was heated in benzene (4 mL) at reflux for 10 hours. The solvent was removed in vacuo and the crude product was purified by flash-column chromatography (4/1 petroleum ether/ ether), affording 172 mg (86%) of cycloadduct: IR (CCl$_4$) 2959, 1704, 1695 cm^{-1}; ^{1}H NMR (CDCl$_3$) δ 1.43–1.88 (m, 4 H), 1.94 (m, 1 H), 2.20–2.40 (m, 2 H), 2.67

(m, 1 H), 2.88 (m, 1 H), 3.49 (q, J = 6.5 Hz, 1 H), 5.41–6.05 (m, 6 H). Anal. Calcd for $C_{14}H_{16}O$: C, 83.94; H, 8.06. Found: C, 83.97; H, 8.05.

6-Methylazulene (Electron-Rich Fulvene–Electron-Poor Diene Cycloaddition).[31] 3-Methyl-3,4-dibromotetrahydrothiophene (3.0 g, 10 mmol) was dissolved in benzene (75 mL). The solution was cooled to 0–5°, and triethylamine (3 mL, 20 mmol) was added. This solution was stirred for 2 hours, the precipitate removed by filtration, and the resulting solution added to 6-dimethylaminofulvene (1.5 g, 12.5 mmol). This solution was stirred at room temperature under a gentle stream of nitrogen for 72 hours, at which time the reaction mixture was evaporated to dryness. The crude product was dissolved in 4:1 petroleum ether–chloroform, insoluble material was removed by filtration, and the solution was chromatographed on alumina (petroleum ether). This afforded 0.36 g (25%) of product.

4,7-Diphenyl-5,6-diazaazulene (Electron-Rich Fulvene–Heterodiene Cycloaddition).[33] A solution of 6-dimethylaminofulvene (0.15 g, 1.2 mmol) and 3,6-diphenyltetrazine (0.29 g, 1.2 mmol) in benzene (20 mL) was stirred at 25° under argon in the dark for 5 days. Removal of solvent in vacuo and purification by column chromatography (chloroform) afforded 0.14 g (40%) of product: mp 289–92° (chloroform). Anal. Calcd for $C_{20}H_{14}N_2$: C, 85.08; H, 5.00; N, 9.92. Found: C, 84.96; H, 5.16; N, 9.87.

7α-(Methoxycarbonyl)-10α-methyl-(1Hβ,6Hβ)-bicyclo[4.4.1]undeca-2,4,8-triene (Metal-Promoted Photocycloaddition Employing an Electron-Deficient Diene).[23] A Canrad–Hanovia medium-pressure mercury lamp operating at 450 W was placed in a water-cooled immersion well constructed of Pyrex glass and equipped with a uranium glass sleeve. To this reaction vessel was added a solution of (η^6-1,3,5-cycloheptatriene) tricarbonylchromium(0) (570 mg,

2.5 mmol) and methyl sorbate (284 mg, 2.25 mmol) in hexanes (320 mL). The solution was purged through vigorous bubbling with Ar and irradiated with continuous bubbling for 30 minutes. The resultant orange solution was filtered and the solution reduced in vacuo to a volume of 20–50 mL. At this time, the mixture was stirred with trimethyl phosphite (10 mL) at 25° for 10 hours. Concentration in vacuo and purification by flash-column chromatography (9/1 hexane/ether) afforded 469 mg (96% based on diene) of a colorless oil: IR (film) 3018, 1739, 1435 cm^{-1}; ^{1}H NMR (CDCl$_3$) δ 1.22 (d, J = 7.2 Hz, 3 H), 2.23 (m, 3 H), 2.55 (m, 1 H), 2.69 (m, 1 H), 3.20 (m, 1 H), 3.60 (m, 1 H), 3.74 (s, 3 H), 4.50 (m, 1 H), 5.78 (m, 4 H), 5.92 (m, 1 H). Anal. Calcd for C$_{14}$H$_{18}$O$_2$: C, 77.03; H, 8.31. Found: C, 77.08; H, 8.21.

7α-Trimethylsilyloxy-(1Hβ,6Hβ)-bicyclo[4.4.1]undeca-2,4,8-triene (Metal-Promoted Photocycloaddition Employing an Electron-Rich Diene).[23]

A solution of (η^6-1,3,5-cycloheptatriene) tricarbonylchromium(0) (310 mg, 1.36 mmol) and 1-trimethylsilyloxy-1,3-butadiene (1.21 g, 8.5 mmol) in hexanes (350 mL) was irradiated (Pyrex) under conditions described above for 30 minutes. The mixture was saturated with carbon monoxide and stirred under a blanket of this gas for 15 hours. The resultant solution was concentrated to 50 mL in vacuo and stirred with trimethyl phosphite (10 mL) at 25° for 10 hours. The mixture was then further concentrated in vacuo, and purification by flash-column chromatography (19/1 hexanes/ethyl acetate) afforded 273 mg (86%) of a colorless oil: ^{1}H NMR δ 0.16 (s, 9 H), 2.13–2.49 (m, 4 H), 2.72 (br s, 1 H), 2.82 (br s, 1 H), 4.67 (m, 1 H), 5.55 (m, 2 H), 5.84 (m, 3 H), 6.05 (m, 1 H). Anal. Calcd for C$_{14}$H$_{22}$OSi: C, 71.73; H, 9.45. Found: C, 72.05; H, 9.37.

7α, 10α-Dimethyl-(1Hβ,6Hβ)-bicyclo[4.4.1]undeca-2,4,8-triene (Thermal, Metal-Promoted Cycloaddition Employing a Hydrocarbon Diene).[23]

A solution of (η^6-1,3,5-cycloheptatriene) tricarbonylchromium(0) (0.154 g, 0.67 mmol) and (E, E)-2,4-hexadiene (1.5 mL, 13 mmol) in n-butyl ether (10 mL) was refluxed for 36 hours, at which time the reaction mixture was cooled to room temperature and the solvent was removed in vacuo. Flash-column chromatography (pentane) of the crude product afforded 0.115 g (70%) of a colorless oil: ^{1}H NMR δ 1.23 (d, J = 7.3 Hz, 6 H), 2.10–2.30 (m, 2 H), 2.55 (m, 2 H), 2.70–2.85 (m, 2 H), 5.28 (m, 2 H), 5.81 (m, 2 H), 5.91 (m, 2 H). High-resolution mass spectrum: Calcd for C$_{13}$H$_{18}$: 174.1408. Found: 174.1410.

TABULAR SURVEY

[6 + 4] Cycloaddition reactions of cyclic trienes, fulvenes, miscellaneous 6π reactants, and metal complexes are grouped in Tables I–IV and follow the order of topics discussed in the Scope and Limitations section. Tables I and II are further divided into subcategories related to the types of 4π reaction partners involved. Table I has three subsections: (A) All-Carbon Trienophiles, (B) 1,3-Dipolar Trienophiles, including TMM equivalents, and (C) Intramolecular Reactions. Table II has four subcategories: (A) All-Carbon Trienophiles, (B) 1,3-Dipolar Trienophiles, (C) Heterodienes, and (D) Intramolecular Reactions. Miscellaneous substrates that could not be placed in either Table I or II are collected in Table III, and Table IV presents metal-mediated [6 + 4] cycloadditions.

Within each table, the reactions are listed according to increasing carbon number in the triene (6π) substrate and the count is based on the total number of carbon atoms in these reactants. Table IV is an exception to this arrangement in that spectator ligands around the metal center are **not** included in the total carbon count.

Within each carbon-number group, the trienes are listed in order of increasing hydrogen count. Yields are given in parentheses and a dash indicates that no yields or experimental conditions were given in the original reference.

Some entries involve structurally large reactant substructures. In these cases, the explicit substructure is provided in the relevant reactant location and indicated in condensed form in the product(s).

The literature has been reviewed from 1952 through mid-1995.

The following abbreviations are used in the tables:

C_2H_3	Vinyl
C_3H_5	Cyclopropyl
C_5H_4N	2-Pyridyl
Cp	Cyclopentadienyl
Et_2O	Diethyl ether
DME	1,2-Dimethoxyethane
rt	Room temperature
THF	Tetrahydrofuran
TBDMS	*tert*-Butyldimethylsilyl
Ts	*p*-Toluenesulfonyl

TABLE I. [6+4] CYCLOADDITIONS OF 2,4,6-CYCLOHEPTATRIEN-1-ONE AND DERIVATIVES

A. All-Carbon Trienophiles

Triene	Trienophile	Conditions	Product(s) and Yield(s) (%)	Refs.
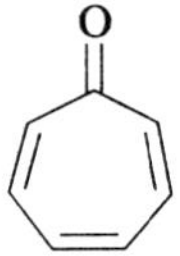 C₇		Xylene, 130°, 10 h	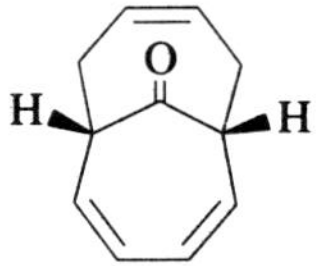 (75) 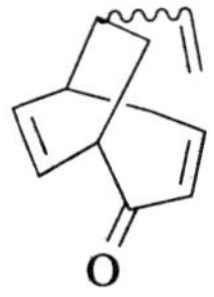 + (8)	62
		Xylene, 130°, 10 h	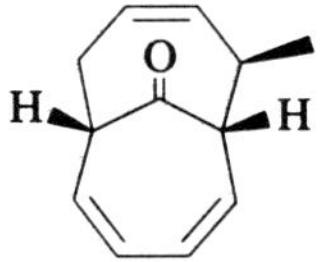 (60)	62
	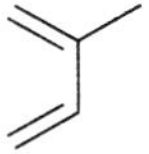	Xylene, 130°, 10 h	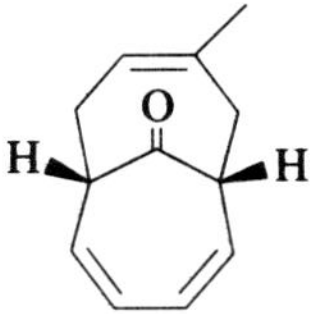 (86)	62
		C₆H₆, 80°, 5 d	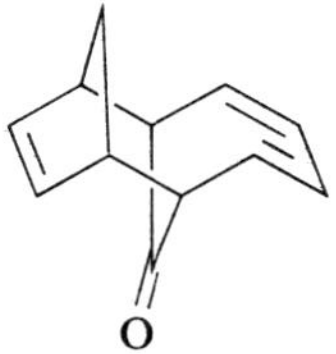 (—)	5, 6

TABLE I. [6+4] CYCLOADDITIONS OF 2,4,6-CYCLOHEPTATRIEN-1-ONE AND DERIVATIVES (*Continued*)

A. *All-Carbon Trienophiles*

Triene	Trienophile	Conditions	Product(s) and Yield(s) (%)	Refs.
	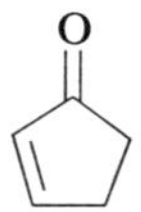	Toluene, 10 Kbar, 100°, 10 h	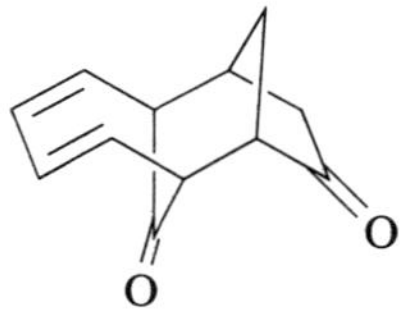 (7)	63
	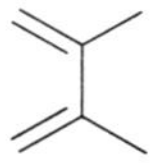	Xylene, 140°, 12 h	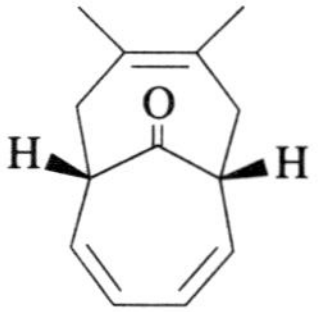 (75)	16
		Xylene, 140°, 5 d	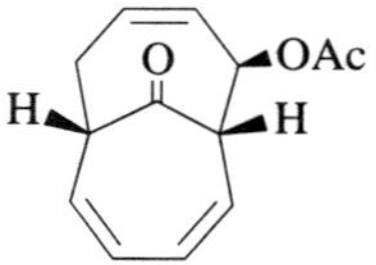 (59)	61, 27, 16, 64, 65
	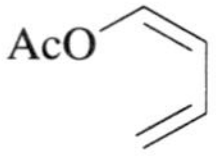	Xylene, 140°, 24 h	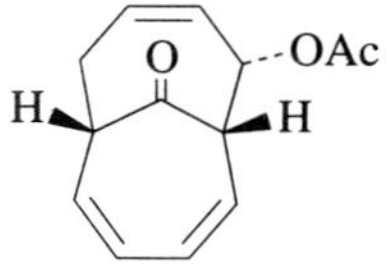 (12)	16, 65

356

TABLE I. [6+4] CYCLOADDITIONS OF 2,4,6-CYCLOHEPTATRIEN-1-ONE AND DERIVATIVES (*Continued*)

A. *All-Carbon Trienophiles*

Triene	Trienophile	Conditions	Product(s) and Yield(s) (%)	Refs.
		—	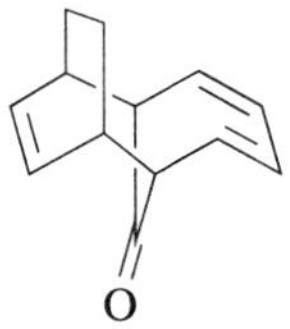 (30)	66, 5, 67
	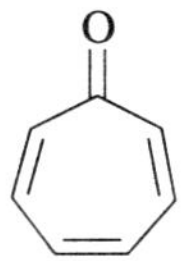	MeCN, 28 h, *hv*	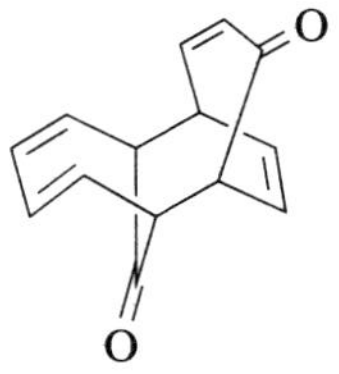 (17)	68, 69
	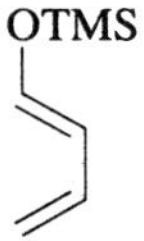	C_6H_6, 80°, 4 d	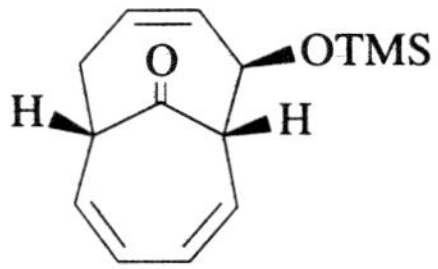 (80)	16
	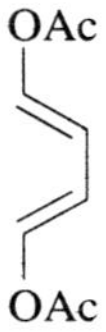	Xylene, 140°, 27 h	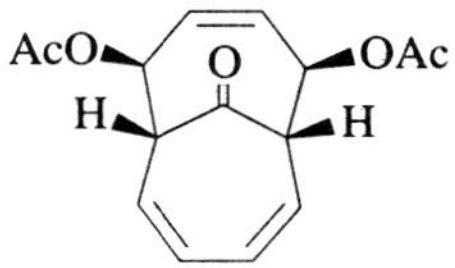 (25)	16

TABLE I. [6+4] CYCLOADDITIONS OF 2,4,6-CYCLOHEPTATRIEN-1-ONE AND DERIVATIVES (*Continued*)

A. *All-Carbon Trienophiles*

Triene	Trienophile	Conditions	Product(s) and Yield(s) (%)	Refs.
	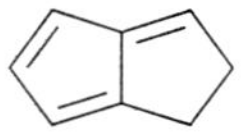	Pentane, rt, 30 d	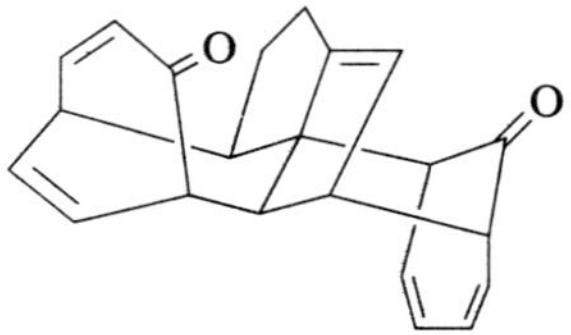 (68)	70
	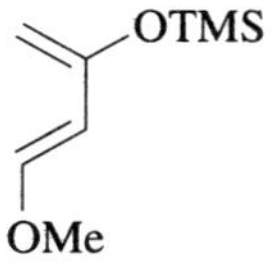	Toluene, 80°, 96 h	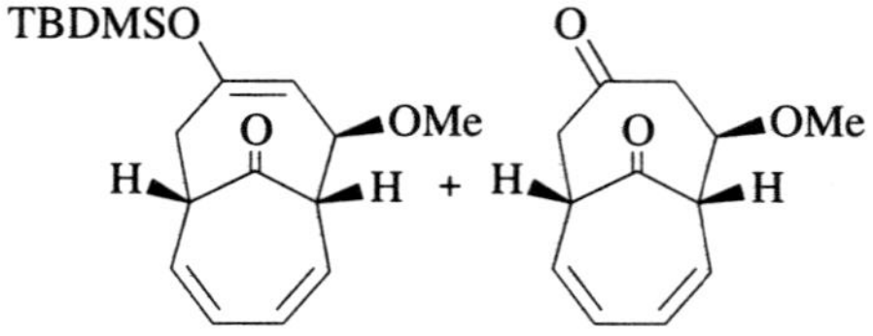 (32) + (8)	16
	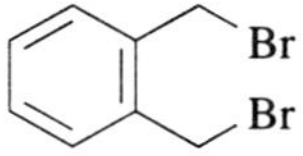	Zn, DMF, rt	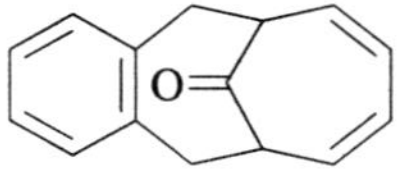 (29)	71
	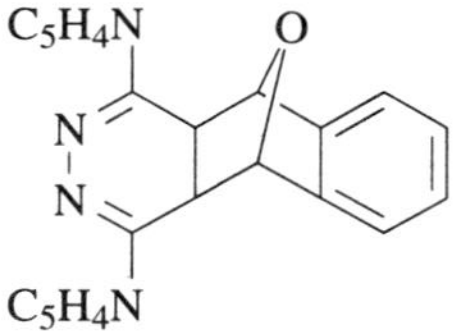	rt	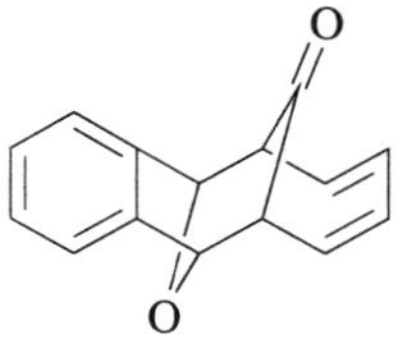 (12)	72

358

TABLE I. [6+4] CYCLOADDITIONS OF 2,4,6-CYCLOHEPTATRIEN-1-ONE AND DERIVATIVES (*Continued*)

A. All-Carbon Trienophiles

Triene	Trienophile	Conditions	Product(s) and Yield(s) (%)	Refs.
	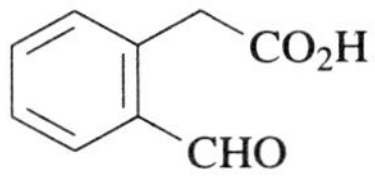	Ac$_2$O, 140°, 1 h	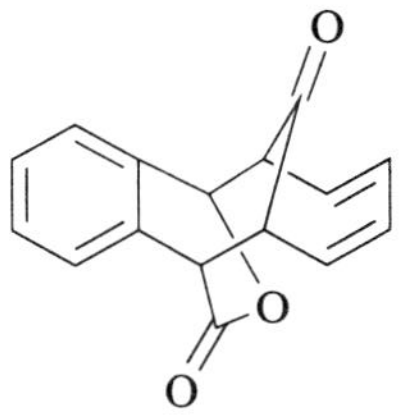 (28)	73
	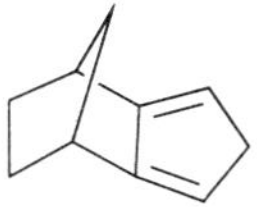	C$_6$H$_6$, rt, 9 d	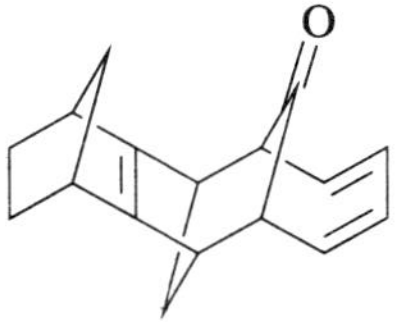 (37) 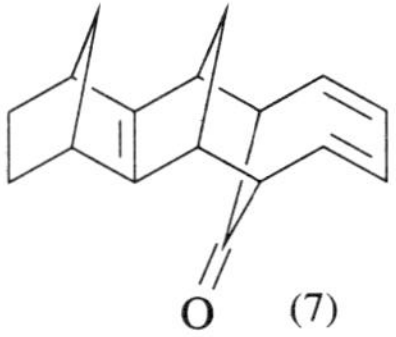 + (7)	74
	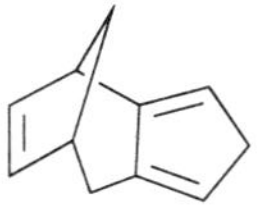	rt	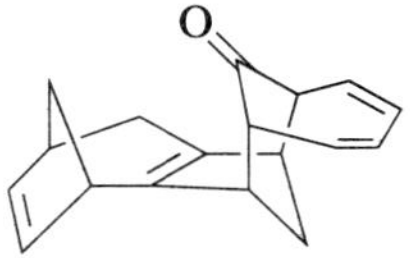 (—)	75
	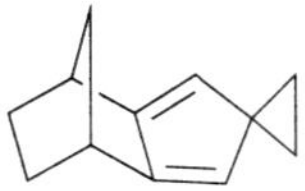	C$_6$H$_6$, 80°, 6 d	 (17)	76

TABLE I. [6+4] CYCLOADDITIONS OF 2,4,6-CYCLOHEPTATRIEN-1-ONE AND DERIVATIVES (*Continued*)

A. *All-Carbon Trienophiles*

Triene	Trienophile	Conditions	Product(s) and Yield(s) (%)	Refs.
		C₆H₆, 80°, 4 d	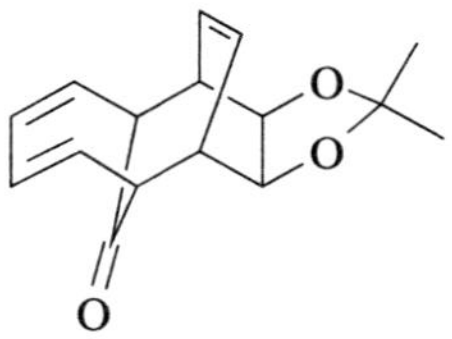 (62)	77
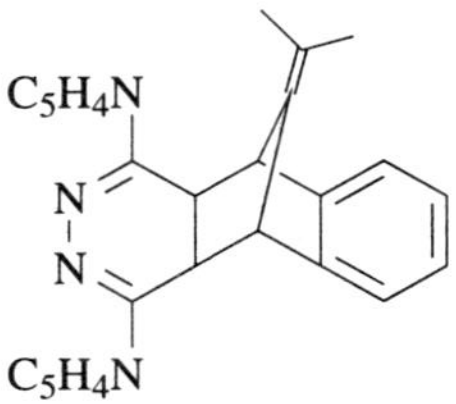	rt		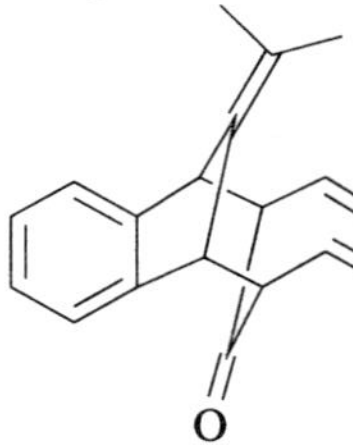 (100)	78, 79
	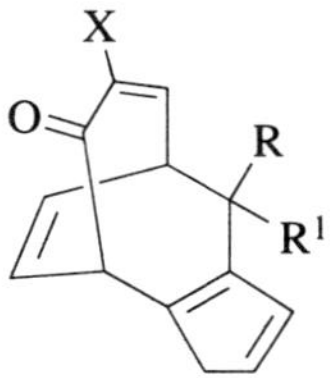		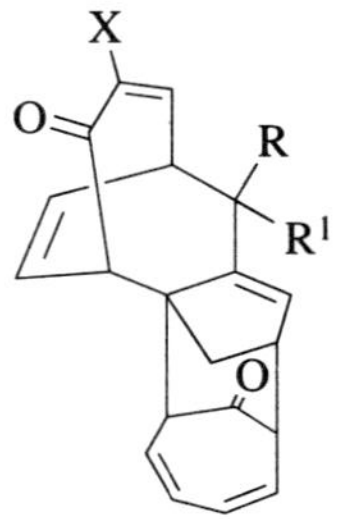	

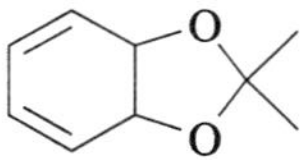

R	R¹	X		
Me	Me	H	THF, 50°, 2 d	(—)	80
Me	H	H	THF, 60°, 12 h	(61)	81
Ph	H	H	MeOH, 60°	(86)	81
Me	Me	Cl	MeOH, 50°, 3 d	(42)	82

360

TABLE I. [6+4] CYCLOADDITIONS OF 2,4,6-CYCLOHEPTATRIEN-1-ONE AND DERIVATIVES (*Continued*)

A. *All-Carbon Trienophiles*

Triene	Trienophile	Conditions	Product(s) and Yield(s) (%)	Refs.
	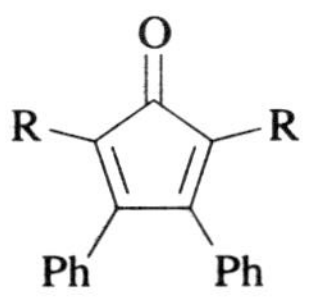		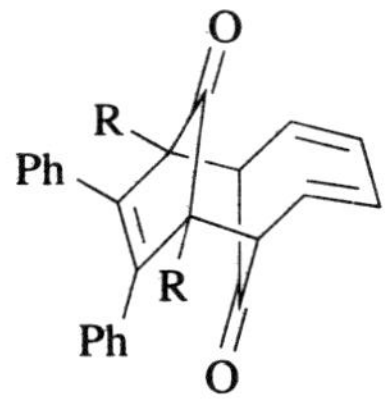	
	R			
	Me	Acetone, 60°, 8 h	(95)	80
	Et	Acetone, 60°, 7 h	(88)	83
	CO_2Me	C_6H_6, 80°, 10 h	(88)	83
	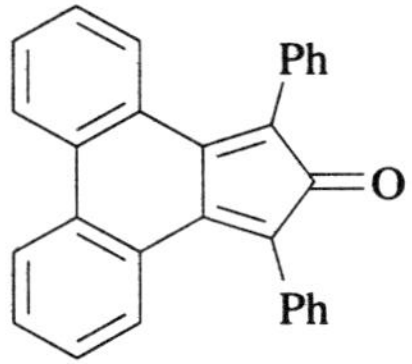	C_6H_6, 80°, 20 h	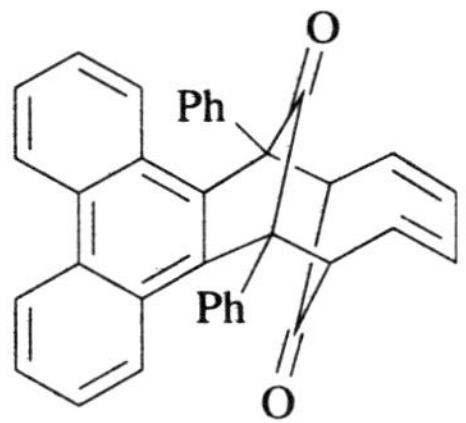 (70)	84
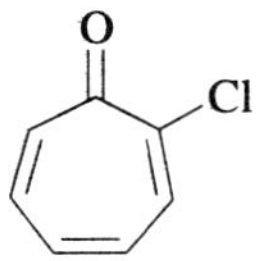		—	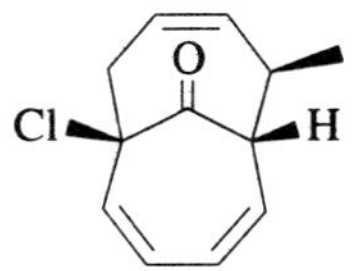 (—)	85

TABLE I. [6+4] CYCLOADDITIONS OF 2,4,6-CYCLOHEPTATRIEN-1-ONE AND DERIVATIVES (*Continued*)

A. All-Carbon Trienophiles

Triene	Trienophile	Conditions	Product(s) and Yield(s) (%)	Refs.
		105°, 3 h	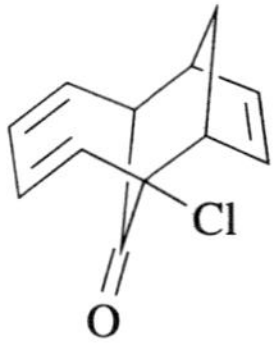 (11)	28, 85
		100°, 9 h	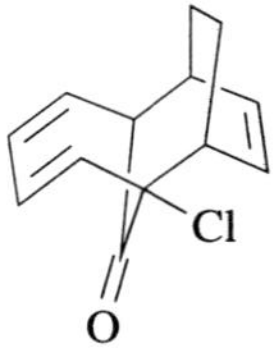 (58)	85
	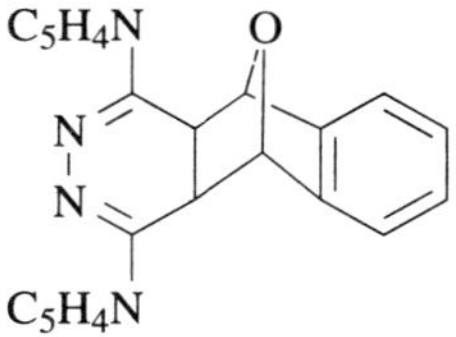	rt	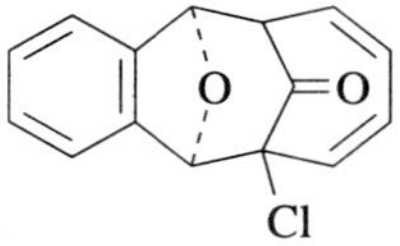 (25)	72
C₈		C₆H₆, 80°, 6 d	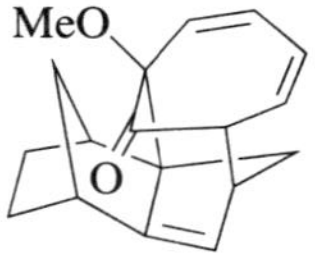 (25)	76

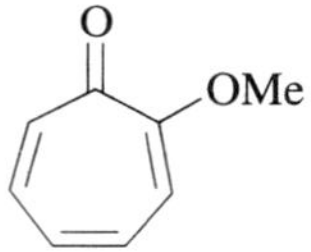

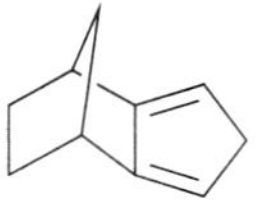

TABLE I. [6+4] C YCLOADDITIONS OF 2,4,6-CYCLOHEPTATRIEN-1-ONE AND DERIVATIVES (*Continued*)

A. All-Carbon Trienophiles

Triene	Trienophile	Conditions	Product(s) and Yield(s) (%)	Refs.
	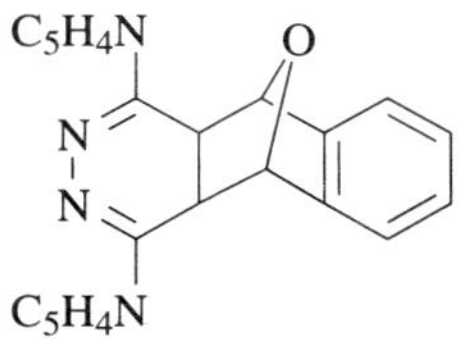	rt	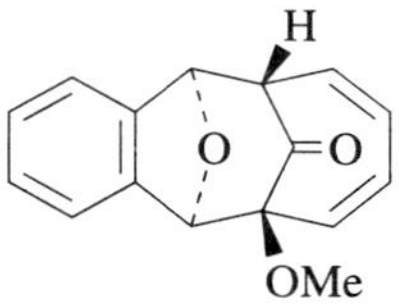 (17)	72
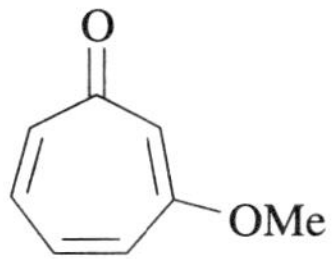	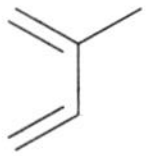	C_6H_6, 160°, 140 h	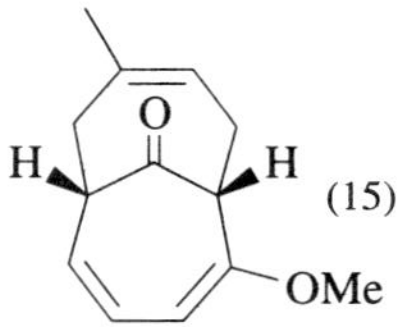 (15) + 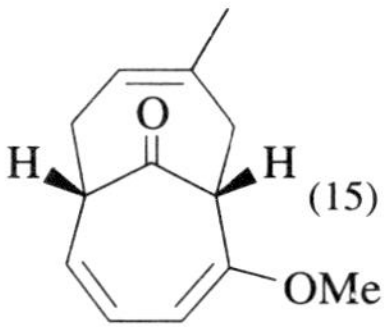 (15)	29
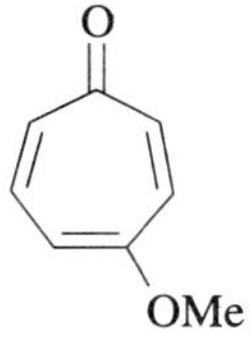	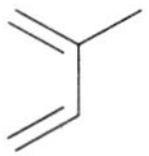	C_6H_6, 175°, 120 h	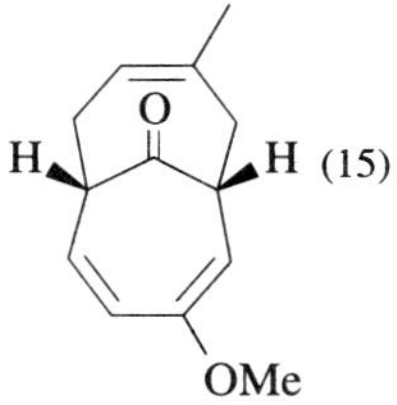 (15) + 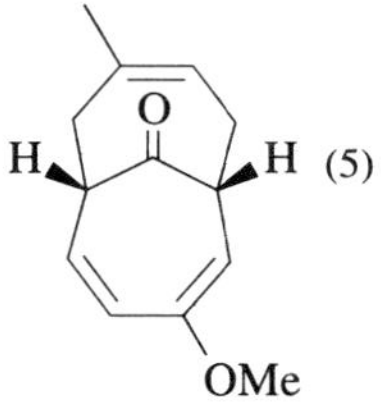 (5)	29
C_{10} 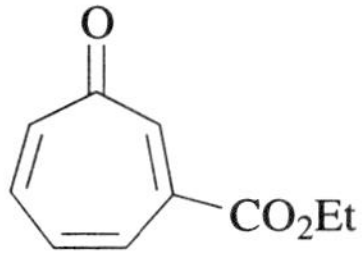		C_6H_6, 120°, 16 h	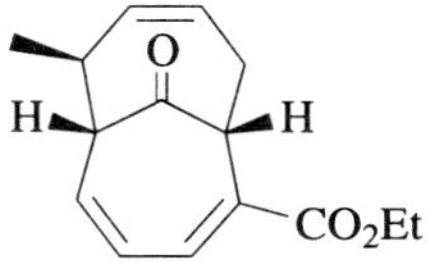 (21)	29

TABLE I. [6+4] CYCLOADDITIONS OF 2,4,6-CYCLOHEPTATRIEN-1-ONE AND DERIVATIVES (*Continued*)

A. *All-Carbon Trienophiles*

Triene	Trienophile	Conditions	Product(s) and Yield(s) (%)	Refs.
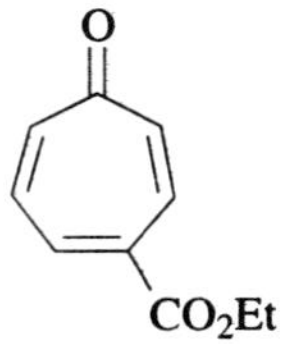	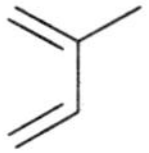	C_6H_6, 110°, 15 h	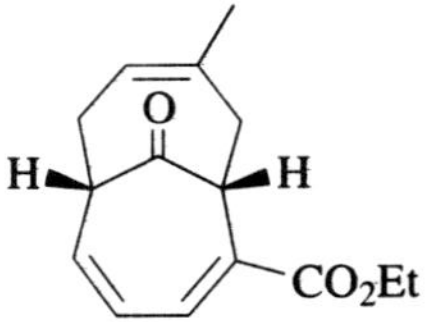 (38)	29
	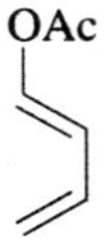	Toluene, 110°, 16 h	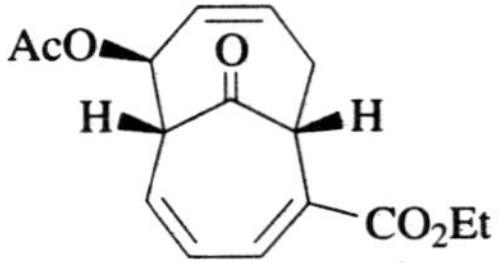 (20)	29
		C_6H_6, 110°, 62 h	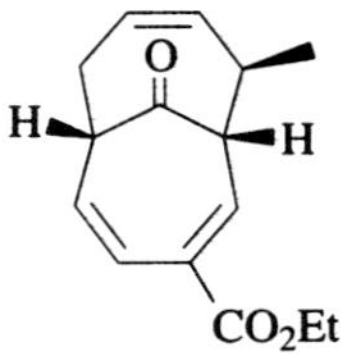 (28)	29
	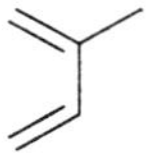	C_6H_6, 110°, 40 h	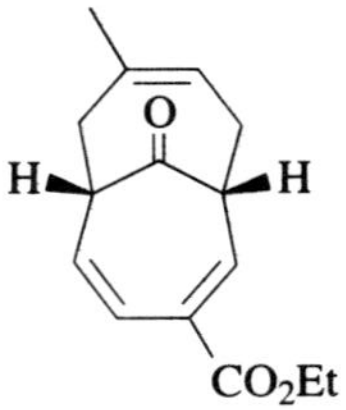 (25)	29

TABLE I. [6+4] CYCLOADDITIONS OF 2,4,6-CYCLOHEPTATRIEN-1-ONE AND DERIVATIVES (*Continued*)

A. *All-Carbon Trienophiles*

Triene	Trienophile	Conditions	Product(s) and Yield(s) (%)	Refs.
	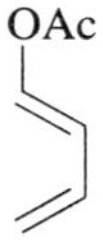	C$_6$H$_6$, 80°, 48 h	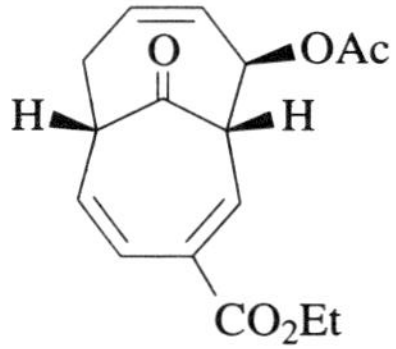 (22)	29
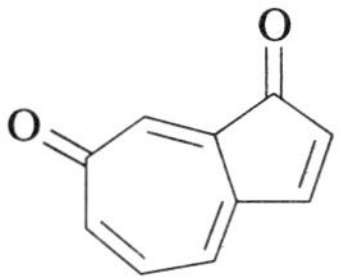	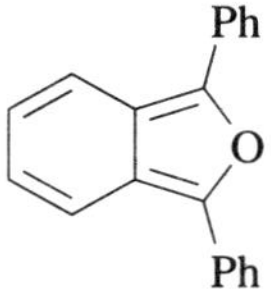	C$_2$H$_4$Cl$_2$, rt, 1 h	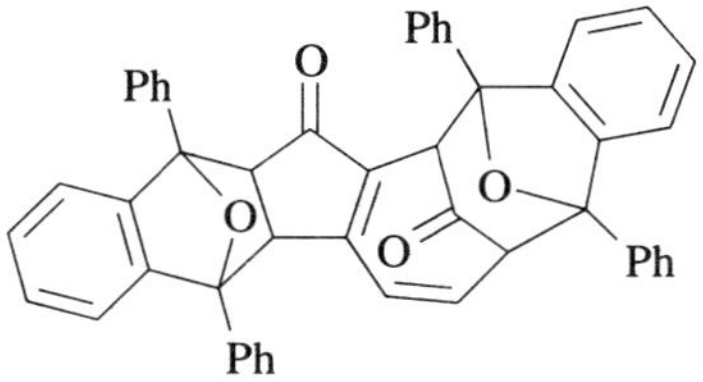 (14)	86
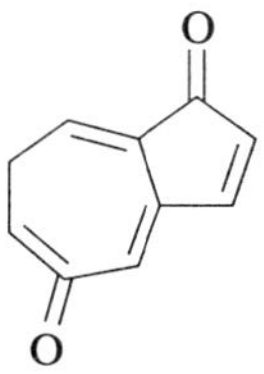	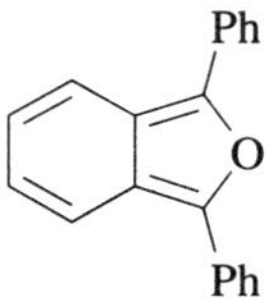	C$_2$H$_4$Cl$_2$, rt, 1 h	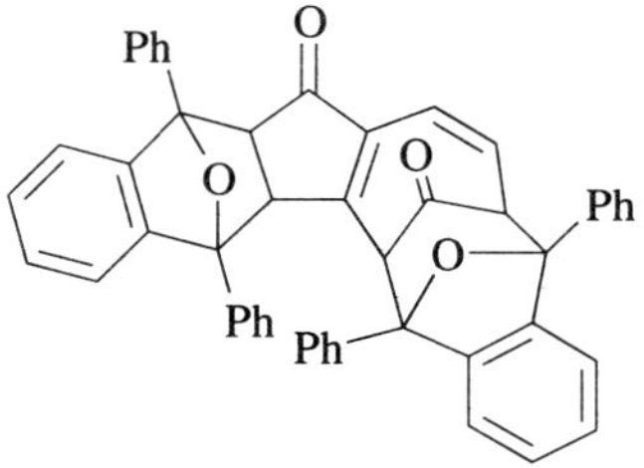 (40)	86

365

TABLE I. [6+4] CYCLOADDITIONS OF 2,4,6-CYCLOHEPTATRIEN-1-ONE AND DERIVATIVES (*Continued*)

A. *All-Carbon Trienophiles*

Triene	Trienophile	Conditions	Product(s) and Yield(s) (%)	Refs.

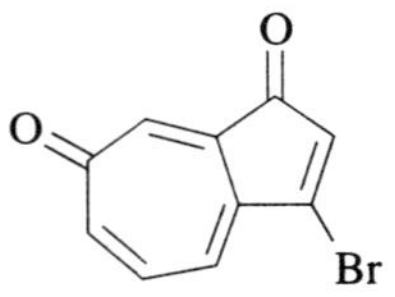

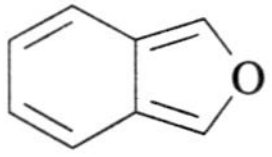

(30) + 87

(11)

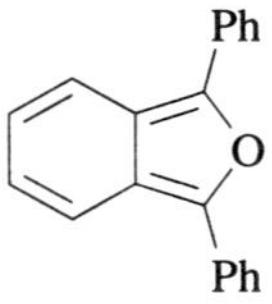

Chlorobenzene, 130°, 4 h

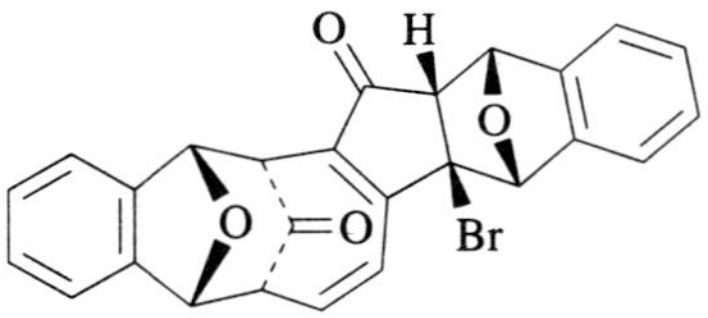

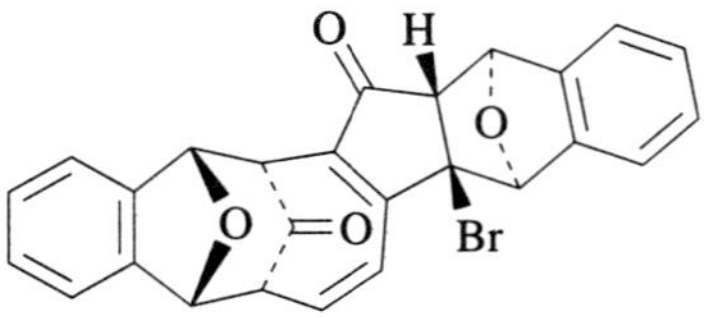

(11) + 88

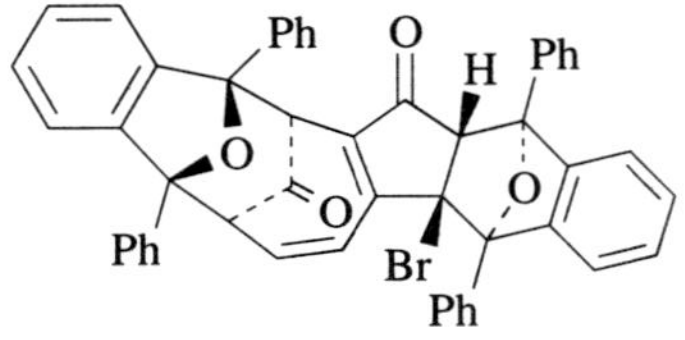

(30)

Chlorobenzene, 130°, 4 h

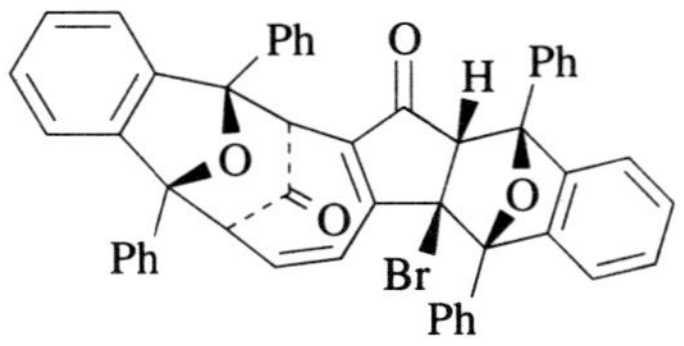

TABLE I. [6+4] CYCLOADDITIONS OF 2,4,6-CYCLOHEPTATRIEN-1-ONE AND DERIVATIVES (*Continued*)

A. *All-Carbon Trienophiles*

Triene	Trienophile	Conditions	Product(s) and Yield(s) (%)	Refs.
		Chlorobenzene, 130°, 4 h	(25) + (28) + (9) + (11)	87

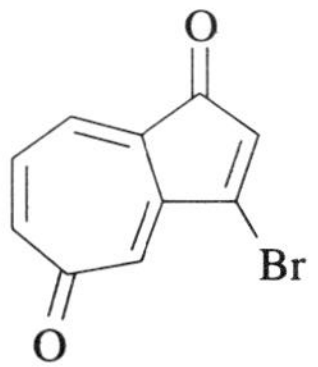

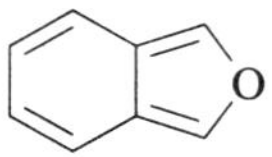

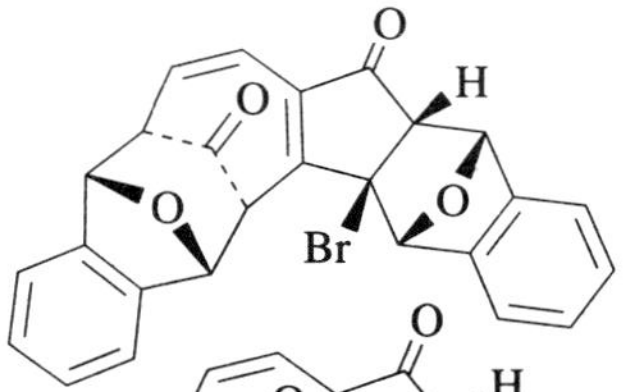

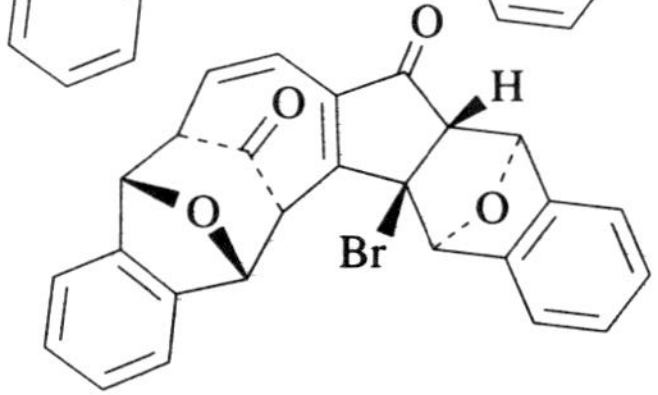

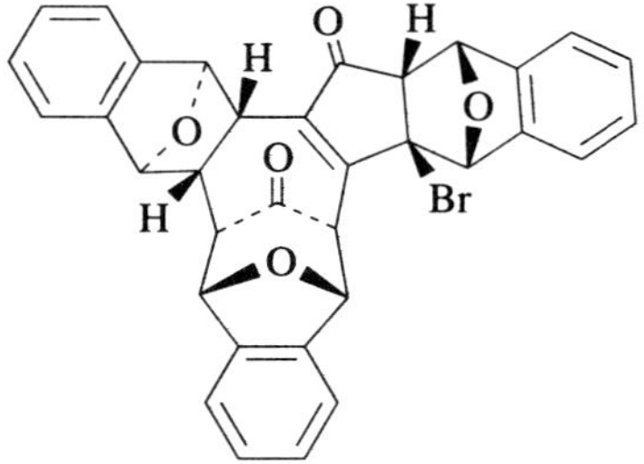

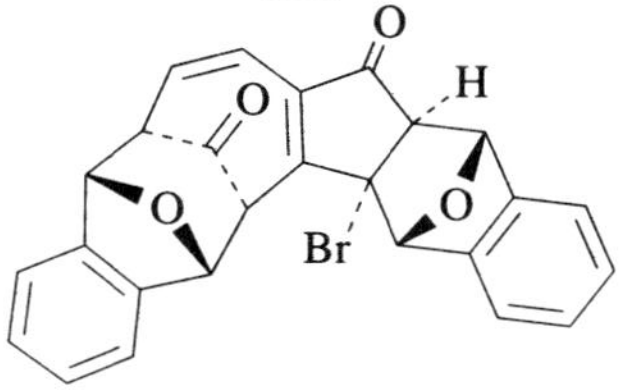

TABLE I. [6+4] CYCLOADDITIONS OF 2,4,6-CYCLOHEPTATRIEN-1-ONE AND DERIVATIVES (*Continued*)

A. All-Carbon Trienophiles

Triene	Trienophile	Conditions	Product(s) and Yield(s) (%)	Refs.
C$_{13}$		CH$_2$Cl$_2$, Et$_3$N, 0°	(20)	89
	$\dfrac{R}{\text{Me} \;\; \text{CO}_2\text{Me}}$	C$_6$H$_6$, 80°, 96 h	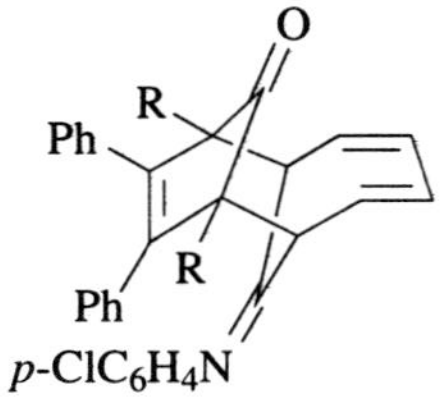 (50) (51)	90
C$_{14}$	$\dfrac{R}{\text{Me} \;\; \text{Et} \;\; \text{CO}_2\text{Me}}$	C$_6$H$_6$, 80°, 96 h	(45) (50) (61)	90

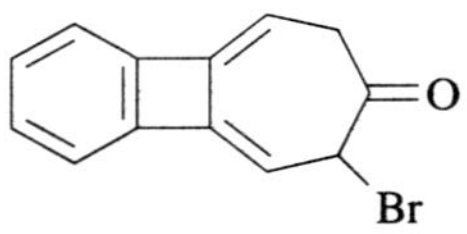

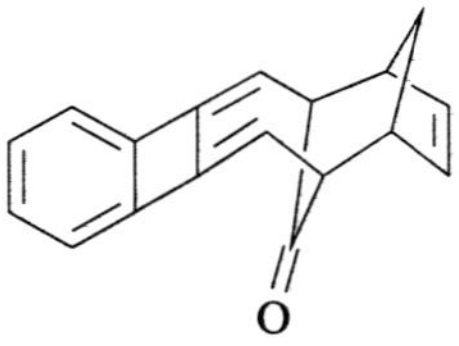

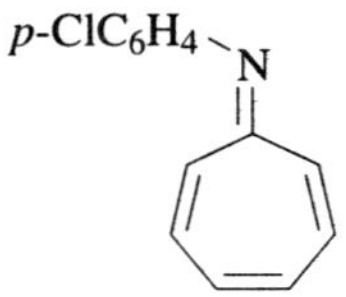

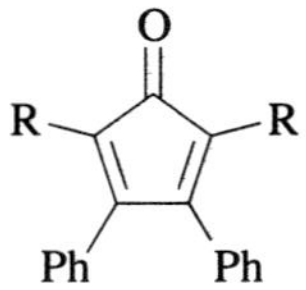

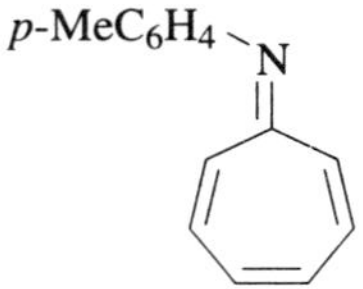

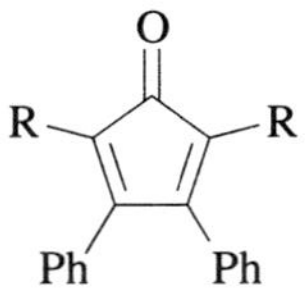

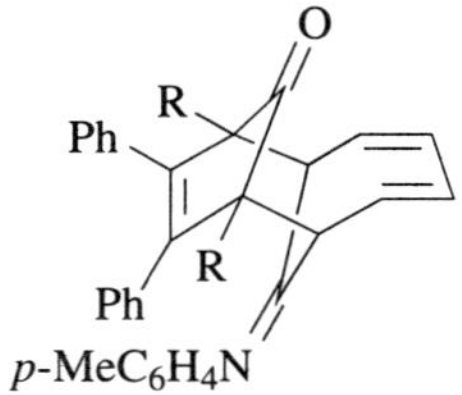

TABLE I. [6+4] CYCLOADDITIONS OF 2,4,6-CYCLOHEPTATRIEN-1-ONE AND DERIVATIVES (*Continued*)

B. 1,3-Dipolar Trienophiles

Triene	Trienophile	Conditions	Product(s) and Yield(s) (%)	Refs.
C$_7$ 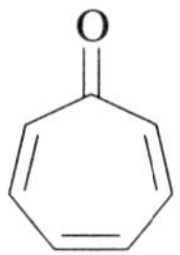	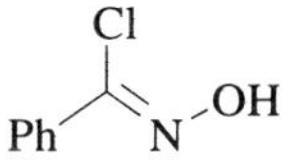 Ph Cl N-OH	C$_6$H$_6$, Et$_3$N, rt, 48 h	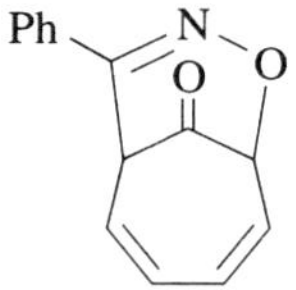 (3.5)	91
	Mesityl C≡N$^+$-O$^-$	C$_6$H$_6$, 15 d	(0.6) + (57)	91
			R = 2,4,6-Me$_3$C$_6$H$_2$	
	TMS OAc	Toluene, Pd(OAc)$_2$, (i-C$_3$H$_7$-O)$_3$P, 80°, 3 h	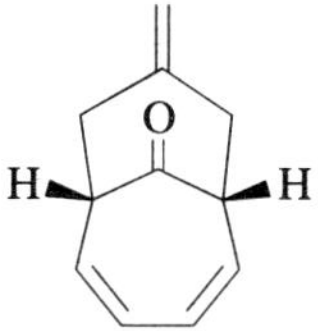 (68)	92
	TMS OCO$_2$Me	Toluene, Pd(OAc)$_2$, (i-C$_3$H$_7$-O)$_3$P, 80°, 6 h	H (23) + H (23)	92

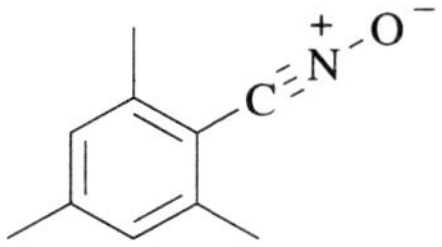

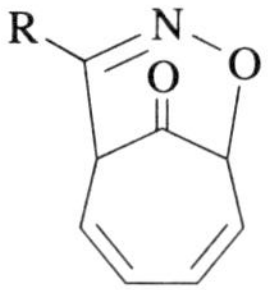

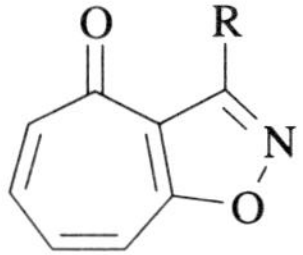

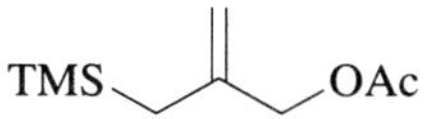

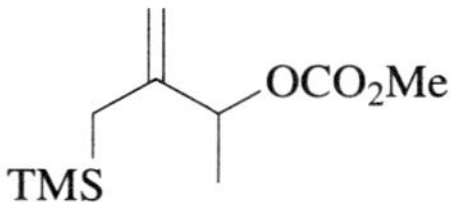

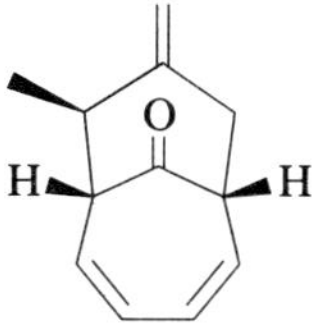

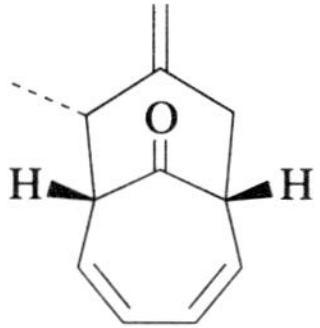

TABLE I. [6+4] CYCLOADDITIONS OF 2,4,6-CYCLOHEPTATRIEN-1-ONE AND DERIVATIVES (Continued)

B. 1,3-Dipolar Trienophiles

Triene	Trienophile	Conditions	Product(s) and Yield(s) (%)	Refs.
	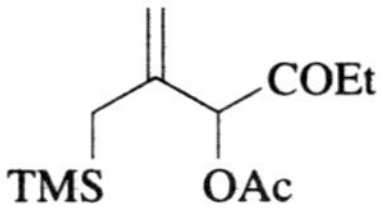 TMS, COEt, OAc	Toluene, Pd(OAc)$_2$, $(i$-C$_3$H$_7$O)$_3$P, 80°, 3 h	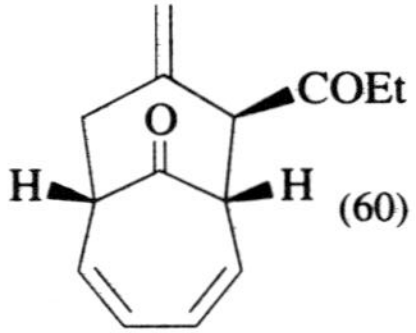 (60) 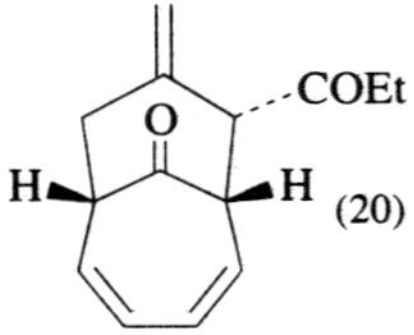 + (20)	92
	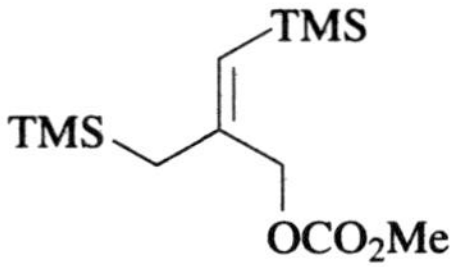 TMS, TMS	Toluene, Pd(OAc)$_2$, $(i$-C$_3$H$_7$O)$_3$P, 80°, 6 h	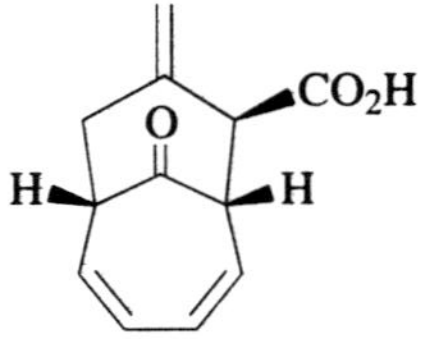 (84)	92
	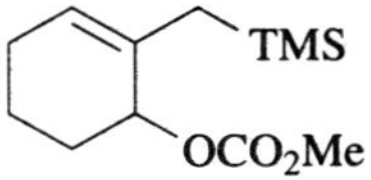 TMS, OCO$_2$Me	Toluene, Pd(OAc)$_2$, $(i$-C$_3$H$_7$O)$_3$P, 80°, 3 h	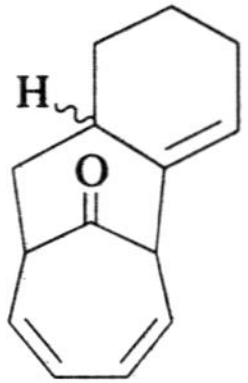 (81)	92
	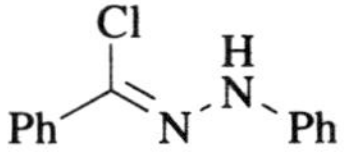 Cl, Ph, N, N-Ph, H	C$_6$H$_6$, Et$_3$N, rt, 12 h	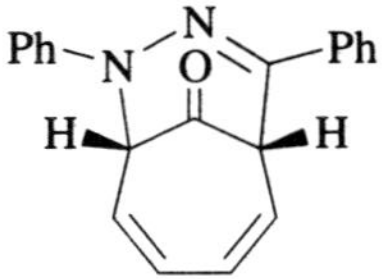 (5)	93, 8, 94

TABLE I. [6+4] CYCLOADDITIONS OF 2,4,6-CYCLOHEPTATRIEN-1-ONE AND DERIVATIVES (*Continued*)

B. 1,3-Dipolar Trienophiles

Triene	Trienophile	Conditions	Product(s) and Yield(s) (%)	Refs.
	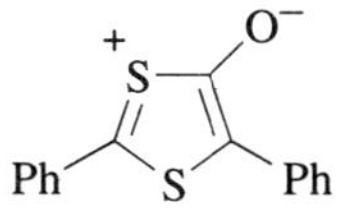	Toluene, 110°, Cu(OAc)$_2$	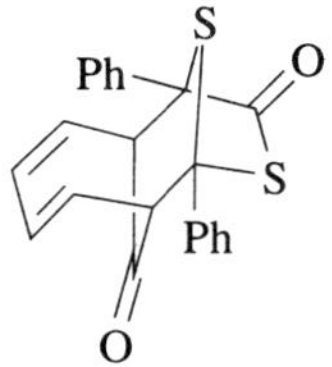 (19)	95
	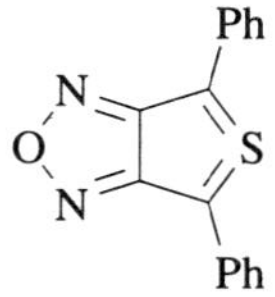	C$_6$H$_6$, 80°, 48 h	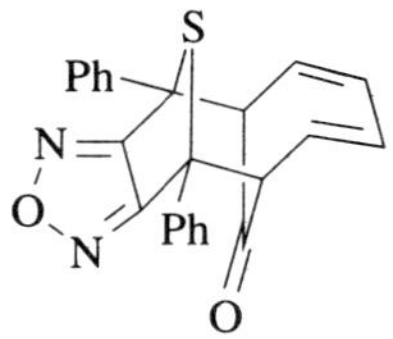 (30)	96
	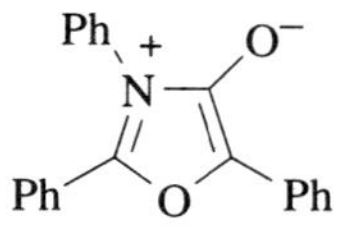	rt	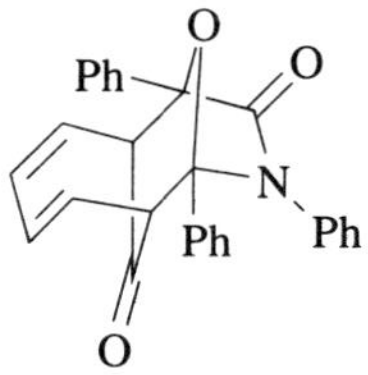 (49)	95

371

TABLE I. [6+4] CYCLOADDITIONS OF 2,4,6-CYCLOHEPTATRIEN-1-ONE AND DERIVATIVES (*Continued*)

B. *1,3-Dipolar Trienophiles*

Triene	Trienophile	Conditions	Product(s) and Yield(s) (%)	Refs.
C$_8$				
	TMS — OAc	Toluene, Pd(OAc)$_2$, (i-C$_3$H$_7$O)$_3$P, 80°, 10 h	(74)	92
	TMS, TMS, OCO$_2$Me	Toluene, Pd(OAc)$_2$, (i-C$_3$H$_7$O)$_3$P, 80°, 12 h	HO$_2$C (62)	92
OMe	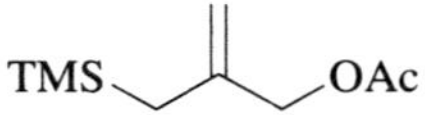 TMS — OAc	Toluene, Pd(OAc)$_2$, (i-C$_3$H$_7$O)$_3$P, 80°, 6 h	OMe (41)	92

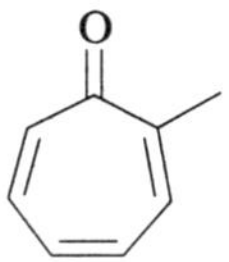

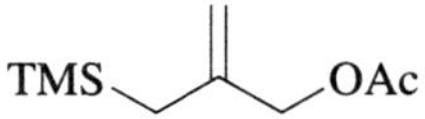

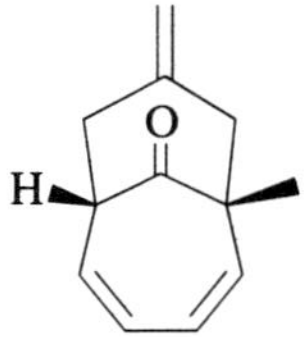

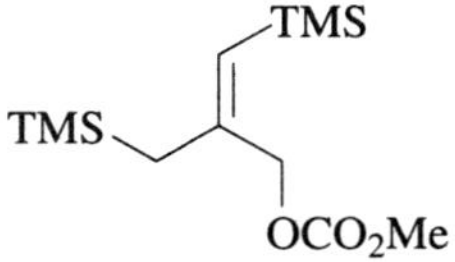

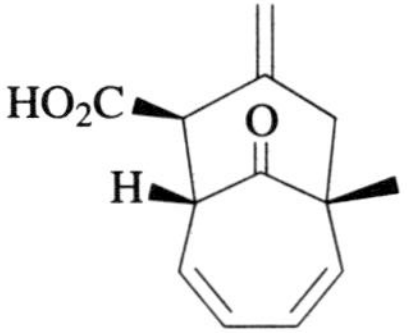

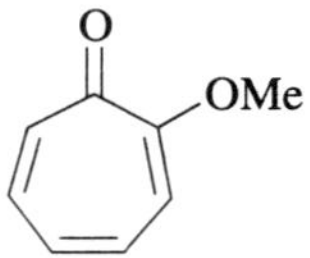

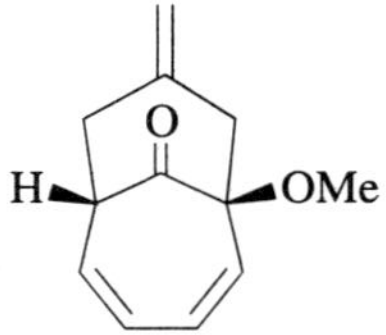

TABLE I. [6+4] CYCLOADDITIONS OF 2,4,6-CYCLOHEPTATRIEN-1-ONE AND DERIVATIVES (*Continued*)

B. 1,3-Dipolar Trienophiles

Triene	Trienophile	Conditions	Product(s) and Yield(s) (%)	Refs.
C$_{13}$ 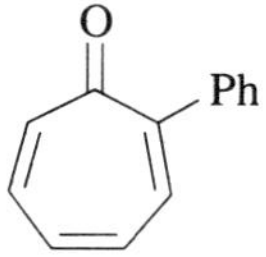	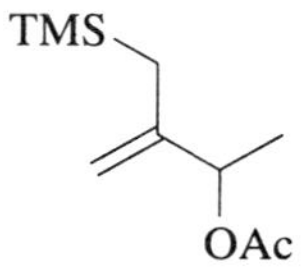	Toluene, Pd(OAc)$_2$, (*i*-C$_3$H$_7$O)$_3$P, 110°	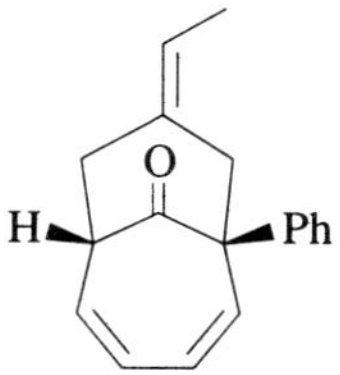 (7) 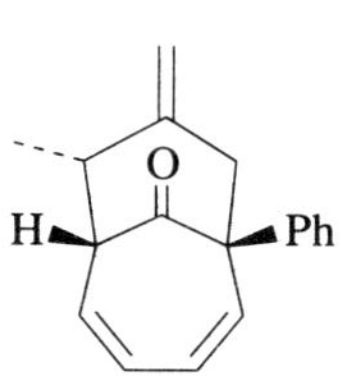 + (36) 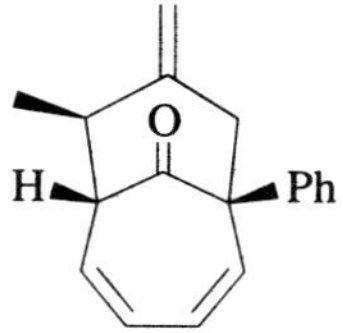 + (11)	97
	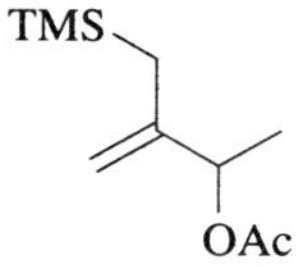	Toluene-C$_6$H$_6$, 15 kbar, (η^3-C$_3$H$_5$PdCl)$_2$, (RO)$_3$P, 50°	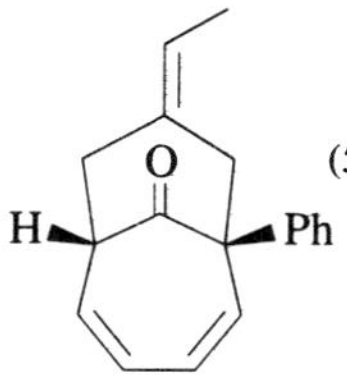 (57) 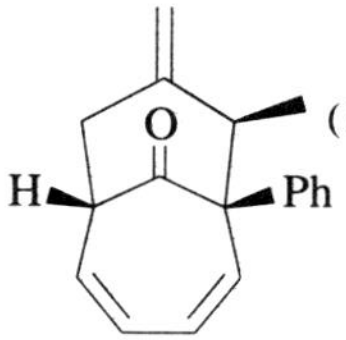 + (16) 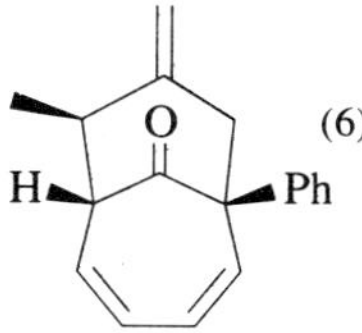 + (6) 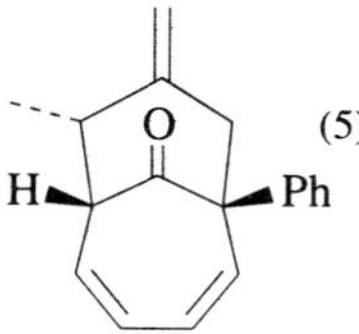 + (5)	97
	(RO)$_3$P ≡ 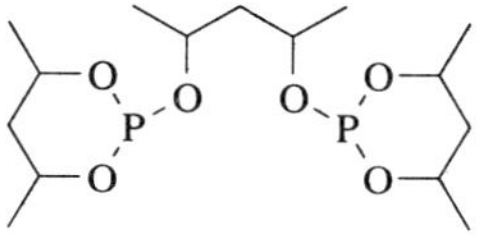			

TABLE I. [6+4] CYCLOADDITIONS OF 2,4,6-CYCLOHEPTATRIEN-1-ONE AND DERIVATIVES (*Continued*)

C. *Intramolecular Reactions*

Triene	Trienophile	Conditions	Product(s) and Yield(s) (%)	Refs.
C_{14}	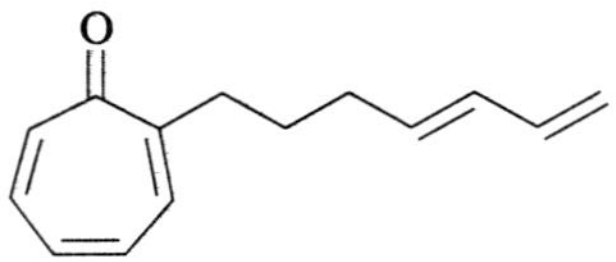	Xylene, 140°, 6 h	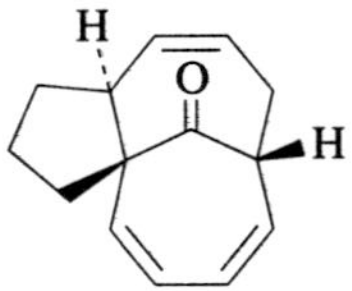 (81)	27
C_{15}	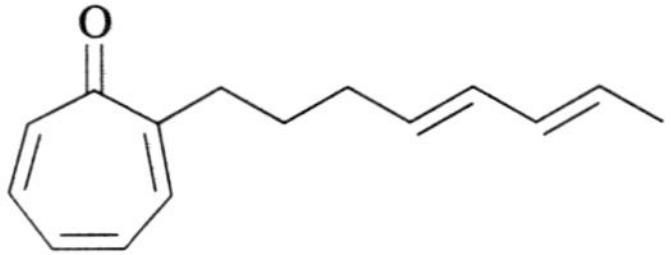	Toluene, 150°, 36 h	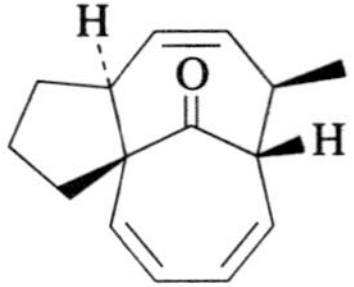 (88)	30
C_{20}	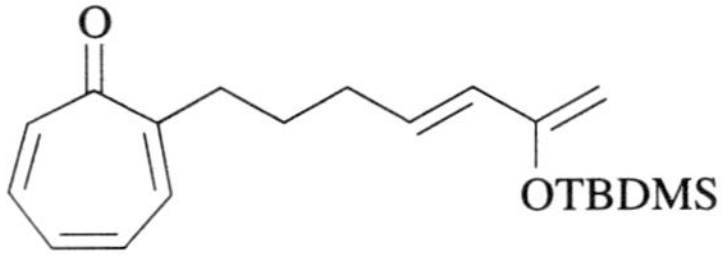 OTBDMS	C_6H_6, 80°, 6 h	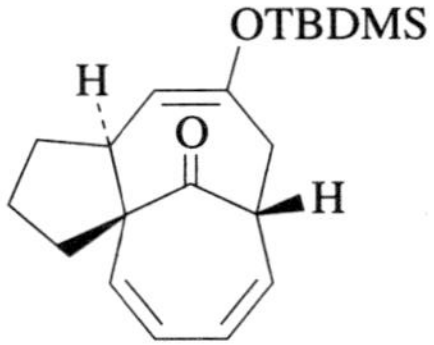 OTBDMS (92)	30

TABLE II. [6+4] CYCLOADDITIONS OF FULVENE DERIVATIVES

A. All-Carbon Trienophiles

Triene	Trienophile	Conditions	Product(s) and Yield(s) (%)	Refs.
C$_6$ 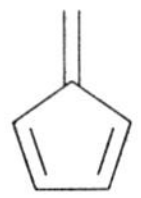		rt, 5 d	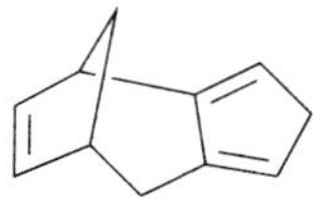 (—)	98
C$_7$ 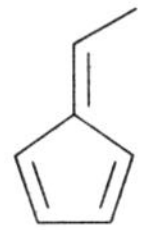	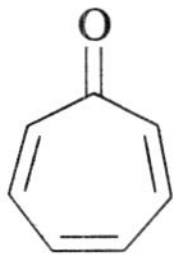	60°, 12 h	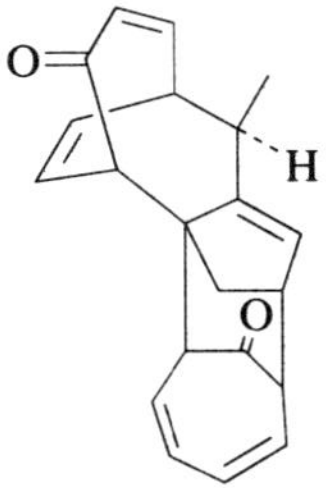 (61)	81, 82
	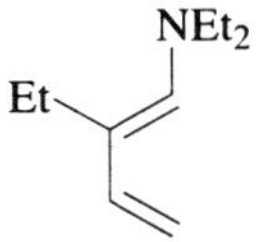	rt, 24 h	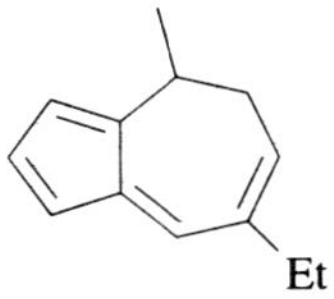 (52)	99
	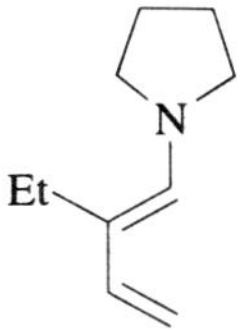	Et$_2$O, rt	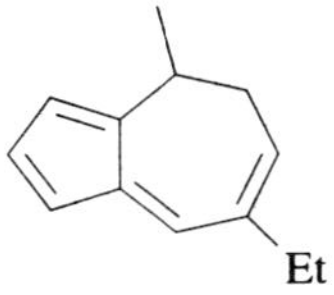 (52)	100

TABLE II. [6+4] CYCLOADDITIONS OF FULVENE DERIVATIVES (*Continued*)

A. All-Carbon Trienophiles

Triene	Trienophile	Conditions	Product(s) and Yield(s) (%)	Refs.
		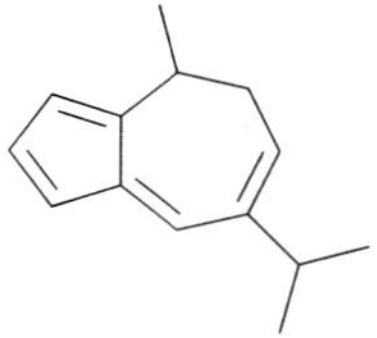 Et₂O, rt	(50)	100
	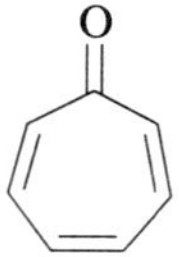	THF, 50°, 2 d	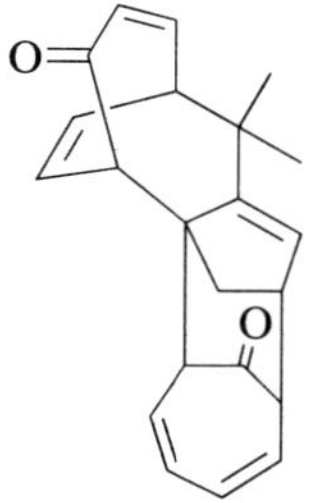 (—)	101, 37, 82
	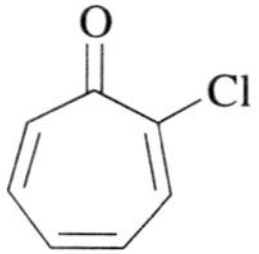	50°, 3 d	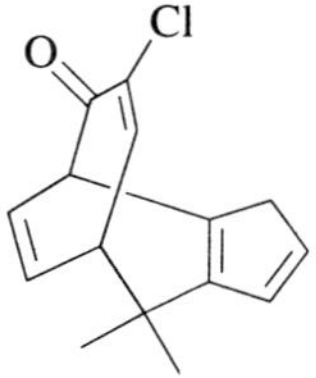 (—)	102
	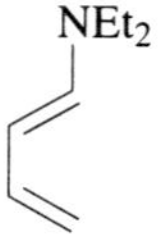	rt, 2 d	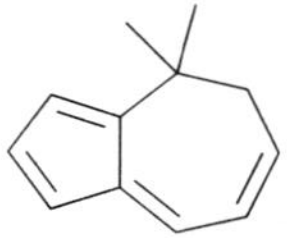 (65)	19, 99

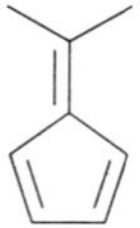

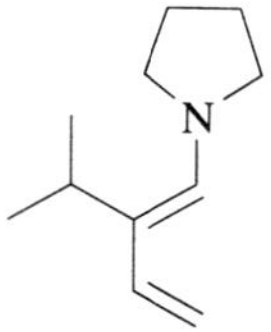

C₈

TABLE II. [6+4] CYCLOADDITIONS OF FULVENE DERIVATIVES (*Continued*)

A. *All-Carbon Trienophiles*

Triene	Trienophile	Conditions	Product(s) and Yield(s) (%)	Refs.
	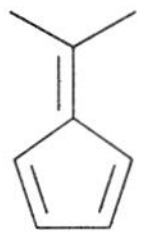	60°, 6 mon.	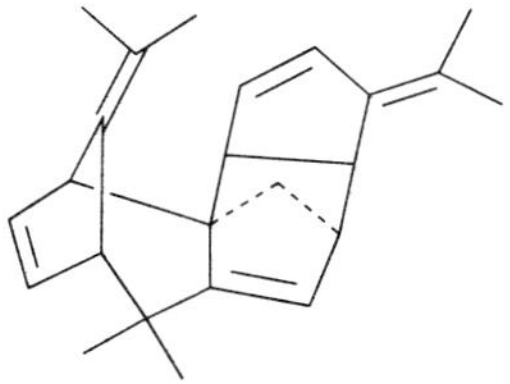 (35)	103, 104
	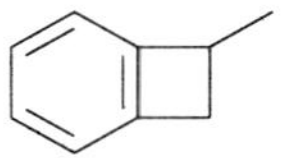	Toluene, 185°, 5 h	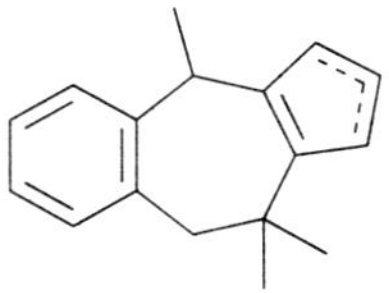 (—)	34
	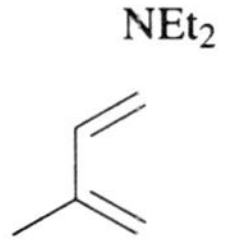 NEt$_2$	rt, 48 h	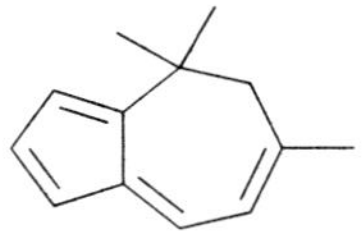 (65)	99
	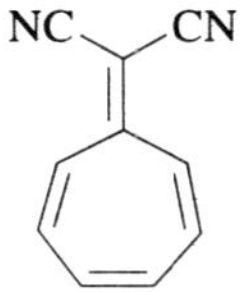 NC CN	CHCl$_3$, rt, 1 d	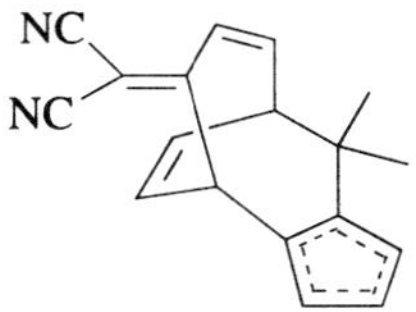 (—)	105

TABLE II. [6+4] CYCLOADDITIONS OF FULVENE DERIVATIVES (*Continued*)

A. All-Carbon Trienophiles

Triene	Trienophile	Conditions	Product(s) and Yield(s) (%)	Refs.
	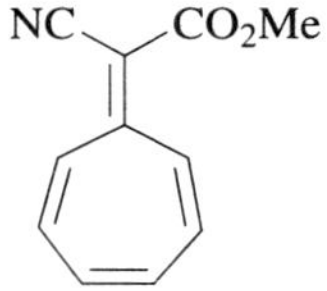 NC CO$_2$Me (cycloheptatriene)	CHCl$_3$, rt, 7 d	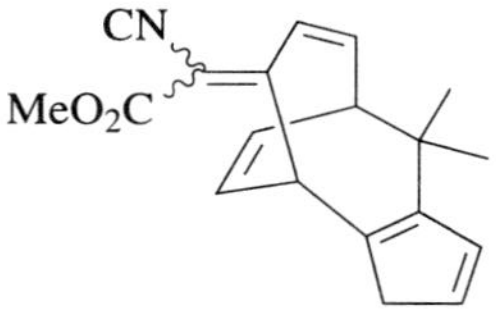 (—)	106
	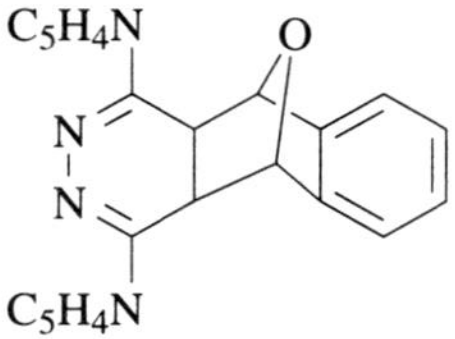 C$_5$H$_4$N / O / N—N / C$_5$H$_4$N	160°, 4 h	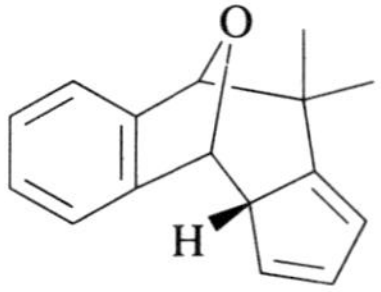 (—)	32
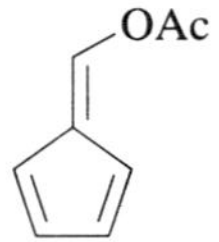 OAc (fulvene)	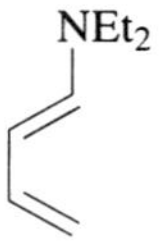 NEt$_2$ (diene)	CHCl$_3$, rt	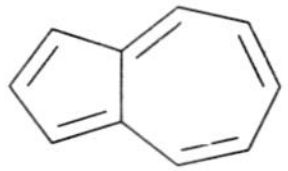 (12)	107
	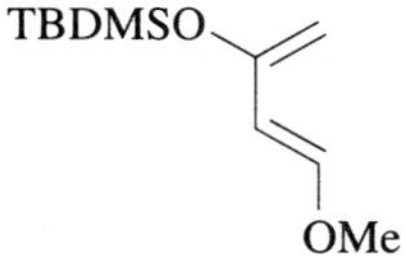 TBDMSO OMe	—	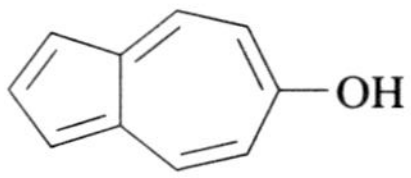 OH (—)	107

TABLE II. [6+4] CYCLOADDITIONS OF FULVENE DERIVATIVES (Continued)

A. All-Carbon Trienophiles

Triene	Trienophile	Conditions	Product(s) and Yield(s) (%)	Refs.

Triene: (fulvene with NMe$_2$ substituent)

Trienophile: (thiophene-S,S-dioxide with R^1, R^2, R^3, R^4 and O_2)

Products: **I** and **II** (azulene derivatives)

R^1	R^2	R^3	R^4	Conditions	I	II	Refs.
H	H	H	H	THF, 67°, 3 h	(33)	(0)	108
H	H	H	H	CHCl$_3$, rt, 12 h	(10)	(0)	31
Me	H	H	H	CHCl$_3$, 50°, 72 h	(5)	(0)	31
H	Me	H	H	CHCl$_3$, rt, 72 h	(25)	(0)	31
H	Et	H	H	CHCl$_3$, rt, 48 h	(12)	(0)	31
H	Ph	H	H	CHCl$_3$, rt, 12 h	(27)	(0)	31
H	Me	Me	H	CHCl$_3$, 50°, 48 h	(8)	(0)	31
H	Cl	Cl	H	CHCl$_3$, 25°, 12 h	(60)	(0)	31
H	Cl	Cl	H	THF, 67°	(46)	(0)	108
H	Ph	Ph	H	CHCl$_3$, rt, 24 h	(13)	(0)	31
Me	H	Me	H	CHCl$_3$, 50°, 48 h	(0)	(10)	31
Cl	Cl	Cl	Cl	C$_6$H$_6$, rt,	(11)	(0)	109

Trienophile: (2-Et, 5-Me thiophene-S,S-dioxide, O_2)

Conditions: C$_5$H$_5$N, 115°

Products:

(4-Et, 7-Me azulene) (16) + (4-Me, 7-Et azulene) (4)

Refs. 100

TABLE II. [6+4] CYCLOADDITIONS OF FULVENE DERIVATIVES (*Continued*)

A. All-Carbon Trienophiles

Triene	Trienophile	Conditions	Product(s) and Yield(s) (%)	Refs.
	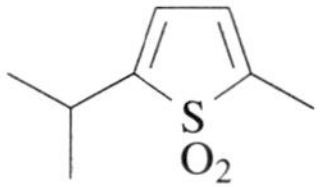	(—)	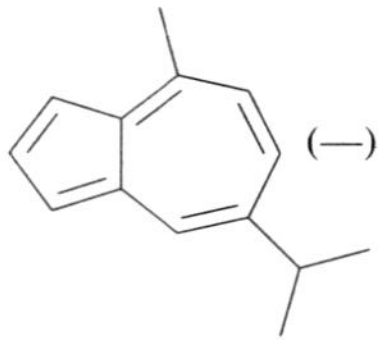 (—) + 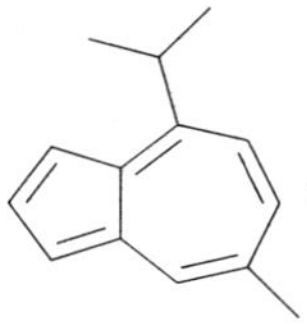 (—)	100
	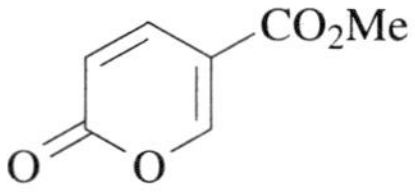 CO2Me	C_6H_6, rt, 5 d	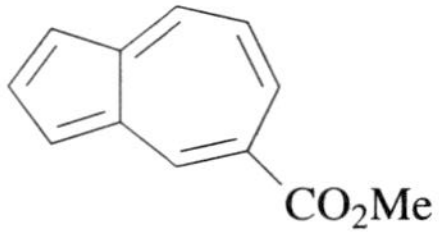 CO2Me (9)	110
	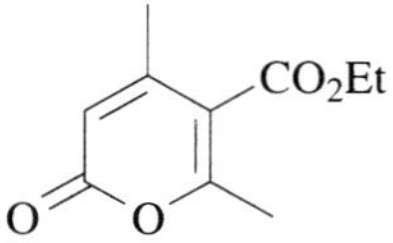 CO2Et	C_6H_6, rt	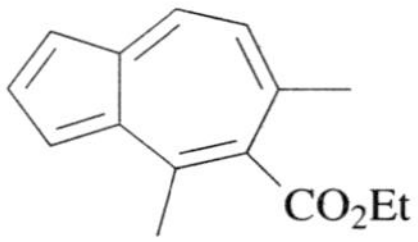 CO2Et (—)	110
C_9 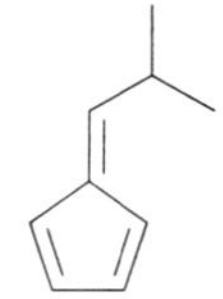	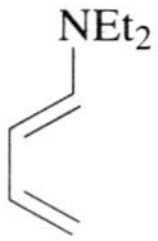 NEt2	rt, 24 h	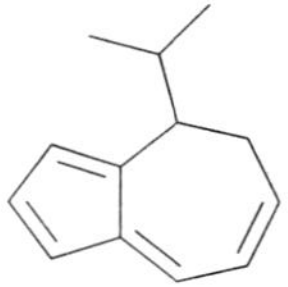 (23)	99
	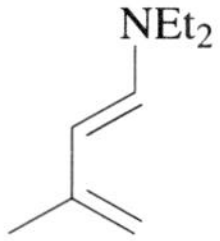 NEt2	rt, 48 h	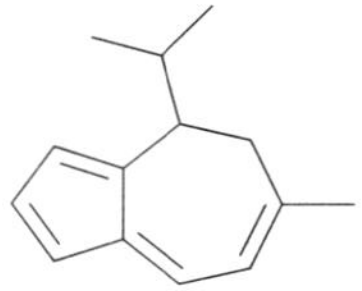 (21)	99

TABLE II. [6+4] CYCLOADDITIONS OF FULVENE DERIVATIVES (*Continued*)

A. *All-Carbon Trienophiles*

Triene	Trienophile	Conditions	Product(s) and Yield(s) (%)	Refs.
		rt, 24 h	(34)	99
		rt, 14 d	(9)	99
		40°, 4 d	(12)	99

TABLE II. [6+4] CYCLOADDITIONS OF FULVENE DERIVATIVES (*Continued*)

A. All-Carbon Trienophiles

Triene	Trienophile	Conditions	Product(s) and Yield(s) (%)	Refs.
NMe_2 fulvene	thiophene dioxide with R^2, R^1 (SO_2) R^1 R^2 H H Me H Et H Ph H Me Me Cl Cl	 CHCl$_3$, rt, 12 h CHCl$_3$, rt, 72 h CHCl$_3$, rt, 96 h CHCl$_3$, rt, 48 h CHCl$_3$, 50°, 72 h CHCl$_3$, rt, 12 h	product with R^1, R^2 R^1 R^2 H H (4) Me H (10) Et H (12) Ph H (5) H Ph (1) Me Me (7) Cl Cl (15)	31
C$_{10}$ morpholine fulvene	CO_2Me pyranone	C$_6$H$_6$, rt, 5 d	azulene-CO_2Me (17)	110
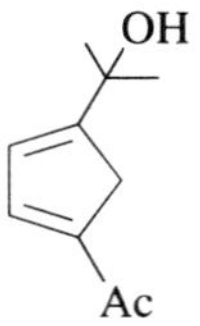 OH, Ac fulvene	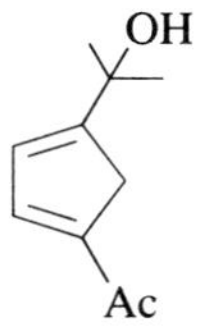 OH, Ac fulvene	CHCl$_3$, 1 h	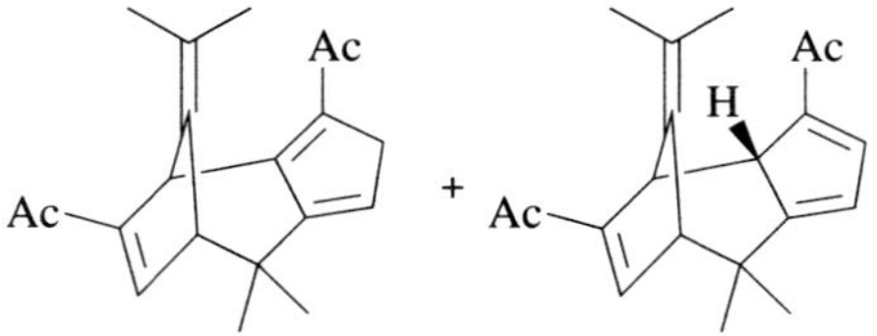 (—) + (—)	111

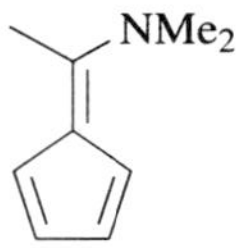

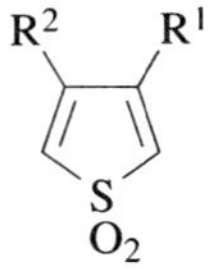

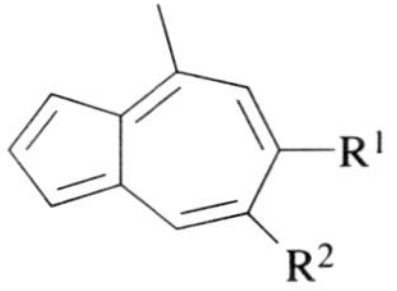

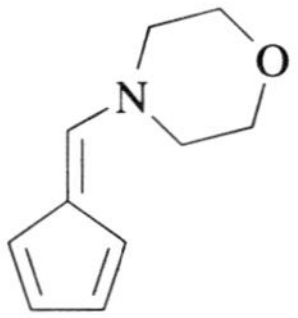

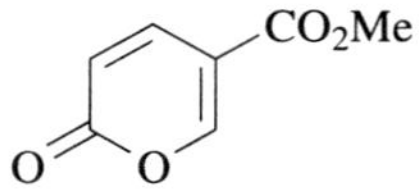

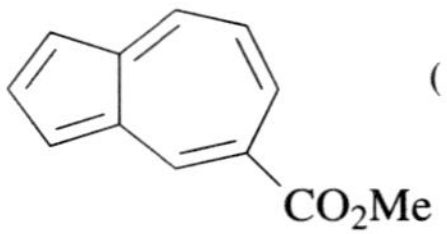

TABLE II. [6+4] CYCLOADDITIONS OF FULVENE DERIVATIVES (*Continued*)

A. All-Carbon Trienophiles

Triene	Trienophile	Conditions	Product(s) and Yield(s) (%)	Refs.
	R^1 R^2 H H Me H Et H Ph H Ph Ph	CHCl$_3$, rt, 12 h CHCl$_3$, rt, 24 h CHCl$_3$, rt, 48 h CHCl$_3$, rt, 12 h CHCl$_3$, rt, 48 h	R^1 R^2 H H (10) Me H (5) Et H (8) Ph H (6) H Ph (1) Ph Ph (10)	31
C$_{11}$ 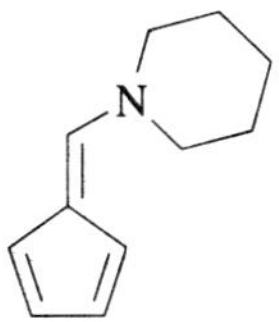	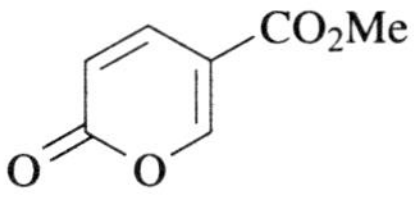	C$_6$H$_6$, rt, 5 d	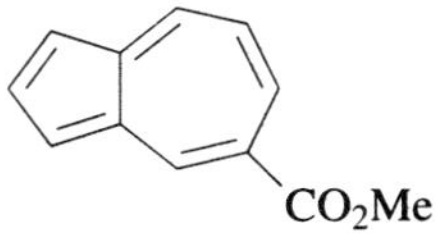 (12)	110
C$_{12}$ 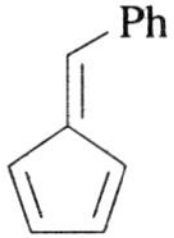	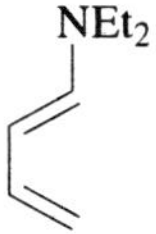	CCl$_4$, 1 d	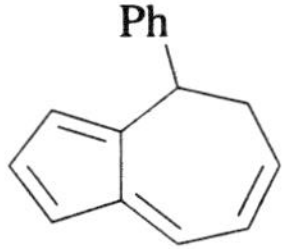 (62)	19, 99

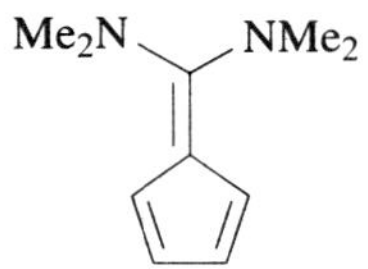

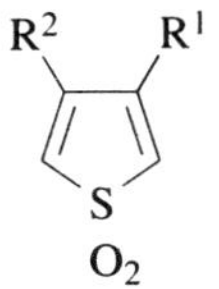

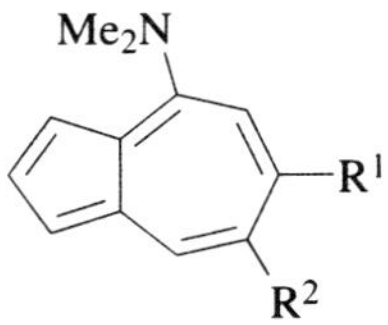

383

TABLE II. [6+4] CYCLOADDITIONS OF FULVENE DERIVATIVES (*Continued*)

A. *All-Carbon Trienophiles*

Triene	Trienophile	Conditions	Product(s) and Yield(s) (%)	Refs.
	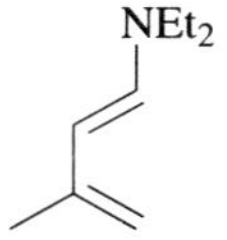 NEt₂ (with methyl)	rt, 24 h	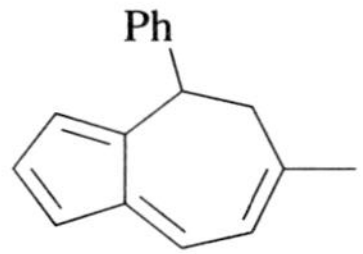 (21)	99
	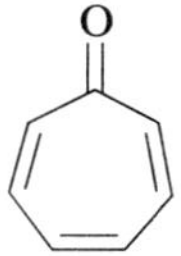 (tropone, O)	MeOH, 60°	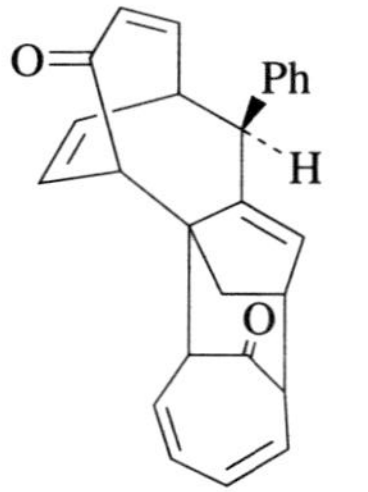 (86)	81
	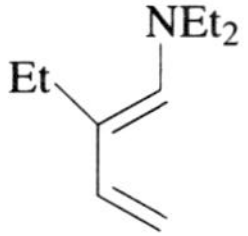 NEt₂, Et	rt, 12 h	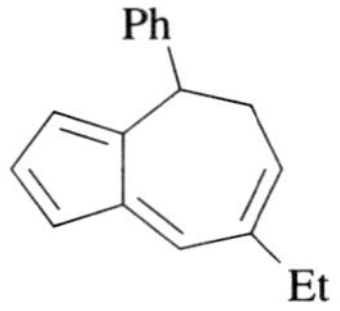 (46)	81
	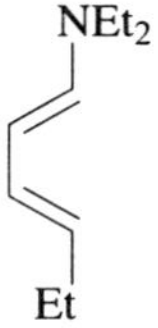 NEt₂, Et	rt, 48 h	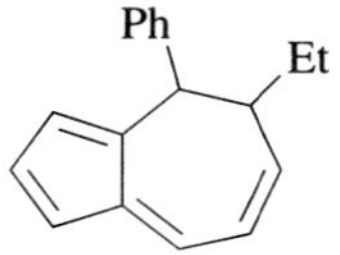 (43)	81

384

TABLE II. [6+4] CYCLOADDITIONS OF FULVENE DERIVATIVES (*Continued*)

A. All-Carbon Trienophiles

Triene	Trienophile	Conditions	Product(s) and Yield(s) (%)	Refs.
C_{13} 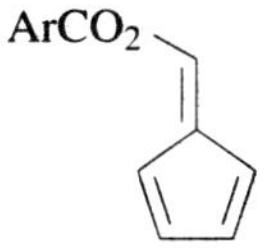 $ArCO_2$—⟨fulvene⟩ $Ar = p\text{-}O_2NC_6H_4$	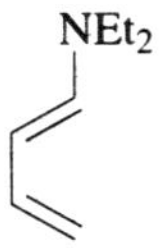 ⟨NEt_2 butadiene⟩	C_6H_6, rt	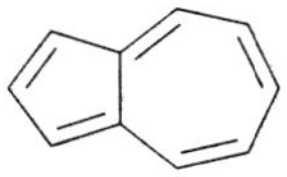 ⟨azulene⟩ (68)	107
	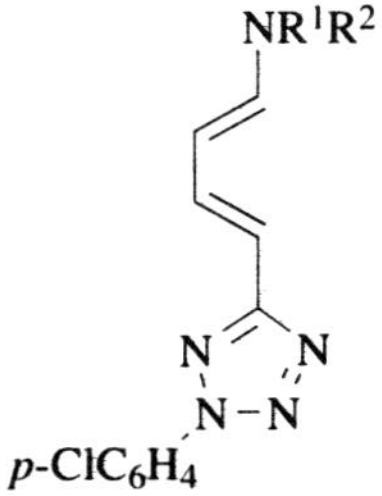 ⟨NR^1R^2-dienyl triazole, $p\text{-}ClC_6H_4$⟩		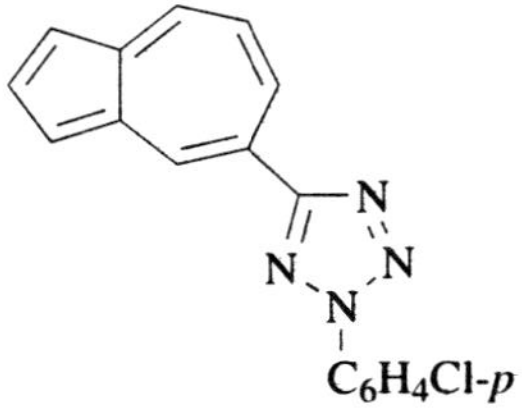 ⟨azulenyl triazole, $C_6H_4Cl\text{-}p$⟩	112

R^1	R^2	Conditions	Yield
Me	Me	C_6H_6, rt, 7 h	(50)
(CH$_2$)$_4$		C_6H_6, rt, 3 h	(33)
(CH$_2$)$_2$O(CH$_2$)$_2$		C_6H_6, 50°, 6 h	(14)
Et	Et	C_6H_6, rt, 7 h	(65)
(CH$_2$)$_5$		C_6H_6, 50°, 6 h	(14)

385

TABLE II. [6+4] CYCLOADDITIONS OF FULVENE DERIVATIVES (*Continued*)

A. All-Carbon Trienophiles

Triene	Trienophile	Conditions	Product(s) and Yield(s) (%)	Refs.
	$p\text{-BrC}_6\text{H}_4$ NEt$_2$ thiazole trienophile	C$_6$H$_6$, 50°, 8 h	azulene-thiazole-C$_6$H$_4$Br-p (46)	112
	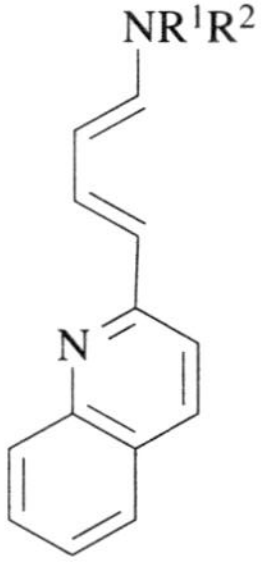 NR^{1}R^2 quinoline trienophile	C$_6$H$_6$, rt, 8 h C$_6$H$_6$, rt, 10 h	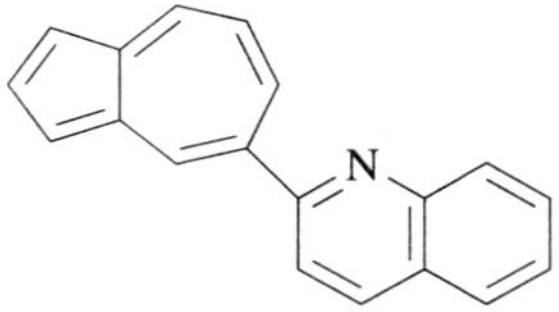 azulene-quinoline (10) (12)	112
	R^{1}R^2 = (CH$_2$)$_4$, (CH$_2$)$_5$			
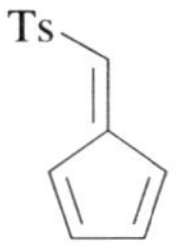 Ts-fulvene	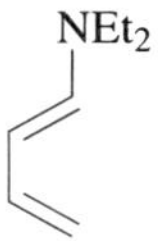 NEt$_2$ butadiene	CCl$_4$, rt	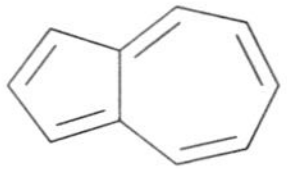 azulene (14)	107

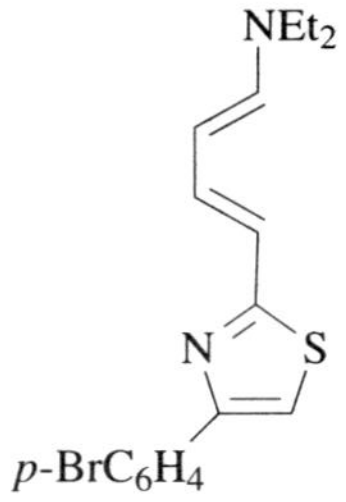

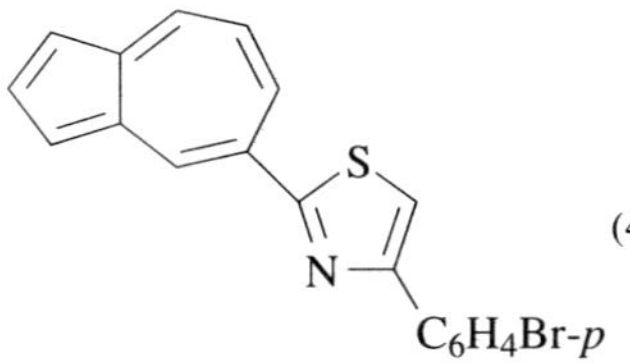

TABLE II. [6+4] CYCLOADDITIONS OF FULVENE DERIVATIVES (*Continued*)

A. *All-Carbon Trienophiles*

Triene	Trienophile	Conditions	Product(s) and Yield(s) (%)	Refs.
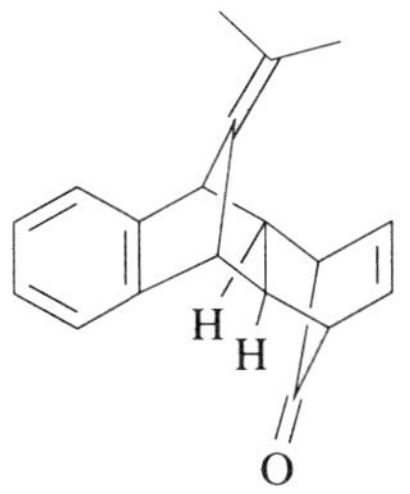 C₁₉	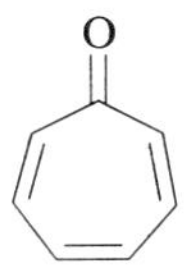	120°	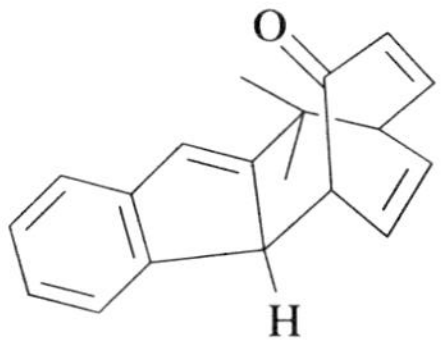 (100)	113, 78
C₃₀		rt	(—)	79

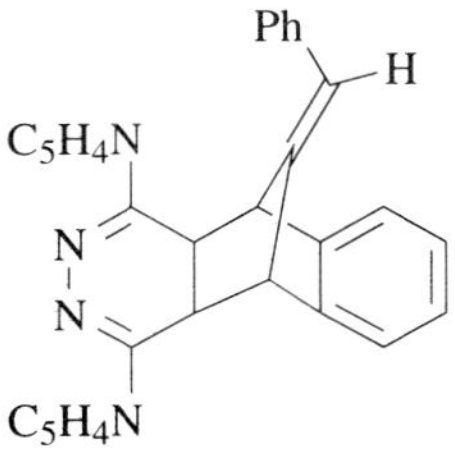

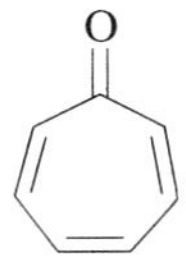

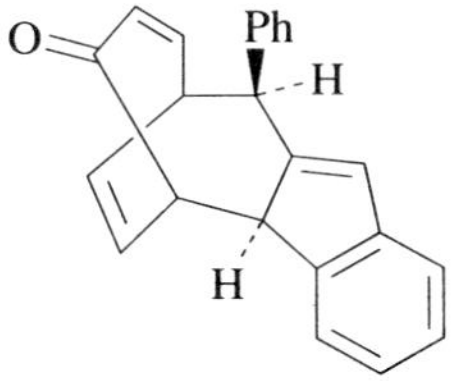

387

TABLE II. [6+4] CYCLOADDITIONS OF FULVENE DERIVATIVES (*Continued*)

B. *1,3-Dipolar Trienophiles*

Triene	Trienophile	Conditions	Product(s) and Yield(s) (%)	Refs.
C$_8$ 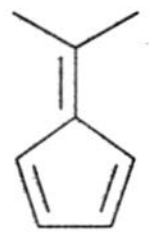	CH$_2$N$_2$	Et$_2$O, 0°, 7 d	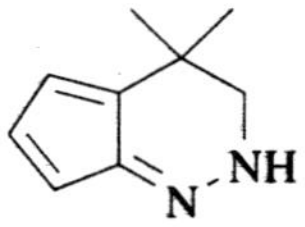 (—)	37, 114
	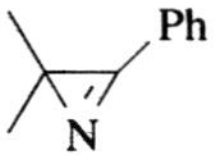	Cyclohexane, *hv*	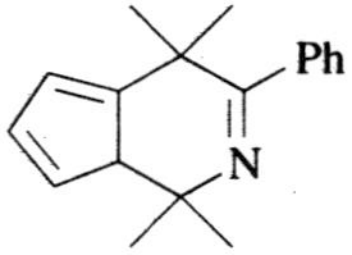 (—)	115
	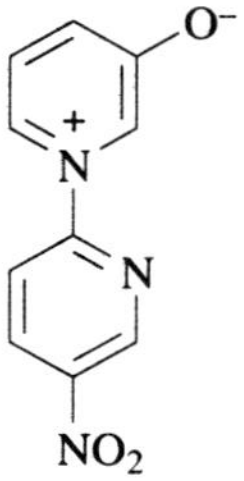	(—)	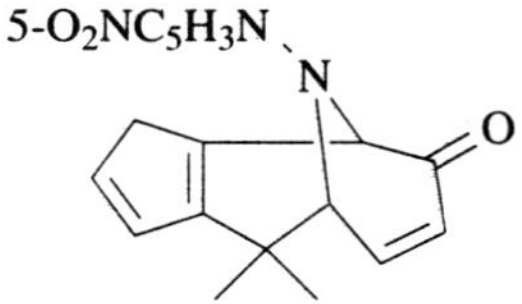 (—)	116
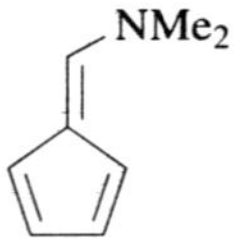	Ph–C≡N–O⁻	Et$_2$O	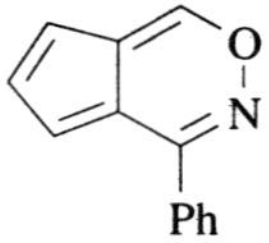 (60)	117

TABLE II. [6+4] CYCLOADDITIONS OF FULVENE DERIVATIVES (*Continued*)

B. 1,3-Dipolar Trienophiles

Triene	Trienophile	Conditions	Product(s) and Yield(s) (%)	Refs.
C$_{12}$ 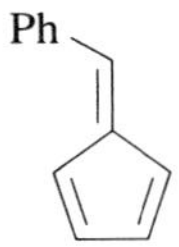	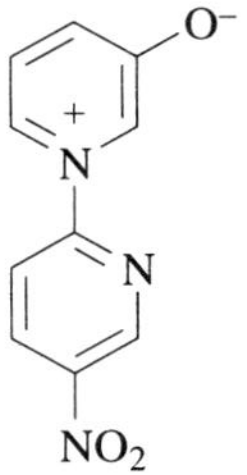	(—)	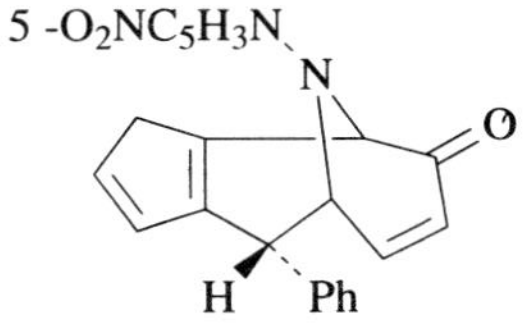 (—)	116
C$_{13}$ 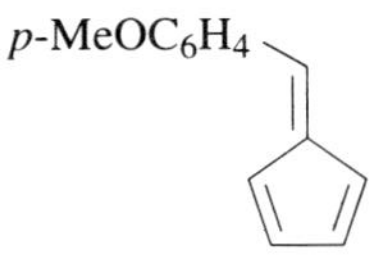	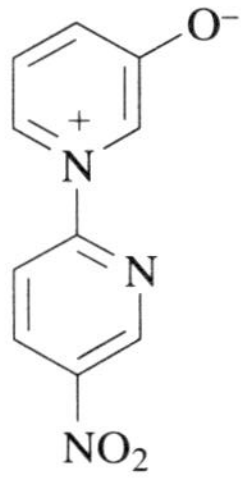	(—)	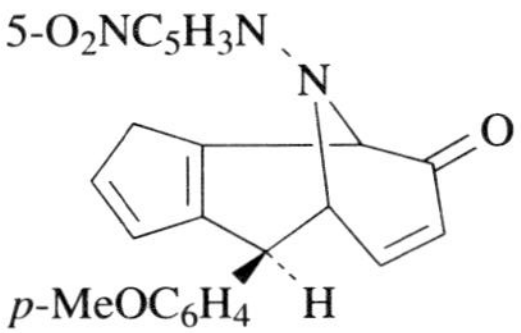 (—)	116

TABLE II. [6+4] CYCLOADDITIONS OF FULVENE DERIVATIVES (*Continued*)

C. Heterodienes

Triene	Trienophile	Conditions	Product(s) and Yield(s) (%)	Refs.
C$_8$ 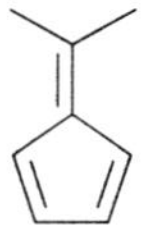	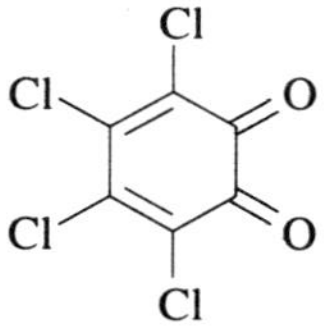	THF, rt, 3 h	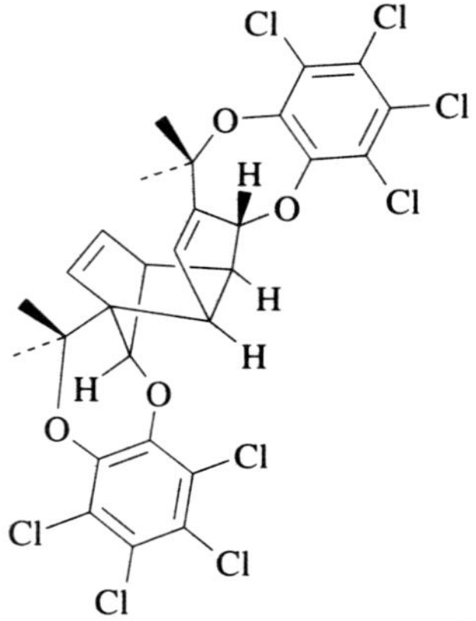 (23) 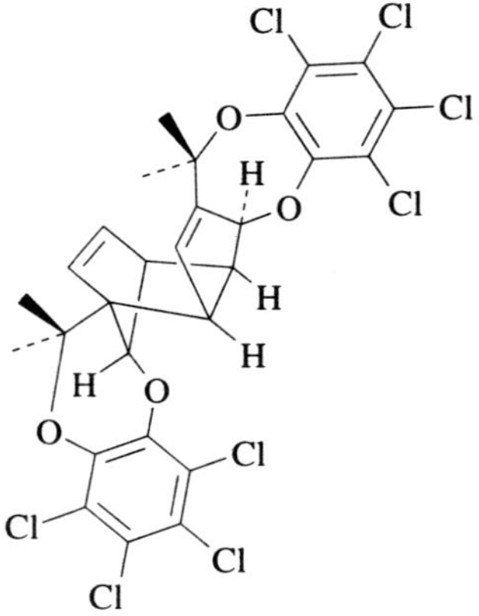 + (36)	118

TABLE II. [6+4] CYCLOADDITIONS OF FULVENE DERIVATIVES (*Continued*)

C. Heterodienes

Triene	Trienophile	Conditions	Product(s) and Yield(s) (%)	Refs.
NMe$_2$ (fulvene deriv.)	SO$_2$Ph / N=, =N–SO$_2$Ph quinone diimide	C$_6$H$_6$, rt	PhSO$_2$N / N–PhSO$_2$ fused ring (93)	119
	ROS–S(O$_2$)–SOR thiadiazole $\dfrac{R}{Me}$ / Et	Acetone, rt, 0.5 h Acetone, rt, 15 h	SONMe$_2$ / N=N / SOR (60) (—)	120
	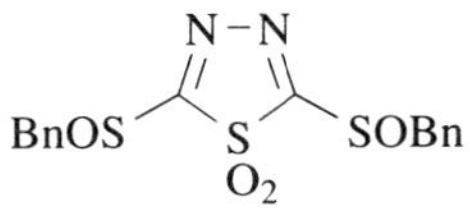 BnOS–S(O$_2$)–SOBn thiadiazole	CHCl$_3$, 50°, 15 h	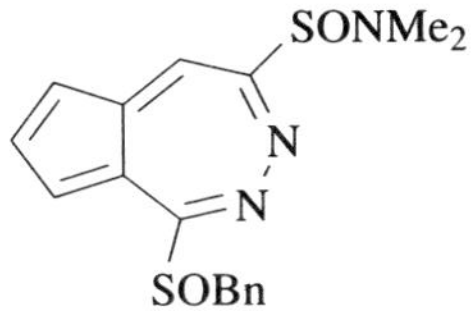 SONMe$_2$ / N–N / SOBn (—)	120
	Ph–(N=N)–Ph tetrazine	C$_6$H$_6$, rt, 5 d	Ph / N=N / Ph (40)	33

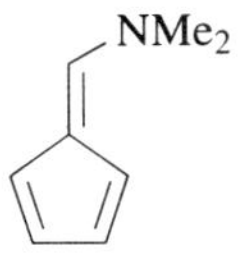

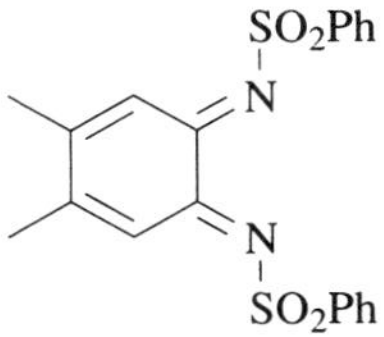

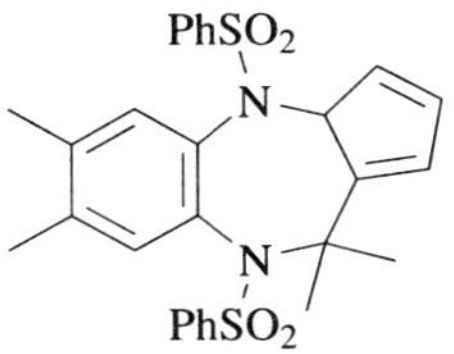

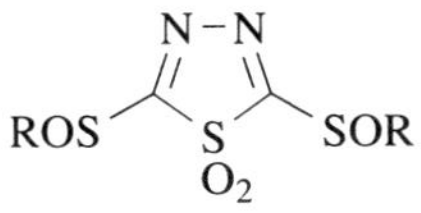

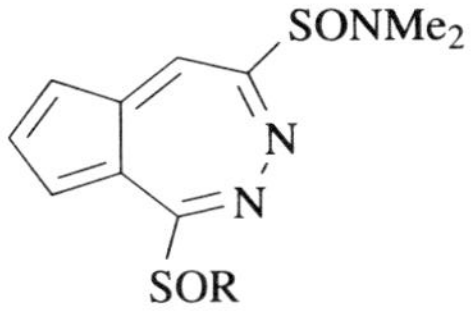

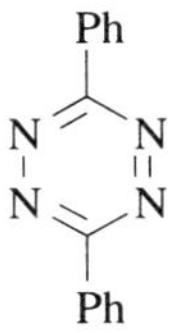

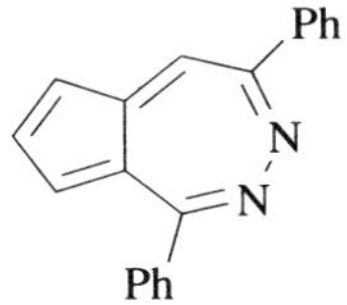

391

TABLE II. [6+4] CYCLOADDITIONS OF FULVENE DERIVATIVES (*Continued*)

C. Heterodienes

Triene	Trienophile	Conditions	Product(s) and Yield(s) (%)	Refs.
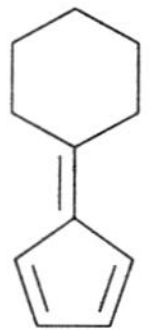 C₁₁	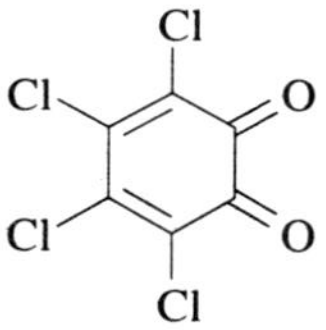	DME, rt, 30 h	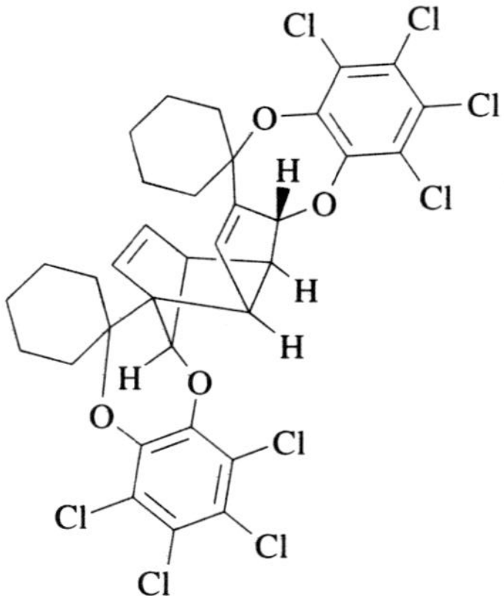 (18) 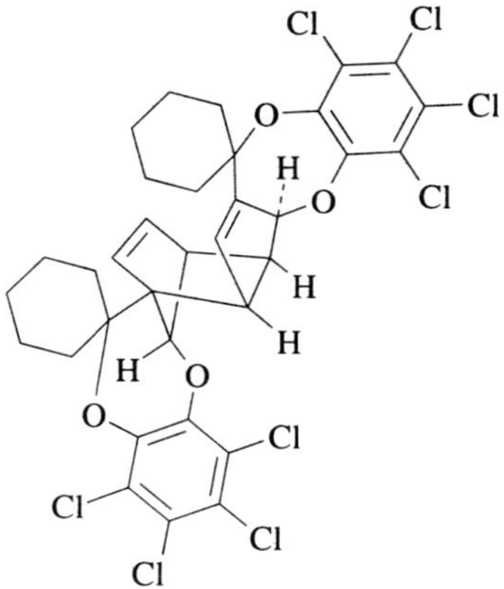 + (25)	118

TABLE II. [6+4] CYCLOADDITIONS OF FULVENE DERIVATIVES (*Continued*)

C. *Heterodienes*

Triene	Trienophile	Conditions	Product(s) and Yield(s) (%)	Refs.
C$_{12}$ 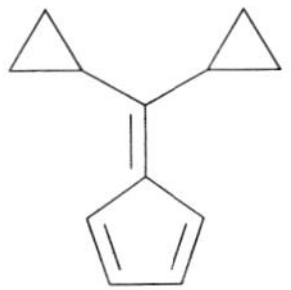	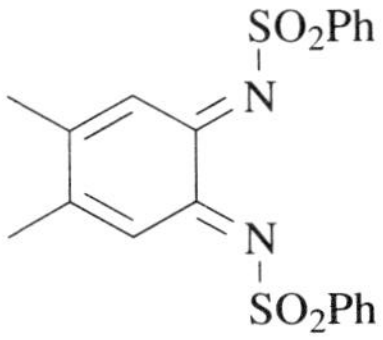	C$_6$H$_6$, rt	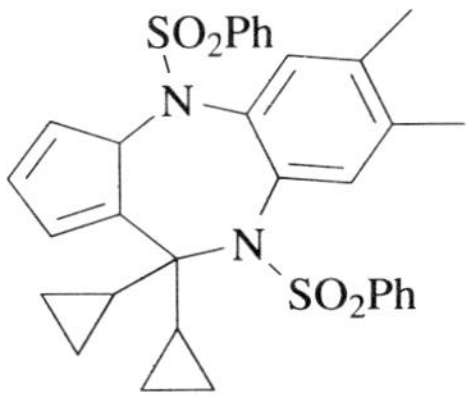 (69)	119, 121
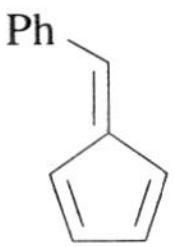	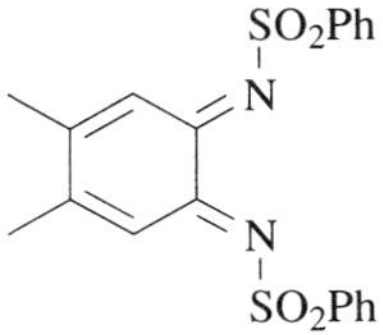	MeCN, rt, 24 h	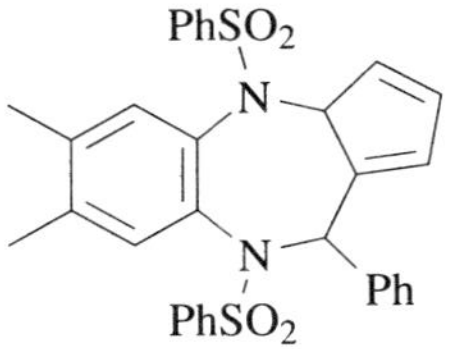 (71)	122
C$_{13}$ 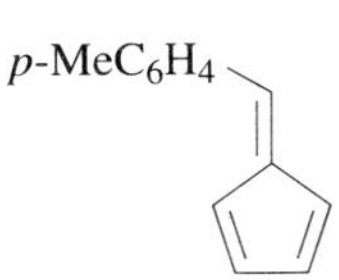 p-MeC$_6$H$_4$	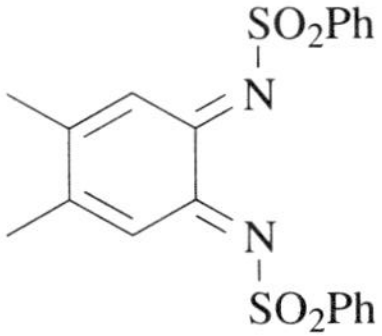	MeCN, rt, 24 h	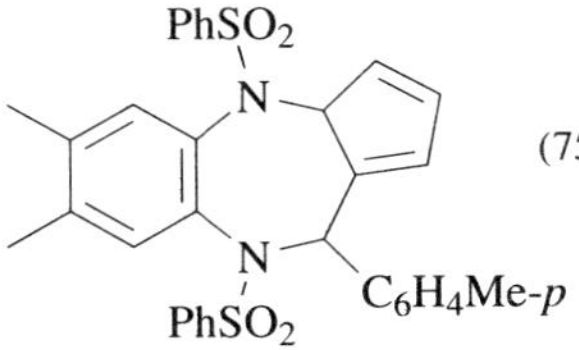 (75)	122
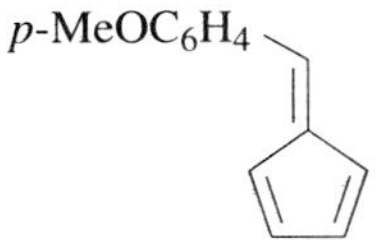 p-MeOC$_6$H$_4$	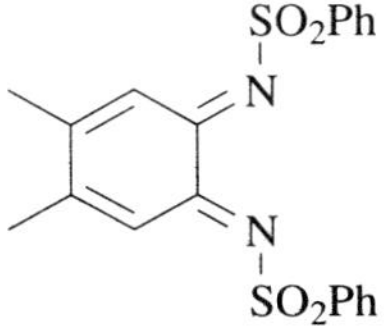	MeCN, rt, 24 h	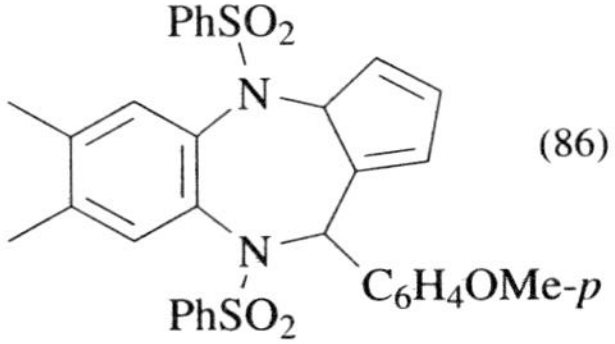 (86)	122

393

TABLE II. [6+4] CYCLOADDITIONS OF FULVENE DERIVATIVES (*Continued*)

C. Heterodienes

Triene	Trienophile	Conditions	Product(s) and Yield(s) (%)	Refs.
C_{14} 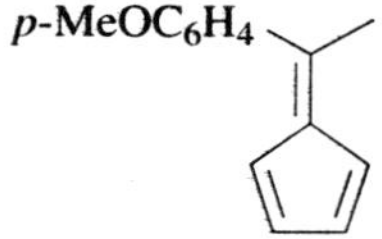 *p*-MeOC$_6$H$_4$ (fulvene)	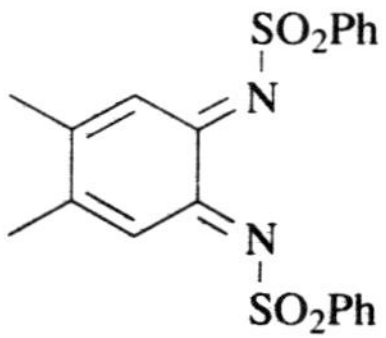 (diimine, SO$_2$Ph)	C$_6$H$_6$, rt	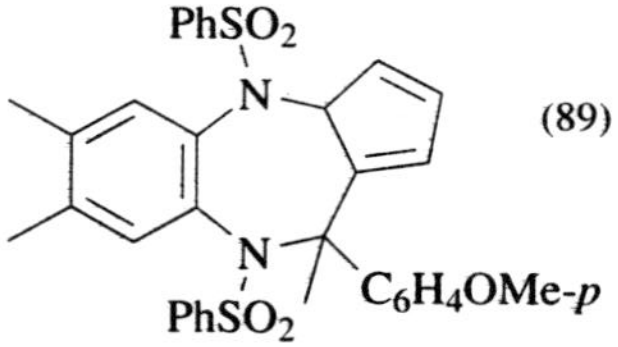 (89)	119
C_{18} Ph, Ph (fulvene)	SO$_2$Ph $\dfrac{R}{\text{Me}}$ $\dfrac{}{\text{Cl}}$	C$_6$H$_6$, rt C$_6$H$_6$, rt	(49) (62)	121, 119
R = *p*-ClC$_6$H$_4$	SO$_2$R^2 R^1 R^2 Me Ph Me *p*-O$_2$NC$_6$H$_4$ Me *p*-MeOC$_6$H$_4$	C$_6$H$_6$, rt C$_6$H$_6$, rt C$_6$H$_6$, rt	(83) (50) (55)	119, 121

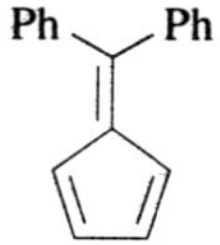

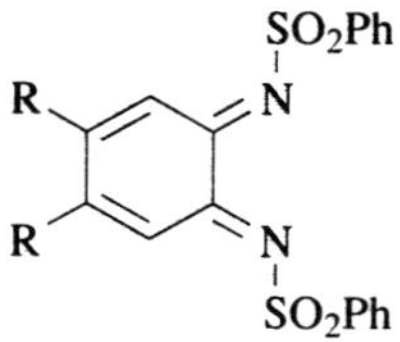

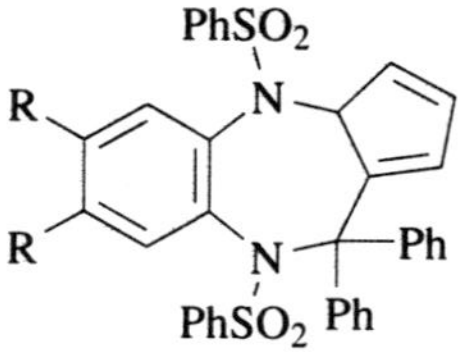

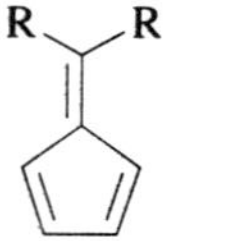

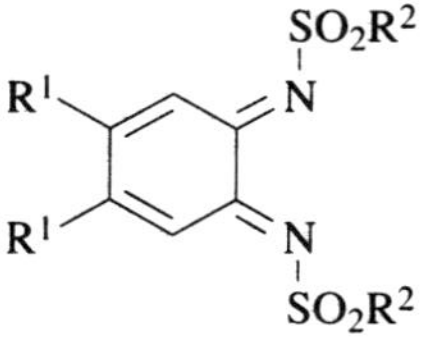

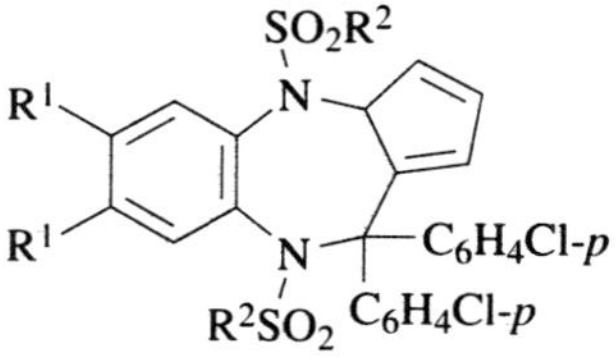

Product structures:

- PhSO$_2$–N ... N–C$_6$H$_4$OMe-*p*, PhSO$_2$ (89)
- PhSO$_2$–N ... N, Ph, PhSO$_2$, Ph
- SO$_2$R^2 ... N, C$_6$H$_4$Cl-*p*, C$_6$H$_4$Cl-*p*, R^{2}SO$_2$, R^1

TABLE II. [6+4] CYCLOADDITIONS OF FULVENE DERIVATIVES (*Continued*)

C. *Heterodienes*

Triene	Trienophile	Conditions	Product(s) and Yield(s) (%)	Refs.
R = p-O₂NC₆H₄		C₆H₆, rt C₆H₆, rt	(92) (70)	119, 121
			R¹ = Me, Cl	
C₂₀ R = p-MeC₆H₄		C₆H₆, rt	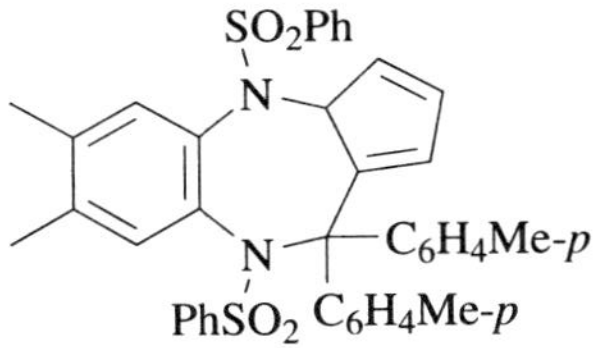 (60)	119, 121
C₃₀		C₆H₆, rt, 10 h	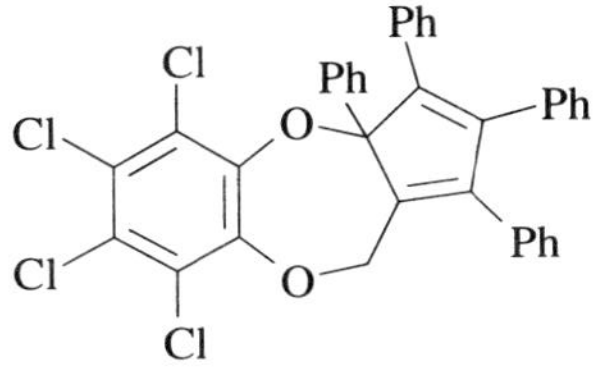 (24)	122a

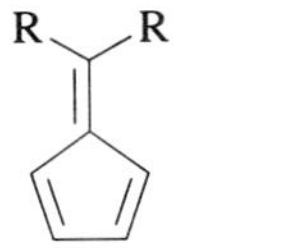

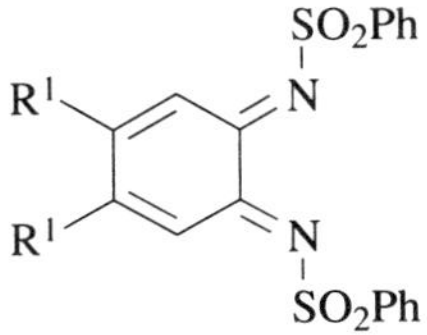

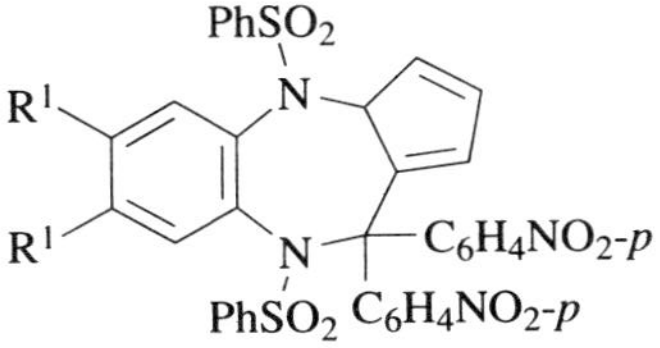

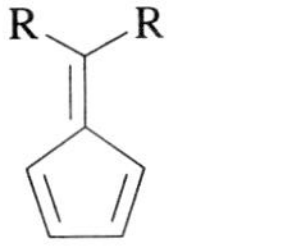

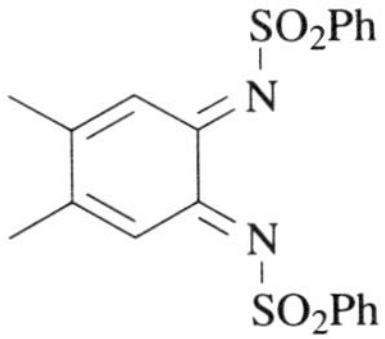

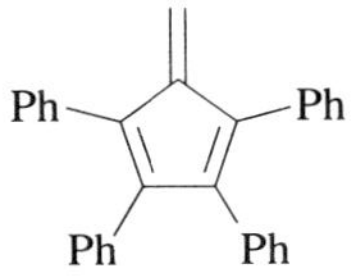

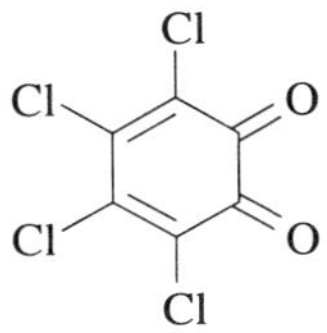

395

TABLE II. [6+4] CYCLOADDITIONS OF FULVENE DERIVATIVES (*Continued*)

C. *Heterodienes*

Triene	Trienophile	Conditions	Product(s) and Yield(s) (%)	Refs.
C_{32}				
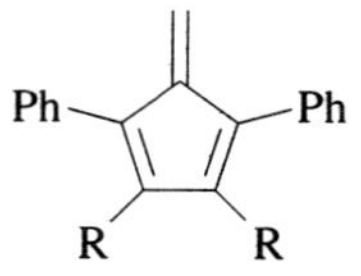 R = p-MeC_6H_4	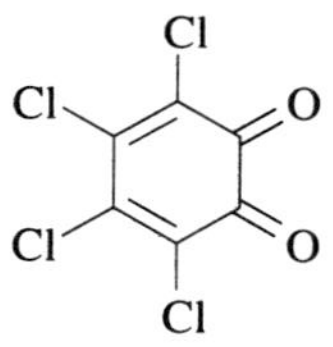	C_6H_6, rt	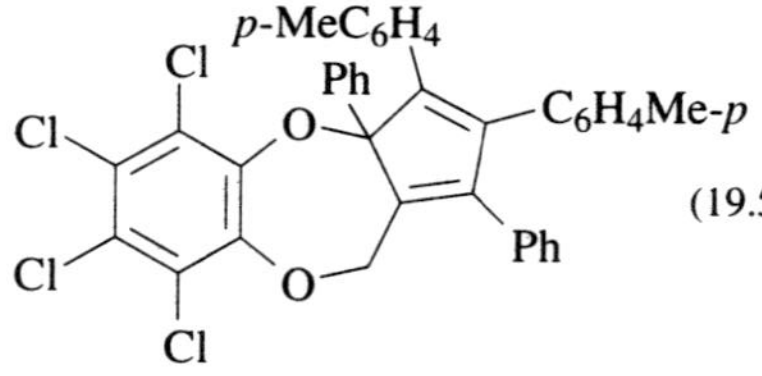 (19.5)	122a
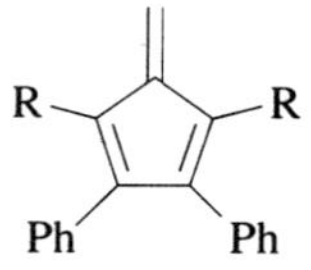 R = p-MeC_6H_4	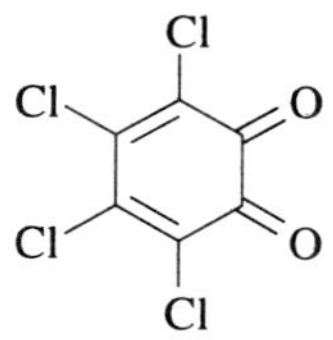	C_6H_6, rt	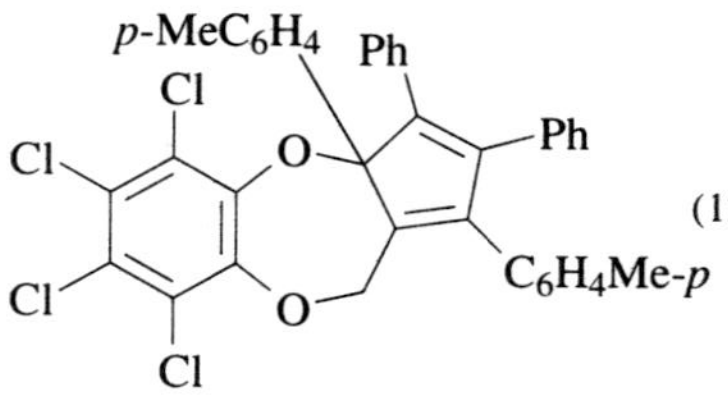 (12.5)	122a

TABLE II. [6+4] CYCLOADDITIONS OF FULVENE DERIVATIVES (*Continued*)

D. *Intramolecular Reactions*

Triene	Trienophile	Conditions	Product(s) and Yield(s) (%)	Refs.
C$_{12}$ 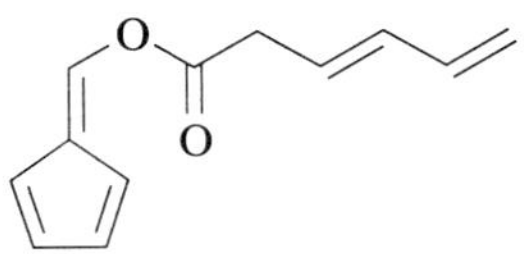		Toluene, 160°	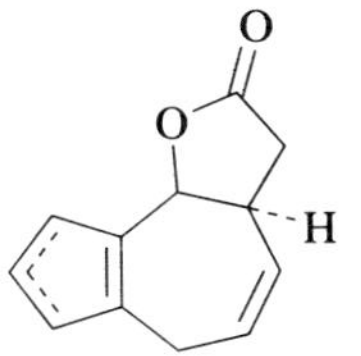 (52)	123
C$_{13}$ 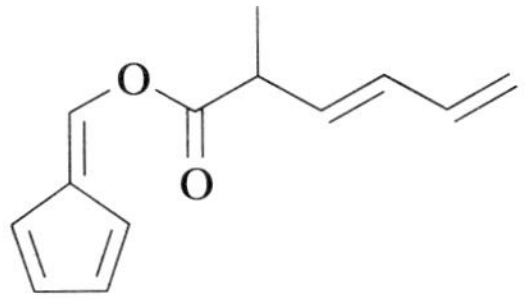		Toluene, 160°	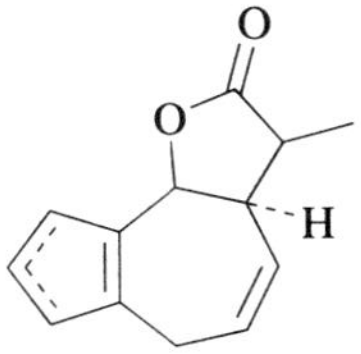 (38)	123
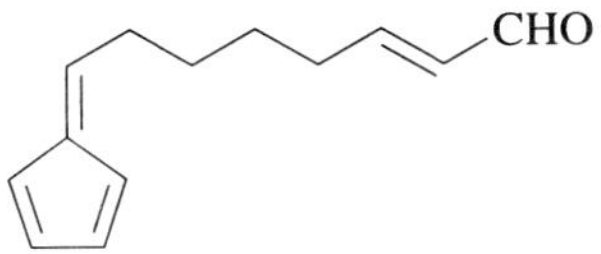		1. Et$_2$NH, K$_2$CO$_3$, C$_6$H$_6$ 2. C$_6$H$_6$, 40°	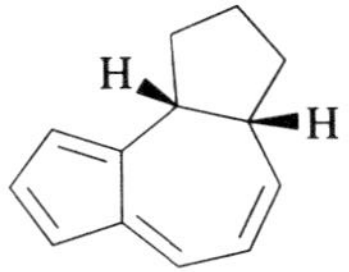 (55)	36
C$_{14}$ 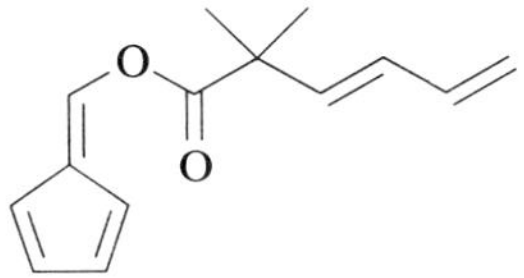		Toluene, 160°, 4 h	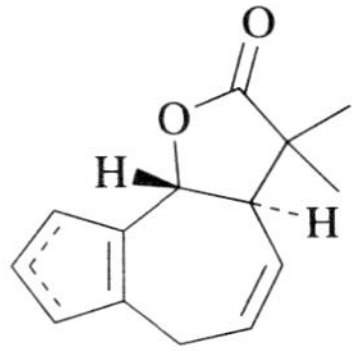 (56)	123

397

TABLE II. [6+4] CYCLOADDITIONS OF FULVENE DERIVATIVES (*Continued*)

D. Intramolecular Reactions

Triene	Trienophile	Conditions	Product(s) and Yield(s) (%)	Refs.
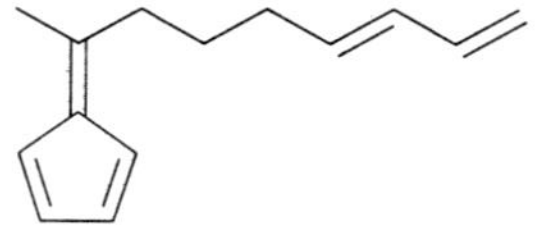		Toluene, 210°, 48 h	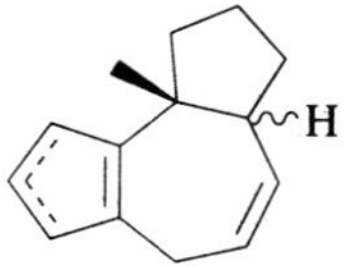 (45)	36
	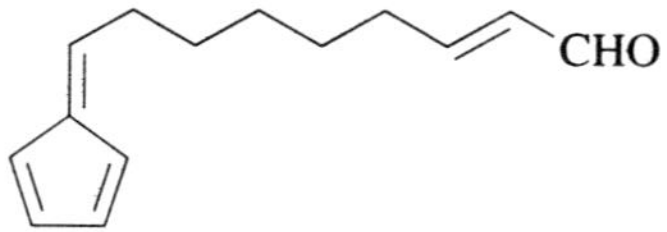 CHO	1. Et$_2$NH, K$_2$CO$_3$, C$_6$H$_6$ 2. C$_6$H$_6$, 40°	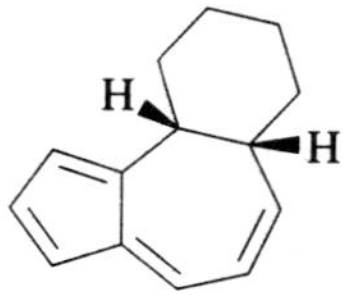 (46)	36
	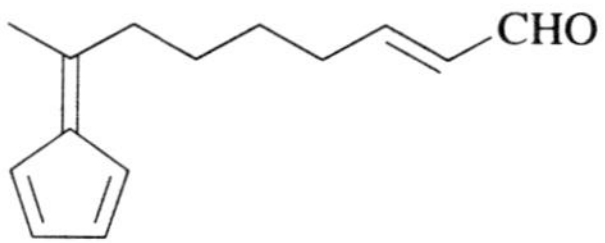 CHO	1. Et$_2$NH, K$_2$CO$_3$, C$_6$H$_6$ 2. C$_6$H$_6$, 40°	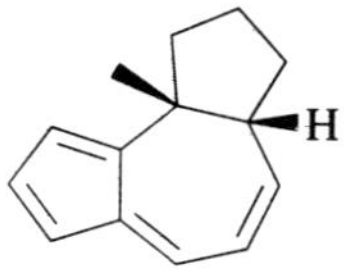 (54)	36
C$_{17}$	OTMS	C$_6$H$_6$, 40°	(—)	36

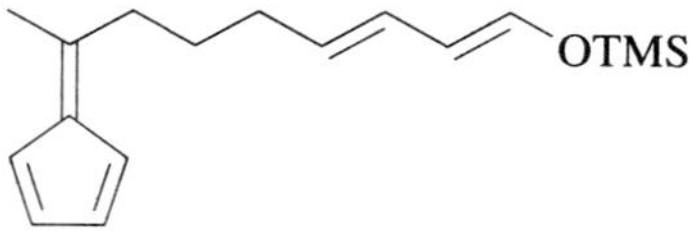

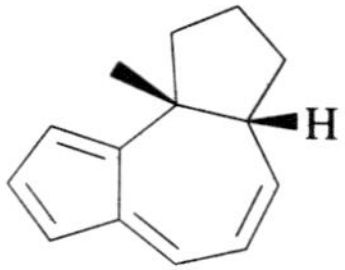

TABLE II. [6+4] CYCLOADDITIONS OF FULVENE DERIVATIVES (*Continued*)

D. Intramolecular Reactions

Triene	Trienophile	Conditions	Product(s) and Yield(s) (%)	Refs.
C_{18} 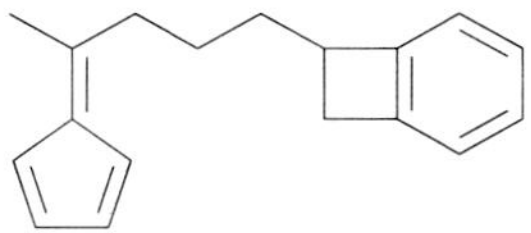		Toluene, 195°, 12 h	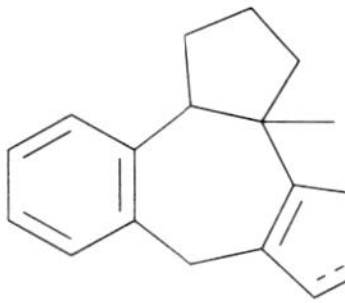 (44)	35
C_{19} 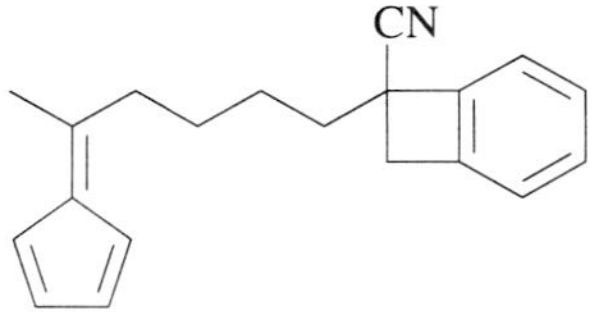		o-Cl$_2$C$_6$H$_4$, 180°, 7 h	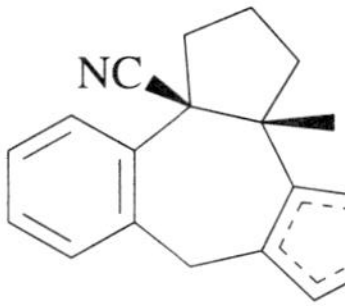 (60)	35

TABLE III. MISCELLANEOUS [6+4] CYCLOADDITION REACTIONS

	Triene	Trienophile	Conditions	Product(s) and Yield(s) (%)	Refs.
C$_5$ 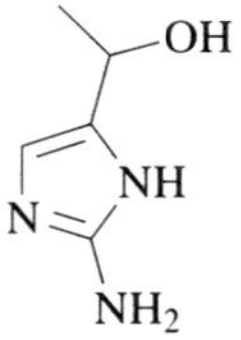			H$_2$SO$_4$, 100°, 12 h	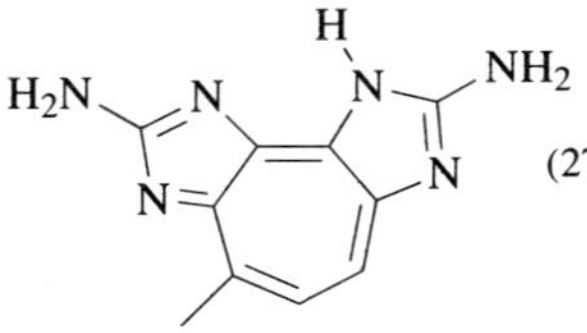 (27)	124
C$_6$ 			C$_6$H$_6$, rt, 1 d	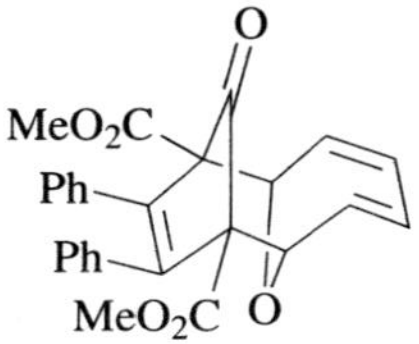 (12)	125

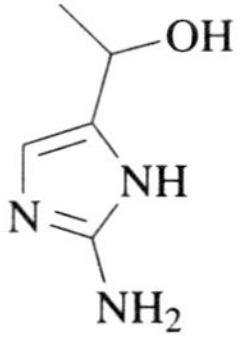

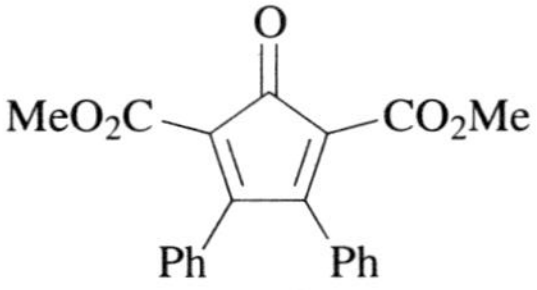

400

TABLE III. MISCELLANEOUS [6+4] CYCLOADDITION REACTIONS (*Continued*)

Triene	Trienophile	Conditions	Product(s) and Yield(s) (%)	Refs.
C$_7$	S$_3$	Sulfolane, pyridine, 70°, 72 h	(21)	126
	CO$_2$H (pyranone)	Xylene, 180°	CO$_2$H (1)	127
	Ph—N=CH$_2$ Ph Li$^+$	Et$_2$O, −45°	HN Ph Ph (47)	128
	(anthracene)	C$_6$H$_6$, rt, *hv*	(34)	129
	CN (anthracene)	C$_6$H$_6$, rt, *hv*	CN (4)	130

TABLE III. MISCELLANEOUS [6+4] CYCLOADDITION REACTIONS (Continued)

Triene	Trienophile	Conditions	Product(s) and Yield(s) (%)	Refs.
	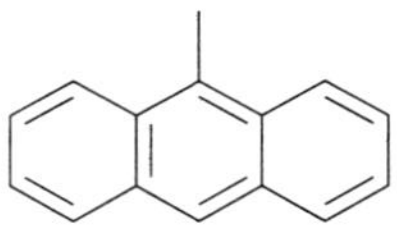	C$_6$H$_6$, rt, hv	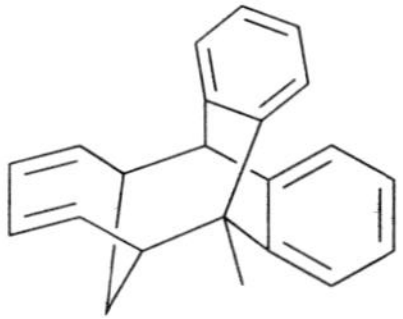 (9)	131
	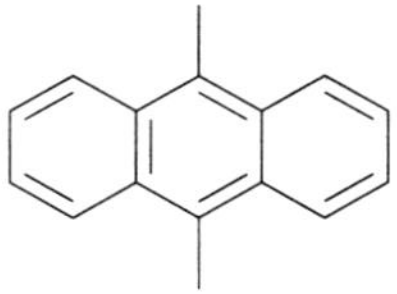	C$_6$H$_6$, rt, hv	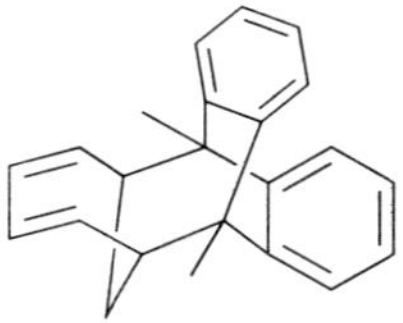 (—)	131
	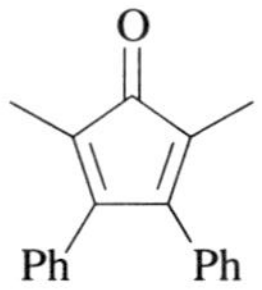	170°, 36 h	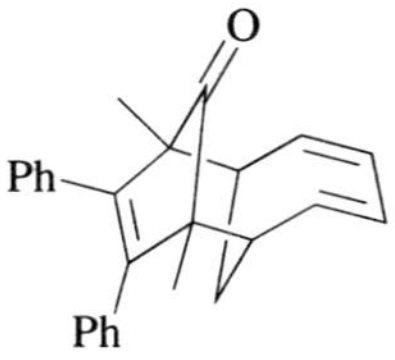 (33)	132
		80°, 39 h rt, 108 h	(21) (57)	133

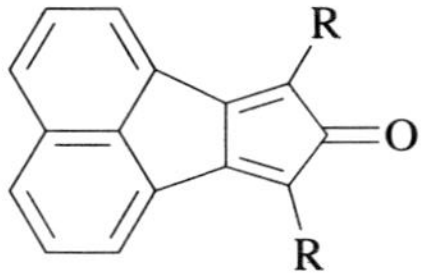

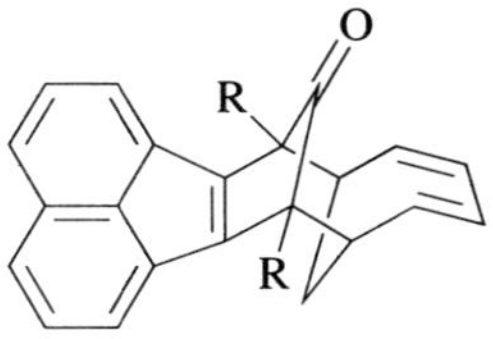

$$
\begin{array}{c}
\text{R} \\ \hline
\text{Et} \\
\text{CO}_2\text{Et}
\end{array}
$$

TABLE III. MISCELLANEOUS [6+4] CYCLOADDITION REACTIONS (*Continued*)

Triene	Trienophile	Conditions	Product(s) and Yield(s) (%)	Refs.
	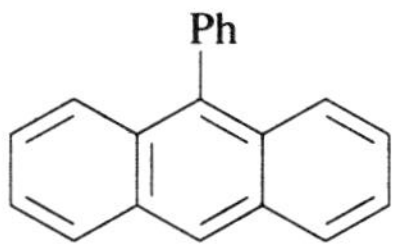	C$_6$H$_6$, rt, *hv*	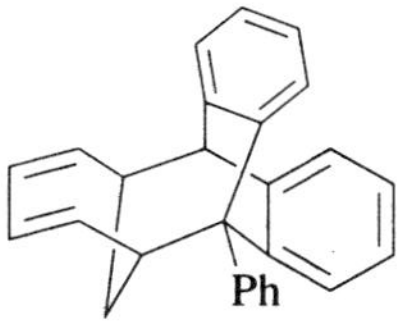 (38)	131
	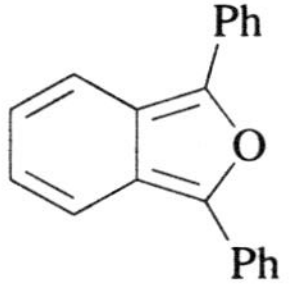	EtOH, *hv*	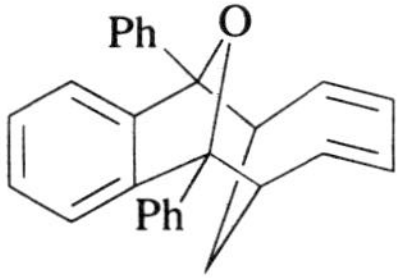 (16) + 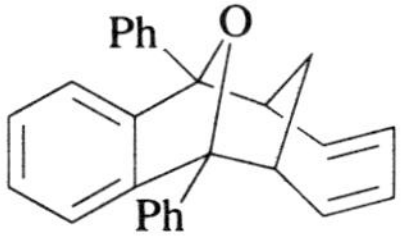 (20)	134, 135
	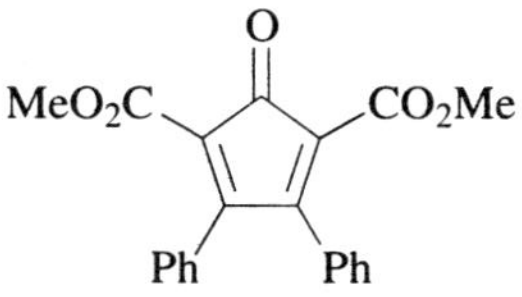	rt, 10 d	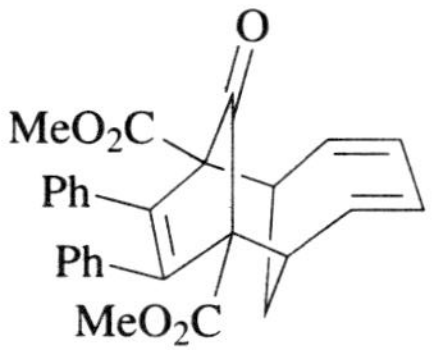 (57)	136
C$_8$ 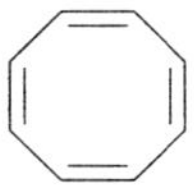	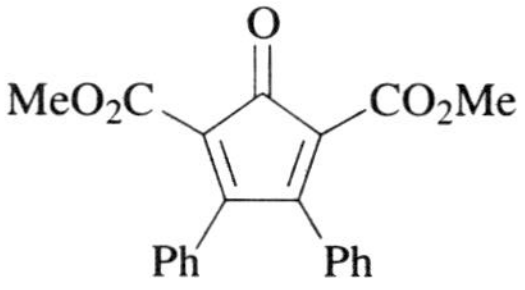	C$_6$H$_6$, 80°	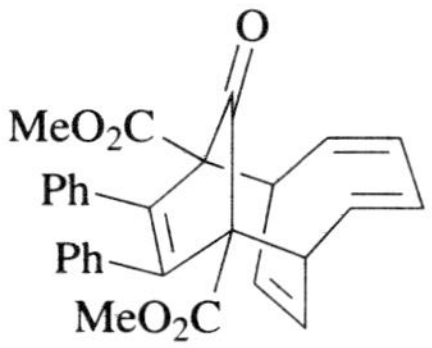 (46)	137, 138

TABLE III. MISCELLANEOUS [6+4] CYCLOADDITION REACTIONS (Continued)

Triene	Trienophile	Conditions	Product(s) and Yield(s) (%)	Refs.
C₉ (N-CO₂Et azepine)	(acenaphthylene-fused cyclopentenone diester, CO₂Et / O / CO₂Et)	C₆H₆, 80°, 8 h	(EtO₂C, EtO₂C, O product) (16)	136
	(tetrachloro-o-benzoquinone, Cl Cl Cl Cl, O O)	C₆H₆, rt, 70 min	(NCO₂Et, O O, Cl Cl Cl Cl product) (46)	139
	(2-pyranone, CO₂Me, O O)	C₆H₆, 80°, 57 h	(MeO₂C, NCO₂Et, O lactone product) (20)	140, 141
	(N-CO₂Et azepine, CO₂Et–N)	125°	(EtO₂CN, NCO₂Et product) (—)	142

404

TABLE III. MISCELLANEOUS [6+4] CYCLOADDITION REACTIONS (*Continued*)

Triene	Trienophile	Conditions	Product(s) and Yield(s) (%)	Refs.
		Xylene, 140°, 6 d	(7)	143
		Xylene, 140°, 6 d	(30)	143
		DMSO, 120°	(87)	144
		DMSO, 120°	(92)	144
C$_{10}$		DMSO, 120°	(91)	144

TABLE III. MISCELLANEOUS [6+4] CYCLOADDITION REACTIONS (*Continued*)

Triene	Trienophile	Conditions	Product(s) and Yield(s) (%)	Refs.
(C$_{11}$)	$Ph\!-\!N\!-\!Ph$ (H)	DMSO, 120°	(82)	144
	t-Bu	CH$_2$Cl$_2$, rt	(40)	145
C$_{13}$	NEt$_2$	C$_6$H$_6$	(31)	146
	NEt$_2$	C$_6$H$_6$	(66)	146

406

TABLE III. MISCELLANEOUS [6+4] CYCLOADDITION REACTIONS (*Continued*)

Triene	Trienophile	Conditions	Product(s) and Yield(s) (%)	Refs.
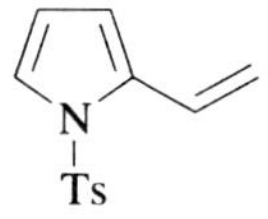 (vinyl pyrrole N-Ts structure)	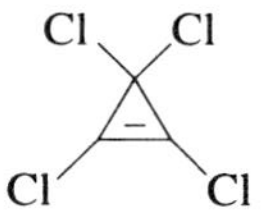 (tetrachlorocyclopropene structure)	CCl₄, 69°	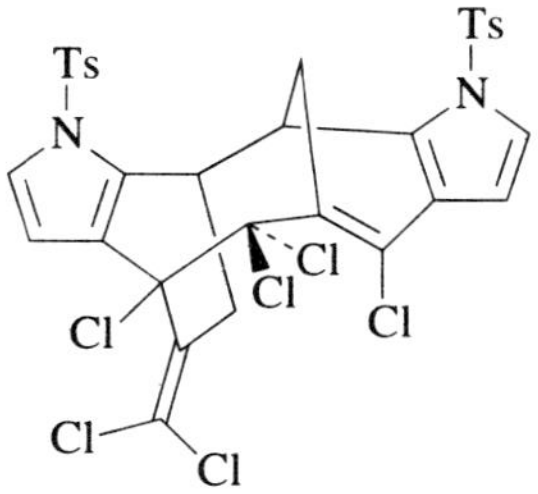 (—)	147
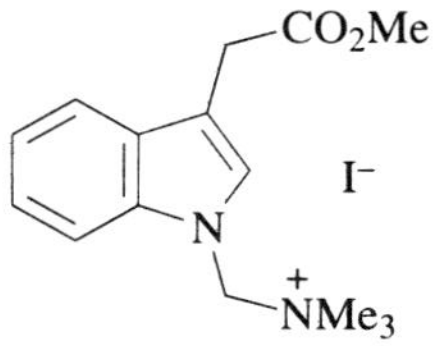 C₁₅ (indole CO₂Me / NMe₃⁺ I⁻ structure)	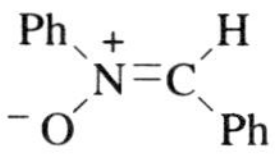 (nitrone Ph-N⁺=C structure)	DMSO, 120°	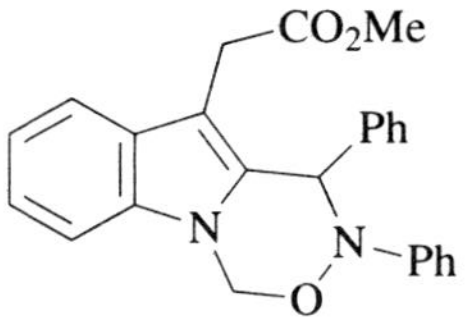 (22)	144

407

TABLE IV. METAL-MEDIATED [6+4] CYCLOADDITIONS

Triene	Trienophile	Conditions	Product(s) and Yield(s) (%)	Refs.
C$_6$				
		1. C$_6$H$_{14}$, CH$_2$Cl$_2$, $h\nu$ 2. O$_2$, Et$_2$O	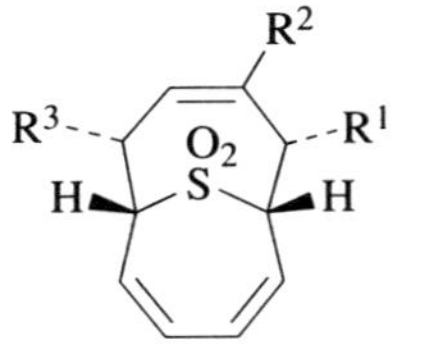	23, 42
	R^1 R^2 R^3 H Me H (77) OAc H H (78) CO$_2$Me H H (38) CO$_2$Me H Me (21) OTMS H H (65)			
		CH$_2$Cl$_2$, rt	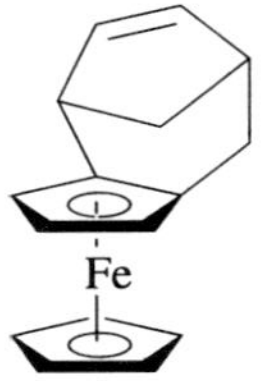 　(—)	148

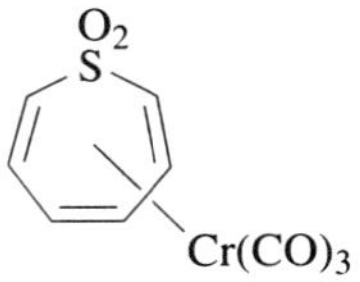

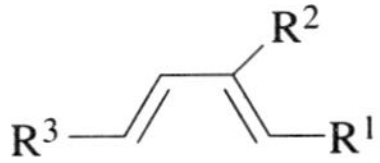

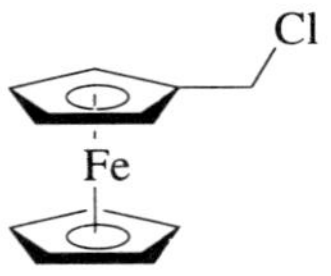

TABLE IV. METAL-MEDIATED [6+4] CYCLOADDITIONS (Continued)

Conditions legend:

A. 1. C_5H_{12}, hv
   2. Me_3P
B. 1. C_6H_{14}, hv
   2. $(MeO)_3P$
C. $n\text{-}Bu_2O$, 142°

Triene (C7): cycloheptatriene–$Cr(CO)_3$

Trienophile: R^3, R^2, R^1, R^4 substituted diene

Product: bicyclic adduct with R^1, R^2, R^3, R^4 and two H

R^1	R^2	R^3	R^4	Conditions	Product(s) and Yield(s) (%)	Refs.
H	H	H	H	A	(21)	149, 150
Me	H	H	H	A	(38)	150
H	Me	H	H	A	(59)	150
Me	H	H	Me	B	(86)	150
Me	H	H	Me	C	(70)	23
H	Me	Me	H	A	(32)	23
CH=CH2	H	H	H	A	(41)	150
OMe	H	H	H	B	(64)	150
OAc	H	H	H	B	(67)	22
OAc	H	H	H	C	(59)	22
CO2Me	H	H	H	B	(83)	23
CO2Me	H	H	H	C	(55)	22
OTMS	H	H	H	B	(86)	23
H	OTMS	H	H	B	(82)	23
CO2Me	H	H	Me	B	(96)	23
CO2Me	H	H	Me	C	(60)	23
OAc	H	H	O Ac	B	(65)	22, 23
CO2Me	H	H	CO2Me	B	(89)	23

409

TABLE IV. METAL-MEDIATED [6+4] CYCLOADDITIONS (Continued)

Triene	Trienophile	Conditions	Product(s) and Yield(s) (%)	Refs.
		1. C$_6$H$_{14}$, hv 2. (MeO)$_3$P	(15) +	22
			(45)	
		1. C$_5$H$_{12}$, hv 2. (MeO)$_3$P	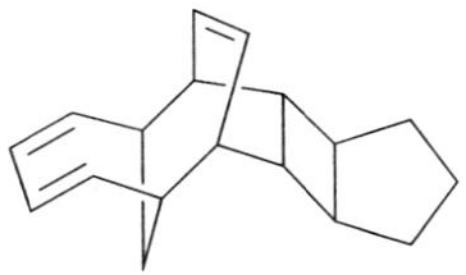 (21)	151
		1. C$_6$H$_{14}$, hv 2. (MeO)$_3$P	(63)	22

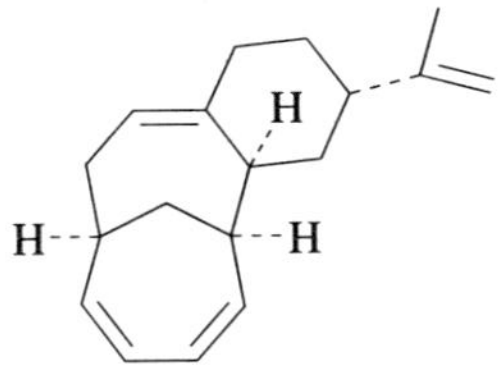

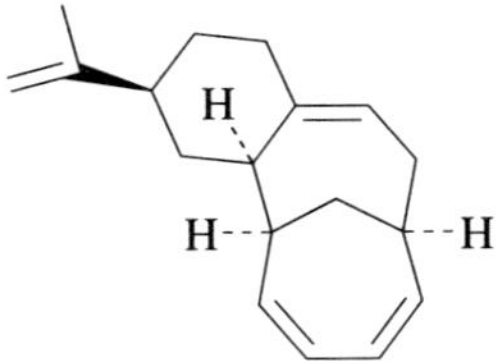

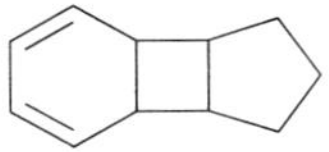

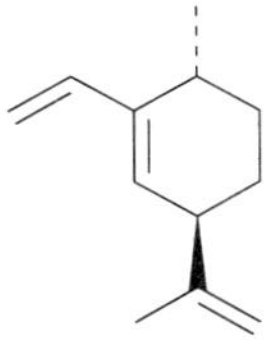

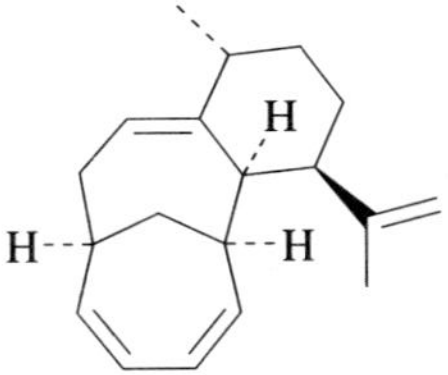

TABLE IV. METAL-MEDIATED [6+4] CYCLOADDITIONS (*Continued*)

Triene	Trienophile	Conditions	Product(s) and Yield(s) (%)	Refs.
		1. C_6H_{14}, *hv* 2. $(MeO)_3P$	(50)	23
			(75)	
			(74)	

TABLE IV. METAL-MEDIATED [6+4] CYCLOADDITIONS (*Continued*)

Triene	Trienophile	Conditions	Product(s) and Yield(s) (%)	Refs.
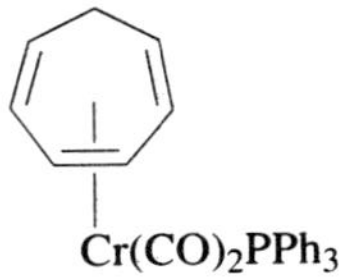 Cr(CO)$_2$PPh$_3$	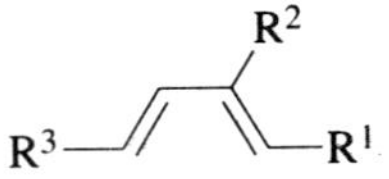 R^2 — R^1 (R^3) R^1 R^2 R^3 CO$_2$Me H Me H OTMS H	1. C$_6$H$_{14}$, *hv* 2. (MeO)$_3$P	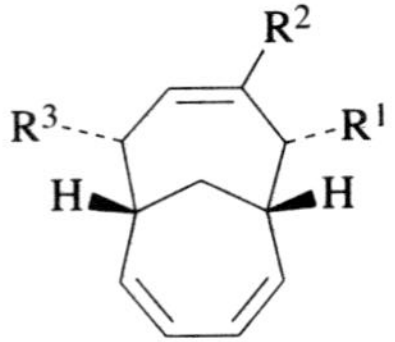 R^2 R^1 H / R^3 H (72) (53)	23
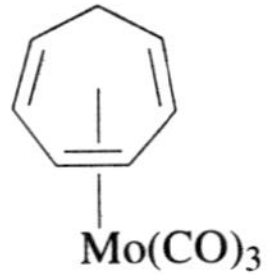 Mo(CO)$_3$	 R^2 — R^1 R^1 R^2 OTMS H Me H Me CO$_2$Me	1. C$_6$H$_{14}$, *hv* 2. (MeO)$_3$P	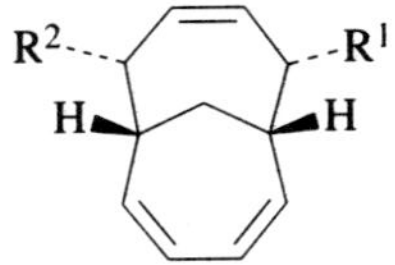 R^1 H / R^2 H (51) (32) (27)	23
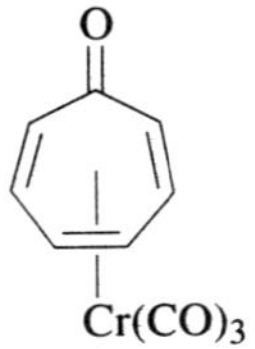 O= Cr(CO)$_3$	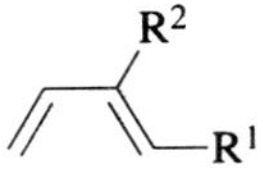 R^2 — R^1 R^1 R^2 Me H H Me	1. C$_6$H$_{14}$, *hv* 2. (MeO)$_3$P	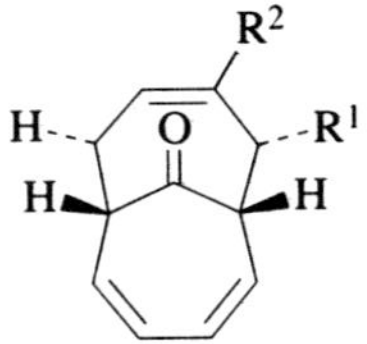 R^2 R^1 H / H H (20) (40)	23

TABLE IV. METAL-MEDIATED [6+4] CYCLOADDITIONS (Continued)

Triene	Trienophile	Conditions	Product(s) and Yield(s) (%)	Refs.
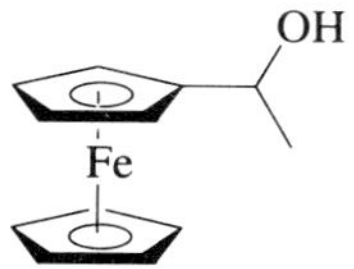		CH_2Cl_2, Ph_3CBF_4	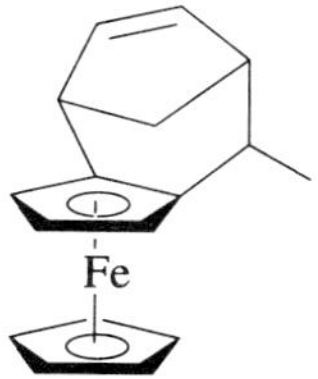 (—)	148
C_8	 R¹ / R² : OMe / H ; Me / CO₂Me	1. C_6H_{14}, $h\nu$ 2. $(MeO)_3P$	(47) (97)	22 23
	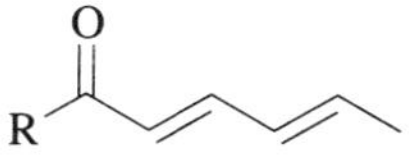 R =	1. C_6H_{14}, $h\nu$ 2. $(MeO)_3P$	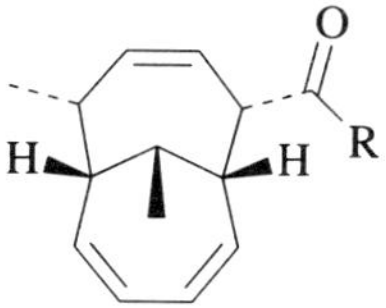 (75)	23

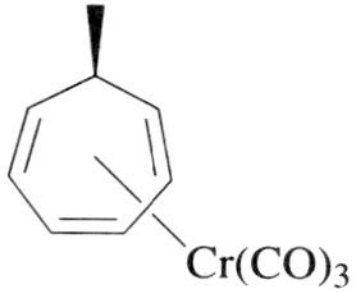

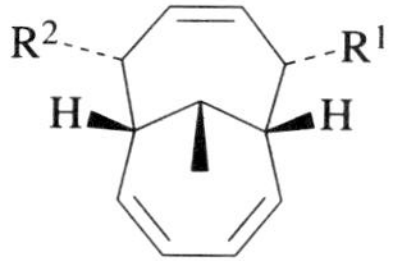

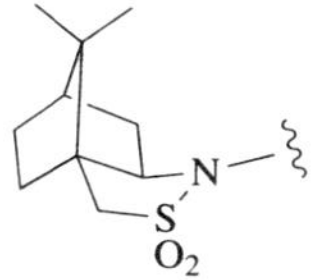

TABLE IV. METAL-MEDIATED [6+4] CYCLOADDITIONS (*Continued*)

Triene	Trienophile	Conditions	Product(s) and Yield(s) (%)	Refs.
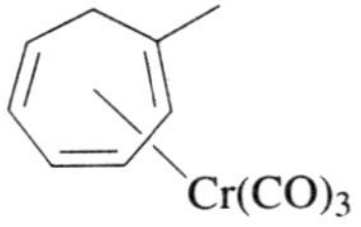	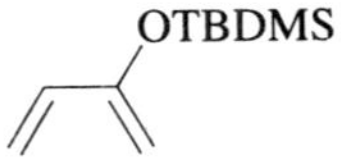 OTBDMS	1. C_6H_{14}, hv 2. $(MeO)_3P$	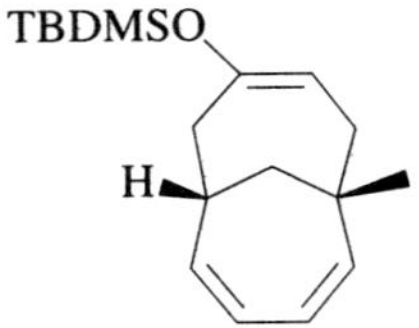 (35)	23
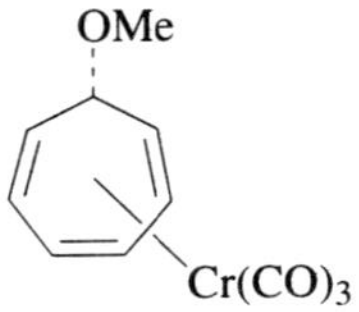	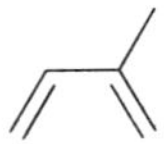	1. C_6H_{14}, hv 2. $(MeO)_3P$	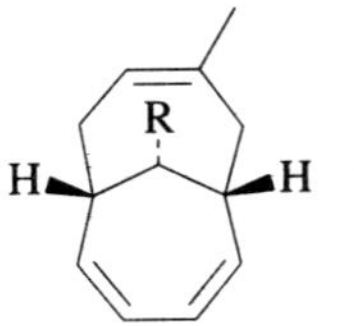 (66) R = OMe	22
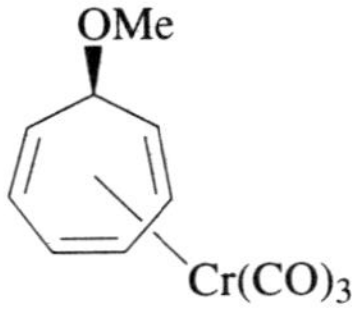	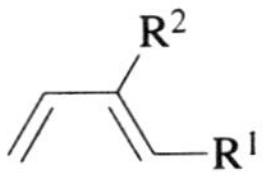 $\dfrac{R^1 \quad R^2}{OAc \quad H}$ $H \quad Me$	1. C_6H_{14}, hv 2. $(MeO)_3P$	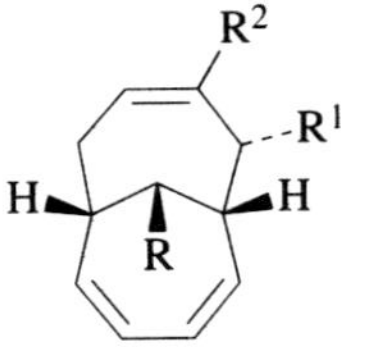 R = OMe (60) (60)	22

TABLE IV. METAL-MEDIATED [6+4] CYCLOADDITIONS (*Continued*)

Triene	Trienophile	Conditions	Product(s) and Yield(s) (%)	Refs.
OMe (cycloheptatriene) Cr(CO)$_3$	R^2, R^1 diene 	R^1	R^2
---	---
H	OTBDMS
Me	H
OTMS	H
H	Me		1. C$_6$H$_{14}$, *hv* 2. (MeO)$_3$P	bicyclic product (R^2, R^3, H, OMe, R^1) | R^1	R^2	R^3	
---	---	---	---
H	OTBDMS	H	(52)
Me	H	H	(45)
OTMS	H	H	(38)
H	Me	H	(36)
H	H	Me	(12)		23
OMe (cycloheptatriene) Cr(CO)$_3$	R^2, R^1 	R^1	R^2
---	---
Me	CO$_2$Me
OTMS	H		1. C$_6$H$_{14}$, *hv* 2. (MeO)$_3$P	bicyclic product (R^2, H, OMe, R^1) 	R^1	R^2	
---	---	---
Me	CO$_2$Me	(44)
CO$_2$Me	Me	(45)
OTMS	H	(45)
H	OTMS	(44)		23

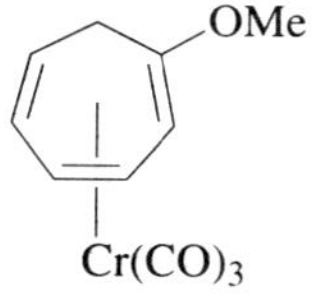

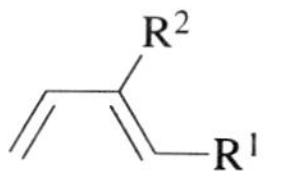

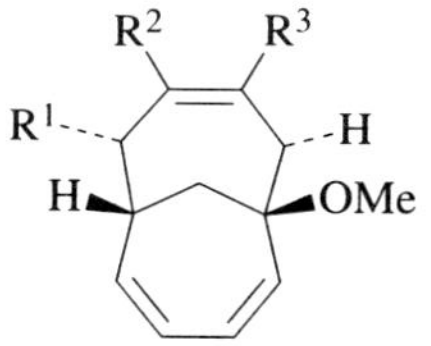

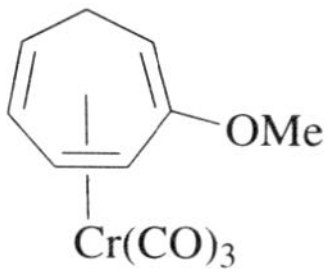

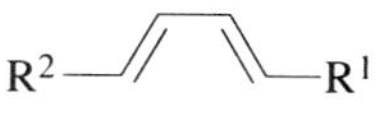

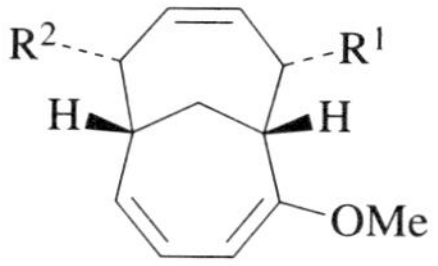

415

TABLE IV. METAL-MEDIATED [6+4] CYCLOADDITIONS (*Continued*)

Triene	Trienophile	Conditions	Product(s) and Yield(s) (%)	Refs.
MeO—Cr(CO)$_3$ (methoxycycloheptatriene chromium tricarbonyl)	diene with R^1, R^2, R^3 R^1 | R^2 | R^3 H | Me | Me OTMS | H | H	1. C$_6$H$_{14}$, *hv* 2. (MeO)$_3$P	bicyclic product with R^1, R^2, R^3, R^4, MeO R^1 | R^2 | R^3 | R^4 H | Me | Me | H (93) OTMS | H | H | H (40) H | H | H | OTMS (39)	23
2-methyltropone chromium tricarbonyl (O=, Cr(CO)$_3$)	diene with R^1, R^2 R^1 | R^2 H | Me	1. C$_6$H$_{14}$, *hv* 2. (MeO)$_3$P	two bicyclic products (O, R^1, R^2) (32.5) + (32.5) (0) + (23)	23
CO$_2$Me—N azepine chromium tricarbonyl, Cr(CO)$_3$	diene with R^1, R^2, R^3 R^1 | R^2 | R^3 H | OTMS | H OAc | H | H OTMS | H | H Me | H | CO$_2$Me	1. C$_6$H$_{14}$, Et$_2$O, *hv* 2. O$_2$, Et$_2$O	N–R bicyclic product, R = CO$_2$Me (87) (75) (79) (83)	23, 42

TABLE IV. METAL-MEDIATED [6+4] CYCLOADDITIONS (*Continued*)

Triene	Trienophile	Conditions	Product(s) and Yield(s) (%)	Refs.
		CH$_2$Cl$_2$, Ph$_3$C$^+$BF$_4^-$	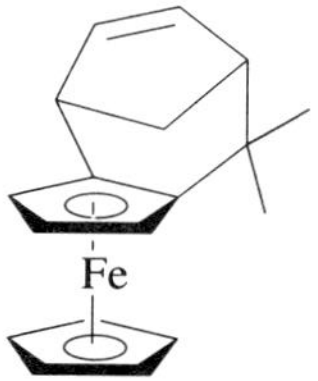 (—)	148
C$_9$	$\begin{array}{cc} R^1 & R^2 \\ \hline Me & H \\ H & Me \end{array}$	1. C$_6$H$_{14}$, hv 2. (MeO)$_3$P	R = OMe (50) (67)	22
		1. C$_6$H$_{14}$, hv 2. (MeO)$_3$P	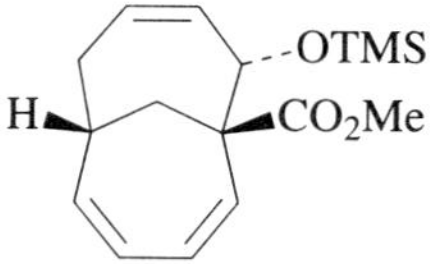 (11) 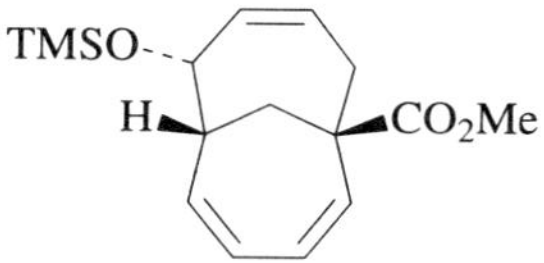 + (11)	23

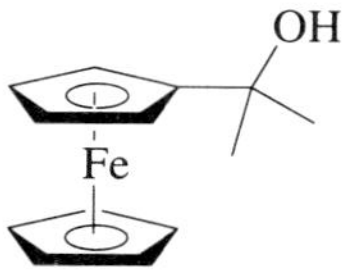

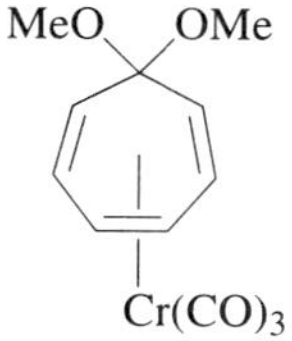

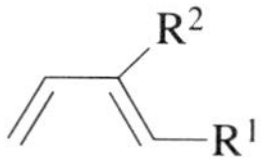

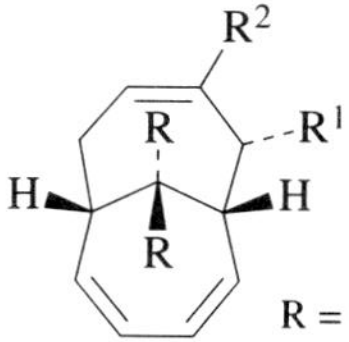

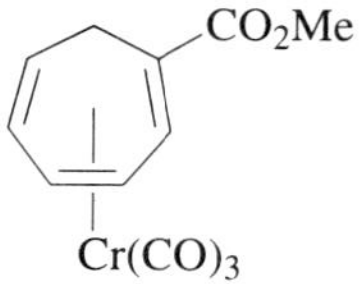

TABLE IV. METAL-MEDIATED [6+4] CYCLOADDITIONS (Continued)

Triene	Trienophile	Conditions	Product(s) and Yield(s) (%)	Refs.

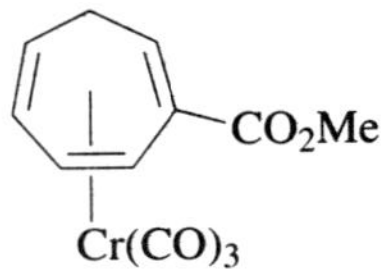

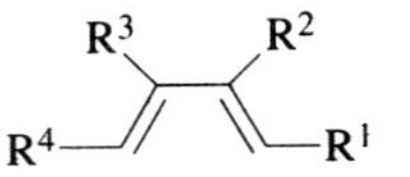

Row 1

Triene: cycloheptatriene with CO_2Me and $Cr(CO)_3$

Trienophile: diene with R^3, R^2, R^4, R^1

R^1	R^2	R^3	R^4
OTMS	H	H	H
Me	H	H	Me

Conditions: 1. C_6H_{14}, hv 2. $(MeO)_3P$

Product with R^3, R^2, R^1, H, R^4, H, CO_2Me

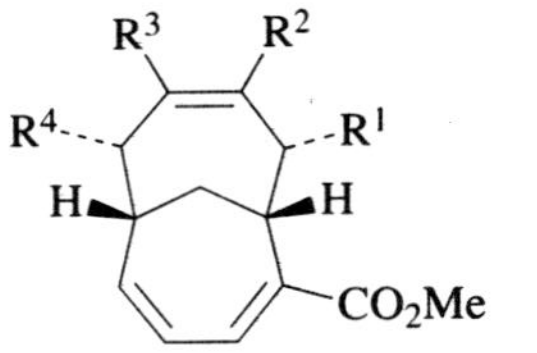

R^1	R^2	R^3	R^4	
OTMS	H	H	H	(37)
H	H	H	OTMS	(38)
H	Me	Me	H	(74)

Refs: 23

Row 2

Triene: cycloheptatriene with $(OC)_3Cr$ and CO_2Me

Trienophile: diene with R^2, R^1

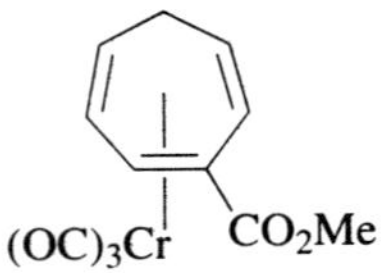

R^1	R^2
Me	Me
Me	CO_2Me

Conditions: 1. C_6H_{14}, hv 2. $(MeO)_3P$

Product with R^1, H, R^2, H, CO_2Me

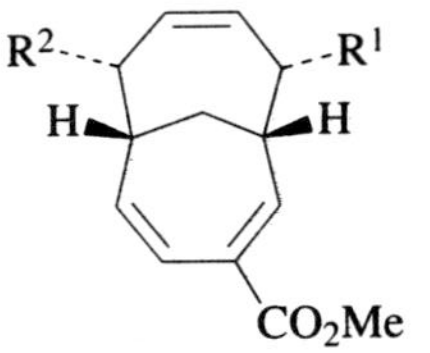

R^1	R^2	
Me	Me	(90)
Me	CO_2Me	(45)
CO_2Me	Me	(45)

Refs: 23

418

TABLE IV. METAL-MEDIATED [6+4] CYCLOADDITIONS (*Continued*)

Triene	Trienophile	Conditions	Product(s) and Yield(s) (%)	Refs.
C$_{10}$				

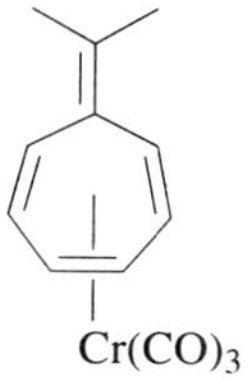

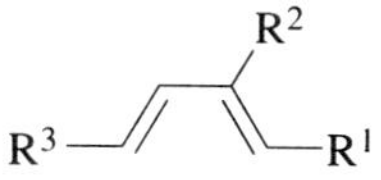

Triene structure (C$_{10}$): isopropylidene-substituted cycloheptatriene–Cr(CO)$_3$ complex.

Trienophile:

R^1	R^2	R^3
H	H	H
Me	H	H
Me	H	Me
H	Me	H

Conditions:
1. C$_6$H$_{14}$, *hv*
2. (MeO)$_3$P

Product:

(49)
(62)
(61)
(50)

Refs.: 24

Trienophile:

Conditions:
1. C$_6$H$_{14}$, *hv*
2. (MeO)$_3$P

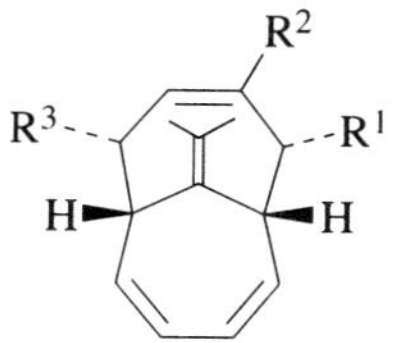

Product:

(74)

Refs.: 24

Trienophile:

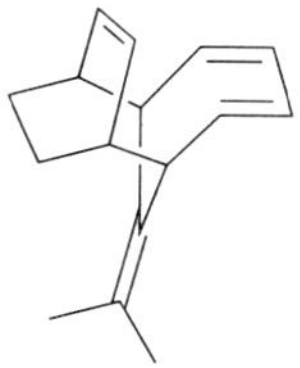

Conditions:
1. C$_6$H$_{14}$, *hv*
2. CO

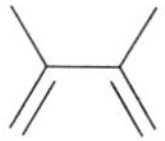

Product:

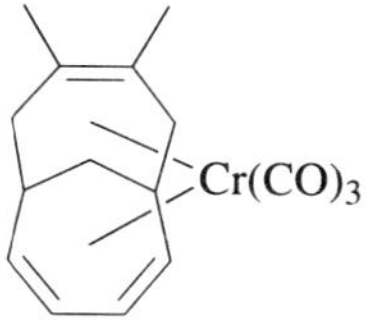

(68)

Refs.: 152

TABLE IV. METAL-MEDIATED [6+4] CYCLOADDITIONS (*Continued*)

Triene	Trienophile	Conditions	Product(s) and Yield(s) (%)	Refs.
C₁₁		1. C₆H₁₄, *hv* 2. (MeO)₃P	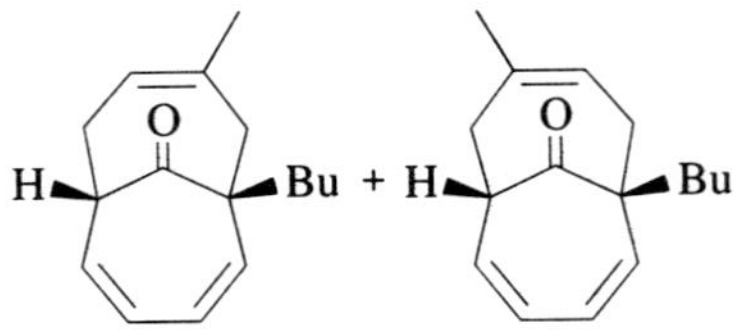 (33) + (32)	23
C₁₄		1. C₆H₁₄, *hv* 2. (MeO)₃P	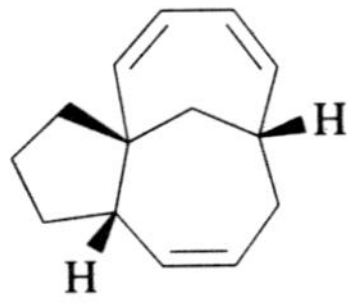 (85)	153
		1. C₆H₁₄, *hv* 2. (MeO)₃P	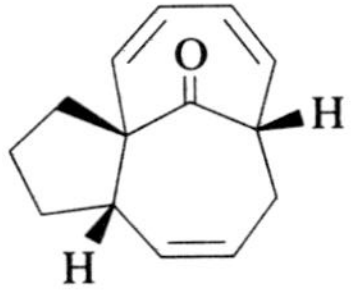 (52)	153

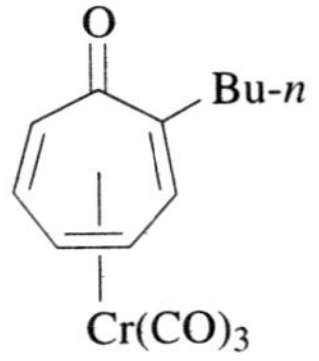

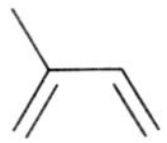

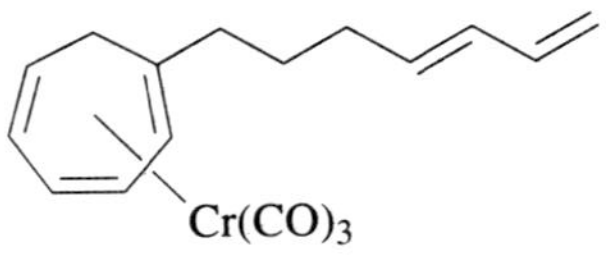

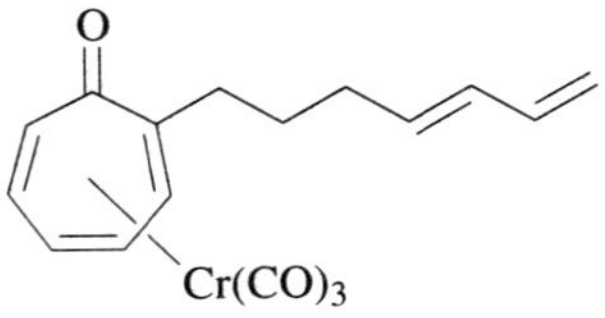

420

TABLE IV. METAL-MEDIATED [6+4] CYCLOADDITIONS (*Continued*)

Triene	Trienophile	Conditions	Product(s) and Yield(s) (%)	Refs.
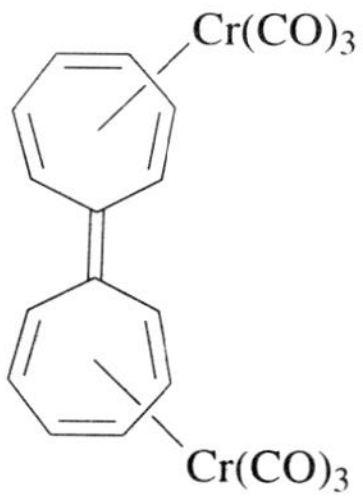		THF, *hv*	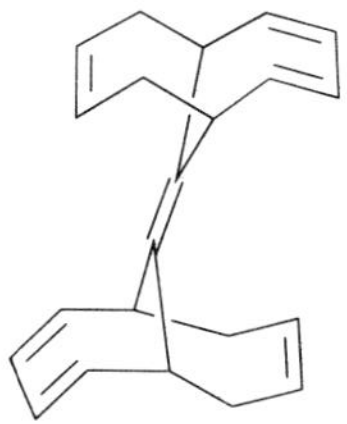 (34)	154
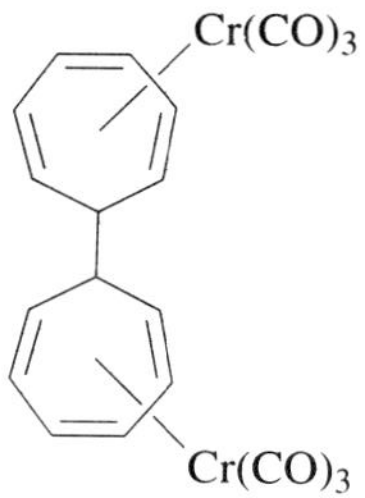	 R¹ R² H H Me H Me Me	1. THF, *hv* 2. (MeO)₃P	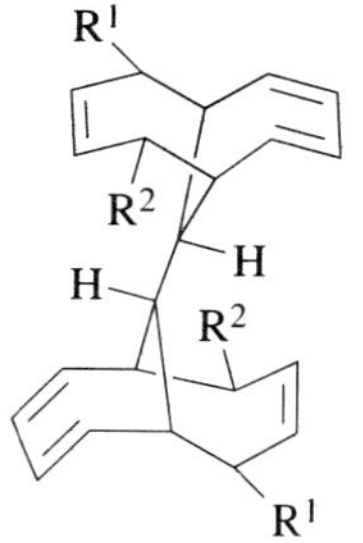 (48) (39) (42)	155
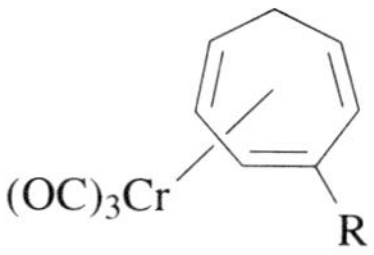 R = (+)-2,10-camphor sultam		1. C₆H₁₄, Et₂O, *hv* 2. O₂, Et₂O	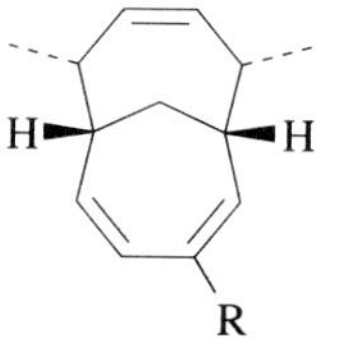 (82) >98% de	41a

421

REFERENCES

[1] Rigby, J. H. in *Comprehensive Organic Synthesis*, Trost, B. M.; Fleming, I., Eds., Vol. 5, Pergamon, Oxford, 1991, pp. 617–643.

[2] Rigby, J. H. *Acc. Chem. Res.* **1993**, *26*, 579.

[3] Kreiter, C. G. *Adv. Organomet. Chem.* **1986**, *26*, 297.

[4] Rigby, J. H.; Short, K. M.; Ateeq, H. S.; Henshilwood, J. A. *J. Org. Chem.* **1992**, *57*, 5290.

[5] Cookson, R. C.; Drake, B. V.; Hudec, J.; Morrison, A. *J. Chem. Soc., Chem. Commun.* **1966**, 15.

[6] Ito, S.; Fujise, Y.; Okuda, T.; Inoue, Y. *Bull. Chem. Soc. Jpn.* **1966**, *39*, 1351.

[7] Hoffmann, R.; Woodward, R. B. *J. Am. Chem. Soc.* **1965**, *87*, 4388.

[8] Mukherjee, D.; Watts, C. R.; Houk, K. N. *J. Org. Chem.* **1978**, *43*, 817.

[9] Woodward, R. B.; Hoffmann, R. *Angew Chem., Int. Ed. Engl.* **1969**, *8*, 781.

[10] LeNoble, W. J.; Ojosipe, B. A. *J. Am. Chem. Soc.* **1975**, *97*, 5939.

[11] Tanida, H.; Pfaendler, H. R. *Helv. Chim. Acta* **1972**, *55*, 3062.

[12] Takeshita, H.; Sugiyama, S.; Hatsui, T. *J. Chem. Soc., Perkin Trans. 2* **1986**, 1491.

[13] Sugiyama, S.; Takeshita, H. *Bull. Chem Soc. Jpn.* **1987**, *60*, 977.

[14] Ito, S.; Sakan, K.; Fujise, Y. *Tetrahedron Lett.* **1970**, 2873.

[15] Houk, K. N.; Woodward, R. B. *J. Am. Chem. Soc.* **1970**, *92*, 4145.

[16] Garst, M. E.; Roberts, V. A.; Prussin, C. *Tetrahedron* **1983**, *39*, 581.

[17] Franck–Neumann, M.; Martina, D. *Tetrahedron Lett.* **1977**, 2293.

[18] Houk, K. N.; Luskus, L. J. *J. Org. Chem.* **1973**, *38*, 3836.

[19] Dunn, L. C.; Chang, Y. -M.; Houk, K. N. *J. Am. Chem. Soc.* **1976**, *98*, 7095.

[20] Houk, K. N.; George, J. K.; Duke, R. E. *Tetrahedron* **1974**, *30*, 523.

[21] Houk, K. N.; Sims, J.; Watts, C. R.; Luskus, L. J. *J. Am. Chem. Soc.* **1973**, *95*, 7301.

[22] Rigby, J. H.; Ateeq, H. S. *J. Am. Chem. Soc.* **1990**, *112*, 6442.

[23] Rigby, J. H.; Ateeq, H. S.; Charles, N. R.; Cuisiat, S. V.; Ferguson, M. D.; Henshilwood, J. A.; Krueger, A. C.; Ogbu, C. O.; Short, K. M.; Heeg, M. J. *J. Am. Chem. Soc.* **1993**, *115*, 1382.

[24] Michels, E.; Sheldrick, W. S.; Kreiter, C. G. *Chem. Ber.* **1985**, *118*, 964.

[25] VanHouwelingen, T.; Stufkens, D. J.; Oskam, A. *Organometallics* **1992**, *11*, 1146.

[26] Harano, K.; Ban, T.; Yasuda, M.; Kanematsu, K. *Tetrahedron Lett.* **1979**, 1599.

[27] Rigby, J. H.; Moore, T. L.; Rege, S. *J. Org. Chem.* **1986**, *51*, 2398.

[28] Ito, S.; Sakan, K.; Fujise, Y. *Tetrahedron Lett.* **1969**, 775.

[29] Garst, M. E.; Roberts, V. A.; Houk, K. N.; Rondan, N. G. *J. Am. Chem. Soc.* **1984**, *106*, 3882.

[30] Funk, R. L.; Bolton, G. L. *J. Am. Chem. Soc.* **1986**, *108*, 4655.

[31] Reiter, S. E.; Dunn, L. C.; Houk, K. N. *J. Am. Chem. Soc.* **1977**, *99*, 4199.

[32] Takeshita, H.; Mori, A.; Sano, S.; Fujise, Y. *Bull. Chem. Soc. Jpn.* **1975**, *48*, 1661.

[33] Sasaki, T.; Kanematsu, K.; Kataoka, T. *J. Org. Chem.* **1975**, *40*, 1201.

[34] Bimanand, A. Z.; Gupta, Y. N.; Doa, M. J.; Eaton, T. A.; Houk, K. N.; Fronczek, F. R. *J. Org. Chem.* **1983**, *48*, 403.

[35] Gupta, Y. N.; Doa, M. J.; Houk, K. N. *J. Am. Chem. Soc.* **1982**, *104*, 7336.

[36] Wu, T. -C.; Mareda, J.; Gupta, Y. N.; Houk, K. N. *J. Am. Chem. Soc.* **1983**, *105*, 6996.

[37] Houk, K. N.; Luskus, L. J. *Tetrahedron Lett.* **1970**, 4029.

[38] Houk, K. N.; Luskus, L. J.; Bhacca, N. S. *J. Am. Chem. Soc.* **1970**, *92*, 6392.

[39] Tate, D. P.; Knipple, W. R.; Augl, J. M. *Inorg. Chem.* **1962**, *1*, 433.

[40] Pauson, P. L.; Smith, G. H.; Valentine, J. H. *J. Chem. Soc. (C)* **1967**, 1057.

[41] Munro, J. D.; Pauson, P. L. *J. Chem. Soc.* **1961**, 3475.

[41a] Rigby, J. H.; Sugathapala, P.; Heeg, M. J. *J. Am. Chem. Soc.* **1995**, *117*, 8851.

[42] Rigby, J. H.; Ateeq, H. S.; Krueger, A. C. *Tetrahedron Lett.* **1992**, *33*, 5873.

[43] Rigby, J. H.; Short, K. M.; Ateeq, H. S.; Henshilwood, J. A. *J. Org. Chem.* **1992**, *57*, 5290.

[44] Vogel, E. *Pure Appl. Chem.* **1982**, *54*, 1015.

[45] Rigby, J. H. in *Studies in Natural Products Chemistry*, Rahman, A.-u., Ed., Vol. 12 (Part H), Elsevier, Amsterdam, 1993, pp. 233–274.

[46] Marshall, J. A.; Conrow, R. E. *J. Am. Chem. Soc.* **1983**, *105*, 5679.

[47] Barrett, D. G.; Liang, G.-B.; McQuade, D. T.; Desper, J. M.; Schladetzky, K. D.; Gellman, S. H. *J. Am. Chem. Soc.* **1994**, *116*, 10525.

[48] Paquette, L. A.; Nitz, T. J.; Ross, R. J.; Springer, J. P. *J. Am. Chem. Soc.* **1984**, *106*, 1446.

[49] Mehta, G.; Pathak, V. P. *J. Chem. Soc., Chem. Commun.* **1987**, 876.

[50] Funk, R. L.; Olmstead, T. A.; Parvez, M. *J. Am. Chem. Soc.* **1988**, *110*, 3298.

[51] Winkler, J. D.; Henegar, K. E.; Williard, P. G. *J. Am. Chem. Soc.* **1987**, *109*, 2850.

[52] Heathcock, C. H.; Del Mar, E. G.; Graham, S. L. *J. Am. Chem. Soc.* **1982**, *104*, 1907.

[53] Heathcock, C. H.; Graham, S. L.; Pirrung, M. C.; Plavac, F.; White, C. T. in *The Total Synthesis of Natural Products*, Apsimon, J., Ed., Vol. 5, Wiley, New York, 1983.

[54] Rigby, J. H. in *Studies in Natural Products Chemistry*, Rahman, A.-u., Ed., Vol. 1 (Part A), Elsevier, Amsterdam, 1988, pp. 545–576.

[55] Jenniskens, L. H. D.; Wijnberg, J. B. P. A.; DeGroot, A. in *Studies in Natural Products Chemistry*, Rahman, A.-u., Ed., Vol. 14 (Part I), Elsevier, Amsterdam, 1994, pp. 355–387.

[56] Rigby, J. H.; Wilson, J. Z. *J. Am. Chem. Soc.* **1984**, *106*, 8217.

[57] Anderson, N. H.; Uh, H. S. *Tetrahedron Lett.* **1973**, 2079.

[58] Marshall, J. A.; Ruth, J. A. *J. Org. Chem.* **1974**, *39*, 1971.

[59] Lautens, M.; Kumanovic, S. *J. Am. Chem. Soc.* **1995**, *117*, 1954.

[60] Wender, P. A.; Takahashi, H.; Witulski, B. *J. Am. Chem. Soc.* **1995**, *117*, 4720.

[60a] Dasler, W; Bauer, C. D. *Ind. Eng. Chem., Anal. Ed.* **1946**, *18*, 52.

[60b] Burfield, D. R. *J. Org. Chem.* **1982**, *47*, 3281.

[61] Rigby, J. H.; Cuisiat, S. V. *J. Org. Chem.* **1993**, *58*, 6286.

[62] Ito, S.; Ohtani, H.; Narita, S. -i.; Honma, H. *Tetrahedron Lett.* **1972**, 2223.

[63] Li, Z. -H.; Mori, A.; Kato, N.; Takeshita, H. *Bull. Chem Soc. Jpn.* **1991**, *64*, 2778.

[64] Fujise, Y.; Shiokawa, T.; Mazaki, Y.; Fukazawa, Y.; Fujii, M.; Ito, S. *Tetrahedron Lett.* **1982**, *23*, 1601.

[65] Roberts, V. A.; Prussin, C.; Garst, M. E. *J. Org. Chem.* **1982**, *47*, 3969.

[66] Nozoe, T.; Mukai, T.; Takase, K.; Nagase, T. *Proc. Jpn. Acad.* **1952**, *28*, 477 (*Chem. Abstr.* **52**, 2676f).

[67] Takeshita, H.; Sugiyama, S.; Toshihide, H. *J. Chem. Soc., Perkin Trans. 2* **1986**, 1491.

[68] Kende, A. S. *J. Am. Chem. Soc.* **1966**, *88*, 5026.

[69] Kende, A. S.; Lancaster, J. E. *J. Am. Chem. Soc.* **1967**, *89*, 5283.

[70] Meier, H.; Pauli, A.; Kolshorn, H. *Chem. Ber.* **1989**, *122*, 101.

[71] Fujise, Y.; Saito, H.; Ito, S. *Tetrahedron Lett.* **1976**, 1117.

[72] Takeshita, H.; Wada, Y.; Mori, D.; Hatsui, T. *Chem. Lett.* **1973**, 335.

[73] Jones, P. W.; Kneen, G. *J. Chem. Soc., Perkin Trans. 1* **1976**, 1647.

[74] Paquette, L. A.; Hathaway, S. J.; Schirch, P. F. T. *J. Org. Chem.* **1985**, *50*, 4199.

[75] Ipaktschi, J.; Herber, J.; Kalinowski, H.-O.; Boese, R. *Tetrahedron Lett.* **1987**, *28*, 3467.

[76] Paquette, L. A.; Hathaway, S. J.; Gallucci, J. C. *Tetrahedron Lett.* **1984**, *25*, 2659.

[77] Mahon, M. F.; Molloy, K.; Pittol, C. A.; Pryce, R. J.; Roberts, S. M.; Ryback, G.; Sik, V.; Williams, J. O.; Winders, S. A. *J. Chem. Soc., Perkin Trans. 1* **1991**, 1255.

[78] Paddon-Row, N. M.; Warrener, R. N. *Tetrahedron Lett.* **1974**, 3797.

[79] Tegmo-Larsson, I. -M.; Houk, K. N. *Tetrahedron Lett.* **1978**, 941.

[80] Houk, K. N.; Luskus, L. J.; Bhacca, N. S. *J. Am. Chem. Soc.* **1970**, *92*, 6392.

[81] Houk, K. N.; Luskus, L. J.; Bhacca, N. S. *Tetrahedron Lett.* **1972**, 2297.

[82] Pfaendler, H. R.; Tanida, H. *Helv. Chim. Acta* **1973**, *56*, 545.

[83] Mori, M.; Hayamizu, A.; Kanematsu, K. *J. Chem. Soc., Perkin Trans. 1* **1981**, 1259.

[84] Sasaki, T.; Kanematsu, K.; Iizuka, K. *J. Org. Chem.* **1976**, *41*, 1105.

[85] Fujise, Y.; Mazaki, Y.; Shiokawa, T.; Fukazawa, Y.; Ito, S. *Tetrahedron Lett.* **1984**, *25*, 3611.

[86] Scott, L. T.; Adams, C. M. *J. Am. Chem. Soc.* **1984**, *106*, 4857.

[87] Takeshita, H.; Yan, Y. Z.; Kato, N.; Mori, A.; Nozoe, T. *Tetrahedron Lett.* **1995**, *36*, 5195.

[88] Takeshita, H.; Yan, Y. Z.; Kato, N.; Mori, A.; Wakabayashi, H.; Nozoe, T. *Tetrahedron Lett.* **1995**, *36*, 5199.

[89] McCulloch, R. K.; Wege, D. *Tetrahedron Lett.* **1976**, 3213.

[90] Gamba, A.; Gandolfi, R.; Oberti, R.; Sardone, N. *Tetrahedron* **1993**, *49*, 6331.

[91] DeMicheli, C.; Gandolfi, R.; Grünanger, P. *Tetrahedron* **1974**, *30*, 3765.

[92] Trost, B. M.; Seoane, P. R. *J. Am. Chem. Soc.* **1987**, *109*, 615.

[93] Bonadeo, M.; DeMicheli, C.; Gandolfi, R. *J. Chem. Soc., Perkin Trans. 1* **1977**, 939.

[94] Houk, K. N.; Watts, C. R. *Tetrahedron Lett.* **1970**, 4025.

[95] Kato, H.; Kobayashi, T.; Tokue, K.; Shirasawa, S. *J. Chem. Soc., Perkin Trans. 1* **1993**, 1617.

[96] Tsuge, O.; Takata, T.; Noguchi, M. *Chem. Lett.* **1980**, 1031.

[97] Trost, B. M.; Parquette, J. R.; Marquart, A. L. *J. Am. Chem. Soc.* **1995**, *117*, 3284.

[98] Niggli, U.; Neuenschwander, M. *Helv. Chim. Acta* **1990**, *73*, 2199.

[99] Dunn, L. C.; Houk, K. N. *Tetrahedron Lett.* **1978**, 3411.

[100] Mukherjee, D.; Dunn, L. C.; Houk, K. N. *J. Am. Chem. Soc.* **1979**, *101*, 251.

[101] Bhacca, N. S.; Luskus, L. J.; Houk, K. N. *J. Chem. Soc., Chem. Commun.* **1971**, 109.

[102] Pfaendler, H. R.; Tanida, H. *Helv. Chim. Acta* **1973**, *56*, 543.

[103] Uebersax, B.; Neuenschwander, M. *Chimia* **1981**, *35*, 400.

[104] Uebersax, B.; Neuenschwander, M.; Engel, P. S. *Helv. Chim. Acta* **1982**, *65*, 89.

[105] Liu, C. -Y.; Ding, S. T. *J. Org. Chem.* **1992**, *57*, 4539.

[106] Liu, C. -Y.; Ding, S. -T.; Chen, S. -Y.; You, C. -Y.; Shie, H. -Y. *J. Org. Chem.* **1993**, *58*, 1628.

[107] Gupta, Y. N.; Mani, S. R.; Houk, K. N. *Tetrahedron Lett.* **1982**, *23*, 495.

[108] Copland, D.; Leaver, D.; Menzies, W. B. *Tetrahedron Lett.* **1977**, 639.

[109] Kanematsu, K.; Harano, K.; Dantsuji, H. *Heterocycles* **1981**, *16*, 1145.

[110] Sato, M.; Ebine, S.; Tsunetsugu, J. *Tetrahedron Lett.* **1974**, 2769.

[111] Thomas, A. F.; Perret, C. *Tetrahedron* **1986**, *42*, 3311.

[112] Messmer, A.; Hajos, G.; Timari, G. *Monatsh. Chem.* **1988**, *119*, 1113.

[113] Tanida, H.; Irie, T.; Tori, K. *Bull. Chem. Soc. Jpn.* **1972**, 1999.

[114] Alder, K.; Braden, R.; Flock, F. H. *Chem. Ber.* **1961**, *94*, 456.

[115] Padwa, A.; Nobs, F. *Tetrahedron Lett.* **1978**, 93.

[116] Dennis, N.; Ibrahim, B.; Katritzky, A. R. *J. Chem. Soc., Chem. Commun.* **1975**, 425.

[117] Caramella, P.; Frattini, P.; Grünanger, P. *Tetrahedron Lett.* **1971**, 3817.

[118] Allmann, R.; Debaerdemaeker, T.; Friedrichsen, W.; Jürgens, H. J.; Betz, M. *Tetrahedron* **1976**, *32*, 147.

[119] Friedrichsen, W.; Oeser, H.-G. *Justus Liebigs Ann. Chem.* **1978**, 1146.

[120] Mori, M.; Kanematsu, K. *J. Chem. Soc., Chem. Commun.* **1980**, 873.

[121] Friedrichsen, W.; Oeser, H.-G. *Tetrahedron Lett.* **1975**, 1489.

[122] Friedrichsen, W.; Oeser, H.-G. *Justus Liebigs Ann. Chem.* **1978**, 1161.

[122a] Friedrichsen, W.; Schröer, W.-P.; Schmidt, R. *Justus Liebigs Ann. Chem.* **1976**, 793.

[123] Gupta, Y. N.; Patterson, R. T.; Bimanand, A. Z.; Houk, K. N. *Tetrahedron Lett.* **1986**, *27*, 295.

[124] Braun, M.; Büchi, G.; Bushey, D. F. *J. Am. Chem. Soc.* **1978**, *100*, 4208.

[125] Ban, T.; Wakita, Y.; Kanematsu, K. *J. Am. Chem. Soc.* **1980**, *102*, 5416.

[126] Fritz, H.; Weis, C. D. *Tetrahedron Lett.* **1974**, 1659.

[127] Sasaki, T.; Kanematsu, K.; Yukimoto, Y.; Hiramatsu, T. *J. Am. Chem. Soc.* **1974**, *96*, 2536.

[128] Bower, D. J.; Howden, M. E. H. *J. Chem. Soc., Perkin Trans. 1* **1980**, 672.

[129] Sasaki, T.; Kanematsu, K.; Hayakawa, K. *J. Am. Chem. Soc.* **1973**, *95*, 5632.

[130] Yang, N. C.; Srinivasachar, K. *J. Chem. Soc., Chem. Commun.* **1976**, 48.

[131] Kondo, H.; Mori, M.; Kanematsu, K. *J. Org. Chem.* **1980**, *45*, 5273.

[132] Houk, K. N.; Woodward, R. B. *J. Am. Chem. Soc.* **1970**, *92*, 4143.

[133] Harano, K.; Yasuda, M.; Kanematsu, K. *J. Org. Chem.* **1982**, *47*, 3736.

[134] Sasaki, T.; Kanematsu, K.; Hayakawa, K. *Tetrahedron Lett.* **1974**, 343.

[135] Sasaki, T.; Kanematsu, K.; Hayakawa, K.; Sugiura, M. *J. Am. Chem. Soc.* **1975**, *97*, 355.

[136] Yasuda, M.; Harano, K.; Kanematsu, K. *J. Am. Chem. Soc.* **1981**, *103*, 3120.

[137] Harano, K.; Hisano, T. *J. Chem. Res. (S)* **1993**, *9*, 356.

[138] .Yasuda, M.; Harano, K.; Kanematsu, K. *Tetrahedron Lett.* **1980**, 627.

[139] Saito, K.; Mukai, T.; Iida, S. *Bull. Chem. Soc. Jpn.* **1986**, *59*, 2485.

[140] Iida, S.; Mukai, T.; Saito, K. *Heterocycles* **1978**, *11*, 401.

[141] Saito, K.; Iida, S.; Mukai, T. *Bull. Chem Soc. Jpn.* **1984**, *57*, 3483.

142 Paul, I. C.; Johnson, S. M.; Barrett, J. H.; Paquette, L. A. *J. Chem. Soc., Chem. Commun.* **1969**, 6.

143 Kurita, J.; Kojima, H.; Tsuchiya, T. *Chem. Pharm. Bull.* **1986**, *34*, 4866.

144 Burger, U.; Bringhen, A. O. *Tetrahedron Lett.* **1988**, *29*, 4415.

145 Szechner, B.; Reg, M.; Dreiding, A. S.; Grieb, R. *Helv. Chim. Acta* **1984**, *67*, 1386.

146 Wu, T. -C.; Houk, K. N. *J. Am. Chem. Soc.* **1985**, *107*, 5308.

147 Keil, J. -M.; Massa, W.; Riedel, R.; Seitz, G.; Wocadlo, S. *Tetrahedron Lett.* **1994**, *35*, 7923.

148 Turbitt, T. D.; Watts, W. E. *J. Chem. Soc., Chem. Commun.* **1971**, 631.

149 Özkar, S.; Kurz, H.; Neugebauer, D.; Kreiter, C. G. *J. Organomet. Chem.* **1978**, *160*, 115.

150 Kreiter, C. G.; Kurz, H. *Chem. Ber.* **1983**, *116*, 1494.

151 Özkar, S.; Kreiter, C. G. *J. Organomet. Chem.* **1985**, *293*, 229.

152 Michels, E.; Kreiter, C. G. *J. Organomet. Chem.* **1983**, *252*, C1.

153 Rigby, J. H.; Sandanayaka, V. P. *Tetrahedron Lett.* **1993**, *34*, 935.

154 Kreiter, C. G.; Michels, E. *J. Organomet. Chem.* **1986**, *312*, 59.

155 Kreiter, C. G.; Michels, E.; Kaub, J. *J. Organomet. Chem.* **1986**, *312*, 221.

CHAPTER 3

CARBON–CARBON BOND-FORMING REACTIONS PROMOTED BY TRIVALENT MANGANESE

Gagik G. Melikyan

Department of Chemistry, California State University, Northridge, California

CONTENTS

Organic Reactions, Vol. 49, Edited by Leo A. Paquette et al.
ISBN 0-471-15655-8   © 1997 Organic Reactions, Inc. Published by John Wiley & Sons, Inc.

INTRODUCTION

In the past decade, major advances in radical chemistry have been made by the use of transition metals. This field has witnessed impressive accomplishments, and tremendous potential lies ahead. There are many transition metal-mediated methods for producing radicals, including (a) oxidation of C—H bonds or unsaturated fragments by transition metals in higher oxidation states, (b) reduction of C—X bonds or unsaturated moieties by transition metals in lower oxidation states, and (c) homolysis of metal–carbon σ bonds. Redox methods for generating radicals are well elaborated and utilize transition metals in different oxidation states,[1-6] such as Mn(III), Ti(IV), Co(III), Cu(II), Fe(II), Ag(II), Pb(IV), Ce(IV), Mn(IV), V(V), Ag(I), Cu(I), Co(II), Fe(II), V(II), Cr(II), Nb(IV), and Ru(II). In the vast majority of these reactions, transient organometallic species have not been either isolated or identified. Their tentative structures have been proposed, in some cases based solely on chemical logic. Accordingly, the mechanisms of these multistep interactions have not been fully established. Particularly lacking is a clear recognition of those elementary steps that occur inside the ligand sphere of the transition metal.

In comparison with traditional methods of radical generation,[5-8] redox initiators demonstrate remarkable regioselectivity and are especially efficient in polyfunctional organic compounds. Furthermore, new types of radicals, inaccessible by traditional approaches, can be successfully generated. The main difference lies in the multiple roles that metal oxidants play during the reaction, namely, one-electron transfer between proradical and transition metal to produce radical species, followed by redox interaction with intermediate adduct radicals. For this reason, metal-mediated reactions differ significantly from those of peroxide- or light-initiated processes.

Trivalent manganese occupies a rather unusual place among metal oxidants in higher oxidation states and is particularly useful in this field. Numerous novel regio-, chemo-, and stereoselective synthetic methods have been developed in both inter- and intramolecular reactions, and their applicability to the construction of complex natural and biologically active compounds has been demonstrated. Despite its growing significance for synthetic chemistry, manganese(III)-mediated reactions have not been comprehensively reviewed in recent years. Reviews by de Klein,[9] Snider,[10] and Melikyan[11] have discussed selected aspects of Mn(III) chemistry; limited coverage is also available in other papers as part of larger topics.[3-6]

The subject of this chapter is the radical carbon–carbon bond-forming reaction induced by trivalent manganese derivatives such as Mn(OAc)₃, Mn(acac)₃, and Mn(pic)₃. It includes the oxidative generation of α- or α,α-dioxoalkyl or alkyl radicals and their subsequent addition to unsaturated moieties. Both inter- and intramolecular processes are discussed, with special emphasis on the regio-, chemo-, and stereoselectivity issues, as well as natural products syntheses. A comprehensive representation of experimental data is accompanied by critical analyses to give a reader adequate ideas of the current status of this field, of what and can be achieved, of what can be anticipated in any new reaction or in any new application of a known process, and of predictions that can be made based on the collective accumulated knowledge. Discussions of oxidations of unsaturated compounds[9,12,13] such as arenes and alkenes, of the α-acetoxylation[14,15] and α-chlorination[16,17] of ketones, and of chlorination of alkenes[18,19] are beyond the scope of this review.

Throughout the chapter the following abbreviations are used: LTR–ligand transfer reaction,[20,21] transfer of an atom or group to a radical center of an adduct radical, presumed to be proceeding in the transition metal ligand sphere; ETR–electron transfer reaction,[20,21] transfer of an electron from a radical center to a transition metal ion; HAA–hydrogen atom abstraction,[7] a propagation step in traditionally initiated radical reactions.

MECHANISM

Generation of Radicals

The initial step of a radical bond-forming reaction is considered to be generation of a carbon-centered radical by one-electron oxidation of the carbonyl component (Eq. 1). The stoichiometry of the process requires an equimolar amount

$$X \underset{Y}{\overset{H}{\underset{|}{\bigwedge}}} Z + Mn(OAc)_3 \longrightarrow X \underset{Y}{\overset{\bullet}{\bigwedge}} Z + Mn(OAc)_2 + AcOH \qquad \text{(Eq. 1)}$$

X, Y, Z = H, Alk, Ar, CHO, COR, CO$_2$H, CO$_2^-$, CO$_2$R, CONH$_2$,
CONR$_2$, CO$_2$COR, NO$_2$

of metal oxidant. The ease of oxidation depends upon the nature of the substituents X, Y, and Z, and is greatly facilitated by carbonyl or nitro groups in the α position. Radicals with one activating group (aldehydes, ketones, monocarboxylic acids, carboxylic acid anhydrides, and nitroalkanes), two activating groups (β-diketones, β-ketoesters, β-ketocarboxylic acids, malonic acid and its half- and diesters and diamides, cyanoacetic acid, cyanoacetamide, α-acyl-γ-lactones, β-ketophosphonates, β-ketosulfoxides, and β-ketosulfones) or three activating groups (ortho esters) are generated efficiently from the corresponding C—H precursor. The greater the number of activating groups, the faster is the radical generation process. For example, the generation of radicals from β-diketones and β-ketoesters occurs smoothly at room temperature, whereas monoketones and monocarboxylic acids require temperatures up to 120–140°.[11] The mechanism of radical generation is not well understood.[9,11] In recent years, novel methods for generating alkyl radicals with Mn(III) have been developed, utilizing as radical precursors cyclopropanols[22,23] and cyclobutanols[24] or Cr(0) complexes (Eq. 2).[25]

n = 0, 1; R = H, Alk, OR

$$(CO)_5Cr = C \underset{R^1}{\overset{O^- NMe_4^+}{\underset{R^2}{\bigvee}}} R^3 + Mn^{III} \longrightarrow R^1 \underset{R^2}{\overset{R^3}{\overset{\bullet}{\bigvee}}} + Mn^{II}$$

$$\text{(Eq. 2)}$$

R^1, R^2, R^3 = Alk, H

Direct experimental proof for the formation of radicals in Mn(III)-induced oxidations of carbonyl compounds is lacking. Indirect evidence for radical intermediates is the formation of dimers and the regioselectivity of aromatic substitution.

Dimers from mono- and dicarbonyl compounds attributable to radical couplings have been isolated in several cases. Ketones[26] and carboxylic acids[27,28] tend to produce C—C dimers, whereas with dicarbonyl compounds both C—C[29-31] and C—O[31] dimers have been isolated (Eq. 3).

The regioselectivity of aromatic substitution reactions, (i.e., the high level of *ortho*-substituted products), is consistent with the formation of free radicals.[32] In particular, carboxymethylation of chlorobenzene and toluene affords substitution products containing 30–46% and 39–58% of *ortho* isomers, respectively.[27] In the analogous reaction, anisole produces 78% of the *ortho*-substituted product.[33] Acetonylation of monosubstituted benzenes[34,35] and nitromethylation reactions[36] show a similar pattern. For steric reasons, this is not the case with di- and tricarbonyl compounds.[37-40]

The regioselectivity of radical generation is crucial with unsymmetrical carbonyl compounds. One of the major advantages of transition metals, in particular trivalent manganese, is the highly selective formation of radicals that are not accessible under "traditional" conditions, for example, with peroxides.[7,8]

Oxidation of aldehydes with Mn(OAc)₃ under homogeneous conditions generates α-formyl alkyl radicals **1**, which can add to olefins (vide infra). In the absence of solvent in a heterogeneous process, acyl radicals **2** are produced from α-formyl alkyl radicals by intermolecular H-atom transfer.[41-43]

Unsymmetrical ketones can generate isomeric α-oxoalkyl radicals by competing oxidation of the primary, secondary, or tertiary C—H bonds. Ketones **3–7**, containing methyl and methylene groups, react with alkenes with low selectivity.[44-47] The exclusive reaction of the methyl group in ketone **8** may reflect the steric inaccessibility of the methylene group.[45] The unreactivity of the methyne group in ketone **9** toward the metal oxidant may also be due to steric hindrance.[46] The highly regioselective oxidation of the methylene group in ketone **10** might be due to the relative instability of the isomeric bridgehead radical.[46a]

3 25%   75%   R = Me   100%   0%
4 30%   70%   R = (CH₂)₂OAc   **8**
5 40%   60%   R = CH₂CO₂Me   100%   0%
6 55%   45%   R = OAc   **9**
7 60%   40%   R = n-C₅H₁₁

R = H, Me
10

It is important to note that the ratio of addition products is not necessarily the same as the relative rates of generation of isomeric radicals, since the rates of their addition to substrates may be different for different radicals. For example, the oxidation of 2-butanone with Mn(OAc)₃ in the absence of alkenes produces 1- and 3-acetoxybutan-2-ones in a ratio of 2.5:1, indicating preferential oxidation of the methyl group,[48] as opposed to the 1:3 ratio of addition of isomeric radicals to an olefin.[44]

Oxidation of β-dicarbonyl compounds and their analogues occurs regioselectively at internal methylene or methyne groups because of their higher acidity and enolizability compared with competing methyl groups.

Adduct Radicals: Formation and Reactivity Patterns

Educt radicals generated in the presence of "matching" unsaturated substrates may attack across multiple bonds to produce adduct radicals **11**.[11] These are short-lived transient intermediates in radical (cyclo)addition reactions, and their direct detection and spectral characterization still remain a challenge for "radical" chemists. The synthetic result and selectivity of the process depend upon the reactivity patterns of adduct radicals; their transition to stable organic products can occur in several ways (Eq. 4). The normal pathway[7,8] is represented by H-atom abstraction, which results in the formation of the more saturated derivatives **12**. For both alkenes (Tables I and II) and alkynes (Table XII), the corresponding HAA-products **12** have been isolated, resulting from atom transfer to alkyl and vinyl adduct radicals, respectively.[44,49,50]

Adduct radicals **11** can be trapped with molecular and redox radical scavengers, enhancing the synthetic potential of the reaction as well as providing experimental proof for the existence of free adduct radicals. Trapping experiments have been accomplished with molecular oxygen as a scavenger and β-dicarbonyl compounds as reagents.[51-56] In most cases, peroxy radicals **13** attack acetyl groups intramolecularly to produce cyclic peroxides (Eq. 5). In the absence of

(Eq. 4)

$R^1 = Ar, Alk;$
$R^2 = Ar, NH_2, RNH, R_2N$

(Eq. 5)

acetyl groups, the corresponding hydroperoxides have been isolated.[54] An alternative radical trap is carbon monoxide, which is highly efficient in both inter- and intramolecular processes (Table XXXIX).

Adduct radicals **11** can be oxidized by Mn(III) or Cu(II) ions to the corresponding carbocations **14**. The ease of oxidation depends dramatically upon the nature of substituents and functional groups α to the radical center. Carbocations **14** undergo β elimination to produce β,γ- and/or γ,δ-unsaturated derivatives **15**, or intramolecular cyclizations to carbonyl-containing groups such as acyl, carboxy, alkoxycarbonyl, or unsaturated fragments like double bonds. The cyclization produces cyclic carbocation **16**, which forms β-elimination products (dihydrofurans, furans, or γ-lactones), undergoes tandem (poly)cyclizations, or reacts with nucleophiles.[57]

Alternative pathways for adduct radicals **11** involve ligand transfer from the Mn(III) complex to alkyl or vinylic radical centers. Numerous examples of AcO group[48,58–62] and Cl atom[50,63,64] transfers have been reported; the formation of LTR products **17** and **18** might occur via carbocations **14**.

Cyclizations on multiple bonds and aromatic rings represent an additional and synthetically useful pattern for adduct radicals **11**. This pattern has been demonstrated by numerous intramolecular (Tables XXIV–XXXI) and tandem cyclizations (Tables XXXII–XXXVIII), as well as by addition–cyclization processes (Table XXXIII).

One of the most crucial points in the design of Mn(III)-mediated reactions is the choice of metal oxidant, in particular, the use of either Mn(III) complexes alone or in combination with catalytic or equimolar amounts of Cu(OAc)$_2$. The latter is widely utilized in intermolecular functionalization of unsaturated substrates with aldehydes,[65] ketones,[44,45,49,59,66,67] carboxylic acids,[68] β-dicarbonyl compounds,[45, 57, 69–78] dicarboxylic acid derivatives,[64] and nitroalkanes,[79] as well as in addition–cyclization reactions,[80–82] intramolecular[83–97] and tandem cyclizations,[56,86–88,91,94,96,98–103] and polycyclizations.[104] The major incentive for using a Cu(II) salt as a cooxidant is to improve the selectivity of the reaction or to redirect it toward the formation of new products. This approach is exemplified by the interaction of 1-heptene with diethyl malonate.[64] In the absence of Cu(OAc)$_2$, addition of the bis(ethoxycarbonyl)methyl radical to the double bond produces adduct radical **19**, which undergoes H-atom abstraction to afford saturated diester **20**. To the contrary, Cu(II) ions efficiently trap transient radicals **19** by oxidation to γ,δ-unsaturated derivative **21**.

In combination with trapping experiments utilizing molecular oxygen,[51–56] Cu(II)-modified reactions provide sound experimental evidence for the existence of free adduct radicals that break away from the manganese cluster prior to conversion to final products. The striking effect of Cu(II) acetate is caused by its ability to oxidize alkyl radicals to carbocations;[21,58] the rate of this oxidation is 350 times greater than that with Mn(OAc)$_3$.[58] This is the reason that Cu(OAc)$_2$ is used in reactions where Mn(III) ions are not sufficiently powerful to oxidize intermediate radicals. In some cases, catalytic amounts of Cu(II) ions are sufficient to achieve the desired result, but the rate of oxidation with Cu(OAc)$_2$ also changes, depending on the structures of adduct radicals. In these cases, the use of equimolar amounts of cooxidant, or even a several-fold excess, might be necessary to obtain good yields.

The choice of metal oxidant still remains empirical, although a large number of examples allow one to draw some conclusions from the structures of adduct radicals as to which might or might not require the use of an oxidant stronger than $Mn(OAc)_3$. Such generalizations are useful for the design of new reactions.

The most common types of adduct radicals are shown in Eq. 6. Intermediates **22–26** constitute a first group of adduct radicals that are oxidized by $Mn(OAc)_3$. Synthetically this means that in the presence of $Mn(OAc)_3$ an oxidative cyclization, elimination, or ligand-transfer reaction can occur, but not H-atom abstraction. Another distinctive feature of these reactions is that the addition of $Cu(OAc)_2$ does not change the distribution of products or appreciably affect yields. The common structural feature of adduct radicals **22–26** is the effective stabilization of their corresponding carbocations by conjugation with unsaturated moieties (**22–24**), by interaction with unshared electron pairs of an α-substituent (**25**), or with a π-bonded metal cluster (**26**). This stabilization decreases the ionization potential of the radicals, thus enabling Mn(III) ion oxidation to carbocations.

The second group of adduct radicals (**27–31**) are oxidized by $Cu(OAc)_2$ but not by $Mn(OAc)_3$. These intermediates are stabilized less than those in the first group. In the absence of Cu(II) ions, they produce either H-atom abstraction products (**27, 28**) or undergo polymerization (**29–31**). The introduction of a more powerful metal oxidant into the reaction results in the formation of oxidative elimination, ligand transfer, or cyclization products.

(Eq. 6)

Most secondary alkyl (**32**) and allylic (**33**) radicals require the use of $Cu(OAc)_2$, although there are some reports on their partial[58,105,106] or even complete[107–109] oxidation with trivalent manganese. Tertiary alkyl radicals **34** are on the borderline between more- and less-stabilized adduct radicals, although they are closer to the first group. There are several reports of their effective oxidation by Mn(III) ions alone,[31,65,70,82,106,110,111] as well as single reports on either partial oxidation[110] or the formation of H-atom abstraction products.[49]

The large number of intermolecular reactions, including the composition of metal oxidants used and product distribution, can be found in Tables I–XXIII. Additional examples are provided by mono- and tandem cyclizations, where Cu(OAc)$_2$ is widely and effectively used to improve the selectivity of the reactions (Tables XXIV–XXXVIII).

A word of warning is relevant here. Although experimental data on the formation of free radicals and free-adduct radicals are well documented, any generalizations should be made with extreme caution. First, whatever is shown for a certain type of unsaturated substrate and carbonyl compound might be incorrect for even a closely related type of interaction. Second, the formation of free educt radicals in the absence of unsaturated substrates does not necessarily mean that this is the case when a substrate is present in the reaction mixture, since C—C bond formation can still occur within the metal–ion ligand sphere.[12,31] Third, dimerization or trapping reactions indicate only that there is a certain fraction of scavengable radicals, but they do not prove that all products are formed via radicals.

Kinetics

Kinetic studies of Mn(III)-mediated reactions are directed toward acquiring in-depth understanding of the process and resolution of the most crucial issues in the multistep mechanism. Among the latter are the following. Are keto or enol forms oxidized by metal oxidant? Does the interaction of educt radicals with unsaturated substrates occur within the ligand sphere of the metal? What is the rate-determining step—enolization, formation of the metal-complexed (or metal free) educt radical, or C—C bond formation? Are products derived from metal-complexed or kinetically free adduct radicals? The available kinetic data shed light on only a few of these issues.

The oxidation of aldehydes with Mn(OAc)$_3$ produces α-formyl alkyl radicals, which may be converted into acyl radicals by intermolecular H-atom transfer.[42,43] The rate-determining step in the oxidation is homolysis of a C—H bond, as established by the high isotopic effect obtained with CD$_3$CHO.[43] Free radicals are in equilibrium with oxallylic complexes of the metal, thus affecting the regioselectivity of the process.[112] In particular, the formation of α-alkyl substituted aldehydes is attributed to the reaction proceeding in the coordination sphere of the metal. Enolization is believed not to precede the oxidation, since oxidation occurs for nonenolizable aldehydes.

The oxidation of ketones by Mn(OAc)$_3$ occurs via the keto form since its rate is a factor of 10 higher than racemization[43,113] or isotope exchange.[114] The kinetic isotopic effect for CD$_3$COCD$_3$ is 5.8, indicating that α-C—H bond cleavage is the slowest step.[114] Somewhat contradictory results have been obtained in the oxidation of cyclohexanone, where enolization is proposed to precede the oxidation step.[115] The rate-determining step in the interaction of ketones with mono- and binuclear aromatic compounds is complexation of the Mn(III) enolate with the arene and subsequent C—C bond formation within the coordination sphere of the metal.[116]

The oxidation of carboxylic acids is independent of alkene concentration and is directly proportional to their acidity over a range of 16 pK_a units.[117] The rate-determining step is the abstraction of a proton from the bridged acetate followed by electron transfer from the enolate ion to the Mn(III) atom.[12,117] This mechanism explains the accelerating effect of acetate anions on the rate of alkene annulation. Control experiments indicate that if the oxidation of acetic acid with Mn(OAc)$_3$ is performed in the absence of alkenes, no succinic acid is formed.[12] The authors concluded that no free carboxymethyl radicals are released and that the formation of the C—C bond occurs within the metal–oxidant ligand sphere.[12] The issue of whether radical dissociation from the metal occurs prior to cyclization is not resolved.

Lactonization of alkenes does not require cooxidant to produce γ-lactones in good to high yields.[11] This observation can be rationalized in terms of an intramolecular cyclization taking place within the Mn(III)–ligand sphere, since free adduct radicals (e.g., 27–29), and most secondary alkyl (32) and allylic (33) radicals are not oxidized by Mn(III) ions and thus cannot cyclize on a carboxy group. Although both cis- and trans-4-octene produce the same ratio of the isomeric γ-lactones (trans:cis = 3.3:1),[117] this result does not resolve the issue of metal binding to the adduct radicals, since stereomutation could also occur with a Mn(III)-bound adduct radical.

The oxidation of β-ketoesters with Mn(OAc)$_3$ is dramatically accelerated in the presence of unsaturated substrates,[31,72,118] indicating that the generation of radicals and C—C bond formation occur within the ligand sphere of the metal. Recent investigation suggests that there is a substantial difference in the behavior of unsubstituted and 2-methyl-substituted acetoacetates.[31] For the former, C—C bond formation within the Mn(III)–ligand sphere is proposed to be the rate-determining step, whereas for the latter the Mn(III) enolate may be produced in the slowest step. In a recent and well-designed comparative study, it has been demonstrated that the nature of the radicals generated by either Mn(III)-mediated oxidation or TBTH-induced atom transfer is the same,[119] thus supporting the existence of free radicals.

SCOPE AND LIMITATIONS

Intermolecular Reactions

Compounds With One Multiple Bond—Regio- and Stereoselectivity.
Manganese(III)-mediated reactions of alkenes with aldehydes suffer from low selectivity, separation problems, and moderate yields. The initial step gives rise to two types of adduct radicals formed by the addition of α-formyl alkyl and acyl radicals across the double bond. Their conversion to products occurs by different pathways, that is, oxidative β-deprotonation, H-atom transfer, or AcO-group transfer, depending on the alkene structure.[43] Isobutene-derived tertiary adduct

radicals tend more toward elimination, producing β,γ- and γ,δ-alkenals **35**. Their proportion in the product mixture is 50–70%, in addition to the saturated aldehyde **36** and ketone **37**.[110] The introduction of Cu(OAc)$_2$ into the reaction mixture as cooxidant[21,58,120] improves both the yield (up to 46%) and selectivity of the process (up to 93% **35**).[110] In contrast to the more easily oxidizable tertiary adduct radicals, secondary radicals favor H-atom abstraction, producing saturated aldehydes and ketones.[41,42,121] This method has been made synthetically attractive by optimization to selectively produce α-alkylated aldehydes **38**.[41] In the presence of Cu(OAc)$_2$, secondary adduct radicals undergo regioselective β elimination yielding γ,δ-alkenals **39** on a preparative scale.[122]

The reactions of alkenes with ketones are mechanistically analogous to those of aldehydes, and include generation of α-oxoalkyl radicals, their regioselective addition to the terminal double bond of alkenes, and subsequent transformation of adduct radicals to afford H-atom abstraction products, isomeric alkenones, and γ-acetoxy ketones (Table II). Both acyclic and cyclic ketones have been used as reagents; terminal alkenes are the most investigated substrates, along with single reports on cycloalkenes. Of practical importance is the telomerization of ethylene when used as an unsaturated substrate.[66,67,123] The chain length and terminal functionality depend on the reaction conditions and the presence of Cu(II) additives, producing either saturated[123] or unsaturated[66,67] telomeric ketones. The nature of the products reflects the reactivity patterns of primary adduct radicals, that is, H-atom abstraction in the presence of Mn(OAc)$_3$ and β elimination or AcO-group transfer mediated by Cu(OAc)$_2$.[120] From the synthetic viewpoint, one of the major achievements of this area is a novel approach to 1,4-diketones using enol acetates as substrates.[46a] The generality of the reaction has been well demonstrated by using acyclic, cyclic, and terpenoid ketones. As shown in the specific example, the addition of an educt radical to the double bond of enol acetate **40** produces adduct radical **41**; the latter is easily oxidized by Mn(III) ions owing to effective stabilization of secondary carbocation **42** by the α-oxygen atom. Subsequent release of the acetyl group yields 1,4-diketone **43**.[46]

The interaction of alkenes with ketones can be converted into conjugated addition by using protonated pyridines as nucleophilic radical traps.[124] Adduct radicals **44** alkylate electron-deficient aromatic rings with high positional α selectivity (9:1), producing pyridine **45** as a major isomer. In a similar manner, but in higher yield, the reaction proceeds with protonated isoquinoline as a radical trap.[124] Although limited success has been achieved with cyclohexene, overall this reaction looks promising and deserves more attention to define further the scope, synthetic utility, and stereoselectivity.

Carboxylic acids, in particular acetic acid, react with alkenes to produce γ-lactones **46** as the major products (Eq. 7).[28,125] Alternative procedures include the

(Eq. 7)

use of acetic anhydride[28] and potassium acetate.[125] A good level of understanding has been achieved concerning separate steps of the mechanism (see also Kinetics section).[12,117] In particular, potassium acetate acts as a base, facilitating the rate-determining deprotonation of bridging acetate ligands, and acetic anhydride is oxidized prior to acetic acid because of its higher acidity (pK_a 18 vs. 25). Yields of γ-lactones are systematically higher with potassium acetate because of the formation of unsaturated and γ-acetoxy carboxylic acids in the presence of acetic anhydride.[117] In some cases, allylic acetates are formed as byproducts,[28,125] although the selectivity for lactonization is usually very high, that is, the ratio γ-lactone:allylic acetate is 30:1 and 50:1 with methylstyrene and 1-octene, respectively.[125]

The regioselectivity of lactonization is high with terminal alkenes, which are attacked by carboxymethyl radicals at the C_1 carbon atom (Table III). Regioselectivity becomes a critical issue with unsymmetrical di- and trisubstituted alkenes, being governed by the relative stabilities of adduct radicals as well as by steric factors. β-Methylstyrene (**47**) reacts regioselectively, demonstrating the higher stability of the benzylic radical compared to the secondary radical (Eq. 8).[117]

(Eq. 8)

Analogously, the greater stability of tertiary versus secondary alkyl radicals determines the regioselectivity with 2-methyl-2-pentene[126] and 3-ethyl-2-pentene.[117] Remote differences in the alkyl chains do not substantially affect the addition of educt radicals.[127] In α,β-unsaturated esters, the regioselectivity depends upon the nature of the β substituent. The relative stability of the adduct radicals with α-methyl and α-tert-butoxycarbonyl groups appears to be close, resulting in low regioselectivity in the lactonization of tert-butyl crotonate.[117] To the contrary, the powerful stabilizing effect of a phenyl group directs the addition of both acetic[117,128] and chloroacetic[129] acids to ester 48 (Eq. 8). Analogous regioselectivity has been observed with coumarins, although the products are α-acetoxymethyl and α-diacetoxymethyl derivatives.[130]

Stereoselectivity is one of the most critical aspects of the annulation reaction. It includes, first, the relative configuration of the "double-bond" carbons, and second, the stereochemical relationship of the latter to the α substituent of the lactone ring. Alkenes and cycloalkenes produce cis- and trans-γ-lactones in different ratios (Tables III and IV). In particular, substrates 47 and 48 (Eq. 8) represent two examples of highly stereoselective formation of trans-γ-lactones (up to 98.5%).[117] Trans stereoselectivity might be even higher in the reactions of cis-4,5-disubstituted cyclohexene 49 (Eq. 9) and norbornene.[131] It is noteworthy that un-

$$\text{(Eq. 9)}$$

substituted cyclohexene undergoes lactonization with preponderant formation of the cis isomer.[117] Exclusive cis annulation has been observed with benzofuran,[132] bornene,[133] and indenes,[117,134] (e.g., 50).[134] One of the critical points in stereochemical studies is the stereomutation observed in lactonization of cis- and trans-4-octenes, which results in the formation of the same mixture of isomeric γ-lactones (trans:cis = 3.3:1).[117] This result excludes a "concerted" mechanism as an option, although it does not address the issue of in- or out-of-ligand-sphere formation of products. Substituted acetic acids and their homologues produce γ-lactones with two (α,γ) or three (α,β,γ) stereocenters.[126,128,129] In all cases where stereochemistry is established, the reactions are nonstereoselective, producing mixtures of two or four stereoisomers in similar amounts;[129] the only exception is the lactonization of methyl cinnamate with chloroacetic acid, which forms a single stereoisomer.[129]

Reactions of alkenes with β-dicarbonyl compounds represent a general method for the synthesis of polysubstituted dihydrofurans **51** (Eq. 10, Table VII).

R^1, R^2, R^3 = H, Alk, Ar; R^4, R^5 = Me, OMe, OEt, —(CH$_2$)$_3$—, Ph

(Eq. 10)

Regioselective generation of the α,α-dioxoalkyl radicals in the presence of alkenes produces adduct radicals **52**, which undergo intramolecular cyclization with acyl or benzoyl groups to form dihydrofurans **51**. The reaction is especially facile with aromatic substituents at the double bond, since benzylic radicals are easily oxidized by Mn(III) ions to carbocations **53**. The latter have not been directly observed, but their formation along the reaction coordinate has been proposed based on overwhelming indirect evidence (see Adduct Radicals section). With alkyl adduct radicals, Cu(OAc)$_2$ is used to effect fast oxidation. Selectivity drops with Mn(OAc)$_3$ alone, since the reactivity patterns of alkyl radicals include H-atom transfer,[70,105] β-elimination,[31,70] and acetoxy group transfer.[31] Most of the reactions are mediated by equimolar amounts of Mn(OAc)$_3$, but there has been occasional use of Mn(acac)$_3$.[45,55,135] Both Mn(III) and Mn(II) acetates have been used in combination with molecular oxygen, which acts as both a regenerating and oxidizing agent.[51,53,54] A synthetically useful modification is the introduction of lithium chloride into the reaction mixture, whereby chlorine ligand transfer becomes a new reaction pathway for adduct radicals.[63] Besides commonly used β-diketones and β-ketoesters, new types of carbonyl components such as β-keto phosphates, β-keto sulfoxides, and β-keto sulfones have recently been used to produce dihydrofurans.[53]

Reactions of terminal and 1,1-disubstituted alkenes are highly regioselective, with the attack of electrophilic educt radicals occurring at the double-bond terminus (Table VII). 1,2-Disubstituted substrates like β-alkyl- and β-alkoxycarbonylstyrenes also react regioselectively owing to the effective stabilization of radical intermediates by α-aryl groups.[136] The scope of the reaction has been further expanded by involving a large number of styrene derivatives and β-aroyl esters (Eq. 11). Nonconjugated alkadienes (Table XV) can be either bis-[109] or monofunctionalized,[72] with faster reaction at the electron-rich double bond.[82]

$$R = H, Me, CH_2OAc, CO_2Et;$$
$$Ar^1, Ar^2 = Ph, 4\text{-}MeOC_6H_4, 3,4\text{-}(MeO)_2C_6H_3$$

54 (27-71%)

The stereochemistry of the process remains mostly unknown because cycloalkenes and 1,2-disubstituted alkenes have not been investigated in detail (Table VII). Only in the reaction of styrenes with β-aroyl esters has the stereochemistry of dihydrofurans 54 been established to be *trans*.[136]

A novel method of isoprenylation of mono- and β-dicarbonyl compounds has been developed by using allyl sulfides, sulfoxides, and sulfones as unsaturated substrates (Table VII).[71] Adduct radicals 55 undergo homolytic β scission to release sulfur-centered radicals. The generality of the method is demonstrated by the use of a large number of acyclic and cyclic β-dicarbonyl compounds, although use of the strong oxidant PbO_2 severely limits the types of prospective substrates.[71]

$$X = SPh, SOPh, SO_2Ph, t\text{-}BuS, t\text{-}BuSO_2$$

Dihydrofuran synthesis by reaction of alkenes with β-dicarbonyl compounds has been brought to a new level by involving endo- and exocyclic enol ethers and enol lactones as substrates (Table XI). An extensive exploration of these reactions has led to the synthesis of a large number of oxaspirolactones and *cis*-fused di- and tricyclic systems.[137-143] Selected examples shown in Eq. 12 feature regioselective addition of α,α-dioxoalkyl radicals to a β carbon producing α-oxygen-substituted adduct radicals. Their oxidation by $Mn(OAc)_3$ is greatly facilitated by the oxygen atom and leads to spiro and fused systems 56–58. The attractive feature of these reactions is their high *cis* stereoselectivity, exemplified by polycyclic fused systems 57 and 58. From the standpoint of synthetic chemistry, these reactions are very attractive for constructing complex polycyclic compounds in one step from readily available starting materials.

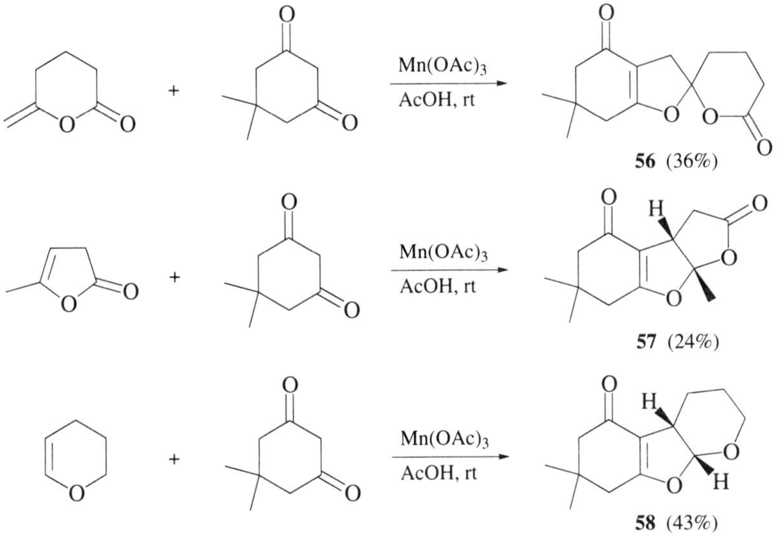

(Eq. 12)

Malonic acid, with *gem*-carboxylate moieties, provides a new dimension in Mn(III)-mediated reactions: sequential lactonization with the participation of two molecules of alkene (Table IX).[144,145] Thus, initially formed γ-lactone **59** is oxidized again to generate educt radical **60**, which reacts with a second molecule of substrate to produce spiro compound **61**. A number of alkenes and cyclo-

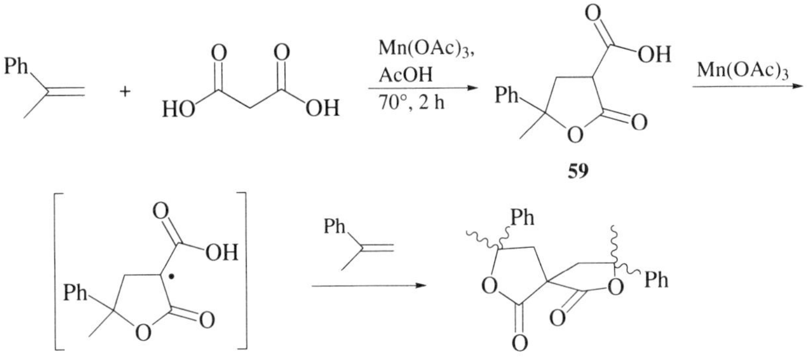

alkenes have been bis-annulated with malonic acid, demonstrating the generality of the reaction and providing facile access to the synthetically useful 2,7-dioxa[4.4]nonane-1,6-diones.[144,145] The lack of stereoselectivity seems to be a major drawback since mixtures of unsymmetrical and symmetrical *syn* and *anti* stereoisomers are usually formed. The highest selectivity is observed in the reaction of cyclopentene (92% of the unsymmetrical isomer).[145] Synthetic utility of the reaction has been enhanced by the use of nonconjugated alkadienes such as 1,5-hexadienes as unsaturated substrates.[145] Although yields are low to moderate, topologically unusual bridged tricyclic systems have been successfully constructed in one step (Table XV). Overall, reactions of malonic acid with alkenes have not received the attention they deserve; major advances would be complete control of the stereoselectivity as well as further expansion of scope. Halide and methyl-substituted malonic acids produce monoannulated products such as saturated and α,β-unsaturated γ-lactones.[146]

Dicarboxylic acid derivatives **62–70** have been widely used as carbonyl components (Eq. 13, Table X). The synthetic outcome of the reaction depends upon

$$
\begin{array}{ccccc}
\underset{\textbf{62}}{RO-\overset{O}{C}-CH_2-\overset{O}{C}-OR} & & \underset{\textbf{63}}{RO-\overset{O}{C}-\underset{Cl}{CH}-\overset{O}{C}-OR} & & \underset{\textbf{64}}{HO-\overset{O}{C}-CH_2-\overset{O}{C}-OR}
\end{array}
$$

$$
\begin{array}{ccccc}
\underset{\textbf{65}}{HO-\overset{O}{C}-\underset{Cl}{CH}-\overset{O}{C}-OR} & & \underset{\textbf{66}}{HO-\overset{O}{C}-CH_2-CN} & & \underset{\textbf{67}}{MeO-\overset{O}{C}-CH_2-\overset{O}{C}-O^- K^+} \qquad \text{(Eq. 13)}
\end{array}
$$

$$
\begin{array}{ccccc}
\underset{\textbf{68}}{H_2N-\overset{O}{C}-CH_2-\overset{O}{C}-NH_2} & & \underset{\textbf{69}}{H_2N-\overset{O}{C}-CH_2-CN} & & \underset{\textbf{70}}{RO-\overset{O}{C}-\underset{\underset{R}{\overset{|}{N}}-COR}{CH}-\overset{O}{C}-OR}
\end{array}
$$

the nature of the functional groups in the carbonyl components and on their ability to interact with radical/cationic centers in the adduct radicals. Malonic diesters **62** and α-chloro diesters **63** produce linear products,[50,63,64,111] since cyclization on the alkoxycarbonyl group is a slow process compared with that of acyl and carboxy groups. Major reactivity patterns of adduct radicals appear to be H-atom abstraction,[64] regioselective β-deprotonation,[64,111] and Cl-atom transfer if lithium chloride is used as an additive.[50,63] Malonic acid derivatives that contain carboxy (**64**,[146,147] **65**,[147] **66**[128,129,147]) or carboxylate (**67**[129,148,149]) groups all produce corresponding γ-lactones analogous to those from monocarboxylic acids (Eq. 7, Tables III–V). Malonic diamides **68** undergo partial hydrolysis to generate carboxy groups and thereby α-aminocarbonyl-γ-lactones.[150] A novel cyclization pathway for aminocarbonyl fragmentation has been observed that gives rise to α,β-unsaturated γ-lactams.[150] This is a new reactivity pattern for benzylic adduct radicals, and although only two examples have been reported, both utilizing heavily substituted alkenes, this reaction might be of general use. Cyanoacetamide **69** produces a variety of structures arising from acetoxy group transfer, cyclizations on carboxy and aminocarbonyl groups, and secondary transformations of γ-lactams.[151] N-Acyl substituted malonic diester **70** is an exotic type of carbonyl compound, the chemistry of which does not include participation of the pendant *N*-fragment.[107]

The reactions of malonic acid derivatives with unsymmetrical alkenes are highly regioselective (Table X). Educt radicals attack double bonds to generate more stable intermediates; phenyl groups have a strong directing effect in β-substituted styrenes,[128,147,148] as does the α-oxygen atom in enol ethers.[149] *tert*-Butyl crotonate also reacts regioselectively with cyanoacetic acid with exclusive attack of the educt radicals on the β carbon.[129] Regioselective functionalization of acyclic[109] and cyclic[147] nonconjugated alkadienes has also been described (Table XV).

The stereoselectivity has not been fully established for every reaction (Table X). *cis* Stereoselection has been reported for the lactonization of norbornene with monoethyl malonate, its chloro derivative, and cyanoacetic acid.[147] To the contrary, the formation of isomeric mixtures has been observed in reactions of 1-octene,[129] 1-decene,[129,147] α,β-unsaturated esters,[129] cyclohexene,[129,147] cyclooctene,[129] and α- and β-substituted styrenes.[129,147]

Alkynes remain one of the least investigated substrates (Table XII). Their interactions with mono- and dicarbonyl compounds have been reported, although apparently unoptimized procedures were used.[44,49,50,152] The major theoretical issue here is the reactivity pattern of vinylic radicals (**28**, Eq. 6), which affects both the yield and synthetic outcome of the processes. Based on a limited number of examples, the present level of knowledge does not permit reliable prediction of the reaction outcome.

Conjugated Systems—Chemo-, Regio-, and Stereoselectivity.   The interaction of 1,3-alkadienes with ketones is represented by a single reaction between 1,3-butadiene and acetone.[61] 1,2-Conjugate addition to the double bond results in 5-acetoxy-6-hepten-2-one (**71**), which undergoes a nonregioselective oxidation of both the methylene and methyl groups. Adduct radicals **72** and **73** undergo a sequential five-membered ring annulation with the formation of isomeric 2,8-dioxa-*cis*-bicyclo[3.3.0]octanes (**74**, 1:1:2) and 1,6-dioxaspiro[4.4]nonanes (**75**,

3:3:1). Cyclization on the acyl group is an unexpected mode of behavior for secondary adduct radicals (Table II), although it is conceivable that the second annulation makes the overall process irreversible. Phenomenologically this is one of the most intriguing reactions in Mn(III) chemistry, producing complex structures from two parent compounds in a single step. The configuration of the products needs to be fully established, and more conjugated dienes and ketones need to be studied to define the scope and synthetic utility of the method.

γ-Lactones **76** are uniformly produced if 1,3-alkadienes are treated with acetic acid in the presence of potassium acetate (Table XIV). The regioselectivity issue arises with unsymmetrical substrates such as isoprene,[128] (1,3-Z)-dodecadiene,[153] 1,1-dichloro-4,4-dimethyl-1,3-butadiene,[154] and methyl sorbate.[108] More stable adduct radicals are preferentially formed if both termini of the substrate are equally accessible,[128] otherwise the less-substituted double bond is the exclusive reaction site.[153] The electrophilic character of the carboxymethyl radical results in selective addition at the more electron-rich double bond,[154] producing a key intermediate in pyrethroid synthesis.[155] For the same reason, the γ,δ-double bond in methyl sorbate is the principal point of attack by the educt radical, with less attack at the α,β-moiety (68 vs. 29%); both regioisomeric γ-lactones are formed with low stereoselectivity.[108] Partial stereomutation of the *cis* allylic adduct radical has been observed in the lactonization reaction that produces the sex pheromone of the Japanese beetle (Eq. 26).[153] Formal *cis*-stereoselective lactonization of 1,3-cyclohexadiene into bicycle **77** has been accomplished in three steps, although in low overall yield.[11,156] Nonconjugated dienes can be functionalized selectively at one of the double bonds,[128] with the more electron-rich olefin reacting preferentially.[82]

1,3-Alkenynes afford γ-alkynyl-γ-lactones **78** in moderate yields if lactonization is carried out in the presence of potassium acetate (Table XVII). The synthetic usefulness of this method has been demonstrated by the syntheses of natural sex pheromones.[153,157-159] Acetic anhydride as a cosolvent makes ligand transfer a major reaction pathway for propargylic radicals (**29**, Eq. 6), producing γ-acetoxy acids **79**.[62] The phenomenon of "critical carbon chain length" has been observed in the selective formation of LTR products **79** vs. γ-lactones **78** if R > C_7H_{15}. This result implies a relationship between the reactivity of the adduct radical and carbon chain length in substituent R, although this has not been established.[62]

76   77   78   79

R¹, R² = H, Cl, Alk,   R¹ = C_3-C_8 R = C_2-C_{10}
CO₂Me   R² = H, Me

Dihydrofurans with vinyl or alkenyl side chains are formed in radical cycloaddition reactions of β-dicarbonyl compounds to 1,3-alkadienes (Table XVI). The mechanism is analogous to that of simple alkenes (Eq. 10), and includes the formation of allylic radicals and their subsequent oxidation by metal oxidants. In most cases, Cu(OAc)$_2$ is used to ensure effective oxidation, although phenyl groups α to the radical center greatly facilitate oxidation, enabling Mn(III) ions alone to provide moderate to high yields of cyclization products.[109] Regioselectivity at C$_1$ vs. C$_2$ is usually high because educt radicals selectively add to either C$_1$ or C$_4$ carbon atoms of the 1,3-butadiene moiety to produce isomeric allylic radicals (Table XVI). Inverse regioselectivity has been observed for diene **80**, heavily substituted with aryl groups.[109] Selective attack at the C$_2$ carbon affords dihydrofuran **81**, and indicates a higher stability of α,α-diarylalkyl radicals (**23**, Eq. 6)

80

Ar = Ph, p-ClC$_6$H$_4$, p-MeC$_6$H$_4$

81 (11-46%)

in comparison with allylic radicals (C-1 attack). Regioselectivity at C$_1$ vs. C$_4$ in 2-substituted 1,3-alkadienes is directed by the relative stabilities of isomeric allylic radicals; ratios are 70:10 and 43:8 for isoprene and myrcene, respectively.[72] With 4-substituted dienes, reactions proceed selectively at C$_1$ directed by steric effects.[73] Low regio-(C$_1$ vs. C$_4$) and stereoselectivity has been reported for 1,4-substituted dienes such as methyl sorbate[72,108] and conjugated dienones.[72] A partial stereomutation of *cis* allylic intermediates was observed, whereas *trans* analogues were configurationally stable.[73]

Lactonization of 1,3-alkadienes and their cyclic analogues with malonic acid derivatives is a practical way to produce highly functionalized and polysubstituted γ-lactones containing unsaturation in a γ side chain (Table XVI).[107,108,128,147] Some noteworthy features are: (1) high C$_1$ vs. C$_2$ regioselectivity; (2) preferential participation of the electron-rich double bond in 2-substituted substrates—a ratio of 39:5 was obtained with isoprene and cyanoacetic acid;[128] (3) low C$_1$ vs. C$_4$ regioselectivity in the reaction of methyl sorbate and potassium methyl malonate [**82**:(**83** + **84**) = 21:59];[108] (4) high stereoselectivity in the lactonization of an α,β double bond, producing lactone **82** as a single stereoisomer, and to the contrary, low stereoselectivity if the γ,δ reactive site participates (**83**:**84** = 44:15).[108]

82

83

84

82 + **83** + **84** = (63%)

Chemoselectivity is the most critical aspect in Mn(III)-mediated reactions of 1,3-alkenynes, and it depends upon the type and degree of substitution in the substrate (Table XVIII).[74–77,160,161] When grouped according to selectivity, 1,3-butenyne and its derivatives form three pairs that react nonchemoselectively (0-, 2-substituted) and chemoselectively at the triple (1-substituted, 1,2-disubstituted) or double bond (4-substituted, 2,4-disubstituted). Thus, interaction of isopropenyl-acetylene with ethyl acetoacetate gives rise to propargylic **85** and vinylic **86** adduct radicals. Both are resistant to Mn(III)-induced oxidation (Eq. 6), thus requiring the use of the stronger oxidant Cu(OAc)$_2$. The corresponding cations undergo intramolecular cyclization to the acetyl group to produce dihydrofuran **87** and furan **88**. The latter undergoes a secondary transformation to afford furan **89** via the easily oxidized α-(2-furyl)alkyl radical **90**, an analogue of benzylic radical **22** (Eq. 6).[76]

A novel strategy has recently been developed to effect a chemoselective reaction between 1,3-alkenynes and β-dicarbonyl compounds.[57,162] A three-step sequence includes protection of the triple bond with the Co$_2$(CO)$_6$ group, Mn(III)-mediated radical reaction with selective participation of the double bond, and oxidative demetalation. This sequence leads to reversal of the chemoselectivity for 1,2-substituted derivatives like cyclohexenylacetylene, which otherwise reacts chemoselectively at the triple bond.[77] The efficiency of the method has been demonstrated for both acyclic and cyclic enynes and carbonyl compounds (Table XIX). Thus, Co$_2$(CO)$_6$-protected cyclopentenylacetylene **91** reacts with 1,3-cyclohexanedione to produce propargyl adduct radical **92**, which can be oxidized by a Mn(III) salt to the corresponding Co$_2$(CO)$_6$-stabilized carbocation,[57] in contrast to its uncomplexed counterpart (**26** vs. **29**, Eq. 6). Tricyclic product **93** is formed with high *cis* stereoselectivity, and can be oxidatively decomplexed to give a pure organic product. Metal-protected cyclohexenylacetylene also produces *cis*-fused systems in moderate yields.[57]

91 **92** **93** (28%)

1,3-Alkadiynes comprise a relatively uninvestigated class of conjugated systems (Table XX). In monosubstituted compounds, both triple bonds are attacked by electrophilic educt radicals (C_1–C_4 addition) to produce isomeric vinylic radicals of type **30** (Eq. 6). The latter undergo polymerization even in the presence of catalytic amounts of $Cu(OAc)_2$. Cyclization products **94** and **95** can be isolated in moderate yields only when a four-fold excess of cooxidant is used.[78,163] With a straight-chain substituent R, the process is nonregioselective because furan **94** is inactive under the reaction conditions. To the contrary, high regioselectivity is achieved with a *tert*-butyl-like substituent R, effectively protecting one of the triple bonds.[78]

R = C_2-C_4

94 (29-32%) **95** (28-30%)

Aromatics. Arenes were among the first substrates investigated in Mn(III)-mediated reactions (Table XXIA).[27,33,164] The interaction of benzene and its monosubstituted derivatives with acetic acid suffers from low selectivity because initial carboxymethylation is followed by stepwise overoxidation. Regioselectivity is also lacking, although there is preponderant formation of *ortho* isomers (30–78%)—typical for aromatic radical substitutions.[32] With acetone, substituted methyl-benzyl ketones are formed (30–80%) containing 52–85% of *ortho* isomers.[34,35] With β-dicarbonyl compounds and diethyl malonates, the regioselectivity depends upon the type of substitution: 1,2- and 1,4-dimethoxybenzene and 1-methoxy-2,3-dimethoxynaphthalene produce single regioisomers, whereas anisole and naphthalene and its 2-methoxy derivative can react nonregioselectively depending on the nature of the carbonyl component.[38,165] Use of malonic acid provides a new synthetic method for one-step regioselective formylation and carboxylation of naphthalene derivatives.[39]

A novel type of educt radical, tris(ethoxycarbonyl)methyl, has been successfully generated by oxidation of *ortho* ester **96**.[37] It has proven to be a viable supplement to α-oxoalkyl radicals in reactions with arenes and heterocycles (Tables XXIA and B). Thus, 2,3-dimethoxybenzene and furan are selectively functionalized to produce **97** and **98**, respectively. More "conventional" carbonyl components, like acetone and dicarbonyl compounds, also react with heterocycles (furan, thiophene, pyrrole) selectively at the α position.[166,167]

A related species is the nitromethyl radical, which can be generated by oxidation of nitromethane with Mn(OAc)$_3$ (Table XXII). Its reactions with arenes[36,168,169] and alkenes[79] have the same characteristics as those of α-oxoalkyl radicals.

Addition–Cyclization Reactions

Addition–cyclization reactions require that three functional groups, such as a carbonyl moiety and unsaturated fragments (double or triple bond, aromatic ring, nitrile group) be present in the two components. Topologically numerous combinations are possible, differing in connectivities between major constituent parts. By function, a carbonyl moiety gives rise to an educt radical, and unsaturated sites undergo sequential radical–radical (or radical–electrophilic) attack. Literature data are summarized in Table XXIII, arranged according to the class of substrate (A, alkene; B, alkyne; and C, alkadiene). Structural analysis reveals their common features and major topological differences (Eq. 14). Combinations **99**, **100**, and **102** have a carbonyl and one of the unsaturated moieties in the same molecule, whereas unsaturated groups are disposed of together in combinations **101** and **103**. The specific feature of **99** is the location of a carbon-centered radical on one side of both functions; in contrast, the oxidation site is located between **100** and **102**.

Addition–cyclization remains a relatively unexplored dimension in Mn(III) chemistry, although the first examples of type **99** and **102** were reported quite early (Table XXIIIA).[170] The synthetic potential of this reaction is truly outstanding, and includes novel approaches to the synthesis of monocyclic, bicyclic,

99   **100, X = CH, N**   **101**

(Eq. 14)

102, X = CH, N   **103**

fused, and spiro compounds,[80–82,117,171] as well as derivatives of tetralone, tetralin, indane,[170] quinoline, isoquinoline,[172] and naphthalene.[173,174]

Regioselectivity of the addition to unsymmetrical alkenes and alkynes is normally high and the addition occurs according to the Markovnikov rule (Table XXIIIA–C). Selectivity in the second, intramolecular step is most crucial for the synthetic outcome of the overall process. With a double bond as the second reaction site, 5-exo and 6-endo cyclizations are the alternatives.[175,176] In most cases (with one exception[81]) the corresponding cyclopentane derivatives have been isolated, indicating that intramolecular cyclizations occur under kinetic control. Selected examples demonstrate the formation of cyclopentanes **104** and **105** by regioselective 5-exo cyclization of intermediate radicals such as **106** (Eq. 15).[80,82] A second annulation of the alkyl adduct radicals on benzene or

Mn(OAc)$_3$
Cu(OAc)$_2$,
AcOH,
110°, 15 min

106   **104**

+   Mn(OAc)$_3$,
Cu(OAc)$_2$
AcOH,
25°, 16 h

CO$_2$Me

MeO$_2$C

105 (75%)

(Eq. 15)

pyridine rings proceeds selectively at the ortho position.[170,172,173,177] In contrast, alkyne-derived vinylic adduct radicals suffer from low regioselectivity (Table XXIIIB). Thus, vinylic intermediate **107** undergoes both 6-endo and 5-exo cyclization to afford dihydronaphthalene **108** and spiro compound **109**, with loss of aromaticity in the latter (Eq. 16).[174] In unsymmetrical aromatics the regioselectivity of the second annulation has a double sense, involving both 6-endo vs. 5-exo cyclization modes and competitive participation of unequivalent *ortho* positions. Thus, 3-substituted pyridines cyclize nonselectively at positions 2 and 4 with both alkyl and vinylic adduct radicals.[172]

108 (10%)   **109** (79%)

(Eq. 16)

Intramolecular Reactions

Manganese(III)-mediated intramolecular reactions constitute an important area in the field of radical chemistry. They have led to better understanding of the driving forces in radical cyclization processes and of the reactivity patterns of different types of adduct radicals. They have also yielded a bewildering array of organic compounds not otherwise easily accessible. The accumulated collective experience provides a solid basis for the optimal design of new substrates for intramolecular reactions to effect a given type of cyclization, or for constructing a certain type of complex target.

Starting materials for intramolecular cyclization can be designed by incorporating both carbonyl and unsaturated moieties in the same molecule. Not only can the nature of active fragments be varied, but also their disposition, in particular the location of a side chain bearing the unsaturated fragment (UF). Known types of substrates (Eq. 17) include 2- and 4-substituted 3-ketoesters **110**, **111**, O-substituted 3-ketoesters **112**, 4- and 2-substituted 1,3-diketones **113**, **114**, O- and C-substituted malonic ester derivatives **115**, **116**, and N-substituted 3-ketoamides **117** (Tables XXIV–XXXI). Monocyclization products are often accompanied by products of tandem cyclizations (Tables XXXII–XXXVII). To avoid duplication in the Tabular Survey, the major product is given a higher priority in determining the location of certain reactions.

(Eq. 17)

Regioselectivity.   Regioselectivity is one of the most critical aspects of intramolecular cyclizations. The competing formation of different ring sizes occurs under kinetic or thermodynamic control, and is best understood for the classical 5-hexenyl radical cyclizations.[175,176] Since both steric and electronic effects contribute to the reversibility or irreversibility of the addition step, the nature of carbonyl-containing groups, unsaturated moieties, and substituents, as well as the location of the latter at vital positions of the substrates, are of primary importance. In particular, terminal double or triple bonds are preferentially attacked at their termini (*endo* mode), whereas introduction of alkyl substituents activates the *exo* mode.

Cyclopentane vs. Cyclohexane (5-exo vs. 6-endo mode).   The largest number of experimental data deal with the formation of five- and/or six-membered rings by cyclizations of polysubstituted 5-hexenyl adduct radicals. All types of substrates (except **114**) have been subjected to this reaction, although to varying degrees. The regioselective 5-*exo* mode leading to cyclopentanes clearly dominates, producing a variety of structures from substrates **110**,[83] **111**,[85,89–91,99] **112**,[92] **115**,[83,92,94–96] **116**,[97,99,178] and **117**.[179] Cyclohexanes have been obtained as single regioisomers from substrates **111**[56,180] and **113**.[83] A synthetically useful method is the novel and general approach to salicylic acid derivatives employing a large number of substrates **111**[80,86,90] and **113**.[84,89] Nonregioselective reactions are rather uncommon, although lack of regiocontrol has been observed for substrate **111** in cyclizations upon double[83] and triple bonds.[99] The benzene ring directs radical attack selectively at *ortho* positions, producing fused bi- and tricyclic systems.[29,181,182] Examples are shown in Eq. 18, including regioselective formation of cyclopentane **118**[90] and salicylic acid derivative **119**.[84]

Cyclobutane vs. Cyclopentane (4-exo vs. 5-endo mode).   This type of cyclization is represented by a single reaction that produces a *spiro* system by regioselective 5-*endo* annulation (Table XXVIII).[93]

118 (50%)

(Eq. 18)

119 (46%)

Cyclohexane vs. Cycloheptane (6-exo vs. 7-endo mode). Competing formation of six- and seven-membered rings has been studied for a limited number of
4-substituted 3-ketoesters **111** (Table XXV). Regioselective formation of cyclohexanes[83,85,87] and cycloheptanes,[87,88] as well as their mixtures, has been observed.[87,88,180] Thus, the effective construction of bridged compound **120** has been
achieved by the 6-*exo* regioselective addition of cyclopentyl radical to the pendant double bond (Eq. 19).[83] In contrast, the formation of cycloheptane **121** as a
single regioisomer has been reported in the cyclization of an α-chloroacetoacetic
ester on an isopropenyl group (Eq. 19).[87]

120 (78%)

(Eq. 19)

121 (n = 1, R = Et, 69%)
122 (n = 2, R = Me, 69%)

Cycloheptane vs. Cyclooctane (7-exo vs. 8-endo mode). Several examples of
regioselective intramolecular cyclizations producing eight-membered rings have
also been described.[87,88] Thus, cyclooctene **122** is formed as a mixture of two regioisomers by the selective 8-*endo* addition of a highly electrophilic educt radical
to the terminus of the double bond.[87]

Stereoselectivity. Stereocontrol is another critical aspect of monocyclization reactions from the viewpoint of synthetic utility. Although a large number of

annulations have been reported, the stereochemistry of the products remains un-elucidated in many reactions (Tables XXIV–XXXI). Especially deficient are data on the configuration of seven- and eight-membered rings. For cyclopentane and cyclohexane derivatives, the relative configurations have been elucidated fully. There are several aspects to the stereochemical outcome; the most important is the spatial relationship of the substituents at a newly formed C — C bond. Of secondary importance is the stereochemistry of double bonds in the side chains, which are usually formed in the *exo* cyclization mode.

Two examples are given above for the stereoselective formation of a five-membered ring (Eq. 18) and a bridged system (Eq. 19). Tables XXVA, XXIX, and XXXA contain more examples of stereoselective annulations that produce five-membered carbocycles or lactones,[83,90,94,178] as well as six-membered rings[85] and bridged systems.[83,180] A nonstereoselective pathway has been reported for the cyclizations of *O*-substituted malonic esters (Table XXIX) to yield γ-lactones[92,95,96] and 4-substituted 3-ketoesters (Table XXVA) resulting in cyclopentanes[83,85,90,91] or cyclohexanes.[83] Some additional highly stereoselective cyclizations used as key steps in directed syntheses can be found in the section Applications to Natural Product Synthesis.

Tandem and Polycyclization Reactions

Tandem cyclizations include as key steps two consecutive intramolecular additions to unsaturated fragments. The nature of the dicarbonyl and unsaturated moieties can be varied, along with their relative disposition, thus creating an array of starting materials with different topology. According to the type of radical (carbocation) recipients, six modes of tandem cyclizations are designated DD-mode (double bond–double bond, Table XXXII), DB-mode (double bond–benzene ring, Table XXXIII), TD-mode (triple bond–double bond, Table XXXIV), TB-mode (triple bond–benzene ring, Table XXXV), DC-mode (double bond–carbalkoxy/carboxy/carboxylate group, Table XXXVI), and DN-mode (double bond–nitrile group, Table XXXVII), as well as DDDD-mode (Table XXXVIII) for multiple cyclizations. The types of substrates involved are given in Eq. 20, each bearing two UFs and representing 4-, 2,4-, and 4,4-substituted 3-ketoesters (**123**, **124**, **125**); β-diketones **126**; *C*-substituted malonates with one or two unsaturated fragments (**127**, **128**); *O*-substituted malonate **129**; *C,O*- and *O,O*-disubstituted malonates (**130**, **131**) and their analogues; and 4-substituted 3-ketoacid **132**. The first step of a tandem cyclization is analogous to that of a monocyclization, involving regioselective generation of α,α-dioxoalkyl radicals and their addition across multiple bonds. The thus-generated alkyl and vinyl intermediates can attack intramolecularly an unsaturated fragment, that is, double bond, benzene ring, carbalkoxy/carboxy, or nitrile group. Copper(II) acetate can be used to promote the second cyclization step when the oxidizing power of Mn(OAc)₃ appears to be insufficient. Copper(II) acetate is especially useful for cyclizations on carbalkoxy groups, although it can affect the product pattern in many other reactions as well.

123

124

125

(Eq. 20)

126

127, 128

129

130

131

132

Regioselectivity.   The synthetic result of a tandem reaction depends upon the regioselectivity of the first and second annulations, either of which can occur by an *exo* or *endo* mode. The disposition of unsaturated fragments and the positions of substituents are major structural factors that determine the direction of initial attack, the conformation of the intermediate radical, and the ease and selectivity of the second cyclization. A large variety of highly substituted and functionalized bridged and fused systems become available from cleverly designed starting materials (Tables XXXII–XXXVII). Among different modes of tandem cyclizations, those with participation of two double bonds are the most common (DD-mode, Table XXXII). Thus, regioselective 6-*endo* and 5-*exo* cyclizations in substrate **133** afford bridged compound **134** via intermediate cyclohexyl adduct radical **135** (Eq. 21).[100] It is noteworthy that the same cyclization mode can pro-

133

135

134 (73%)

(Eq. 21)

duce different structures depending on the starting material, and in contrast, different modes can afford the same structure by utilizing isomeric substrates. Additional examples of the mode of cyclization vs. type of structure are the following: (a) 6-*endo*, 5-*exo*: bicyclo[4.3.0]nonanes[97,98,101] or bicyclo-

[3.2.1]octanes;[86,99–102] (b) 5-*exo*, 6-*endo*: bicyclo[4.3.0]nonane;[103] (c) 6-*endo*, 6-*endo*: bicyclo[4.4.0]decane;[56,91] (d) 6-*exo*, 5-*exo*: bicyclo[4.3.0]nonane;[100,103] (e) 7-*endo*, 5-*exo*: bicyclo[4.2.1]nonane[87,88] and (f) 8-*endo*, 5-*exo*: bicyclo-[5.2.1]decane.[87,88]

Tandem cyclizations by the DB-mode include an initial addition across a double bond and subsequent cyclization upon a benzene ring (Table XXXIII). In most cases, the first step proceeds regioselectively, that is, either 5-*exo*[90,94,103] or 6-*endo*[85,86,103] annulations have been observed for 5-hexenyl radicals, and exclusively the 6-*exo*[90,103] mode for 6-heptenyl radicals. There is a single report of a nonregioselective reaction of a 5-hexenyl radical to produce 5-*exo* and 6-*endo* products in a ratio of 1:3.[90] The cyclization on a benzene ring is selectively directed to the *ortho* position, formally representing the 6-*endo* mode (Table XXXIII). A doubly regioselective 6-*endo*, 6-*endo* mode is exemplified by the transformation of 3-ketoester **136** to phenanthrene derivative **137** via tertiary adduct radical **138**.

(Eq. 22)

Few examples of TD- and TB-mode tandem cyclizations have been reported (Tables XXXIV and XXXV). For the former, 2,4-substituted 3-ketoesters (Type **124**, Eq. 20) with tethered triple and double bonds undergo regioselective cyclization on a disubstituted triple bond (5-*exo*), whereas in terminal alkynes both acetylenic carbons undergo radical attack (5-*exo*:6-*endo* ca. 2:1). Subsequent cyclization of vinylic radicals on a double bond proceeds selectively in the 5-*exo* mode to produce bicyclo[3.3.0]octanes.[99] β-Diketones of type **126** (Eq. 20) were the first substrates for TB-type tandem cyclizations (Table XXXV). The first cyclization is 6-*exo* selective, the second one involves attack of vinylic radicals on the benzene ring (6-*endo* mode).[103]

DC-mode tandem cyclizations represent the second widely investigated group of reactions (Table XXXVI). The principal types of substrates employed are *C*-substituted malonate **127**, *O*-substituted malonate **129**, *C,O*-disubstituted malonate **130**, and 4-substituted 3-ketoacid **132** (Eq. 20). Each substrate contains a tethered double bond that reacts in the first step, and a carbalkoxy/carboxy/carboxylate group that undergoes intramolecular attack by the adduct radical. In most cases the initiation step generates 5-hexenyl educt radicals, which undergo 5-*exo* selective cyclization to afford cyclopentane or γ-lactone rings.[94,96,99,178,183]

The formation of a cyclohexane ring from both 5-hexenyl[91] and 6-heptenyl[178] radicals has been also reported. The second step involves cyclization of alkyl adduct radicals upon carbalkoxy, carboxy, or carboxylate groups. The nature of these closely related moieties is nevertheless different, as is the mechanism of cyclization. Thus, addition on a carbalkoxy group requires the use of Cu(OAc)$_2$ as a cooxidant,[91,94,96,99] whereas both carboxy[91,183] and carboxylate[178] fragments produce γ-lactones with Mn(OAc)$_3$ alone. It is noteworthy that the species that cyclize on carbalkoxy groups are mostly primary radicals,[94,96,99] although secondary radicals also afford γ-lactones even in the absence of Cu(OAc)$_2$, although in lower yield.[91] One of the pioneering DC-mode tandem reactions is the Mn(III)-mediated cyclization of the appropriately designed ketoacid **139**.[183] Angularly *cis*-fused ketolactones **140**, as well as related dilactones, are key intermediates for the construction of naturally occurring polycyclic compounds.

(Eq. 23)

139   **140** (63%)

The nitrile group is a novel type of radical trap recently utilized in intramolecular reactions. Its combination with a double bond creates a new dimension in tandem cyclizations, a DN-mode (Table XXXVII). Thus, secondary adduct radical **141** adds regioselectively across the C≡N bond to produce intermediate imine **142** and subsequently bicyclic ketone **143** (Eq. 24).[171]

141   **142** X = NH
                           143 X = O, (57%)

(Eq. 24)

The polycyclization reaction upon multiple double bonds (DDDD-mode) is represented by a single example,[104] producing a tetracyclic skeleton by regioselective all-6-*endo* cyclizations (Table XXXVIII).

Stereoselectivity.   The stereochemistry of tandem cyclizations is one of the most important aspects of the reaction, which determines its significance for organic synthesis, and also its place among other stereoselective methods for the construction of cyclic assemblies. As substantiated by examples given in Tables

XXXII–XXXIX, Mn(III) chemistry provides a powerful and highly versatile tool for stereoselective generation of complex organic molecules. There is no complete understanding of the observed selectivity for every reported reaction; even more, in many cases the reasons for the selectivity remain beyond the scope of discussion. Nevertheless the development of such knowledge is critical to predicting the stereochemical outcome of tandem cyclizations. We largely limit our coverage of this topic to two selected reactions where the observed stereoselectivity has become a subject of special consideration. Other examples of highly stereoselective tandem reactions of DD-, DB-, DC-, and DN-modes are shown in Eqs. 21–24, and can also be found in the tables.

Detailed investigation of the tandem cyclization of *O*-allyl malonate **144** has been undertaken recently.[94] The first cyclization converts the starting compound into diastereomeric primary alkyl radicals **145** and **146**. The reaction is regioselective (5-*exo* mode), although stereoselectivity is lacking. In the presence of Cu(OAc)$_2$, the cyclization of adduct radicals **145** and **146** proceeds stereoselectively upon molecular fragments that occupy positions *cis* to the radical center, that is carbethoxy and phenyl groups. *Cis*-fused products **147** and **148** are formed in a ratio of 2:1. In the absence of cooxidant, primary alkyl radical **145** attacks the phenyl group, even though it is in a *trans* position on the five-membered ring. Isomeric lactones **148** and **149** are isolated almost in equal amounts, clearly indicating that cyclization on the benzene ring is purely radical and thus not affected by Cu(OAc)$_2$.

(Eq. 25)

A tandem cyclization of the DB-mode produces *trans*-fused phenanthrene derivative **137** with a *cis* relationship between the methoxycarbonyl group and the bridgehead proton (Eq. 22).[85] The observed stereochemistry might not be genuine if the initially formed stereoisomer undergoes inversion at one of the stereocenters. In particular, enolization could intervene to cause stereomutation at the α carbon. To check this hypothesis, the methyl-substituted analogue **150** has been used in the reaction under identical conditions. Product **151** also has a *trans* junction between cyclohexyl rings, although the methoxycarbonyl group and bridgehead proton are *trans* to each other. This represents the initial stereochemical relationship between these groups, which remains fixed with the methyl substituent, but might also isomerize to the thermodynamically more stable isomer by subsequent enolization.[91]

150   **151** (50%)

Enantioselectivity.   Complete stereocontrol in radical C — C bond-forming reactions remains a major challenge for synthetic chemists. The first success in this area was reported in 1991; the utilization of an enantiomerically pure sulfoxide moiety provides effective asymmetric induction in DD-mode tandem cyclizations.[184] Thus, β-keto sulfoxide **152** undergoes 6-*endo*, 5-*exo* double annulation to afford bicyclo[3.2.1]octanone **153** as a single enantiomer. The chiral auxiliary is removed in two steps to produce the enantiomerically pure target **154**. This strategy of using chiral auxiliaries in the β-dicarbonyl site of the substrate has been further developed by utilizing optically active alcohols (phenylmenthol, *trans*-2-phenylcyclohexanol, naphthylborneol) or amines (i.e., 2,5-dimethylpyrrolidine).[185,186] The level of asymmetric induction is 23–92%, with the best result obtained with phenylmenthol (90% yield, 92% de).

152   **153** (44%)   **154**

Oppolzer's D-camphorsultam has been used to effect asymmetric induction in β-ketoamide **155**.[187] Manganese(III)-promoted tandem cyclization (DD-mode) produces diastereomers **156** and **157** in the ratio 3:1 (50% de). The opposite stereochemical result has been obtained with the L form of the chiral auxiliary.

Reductive liberation of separable diastereomers **156** and **157** has also been described.

155　　　　　　　　　　　　　**156** (49%)　　　**157** (17%)

R = D-Camphor sultam

NEW DIRECTIONS IN MANGANESE(III) CHEMISTRY

There have been many recent advances in the field of Mn(III)-mediated reactions. Among them are the use of carbon monoxide as a trapping agent for adduct radicals (Table XXXIX), electrochemical generation of trivalent manganese (Table XL), the use of ultrasound as a promoter (Table XLI), and novel methods for alkyl radical generation by oxidation of cycloalkanols (Table XLII) or Cr(0) complexes (Table XLIII) by Mn(III) salts.

Carbon Monoxide Trapping Reactions. Mn(III)-mediated inter- and intramolecular reactions, as well as addition–cyclizations, are altered in the presence of carbon monoxide owing to its effective interaction with intermediate adduct radicals.[188] Thus, 5-*exo* cyclization of alkenyl malonate **158** initially produces primary radical **159** and subsequently acyl radical **160**, which oxidatively converts to carboxylic acid **161**. This novel modification is highly useful for synthetic chemistry, and also valuable from a mechanistic point of view as another proof for the formation of free-adduct radicals.

158　　　　　　　　　　　　　　**159**

160　　　　　　　　　　　　**161** (50%)

Electrochemical generation of Mn(OAc)₃ is an environmentally benign procedure because it allows use of catalytic amounts of Mn(OAc)$_2$. The results of intermolecular reactions (Table XLA), addition–cyclizations (Table XLB), as well as intramolecular processes[189] are virtually the same as those of the nonelectrochemical variants, although some new types of proradicals, such as $BrCCl_3$, CBr_4, CF_2Br_2, and alkyl iodides and dibromoacetates, have been successfully employed.[190] The stereoselectivity is remarkably high in the addition–cyclization reaction of 1,5-cyclooctadiene, affording cis-bicyclo[3.3.0]octane 162 in high yield.[190]

162 (83%)

Sonochemical reactions between alkenes and cyanoacetic acid and potassium methyl malonate produce γ-lactones (Table XLI).[191] Noteworthy features are lower reaction temperature (0°) and high stereoselectivity in the annulation of cycloalkenes and enol ethers. Thus, cyclohexene is lactonized with both carbonyl compounds producing cis-fused bicyclics with high stereoselectivity.[191]

Cycloalkanols and Cr(0)-Complex-Derived Alkyl Radicals. Oxidation of cyclopropanols, cyclobutanols, and Cr(0) complexes with Mn(III) salts constitute new methods of alkyl radical generation (Eq. 2).[22,23,25] Trapping the alkyl radicals with a large number of enol ethers and silyl enol ethers, as well as with electron-deficient alkenes, results mostly in the formation of H-atom transfer products (Tables XLIIA and XLIII). Intramolecular additions of the cyclobutanol-derived alkyl radicals produce polycyclic compounds with *cis* stereoselectivity (Table XLIIC).[24] In contrast, analogous cyclizations of cyclopropanol-derived alkyl radicals afford *trans*-fused bicyclic systems (Table XLIIB).[23] Thus, substrate 163 undergoes oxidative ring enlargement with the formation of cycloheptenyl radical 164; the latter attacks the double bond in 5-*exo* mode to produce *trans*-fused bicycle 165. Trapping of adduct radicals has also been accomplished with silyl enol ethers, tributyltin hydride, and diphenyl diselenide.[192]

163

164

165 (81%)

APPLICATIONS TO NATURAL PRODUCT SYNTHESIS

The maturity and significance of any field of organic chemistry can be assessed by its applications to the synthesis of natural and biologically active compounds. Manganese(III) chemistry is at the initial stage of its "practical" utilization, which at this point looks very promising. The common approach is the use of the original intra- and intermolecular model reactions between certain types of substrates and reagents as key steps in the construction of naturally occurring molecules or their analogues and precursors. As the examples of this section demonstrate, manganese(III) mediation is highly efficient in the construction of the diverse structural types ranging from straight-chain acyclic compounds to architecturally complex polycyclic multifunctional derivatives.

Pheromones

Pheromones constitute a group of naturally occurring compounds that effect chemical communication among insects and animals. In particular, sex pheromones are highly efficient in field trials for monitoring purposes as well as for plant protection. In the past two decades, major efforts have been directed toward the elaboration of novel synthetic methods for supplying sufficient amounts of these compounds for their practical use. Manganese(III)-mediated reactions have made a solid contribution to this field, which is substantiated by examples described in this section.

Queen Substance.[193]   Queen bee pheromone **166** plays an important role in the metabolism of bees. An alkene–ketone reaction (Table II) has been used for three-carbon homologation of the starting material **167**. The useful yield of **168**, as well as its easy conversion to the final product, make this approach one of the most effective among those reported in literature.[193]

Popillia Japonica.[153]   Manganese(III)-mediated lactonization of conjugated enynes (Table XVII) has been used as a key step in the synthesis of (5Z)-tetradecen-4-olide (**169**), the sex pheromone of the Japanese beetle *Popillia japonica*. Stereoselective hydrogenation of the triple bond and subsequent lactonization with acetic acid have been used in both direct and reversed order, providing different stereochemical results. In the former sequence, enyne **170** is reduced stereoselectively to (1,3Z)-dodecadiene (**171**), which reacts with acetic acid to afford an isomeric mixture of the natural compound (**169**:**172** = 56:44).

The observed partial inversion of the Z double bond occurs in Z-allylic adduct radical **173**, analogous to that in dihydrofuran synthesis.[73] In contrast, lactonization–hydrogenation sequence **170** → **174** → **169** produces the sex pheromone **169** with high stereoselectivity (Eq. 26).[153]

(Eq. 26)

Acanthoscelides Obtectus.[157]   Acetylenic lactone **174** has also been used as an intermediate in the synthesis of methyl (2E,4,5)-tetradecatrienoate (**175**), the sex pheromone of the bean weevil moth *Acanthoscelides obtectus*. The method effectively produces multigram quantities of racemic **175**, although its potential for enantioselective synthesis remains to be explored (Eq. 27).[157]

(Eq. 27)

Scotia Exclamationis.[159]   The "magic" lactone **174** has become the source of (5Z)-tetradecenyl acetate (**176**), the sex pheromone of the butterfly *Scotia exclamationis*. Along with Mn(OAc)₃-promoted lactonization, another crucial step in this approach is the novel regioselective reduction of the secondary propargyl chloride **177** effected by the Co₂(CO)₆ group (Eq. 28).[159]

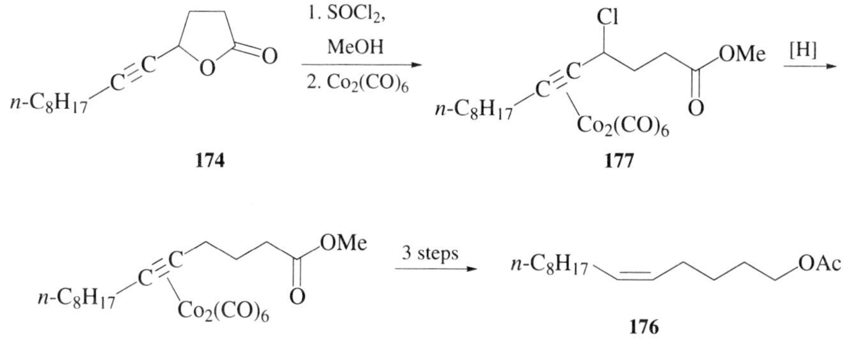

174 177 176

(Eq. 28)

Keiferia Lycopersicella.[158] Lactonization of 1,3-undecenyne with acetic acid (Table XVII) provides lactone **178**, a key intermediate in the synthesis of (4*E*)-tridecenyl acetate (**179**), the sex pheromone of the tomato pinworm *Keiferia lycopersicella*. This scheme features the highly regio- and *E*-stereoselective one-step conversion of lactone **178** into carboxylic acid **180** (hydrogenolysis–migration reaction, Eq. 29).[158]

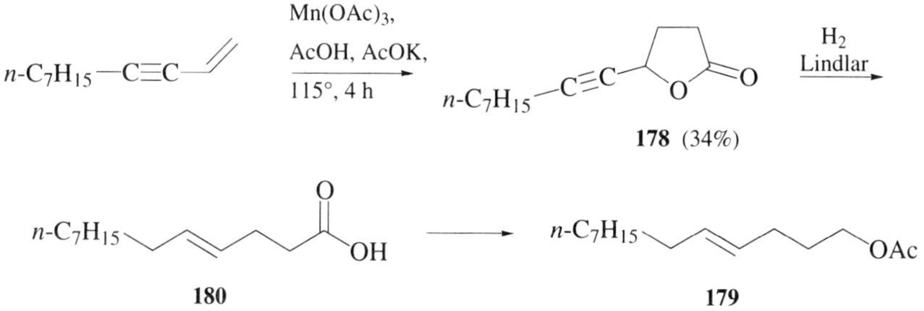

178 (34%) 180 179

(Eq. 29)

Hylecoetus Dermestoides.[156] The manganese(III)-mediated reaction of 1,3-alkadienes and β-dicarbonyl compounds (Table XVI) has been used as an initial step in the synthesis of 2-isopropyl-2,5-dimethyl-2,3-dihydrofuran (**181**), the sex pheromone of the beetle *Hylecoetus dermestoides* (Eq. 30).[156]

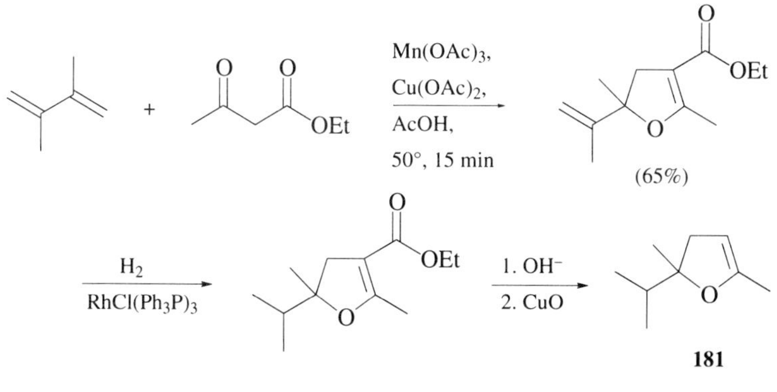

(65%) 181

(Eq. 30)

Terpenes and Terpenoids

Himasecolone.[194]   An alkene–ketone reaction (Table II) has been utilized in the short synthesis of *Himasecolone* **182**, a phenolic sesquiterpenoid isolated from plant sources (Eq. 31).[194] Secondary adduct radicals formed in the addition step abstract a proton from acetone or acetic acid, since $Mn(OAc)_3$ does not oxidatively interact with these transient species (**32**, Eq. 6). Subsequent hydrolysis of the HAA product releases target molecule **182**.

182 (50%)

(Eq. 31)

β-Bisabolene.[195]   Carboxymethylation of limonene with acetic acid and acetic anhydride (Tables III and XV) has produced 4-alkenoic acid **183** in low yield. Other products formed remain undescribed, although based on related studies their nature could be anticipated. Compound **183** was converted to the natural sesquiterpene β-bisabolene **184** in three conventional steps (Eq. 32).[195]

Norbisabolide.[195,196]   The C_{12} terpene lactone **185**, isolated from plants, was first synthesized from limonene in one step and in low yield.[195] In an optimized protocol, the same composition $Mn(OAc)_3$–AcOH–Ac_2O produces 4-alkenoic acid **183**, which undergoes acid-catalyzed cyclization to the target molecule **185** in 43% overall yield (Eq. 33).[196]

3 steps   (Eq. 32)

$Mn(OAc)_3$,
AcOH, AcOK
Ac_2O, 120°, 1 h

184

183

HCO_2H (Eq. 33)

185 (43%)

Dihydropallescensin D.[197] Intramolecular cyclization of olefinic β-ketoester **186** (Table XXVA) has been used as a key step in the synthesis of dihydropallescensin D, the natural furanosesquiterpene **187**.[197] The cyclization proceeds with high regio- and stereoselectivity and affords bicyclo[4.3.1]decane **188** as a single stereoisomer with a *cis* relationship between bridgehead hydrogens and the carbomethoxy group. Copper(II) acetate is used as a cooxidant to promote an oxidative deprotonation of the intermediate tertiary radical adduct. The regioselectivity of the elimination step is determined by the higher thermodynamic stability of regioisomer **188** with the exocyclic double bond (Eq. 34).[197]

186

Mn(OAc)$_3$,
Cu(OAc)$_2$
AcOH,
25°, 3 h

188 (62%)

1. LiCl, DMSO
2. Me$_3$SiCl
3. MCPBA
4. HCl

1. TMSC≡CLi
2. K$_2$CO$_3$
3. H$^+$

187

(Eq. 34)

Bioactive Compounds

(±)-Paeoniflorigenin.[198] Paeoniflorigenin (**189**) is the nonglucoside part of Paeoniflorin, an active ingredient of the essential oils from *Paeonia lactiflora* that have diverse biological activity.[198] Their syntheses remained an unsolved problem until recently, when both target molecules were successfully constructed. The first step is of immediate relevance, and consists of the lactonization of unsymmetrically substituted 1,4-cyclohexadiene **190** with cyanoacetic acid (Tables X and XV). Lactone **191** is formed chemo-, regio-, and stereoselectively, that is, the electron-rich double bond is exclusively attacked by an electrophilic educt radical, and the bicyclic[4.3.0] system with *trans* configuration is formed. It is noteworthy that lactonization of the unsubstituted cyclohexene with cyanoacetic acid at elevated temperatures produces a mixture of four stereoisomers.[129] The observed *trans* stereoselectivity may be produced by the bulky substituent at the double bond (Eq. 35).[198]

190

Mn(OAc)$_3$
AcOK, MeCN,
rt, 15 h

191 (48%)

12 steps

189

(Eq. 35)

Podophyllotoxin Analogue.[136]   Reactions between styrenes and β-ketoesters have long been known to produce 4,5-dihydrofurans.[33] Styrenes with activated double bonds and aromatic 3-ketoesters react analogously (Eq. 11).[136] In particular, styrene **192** and carbonyl compound **193** produce dihydrofuran **194** by intramolecular cyclization of the intermediate benzylic cation on the benzoyl group. This reaction proceeds with high regioselectivity in the addition step and high stereoselectivity in the intramolecular cyclization. Under acidic conditions, *trans* isomer **194** is rearranged into 4-aryltetralone **195**, a structural analogue of the naturally occurring lignane podophyllotoxin (Eq. 36).[136]

(±)-14-Epiupial.[199]   Racemic 14-epiupial (**196**) represents one of the most sophisticated natural molecules that has been synthesized using Mn(III) methodology. Cyclohexene derivative **197**, containing a malonic ester moiety, undergoes partial hydrolysis and subsequent tandem intramolecular cyclization with the sequential formation of six-membered and γ-lactone rings (Eq. 37). The lactone-bridged bicyclo[3.3.1]nonane **198** has been converted by a sequence of steps into target molecule **196**, which appeared to be isomeric with the natural compound.

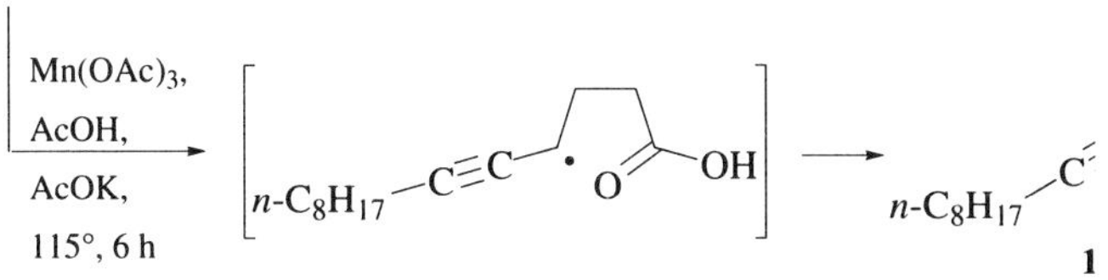

(Eq. 37)

Further attempts to construct the "right" isomer clearly revealed the high sensitivity of Mn(III)-mediated cyclizations toward the configuration of stereogenic centers. Thus, the reaction fails completely after epimerization of the MOMO-bearing carbon atom (a) and proceeds with low yield (9%) if the stereochemistry of carbon (b) is inverted (Eq. 38).[199]

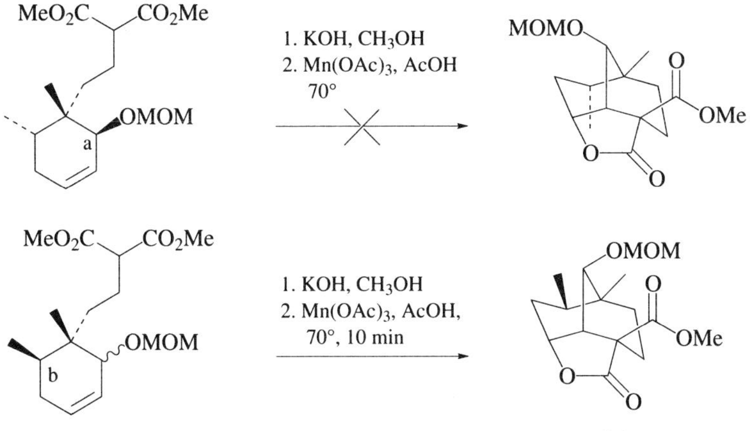

(Eq. 38)

(9%)

Sorbic Acid.[200] Sorbic acid (**199**) is a food preservative that is produced on a large commercial scale. An efficient practical method for its synthesis has been developed, including (a) 1,2 and 1,4 conjugate addition to butadiene mediated by electrochemically generated Mn(OAc)$_3$; (b) Cu(I)-promoted conversion of the isomeric acetoxy acids into γ-vinyl-γ-butyrolactone (**200**); and (c) acid-catalyzed rearrangement of the latter into sorbic acid (**199**, Eq. 39).[200]

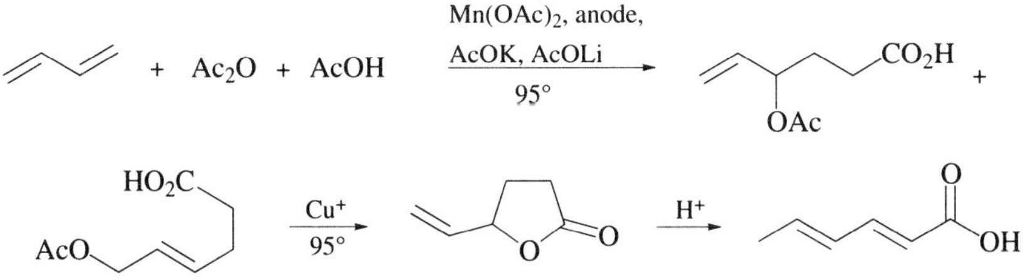

(Eq. 39)

Formal Syntheses

Pyrenophorin.[201] The model reaction between alkenes and ketones mediated by Mn(OAc)₃ and producing saturated ketones (Table II) has been modified, first by using Cu(OAc)₂ to effect oxidative deprotonation of adduct radicals, and second by introducing a sulfonyl group into substrate **201**.The latter allegedly directs regioselective deprotonation in adduct radicals, producing *exo*-methylene derivative **202**. The scope of the reaction was later expanded[59] by involving both acyclic and cyclic ketones. γ,δ-Enone **202**,isolated in moderate yield (27%), was converted into the known precursor **203** in four steps. This method constitutes a facile access to functionalized γ-keto acrylates, and a formal synthesis of the natural fungicide pyrenophorin (**204**) (Eq. 40).[201]

(Eq. 40)

Podocarpic Acid.[85] Intramolecular tandem cyclization in the DB-mode (Table XXXIII) has been used to construct stereoselectively the tricyclic core of the phenanthrene derivative **151**. The latter has been reduced to **205**, a racemic precursor in the synthesis of podocarpic acid (Eq. 41).[85,91]

(Eq. 41)

EXPERIMENTAL CONDITIONS

Manganese(III) Salts

Mn(OAc)₃. Manganese(III) acetate has two modifications, a dihydrate form Mn(OAc)₃ · 2H₂O and an anhydrous form Mn(OAc)₃. The dihydrate is commercially available (Fluka). It can be synthesized in the laboratory by oxidation of $Mn(OAc)_2$ with $KMnO_4$,[164,202] or in situ by anodic oxidation.[169,190,203–205] Anhydrous $Mn(OAc)_3$ can be prepared by oxidation of $Mn(OAc)_2$ with $KMnO_4$ in glacial AcOH in the presence of Ac_2O,[9,27] or by treating $Mn(NO_3)_2 \cdot 6H_2O$ with Ac_2O.[164] The structure of the anhydrous form has been elucidated by X-ray diffraction,[206] and is polymeric with repeating units consisting of three molecules of $Mn(OAc)_3$. Manganese atoms form an equilateral oxo-centered triangle surrounded by six bridging acetate ligands and another two acetate moieties that link repeating units to each other. The trimer of $Mn(OAc)_3$ might be capable of oxidizing up to three radical precursors, although the fact that an excess is often required to bring reactions to completion indicates that its oxidizing power and mode of performance depends on the nature of the process. The hydrated form has been used in the vast majority of Mn(III)-mediated reactions because of its availability and because the anhydrous form has no apparent advantages. In one comparative study on lactonization of alkenes,[117] the hydrated form provided higher yields of γ-lactones.

The thermal decomposition of $Mn(OAc)_3$ has been investigated by TG, DTG, and DTA methods,[207] as well as in AcOH[27] and Ac_2O.[208] Thermogravimetric analysis reveals its stability up to 110° in air and, surprisingly, only up to 40° under anaerobic conditions.[207] At 110° it undergoes chemical degradation by releasing carboxymethyl radical, which produces acetoxyacetic (20%) and succinic (2%) acids along with carbon dioxide (15%) and methane (2%).[27] The reduction reaction is first order in trivalent manganese with an energy of activation of ~28 kcal mol^{-1}.[27] Succinic anhydride and acetoxyacetic acid are the major products when it is treated at 120° in Ac_2O containing 10% AcOH.[208] This reaction is zero order in Mn(III) salt and is accounted in terms of enolization of Ac_2O as the rate-determining step. Trifluoroacetic acid is suggested to accelerate the enolization step, and thus the reduction overall.[208]

Mn(acac)₃. Tris(acetylacetonato)manganese(III) has been used in some reactions to produce acetylacetonyl radicals.[38,45,55,109,135] The chemistry of the process is virtually the same as that of acetylacetone oxidation by $Mn(OAc)_3$, although the former appears to be a milder oxidant.[13]

Mn(pic)₃. ($PyCO_2)_3Mn$, or Mn(III) tris(2-pyridinecarboxylate), has a short history as a mediator of radical reactions between unsaturated systems and carbonyl compounds. Japanese authors have recently described its use under mild conditions in DMF.[22,23,25,118,209] Analogous to $Mn(OAc)_3$, it is able to oxidize carbonyl compounds,[118,209] as well as cyclopropanols,[22,23] cyclobutanols,[24] and Cr(0) complexes.[25]

Stoichiometry

The initial step of the reaction is a one-electron oxidation of a carbonyl compound with a Mn(III) salt, requiring 1 equivalent of the latter. If manganese(III) ions participate in the oxidation of the intermediate radical-adducts to cations, overall 2 equivalents of metal oxidant will be required, unless regeneration of Mn(III) ions occurs during the reaction. It does occur in the presence of oxygen, which enables the use of catalytic amounts of $Mn(OAc)_3$ or $Mn(OAc)_2$,[52] although this protocol has serious limitations because of the sensitivity of many unsaturated organic compounds to oxygen. One of the latest developments is the electrochemical generation of $Mn(OAc)_3$ by anodic oxidation of equimolar or even catalytic amounts of $Mn(OAc)_2$.[169,189,190,203-205] The use of $Cu(OAc)_2$ as a cooxidant requires doubling the amount of Mn(III) salt, since the latter regenerates Cu(II) from Cu(I) ions. A separate issue is the molar ratio of substrate to carbonyl compound. Although equimolar amounts are normally required by the stoichiometry of the reaction, the carbonyl components have often been used in excess or even as solvent. The optimization of any reaction remains empirical, and not every reported reaction has been optimized. Even if a closely related type of process has been described, one has to realize that structural changes, even subtle ones either in substrate or carbonyl component, might require some additional optimization.

Solvents

The most common solvent in Mn(III)-mediated reactions is glacial acetic acid. It was first described in the pioneering papers,[28,125] and since then has been heavily used. Although $Mn(OAc)_3$ is not soluble in acetic acid at room temperature, reactions can be effectively executed with the suspension. At elevated temperatures (50–140°), the system becomes homogeneous. The intermediate case consists of initial homogenization of the solution by short heating, and subsequent cooling to room temperature prior to addition of the reagents. This procedure allows operation with an almost homogeneous system at lower temperatures. Although glacial acetic acid is efficient in a large number of reactions, it is not free of limitations. First, the acidic character makes it incompatible with acid-sensitive substrates. Second, acetic acid can participate in solvolysis or LTR reactions to produce acetoxy derivatives which are not always desired. Alternative solvents used in recent years are aliphatic alcohols (methanol,[57] ethanol[24,87,95,97,99,171,179]), dimethylformamide,[22,23,25,118,209] hexane,[49] dioxane,[82] acetonitrile,[172] and chloroform.[93]

Temperature

The temperature of the process is largely determined by the first step—generation of the α-oxo- and α,α-dioxoalkyl radicals by oxidation of the carbonyl compounds. The rate of the reaction is dependent on the structure of the carbonyl component, in particular, the number of carbonyl groups, enolizability, and C—H acidity. Usually, compounds with more carbonyl groups (more easily enolizable and more acidic) react with the oxidant at lower temperatures. The overview for different types of carbonyl compounds provides an idea of the temperature ranges used for different classes of carbonyl compounds:

Aldehydes	50–70°
Ketones	40–120°
Monocarboxylic acids	100–140°
Anhydrides	120–140°

β-Dicarbonyl compounds	20–100°
β-Ketocarboxylic acids	20°
Dicarboxylic acids	70–120°
Dicarboxylic acid derivatives	23–120°
Ortho esters	60–65°
Nitroalkanes	83–120°

The temperature limit for Mn(III)-promoted reactions has recently been brought down to 0° by using ultrasound[191] or Cr(0) complexes as a source of alkyl radicals.[25]

Concentration

The role of metal oxidant entails the generation of radicals and the oxidative interaction with adduct-radicals formed along the reaction coordinate. The relative rates of the intermediate steps determine the pattern of products and depend upon the concentration of metal oxidant. Thus, concentration becomes an important parameter of the reaction, and the selectivity of the process is dependent on it. This parameter, which has often been chosen by analogy with previous work or selected empirically, remains mostly arbitrary. It has resulted in the use of different concentrations of $Mn(OAc)_3$ even in the same type of reaction. The most striking example is the interaction of alkenes with acetic acid–acetic anhydride, where the concentration of $Mn(OAc)_3$ has varied widely (0.01, 0.067, 0.14, 0.35, 0.50, 0.56, 0.69, 0.81, and 1.69 mol L^{-1}). It is not surprising that this results in a diversity of product distributions, including the formation of saturated carbon acids or alkan-4-olides as single products, simultaneous formation of 4-acetoxy-acids, alkan-4-olides, 3- and 4-alkenoic acids, as well as alkan-4-olides together with saturated carboxylic acids or allylic acetates. The importance of this parameter is not adequately recognized; in most publications it is not considered. The Mn(III) concentration should be kept uniform for given types of reactions so that procedures will become standardized and the results predictable for analogous processes. The arguments above are also germane to $Cu(OAc)_2$, which is the commonly used cooxidant.

EXPERIMENTAL PROCEDURES

n-C$_5$H$_{11}$ ⟍⫽ + ⟍CHO →[Mn(OAc)$_3$, AcOH, 60°, 1 h] n-C$_5$H$_{11}$⟍⟍CHO

206 (28%)

2-Methylnonanal (206).[41] $Mn(OAc)_3 \cdot 2H_2O$ (23.5 g) was dissolved in 105 mL of AcOH, and propionaldehyde (101.5 g) and 1-heptene (34.2 g) were added to the solution. The mixture was then heated in an atmosphere of nitrogen at 60°. After the brown color of the trivalent manganese salt had disappeared (1 hour), the excess of the original reagents and most of the acetic acid were distilled under vacuum. The residue was treated with water, the mixture was extracted with ether, and the extract was dried and submitted to fractional distillation in a stream of nitrogen. 2-Methylnonanal (**206**, 3.8 g, 28%) was obtained, bp 68–70° (7 mm Hg). DNPH-derivative: mp 87–88°; IR 1725 cm^{-1}; ^{1}H NMR (CDCl$_3$) δ 9.4–9.5 (J = 1.6–2.7 Hz). Anal. Calcd for C$_{10}$H$_{20}$O: C, 76.86; H, 12.90. Found: C, 76.62; H, 12.92.

207 (56%)

4,5-Diethoxycarbonyl-(2H)-tetrahydrofuran-2-one (207),[131] In a round-bottomed flask equipped with a condenser were added diethyl maleate (2.0 mmol), glacial AcOH (40 mL), and Mn(OAc)$_3$ · 2H$_2$O (1.08 g, 4.0 mmol). The mixture was heated under reflux in a nitrogen atmosphere until the dark-brown color of the manganese(III) salt disappeared (7 hours), after which most of the AcOH was removed under reduced pressure, and the mixture was allowed to cool to room temperature. The resulting precipitate of Mn(OAc)$_2$ was separated by filtration and washed carefully with a small amount of EtOAc. The washing and filtrate were combined and evaporated under reduced pressure to afford a residue which was chromatographed on a silica gel column using 50% ethyl acetate–hexane as eluent. Lactone **207** was obtained in a 56% yield as a mixture of cis:$trans$ isomers (1:5.7), bp 138–143° (0.25 mm Hg). ^{1}H NMR (CDCl$_3$) cis isomer: δ 3.7–3.8 (m), 5.10 (d, J = 8.4 Hz); $trans$ isomer: 3.4–3.5 (m), 5.15 (d, J = 4.4 Hz). Anal. Calcd for C$_{10}$H$_{14}$O$_6$: C, 52.17; H, 6.13. Found: C, 51.99; H, 6.19.

R^1, R^2 = Ph, Alk; R^1 + R^2 = (CH$_2$)$_4$ **46** (29-82%)

Lactone Annulation (General Procedure).[117] A 100-mL round-bottomed flask equipped with a reflux condenser, nitrogen inlet, and magnetic stirrer was charged with the alkene (5 mmol), Mn(OAc)$_3$ · 2H$_2$O (4.17 mmol), AcOK (25 mmol), and glacial AcOH (50 mL). The mixture was refluxed until the dark brown color disappeared, cooled, and diluted with water (200 mL). The organic products were extracted with ether (5 × 40 mL). The combined ethereal extracts were washed with H$_2$O (2 × 40 mL), saturated NaHCO$_3$ solution (2 × 40 mL), dried (MgSO$_4$), evaporated, and chromatographed. Lactones **46** were isolated in 29–82% yield.

208 (76%)

2-Methyl-3-(phenylsulphenyl)-5,5-diphenyl-4,5-dihydrofuran (208).[53] To a heated solution of phenylsulfenylacetone (1 mmol) and 1,1-diphenylethene (1 mmol) in AcOH (25 mL), Mn(OAc)$_3$ · 2H$_2$O (4 mmol) was added. The mixture was stirred at 80° for 12 minutes. The reaction was quenched by adding H$_2$O (60 mL), and the mixture was then extracted with benzene. After removing the benzene, the product **208** was separated as a pale yellow oil, either on TLC (Wacogel B10) eluting with CHCl$_3$ or on a silica-gel column eluting with benzene. IR (CHCl$_3$) 1642, 1018 cm^{-1}; ^{1}H NMR (60 MHz, CDCl$_3$) δ 2.38 (t, J = 1.6 Hz, 3 H), 2.86 (dq, J = 1.6, 14.4 Hz, 1 H), 3.72 (dq, J = 1.6, 14.4 Hz, 1 H), 7.07–7.73 (m, 15 H); high-resolution MS m/e Calcd for C$_{23}$H$_{20}$O$_2$S 360.1184. Found 360.1273.

61 (81%)

(±)-(3α,5β,8R*)-, (±)-(3α,5α,8S*)- and (±)-(3α,5α,8R*)-3,8-Dimethyl-3,8-diphenyl-2,7-dioxaspiro[4.4]nonane-1,6-dione (61). (Spirodilactonization Reaction).[145] Mn(OAc)$_3$ · 2H$_2$O (2.7 g, 3.35 mmol) was weighed into a 50-mL flask equipped with a magnetic stirrer. Glacial AcOH (25 mL) was added and the flask was placed in a 70° oil bath and stirred. Malonic acid (0.26 g, 2.5 mmol) was added, immediately followed by addition of α-methylstyrene (0.59 g, 5 mmol). The flask was fitted with a reflux condenser and gas-inlet tube (N$_2$) and the reaction was allowed to proceed until the mixture turned colorless (2 hours). After being cooled, the reaction mixture was quenched with 250 mL of H$_2$O and extracted with 50–75 mL of CH$_2$Cl$_2$ or Et$_2$O. The organic extract was washed with saturated NaHCO$_3$ until the washings were no longer acidic, subsequently washed twice with H$_2$O, dried (MgSO$_4$), and concentrated to give spirodilactones **61**. MPLC separation of the diastereomers employed 20–40% AcOEt–hexane. Diastereomeric ratio ss:u:sa 20:49:31. Total yield 81%. **61-ss** : mp 216.5–217°; IR (KBr) 1775, 1753, 1600, 1495 cm^{-1}; ^{1}H NMR (CDCl$_3$, 300 MHz) δ 1.78 (s, 6 H), 2.77 (d, J = 13.3 Hz, 2 H), 3.25 (d, J = 13.3 Hz, 2 H), 7.5–7.2 (m, 10 H); **61-u** : mp 96.5–97°; IR (KBr) 1775, 1753, 1595, 1495 cm^{-1}. ^{1}H NMR (CDCl$_3$, 300 MHz) δ 1.65 (s, 3 H), 1.90 (s, 3 H), 2.10 (d, J = 13.6 Hz, 1 H), 2.80 (d, J = 13.2 Hz, 1 H), 2.80 (d, J = 13.6 Hz, 1 H), 3.15 (d, J = 13.2 Hz, 1 H), 7.5–

7.1 (m, 10 H); **61-sa** : mp 141–143°; IR (KBr) 1778, 1754, 1600, 1495 cm^{-1}; ^{1}H NMR (CDCl$_3$, 300 MHz) δ 1.75 (d, 6 H), 2.17 (d, J = 13.3 Hz, 2 H), 2.67 (d, J = 13.3 Hz, 2 H), 7.5-7.2 (m, 10 H).

209 (64%)

6,6-Dimethyl-4,5,6,7,4′,5′-hexahydrospiro[benzofuran-2(3H),2′(3′H)-furan-4,5′-dione] (209).[141]   Mn(OAc)$_3$ · 2H$_2$O (1.0 g, 3.7 mmol) was heated in acetic acid (8 mL) under nitrogen at 60–70° until a black homogeneous solution was obtained. Dimedone (0.38 g, 2.7 mmol) and γ-methylene-γ-butyrolactone (0.24 g, 2.5 mmol) were added and the reaction mixture was kept at 60° until the color had disappeared. To the cold mixture, H$_2$O was added and the solution was extracted with CH$_2$Cl$_2$. The combined organic extracts were washed with saturated NaHCO$_3$ solution and evaporated under reduced pressure to give an oil. Recrystallization (EtOAc) afforded as white flakes compound **209** (0.38 g, 64%), mp 116–117° IR (CHCl$_3$) 1815 and 1650 cm^{-1}; ^{1}H NMR δ 1.11 (s, 3 H, CH$_3$), 2.25 (m, 2 H, CH$_2$), 2.36 (m, 2 H, CH$_2$), 2.40–2.90 (m, 4 H, 2CH$_2$), 2.98 (d, J = 15 Hz, 1 H) and 3.20 (d, J = 15 Hz, 2 H, CH$_2$). MS m/z: M$^+$ 236 (37%). Anal. Calcd for C$_{13}$H$_{16}$O$_4$: C, 65.7; H, 6.76. Found: C, 66.0; H, 6.74.

200 (39%)

γ-Vinyl-γ-butyrolactone (200).[200]   Under nitrogen, 115 g (0.43 mol) of Mn(OAc)$_3$ · 2H$_2$O, 24.5 g (0.35 mol) of AcOK, and 385 mL of glacial AcOH were charged into a 760-mL top-stirred stainless steel Parr reactor. The reactor was chilled and 43 g (0.89 mol) of 1,3-butadiene was condensed into the reactor. Stirring was begun and the reactor was heated to 140° over 20 minutes. The reactor was maintained at 140° with stirring for 3 hours, then cooled and vented. The product mixture, consisting of a white solid plus yellow liquid, was taken up in 210 mL of H$_2$O and 100 mL of ether. The water layer was separated and extracted with three 75-mL portions of ether. The combined ether solutions were dried over MgSO$_4$ and stripped on a rotary evaporator to remove ether. The residue was distilled through a short-path microware still at 25–160° (0.8 mm Hg) to give 25.3 g of distillate (trapped in dry ice). The distillate was fractionated through a 6-in. Vigreux column to give 9.3 g (39%) of product **200**, bp 97–100° (15 mm Hg), 99% purity by GC analysis. ^{1}H NMR (60 MHz, CD$_3$CN) δ 1.6–2.7 (m, 4 H), 4.7–6.3 (m, 4 H). MS m/z: 112 (100%).

210 (57%)

4-Acetoxy-5-tetradecynoic Acid (210).[62] In a dried, argon-filled round-bottomed flask fitted with stirrer and thermometer was placed Mn(OAc)$_3$ · 2H$_2$O (5.36 g, 20 mmol) and 1-tetradecen-3-yne (1.64 g, 10 mmol); glacial AcOH (120 mL) and Ac$_2$O (40 mL) were consequently added at room temperature. The suspension was heated to reflux until the brown color disappeared (115°, 10 minutes). The mixture was cooled to room temperature, diluted with H$_2$O (300 mL), and extracted with Et$_2$O (2 × 150 mL). The combined ethereal layers were washed with H$_2$O (2 × 100 mL), and then saturated Na$_2$CO$_3$ was poured in slowly until the solution was basic. The aqueous layer was separated, acidified with 5% HCl to pH 1, and extracted with Et$_2$O (3 × 100 mL). The combined ethereal layers were washed with water (2 × 100 mL) and dried (MgSO$_4$). The ether was evaporated, the residue was chromatographed over silica gel (15 g, hexane–Et$_2$O, 4:1) to give 1.61g (57%) of **210**. R_f = 0.42 in hexane–ether 1:2. IR (neat) 3400–2500, 2254, 1745, 1720 cm^{-1}. ^{1}H NMR (400 MHz, CDCl$_3$) δ 0.88 (t, J = 6.7 Hz, 3 H), 1.17–1.40 (br.s, 10 H), 1.50 (quintet, J = 7.3 Hz, 2 H), 2.06 (s, 3 H), 2.07 (td, J = 7.4, 6.1 Hz, 2 H), 2.19 (td, J = 1.9 Hz, 2 H), 2.54 (t, 2 H), 5.44 (tt, 1 H), 10.6 (s, 1 H). ^{13}C NMR (75 MHz, CDCl$_3$) δ 14.11, 18.67, 20.98, 22.66, 28.41, 28.85, 29.03, 29.16, 29.70, 29.86, 31.83, 63.35, 76.38, 87.23, 169.74, 178.73.

93 (28%)

(*cis*-1,8-Dehydro-3-ethynyl-2-oxatricyclo[6.4.0.0^{3,7}]dodecan-9-one) Dicobalt Hexacarbonyl (93).[57] A reaction flask was charged with Mn(OAc)$_3$ · 2H$_2$O (4.34 g, 16.2 mmol) under nitrogen. After five pump-and-fill cycles, a solution of (cyclopentenylacetylene)Co$_2$(CO)$_6$ complex(1.0 g, 2.7 mmol) and 1,3-cyclohexanedione (3.63 g, 32.4 mmol) in a glacial AcOH (54 mL) was added in one portion [molar ratio substrate: Mn(III):β-dicarbonyl compound 1:6:12]. The mixture was heated for 1 hour at 30° with stirring (TLC monitoring). Aqueous workup and subsequent column chromatography (SiO$_2$, 200 g, PE/E 1:2) gave **93** (370 mg, 28%) as dark-red crystals together with **211** (56 mg, 10%). T$_{decomp.}$130–135°. R_f = 0.48 (PE:E, 1:3). ^{1}H NMR (300 MHz, CDCl$_3$): δ 1.56–1.73 (m, 1 H), 1.75–2.10 (m, 7 H), 2.26–2.38 (m, 2 H), 2.40–2.54 (m, 2 H), 3.25 (d, J = 7.3

Hz, 1 H), 6.05 (s, 1 H). MS m/z 488 (1%), 320 (26%). Anal. Calcd for
$C_{19}H_{14}O_8Co_2$: C, 46.72; H, 2.87. Found: C, 46.60; H, 2.72.

cis-1,8-Dehydro-3-ethynyl-2-oxatricyclo[6.4.0.0^{3,7}]dodecan-9-one (211).[57]
Decomplexation of 0.64 mmol of **93** with $Ce(NH_4)_2(NO_3)_6$ at $-78°$ afforded **211**
(111 mg, 86%), R_f = 0.39 (PE:E, 1:3). ^1H NMR (300 MHz, ac-d_6) δ 1.45–1.57
(m, 1 H), 1.71–1.90 (m, 3 H), 1.93–2.09 (m, 4 H), 2.21–2.29 (m, 2 H), 2.39–
2.50 (m, 2 H), 3.38 (s, 1 H), 3.55 (d, J = 7.9 Hz, 1 H). ^{13}C NMR (ac-d_6): 22.5,
24.2, 24.7, 32.9, 37.3, 43.1, 53.8, 77.1, 84.3, 92.6, 115.7, 176.2, 194.4. MS m/z
M + 202. Anal. Calcd for $C_{13}H_{14}O_2$: C, 77.23; H, 6.93. Found: C, 77.02; H, 6.89.

Triethyl (5-Benzoylpyrrol-2-yl)methanetricarboxylate (212).[37] $KMnO_4$
(1.0 g, 6.3 mmol) was added to a hot stirred solution of $Mn(OAc)_2 \cdot 4H_2O$ (6.13 g,
25 mmol) in AcOH (50 mL). After 0.5 hour, Ac_2O (9.68 g, 75 mmol) was added
cautiously and then the mixture was cooled to room temperature. Triethyl
methanetricarboxylate (2.55 g, 11 mmol), 2-benzoylpyrrole (1.88 g, 11 mmol)
and AcONa (1.64 g, 20 mmol) were added and the resulting mixture was stirred
at 60–65° in a N_2 atmosphere for 1 day. Water (20 mL) was added and the product was extracted into toluene (3 × 100 mL). The extract was washed with H_2O
and evaporated in vacuo. The oily residue was purified by flash-column chromatography on silica gel using hexane/EtOAc to produce **212** (86%), mp 72.5–74°
(MeOH). ^1H NMR (300 MHz, $CDCl_3$) δ 1.32 (t, J = 7.1 Hz, 9 H), 4.35 (q,
J = 7.1 Hz, 6 H), 6.36 (dd, J = 2.7, 3.9 Hz, 1 H), 6.82 (dd, J = 2.7 Hz, 1 H),
7.57 (m, 3 H), 7.88 (m, 2 H), 10.48 (br, 1 H). MS m/z: 401.

1,1-Diethoxycarbonyl-3-hexyl-4-methylenecyclopentane (213).[81]   A mixture of 115 mg (0.58 mmol) diethyl allylmalonate, 1.3 g (11.6 mmol) of 1-octene,

462 mg (1.72 mmol) of Mn(OAc)$_3$ · 2H$_2$O, and 178 mg (0.89 mmol) of Cu(OAc)$_2$ · H$_2$O in 10 mL of glacial AcOH was heated in a 90° oil bath for 6 hours. The reaction mixture was diluted with 50 mL of EtOAc, washed with three 25-mL portions of saturated NaHCO$_3$, dried (Na$_2$SO$_4$) and concentrated in vacuo. The residue was chromatographed over 10 g of silica gel (EtOAc:hexane, 1:15) to give 1.43 mg (80%) of **213** as colorless oil. IR (CHCl$_3$) 1796 cm^{-1}. ^{1}H NMR (CDCl$_3$) δ 0.88 (t, J = 5.1 Hz, 3 H), 1.24 (t, J = 7.1 Hz, 6 H), 1.05–2.8 (m, 13 H), 2.9–3.1 (m, 2 H), 4.18 (q, J = 7.1 Hz, 4 H), 4.7–5.1 (m, 2 H).

214 (71%)

2-Carbomethoxy-3-(1-propenyl)cyclohexanone (214).[83] To a solution of Mn(OAc)$_3$ · 2H$_2$O (1.376 g, 5.1 mmol) and Cu(OAc)$_2$ · H$_2$O (0.51 g, 2.55 mmol) in 18 mL of glacial AcOH was added a solution of methyl 3-keto-7-decenoate (0.505 g, 2.55 mmol) in 7 mL of glacial AcOH to give an opaque brownish green solution. The mixture was stirred for 1 hour at 50° at which time the solution was light blue and contained a white precipitate. Water was added to give a single cloudy phase in which the white precipitate had dissolved. The solution was extracted with five 15-mL portions of CH$_2$Cl$_2$. The combined organic layers were washed with saturated NaHCO$_3$ until neutral and then with water. The aqueous layer was back-extracted with two 15-mL portions of CH$_2$Cl$_2$. The combined organic layers were dried over MgSO$_4$, and the solvent was removed in vacuo to provide crude **214**. Flash chromatography on silica gel (hexane–ether 3:1) gave 0.365g (71%) of **214** as a 1.3:1 mixture of keto and enol tautomers. IR (neat) 1745, 1715 cm^{-1}. Anal. Calcd for C$_{11}$H$_{16}$O$_3$: C, 67.32; H, 8.22. Found: C, 66.90; H, 8.33. Keto tautomer ^{1}H NMR δ 1.54 (d, J = 3.7 Hz, 3 H), 1.57–2.29 (m, 5 H), 2.32–2.35 (m, 1 H), 2.75 (dddd, J = 12.0, 12.0, 8.0, 4.0 Hz, 1 H), 3.12 (d, J = 12.0 Hz, 1 H), 3.63 (s, 3 H), 5.08–5.47 (m, 2 H). Enol tautomer ^{1}H NMR δ 1.54 (d, J = 3.7 Hz, 3 H), 1.57–2.29 (m, 5 H), 2.36–2.41 (m, 1 H), 3.05–3.15 (m, 1 H), 3.63 (s, 3 H), 5.08–5.47 (m, 2 H).

215   **188** (62%)

Methyl 6,6-Dimethyl-10-methylene-2-oxo-5β,9β-[4.3.1]-bicyclodecane-1β-carboxylate (188).[197] To a stirred solution of Mn(OAc)$_3$ · 2H$_2$O (213 mg,

0.795 mmol) in glacial AcOH (5 mL) heated at 58° in an oil bath, a solution of keto ester **215** (100 mg, 0.4 mmol) in AcOH (1 mL) was added. After the solution turned colorless (15 minutes) it was poured into H_2O (25 mL) and the mixture was extracted with ether (3 × 25 mL). The combined ether extract was washed successively with H_2O, saturated $NaHCO_3$ and brine, dried over Na_2SO_4, decanted, and evaporated in vacuo. Column chromatography (1% gradient EtOAc in hexane) afforded 61 mg (62%) of **188** as a colorless solid, mp 91–92° (hexane). IR (neat) 1707, 1645 cm^{-1}. ^1H NMR (400 MHz, CDCl$_3$) δ 0.88 (s, 3 H), 0.90 (s, 3 H), 1.13 (br d, J = 11 Hz, 1 H), 1.38 (br d, J = 9.3 Hz, 1 H), 1.7–1.9 (m, 2 H), 1.78–1.9 (m, 1 H), 1.79–1.9 (m, 1 H), 2.05 (t, J = 9, 6.1 Hz, 1 H), 2.2 (ddd, J = 19, 13.4, 4.3 Hz, 1 H), 2.37 (dt, J = 19, 3.5 Hz, 1 H), 3.1 (br d, J = 8.65 Hz, 1 H), 3.68 (s, 3 H), 4.03 (d, J = 8.65 Hz, 1 H), 4.7 (d, J = 1.97 Hz, 1 H), 4.87 (d, J = 1.97 Hz, 1 H). Anal. Calcd for $C_{15}H_{22}O_3$: C, 71.98; H, 8.89. Found: C, 71.87; H, 9.07.

217   216 (86%)

Methyl 5-Methyl-6-methylene-2-oxobicyclo[3.2.1]octane-1-carboxylate (216).[101]   To a stirred solution of Mn(OAc)$_3$ · 2H$_2$O (0.804 g, 3 mmol) and Cu(OAc)$_2$ · H$_2$O (0.3 g, 1.5 mmol) in 13.5 mL of glacial AcOH was added ketoester **217** (0.307 g, 1.5 mmol) in 4 mL of glacial AcOH. The reaction mixture was stirred at room temperature for 26 hours, at which time 100 mL of H$_2$O was added. A solution of 10% NaHSO$_3$ was added dropwise to the mixture to decompose any residual Mn(OAc)$_3$. The resulting solution was extracted with three 30-mL portions of CH$_2$Cl$_2$. The combined organic extracts were washed with saturated NaHCO$_3$ solution and dried over Na$_2$SO$_4$. Removal of the solvent in vacuo gave 0.301 g (96%) of a yellow solid which was recrystallized from pentane to give pure **216** (86%). A second recrystallization provided an analytical sample, mp 71.8–72.5°. ^1H NMR δ 5.08 (dd, J = 2.3, 3.1 Hz, 1 H, = CH$_2$), 5.01 (dd, J = 2.3, 3.1 Hz, 1 H, = CH$_2$), 3.76 (s, 3 H), 2.94 (dddd, J = 0.9, 1.9, 2.9, 18.4 Hz, 1 H, H-7 endo), 2.83 (br d, J = 18.4 Hz, 1 H, H-7 exo), 2.52 (dddd, J = 1.0, 8.9, 12.5, 17.0 Hz, 1 H, H −3 endo), 2.36 (ddd, J = 2.0, 6.9, 17.0 Hz, 1 H, H-3 *exo*), 2.09 (br s, 2 H, 2 H-8), 1.79 (ddd, J = 6.9, 12.0, 12.5 Hz, 1 H, H-4 exo), 1.68 (ddddd, J = 2.0, 2.0, 2.0, 8.9, 12.0 Hz, 1 H, H-4 *endo*), 1.25 (s, 3 H). Anal. Calcd for $C_{12}H_{16}O_3$: C, 69.21; H, 7.24. Found: C, 68.89; H, 7.88.

218 (55%)

2-Ethoxycarbonyl-4-carboxycyclohexanone (218).[188]

Deaerated AcOH (10 mL), Mn(OAc)$_3$ · 2H$_2$O (0.807 g, 3 mmol) and ethyl 3-keto-6-heptenoate (0.204 g, 1.29 mmol) were placed in a 50-mL stainless steel autoclave containing a glass liner. The autoclave was then pressurized with 600 psi of CO and heated with stirring at 70° for 13 hours. After excess CO was discharged at room temperature, the reaction mixture was filtered through Celite. The filtrate was diluted with ether and washed twice with water, and then the water layer was extracted with ether (3 times). The combined organic extracts were dried over MgSO$_4$. Filtration and solvent removal gave an oil, which was purified by flash chromatography (hexane : EtOAc : EtOH, 8 : 2 : 1), affording 152 mg of crystallized **218** (55%).

197

198 (68%)

Key Intermediate in the Synthesis of 14-Epiupial (198).[199]

A solution of 0.7 g (2.123 mmol) of **219** in 10 mL of anhydrous MeOH was cooled to 0° under argon and 1.12 mL (2.23 mmol) of a 1.99 M solution of KOH in MeOH was added dropwise over a period of 5 minutes. The solution was allowed to warm to room temperature and stirred for 7 days. The solvent was evaporated and 15 mL of glacial AcOH together with 1.1 g (4.46 mmol) of Mn(OAc)$_3$ · 2H$_2$O were added. The brown solution was heated in an oil bath at 70° for 2 hours, cooled, decolorized by the addition of solid NaHCO$_3$ (< 25mg), and poured into 75 mL of H$_2$O. The aqueous phase was extracted with ether (4 × 25 mL) and the combined organic layers were carefully added to 50 mL of saturated NaHCO$_3$ solution. Solid NaHCO$_3$ was added to this mixture in small portions until bubbling ceased and the solution was basic (pH 9). The phases were separated and the aqueous portion was extracted with ether (2 × 25 mL). The combined organic layers were washed with saturated NaHCO$_3$ solution (2 × 25 mL) and brine (25 mL), dried, and evaporated to leave a residue which was purified by MPLC (silica gel, 35% ethyl acetate in petroleum ether). There was isolated 0.45g (68%) of **198**, mp 91.5–92°. IR (CHCl$_3$) 1770, 1740 cm^{-1}. ^{1}H NMR (300 MHz, CDCl$_3$) δ 0.91 (s, 3H), 0.96 (d, J = 6 Hz, 3 H), 1.44–1.76 (m, 5 H), 2.22 (br. t, J = 9 Hz, 1 H), 2.69 (br q, J = 9 Hz, 1 H), 3.34 (s and d, 4 H), 3.56 (dd, J = 4, 7 Hz, 1 H), 3.75 (s, 3 H), 4.55 (AB, J = 7 Hz, 2 H), 4.80 (br q, J = 7 Hz, 1 H). MS m/z M$^+$C$_2$H$_5$O$_2$ calcd 251.1286, obsd 251.1268.

TABULAR SURVEY

The tabular survey includes all examples found in the literature to April 1995. Computer searches of *Chemical Abstracts and Science Citation Index* were conducted using key papers.

Table entries are in order of increasing **total** carbon and hydrogen numbers of the **substrate**. Departures from this criterion are made to group closely related compounds. Intramolecular reactions are further prioritized by the ring size of the products (cyclopentanes, cyclohexanes, etc.); in tandem cyclizations the first step is taken into account. If several reagents react with the same substrate, they are also arranged in order of increasing **total** carbon and hydrogen numbers. The same criterion is used for manganese(III) salts. Reagents with the same count of carbon and hydrogen atoms are further prioritized based on their structural features (i.e., acyclic–cyclic, straight-chain–branched-chain, terminal–internal alkenes, or hydrocarbons–functionalized derivatives). Chronology is used as a last criterion when all other parameters are identical.

Yields in parenthesis are isolated yields; numbers separated by colons are ratios of products or stereoisomers. A dash indicates that no datum for yield is reported.

Mn(OAc)$_3$ refers to the dihydrated form Mn(OAc)$_3$ · 2H$_2$O unless it is specifically mentioned that the anhydrous form was used.

The following abbreviations are used in the tables:

Bs	Benzenesulfonyl
Mn(pic)$_3$	Manganese(III) tris(2-pyridinecarboxylate)
TBDMS	*tert*-Butyldimethylsilyl
TMS	Trimethylsilyl
Ts	*p*-Toluenesulfonyl

TABLE I. ALKENES AND ALDEHYDES

Substrate	Reagent	Conditions	Product(s) and Yield(s) (%)	Refs.
C4 (2-methylpropene structure)	R–CH$_2$–CHO	Mn(OAc)$_3$, AcOH, 60°, 1 - 2 h	I + II + III (aldehyde/ketone products)	110

$$\textbf{I}:\textbf{II}:\textbf{III}$$

R	I : II : III
R = Me	70 : 27 : 3 (15)
R = Et	50 : 45 : 5 (17)
R = i-Pr	67 : 17 : 16 (10)

Conditions: Mn(OAc)$_3$, Cu(OAc)$_2$, AcOH, 50°, 5 - 10 h

R	I : II
R = H	93 : 7 (10)
R = Me	93 : 7 (36)
R = Et	93 : 7 (45)
R = i-Pr	90 : 10 (30)
R = n-C$_7$H$_{15}$	87 : 13 (46)

Substrate	Reagent	Conditions	Product(s) and Yield(s) (%)	Refs.
C$_5$-C$_{10}$ R^1–CH=CH$_2$	R^2–CH$_2$–CHO	Mn(OAc)$_3$, AcOH, 60°, 1 h	aldehyde (R^1, R^2)	41

R^1	R^2
n-Pr	Me (20)
n-Bu	Me (25)
n-C$_5$H$_{11}$	Me (30)
n-C$_6$H$_{13}$	Me (25)
n-C$_8$H$_{17}$	Me (25)

R^1	R^2
n-Bu	Et (20)
n-C$_5$H$_{11}$	Et (25)
n-Bu	i-Pr (20)
n-C$_5$H$_{11}$	i-Pr (20)

TABLE I. ALKENES AND ALDEHYDES (Continued)

Substrate	Reagent	Conditions	Product(s) and Yield(s) (%)	Refs.
C6		Mn(OAc)$_3$, AcOH, 60°, 1 - 2 h	CHO (22)	110
C7 n-C$_5$H$_{11}$	CHO Et	Mn(OAc)$_2$, O$_2$, AcOH, 70°, 5 h	n-C$_7$H$_{15}$ R + n-C$_7$H$_{15}$ Et Et (23) R = H (48), R = OH (14)	121
C7-C9 R^1	R^2 H		I + II + III	72

R^1	R^2	Conditions	I : II : III
n-C$_5$H$_{11}$	H	Mn(OAc)$_3$, 50°, 12 h	20 : 0 : 73 (92)
		Mn(OAc)$_3$, AcOH, 23°, 20 h	51 : 39 : 10 (38)
n-C$_7$H$_{15}$	Me	Mn(OAc)$_3$, AcOH, 20°, 50 h	74 : 17 : 9 (38)
n-C$_5$H$_{11}$	n-Pr	Mn(OAc)$_3$, AcOH, 70° 1.5 h	27 : 22 : 48 (89)
n-C$_7$H$_{15}$	n-Pr	Mn(OAc)$_3$, AcOH, 70°, 12 min	72 : 28 : 0 (44)

486

TABLE I. ALKENES AND ALDEHYDES (*Continued*)

Substrate	Reagent	Conditions	Product(s) and Yield(s) (%)	Refs.
C$_9$ 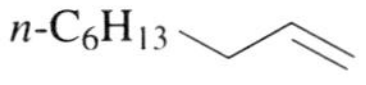 n-C$_6$H$_{13}$ (alkene)	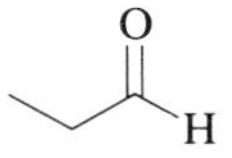 propionaldehyde	Mn(OAc)$_3$, AcOH, 60°, 1 - 2 h	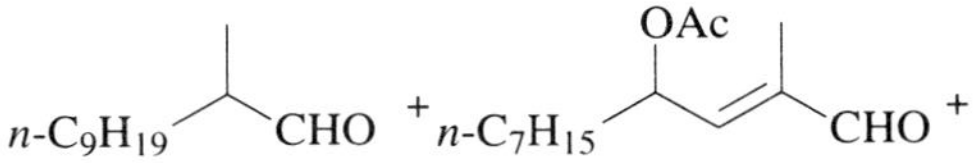 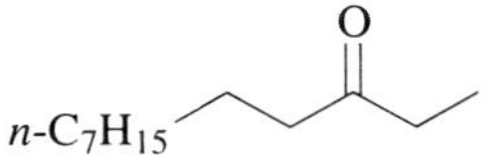 n-C$_9$H$_{19}$–CHO + n-C$_7$H$_{15}$–CHO + (OAc-substituted CHO product) 72 : 17 : 9 (38)	110
C$_{10}$ 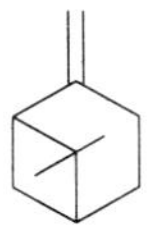 (pinene structure)	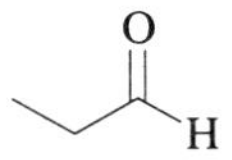 propionaldehyde	Mn(OAc)$_3$, AcOH, 60°	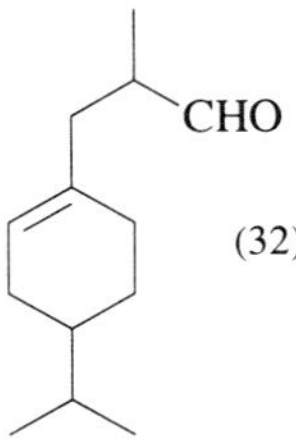 (isopropyl cyclohexenyl CHO product) (32) + (ethyl ketone isopropyl cyclohexenyl product) (6)	65
	R = H, R = Me, R = Et (aldehyde)	Mn(OAc)$_3$, Cu(OAc)$_2$, AcOH, 60°	(R-CHO isopropenyl cyclohexenyl products) (5), (34), (12) + (R-CHO OAc products) (6), (6), (3)	65

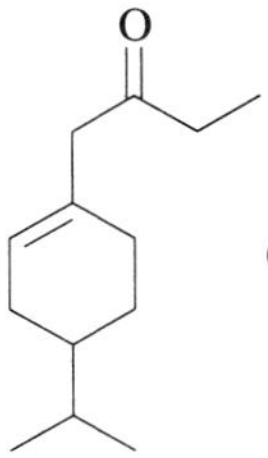

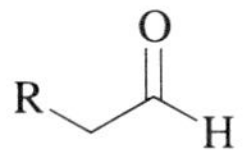

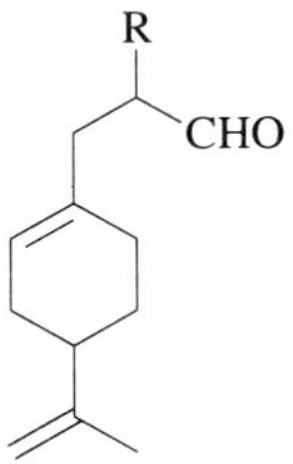

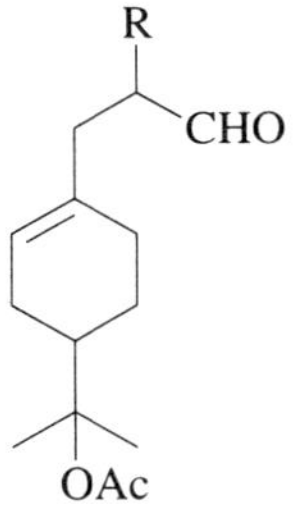

487

TABLE II. ALKENES AND KETONES

Substrate	Reagent	Conditions	Product(s) and Yield(s) (%)	Refs.
C_2 $CH_2{=}CH_2$	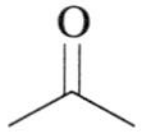	Mn(OAc)$_3$, AcOH, 85°, 6 h, 50 atm	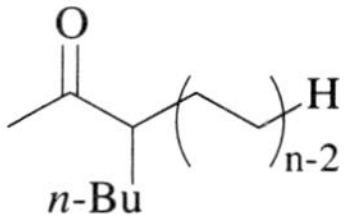 (—) n = 3 - 9	123
	"	Mn(OAc)$_3$, Cu(OAc)$_2$, AcOH, EtOAc, 80°, 2 h, 100 atm	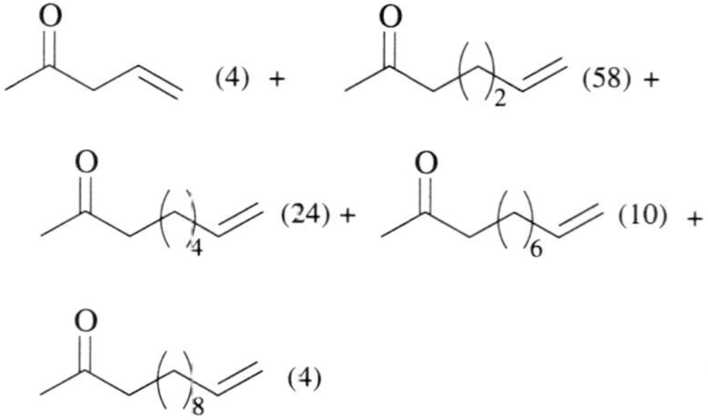 (4) + (58) + (24) + (10) + (4)	67
	"	Mn(OAc)$_3$, Cu(OAc)$_2$, AcOH, 85°, 4 h, 50 atm	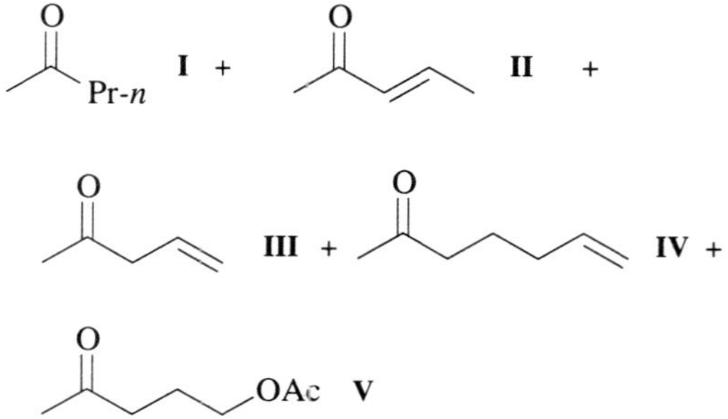 I + II + III + IV + V I:II:III:IV:V 10:18:30:28:1 4 (28)	66

488

TABLE II. ALKENES AND KETONES (*Continued*)

Substrate	Reagent	Conditions	Product(s) and Yield(s) (%)	Refs.
C_4		$Mn(OAc)_3$, C_6H_{14}, 40°, 24 h	(65)	49
C_4-C_8 R^1 R^2		$Mn(OAc)_3$, $Cu(OAc)_2$, AcOH	R^2 R^1	49

R^1	R^2		
Me	H	40°, 9 h	(44)
H	n-Pr	60°, 0.5 h	(43)
H	n-Bu	60°, 0.5 h	(52)
H	n-C$_5$H$_{11}$	80°, 0.5 h	(56)

Substrate	Reagent	Conditions	Product(s) and Yield(s) (%)	Refs.
C_5 OAc		$Mn(OAc)_3$, AcOH, 70°, 10 min	(22)	46
		$Mn(OAc)_3$, AcOH	+ $Z:E$ 6:1 (—) (—)	46

489

TABLE II. ALKENES AND KETONES (*Continued*)

Substrate	Reagent	Conditions	Product(s) and Yield(s) (%)	Refs.
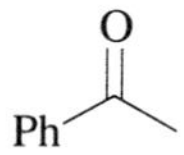 Ph—C(O)CH$_3$	Mn(OAc)$_3$, AcOH	(—)		46
n-C$_5$H$_{11}$ ketone	"	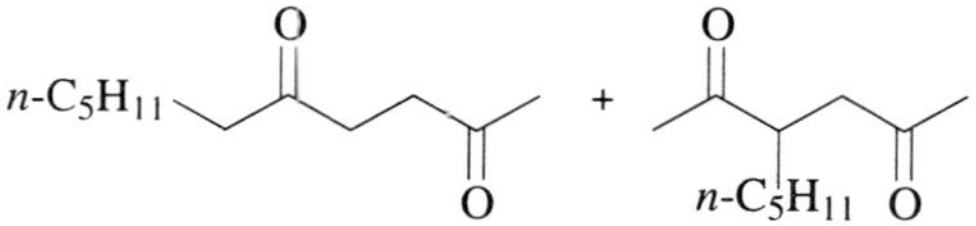 n-C$_5$H$_{11}$ + n-C$_5$H$_{11}$ 1.5 : 1 (—)		46
n-Bu ketone	"	(—)^a		46
pinanone	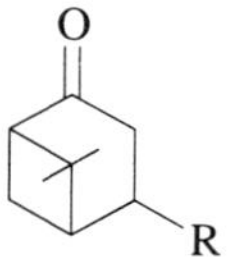 "	(—)	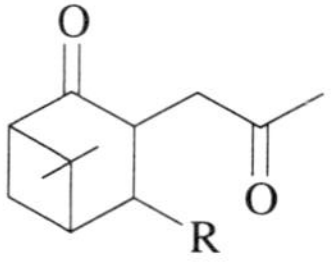	46a
R = H, R = Me	"	(—) (—)		46a

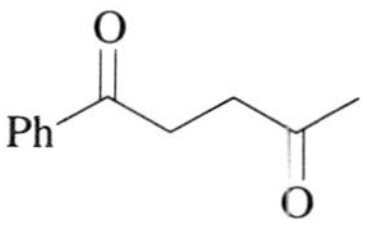

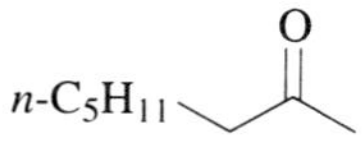

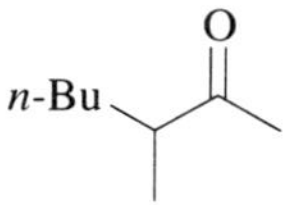

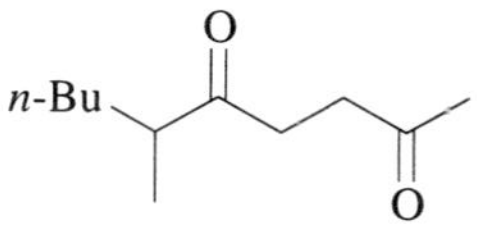

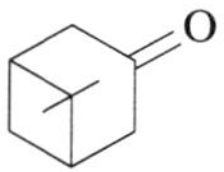

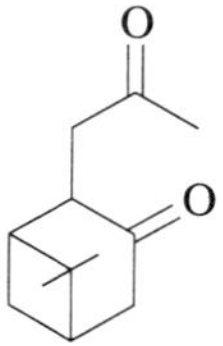

490

TABLE II. ALKENES AND KETONES (*Continued*)

Substrate	Reagent	Conditions	Product(s) and Yield(s) (%)	Refs.

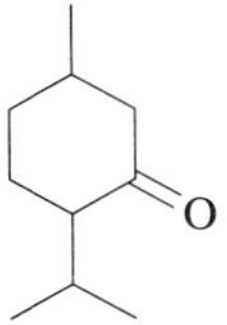

C$_6$

Mn(OAc)$_3$, AcOH

(—) + (—) + Pr-*i*

46a

Mn(OAc)$_3$, AcOH, reflux, 1-3 h

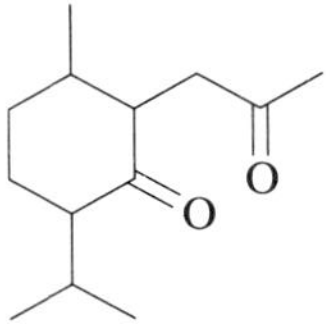

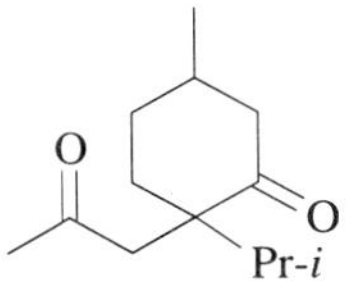

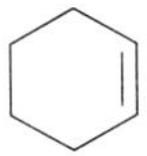

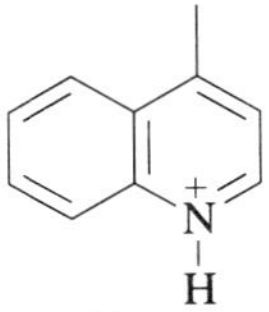

(15)^b

124

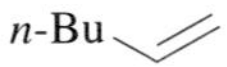

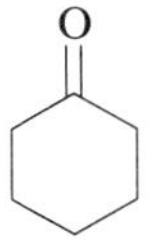

n-Bu

Mn(OAc)$_3$, Cu(OAc)$_2$, AcOH, 50°, 15 h

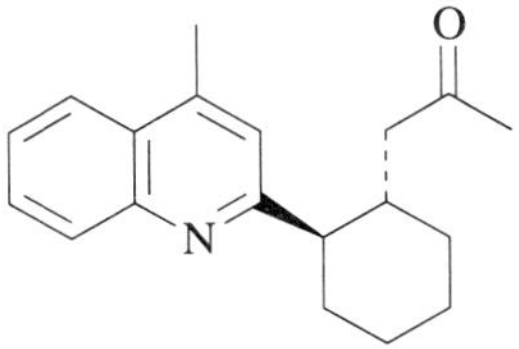

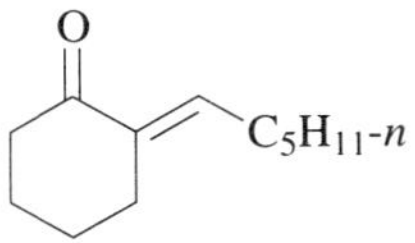

C$_5$H$_{11}$-*n* + Bu-*n* +
I **II**

Pr-*n* + C$_6$H$_{13}$-*n*
III **IV**

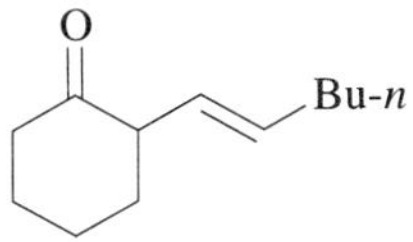

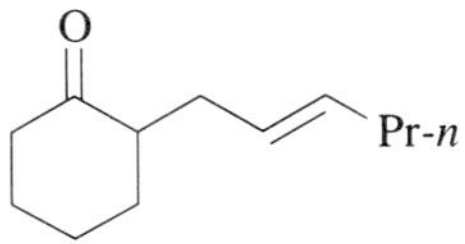

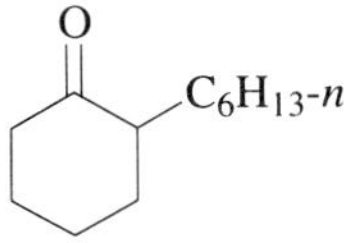

I : II : III : IV 35 : 15 : 35 : 15 (56)

44

491

TABLE II. ALKENES AND KETONES (*Continued*)

Substrate	Reagent	Conditions	Product(s) and Yield(s) (%)	Refs.
	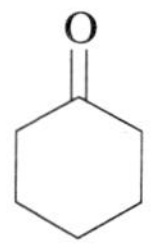	Mn(OAc)$_3$, 75°, 3 h	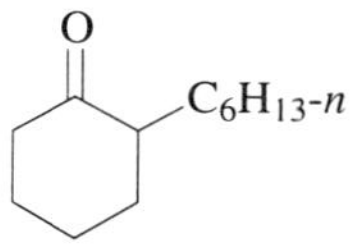 C$_6$H$_{13}$-n (53)	44
		Mn(OAc)$_3$, AcOH, Ac$_2$O, 100°	Bu-n, AcO (16) + Pr-n (12)	48
			O, Pr-n (13) + C$_6$H$_{13}$-n (6)^c	
n-Pr, Pr-n	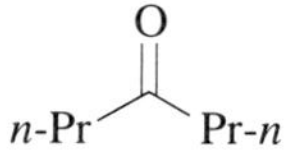	Mn(OAc)$_3$, Cu(OAc)$_2$, AcOH, 70°, 10 h	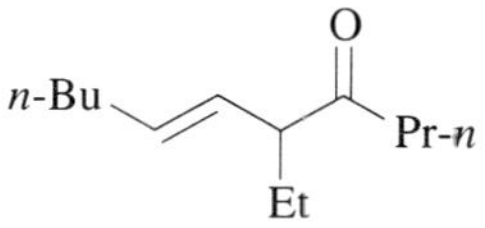 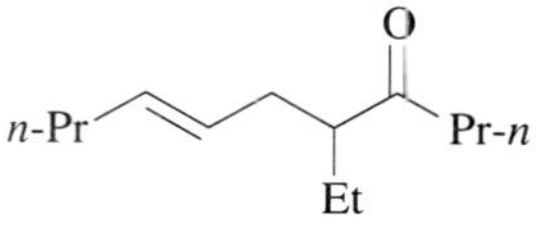 O, Pr-n, Et, n-Bu + O, Pr-n, Et, n-Pr (50) 3 : 1	44

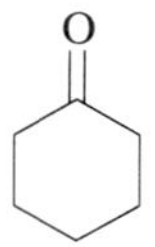

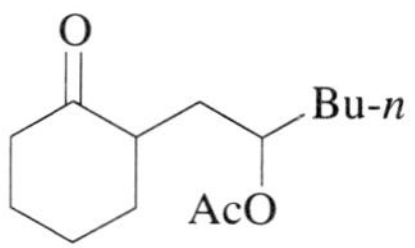

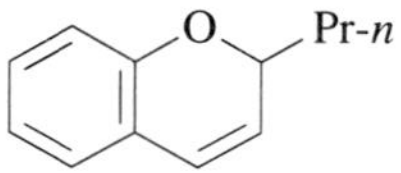

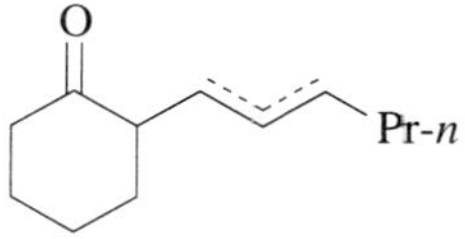

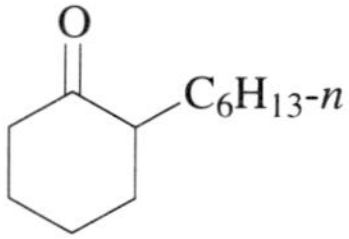

TABLE II. ALKENES AND KETONES (*Continued*)

Substrate	Reagent	Conditions	Product(s) and Yield(s) (%)	Refs.
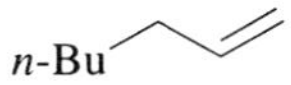 C₇ *n*-Bu (alkene structure)	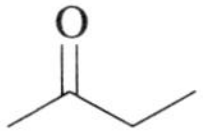 (methyl ethyl ketone structure)	Mn(OAc)₃, Cu(OAc)₂, AcOH, 70°, 10 h	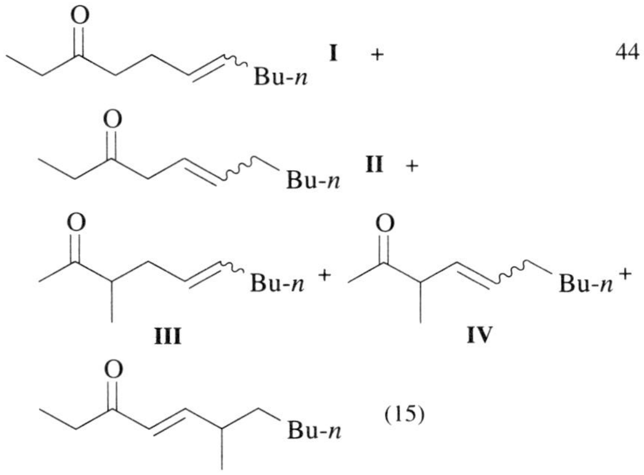 **I** + **II** + **III** + **IV** **I** : **II** : **III** : **IV** 18 : 7 : 25 : 50 (28)	44
	R (ketone structure)	Mn(OAc)₃, Cu(OAc)₂, AcOH, 60°, 2 h	**I** + **II** **I** : **II** 30 : 70 (24) 40 : 60 (35) 55 : 45 (32) 100 : 0 (30)	45

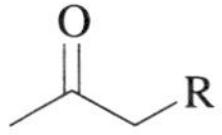

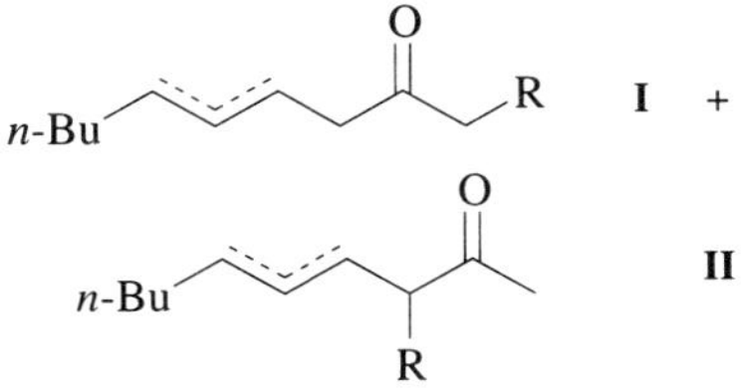

(15) — associated with structure **III**

R = CH₂CH₂OAc
R = CH₂CO₂Me
R = OAc
R = Me₂C(OAc)

TABLE II. ALKENES AND KETONES (*Continued*)

Substrate	Reagent	Conditions	Product(s) and Yield(s) (%)	Refs.
C_8 $n\text{-}C_6H_{13}$	OEt structure	Mn(OAc)$_3$, Cu(OAc)$_2$, AcOH, 60°, 2 h	OEt (24)	45
	acetone	Mn(OAc)$_3$, AcOH, AcOK, 85°	OAc $n\text{-}C_6H_{13}$ (24) + $n\text{-}C_6H_{13}$ (15) O + $n\text{-}C_6H_{13}$ (7) O	58
	X pyridinium, CF$_3$CO$_2^-$, acetone; X = CN, X = COMe	Mn(OAc)$_3$, AcOH, reflux, 1 - 3 h	X $C_6H_{13}\text{-}n$ (40) (36)	124
	quinolinium, CF$_3$CO$_2^-$, acetone	Mn(OAc)$_3$, AcOH, reflux, 1 - 3 h	$C_6H_{13}\text{-}n$ (60)	124

TABLE II. ALKENES AND KETONES (*Continued*)

Substrate	Reagent	Conditions	Product(s) and Yield(s) (%)		Refs.
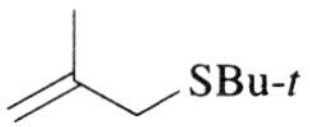 (SBu-*t* alkene)	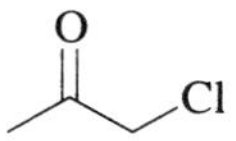 (O Cl acetone chloride)	Mn(OAc)₃, Cu(OAc)₂, PbO₂, AcOH, rt, 2 d	(chloro ketone product)	(43)	71
	(O NO₂ acetophenone)	"	(nitro aryl ketone)	(85)	71
	(O CH₂Cl phenacyl chloride)	"	(Cl aryl ketone)	(84)	71
	(O Cl acetone chloride)	"	(Cl chloro ketone)	(—)	71
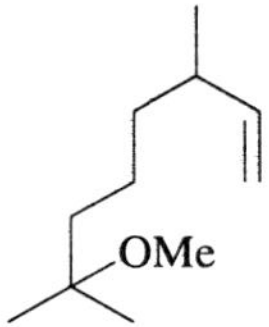 C₁₁ (PhS alkene)	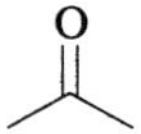 (O acetone)	Mn(OAc)₃, reflux, 24 h	(MeO methyl ketone)	(—)	82
(OMe alkene)	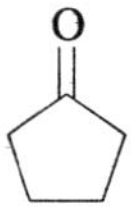 (O cyclopentanone)	Mn(OAc)₃, dioxane	(MeO cyclopentanone product)	(—)	82

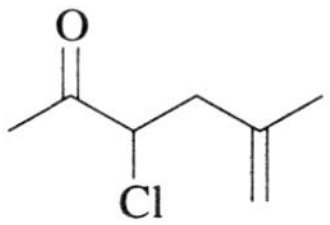

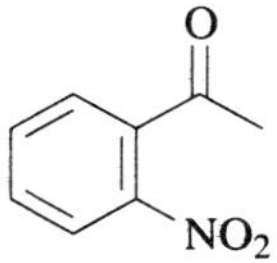

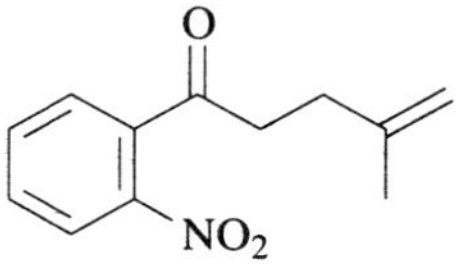

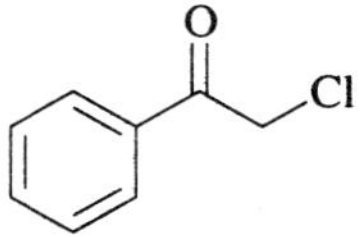

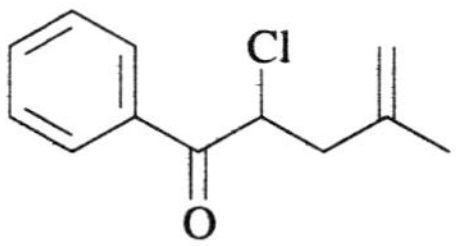

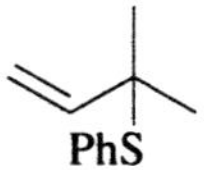

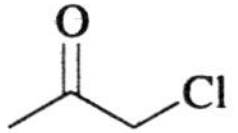

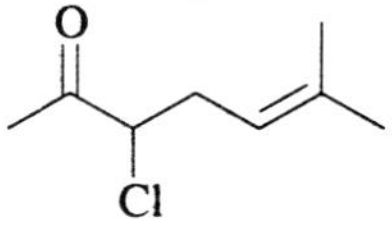

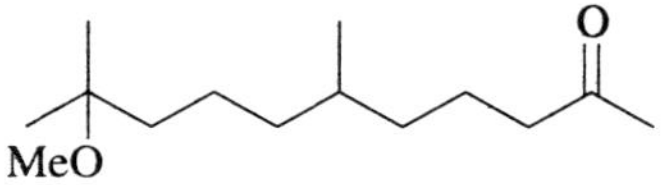

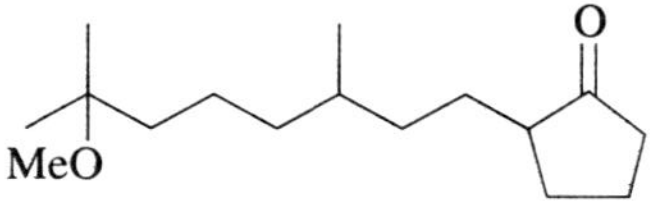

495

TABLE II. ALKENES AND KETONES (*Continued*)

Substrate	Reagent	Conditions	Product(s) and Yield(s) (%)	Refs.
C₁₂		Mn(OAc)₃, Cu(OAc)₂, AcOH, 50°, 1.5 h	90 : 10 (30) (25) (5)	59
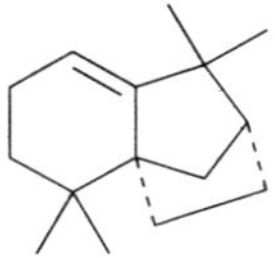 C₁₅	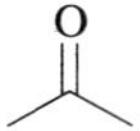	Mn(OAc)₃, AcOH, 110°, 24 h	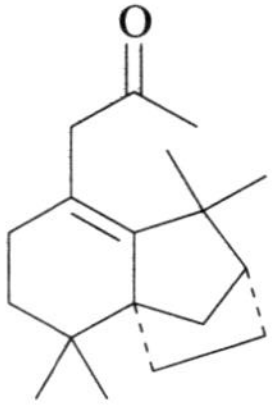 (—)ᵈ	111

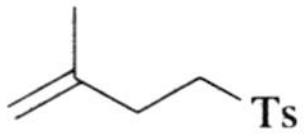

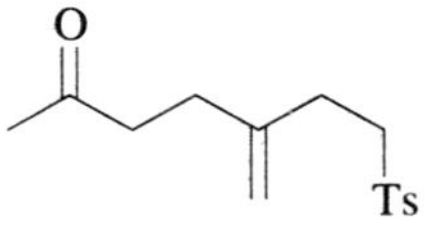

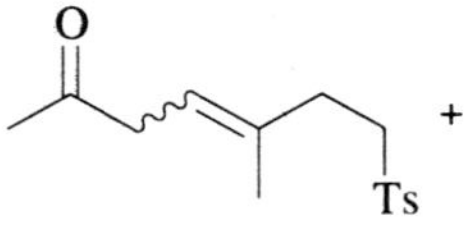

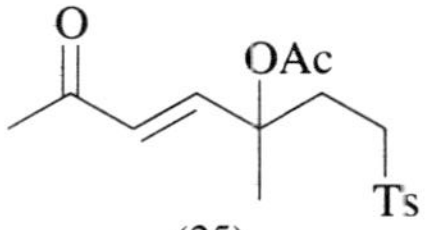

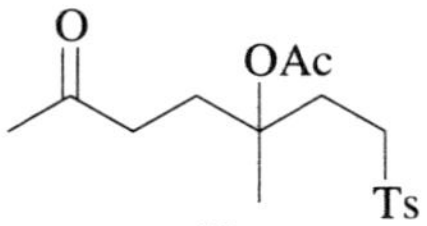

ᵃ This was the major isomer.
ᵇ The content of the cis isomer was 10%.
ᶜ 4-Acetoxyoctanoic acid (4%) and 4-octanolide (12%) were also isolated.
ᵈ Allylic oxidation products were formed.

TABLE III. ALKENES AND ACETIC ACID OR ITS DERIVATIVES

Substrate	Reagent	Conditions	Product(s) and Yield(s) (%)	Refs.
C₂				
$CH_2{=}CH_2$	AcOH	Mn(OAc)₃, AcOH, 120°, 4 h, 15 atm	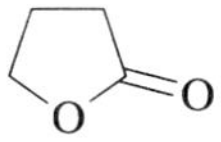 (73)	210

$$C_4{-}C_{14}$$

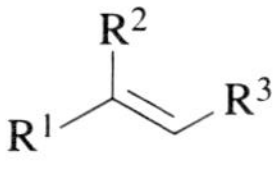

Mn(OAc)₃, AcOH, AcOK, reflux (AcOH reagent)

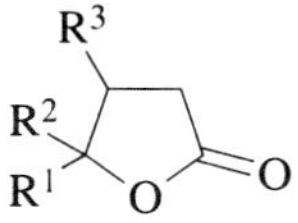

R¹	R²	R³	
$n\text{-}C_6H_{13}$	H	H	(74)
Ph	H	H	(60)
Ph	Me	H	(74)
Me	Me	H	(30)
n-Bu	H	H	(48)
n-Pr	H	n-Pr	(44)
Ph	H	Ph	(16)
Ph	H	Me	(79)
$CH_2C{\equiv}CC_5H_{11}\text{-}n$	H	H	(50)

125
128

C₆

n-Bu (substrate), AcOH

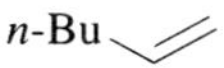

Mn(OAc)₃, Ac₂O, AcOH, 100°

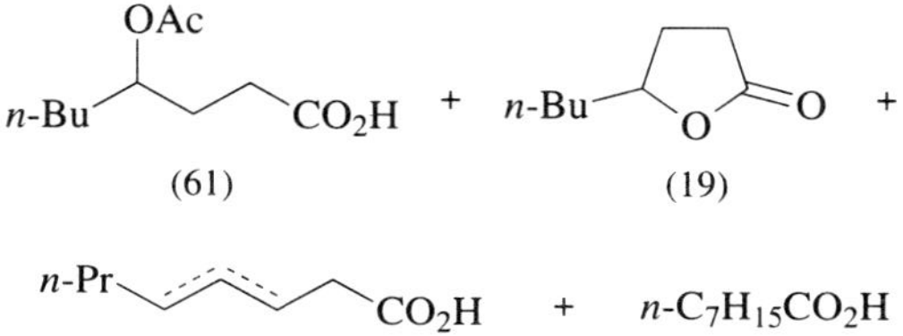

(61) + (19) + (13) + (7)

126

TABLE III. ALKENES AND ACETIC ACID OR ITS DERIVATIVES (*Continued*)

Substrate	Reagent	Conditions	Product(s) and Yield(s) (%)	Refs.
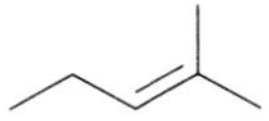 (alkene structure)	AcOH	Mn(OAc)$_3$, Ac$_2$O, AcOH, 100°	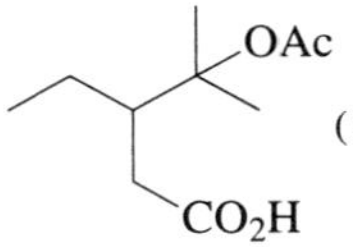 (14) 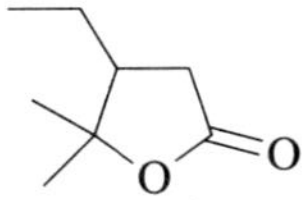 + (55) 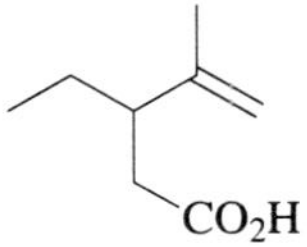 +	126
C$_6$-C$_9$ 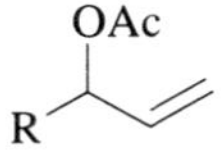 (OAc alkene) R = Me, Et, *i*-Pr, *i*-Bu, *s*-Bu	AcOH	Mn(OAc)$_3$, AcOH, Ac$_2$O, reflux, 15 min	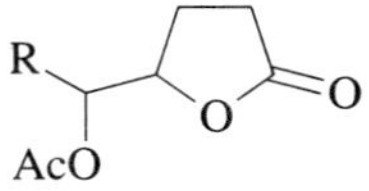 (22) (20 - 40)	211
C$_6$-C$_{12}$ 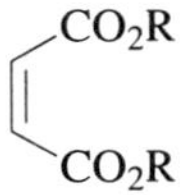 (CO$_2$R alkene)	AcOH	Mn(OAc)$_3$, AcOH, reflux, 7 h	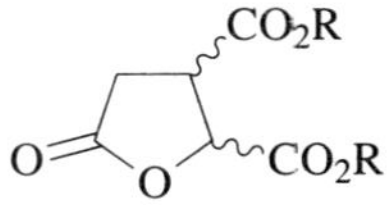	131

R = Me, Et, *i*-Pr, *i*-Bu, *s*-Bu

R	Z : E	
Me	1 : 6.8	(44)
Et	1 : 5.7	(56)
n-Pr	1 : 4.2	(49)
i-Pr	1 : 4.2	(73)
n-Bu	1 : 3.9	(43)
i-Bu	1 : 4.2	(63)

498

TABLE III. ALKENES AND ACETIC ACID OR ITS DERIVATIVES (*Continued*)

Substrate	Reagent	Conditions	Product(s) and Yield(s) (%)	Refs.
C$_6$-C$_{14}$				

C_6-C_{14}

Substrate:

$$\begin{array}{c} R^2 \\ R^1 \end{array}\!\!=\!\!\begin{array}{c} R^3 \\ R^4 \end{array}$$

Reagent: AcOH

Conditions: Mn(OAc)$_3$, AcOH, Ac$_2$O, reflux, 0.5-1 h

Product:

$$\begin{array}{c} R^3\;R^4 \\ R^2 \\ R^1\quad O \end{array}\!\!\!=\!\!O$$

R^1	R^2	R^3	R^4	
Ph	H	H	H	(39)
Ph	Me	H	H	(31)
Ph	H	Me	H	(21)
Bn	H	H	H	(16)
Ph	H	H	Ph	(20)
t-Bu	H	H	H	(12)

Refs.: 28

C$_7$

Substrate: (trisubstituted alkene — 3-ethyl-2-pentene type)

Reagent: AcOH

Conditions: Mn(OAc)$_3$, AcOH, AcOK, reflux

Product: (γ-butyrolactone, methyl and ethyl substituted) (43)

Refs.: 117

C$_8$

Substrate: EtO$_2$C $\diagup\!\!\diagdown$ CO$_2$Et

Reagent: AcOH

Conditions: Mn(OAc)$_3$, AcOH, reflux, 7 h

Product:

(lactone bearing CO$_2$Et groups)

$Z : E$ 1 : 5.7 (83)

Refs.: 131

TABLE III. ALKENES AND ACETIC ACID OR ITS DERIVATIVES (*Continued*)

Substrate	Reagent	Conditions	Product(s) and Yield(s) (%)	Refs.
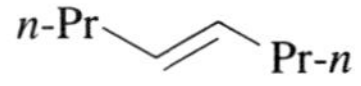 n-Pr⌒Pr-n	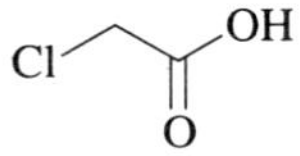 Cl⌒C(O)OH	Mn(OAc)₃, 120°, 2-2.5 h	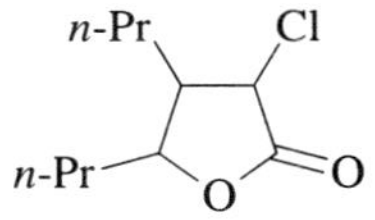 1.0 : 7.3 : 6.3 : 2.0 (33)	129
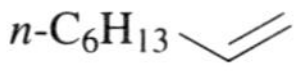 n-C₆H₁₃⌒	AcOH	Mn(OAc)₃, AcOH, Ac₂O, reflux, 20 min	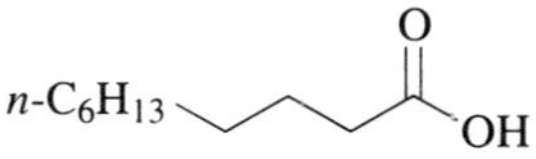 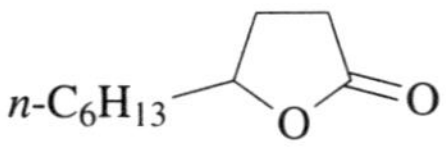 n-C₆H₁₃ ... (—) + n-C₆H₁₃ ... (38)	212
	AcOH	Mn(OAc)₃, Cu(OAc)₂, AcOH : Ac₂O, 1 : 1, 115°, 10 min	n-C₅H₁₁ ... (56) + n-C₅H₁₁ ... (20)	68
	AcOH	Mn(OAc)₃, Cu(OAc)₂, AcOH : Ac₂O, 1 : 4, 110°, 69 min	(56) (0)	68
	AcOH	Mn(OAc)₃, CuCl₂, AcOH : Ac₂O, 1 : 9, 120°, 15 min	(42) (9)[a]	68

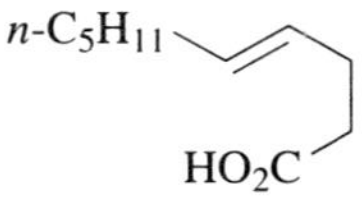

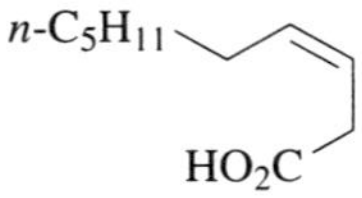

TABLE III. ALKENES AND ACETIC ACID OR ITS DERIVATIVES (*Continued*)

Substrate	Reagent	Conditions	Product(s) and Yield(s) (%)	Refs.
	AcOH	Mn(OAc)$_3$, AcOH, 100°, 95 h	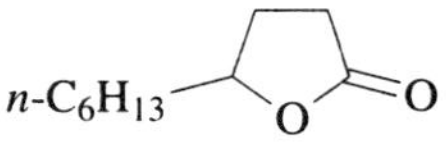 n-C$_6$H$_{13}$ (57)	60
	AcOH		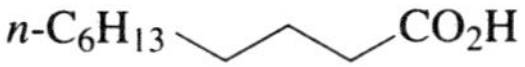  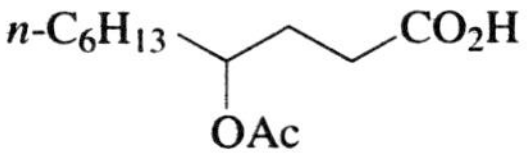 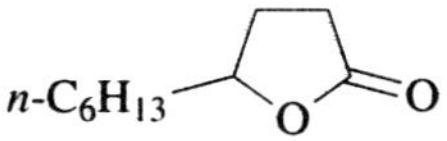 n-C$_6$H$_{13}$ CO$_2$H **I** +	60
			n-C$_5$H$_{11}$ CO$_2$H **II** +	
			n-C$_6$H$_{13}$ CO$_2$H **III** + OAc	
			n-C$_6$H$_{13}$ **IV**	

	I	II	III	IV
Mn(OAc)$_3$, AcOH, Ac$_2$O, 100°, 2 h	(7)	(19)	(54)	(21)
Mn(OAc)$_3$, AcOH, Ac$_2$O, AcOK 100°, 48 min	(16)	(6)	(64)	(14)

TABLE III. ALKENES AND ACETIC ACID OR ITS DERIVATIVES (*Continued*)

Substrate	Reagent	Conditions	Product(s) and Yield(s) (%)	Refs.
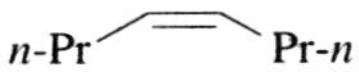 $n\text{-Pr}$—$=$—$\text{Pr-}n$	AcOH	$Mn(OAc)_3$, AcOH, AcOK, reflux	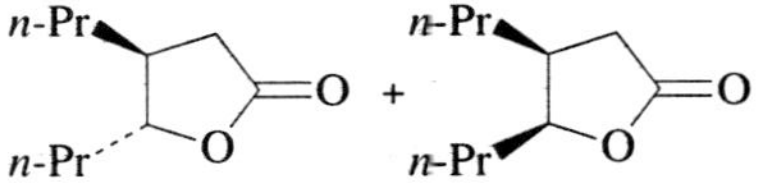 (lactone, n-Pr) + (lactone, n-Pr) 3.3:1 (60)	117
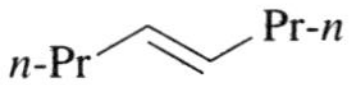 $n\text{-Pr}$—$=$—$\text{Pr-}n$	AcOH	$Mn(OAc)_3$, AcOH, AcOK, reflux	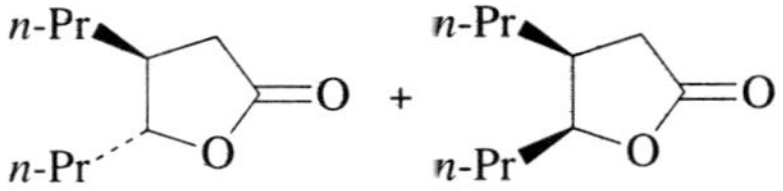 (lactone, n-Pr) + (lactone, n-Pr) 3.4:1 (69)	117
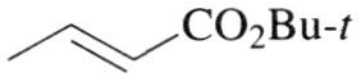 $\diagup\diagdown\diagup^{CO_2Bu\text{-}t}$	AcOH	$Mn(OAc)_3$, AcOH, AcOK, reflux	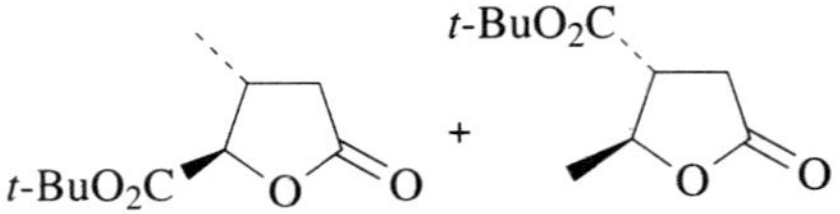 (t-BuO_2C lactone) + (t-BuO_2C lactone) 3.8:1 (57)	117
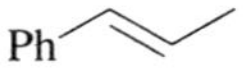 Ph—$=$	AcOH	$Mn(OAc)_3$, AcOH, AcOK, reflux	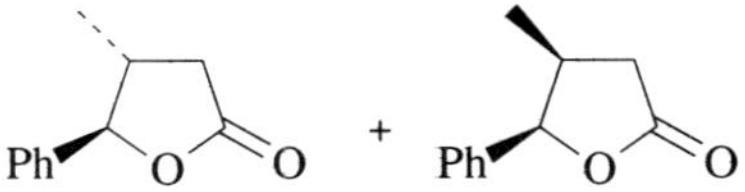 (Ph lactone) + (Ph lactone) 67:1 (68)	117
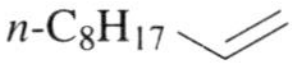 $n\text{-}C_8H_{17}$—$=$	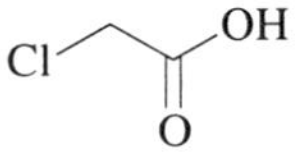 ClCH_2CO_2H	$Mn(OAc)_3$, 120°, 2–2.5 h	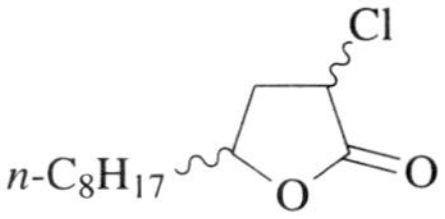 (Cl lactone, $n\text{-}C_8H_{17}$) 1.25:1 (52)	129
$n\text{-}C_8H_{17}$—$=$	AcOH	$Mn(OAc)_3$, AcONa, AcOH, Ac$_2$O, reflux, 2 h	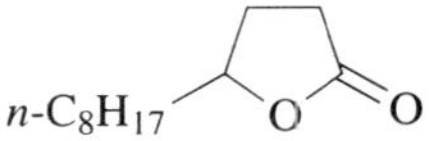 (lactone, $n\text{-}C_8H_{17}$) (85)	213

C_9

C_{10}

502

TABLE III. ALKENES AND ACETIC ACID OR ITS DERIVATIVES (*Continued*)

Substrate	Reagent	Conditions	Product(s) and Yield(s) (%)	Refs.
	AcOH	Mn(OAc)₃, AcOH, AcOK, reflux	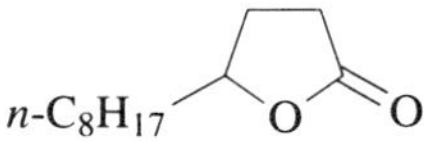 (79) 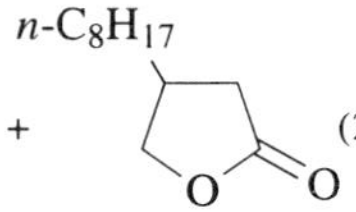 + (2)	117
	AcOH	1. Mn(OAc)₃, AcOH, Ac₂O, reflux, 1.5 h 2. CH₂N₂	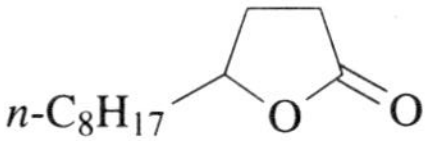 (17) 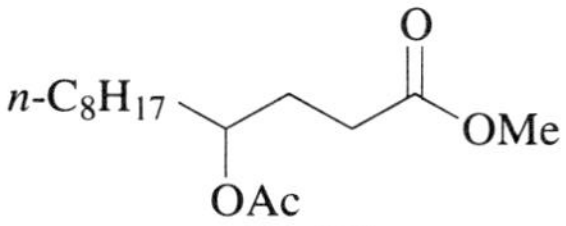 + (20) 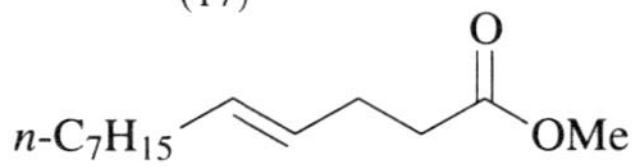 + (10) 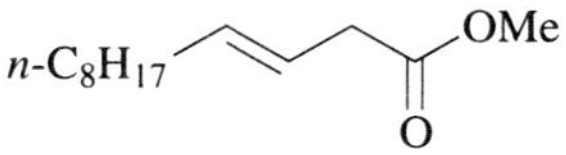 + (3)	117
	AcOH	Mn(OAc)₃, AcOH, Ac₂O, AcOK, reflux, 1.5 h	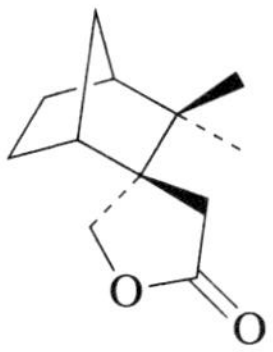 (50) 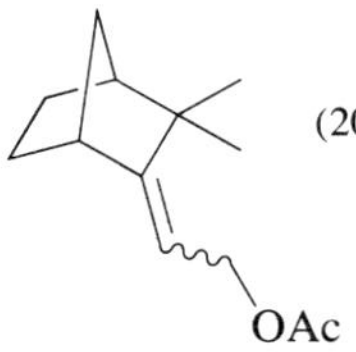 + (20)	214
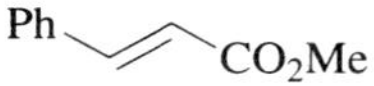 Ph⁓CO₂Me	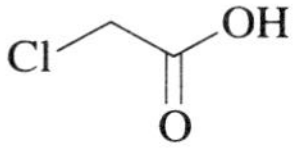 Cl⁓CO‑OH	Mn(OAc)₃, 120°, 2 – 2.5 h	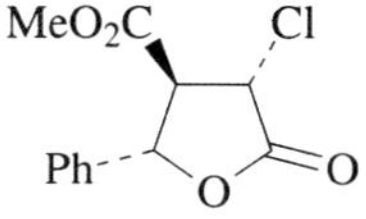 (35)	129

TABLE III. ALKENES AND ACETIC ACID OR ITS DERIVATIVES (*Continued*)

Substrate	Reagent	Conditions	Product(s) and Yield(s) (%)	Refs.
C$_{11}$	AcOH	Mn(OAc)$_3$, AcOH, AcOK, reflux	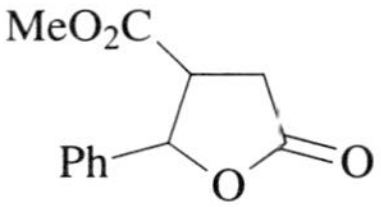 (45)	128
	AcOH	Mn(OAc)$_3$, AcOH, AcOK, reflux	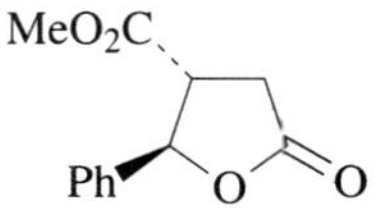 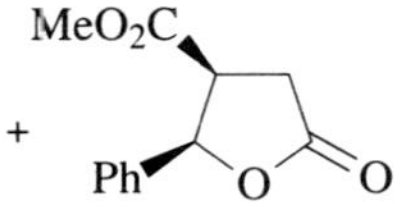 26:1 (82)	117
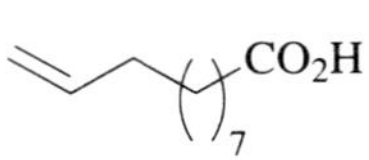	AcOH	Mn(OAc)$_3$, AcOH, Ac$_2$O, AcOK, reflux, 5 h	 (25) 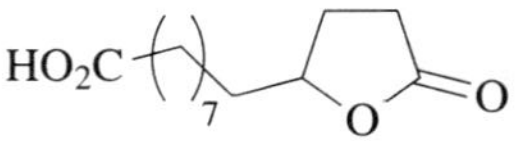 + (65)	127
C$_{15}$ 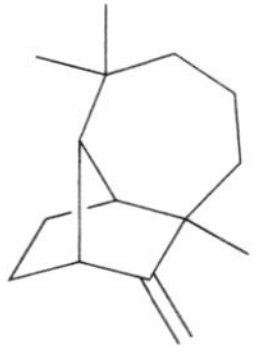	AcOH	Mn(OAc)$_3$, AcOH, Ac$_2$O, AcOK, reflux, 1.5 h	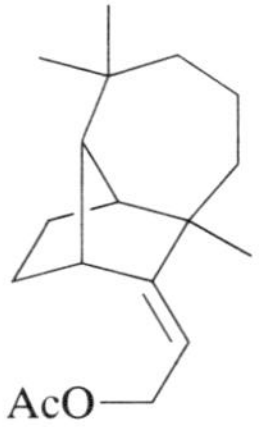 (47)	214

TABLE III. ALKENES AND ACETIC ACID OR ITS DERIVATIVES (Continued)

Substrate	Reagent	Conditions	Product(s) and Yield(s) (%)	Refs.
	AcOH	Mn(OAc)$_3$, AcOH, Ac$_2$O, AcOK, reflux, 2 h	(32)	215
C$_{16}$	AcOH	Mn(OAc)$_3$, AcOH, Ac$_2$O, AcOK, reflux, 2 h	(34) + (12)	215
C$_{18}$	AcOH	Mn(OAc)$_3$, AcOH, Ac$_2$O, AcOK, reflux, 8 h	(20)[b]	127

R^1	R^2
OAc	CH$_2$CO$_2$H[c]
CH$_2$CO$_2$H	OAc[c]

[a] γ-Hexyl-γ-butyrolactone was also formed (7%).

[b] The corresponding allylic acetates were major products (60%).

[c] The structures were not fully established.

505

TABLE IV. CYCLOALKENES AND ACETIC ACID

Substrate	Reagent	Conditions	Product(s) and Yield(s) (%)	Refs.
C_6 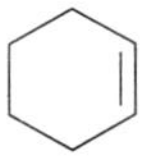	AcOH, Ac$_2$O	Mn(OAc)$_3$, AcOH, Ac$_2$O, reflux, 30-60 min	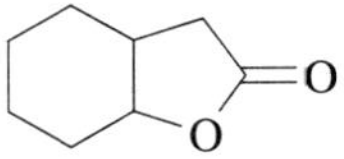 (10)	28
	AcOH	Mn(OAc)$_3$, AcOH, AcOK, reflux	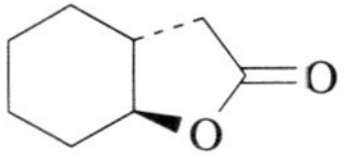 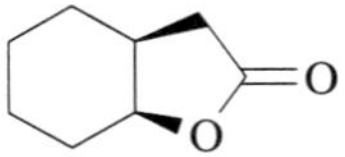 + 1:5.4 (29)	117
C_7 	AcOH	Mn(OAc)$_3$, AcOH, AcOK, reflux	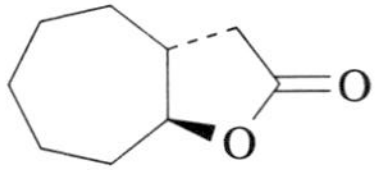 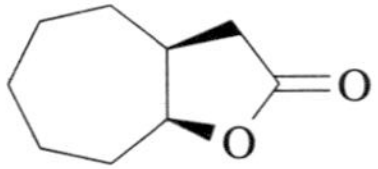 + 1:1.4 (75)	117
	AcOH	Mn(OAc)$_3$, AcOH, AcOK, reflux	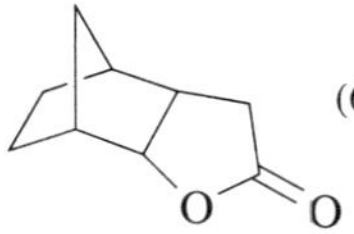 (63) 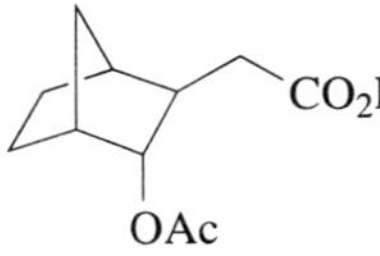 + CO$_2$H (5) OAc	117
C_7 n = 0	Cl—CH$_2$—CO—OH	Mn(OAc)$_3$, 120°, 2 - 2.5 h	1.2 : 1.1 : 1.0 : 1.5 (41)	129
C_8 n = 1			1.0 : 7.3 : 6.3 : 2.0 (53)	

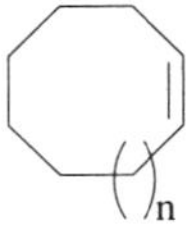

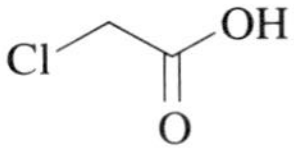

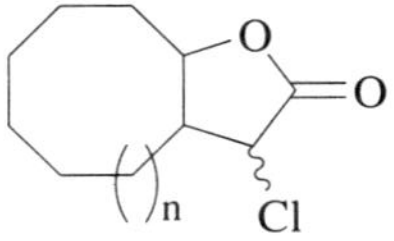

506

TABLE IV. CYCLOALKENES AND ACETIC ACID (*Continued*)

Substrate	Reagent	Conditions	Product(s) and Yield(s) (%)	Refs.
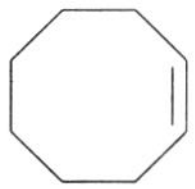 C₈	AcOH	Mn(OAc)₃, AcOH, AcOK, reflux	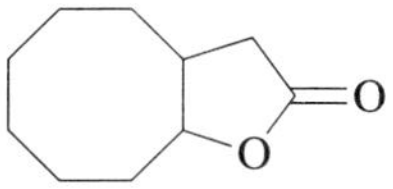 (62)	125 128
	AcOH	Mn(OAc)₃, AcOH, AcOK, reflux	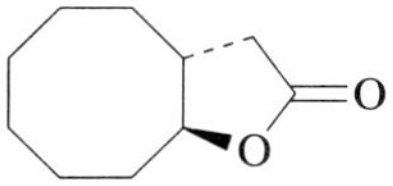 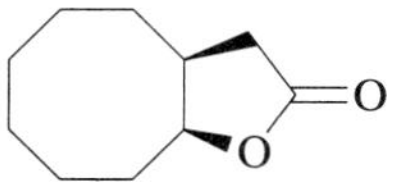 + 2.4:1 (68)	117
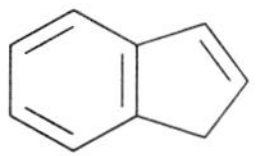 C₉	AcOH	Mn(OAc)₃, AcOH, AcOK, reflux	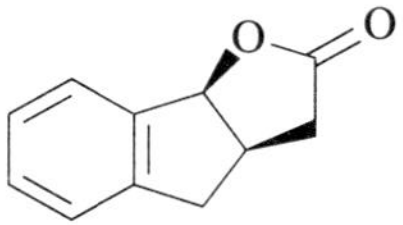 (40)	117
C₁₀	Ac₂O, AcOH	Mn(OAc)₃, AcOH, Ac₂O, 110°, 45 min	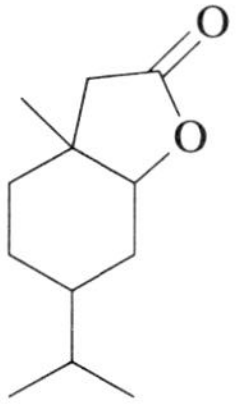 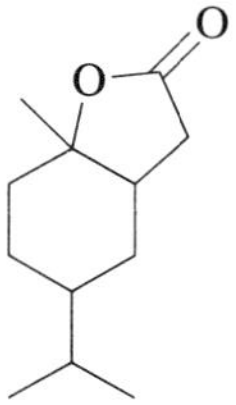 (—) + (—)	216
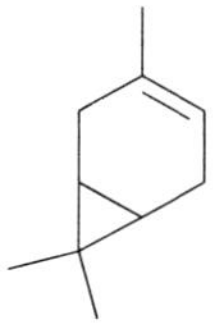	Ac₂O, AcOH	Mn(OAc)₃, AcOH, Ac₂O, 110-130°, 1.5 h	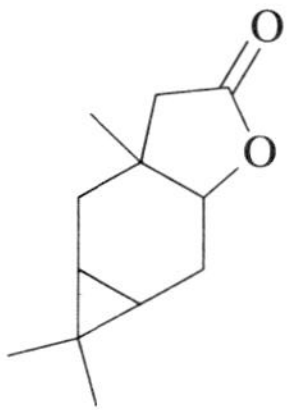 (8)	217

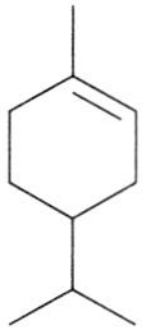

TABLE IV. CYCLOALKENES AND ACETIC ACID (*Continued*)

Substrate	Reagent	Conditions	Product(s) and Yield(s) (%)	Refs.
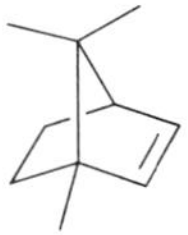	AcOH	Mn(OAc)₃, AcOH, Ac₂O, AcOK, 118°, 5 h	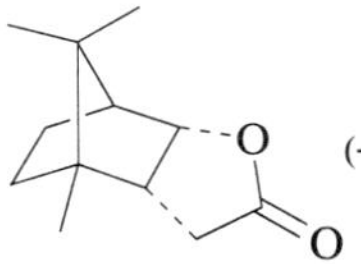 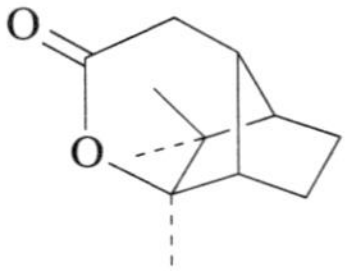 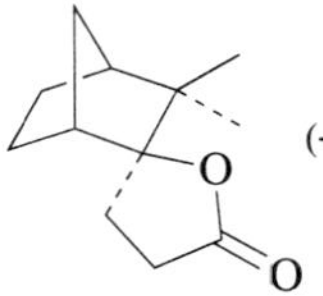 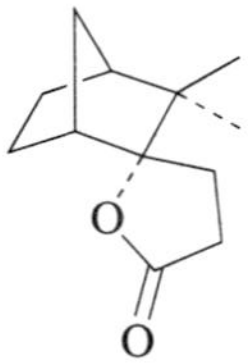 (—) + (—) + (—)	133
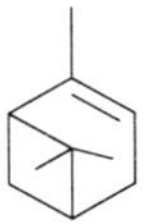 C₁₀-C₁₆	AcOH	Mn(OAc)₃, AcOH, Ac₂O, 110-130°, 45 min	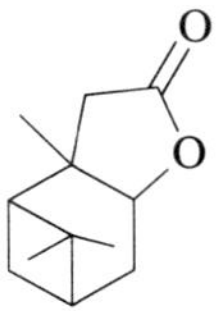 (—)[a]	218
	AcOH	Mn(OAc)₃, AcOH, reflux, 8 - 10 h	R = Me (56) R = Et (26) R = *n*-Pr (32) R = *i*-Pr (40) R = *n*-Bu (35)	131

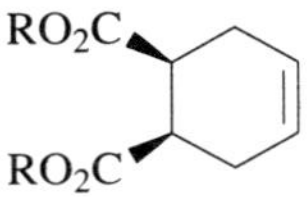

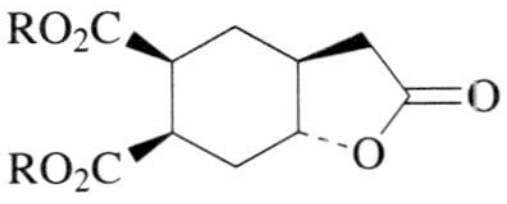

508

TABLE IV. CYCLOALKENES AND ACETIC ACID (*Continued*)

Substrate	Reagent	Conditions	Product(s) and Yield(s) (%)	Refs.
C₁₁–C₁₅	AcOH	Mn(OAc)₃, AcOH, reflux, 9 h	(42) (38) (33)	131
C₁₁	AcOH	Mn(OAc)₃, AcOH, Ac₂O, reflux, 30 min	(68)	134

^{a} α-Terpineol acetate and four isomeric allylic acetates were also formed.

509

TABLE V. ALKENES OR CYCLOALKENES AND MONOCARBOXYLIC ACIDS OTHER THAN ACOH

Substrate	Reagent	Conditions	Product(s) and Yield(s) (%)	Refs.
C$_6$ 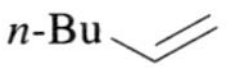 *n*-Bu ⟋⟍	(EtCO)$_2$O, EtCO$_2$H	Mn(OAc)$_3$, (EtCO)$_2$O, EtCO$_2$H, 100°	EtCO$_2$ ⟍ *n*-Bu CO$_2$H (44) + *n*-Bu lactone O (15)	126
	(EtCO)$_2$O, EtCO$_2$H		*n*-C$_6$H$_{13}$ CO$_2$H (41)	
	(EtCO)$_2$O, EtCO$_2$H	Mn(OAc)$_3$, (EtCO)$_2$O, EtCO$_2$H, 100°	lactone (35) + CO$_2$H (38)	126
Cl ⟋⟍ CO$_2$H OH	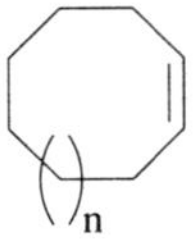 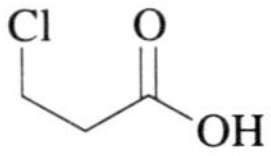 Mn(OAc)$_3$, 100°, 2 - 3 h	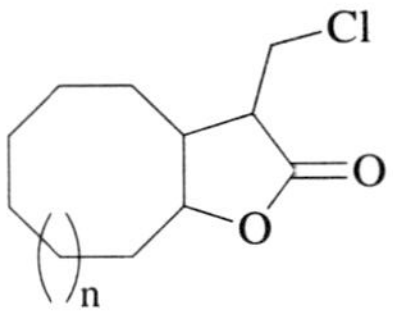 **I**	129
C$_7$ n = 0 C$_8$ n = 1			**I** (30) **I** (50) 1.9:1.7:1.0:1.7	
C$_8$ 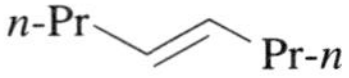 *n*-Pr ⟍⟋ Pr-*n*	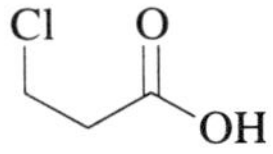 Cl ⟋⟍ CO$_2$H OH	Mn(OAc)$_3$, 100°, 2 - 3 h	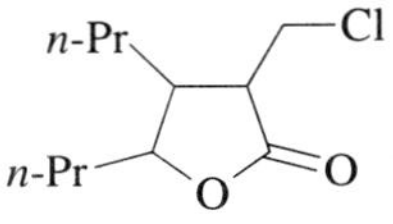 Cl *n*-Pr *n*-Pr lactone (29)	129

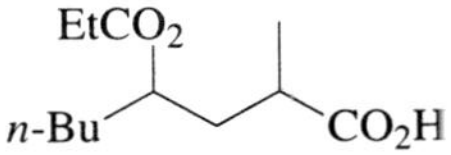

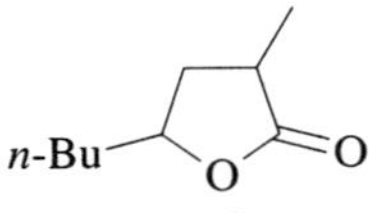

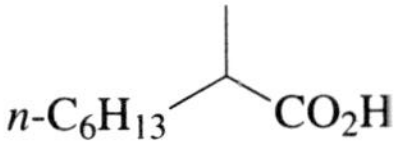

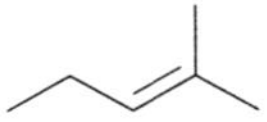

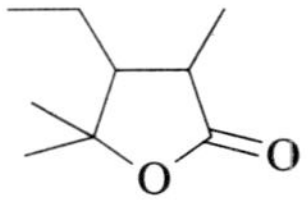

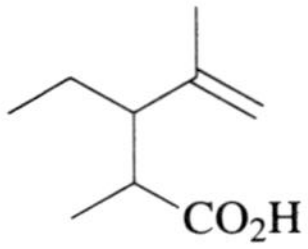

510

TABLE V. ALKENES OR CYCLOALKENES AND MONOCARBOXYLIC ACIDS OTHER THAN ACOH (*Continued*)

Substrate	Reagent	Conditions	Product(s) and Yield(s) (%)	Refs.
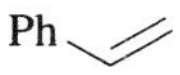 Ph⁀	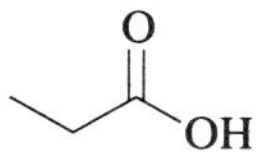 propanoic acid (O, OH)	Mn(OAc)$_3$, AcOH, AcOK, reflux	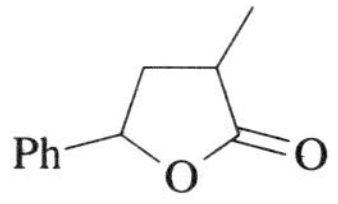 Ph-substituted γ-butyrolactone with methyl (50)	128
C$_{10}$				
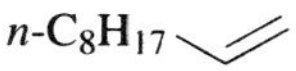 n-C$_8$H$_{17}$⁀	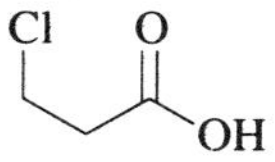 Cl-CH$_2$CH$_2$-COOH (Cl, O, OH)	Mn(OAc)$_3$, 100°, 2 - 3 h	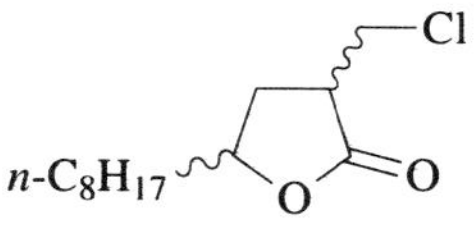 n-C$_8$H$_{17}$ lactone with CH$_2$Cl 1.5 : 1.0 (50)	129

TABLE VI. ALKENES AND CARBOXYLIC ACID ANHYDRIDES

Substrate	Reagent	Conditions	Product(s) and Yield(s) (%)	Refs.
C8				
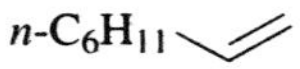 $n\text{-}C_6H_{11}$⎯⧵	Ac_2O	$Mn(OAc)_3$, $Cu(OAc)_2$, Ac_2O, 119°, 40 min	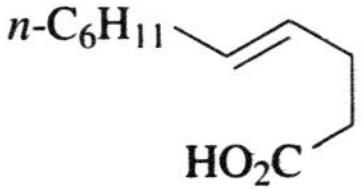 $n\text{-}C_6H_{11}$ ⎯⧵⧸ HO_2C (63) 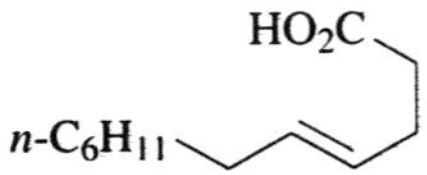 $+$ $n\text{-}C_6H_{11}$ ⎯⧵⧸ CO_2H (8)	68
	Ac_2O	$Mn(OAc)_3$, Ac_2O, 100°, 2.5 h	 $n\text{-}C_6H_{13}$⎯CO_2H (29) 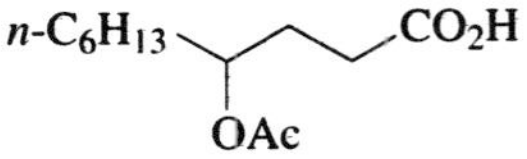 $+$ $n\text{-}C_6H_{13}$⎯CO_2H, OAc (21)	60
C8-C12				
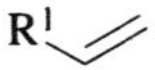 R^1⎯⧵	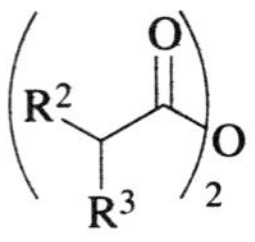 (R²R³CH—C(=O)—O)₂	1. $Mn(OAc)_3$, 120 - 140°, 3 h 2. AcOH, H_2O, 110°, 1 h	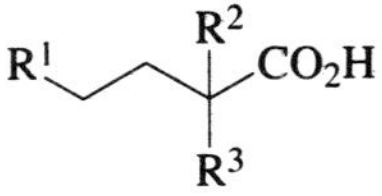 R^1⎯$CH(R^2)(R^3)CO_2H$ (70 - 80)	219

R^1	R^2	R^3
$n\text{-}C_6H_{13}$	H	H
$n\text{-}C_8H_{17}$	H	Me
$n\text{-}C_8H_{17}$	Me	Me
$n\text{-}C_8H_{17}$	H	Et
$n\text{-}C_{10}H_{21}$	H	H

512

TABLE VII. ALKENES OR CYCLOALKENES AND β-DICARBONYL COMPOUNDS

Substrate	Reagent	Conditions	Product(s) and Yield(s) (%)	Refs.
C₂				
$CH_2{=}CH_2$	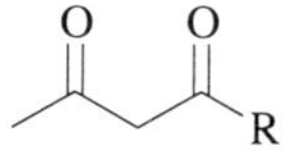 R=Me R=OEt	Mn(OAc)₃, Cu(OAc)₂, AcOH, 60°, 2 h, 50 atm	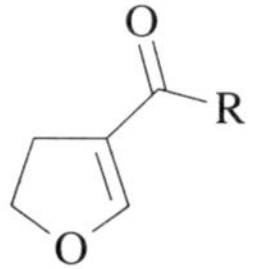 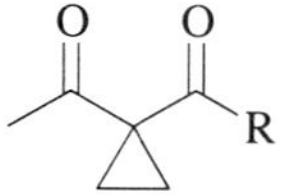 8 : 1 (20) 15 : 1 (20)	69
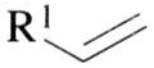	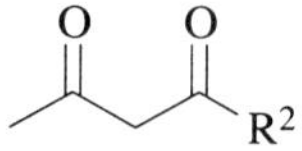	Mn(OAc)₃, LiCl, AcOH	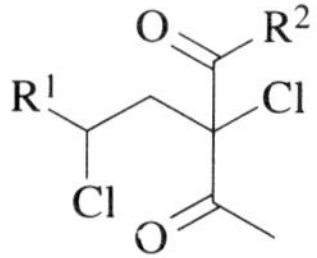	63
	R²			
C₂	Me	55°, 45 min, 45 atm	(30)	
C₆	Me	50°, 40 min	(60)	
C₆	OEt	70°, 40 min	(67)	
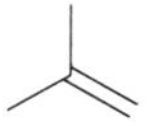 C₄	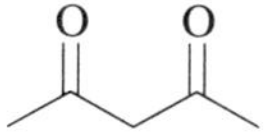	Mn(OAc)₃, Cu(OAc)₂, AcOH, 40°, 5 h	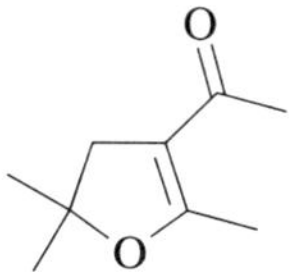 **I** (35) 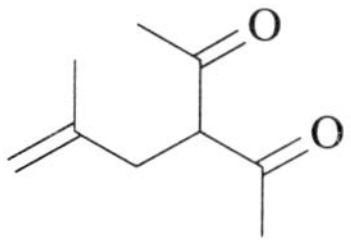 + **II** (24)	70
	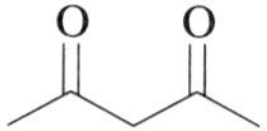	Mn(OAc)₃, AcOH, 40°, 6 h	**I** (35) + **II** (18) 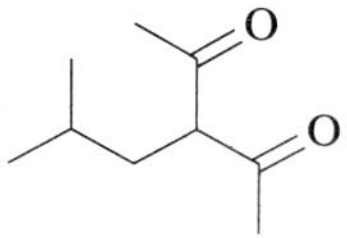 + (3)	70

513

TABLE VII. ALKENES OR CYCLOALKENES AND β-DICARBONYL COMPOUNDS (*Continued*)

Substrate	Reagent	Conditions	Product(s) and Yield(s) (%)	Refs.
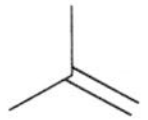	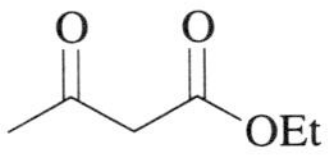	Mn(OAc)$_3$, Cu(OAc)$_2$, AcOH, 60°, 2 h	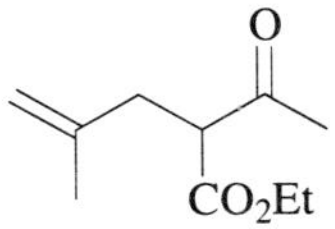 (48)	45
	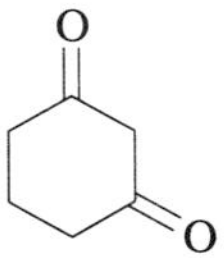	Mn(OAc)$_3$, AcOH, 45°, 10 min	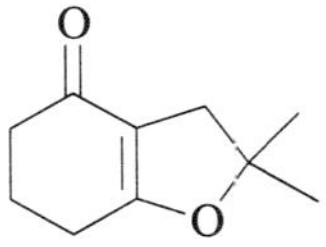 (40)	106
C$_4$–C$_{18}$ 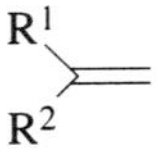	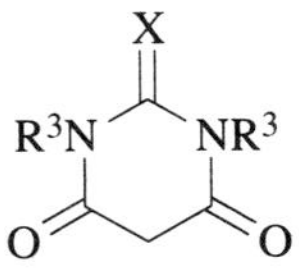	Mn(OAc)$_2$, O$_2$, AcOH, 20–28°	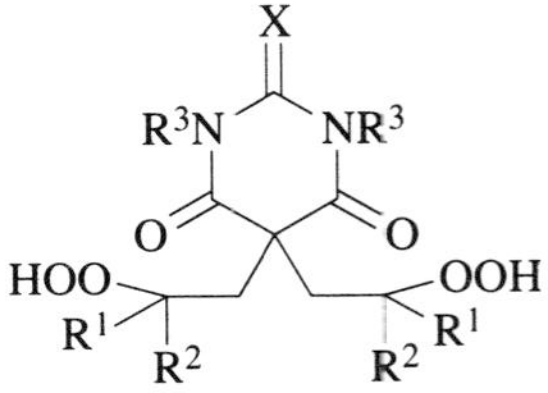	54

R^1	R^2	R^3	X	(h)	
Ph	Ph	Me	O	5	(94)[a]
Ph	Ph	H	O	6	(86)
Ph	Ph	Et	S	4	(62)
Me	Me	Me	O	5	(81)
Et	Et	Me	O	3	(88)
4-MeC$_6$H$_4$	4-MeC$_6$H$_4$	Me	O	3	(97)
4-ClC$_6$H$_4$	4-ClC$_6$H$_4$	Me	O	3	(99)
4-FC$_6$H$_4$	4-FC$_6$H$_4$	Me	O	3	(92)
Me	Ph	Me	O	2	(73)
Ph	4-ClC$_6$H$_4$	Me	O	3	(97)
Ph	4-BrC$_6$H$_4$	Me	O	3	(84)
Ph	1-naphthyl	Me	O	3	(93)

TABLE VII. ALKENES OR CYCLOALKENES AND β-DICARBONYL COMPOUNDS (*Continued*)

Substrate	Reagent	Conditions	Product(s) and Yield(s) (%)	Refs.
C$_6$				
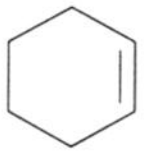 (cyclohexene)	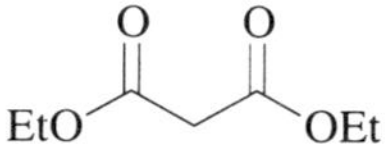 (EtO diethyl malonate)	Mn(OAc)$_3$, LiCl, AcOH, 80°, 50 min	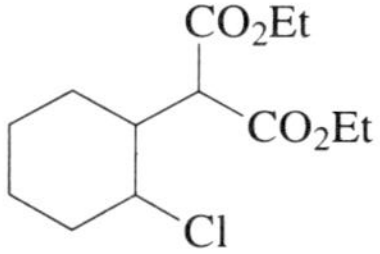 (chlorocyclohexyl diethyl malonate) (25)^b	63
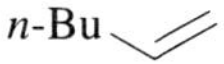 *n*-Bu (1-hexene)	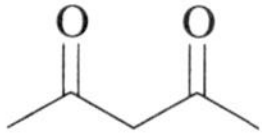 (pentane-2,4-dione)	Mn(OAc)$_3$, AcOH, 60°, 1 h	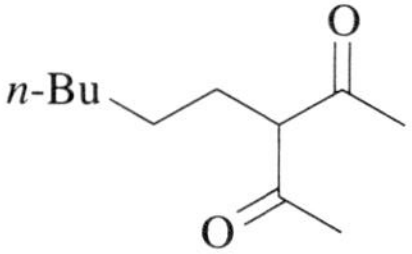 (15) 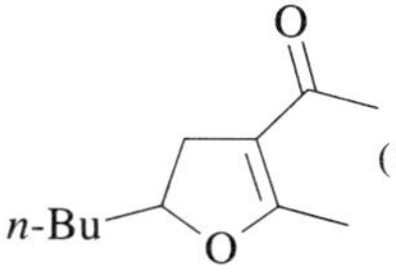 + (10) 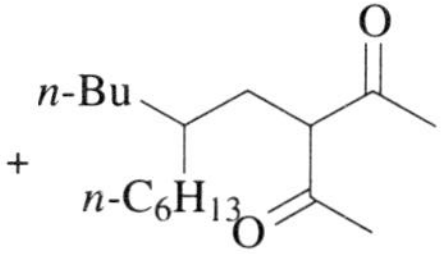 + (6)	105
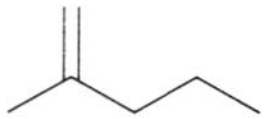	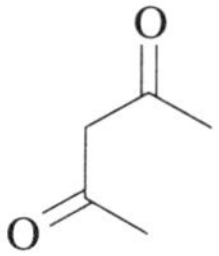	Mn(OAc)$_3$, AcOH, 45°, 10 min	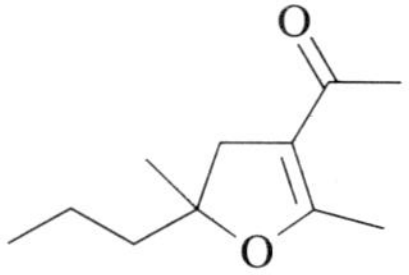 (40)	106
	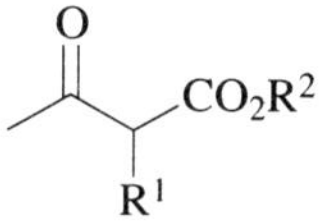 R^1 CO$_2$R^2	Mn(OAc)$_3$, AcOH, 40°, 24 h	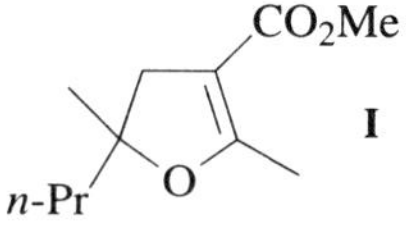 I 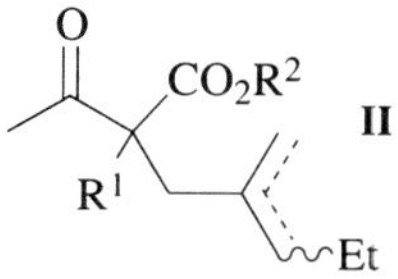 + II +	31

515

TABLE VII. ALKENES OR CYCLOALKENES AND β-DICARBONYL COMPOUNDS (*Continued*)

Substrate	Reagent	Conditions	Product(s) and Yield(s) (%)	Refs.

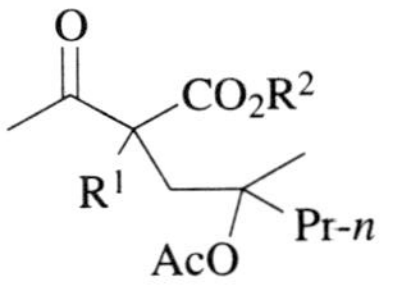

III

$$CO_2R^2$$

$$R^1$$

O / Pr-*n* / AcO

R^1	R^2		I	II	III	
H	Me		(30)	(16)	(13)	
Me	Et		(0)	(35)	(5)	

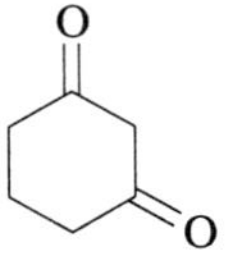

Reagent: 1,3-cyclohexanedione (O=...=O)

Conditions: Mn(OAc)$_3$, AcOH, 45°, 10 min

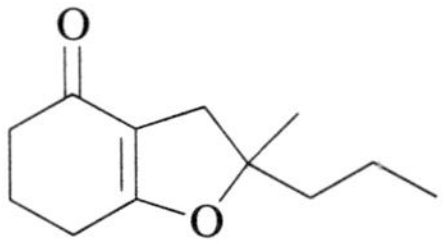

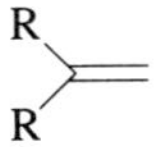

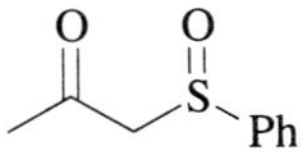

Product: (74)

Refs.: 106

C$_6$-C$_{16}$

Reagent:
O=, S=O, Ph (PhSO–CH$_2$–CO–CH$_3$)

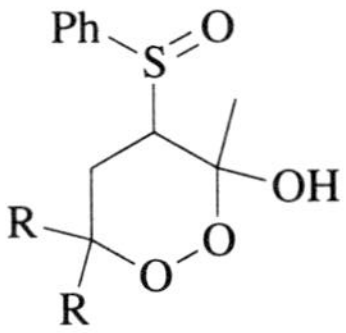

Conditions: Mn(OAc)$_3$, O$_2$, AcOH, 32°, 10 - 12 h

Product: Ph–S(=O)– ... OH / O–O / R

R		
Ph	92 : 8	(63)
p-ClC$_6$H$_4$	65 : 35	(69)
p-MeC$_6$H$_4$	85 : 15	(76)
p-FC$_6$H$_4$	81 : 19	(72)
p-MeOC$_6$H$_4$	88 : 12	(58)
Et	70 : 30	(51)

Refs.: 53

TABLE VII. ALKENES OR CYCLOALKENES AND β-DICARBONYL COMPOUNDS (Continued)

Substrate	Reagent	Conditions	Product(s) and Yield(s) (%)	Refs.
R^1 (alkene)	(β-dicarbonyl, R^2)	Mn(OAc)$_3$, O$_2$, AcOH, 23°, 11–14 h	(structure I, OH, R^2, R^1)	51

R^1	R^2	
4-MeC$_6$H$_4$	4-MeC$_6$H$_4$NH	(99)
4-MeC$_6$H$_4$	PhNH	(98)
4-MeC$_6$H$_4$	2-MeC$_6$H$_4$NH	(100)
4-MeC$_6$H$_4$	4-ClC$_6$H$_4$NH	(99)
4-MeC$_6$H$_4$	2-ClC$_6$H$_4$NH	(100)
4-MeC$_6$H$_4$	NH$_2$	(96)
4-FC$_6$H$_4$	PhNH	(94)
4-FC$_6$H$_4$	Me$_2$N	(59)
4-ClC$_6$H$_4$	2-ClC$_6$H$_4$NH	(87)
4-ClC$_6$H$_4$	Me$_2$N	(89)
4-MeOC$_6$H$_4$	2-ClC$_6$H$_4$NH	(83)
4-MeOC$_6$H$_4$	Me$_2$N	(78)
Et	PhNH	(68)
Et	4-ClC$_6$H$_4$NH	(62)
Ph	4-MeC$_6$H$_4$NH	(92)
Ph	PhNH	(96)
Ph	4-MeOC$_6$H$_4$NH	(82)
Ph	2-MeOC$_6$H$_4$NH	(84)
Ph	2-MeC$_6$H$_4$NH	(94)
Ph	4-ClC$_6$H$_4$NH	(94)
Ph	2-ClC$_6$H$_4$NH	(93)
Ph	NH$_2$	(88)
Ph	MeNH	(85)

517

TABLE VII. ALKENES OR CYCLOALKENES AND β-DICARBONYL COMPOUNDS (*Continued*)

Substrate	Reagent	Conditions	Product(s) and Yield(s) (%)	Refs.
R^1	R^2		**I**	55
Ph	Me_2N		$\overline{(87)}$	
Ph	Et_2N		(89)	
Ph	 morpholino		(88)	

R^1, R^2, R^3 substrate with $Mn(acac)_3, O_2$

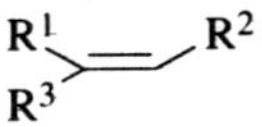

Conditions: $Mn(acac)_3$, AcOH, 23°, 11 h

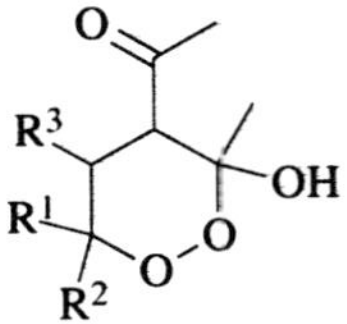

Product:

R^1	R^2	R^3	Yield
Ph	H	Ph	(92)
4-ClC$_6$H$_4$	H	4-ClC$_6$H$_4$	(90)
4-MeOC$_6$H$_4$	H	4-MeOC$_6$H$_4$	(87)
4-MeC$_6$H$_4$	H	4-MeC$_6$H$_4$	(77)
4-FC$_6$H$_4$	H	4-FC$_6$H$_4$	(72)
Ph	H	H	(34)
n-C$_6$H$_{13}$	H	H	(8)
n-C$_7$H$_{15}$	H	H	(35)
(CH$_2$)$_4$			(11)
(CH$_2$)$_6$			(43)

C$_7$

n-C$_5$H$_{11}$ (alkene) + pentane-2,4-dione

Conditions: $Mn(OAc)_3$ or $Mn(acac)_3$, 100°, 30 h

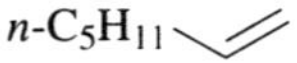

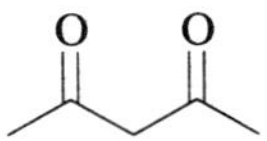

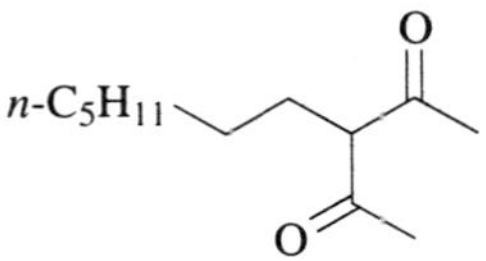

Product: n-C$_5$H$_{11}$–CH(COCH$_3$)$_2$ derivative (60) — Refs. 70

TABLE VII. ALKENES OR CYCLOALKENES AND β-DICARBONYL COMPOUNDS (*Continued*)

Substrate	Reagent	Conditions	Product(s) and Yield(s) (%)	Refs.
$n\text{-}C_5H_{11}$ ⟍	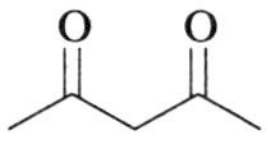 (acetylacetone)	Mn(OAc)$_3$, Cu(OAc)$_2$, AcOH, 60°, 3 h	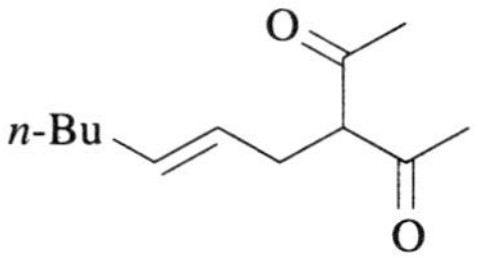 (16) + $n\text{-}C_7H_{15}$ (8)	70
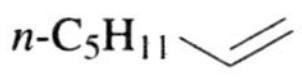	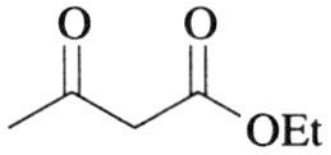 (ethyl acetoacetate, OEt)	Mn(OAc)$_3$, Cu(OAc)$_2$, AcOH, 60°, 2 h	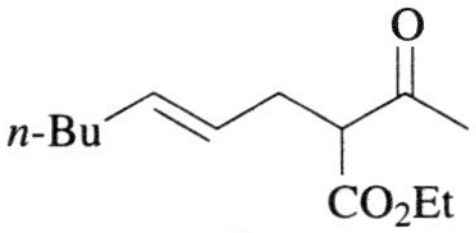 n-Bu CO$_2$Et (33)	45

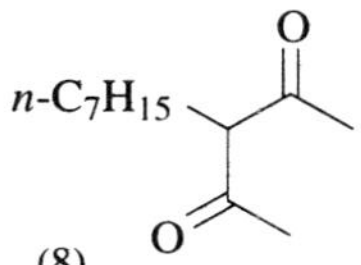

R^1⟍R^3⟍=R^2/H   Mn(acac)$_3$

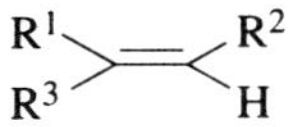

Mn(acac)$_3$, AcOH, reflux, 1-5 min

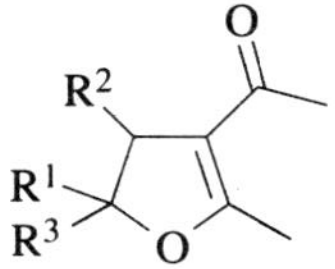

R^1	R^2	R^3	
Ph	H	Ph	(89)
4-MeOC$_6$H$_4$	H	4-MeOC$_6$H$_4$	(88)
Me	H	Ph	(97)
Ph	H	H	(75)
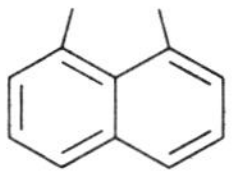		H	(70)
Ph	Ph	Ph	(41)
4-MeOC$_6$H$_4$	Ph	4-MeOC$_6$H$_4$	(29)
4-MeOC$_6$H$_4$	4-MeOC$_6$H$_4$	4-MeOC$_6$H$_4$	(13)
n-C$_5$H$_{11}$	H	H	(12)

135

519

TABLE VII. ALKENES OR CYCLOALKENES AND β-DICARBONYL COMPOUNDS (*Continued*)

Substrate	Reagent	Conditions	Product(s) and Yield(s) (%)	Refs.
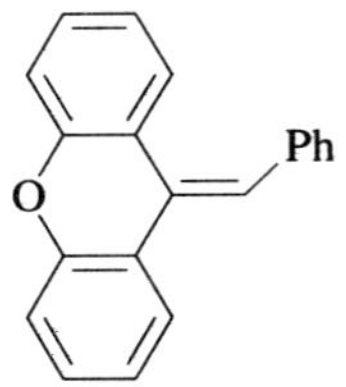	$Mn(acac)_3$	$Mn(acac)_3$, AcOH, reflux, 1-5 min	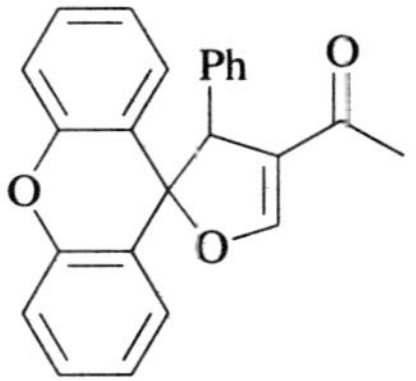 (44)	135
C_8 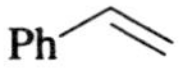	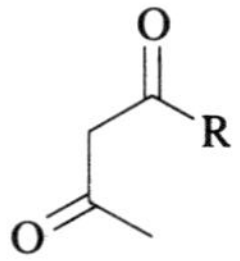 R = Me R =OEt	$Mn(OAc)_3$, AcOH, 45°, 10 min	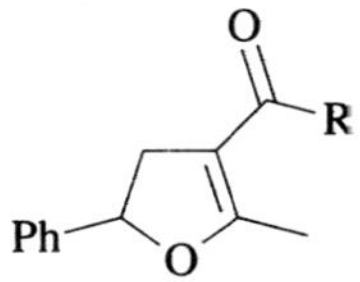 (30) (57)	106
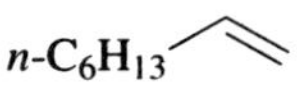	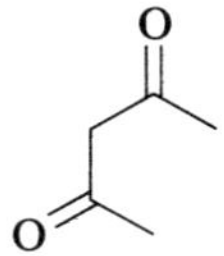	$Mn(OAc)_3$, AcOH, 45°, 10 min	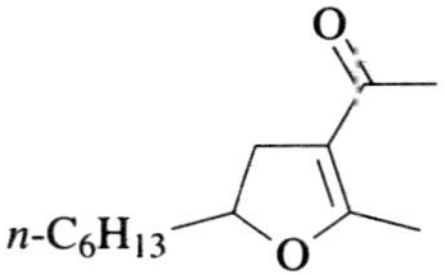 (10)	106
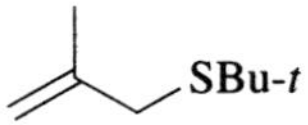	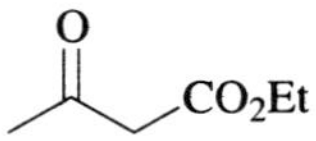	$Mn(OAc)_3$, $Cu(OAc)_2$, PbO_2, AcOH, rt, 2 d	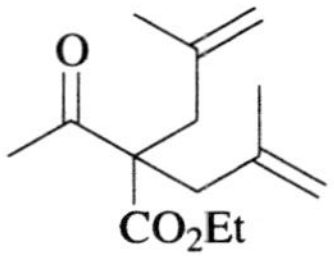 (35)	71
	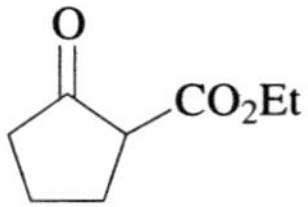	"	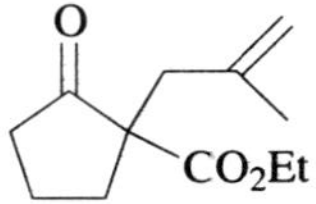 (86)	71

TABLE VII. ALKENES OR CYCLOALKENES AND β-DICARBONYL COMPOUNDS (*Continued*)

Substrate	Reagent	Conditions	Product(s) and Yield(s) (%)	Refs.
	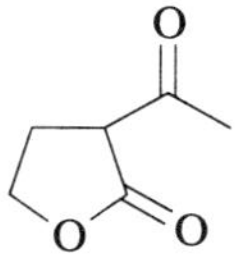	Mn(OAc)$_3$, Cu(OAc)$_2$, PbO$_2$, AcOH, rt, 2 d	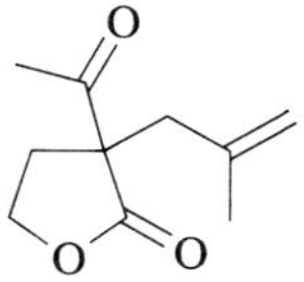 (78)	71
	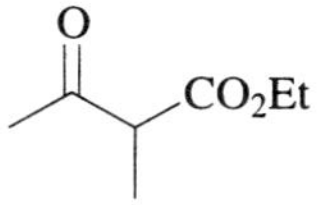	"	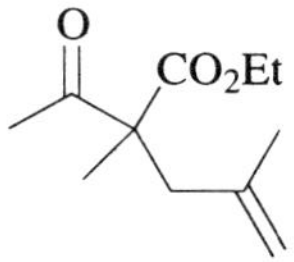 (88)	71
	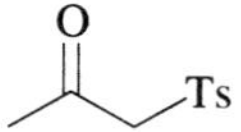	"	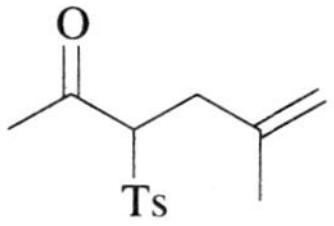 (69)	71
	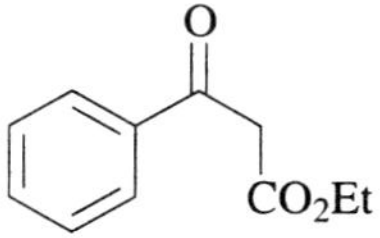	"	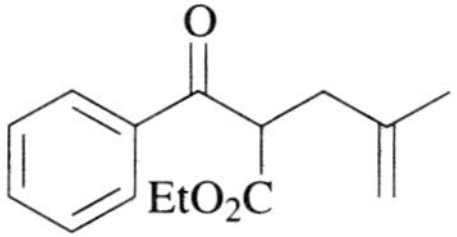 (20)	71
	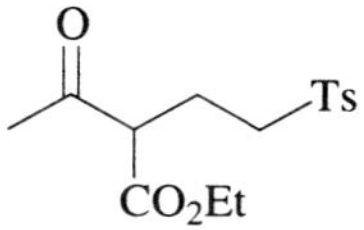	"	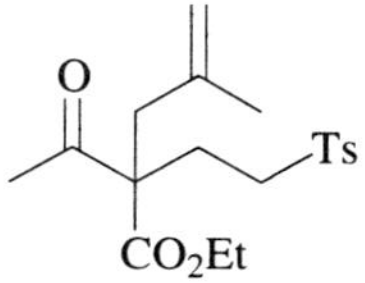 (96)	71

521

TABLE VII. ALKENES OR CYCLOALKENES AND β-DICARBONYL COMPOUNDS (*Continued*)

Substrate	Reagent	Conditions	Product(s) and Yield(s) (%)	Refs.
C$_8$–C$_{10}$				
		Mn(OAc)$_3$, Cu(OAc)$_2$, PbO$_2$, AcOH, rt, 2 d		71

X	
PhSO$_2$	(36)
t-BuSO$_2$	(41)
PhS	(65)
t-BuS	(73)
PhSO	(—)

C$_8$–C$_{14}$

R	Ar^1	Ar^2		
H	Ph	Ph	Mn(OAc)$_3$, AcOH, 75-85°, 10-30 min	(27)	136
CO$_2$Me	Ph	Ph		(40)	
CH$_2$OAc	Ph	Ph		(61)	
Me	4-MeOC$_6$H$_4$	3,4-(MeO)$_2$C$_6$H$_3$		(52)	
CO$_2$Et	3,4-(MeO)$_2$C$_6$H$_3$	3,4-(MeO)$_2$C$_6$H$_3$		(57)	
CH$_2$OAc	3,4-(MeO)$_2$C$_6$H$_3$	3,4-(MeO)$_2$C$_6$H$_3$		(48)	
CO$_2$Et	3,4-CH$_2$OCH$_2$C$_6$H$_3$	3,4-(MeO)$_2$C$_6$H$_3$		(57)	
CO$_2$Et	3,4-CH$_2$OCH$_2$C$_6$H$_3$	3,4-CH$_2$OCH$_2$C$_6$H$_3$		(71)	
CO$_2$Et	3,4,5-(MeO)$_3$C$_6$H$_2$	3,4-CH$_2$OCH$_2$C$_6$H$_3$		(56)	
CH$_2$OAc	3,4-CH$_2$OCH$_2$C$_6$H$_3$	3,4-(MeO)$_2$C$_6$H$_3$	Mn(OAc)$_3$, AcOH, rt, 1-4 d	(58)	
CO$_2$Et	3,4-CH$_2$OCH$_2$C$_6$H$_3$	3,4-CH$_2$OCH$_2$C$_6$H$_3$		(71)	
CH$_2$OAc	3,4-CH$_2$OCH$_2$C$_6$H$_3$	3,4-CH$_2$OCH$_2$C$_6$H$_3$		(55)	
CO$_2$Et	3,4,5-(MeO)$_3$C$_6$H$_2$	3,4-CH$_2$OCH$_2$C$_6$H$_3$		(54)	
CH$_2$OAc	3,4-CH$_2$OCH$_2$C$_6$H$_3$	3,4,5-(MeO)$_3$C$_6$H$_2$		(65)	

522

TABLE VII. ALKENES OR CYCLOALKENES AND β-DICARBONYL COMPOUNDS (*Continued*)

Substrate	Reagent	Conditions	Product(s) and Yield(s) (%)	Refs.

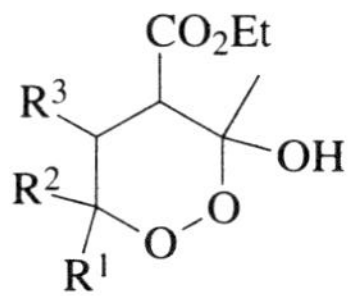

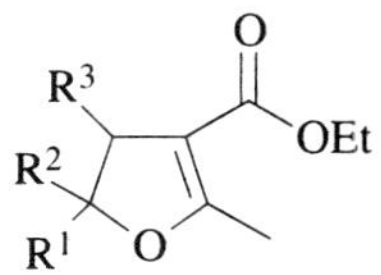

C8–C16

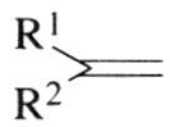

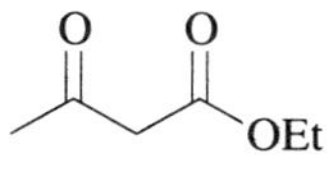

R¹	R²
Ph	Ph
4-ClC₆H₄	4-ClC₆H₄
4-MeOC₆H₄	4-MeOC₆H₄
4-MeC₆H₄	4-MeC₆H₄
4-FC₆H₄	4-FC₆H₄
Ph	H

Mn(OAc)₃, AcOH, rt, 11 h

(82) (27) (54) (55) (50) (50)

(0) (34) (46) (46) (17) (13)

55

C₉

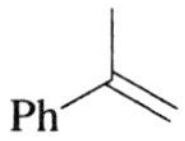

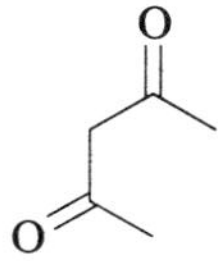

Mn(OAc)₃, AcOH, 45°, 10 min — (100) — 106

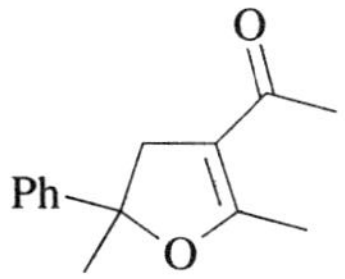

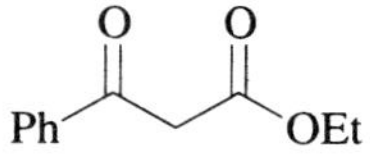

Mn(OAc)₃, AcOH, 75-85°, 10-30 min — (56) — 136

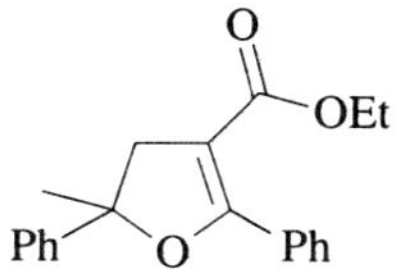

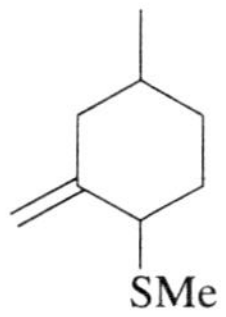

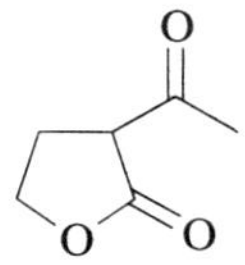

Mn(OAc)₃, Cu(OAc)₂, PbO₂, AcOH, rt, 2 d — (73) — 71

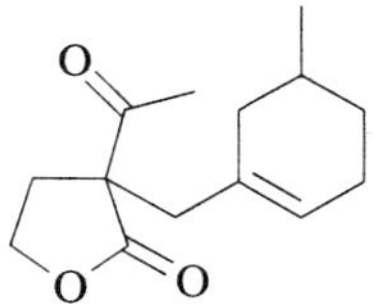

523

TABLE VII. ALKENES OR CYCLOALKENES AND β-DICARBONYL COMPOUNDS (*Continued*)

Substrate	Reagent	Conditions	Product(s) and Yield(s) (%)	Refs.
C_{11} (structure, SPh)	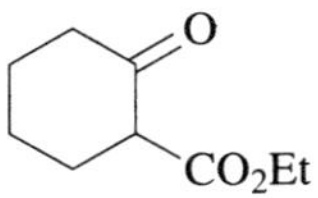 (structure, CO_2Et)	Mn(OAc)$_3$, Cu(OAc)$_2$. PbO$_2$, AcOH, rt, 2 d	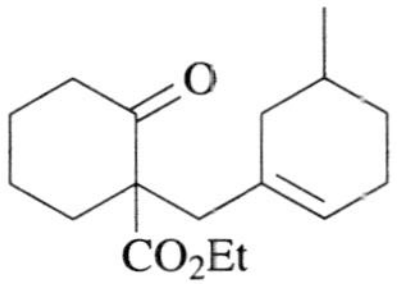 (structure, CO_2Et) (63)	71
	(structure, CO_2Et)	"	(structure, CO_2Et) (—)	71
(furan structure, $O=\!\!<^R$)	(structure, R, O)	Mn(OAc)$_3$, Cu(OAc)$_2$, AcOH, 30°, 3 h	(bicyclic structure) R = Me (70) R = OEt (77)	77
C_{13} C_{14}	R = Me R = OEt			
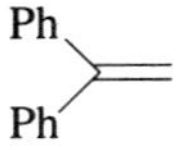 C_{14} Ph / Ph (alkene)	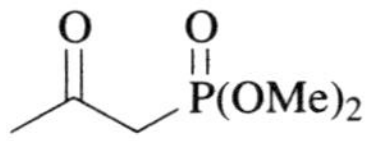 (structure, P(OMe)$_2$)	Mn(OAc)$_3$, AcOH, 80°, 0.5-3 h	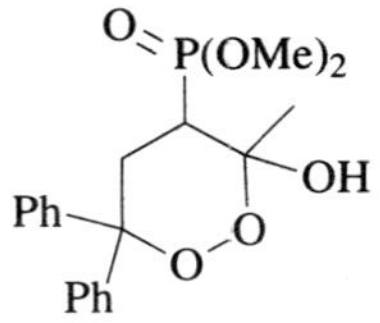 (structure, O=P(OMe)$_2$, OH, Ph, Ph) (46) + (structure, O=P(OMe)$_2$, Ph, Ph) (21)	53

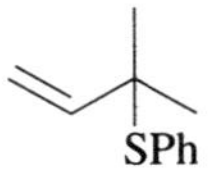

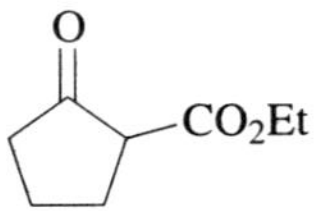

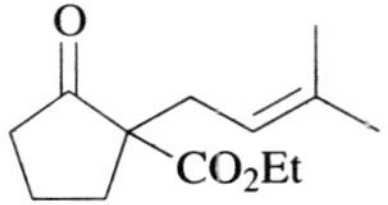

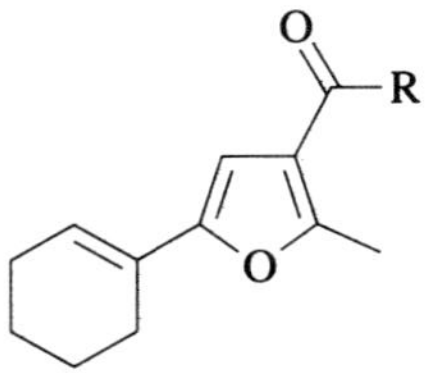

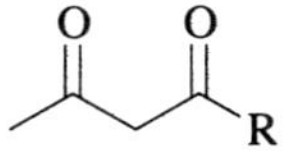

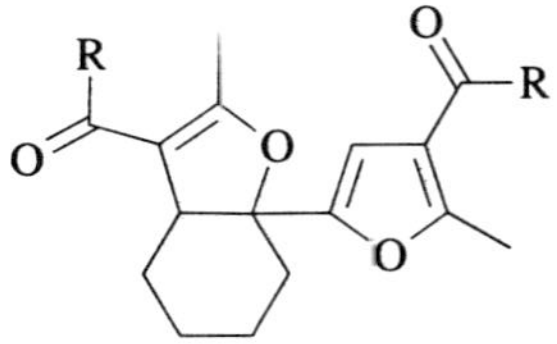

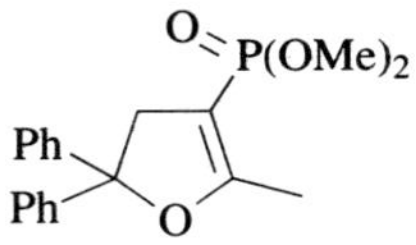

TABLE VII. ALKENES OR CYCLOALKENES AND β-DICARBONYL COMPOUNDS (*Continued*)

Substrate	Reagent	Conditions	Product(s) and Yield(s) (%)	Refs.
C_{14}-C_{16}				
		Mn(OAc)$_3$, O$_2$, AcOH, 40°, 12 h		53

R^1	R^2	
Ph	SOPh	(62)
Ph	SOPh	(76)
Ph	SOPh	(83)
p-ClC$_6$H$_4$	SOPh	(78)
Ph	SO$_2$Ph	(88)
p-ClC$_6$H$_4$	SO$_2$Ph	(80)
p-MeC$_6$H$_4$	SO$_2$Ph	(82)
Ph	SO$_2$C$_6$H$_4$Me-p	(77)
Ph	P(O)(OMe)$_2$	(68)

Substrate	Reagent	Conditions	Product(s) and Yield(s) (%)	Refs.
		Mn(OAc)$_3$, O$_2$, AcOH, 48°, 10-16 h		53

R^1	R^2		
Ph	Me	86 : 14 (58)	(15)
Ph	H	85 : 15 (52)	(17)
p-ClC$_6$H$_4$	H	87 : 13 (47)	(11)
p-MeC$_6$H$_4$	H	89 : 11 (63)	(9)

TABLE VII. ALKENES OR CYCLOALKENES AND β-DICARBONYL COMPOUNDS (*Continued*)

Substrate	Reagent	Conditions	Product(s) and Yield(s) (%)	Refs.
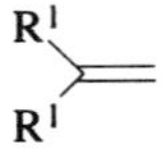 R^1 —(C=CH$_2$)	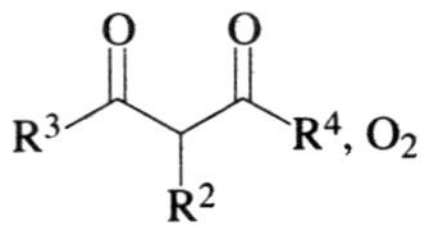 R^3–CO–CH(R^2)–CO–R^4, O$_2$	Mn(OAc)$_2$, O$_2$, AcOH, 23°, 11–39 h	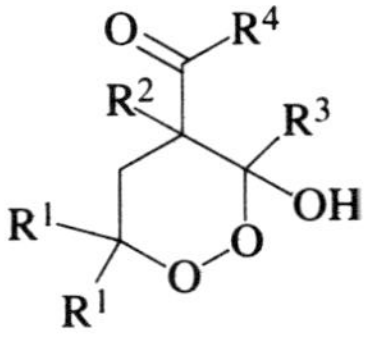 cyclic peroxide product	52

R^1	R^2	R^3	R^4	
Ph	Me	(CH$_2$)$_3$		(75)
Ph	Me	(CH$_2$)$_2$		(93)
Ph	(CH$_2$)$_4$		Me	(67)
Ph	H	Me	OEt	(48)
Ph	Me	Me	Me	(66)
Ph	H	(CH$_2$)$_3$		(79)
p-ClC$_6$H$_4$	Me	(CH$_2$)$_3$		(60)
p-MeOC$_6$H$_4$	Me	(CH$_2$)$_3$		(54)
Ph	H	Me	Me	(7)
p-ClC$_6$H$_4$	H	Me	Me	(6)
p-MeC$_6$H$_4$	H	Me	Me	(11)

[a] The same yield was obtained with Mn(OAc)$_3$.
[b] Cyclohexenyl acetate (10%) was also formed.

526

TABLE VIII. ALKENES AND β-KETOCARBOXYLIC ACIDS

Substrate	Reagent	Conditions	Product(s) and Yield(s) (%)	Refs.
C₄ (SMe,SMe alkene)	(Ph-C(=O)-CH(CH₃)-CO₂H)	(PyCO₂)₃Mn, DMF, rt, 24 h	(SMe,SMe product) (40)	209
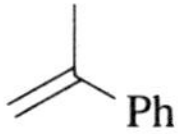 C₉ (Ph alkene)	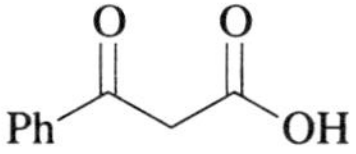 (Ph-C(=O)-CH₂-CO₂H)	Mn(OAc)₃, DMF, rt	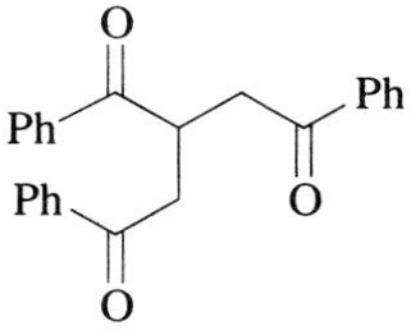 (diketone product) (44) + (Ph diketone) (7) (lactone product) (6)	209
	(Ph-C(=O)-CH₂-CO₂H)	(PyCO₂)₃Mn, DMF, rt	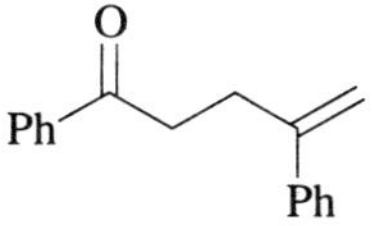 (8) 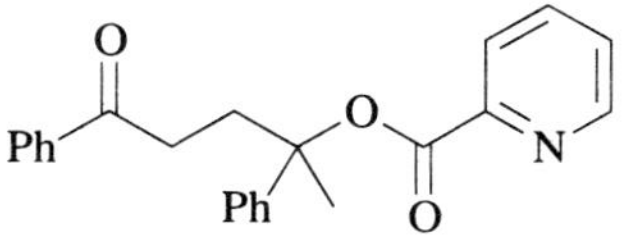 + (14)	209

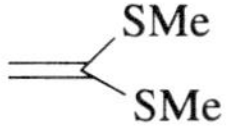

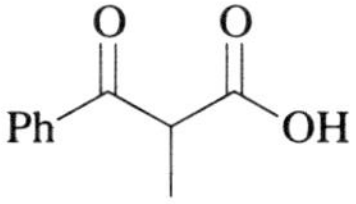

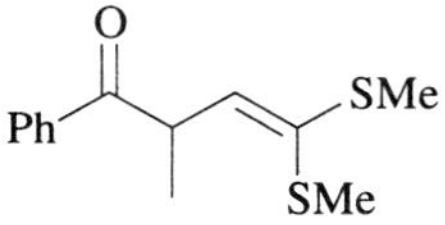

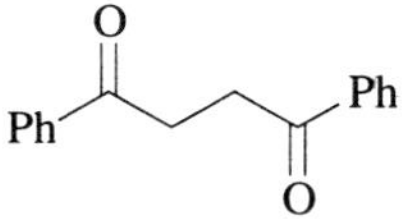

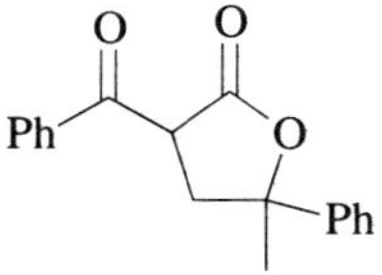

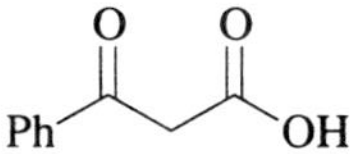

TABLE VIII. ALKENES AND β-KETOCARBOXYLIC ACIDS (Continued)

Substrate	Reagent	Conditions	Product(s) and Yield(s) (%)	Refs.
C₁₁ OTBDMS	β-ketoacid (Ph)	(PyCO₂)₃Mn, DMF, rt	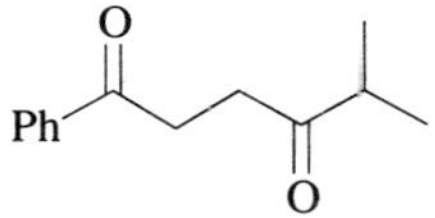 (45)	209
OSiR₃ (Ph); R = Me, R₃ = t-BuMe₂	β-ketoacid (Ph, Me)	(PyCO₂)₃Mn, DMF, rt	(70), (68) + (47), (0)	209
C₁₂ Ph–N(morpholine)	β-ketoacid (Ph, Me)	(PyCO₂)₃Mn, DMF, rt, 24 h	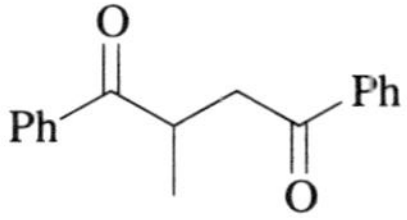 (66)	209
C₁₄ OTBDMS (Ph)	β-ketoacid	(PyCO₂)₃Mn, DMF, rt	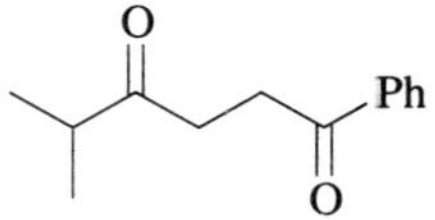 (55)	209
C₂₁ SnPh₃	β-ketoacid (Ph, Me)	(PyCO₂)₃Mn, DMF, rt, 110 h	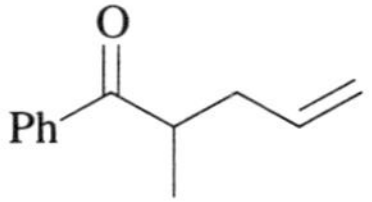 (63)	209

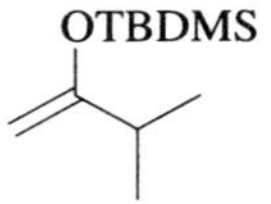

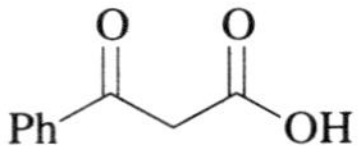

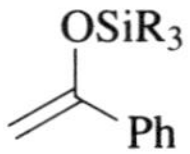

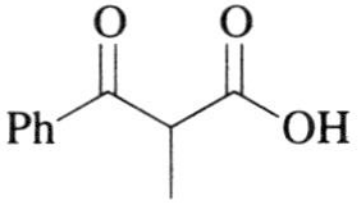

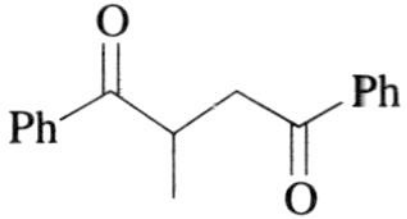

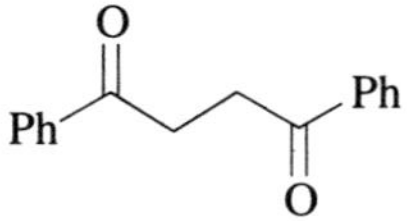

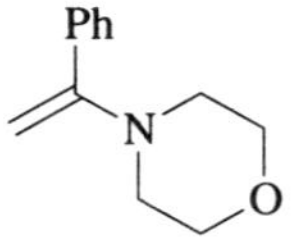

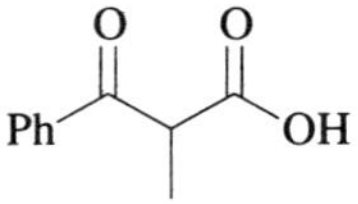

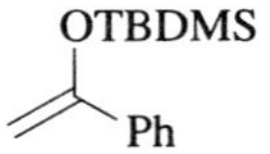

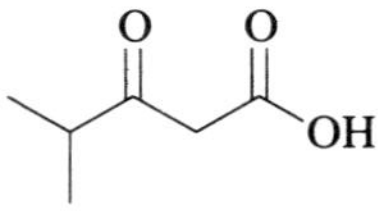

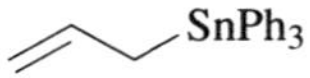

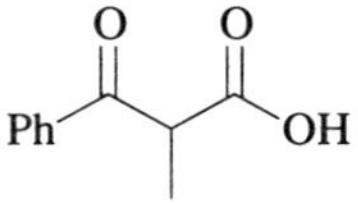

528

TABLE IX. ALKENES OR CYCLOALKENES AND DICARBOXYLIC ACIDS

Substrate	Reagent	Conditions	Product(s) and Yield(s) (%)	Refs.
C_2-C_{14} 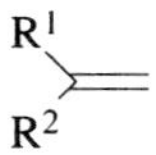	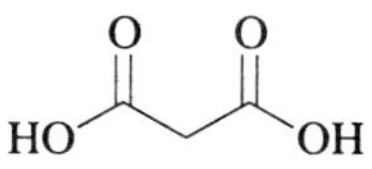	Mn(OAc)$_3$, AcOH	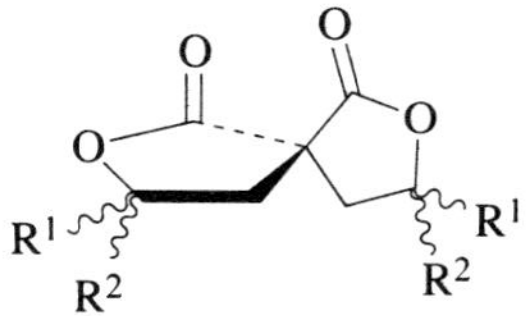	

R^1	R^2			
n-Bu	H	70°, 2 h	9 : 47 : 44[a] (100)	145
n-C$_6$H$_{13}$	H	"	11 : 59 : 30[a] (100)	145
n-C$_6$H$_{13}$	H	reflux, 1 - 5 min	(4) (17) (19)[b]	144
t-Bu	H	70°, 2 h	2 : 48 : 50[a] (42)	145
Ph	Me	"	20 : 49 : 31[a] (81)	145
Ph	Me	reflux, 1 - 5 min	(10) (19) (39)[b]	144
CH$_2$Cl	H	70°, 2 h	9 : 60 : 31[a] (30)	145
Ph	Ph	"	(93)	145
Ph	Ph	reflux, 1 - 5 min	(84)	144
Me	Me	70°, 2 h	(51)	145
H		"	(29)	145
(CH$_2$)$_5$		"	(42)	145
(CH$_2$)$_5$		reflux, 1 - 5 min	(27)	144
Ph	H	"	(9) (30) (40)[b]	144
p-MeOC$_6$H$_4$	H	"	(84)	144

529

TABLE IX. ALKENES OR CYCLOALKENES AND DICARBOXYLIC ACIDS (Continued)

Substrate	Reagent	Conditions	Product(s) and Yield(s) (%)	Refs.
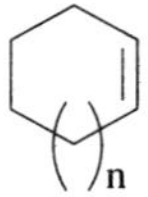	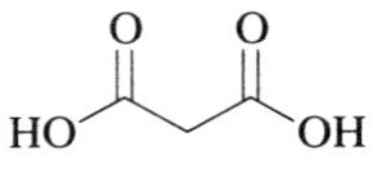	Mn(OAc)$_3$, AcOH 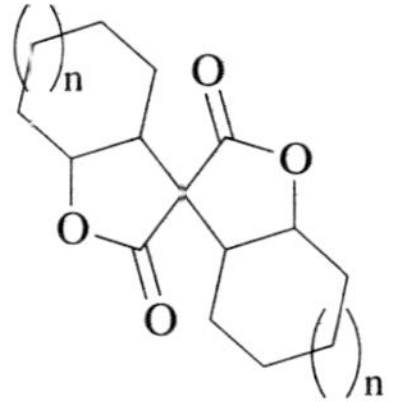		
C$_5$ n = 0		70°, 2 h	0 : 92 : 8^a (40)	145
C$_6$ n = 1		70°, 2 h	(16)	145
n = 1		reflux, 1-5 min	(27)	144
C$_8$ 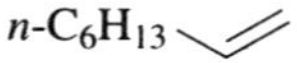 n-C$_6$H$_{13}$	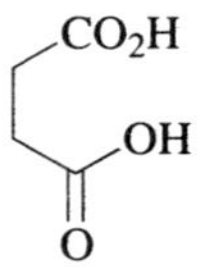	Mn(OAc)$_3$, AcOH, AcOK, reflux	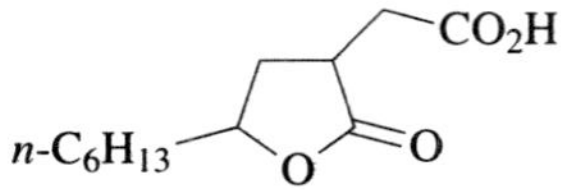 (25)	128
C$_8$-C$_{16}$ 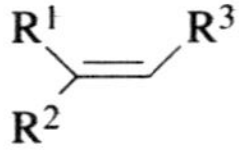	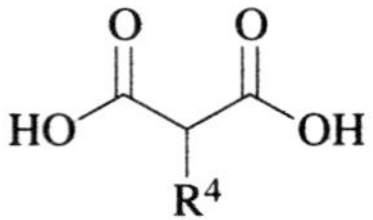	Mn(OAc)$_3$, AcOH, AcOK, 100°, 3-5.5 min	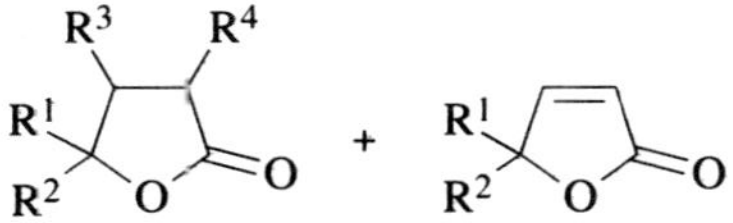	146

R^1	R^2	R^3	R^4		
Ph	Ph	H	Br	(68)	(0)
Ph	Ph	H	Cl	(82)	(0)
4-MeOC$_6$H$_4$	4-MeOC$_6$H$_4$	H	Br	(0)	(82)
4-MeOC$_6$H$_4$	4-MeOC$_6$H$_4$	H	Cl	(19)	(63)
4-MeC$_6$H$_4$	4-MeC$_6$H$_4$	H	Br	(4)	(69)
4-MeC$_6$H$_4$	4-MeC$_6$H$_4$	H	Cl	(74)	(0)
4-MeOC$_6$H$_4$	Ph	H	Br	(0)	(67)

530

TABLE IX. ALKENES OR CYCLOALKENES AND DICARBOXYLIC ACIDS (Continued)

Substrate			Reagent	Conditions	Product(s) and Yield(s) (%)	Refs.

First block

R¹	R³	R²	R⁴			
4-MeOC₆H₄	H	Ph	Cl		(36)	(30)
Ph	H	Me	Br		(54)	(0)
Ph	H	Me	Cl		(46)	(0)
		H	Br		(40)	(0)
		H	Cl		(32)	(0)
Ph	H		Br		(28)	(0)
Ph	H		Cl		(28)	(0)

Substrate: (1,8-dimethylnaphthalene)

Second block

Reagent:

$$\text{HO–C(=O)–CH(Me)–C(=O)–OH}$$

Conditions: Mn(OAc)₃, AcOH, 100°, 2.5 - 10 min

Product:

R^3–$\overset{}{\underset{}{}}$ lactone with CO₂H, R¹, R² (γ-butyrolactone)

Refs.: 146

R¹	R³	R²				
Ph	H	Ph			(78)	
4-MeOC₆H₄	H	4-MeOC₆H₄			(70)	
4-MeC₆H₄	H	4-MeC₆H₄			(88)	
Ph	H	Me			(68)	
(1,8-dimethylnaphthalene)		H	H		(13)	
Ph	H	H			(48)	
(CH₂)₅[c]	H				(56)	
n-C₆H₁₃	H	H			(42)	

[a] Denotes the diastereomeric ratio, *symm-syn* : *unsymm* : *symm-anti*.
[b] These are the yields of the individual diastereomers.
[c] R¹ + R².

531

TABLE X. ALKENES OR CYCLOALKENES AND DICARBOXYLIC ACID DERIVATIVES

Substrate	Reagent	Conditions	Product(s) and Yield(s) (%)	Refs.
C_2 $CH_2=CH_2$	$NC-CH_2-CO_2H$	$Mn(OAc)_3$, AcOH, 70°, 10 - 15 min	(lactone with CN) (4)	129
C_3-C_6 $R-CH=CH_2$ R: n-Bu, CN, CO_2Me	EtO$_2$C-CHCl-CO$_2$Et (diethyl chloromalonate)	$Mn(OAc)_3$, LiCl, AcOH, 70-90°, 0.5 - 3 h	(Cl, Cl, CO$_2$Me, CO$_2$Me product) (65) (32) (29)	50
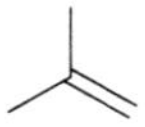 C_4 (isobutylene)	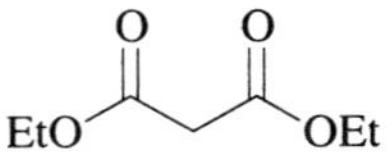 EtO$_2$C-CH$_2$-CO$_2$Et (diethyl malonate)	$Mn(OAc)_3$, $Cu(OAc)_2$, AcOH, 60°	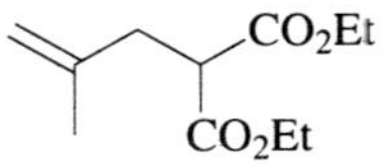 (CO$_2$Et, CO$_2$Et product) (60)	64
C_6 (cyclohexene)	$HO_2C-CH_2-CH_2-CN$	$Mn(OAc)_3$, AcOH, 23°, 15 min	(bicyclic lactone with CN) (79)	147
		$Mn(OAc)_3$, AcOH, 70°, 10 - 15 min	6.8 : 1.0 : 1.0 : 1.9 (46)	129

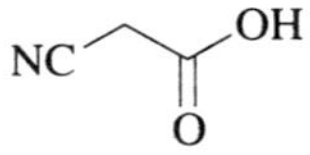

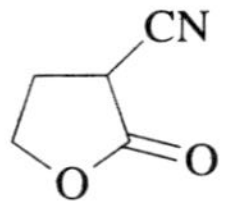

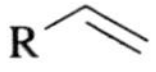

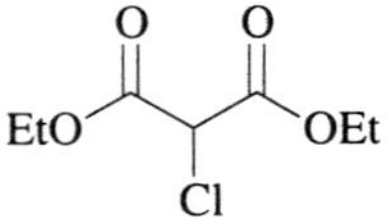

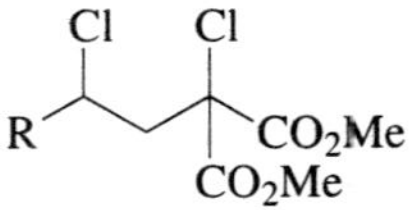

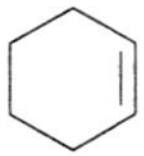

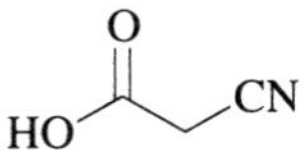

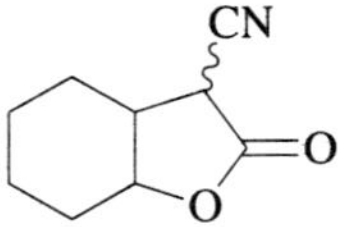

TABLE X. ALKENES OR CYCLOALKENES AND DICARBOXYLIC ACID DERIVATIVES (*Continued*)

Substrate	Reagent	Conditions	Product(s) and Yield(s) (%)	Refs.
(tetramethylethylene)	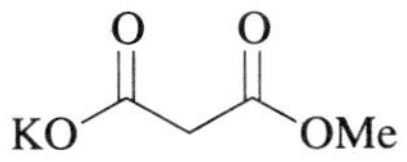 KO $\overset{O}{\!}$ $\overset{O}{\!}$ OMe	Mn(OAc)$_3$, AcOH, 70°, 10 - 12 min	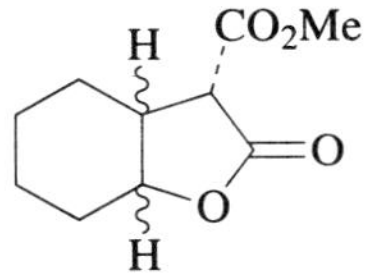 (hexahydrobenzofuranone) CO$_2$Me 1.0 : 1.3 (39)	129
	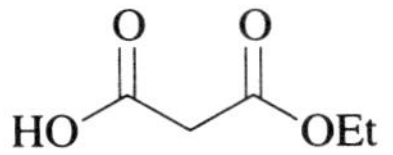 HO $\overset{O}{\!}$ $\overset{O}{\!}$ OEt	Mn(OAc)$_3$, AcOH, 23°, 4 - 6 h	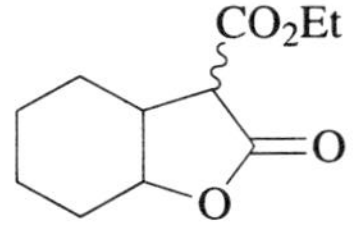 (hexahydrobenzofuranone) CO$_2$Et (75)	147
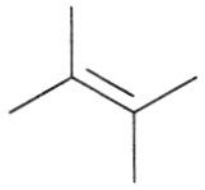	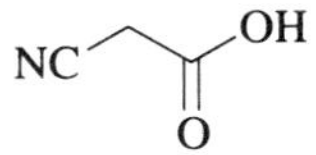 NC $\overset{O}{\!}$ OH	Mn(OAc)$_3$, AcOH, 70°, 10 - 15 min	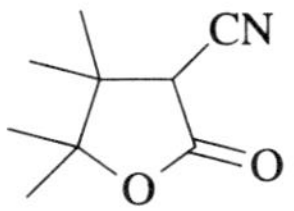 (dimethyl lactone) CN (62)	129
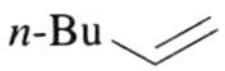 *n*-Bu ⟶	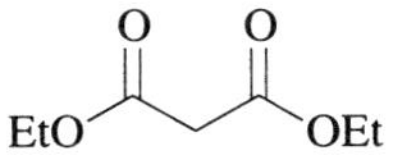 EtO $\overset{O}{\!}$ $\overset{O}{\!}$ OEt	Mn(OAc)$_3$, LiCl, AcOH, 60°, 2 h	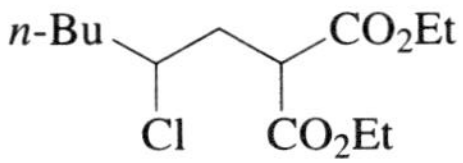 *n*-Bu $\overset{CO_2Et}{\underset{Cl}{\!}}$ CO$_2$Et (61)	63
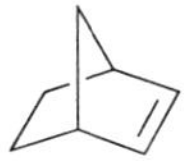 (norbornene)	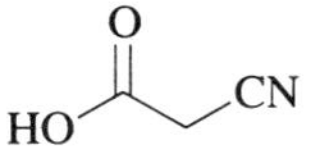 HO $\overset{O}{\!}$ CN	Mn(OAc)$_3$, AcOH, 23°, 15 min	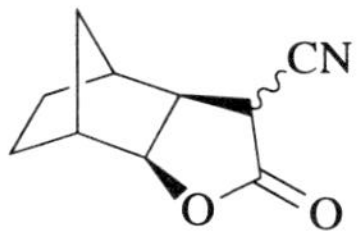 (norbornane lactone) CN (81)	147
	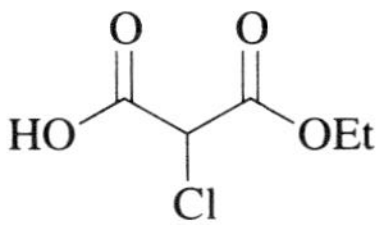 HO $\overset{O}{\!}$ $\overset{Cl}{\!}$ OEt	Mn(OAc)$_3$, AcOH, 23°	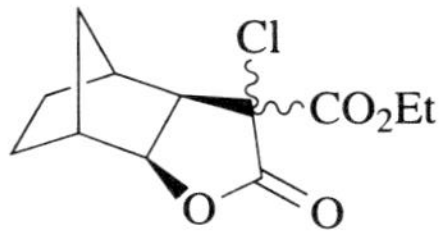 (norbornane lactone) Cl CO$_2$Et (88)	147

C$_7$

TABLE X. ALKENES OR CYCLOALKENES AND DICARBOXYLIC ACID DERIVATIVES (*Continued*)

Substrate	Reagent	Conditions	Product(s) and Yield(s) (%)	Refs.
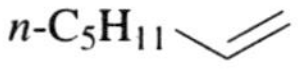	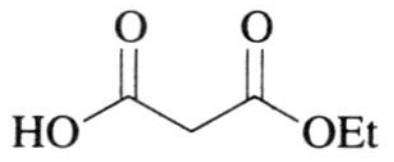 HO–CO–CH₂–CO–OEt	Mn(OAc)₃, AcOH, 23°, 4 - 6 h	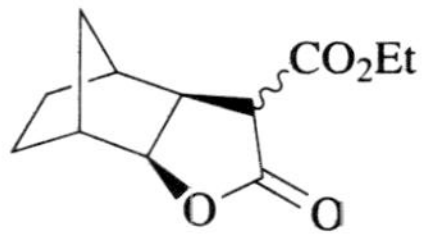 (lactone with CO₂Et) (73)	147
n-C₅H₁₁ (alkene)	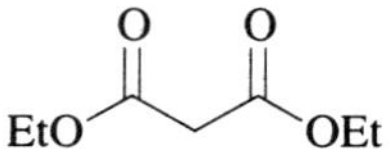 EtO–CO–CH₂–CO–OEt	Mn(OAc)₃, AcOH, 90°	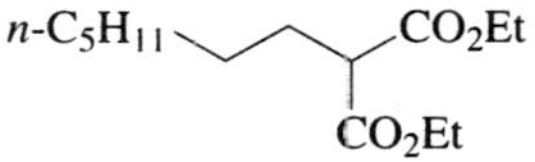 n-C₅H₁₁–CH(CH₂CH₂CO₂Et)CO₂Et (50)	64
"	"	Mn(OAc)₃, Cu(OAc)₂, AcOH	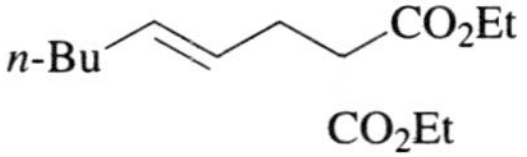 n-Bu–CH=CH–CH₂CH₂CO₂Et (—)	64

C_7–C_{20}

Substrate	Reagent	Conditions	Product(s) and Yield(s) (%)	Refs.
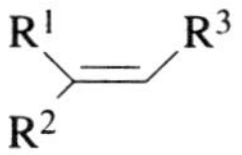 R^1R^2C=CH–R^3	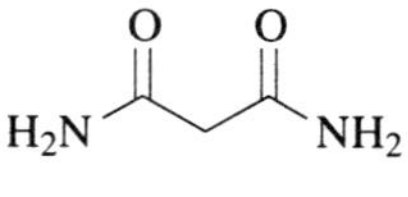 H₂N–CO–CH₂–CO–NH₂	Mn(OAc)₃, AcOH, reflux, 1 - 3 min	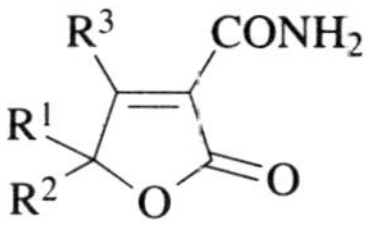 (butenolide: R^3, CONH₂, R^1, R^2, O)	150

R¹	R²	R³	
Ph	Ph	H	(56)
Me	Ph	H	(38)
Ph	Ph	Ph	(69)
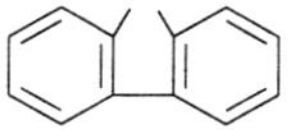 (fluorenylidene)		CO₂H	(48)
(CH₂)₅		H	(26)

534

TABLE X. ALKENES OR CYCLOALKENES AND DICARBOXYLIC ACID DERIVATIVES (*Continued*)

Substrate	Reagent	Conditions	Product(s) and Yield(s) (%)	Refs.
C$_8$				
Ph-CH=CH$_2$	H$_2$NCOCH$_2$CONH$_2$	Mn(OAc)$_3$, AcOH, reflux, 1 - 3 min	(4) + (5) [structures shown]	150
	EtO$_2$C-CH(NR^{1}COR^2)-CO$_2$Et R^1/R^2: Me / H ; C$_6$H$_4$CO / H	Mn(OAc)$_3$, AcOH, 80°, 8 h	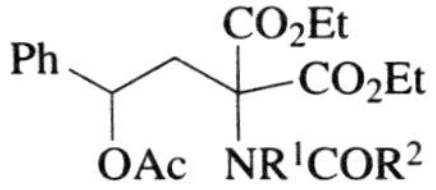 [structure] (88) (87)	107
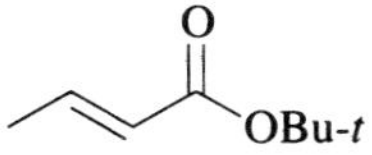 [structure, crotonic acid t-Bu cyanoacetate NC-CH$_2$-CO$_2$H, OBu-t]	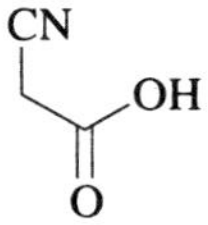	Mn(OAc)$_3$, AcOH, 70°, 10 - 15 min	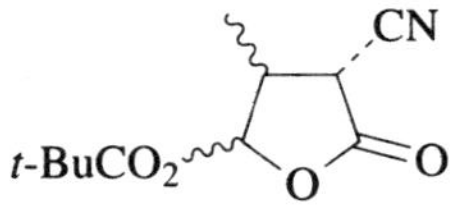 [lactone structure] 1.6 : 1.0 (9)	129
[cyclooctene]		"	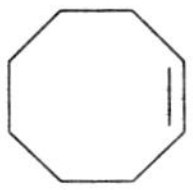 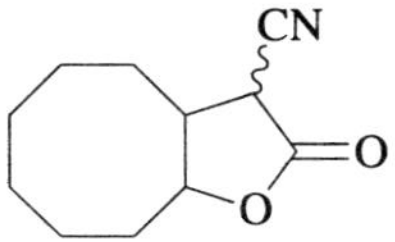 [structure] 1.6 : 1.0 : 9.6 : 2.1 (66)	129
[cyclooctene]	KO$_2$C-CH$_2$-CO$_2$Me	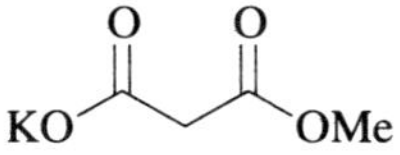 Mn(OAc)$_3$, AcOH, 70°, 10 - 12 min	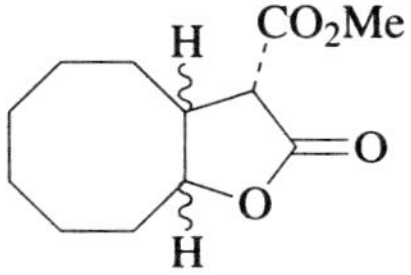 [structure] 5.7 : 1.0 (67)	129

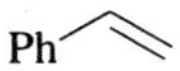

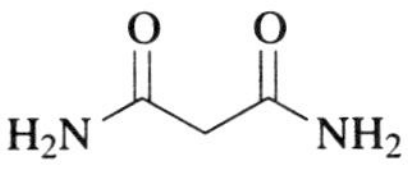

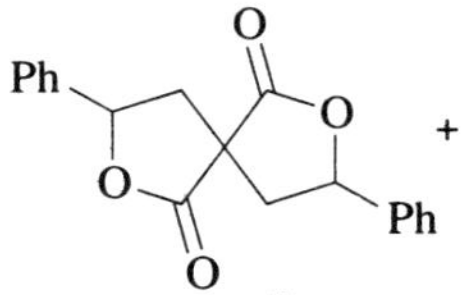

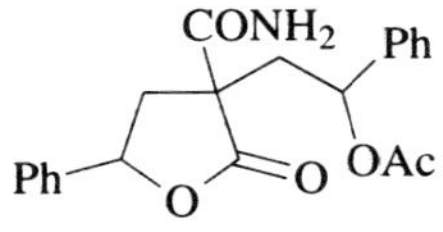

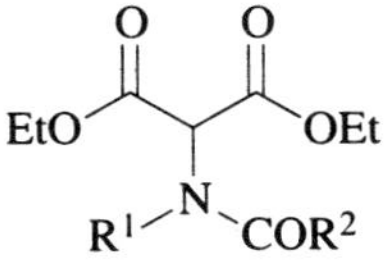

535

TABLE X. ALKENES OR CYCLOALKENES AND DICARBOXYLIC ACID DERIVATIVES (Continued)

Substrate	Reagent	Conditions	Product(s) and Yield(s) (%)	Refs.
$n\text{-}C_6H_{13}$ ⟶ (alkene)	KO–malonate–OMe	Mn(OAc)$_3$, AcOH, 70°, 10 - 12 min	CO$_2$Me lactone, $n\text{-}C_6H_{13}$ 1 : 1 (76)	129

C$_8$–C$_9$

Substrate	Reagent	Conditions	Product(s) and Yield(s) (%)	Refs.
R^2, R^3 alkene (R^1)	NC–CH$_2$–COOH	Mn(OAc)$_3$, AcOH, AcOK, reflux	CN lactone, R^3, R^2, R^1	128

R^1	R^2	R^3	
$n\text{-}C_6H_{13}$	H	H	(60)
Ph	H	H	(41)
Ph	Me	H	(43)
$n\text{-}Pr$	H	$n\text{-}Pr$	(49)
Ph	H	Me	(51)

Substrate	Reagent	Conditions	Product(s) and Yield(s) (%)	Refs.
R ⟶ (alkene) C$_8$ R = $n\text{-}C_6H_{13}$ C$_{10}$ R = $n\text{-}C_8H_{17}$	NC–CH$_2$–COOH	Mn(OAc)$_3$, AcOH, 70°, 10 - 15 min	CN lactone, R α : β 3.7 : 1.0 (61) α : β 3.3 : 1.0 (69)	129

C$_9$

Substrate	Reagent	Conditions	Product(s) and Yield(s) (%)	Refs.
Ph (isopropenyl)	NC–CH$_2$–COOH	Mn(OAc)$_3$, AcOH, 70°, 10 - 15 min	CN lactone, Ph α : β 1.5 : 1.0 (70)	129

536

TABLE X. ALKENES OR CYCLOALKENES AND DICARBOXYLIC ACID DERIVATIVES (*Continued*)

Substrate	Reagent	Conditions	Product(s) and Yield(s) (%)	Refs.
C$_{10}$				
n-C$_8$H$_{17}$ (alkene)	HO$_2$C–CH$_2$–CN	Mn(OAc)$_3$, AcOH 23°, 15 min	NC-substituted lactone, C$_8$H$_{17}$-n (85)	147
	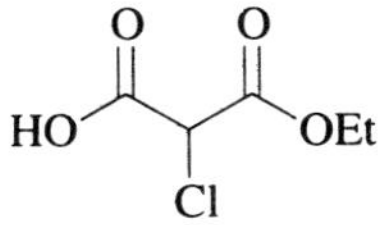 HO$_2$C–CHCl–CO$_2$Et	Mn(OAc)$_3$, AcOH, 23°	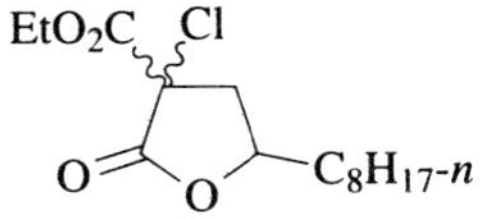 EtO$_2$C, Cl lactone, C$_8$H$_{17}$-n (40)	147
	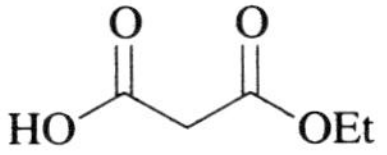 HO$_2$C–CH$_2$–CO$_2$Et	Mn(OAc)$_3$, AcOH, 23°, 4 - 6 h	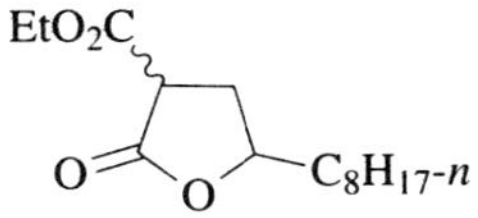 EtO$_2$C lactone, C$_8$H$_{17}$-n (74)	147
C$_{11}$				
AcO–CH$_2$–CH=CH–Ph	HO$_2$C–CH$_2$–CN	Mn(OAc)$_3$, AcOH 23°, 15 min	NC, OAc, Ph lactone (77)	147
	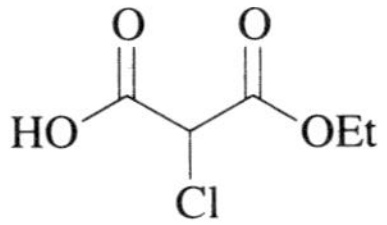 HO$_2$C–CHCl–CO$_2$Et	Mn(OAc)$_3$, AcOH, 23°	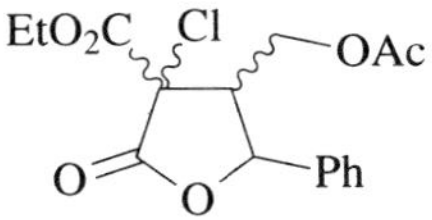 EtO$_2$C, Cl, OAc, Ph lactone (71)	147
	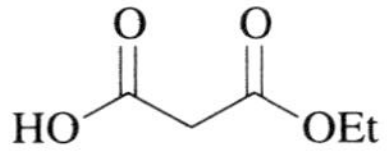 HO$_2$C–CH$_2$–CO$_2$Et	Mn(OAc)$_3$, AcOH, 23°, 4 - 6 h	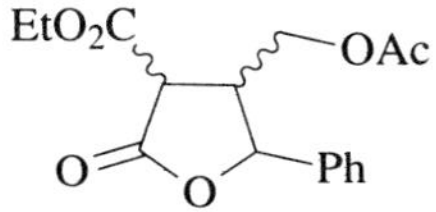 EtO$_2$C, OAc, Ph lactone (78)	147

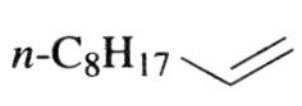

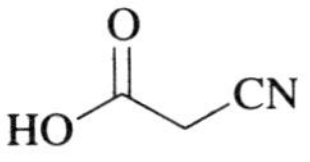

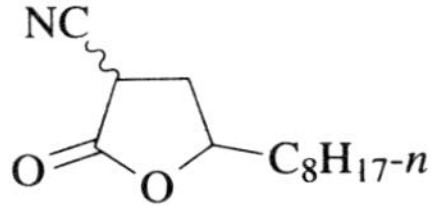

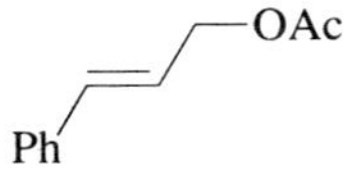

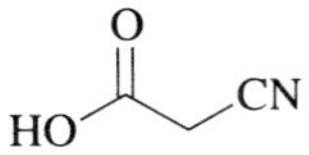

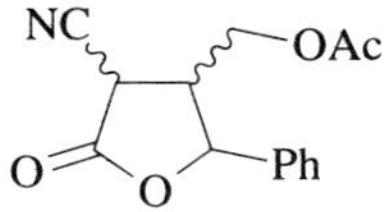

537

TABLE X. ALKENES OR CYCLOALKENES AND DICARBOXYLIC ACID DERIVATIVES (*Continued*)

Substrate	Reagent	Conditions	Product(s) and Yield(s) (%)	Refs.
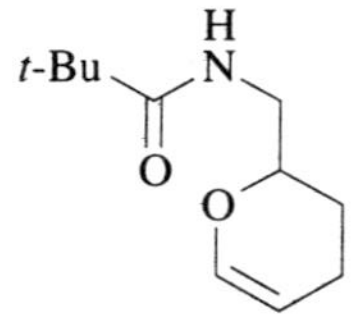	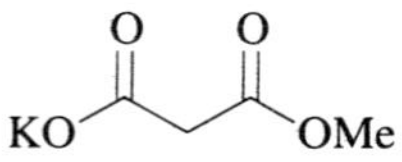	Mn(OAc)$_3$, AcOH, 110°, 10 min	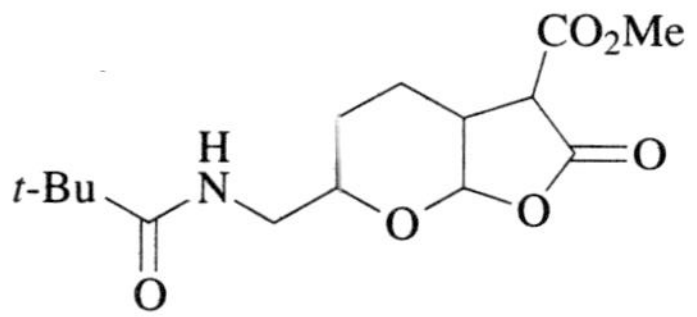 (41)	149
C$_{11}$-C$_{13}$ 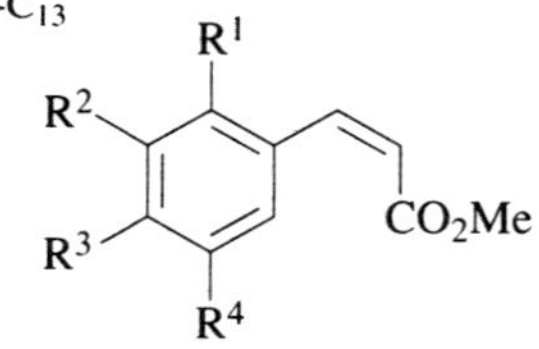	"	Mn(OAc)$_3$, AcOH, 70°, 15 min	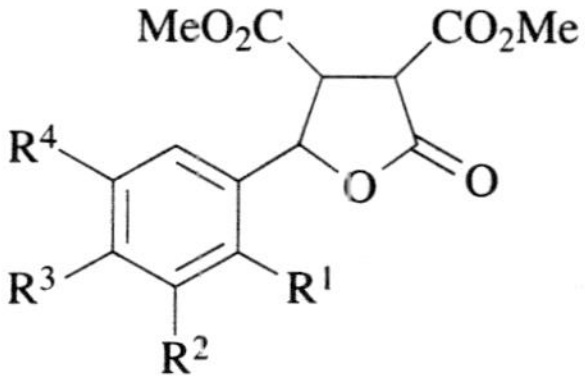	148

R^1	R^2	R^3	R^4	
H	H	OMe	H	(63)
OMe	H	OMe	H	(80)
H	OMe	OMe	H	(66)
OMe	H	H	OMe	(60)
OMe	H	OMe	OMe	(75)
H	OMe	OMe	OMe	(73)
OMe	OMe	OMe	H	(80)

| C$_{14}$ 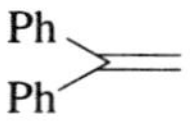 | 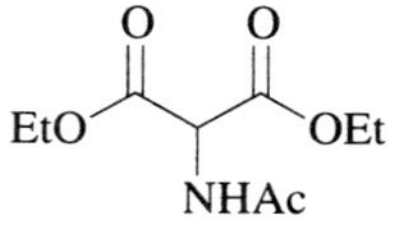 | Mn(OAc)$_3$, AcOH, 80°, 8 h | 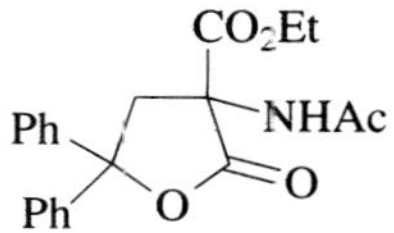 (90) | 107 |

TABLE X. ALKENES OR CYCLOALKENES AND DICARBOXYLIC ACID DERIVATIVES (*Continued*)

Substrate	Reagent	Conditions	Product(s) and Yield(s) (%)	Refs.
C_{14}-C_{21} R^1, R^2, R^3 alkene (R^2–H, R^1–C=C–R^3–H) 	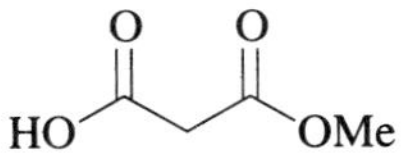 HO–C(=O)–CH_2–C(=O)–OMe	Mn(OAc)_3, AcOH, reflux, 22 min	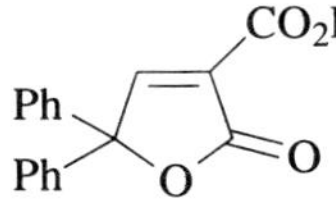 Ph, Ph lactone with CO_2Et (25) + EtO_2C lactone with Ph, Ph (12) + spiro bis-lactone Ph, Ph (11)	146
	NC–CH_2–C(=O)–NH_2 R^1 R^2 R^3 Ph Ph H 4-ClC_6H_4 4-ClC_6H_4 H Ph	Mn(OAc)_3, AcOH, reflux, 1 min	R^1, R^2, R^3 OAc CONH_2 CN (21) (24) (30) + R^3 CN R^1 R^2 O furanone (14) (15) (0)	151
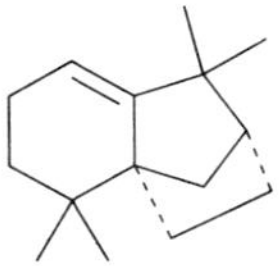 C_{15}	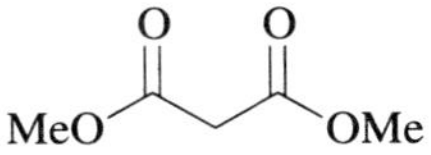 MeO–C(=O)–CH_2–C(=O)–OMe	Mn(OAc)_3, AcOH, 110°, 24 h [a]	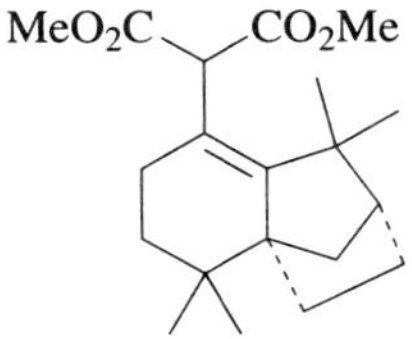 MeO_2C, CO_2Me decalin product (—)	111

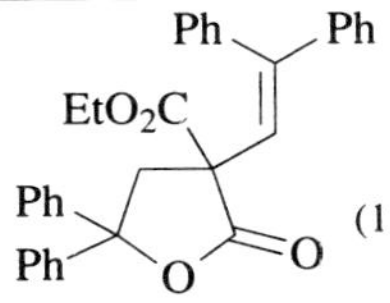

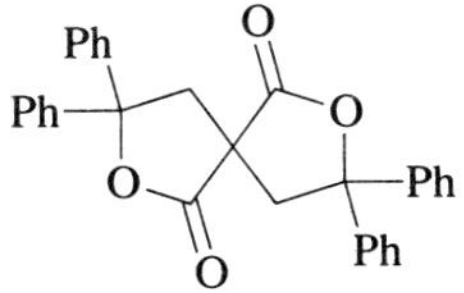

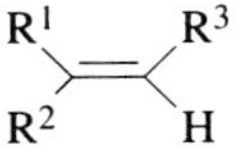

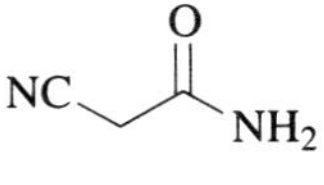

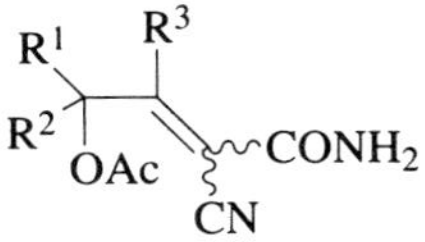

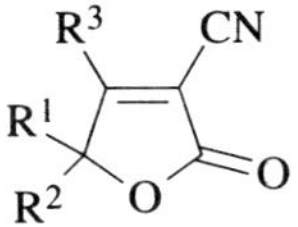

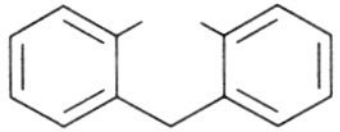

TABLE X. ALKENES OR CYCLOALKENES AND DICARBOXYLIC ACID DERIVATIVES (*Continued*)

Substrate	Reagent	Conditions	Product(s) and Yield(s) (%)	Refs.

$C_{15}-C_{20}$

Substrate (structure with R^1, R^2, R^3, H):

R^1	R^2	R^3		
4-MeOC$_6$H$_4$	4-MeOC$_6$H$_4$	H	(23)/(33)	
Ph	4-MeOC$_6$H$_4$	H	(7)/(37)	
		Ph	(14)/(0)	

Reagent: NC–CH$_2$–C(=O)–NH$_2$

Conditions: Mn(OAc)$_3$, AcOH, reflux, 1 min

Refs. 151

C_{16}

Substrate: 4-MeC$_6$H$_4$–C(=CH$_2$)– ; 4-MeC$_6$H$_4$

Reagent: NC–CH$_2$–C(=O)–NH$_2$

Conditions: Mn(OAc)$_3$, AcOH, reflux, 1 min

Product: (44)

Refs. 151

Substrate (structure with R^1, R^2, R^3):

	R^1	R^2	R^3	
	4-MeOPh	4-MeOPh	H	(9)/(18)
	4-MeOPh	4-MeOPh	Ph	(14)/(17)

Reagent: H$_2$N–C(=O)–CH$_2$–C(=O)–NH$_2$

Conditions: Mn(OAc)$_3$, AcOH, reflux, 1 - 3 min

Refs. 150

	R^1	R^2	R^3
C_{16}	4-MeOPh	4-MeOPh	H
C_{22}	4-MeOPh	4-MeOPh	Ph

TABLE X. ALKENES OR CYCLOALKENES AND DICARBOXYLIC ACID DERIVATIVES (*Continued*)

Substrate	Reagent	Conditions	Product(s) and Yield(s) (%)	Refs.
C$_{20}$ Ph Ph, Ph	NC–CH_2–$CONH_2$	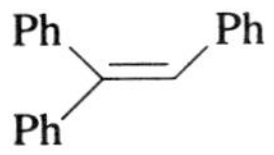 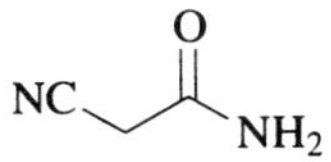 Mn(OAc)$_3$, AcOH, reflux, 1 min	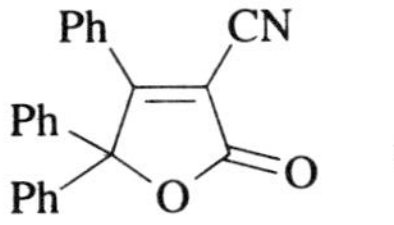 (lactone, CN, Ph, Ph, Ph) (30)	151
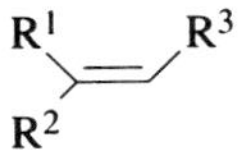 R^1, R^2, R^3 alkene	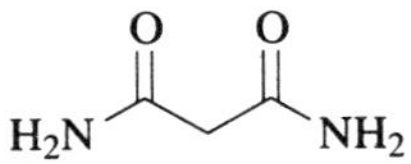 H_2N–CO–CH_2–CO–NH_2	Mn(OAc)$_3$, AcOH, reflux, 1 - 3 min	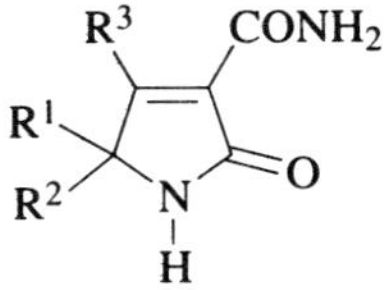 (pyrrolinone: R^3, CONH$_2$, R^1, R^2, N–H)	150

	R^1	R^2	R^3	
C$_{20}$	(diphenyl ether, O)	Ph	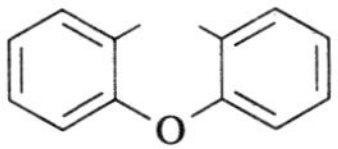 Ph	(59)
C$_{21}$	(diphenylmethane)	Ph	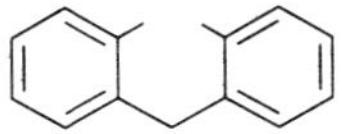 Ph	(90)

C$_{29}$				
(steroid; AcO, C$_8$H$_{17-n}$)	MeO–CO–CH_2–CO–OMe	Mn(OAc)$_3$, AcOH, 4 h	(steroid; AcO, MeO$_2$C, CO$_2$Me) (—)	111

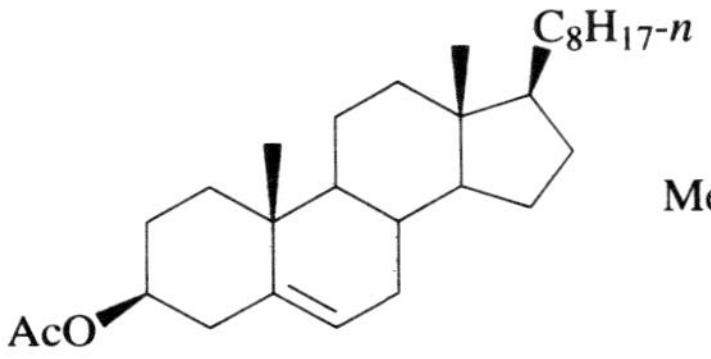

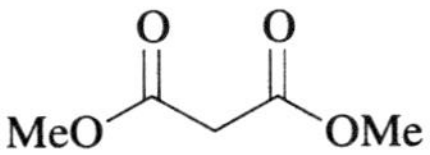

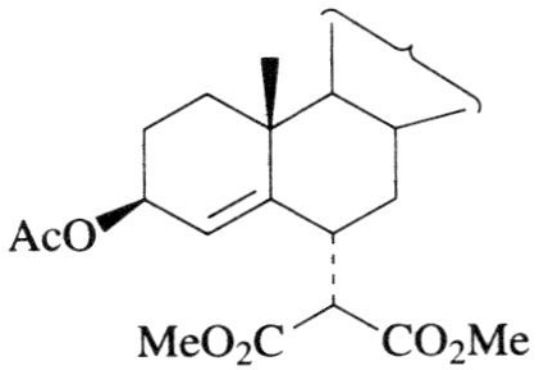

TABLE XI. ENOL ETHERS OR ENOL LACTONES AND β-DICARBONYL COMPOUNDS

Substrate	Reagent	Conditions	Product(s) and Yield(s) (%)	Refs.
C₄				
 (dihydrofuran)	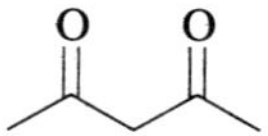 (pentane-2,4-dione)	Mn(OAc)₃, AcOH, 60°	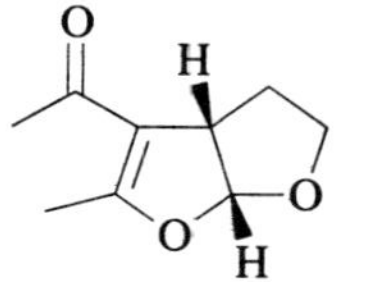 (26)	142
"	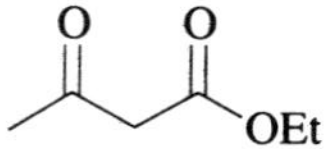 (ethyl acetoacetate, OEt)	"	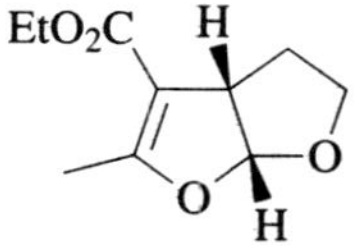 EtO₂C (53)	139, 142
"	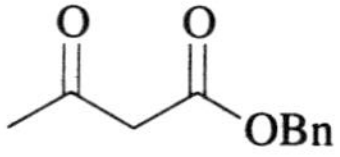 (benzyl acetoacetate, OBn)	"	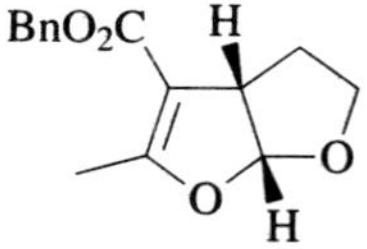 BnO₂C (56)	139, 142
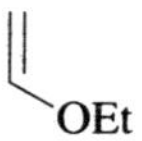 (ethyl vinyl ether, OEt)	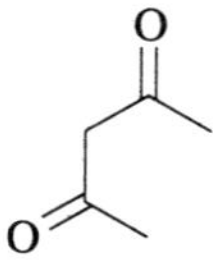 (pentane-2,4-dione)	Mn(OAc)₃, AcOH, 45°, 30 min	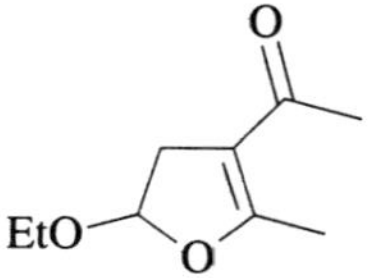 EtO (73)	137
"	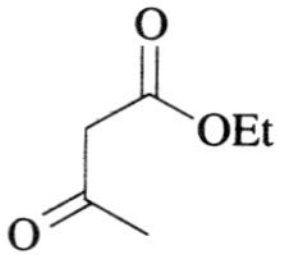 (ethyl acetoacetate, OEt)	Mn(OAc)₃, AcOH, 40°, 10 min	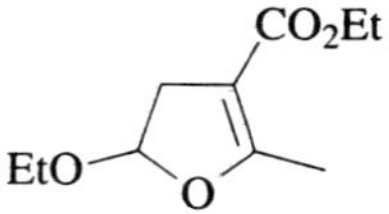 CO₂Et, EtO (89)	137
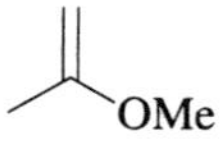 (isopropenyl methyl ether, OMe)	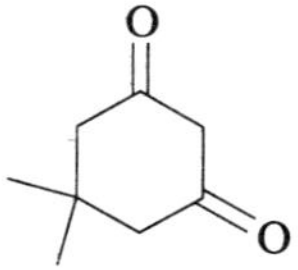 (5,5-dimethyl-1,3-cyclohexanedione)	Mn(OAc)₃, AcOH, 23°, 10 min	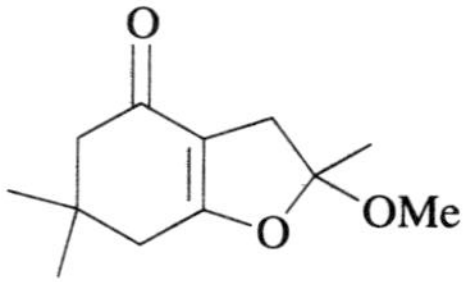 OMe (98)	137

TABLE XI. ENOL ETHERS OR ENOL LACTONES AND β-DICARBONYL COMPOUNDS (*Continued*)

Substrate	Reagent	Conditions	Product(s) and Yield(s) (%)	Refs.
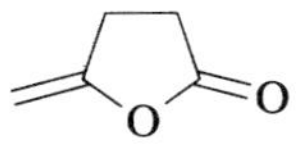 C₅	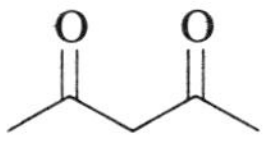	Mn(OAc)₃, AcOH, 65°	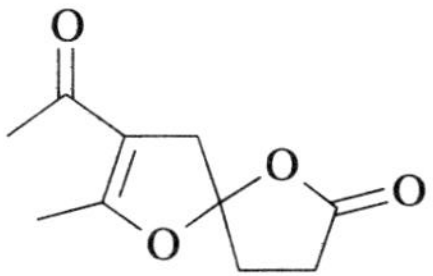 (56)	139, 141
	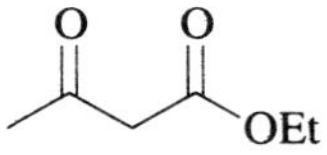 OEt	"	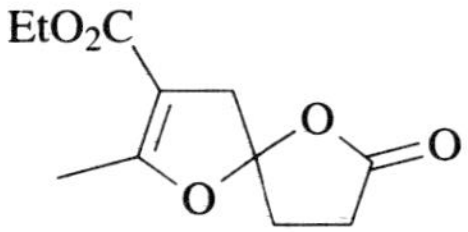 (47)	139, 141
	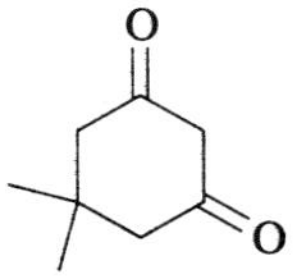	"	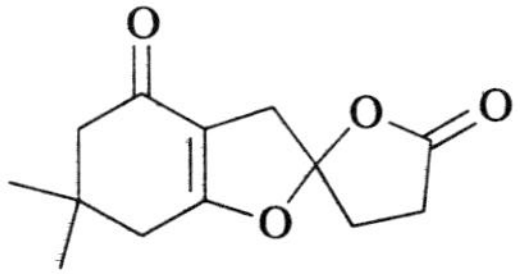 (64)	139, 141
	OBu-*t*	"	(31)	141
	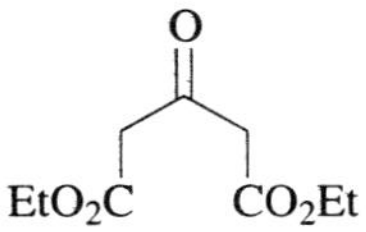 EtO₂C, CO₂Et	"	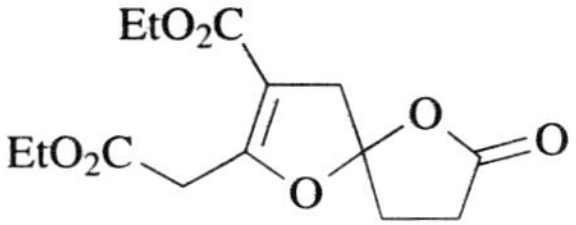 (38)	139, 141
	OBn	"	(52)	141

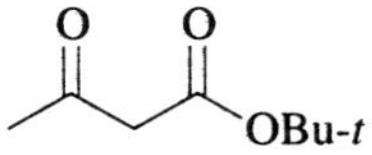

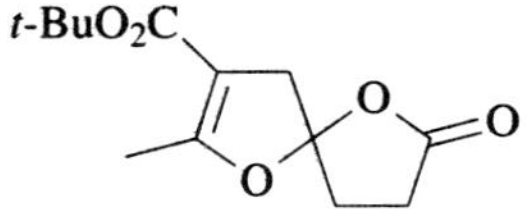

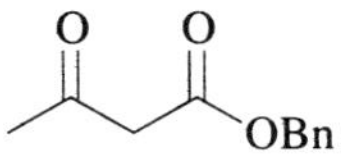

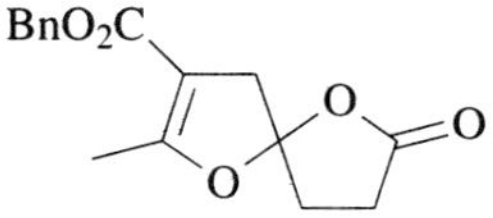

543

TABLE XI. ENOL ETHERS OR ENOL LACTONES AND β-DICARBONYL COMPOUNDS (*Continued*)

Substrate	Reagent	Conditions	Product(s) and Yield(s) (%)	Refs.
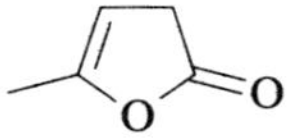	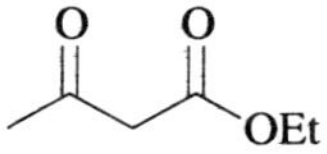	Mn(OAc)₃, AcOH, 60°	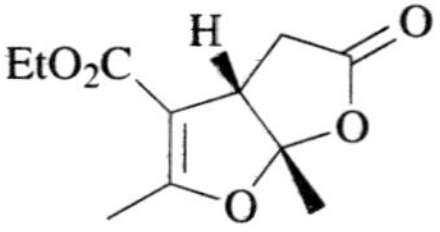 (28)	142
	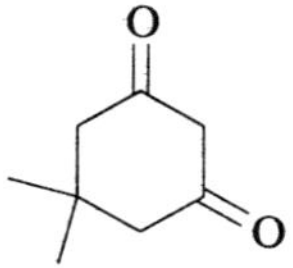	"	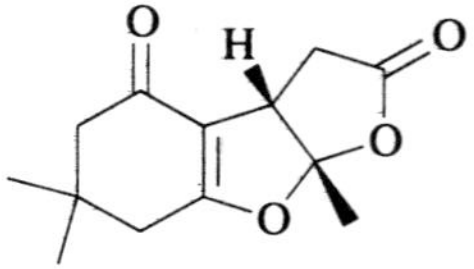 (24)	142
	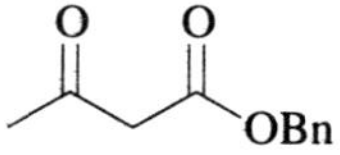	"	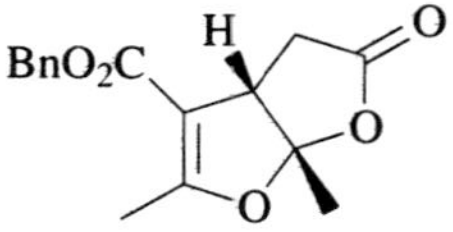 (15)	142
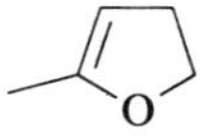	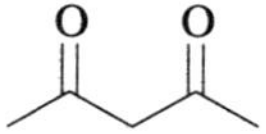	"	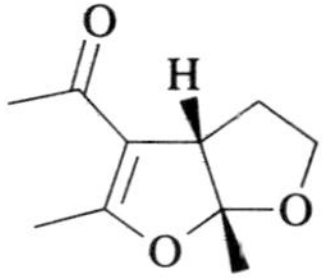 (44)	142
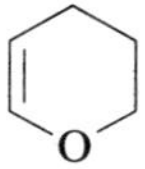	"	"	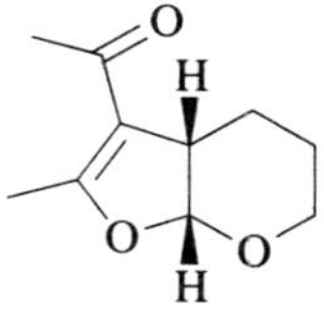 (38)	142

544

TABLE XI. ENOL ETHERS OR ENOL LACTONES AND β-DICARBONYL COMPOUNDS (*Continued*)

Substrate	Reagent	Conditions	Product(s) and Yield(s) (%)	Refs.
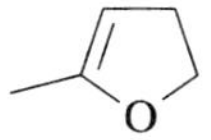	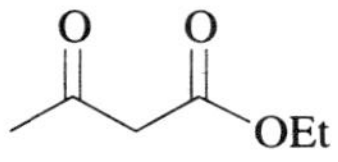	Mn(OAc)$_3$, AcOH, 60°	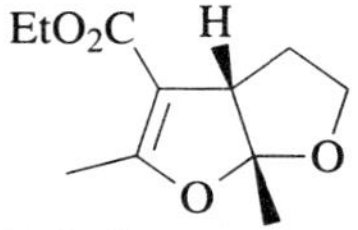 (51)	139, 142
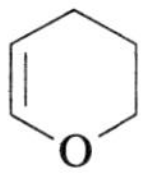	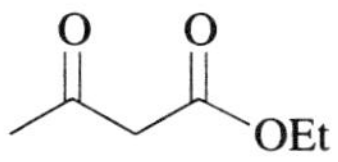	Mn(OAc)$_3$, AcOH, 60°	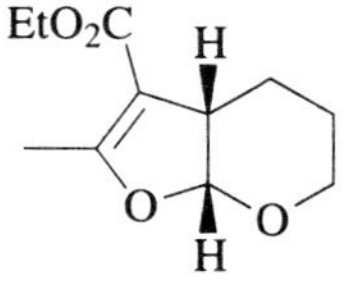 (47)	142
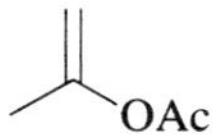	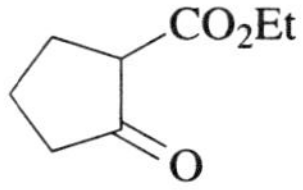	Mn(OAc)$_3$, AcOH, 40°, 30 min	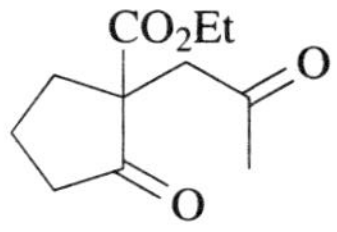 (64)	138
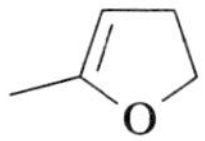	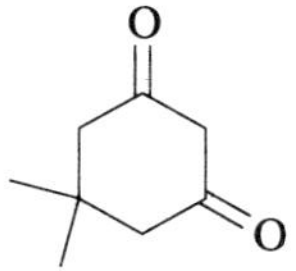	Mn(OAc)$_3$, AcOH, 60°	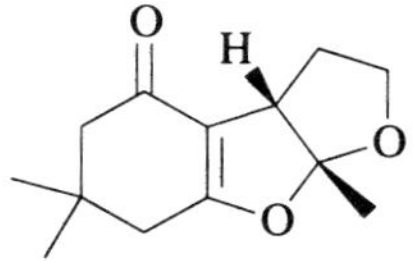 (40)	139, 142
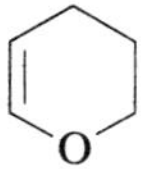	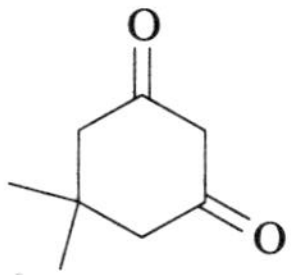	Mn(OAc)$_3$, AcOH, 60°	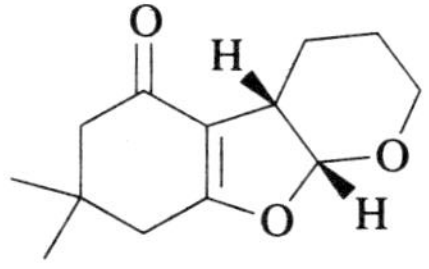 (43)	139, 142
	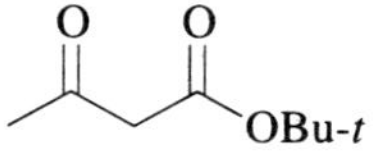	Mn(OAc)$_3$, AcOH, 60°	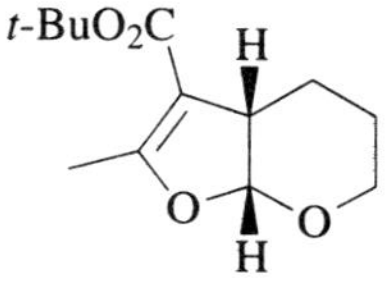 (31)	142

545

TABLE XI. ENOL ETHERS OR ENOL LACTONES AND β-DICARBONYL COMPOUNDS (*Continued*)

Substrate	Reagent	Conditions	Product(s) and Yield(s) (%)	Refs.
		Mn(OAc)$_3$, AcOH, 40°, 30 min	(71)	138
		Mn(OAc)$_3$, AcOH, 60°	(50)	142
		Mn(OAc)$_3$, AcOH, rt	(30)	141
		Mn(OAc)$_3$, AcOH, rt	(36)	141
C$_6$		Mn(OAc)$_3$, AcOH, rt	(19)	141

546

TABLE XI. ENOL ETHERS OR ENOL LACTONES AND β-DICARBONYL COMPOUNDS (*Continued*)

Substrate	Reagent	Conditions	Product(s) and Yield(s) (%)	Refs.
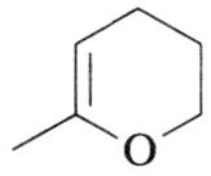 C6–C8 (6-methyl dihydropyran)	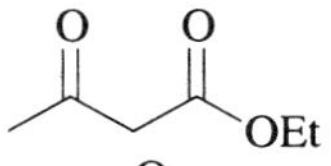 ethyl acetoacetate (EtO₂C type)	Mn(OAc)₃, AcOH, 60°	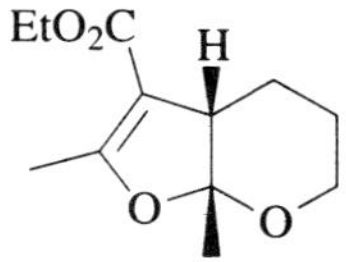 (39)	142
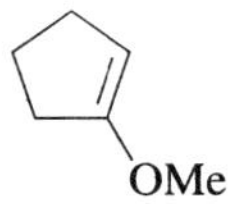 (1-methoxycyclopentene)	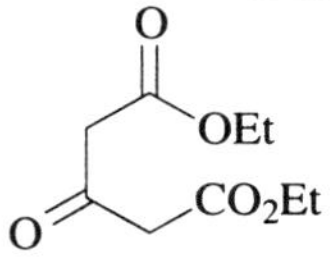 ethyl 3-oxoglutarate type (OEt, CO₂Et)	Mn(OAc)₃, AcOH, 23°, 30 min	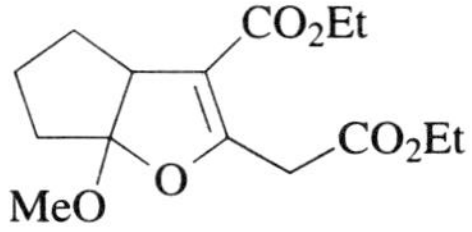 (69)	137
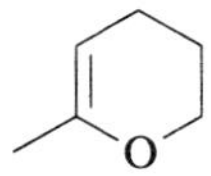 (6-methyl dihydropyran)	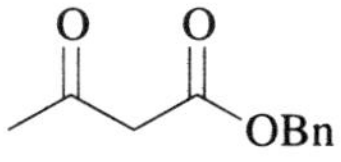 benzyl acetoacetate (OBn)	Mn(OAc)₃, AcOH, 60°	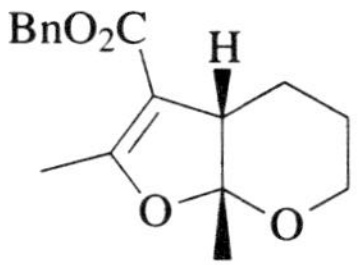 (39)	142
(1-methoxycycloalkene, n = 0, 1, 2)	ethyl acetoacetate (OEt)	Mn(OAc)₃, AcOH, 23°, 10 min	(79) (86) (79)	137
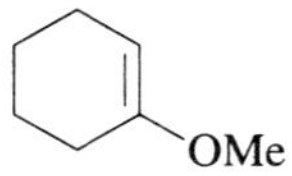 C7 (1-methoxycyclohexene)	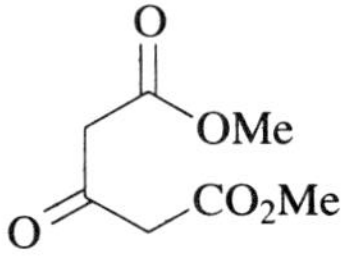 methyl malonate type (OMe, CO₂Me)	Mn(OAc)₃, AcOH, 23°, 20 min	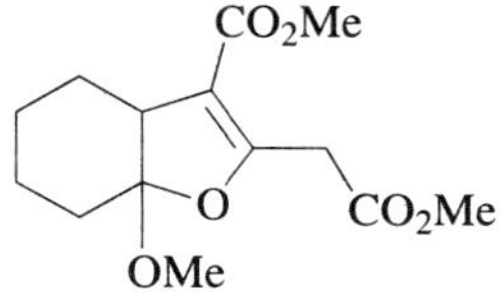 (71)	137

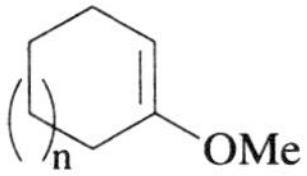

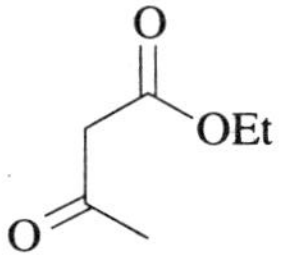

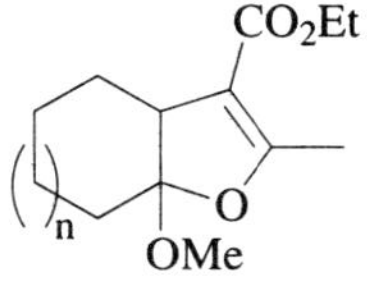

547

TABLE XI. ENOL ETHERS OR ENOL LACTONES AND β-DICARBONYL COMPOUNDS (*Continued*)

Substrate	Reagent	Conditions	Product(s) and Yield(s) (%)	Refs.
	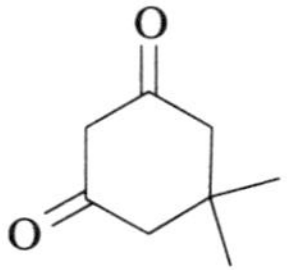	$Mn(OAc)_3$ AcOH, 23°, 10 min	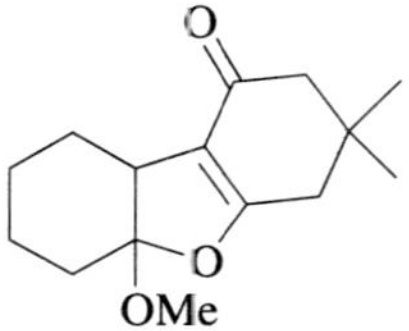 (76)	137
C_8 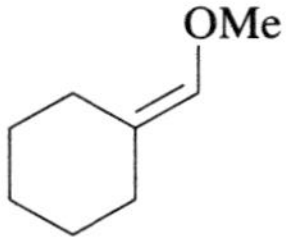	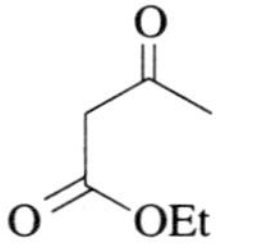	$Mn(OAc)_3$, AcOH, 23°, 25 min	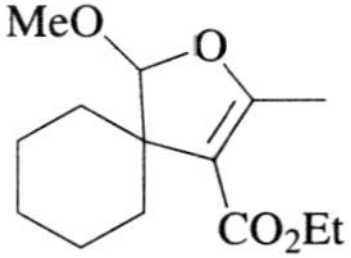 (86)	138
C_9 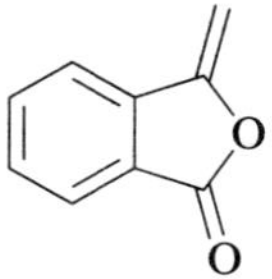	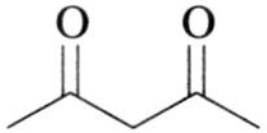	$Mn(OAc)_3$, AcOH, 65°	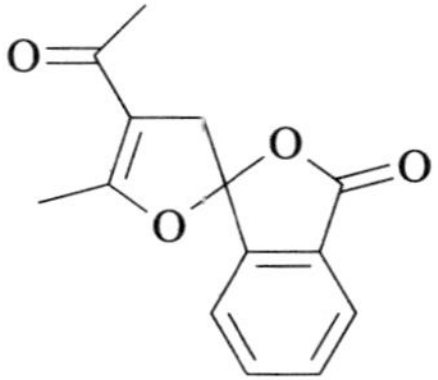 (91)	141
	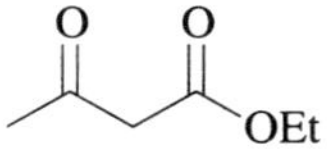	$Mn(OAc)_3$, AcOH, 65°	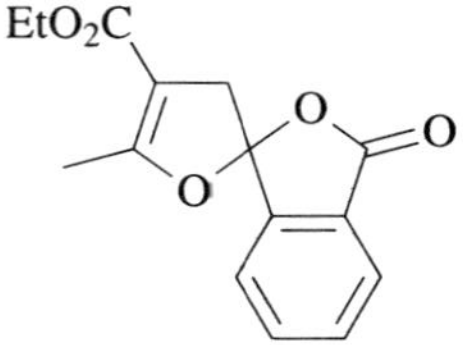 (85)	141

TABLE XI. ENOL ETHERS OR ENOL LACTONES AND β-DICARBONYL COMPOUNDS (*Continued*)

Substrate	Reagent	Conditions	Product(s) and Yield(s) (%)	Refs.
	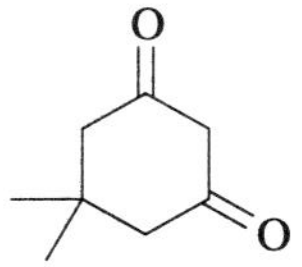	Mn(OAc)$_3$, AcOH, 65°	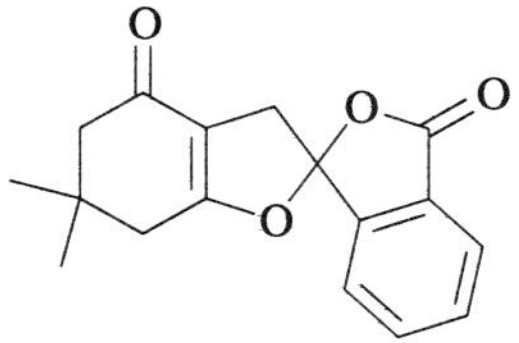 (87)	141
	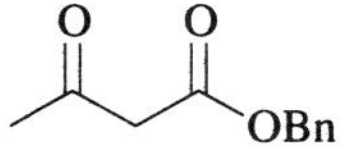	"	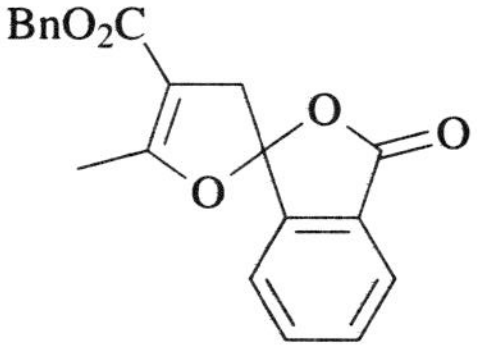 (78)	141
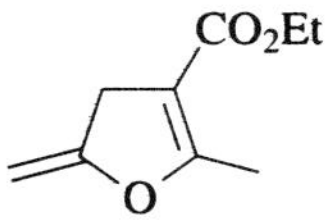	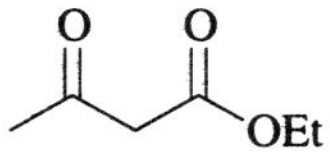	Mn(OAc)$_3$, AcOH, 60°	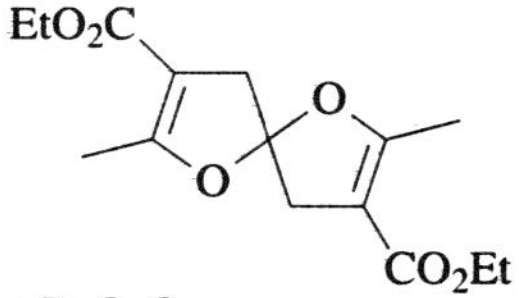 (37)	139, 143
	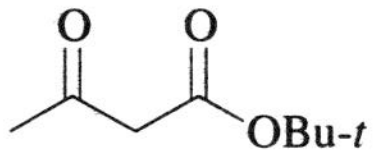	"	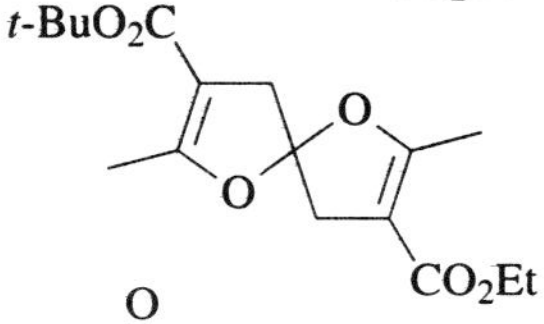 (11)	143
	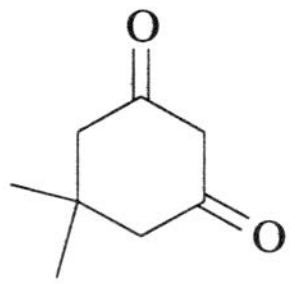	"	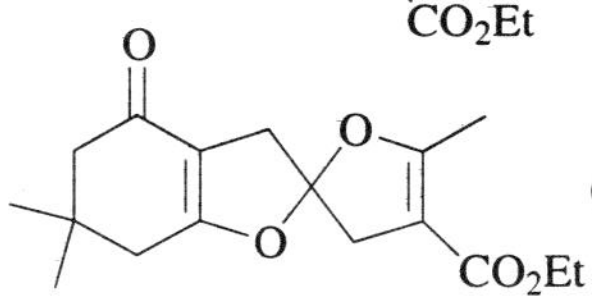 (43)	139, 143

549

TABLE XI. ENOL ETHERS OR ENOL LACTONES AND β-DICARBONYL COMPOUNDS (*Continued*)

Substrate	Reagent	Conditions	Product(s) and Yield(s) (%)	Refs.
	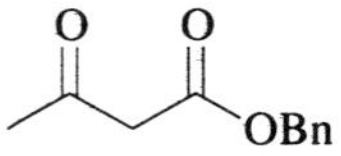	Mn(OAc)$_3$, AcOH, 60°	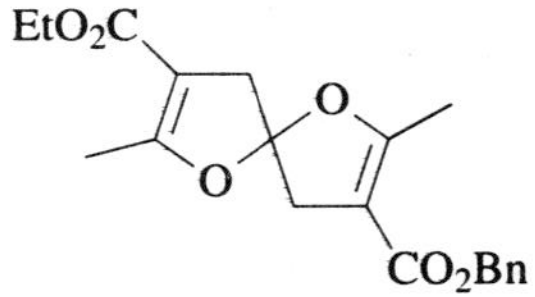 (40)	139, 143
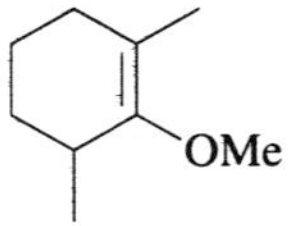	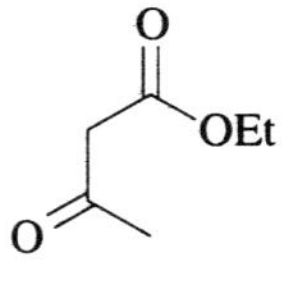	Mn(OAc)$_3$, AcOH, 60°, 30 min	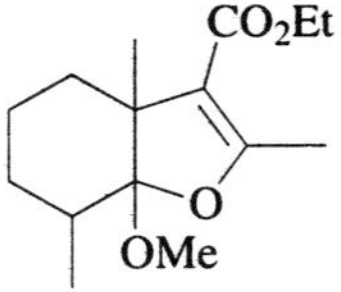 (—)	138
C$_{10}$ 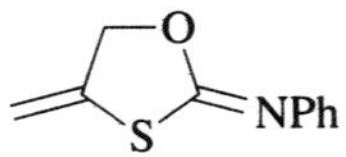	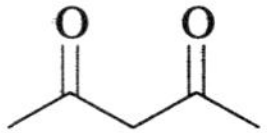	Mn(OAc)$_3$, AcOH, 23°	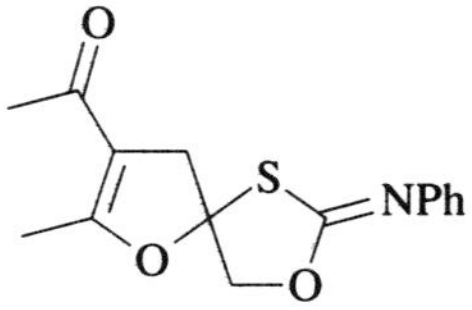 (57)	140
	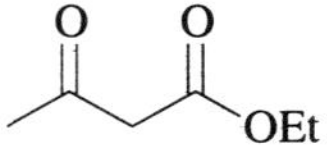	"	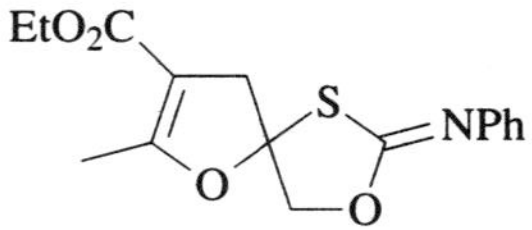 (47)	140
	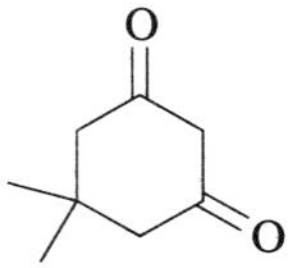	"	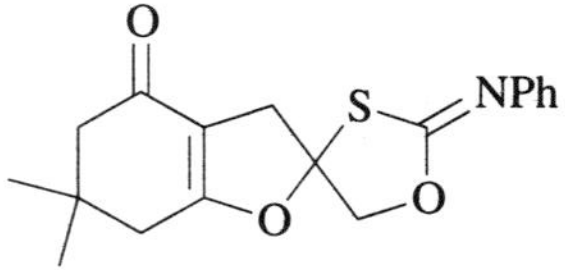 (65)	140

TABLE XI. ENOL ETHERS OR ENOL LACTONES AND β-DICARBONYL COMPOUNDS (*Continued*)

Substrate	Reagent	Conditions	Product(s) and Yield(s) (%)		Refs.
	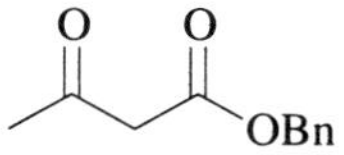	$Mn(OAc)_3$, AcOH, 23°	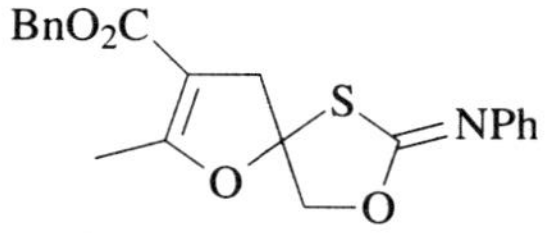	(46)	140
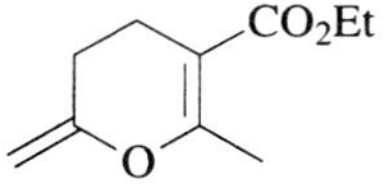	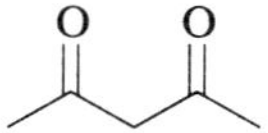	$Mn(OAc)_3$, AcOH, 60°	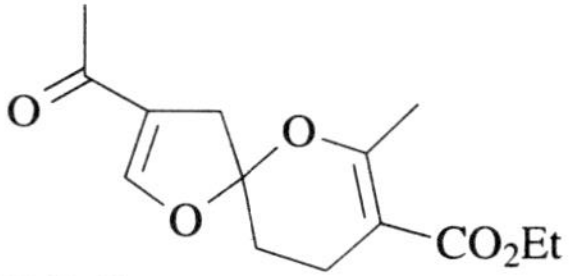	(66)	143
	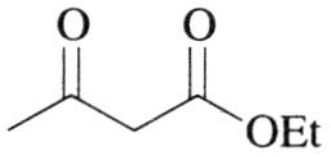	"	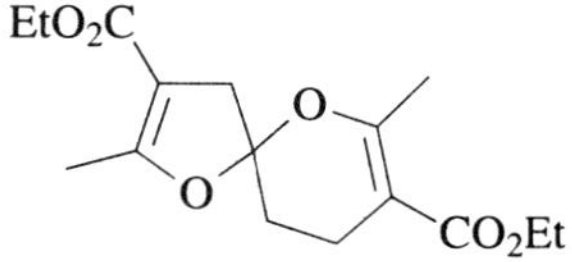	(58)	143
	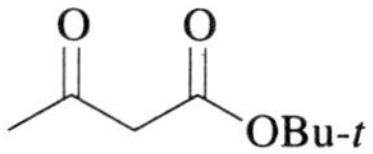	"	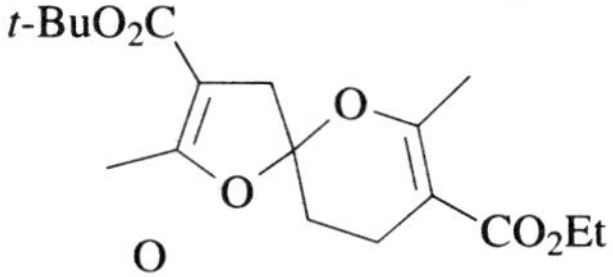	(44)	143
	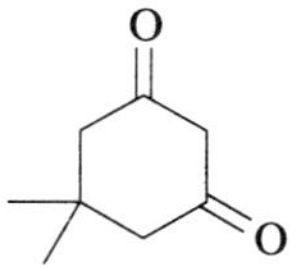	"	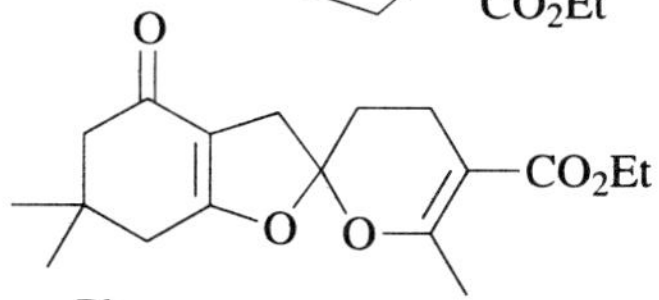	(69)	143
	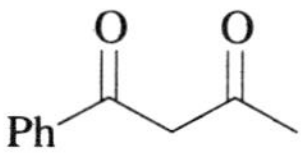	"	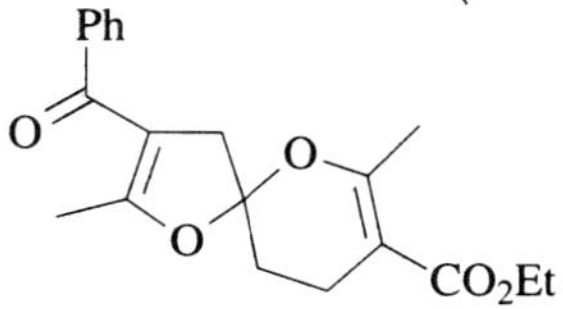	(33)	143

TABLE XI. ENOL ETHERS OR ENOL LACTONES AND β-DICARBONYL COMPOUNDS (*Continued*)

Substrate	Reagent	Conditions	Product(s) and Yield(s) (%)	Refs.
	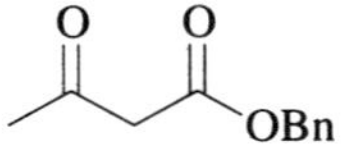	$Mn(OAc)_3$, AcOH, 60°	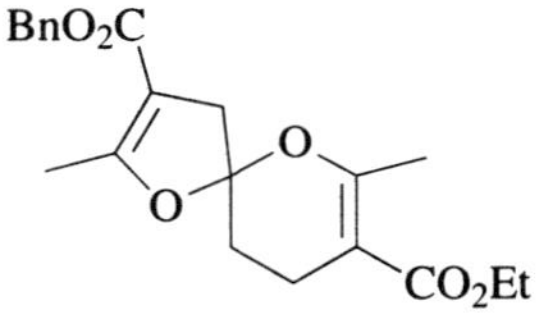 (61)	139, 143
	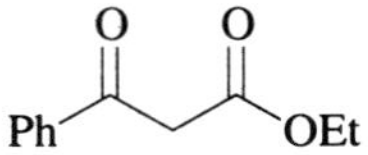	"	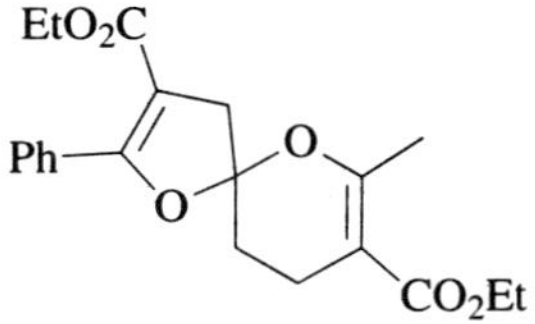 (41)	143
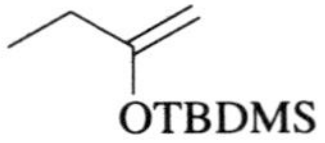	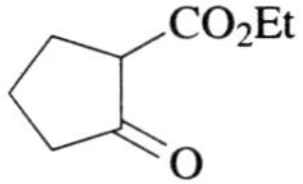	$Mn(OAc)_3$, AcOH, 40°, 30 min	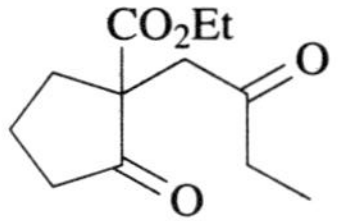 (46)	138
C_{11}				
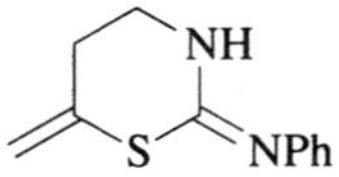	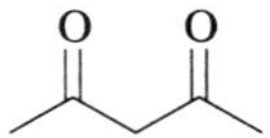	$Mn(OAc)_3$, AcOH, 23°	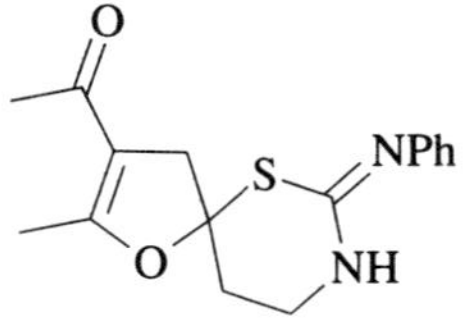 (55)	140
	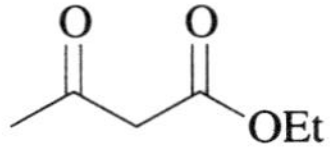	"	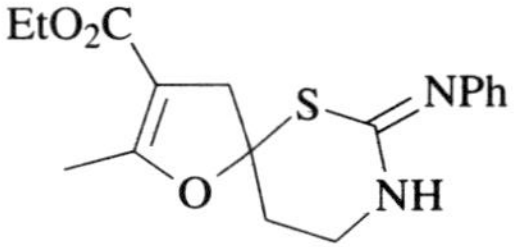 (60)	140

TABLE XI. ENOL ETHERS OR ENOL LACTONES AND β-DICARBONYL COMPOUNDS (*Continued*)

Substrate	Reagent	Conditions	Product(s) and Yield(s) (%)	Refs.
	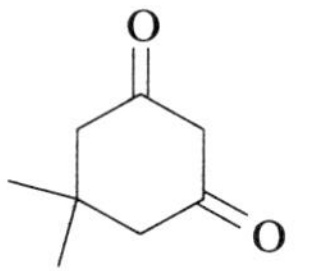	Mn(OAc)$_3$, AcOH, 23°	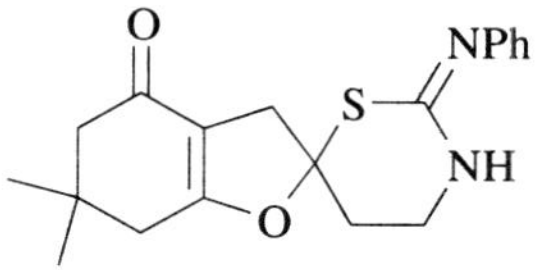 (62)	140
C$_{12}$ 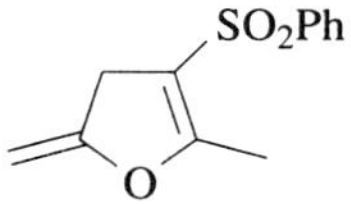	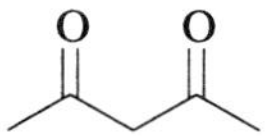	Mn(OAc)$_3$, AcOH, 60°	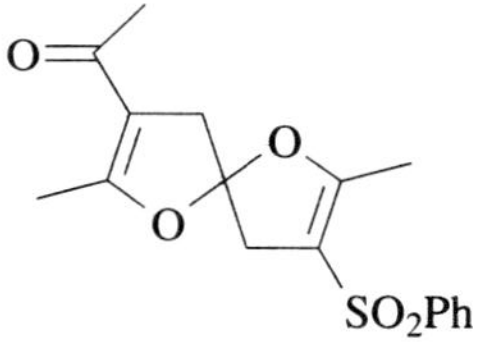 (36)	143
	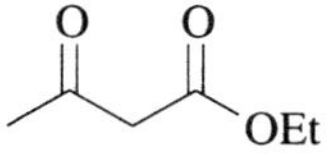	"	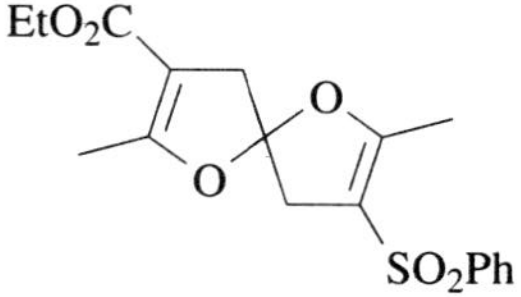 (26)	139, 143
	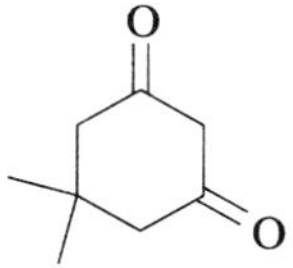	"	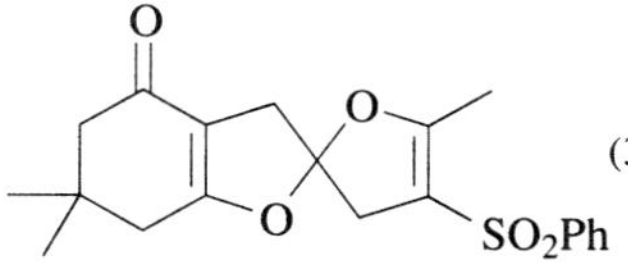 (30)	143
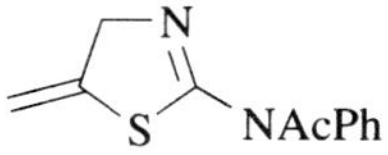	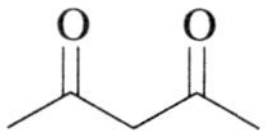	Mn(OAc)$_3$, AcOH, 23°	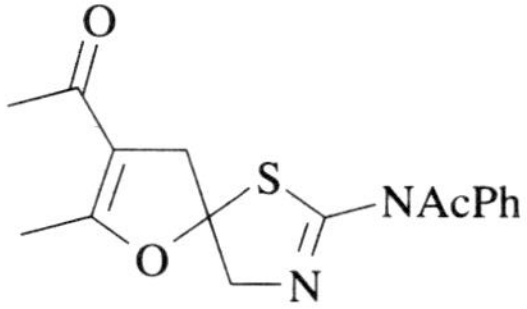 (58)	140

TABLE XI. ENOL ETHERS OR ENOL LACTONES AND β-DICARBONYL COMPOUNDS (*Continued*)

Substrate	Reagent	Conditions	Product(s) and Yield(s) (%)	Refs.
	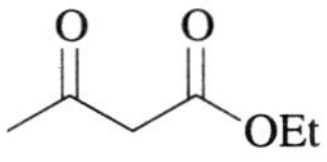	Mn(OAc)$_3$, AcOH, 23°	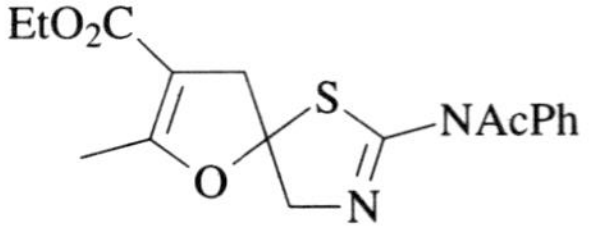 (48)	140
	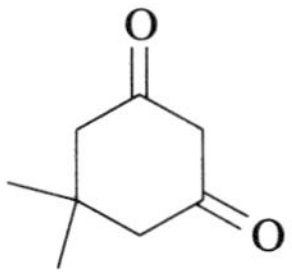	"	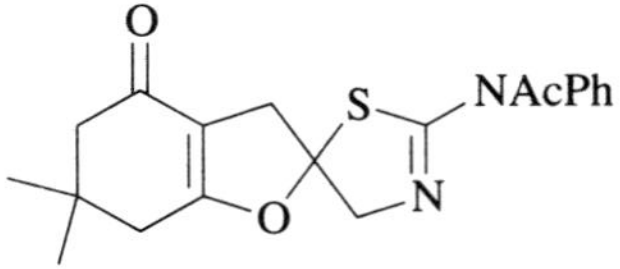 (76)	140
	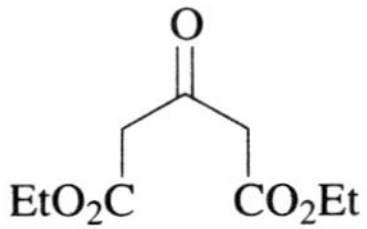	"	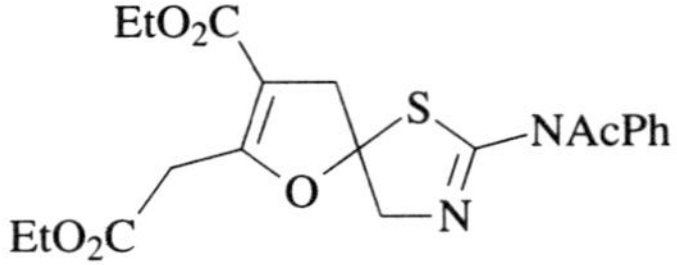 (32)	140
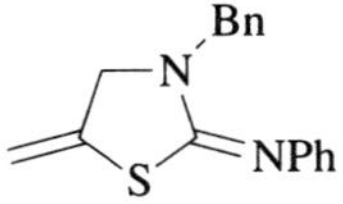	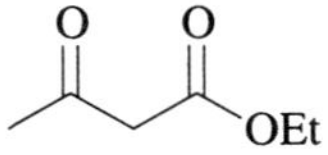	"	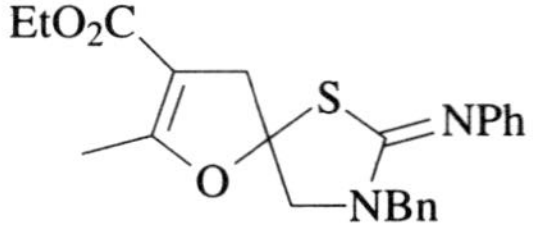 (48)	140
C$_{17}$	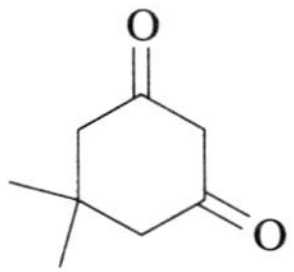	"	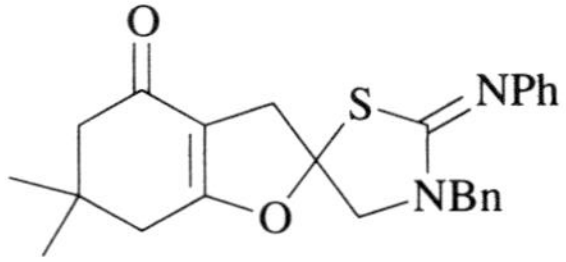 (59)	140

554

TABLE XII. ALKYNES AND CARBONYL COMPOUNDS

	Substrate	Reagent	Conditions	Product(s) and Yield(s) (%)	Refs.
C$_5$	n-PrC≡CH	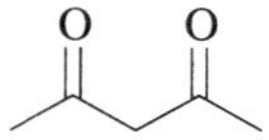 (pentane-2,4-dione)	Mn(OAc)$_3$, AcOH, 70°, 0.5 h	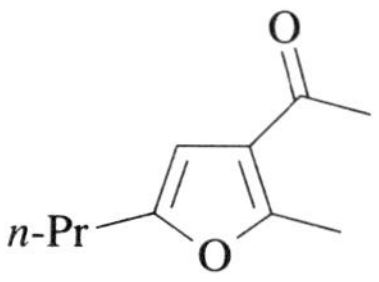 (acetyl methyl propyl furan) (18)	152
C$_6$	n-BuC≡CH	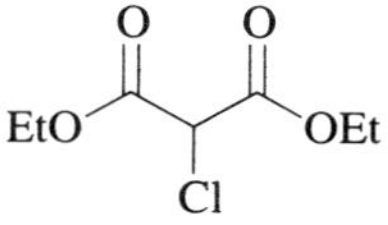 (ethyl ethoxycarbonyl chloro ester)	Mn(OAc)$_3$, LiCl, AcOH	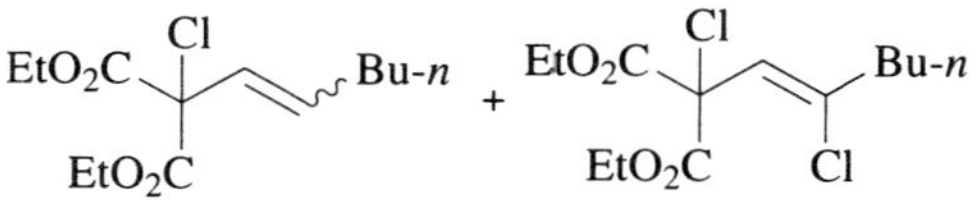 EtO$_2$C / Cl / Bu-n + EtO$_2$C Cl / EtO$_2$C Bu-n Cl 3 : 2 (—)	50
	RC≡CH	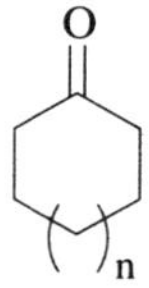 (cyclohexanone, n)	Mn(OAc)$_3$, 80°, 0.5 – 1 h	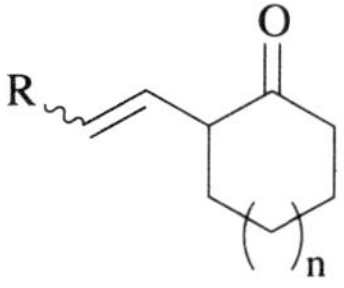 (2-substituted cyclohexanone)	49
	R	n			
C$_6$	n-Bu	0		(46)	
C$_6$	n-Bu	1		(42)	
C$_7$	n-C$_5$H$_{11}$	0		(52)	
C$_8$	n-C$_6$H$_{13}$C≡CH	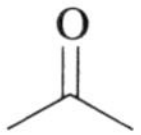 (acetone)	Mn(OAc)$_3$, Cu(OAc)$_2$, AcOH, 60°, 2 h	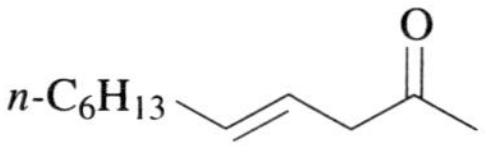 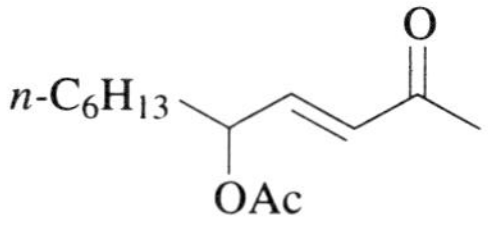 n-C$_6$H$_{13}$ (dienone) (13) + n-C$_6$H$_{13}$ OAc (enone) (18)	44

TABLE XIII. 1,3 - ALKADIENES AND KETONES

Substrate	Reagent	Conditions	Product(s) and Yield(s) (%)	Refs.
C₄		Mn(OAc)₃, Cu(OAc)₂, AcOH, 80°, 3 h	(7) +	61
			1 : 1 : 2[a] (—) +	
			3 : 3 : 1 (—)	
	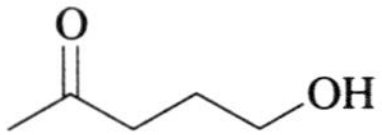	"	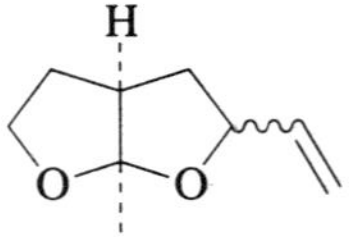 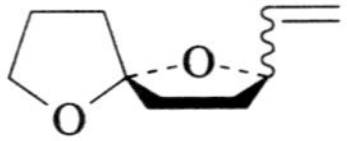 (—) +	61
			(—)	

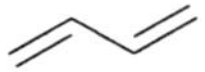

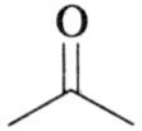

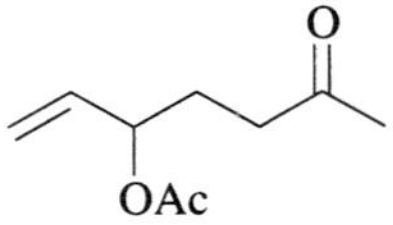

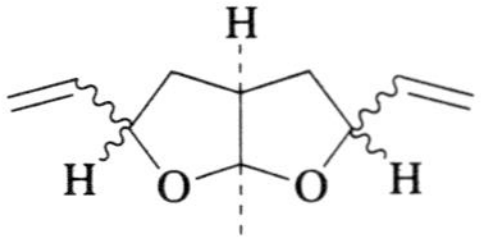

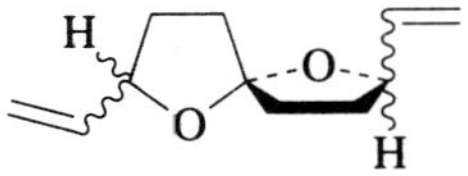

[a] Ratio of exo-exo, exo-endo, and endo-endo stereoisomers.

556

TABLE XIV. 1,3 - ALKADIENES AND ACETIC ACID

Substrate	Reagent	Conditions	Product(s) and Yield(s) (%)	Refs.
C_4 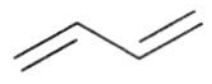	AcOH	Mn(OAc)$_3$, AcOH, AcOK, 140°, 3 h	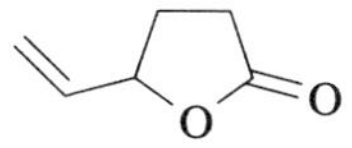 (39)	200
	"	Mn(OAc)$_3$, AcOH, AcOK, reflux	(30)	128
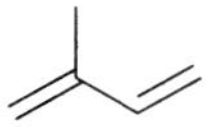	"	"	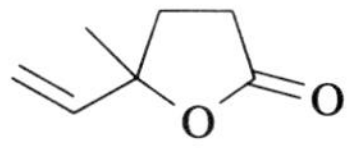 (37) 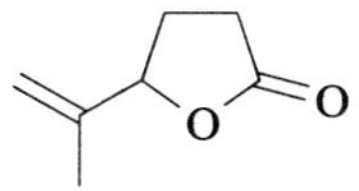 + (13)	128
C_5				
C_6 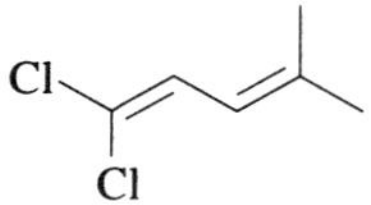	"	Mn(OAc)$_3$, AcOH, AcOK, reflux, 100 h	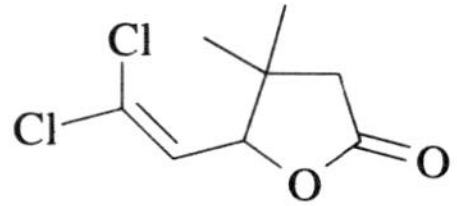 (28)	154
	AcOH, Ac$_2$O	1. Mn(OAc)$_3$ AcOH, Ac$_2$O, reflux, 15 min 2. KOH, EtOH, H$_2$O, 25°, 30 min 3. H$_2$SO$_4$, 60°, 8 h	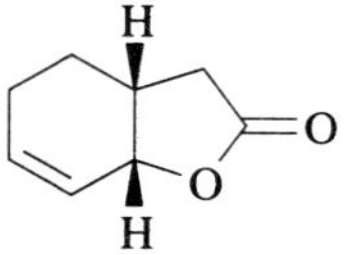 (8)	156

557

TABLE XIV. 1,3 - ALKADIENES AND ACETIC ACID (*Continued*)

Substrate	Reagent	Conditions	Product(s) and Yield(s) (%)	Refs.
C$_7$	AcOH	Mn(OAc)$_3$, AcOH, AcOK, 135°	*E : Z* 20 : 9 + *E : Z* 45 : 23 (60)	108
C$_{12}$	"	Mn(OAc)$_3$ AcOH, AcOK 115°, 6 h	+ (45) 56 : 44	153

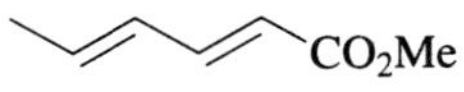

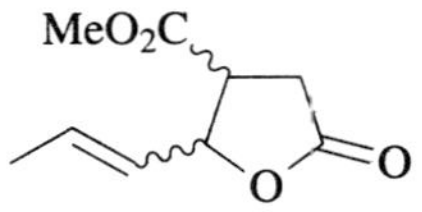

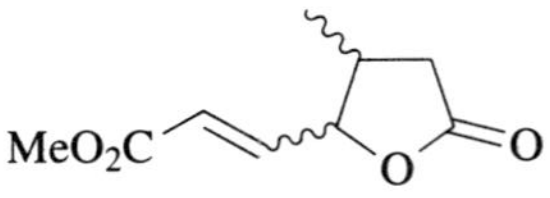

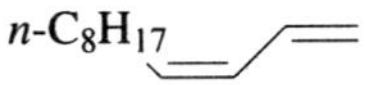

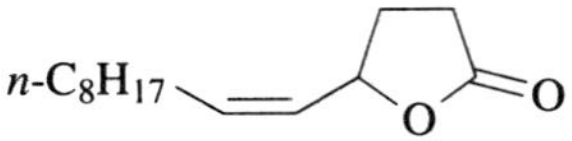

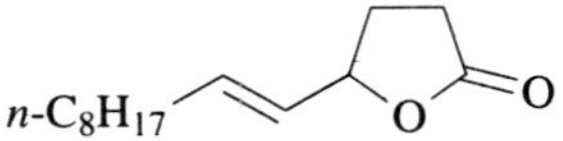

558

TABLE XV. NONCONJUGATED DIENES AND CARBONYL COMPOUNDS

Substrate	Reagent	Conditions	Product(s) and Yield(s) (%)	Refs.
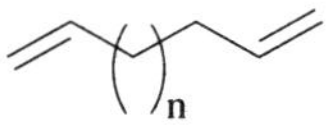 C_6 n = 1 C_8 n = 3	AcOH	Mn(OAc)$_3$, AcOH, AcOK, reflux	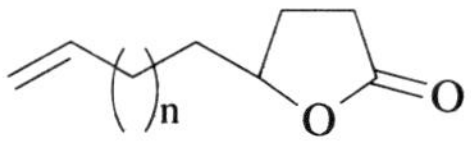 (24) (26)	128
C_8 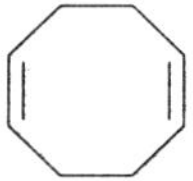	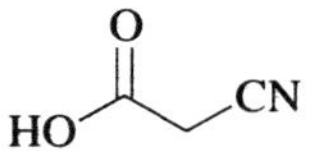	Mn(OAc)$_3$, AcOH 23°, 15 min	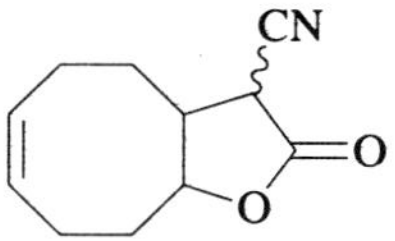 (83)	147
	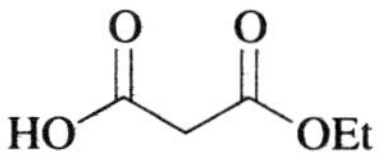	Mn(OAc)$_3$, AcOH, 23°, 4 - 6 h	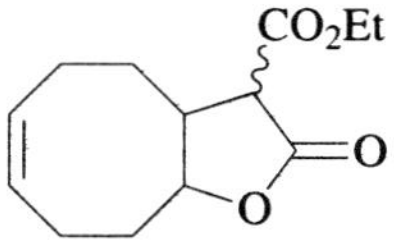 (58)	147
C_{10} 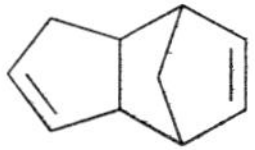	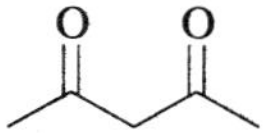	Mn(OAc)$_3$, Cu(OAc)$_2$, AcOH, 60°, 10 - 15 min	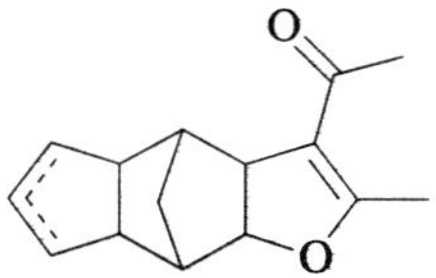 (66)	72

TABLE XV. NONCONJUGATED DIENES AND CARBONYL COMPOUNDS (*Continued*)

Substrate	Reagent	Conditions	Product(s) and Yield(s) (%)	Refs.
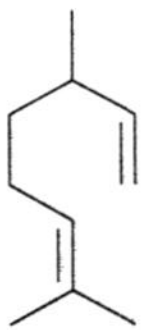	AcOH	Mn(OAc)₃, AcOH, Ac₂O, reflux, 10 min	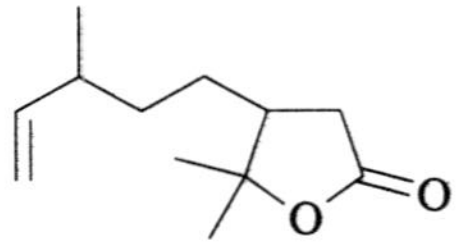 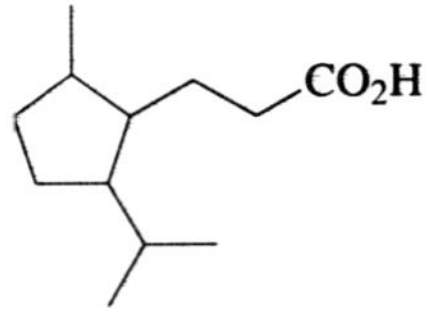 (—) +	82
	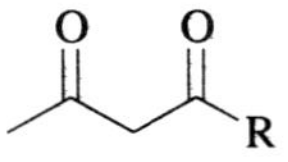 $\begin{array}{c}O\\\parallel\\R\end{array}$ and $\begin{array}{c}O\\\parallel\end{array}$ R = Me R = OEt	Mn(OAc)₃, AcOH, reflux, 10 min	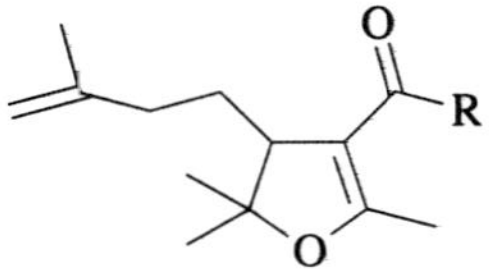 (—)	82
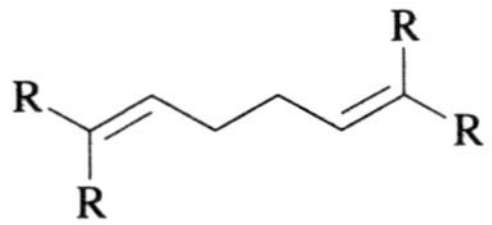 C₁₀ R = Me C₃₀ R = Ph	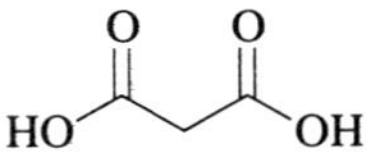 $\begin{array}{c}O\quad O\\\parallel\quad\parallel\\HO\quad OH\end{array}$	Mn(OAc)₃, AcOH, 70°, 2 h	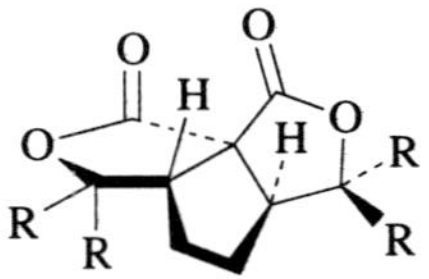 (40) (12)	145

TABLE XV. NONCONJUGATED DIENES AND CARBONYL COMPOUNDS (*Continued*)

Substrate	Reagent	Conditions	Product(s) and Yield(s) (%)	Refs.
C$_{29}$-C$_{34}$				
		Mn(OAc)$_3$, AcOH, reflux, 4 - 9 min		109

n	R^1	R^2		
1	Ph	Ph		(22)
1	*p*-ClC$_6$H$_4$	*p*-ClC$_6$H$_4$		(17)
1	*p*-MeC$_6$H$_4$	*p*-MeC$_6$H$_4$		(14)
2	Ph	Ph		(89)
2	*p*-ClC$_6$H$_4$	*p*-ClC$_6$H$_4$		(71)
2	*p*-MeC$_6$H$_4$	*p*-MeC$_6$H$_4$		(79)
2	*p*-MeOC$_6$H$_4$	*p*-MeOC$_6$H$_4$		(62)

Substrate	Reagent	Conditions	Product(s) and Yield(s) (%)	Refs.
	Mn(acac)$_3$[a]	Mn(acac)$_3$, AcOH, reflux, 1 - 3 min		109

n	R^1	R^2		
1	Ph	Ph		(13)
1	*p*-ClC$_6$H$_4$	*p*-ClC$_6$H$_4$		(23)
1	*p*-MeC$_6$H$_4$	*p*-MeC$_6$H$_4$		(71)
2	Ph	Ph		(71)
2	*p*-ClC$_6$H$_4$	*p*-ClC$_6$H$_4$		(46)
2	*p*-MeC$_6$H$_4$	*p*-MeC$_6$H$_4$		(29)
2	*p*-MeOC$_6$H$_4$	*p*-MeOC$_6$H$_4$		(29)

TABLE XV. NONCONJUGATED DIENES AND CARBONYL COMPOUNDS (*Continued*)

Substrate	Reagent	Conditions	Product(s) and Yield(s) (%)	Refs.
C$_{30}$-C$_{34}$		Mn(OAc)$_3$, AcOH, reflux, 1 - 6 min		109

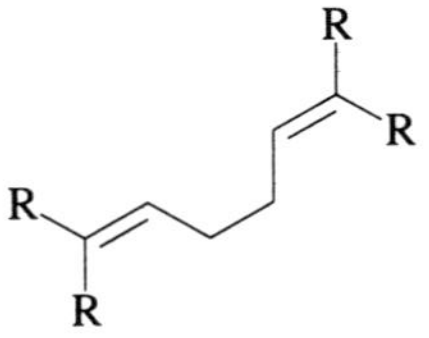

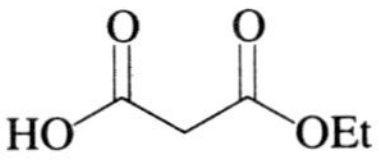

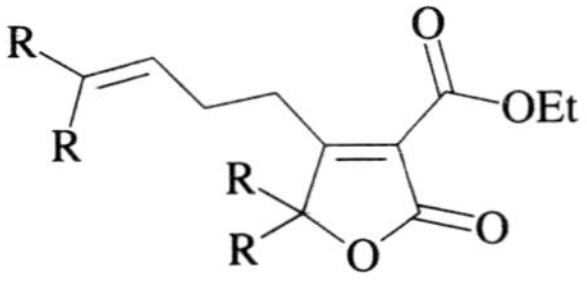

R	
Ph	(81)
p-ClC$_6$H$_4$	(65)
p-MeC$_6$H$_4$	(47)
p-MeOC$_6$H$_4$	(33)

[a] Monodihydrofurans were formed in 11 - 39% yields.

562

TABLE XVI. 1,3-ALKADIENES OR 1,3-CYCLOALKADIENES AND DICARBONYL COMPOUNDS

Substrate	Reagent	Conditions	Product(s) and Yield(s) (%)	Refs.
C₄ 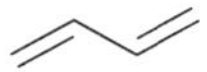	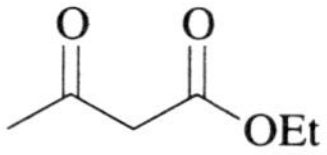	Mn(OAc)₃, Cu(OAc)₂, AcOH, 60°, 10 - 15 min	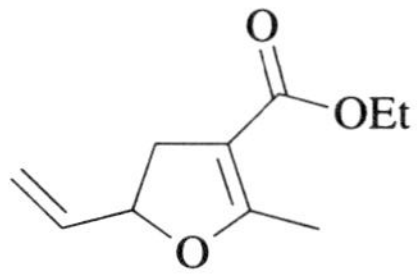 (64)	72
	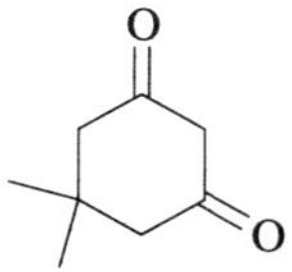	"	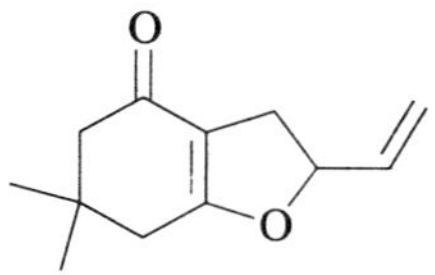 (42)	72
	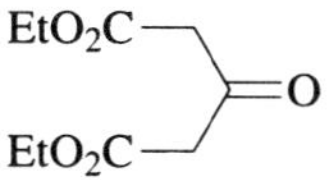	"	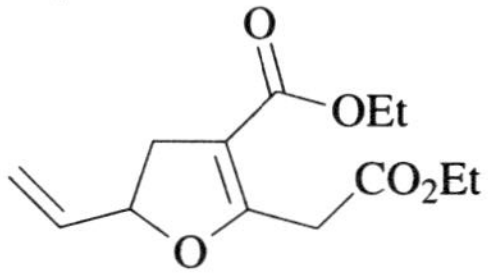 (52)	72
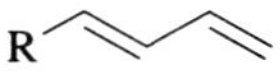	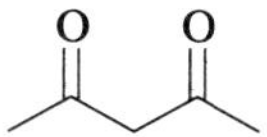	"	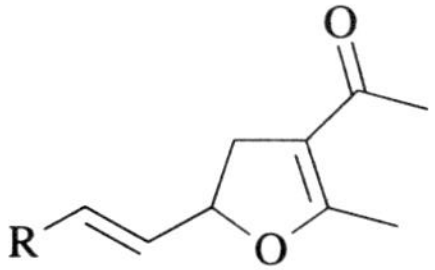 (97) (65)	72, 74
C₄ R = H C₅ R = Me			(20)	
 C₅	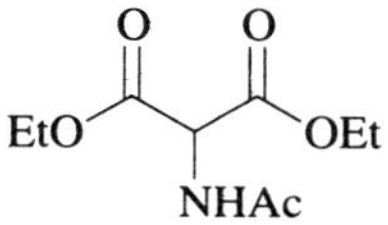	Mn(OAc)₃, AcOH, 80°, 8 h	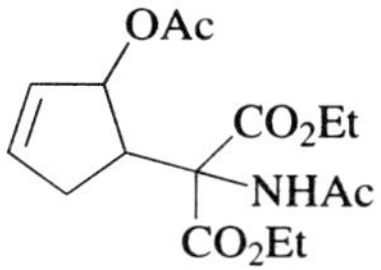 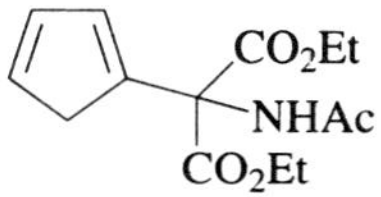 + (10)	107

563

TABLE XVI. 1,3-ALKADIENES OR 1,3-CYCLOALKADIENES AND DICARBONYL COMPOUNDS (*Continued*)

Substrate	Reagent	Conditions	Product(s) and Yield(s) (%)	Refs.
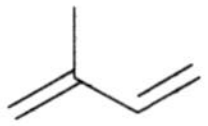 (2-methyl-1,3-butadiene)	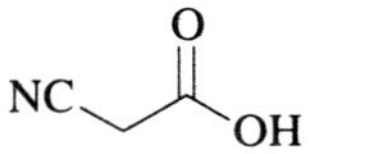 NC–CH₂–CO₂H	Mn(OAc)₃, AcOH, AcOK, reflux	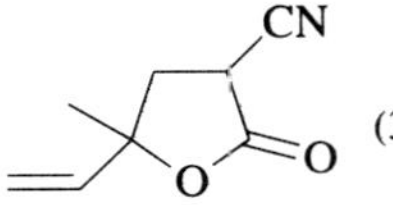 (39) 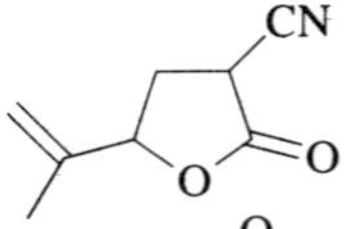 + (5)	128
	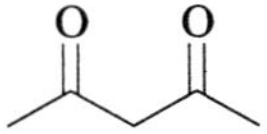 pentane-2,4-dione	Mn(OAc)₃, Cu(OAc)₂, AcOH, 60°, 10 - 15 min	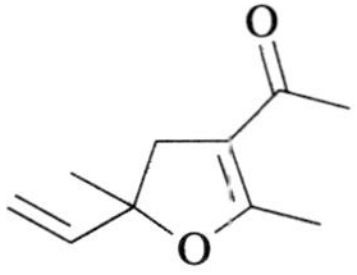 (70) 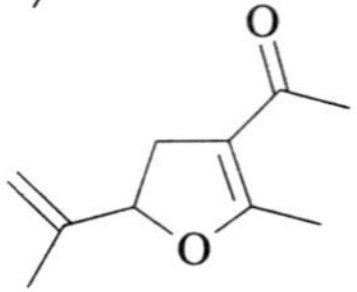 + (10)	72
 C₆ (1,3-cyclohexadiene)	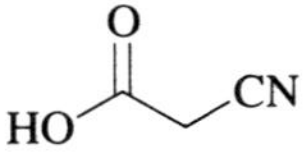 NC–CH₂–CO₂H	Mn(OAc)₃, AcOH, 23°, 15 min	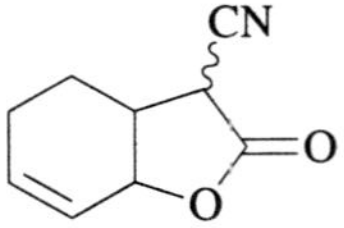 (78)	147
	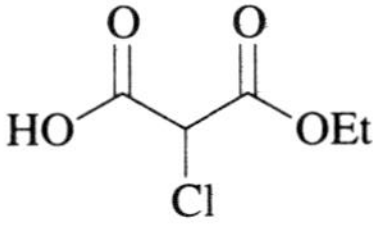 EtO₂C–CHCl–CO₂H	Mn(OAc)₃, AcOH, 23°	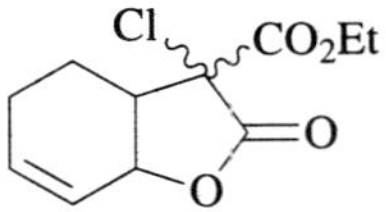 (82)	147
	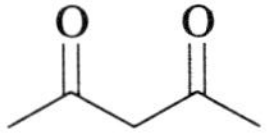 pentane-2,4-dione	Mn(OAc)₃, Cu(OAc)₂, AcOH, 60°, 10 - 15 min	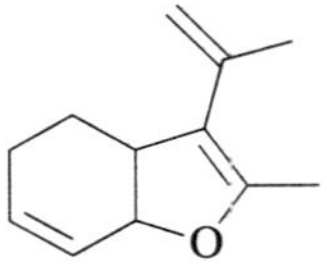 (32)	72
	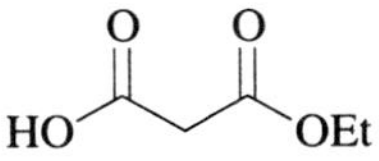 EtO₂C–CH₂–CO₂H	Mn(OAc)₃, AcOH, 23°, 4 - 6 h	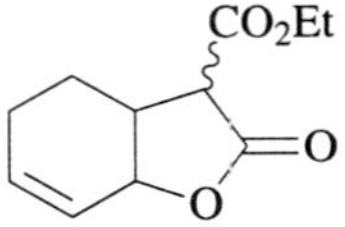 (76)	147

564

Substrate	Reagent	Conditions	Product(s) and Yield(s) (%)	Refs.
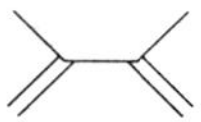	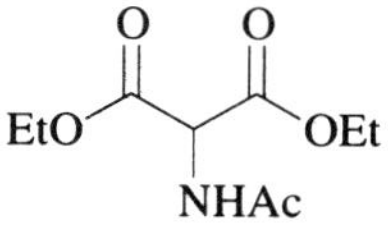	Mn(OAc)$_3$, AcOH, 80°, 8 h	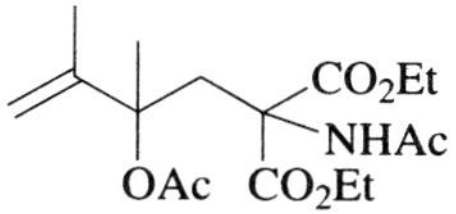 (40) 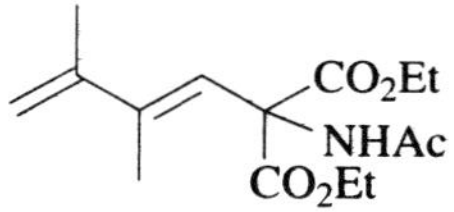 + (28)	107
C$_7$		Mn(OAc)$_3$, AcOH, 70°	I : II : III 21 : 44 : 15 (63)	108

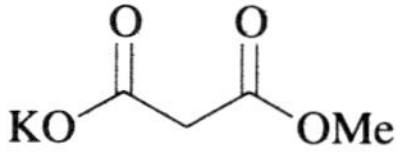

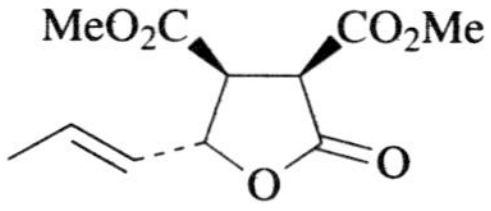

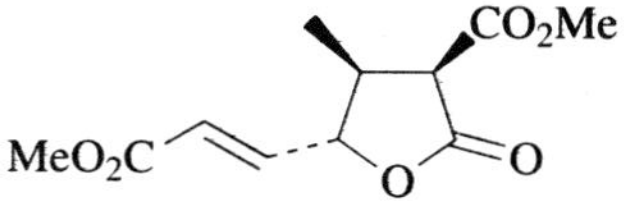

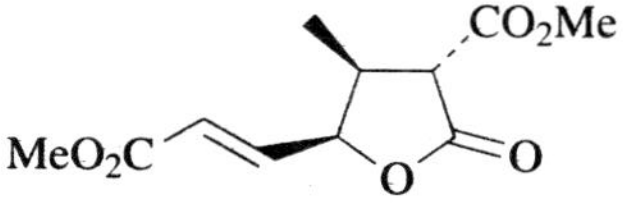

TABLE XVI. 1,3-ALKADIENES OR 1,3-CYCLOALKADIENES AND DICARBONYL COMPOUNDS (*Continued*)

Substrate	Reagent	Conditions	Product(s) and Yield(s) (%)	Refs.
	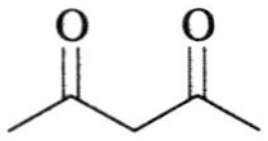	Mn(OAc)$_3$, AcOH, 70°	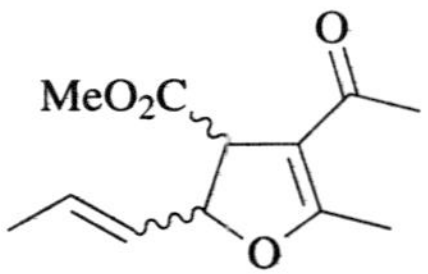 $E:Z$ 36:12 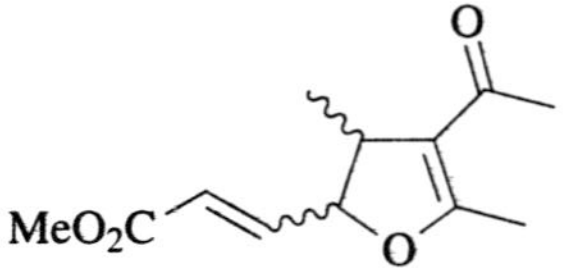 + (57) $E:Z$ 15:19	108
R = OMe R = Me		Mn(OAc)$_3$, Cu(OAc)$_2$, AcOH, 60°, 10 - 15 min	**I** + **II** I (35) + II (7) I (40) + II (8)	72

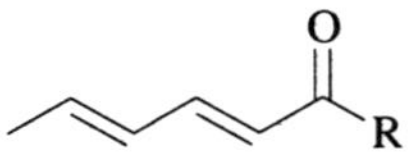

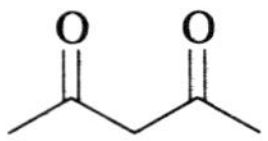

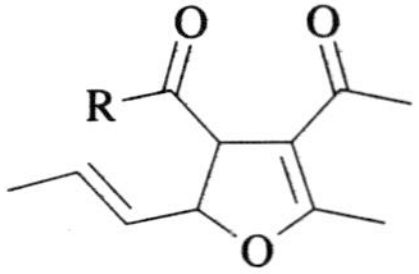

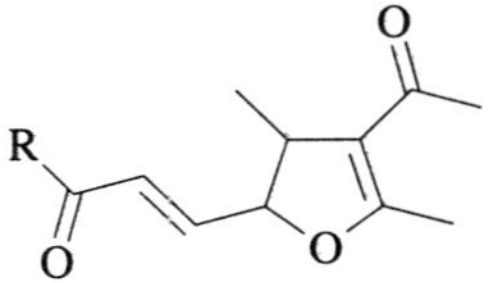

TABLE XVI. 1,3-ALKADIENES OR 1,3-CYCLOALKADIENES AND DICARBONYL COMPOUNDS (*Continued*)

Substrate	Reagent	Conditions	Product(s) and Yield(s) (%)	Refs.
(structure, HO-)	(diketone, O O R)	Mn(OAc)$_3$, Cu(OAc)$_2$, AcOH, 60°, 15 min	I + (structures) R = Me; R = OEt; I + II (56) I:II = 30:70; I + II (60) I:II = 40:60	73
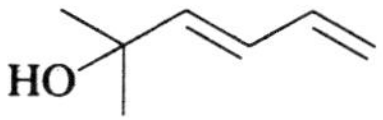 (structure, HO-)	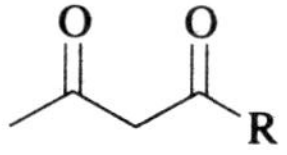 (diketone, O O R)	Mn(OAc)$_3$, Cu(OAc)$_2$, AcOH, 60°, 15 min	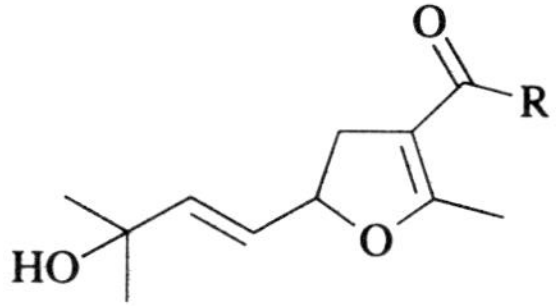 R = Me (53); R = OEt (57)	73
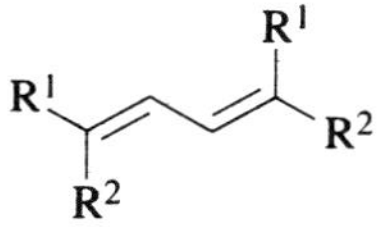 (structure, R^1, R^2; Me/Me; Ph/H)	Mn(acac)$_3$	Mn(acac)$_3$, AcOH, reflux, 2 min	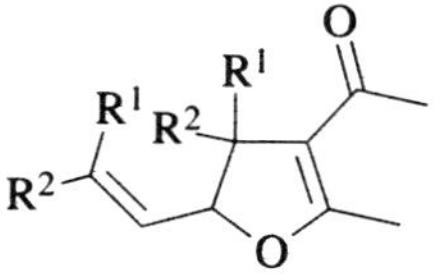 (structure, R^1, R^2) (72); (85)	109

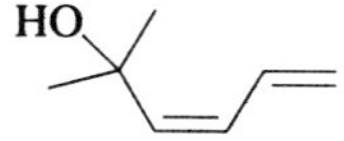

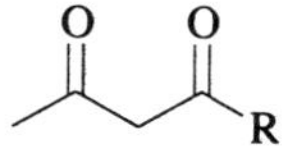

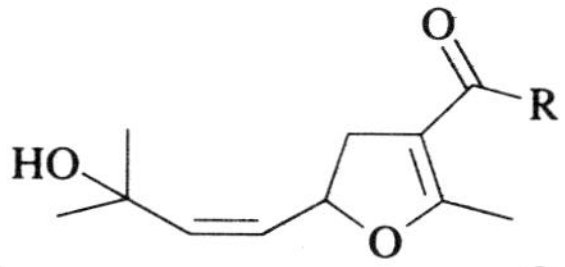

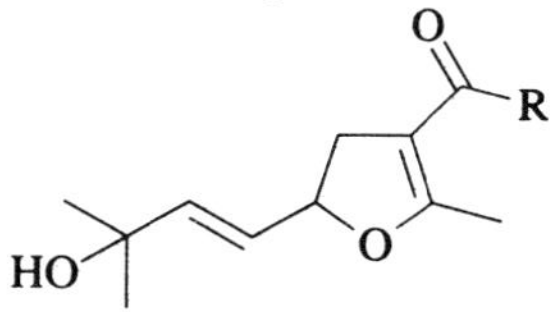

R^1 / R^2: Me / Me; Ph / H

Substrate	Reagent	Conditions	Product(s) and Yield(s) (%)	Refs.
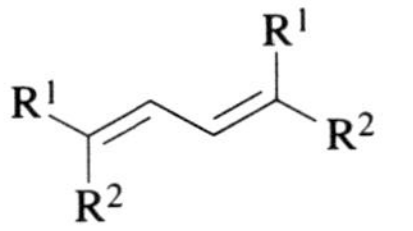 C8 $R^1 = Me$, $R^2 = Me$; C16 $R^1 = Ph$, $R^2 = H$	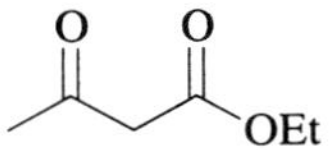 (ethyl acetoacetate)	Mn(OAc)3, AcOH, reflux, 1 - 4 min	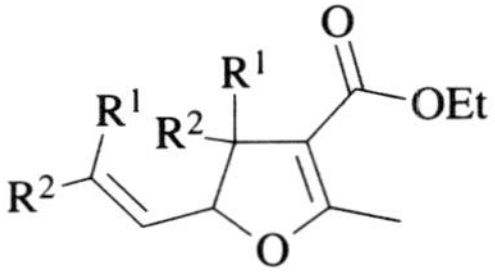 (57) (49)	109
 C10	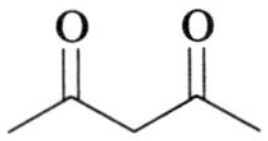 (acetylacetone)	Mn(OAc)3, Cu(OAc)2, AcOH, 60°, 10 - 15 min	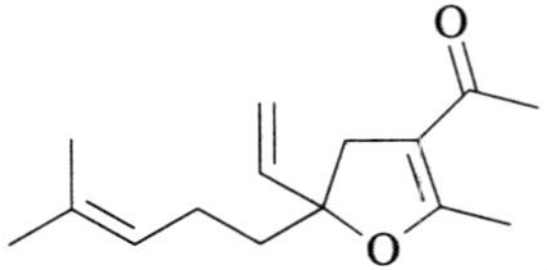 (43) 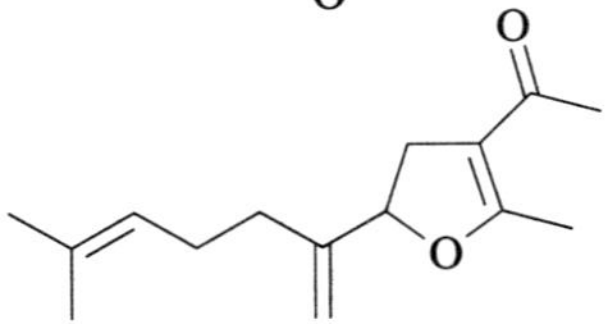 + (8)	72
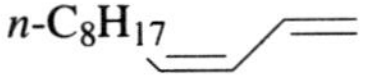 C12 n-C8H17	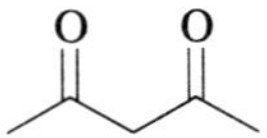 (acetylacetone)	Mn(OAc)3, Cu(OAc)2, AcOH, 60°, 15 min	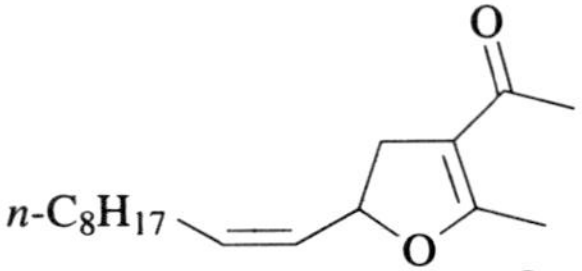 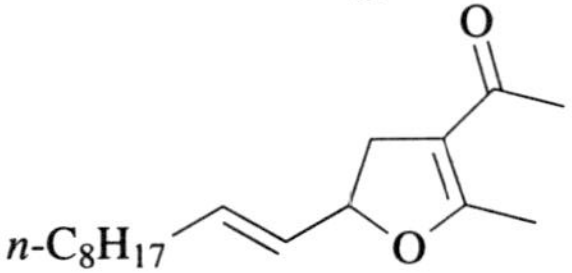 + 80 : 20 (77)	73

568

TABLE XVI. 1,3-ALKADIENES OR 1,3-CYCLOALKADIENES AND DICARBONYL COMPOUNDS (*Continued*)

Substrate	Reagent	Conditions	Product(s) and Yield(s) (%)	Refs.
C_{28}–C_{32}				
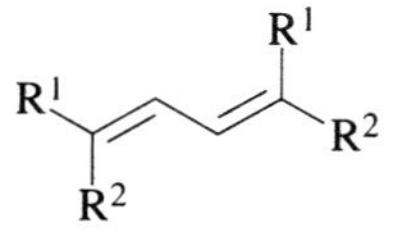	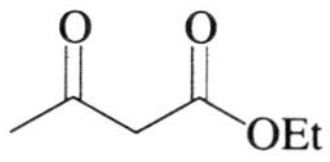	Mn(OAc)$_3$, AcOH, reflux, 3 - 7 min	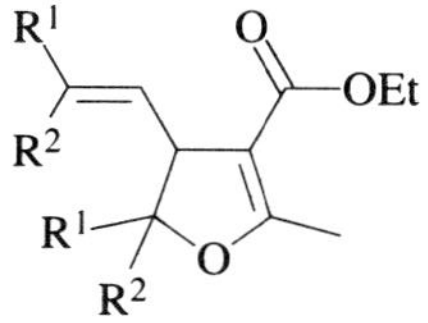	109

R^1	R^2
Ph	Ph	(46)
p-ClC$_6$H$_4$	p-ClC$_6$H$_4$	(29)
p-MeC$_6$H$_4$	p-MeC$_6$H$_4$	(11)

| | Mn(acac)$_3$ | Mn(acac)$_3$, AcOH, reflux, 2 min | 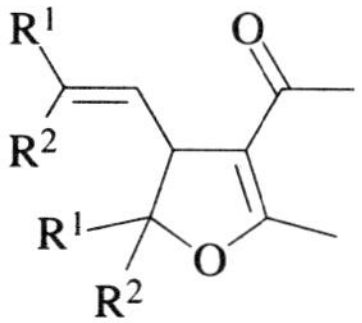 | 109 |

R^1	R^2
Ph	Ph	(19)
p-ClC$_6$H$_4$	p-ClC$_6$H$_4$	(30)
p-MeC$_6$H$_4$	p-MeC$_6$H$_4$	(38)

TABLE XVII. 1,3-ALKENYNES AND ACETIC ACID

Substrate	Reagent	Conditions	Product(s) and Yield(s) (%)		Refs.
C_6-C_{14} R—C≡C— (vinyl)	AcOH	Mn(OAc)₃, AcOH, Ac₂O, 115°, 10 min	(OAc/OH acid)	(lactone, R—C≡C)	62
R					
Et			(27)	(17)	
n-Bu			(40)	(12)	
n-C₇H₁₅			(40)	(0)	
n-C₈H₁₇			(57)	(0)	
n-C₁₀H₂₁			(49)	(0)	
C_7-C_{12} R—C≡C— (vinyl)	"	Mn(OAc)₃, AcOH, AcOK, 115°, 4 h	(lactone, R—C≡C)		158, 153, 156
R					
Me₂C(OMe)			(33)		
n-Pr			(47)		
n-Bu			(50)		
n-C₇H₁₅			(34)		
n-C₈H₁₇			(51)		
C_8-C_9 R—C≡C— (isopropenyl)	"	"	(methyl lactone, R—C≡C)		156
R					
n-Pr			(30)		
n-Bu			(40)		
i-Bu			(52)		

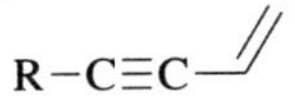

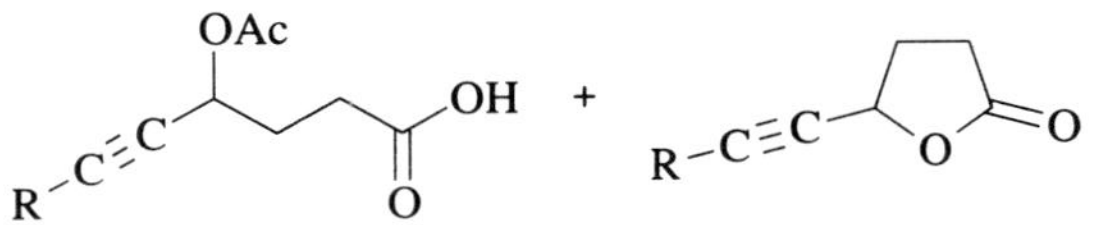

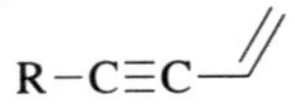

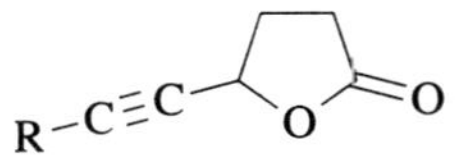

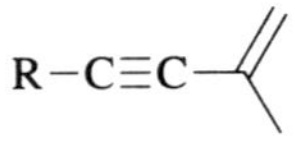

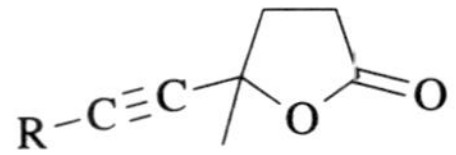

TABLE XVIII. 1,3-ALKENYNES AND β-DICARBONYL COMPOUNDS

Substrate	Reagent	Conditions	Product(s) and Yield(s) (%)	Refs.
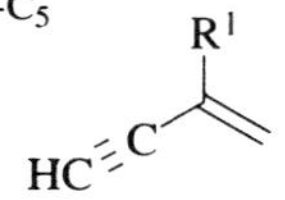 C4-C5	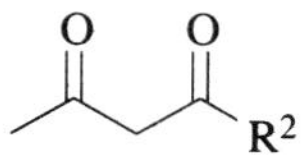	Mn(OAc)₃, Cu(OAc)₂, AcOH, 70°, 15 min	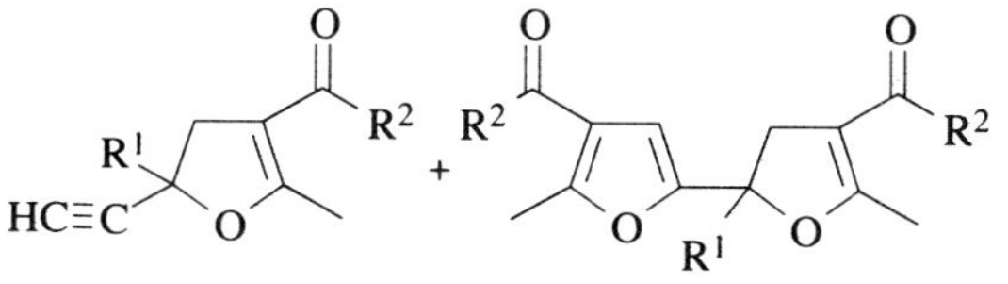	74, 76

R¹	R²		
H	Me	(4)	(14)
H	OEt	(9)	(18)
Me	OEt	(51)	(11)

C5-C12 " 75, 77, 161

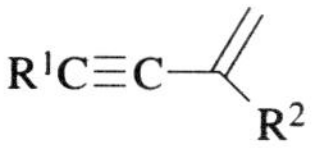

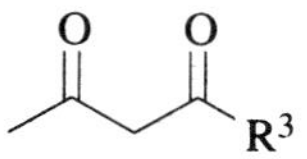

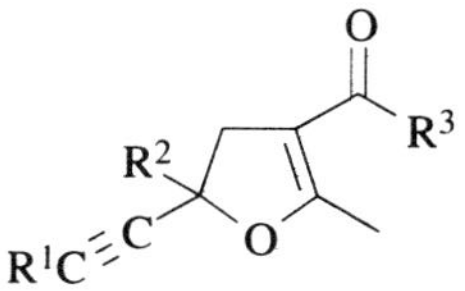

R¹	R²	R³	
n-Bu	H	OEt	(68)
n-Bu	H	Me	(36)
CH₂OH	H	OEt	(50)
i-PrCH(OH)	H	OEt	(59)
i-PrCH(OH)	H	Me	(55)
Me₂C(OH)	H	OEt	(54)
Me₂C(OMe)	H	OEt	(50)
Me₂C(OAc)	H	OEt	(22)
(prenyl-OH)	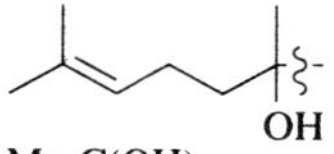 H	OEt	(48)
Me₂C(OH)	Me	OEt	(66)
Me₂C(OH)	Me	Me	(60)
(prenyl-OH)	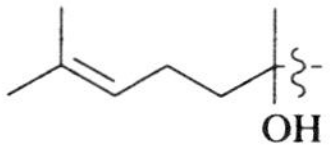 Me	OEt	(50)

TABLE XVIII. 1,3-ALKENYNES AND β-DICARBONYL COMPOUNDS (*Continued*)

Substrate	Reagent	Conditions	Product(s) and Yield(s) (%)	Refs.
C$_7$	(ethyl acetoacetate, OEt)	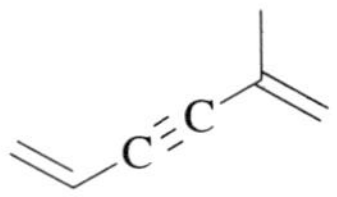 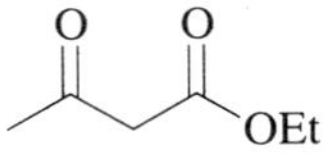 Mn(OAc)$_3$, Cu(OAc)$_2$, AcOH, 70°, 15 min	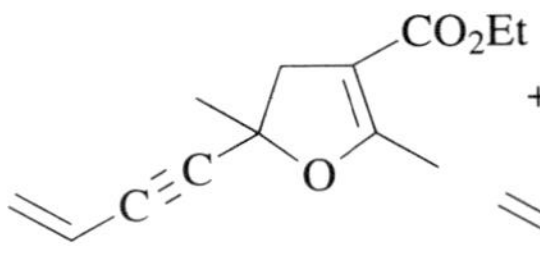 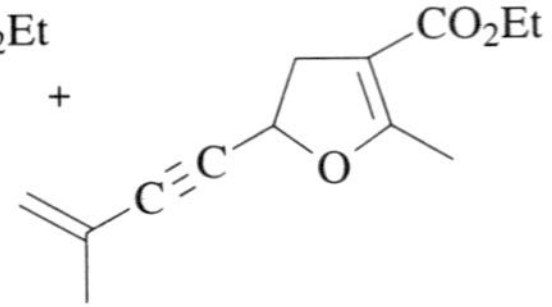 9 : 1 (64)	76
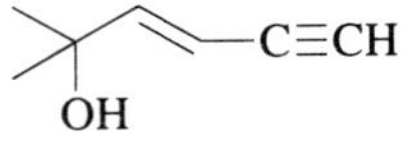	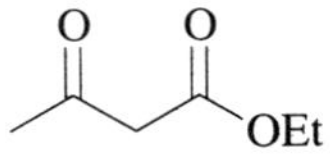 (ethyl acetoacetate, OEt)	Mn(OAc)$_3$, Cu(OAc)$_2$, AcOH, 30°, 3 h	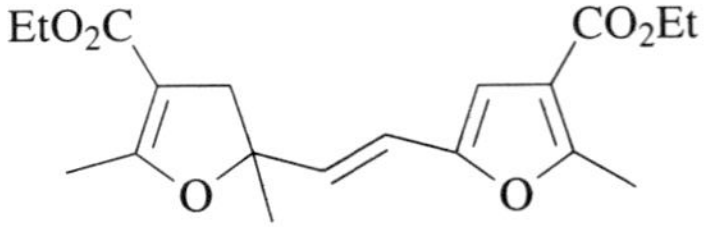 (36)	77
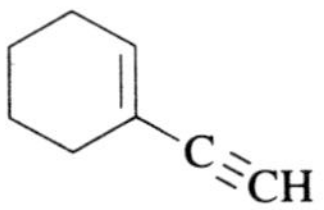 C$_8$	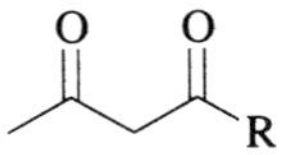	AcOH, 30°, 3 h		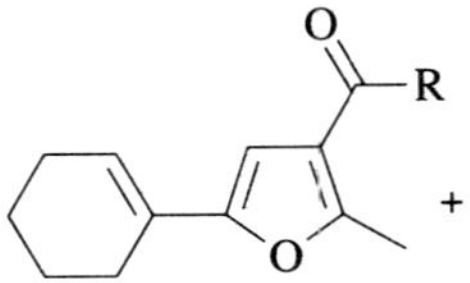 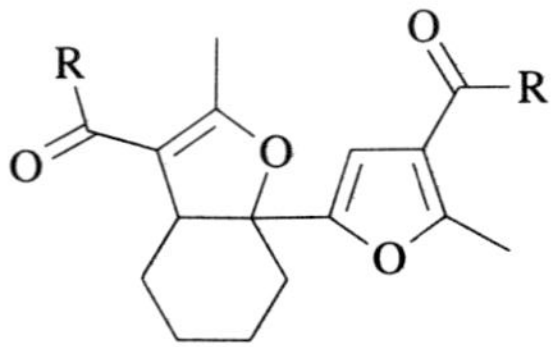 77
	R = Me	Mn(OAc)$_3$, (1 eq) Cu(OAc)$_2$, (1 eq)	(20) (24)	
	R = OEt	"	(23) (19)	
	R = Me	Mn(OAc)$_3$, (3 eq) Cu(OAc)$_2$, (3 eq)	(63) (0)	
	R = OEt	"	(60) (0)	

TABLE XVIII. 1,3-ALKENYNES AND β-DICARBONYL COMPOUNDS (*Continued*)

Substrate	Reagent	Conditions	Product(s) and Yield(s) (%)	Refs.
C_{10} n = 0 C_{11} n = 1		Mn(OAc)$_3$, Cu(OAc)$_2$, AcOH, 30° 1 h 2.5 h	(47) + (25) 89 : 11 (0)	57
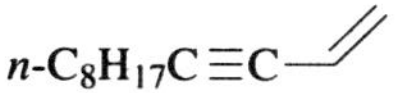 C_{12} $n\text{-}C_8H_{17}C\equiv C$	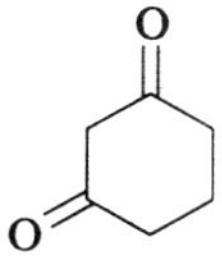	Mn(OAc)$_3$, Cu(OAc)$_2$, AcOH, 30°, 30 min	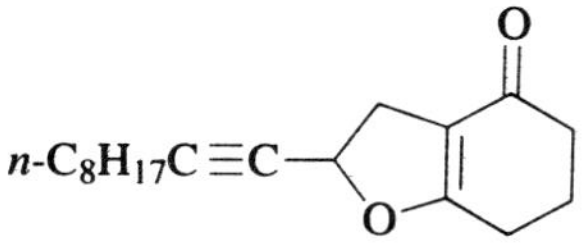 (78)	57

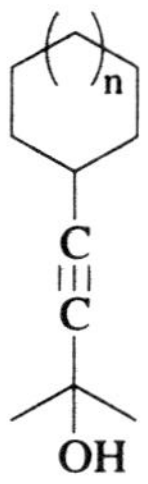

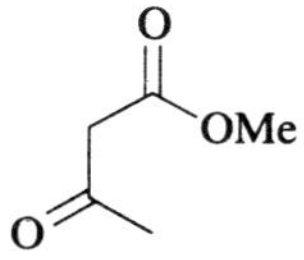

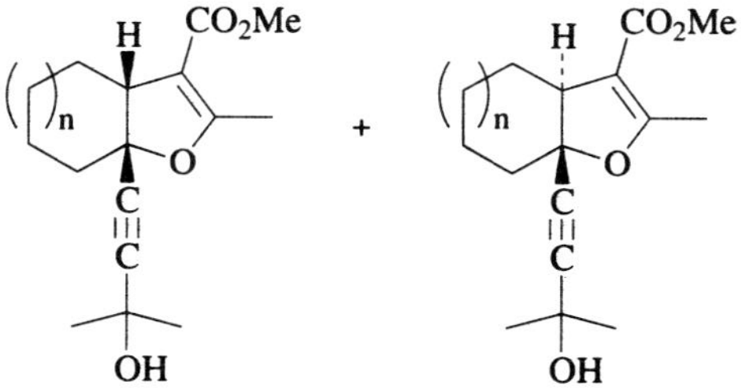

573

TABLE XIX. Co-COMPLEXED 1,3-ALKENYNES AND β-DICARBONYL COMPOUNDS

Substrate	Reagent	Conditions	Product(s) and Yield(s) (%)	Refs.
C₁₁	 	R¹ R² Me OMe Me Me (CH₂)₃	Mn(OAc)₃, AcOH, 30°, 30 min	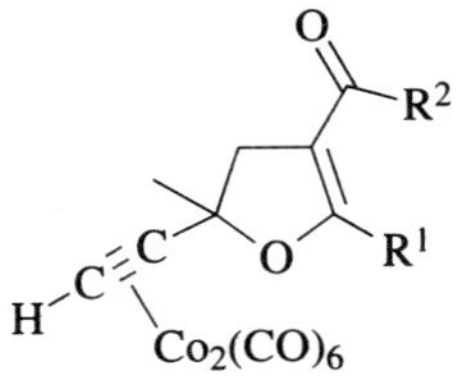 (65) + (8)ᵃ (52) + (6)ᵃ (46) + (10)ᵃ	57
	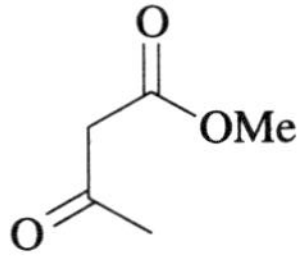	Mn(OAc)₃, MeOH, 30°, 30 min	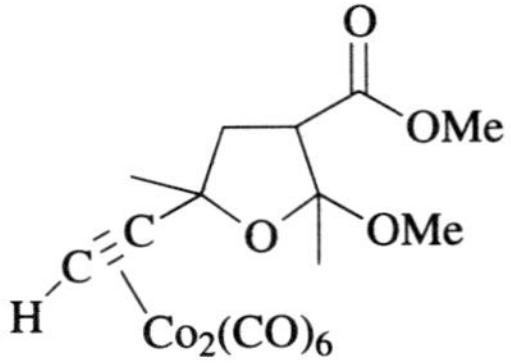 (46)	57
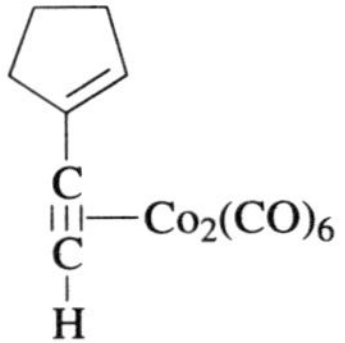 C₁₃	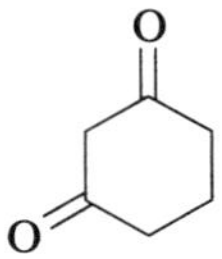	Mn(OAc)₃, AcOH, 30°, 1 h	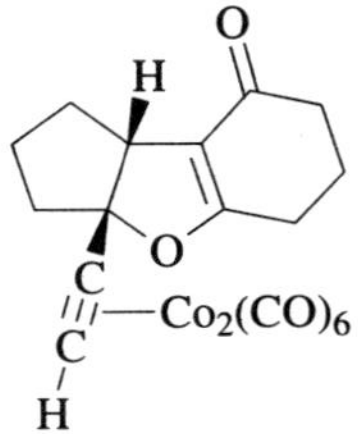 (28) + (10)ᵃ	57

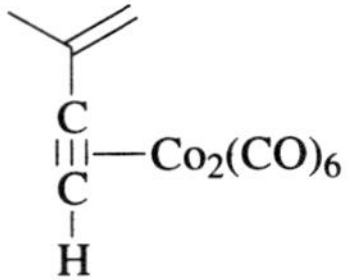

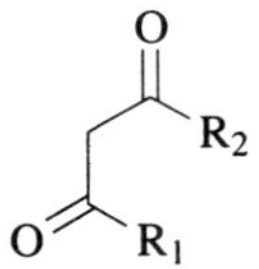

TABLE XIX. Co-COMPLEXED 1,3-ALKENYNES AND β-DICARBONYL COMPOUNDS (*Continued*)

Substrate	Reagent	Conditions	Product(s) and Yield(s) (%)	Refs.
cyclohexenyl–C≡C–R·Co₂(CO)₆ C₁₄ R = H C₁₇ R = Me₂C(OH)	methyl acetoacetate (OMe)	Mn(OAc)₃, AcOH, 30°, 2.5 h	bicyclic furan CO₂Me, C≡C–Co₂(CO)₆, R (22) + (3)[a] (27) + (20)[a]	57
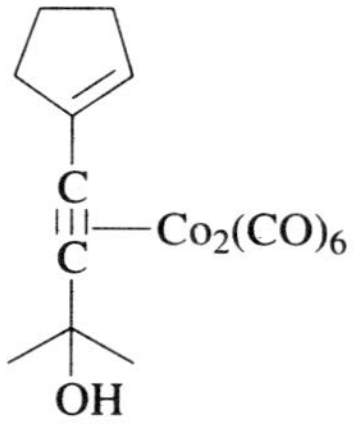 C₁₆ cyclopentenyl–C≡C–C(CH₃)₂OH·Co₂(CO)₆	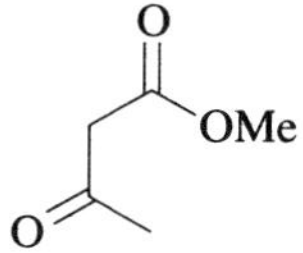 methyl acetoacetate (OMe)	Mn(OAc)₃, Cu(OAc)₂, AcOH, 30°, 1 h	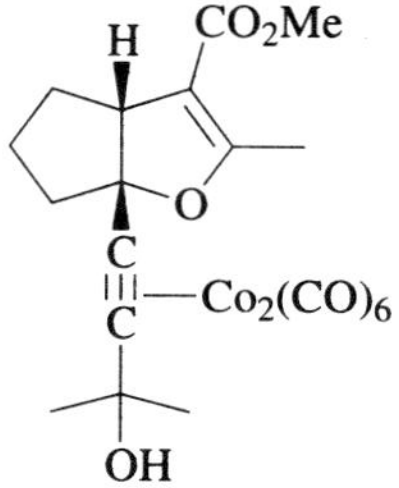 bicyclic furan CO₂Me, C≡C–Co₂(CO)₆, C(CH₃)₂OH (61) + (17)[a]	57
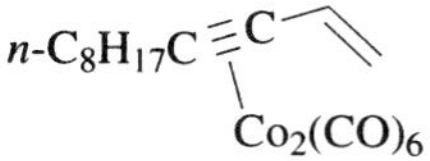 C₁₈ n-C₈H₁₇C≡C–CH=CH₂ Co₂(CO)₆	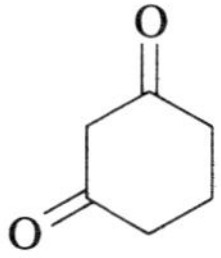 1,3-cyclohexanedione	Mn(OAc)₃, AcOH, 30°, 30 min	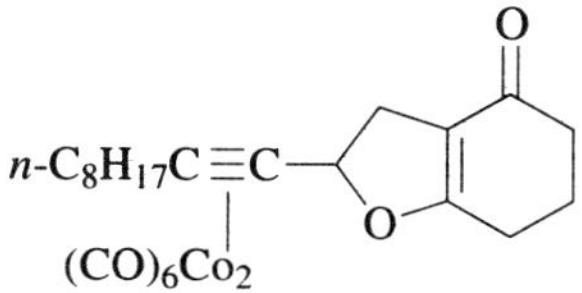 n-C₈H₁₇C≡C–, (CO)₆Co₂ bicyclic dihydrofuranone (41) + (27)[a]	57

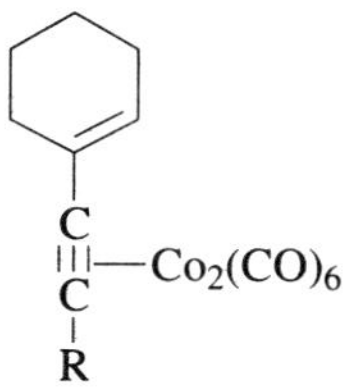

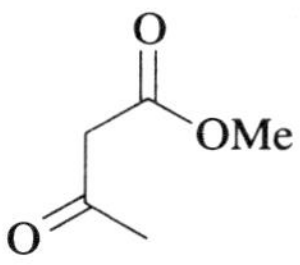

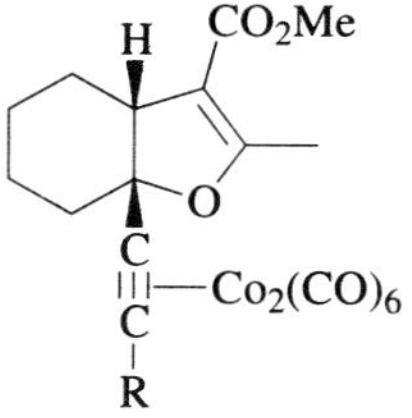

[a] Decomplexation product.

TABLE XX. 1,3-ALKADIYNES AND β-DICARBONYL COMPOUNDS

Substrate	Reagent	Conditions	Product(s) and Yield(s) (%)	Refs.
C_6–C_8				
R¹C≡C—C≡CH	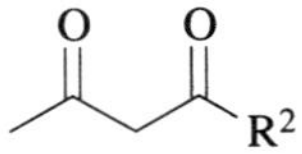	Mn(OAc)₃, Cu(OAc)₂, AcOH, 30°, 3 h		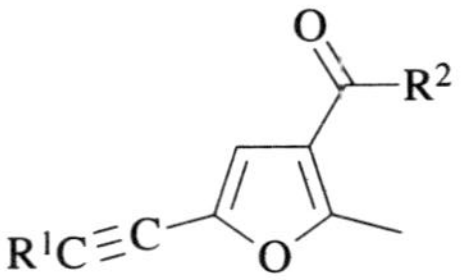 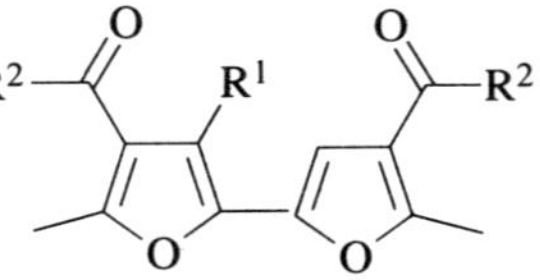 163, 78

R¹	R²		
Et	OEt	(13)	(5)
n-Pr	Me	(29)	(28)
n-Pr	OEt	(31)	(13)
n-Bu	Me	(32)	(30)
n-Bu	OEt	(30)	(14)

C_7				
		"		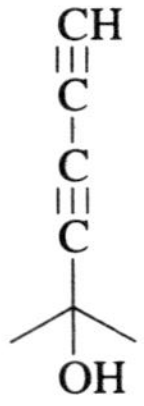 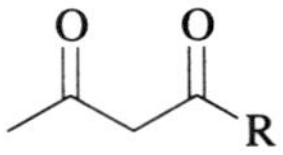 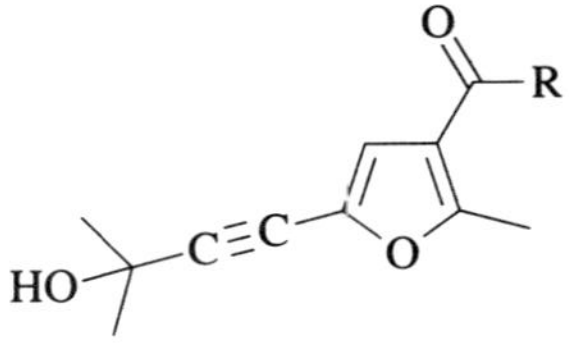 78

R = Me	(37)
R = OEt	(40)

TABLE XXIA. ARENES AND CARBONYL COMPOUNDS

Substrate	Reagent	Conditions	Product(s) and Yield(s) (%)	Refs.
C$_6$ 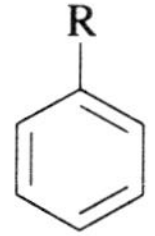 R-C$_6$H$_5$	AcOH	Mn(OAc)$_3$, AcOH, 24 h	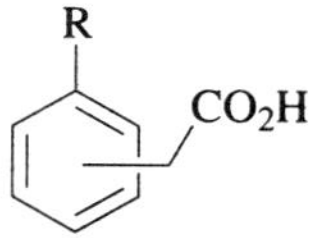 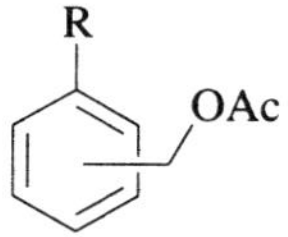 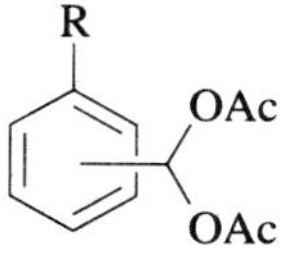 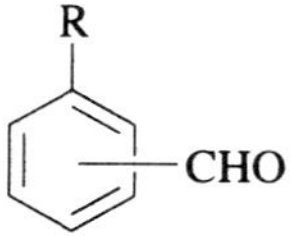 R-CH$_2$CO$_2$H **I** + R-CH$_2$OAc **II** + R-CH(OAc)$_2$ **III** + R-CHO **IV**	27
				I	**II**	**III**	**IV**	
R = H		101°	(8)	(51)	(9)	(18)	
R = Cl		110°	(13) 40^a	(21) 46^a	(17) 30^a	(11) 39^a	
C$_6$-C$_7$ 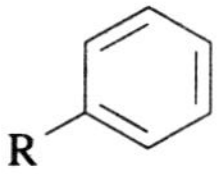 R-C$_6$H$_5$	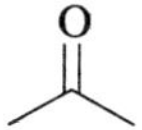 (acetone)	Mn(OAc)$_3$, AcOH, reflux	R-C$_6$H$_4$-COCH$_3$				34, 35
R = H		1.5 h	(40)				
R = Me		1 h	66 : 20 : 14^b (51)				
R = OMe		45 min	84 : 3 : 13^b (74)				
R = Cl		—	72 : 6 : 22^b (25)				
R = F		105 min	71 : 10 : 19^b (29)				
C$_7$ 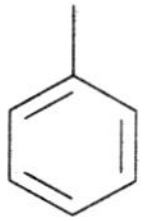 C$_6$H$_5$CH$_3$	AcOH	Mn(OAc)$_3$, AcOH, 110°, 24 h	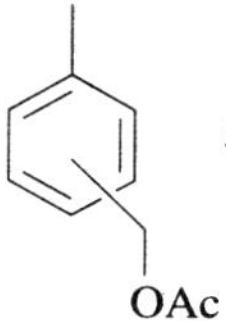 CH$_3$-C$_6$H$_4$-CH$_2$OAc 58 : 23 : 19^b (67) 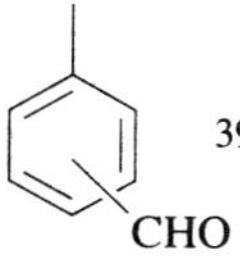 + CH$_3$-C$_6$H$_4$-CHO 39^a (7)	27

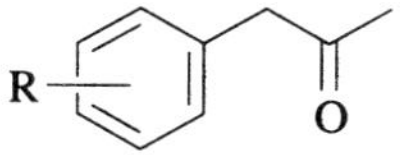

TABLE XXIA. ARENES AND CARBONYL COMPOUNDS (*Continued*)

Substrate	Reagent	Conditions	Product(s) and Yield(s) (%)	Refs.

Row 1:

Substrate: 4-methylanisole (OMe on ring with methyl)

Reagent: AcOH

Conditions: Mn(OAc)$_3$, AcOH

Products: (—) [CH$_2$CO$_2$H derivative] + (—) [CH$_2$OAc derivative]

Refs.: 33

Row 2:

Reagent: AcOH

Conditions: Mn(OAc)$_3$, AcOH

Products: OMe–ring–CO$_2$H + OMe–ring–CH$_2$CO$_2$H   78 : 5 : 17[b] (75)

Refs.: 33

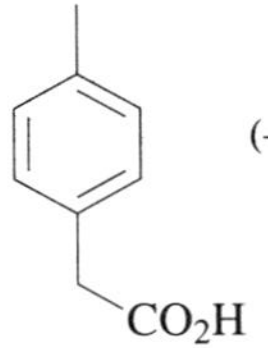

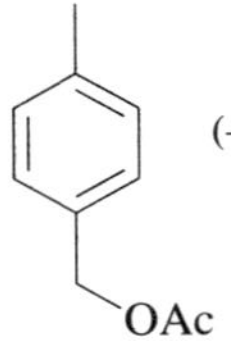

Row 3:

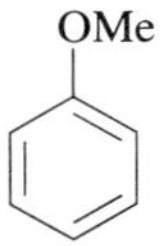

Reagent: EtO–C(=O)–CH(CH$_3$)–C(=O)–OEt (diethyl methylmalonate / 2-methyl diester)

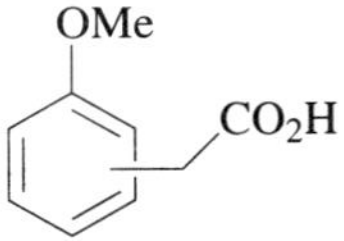

Conditions: Mn(OAc)$_3$, AcOH, 80°, 4 h

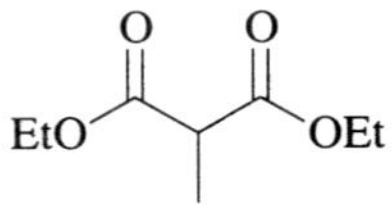

Products: OMe (ortho) –C(CH$_3$)(CO$_2$Et)CO$_2$Et + OMe (para) –C(CH$_3$)$_2$... C(CO$_2$Et)$_2$ 60 : 40 (15)

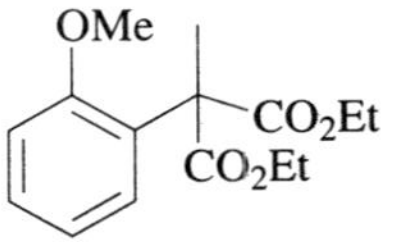

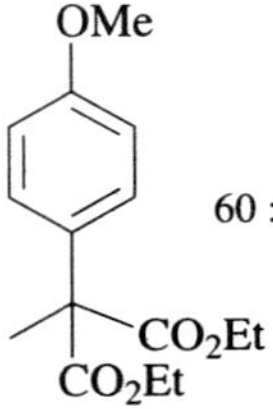

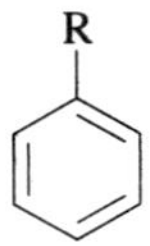

Refs.: 165

Row 4:

Substrate: R–C$_6$H$_5$

Reagent: HC(CO$_2$Et)$_3$

Conditions: Mn(OAc)$_3$, NaOAc, AcOH, 60 - 65°, 2 d

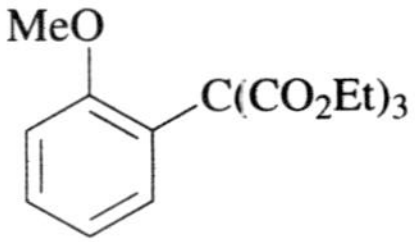

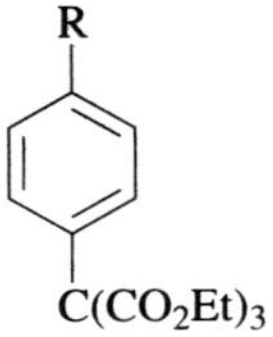

Products: MeO (ortho)–C(CO$_2$Et)$_3$ + R–(para)–C(CO$_2$Et)$_3$

(8)   (12)
(0)   (31)

Refs.: 37

C$_7$ R = OMe

C$_8$ R = NHCOMe

578

TABLE XXIA. ARENES AND CARBONYL COMPOUNDS (*Continued*)

Substrate	Reagent	Conditions	Product(s) and Yield(s) (%)	Refs.
C₈				
(1,4-dimethoxybenzene)	(acetone)	Mn(OAc)₃, AcOH, reflux, 40 min	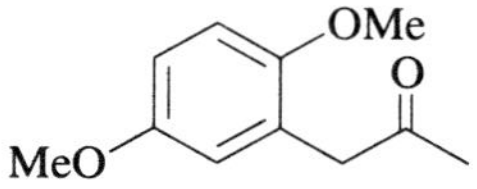 (39)	35
	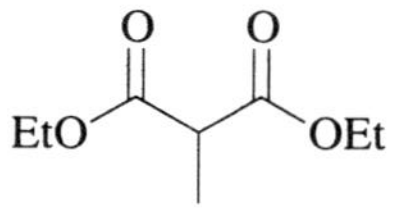 (ethyl 2-methylmalonate, EtO/OEt)	Mn(OAc)₃, AcOH, 80°, 4 h	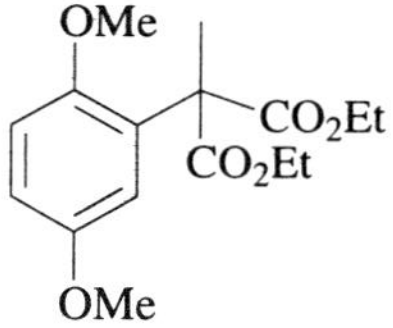 (12)	165
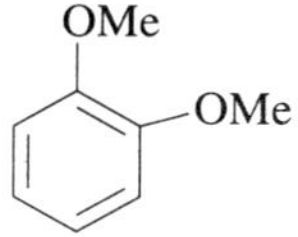 (1,2-dimethoxybenzene)	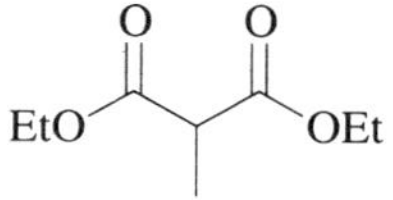	Mn(OAc)₃, AcOH, 80°, 4 h	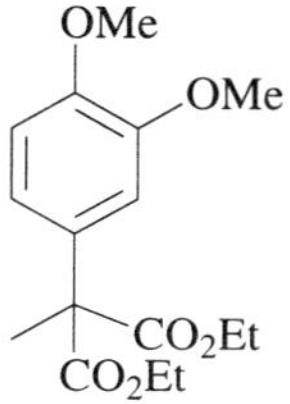 (20)	165
R = 2-OMe, R = 3-OMe	HC(CO₂Et)₃	Mn(OAc)₃, NaOAc, AcOH 60 – 65°, 2d	(35) (28)	37

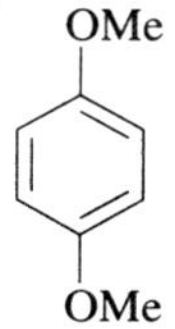

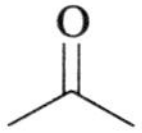

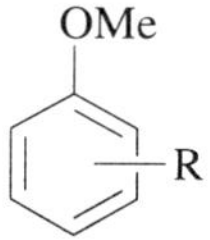

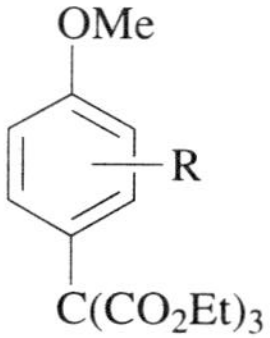

TABLE XXIA. ARENES AND CARBONYL COMPOUNDS (*Continued*)

Substrate	Reagent	Conditions	Product(s) and Yield(s) (%)	Refs.
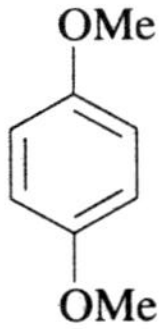 C₉ (1,4-dimethoxybenzene)	HC(CO₂Et)₃	Mn(OAc)₃, NaOAc, AcOH, 60 - 65°, 2d	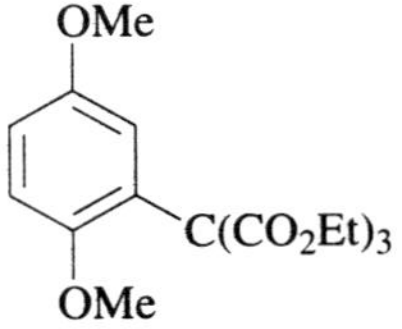 (36) [OMe, OMe, C(CO₂Et)₃ arene]	37
(1,2,3-trimethoxybenzene)	HC(CO₂Et)₃	Mn(OAc)₃, NaOAc, AcOH, 60 - 65°, 2d	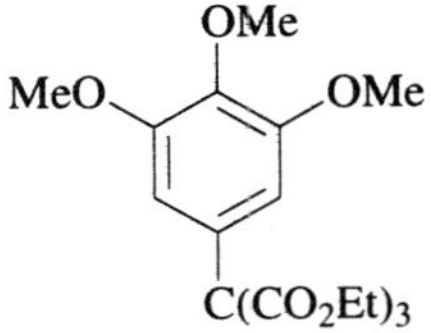 (35) [MeO, OMe, OMe, C(CO₂Et)₃ arene]	37
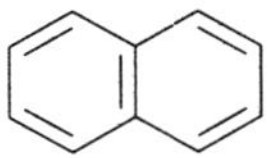 C₁₀ (naphthalene)	 (acetone)	Mn(OAc)₃, AcOH, reflux, 25 min	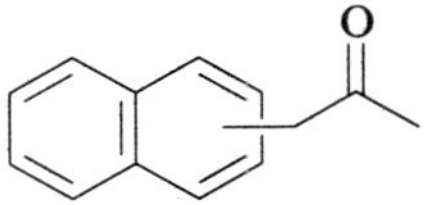 α : β 92 : 8 (77)	35
	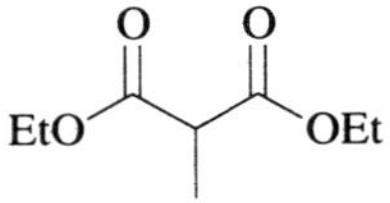 (diethyl methylmalonate)	Mn(OAc)₃, AcOH, 80°, 4 h	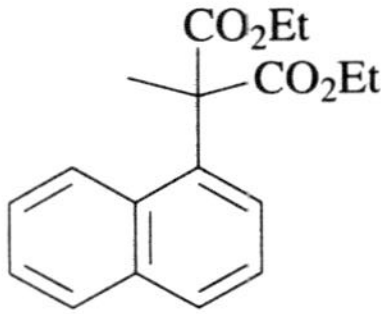 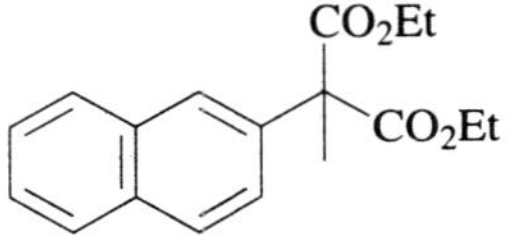 [1-naphthyl C(CH₃)(CO₂Et)₂] + [2-naphthyl C(CH₃)(CO₂Et)₂] 91 : 9 (19)	165

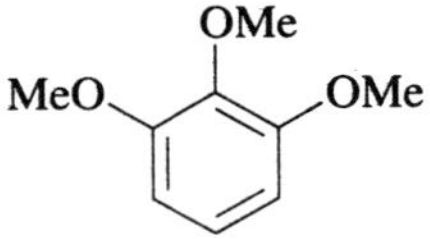

TABLE XXIA. ARENES AND CARBONYL COMPOUNDS (Continued)

Substrate	Reagent	Conditions	Product(s) and Yield(s) (%)	Refs.

C_{10}-C_{12}

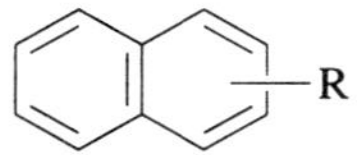

Substrate:

Naphthalene with R substituent

R = 2,3-(MeO)$_2$
R = 1-Me
R = 2-Me
R = 1,8-Me$_2$
R = H

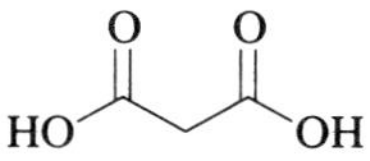

Reagent: malonic acid (HO$_2$C-CH$_2$-CO$_2$H)

Conditions: Mn(OAc)$_3$, AcOH, reflux, 1 min

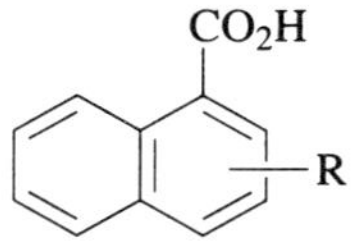

Product: naphthalene with CO$_2$H and R

R = 6,7-(MeO)$_2$
R = 4-Me
R = 2-Me
R = 4,5-Me$_2$
R = H

Refs.: 39

Substrate: naphthalene with R^1, R^2, R^3, R^4

R^1	R^2	R^3	R^4
MeO	H	H	H
MeO	H	H	MeO
MeO	H	MeO	H
MeO	MeO	H	H
H	H	H	H

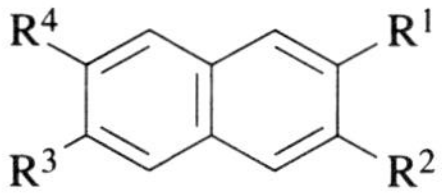

Reagent: Mn(acac)$_3$

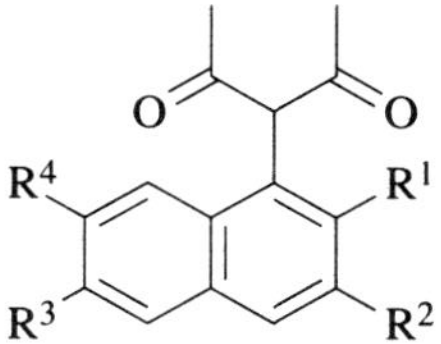

Conditions: Mn(OAc)$_3$, AcOH, 100°, 2-25 min

Product: naphthalene with R^1, R^2, R^3, R^4 and diacetyl group

(54)
(52)
(46)
(41)
(8)

Refs.: 40

C_{11}

Substrate: 1-methoxynaphthalene (OMe)

Reagent: diethyl malonate (EtO$_2$C-CH$_2$-CO$_2$Et)

Conditions: Mn(OAc)$_3$, AcOH, 80°, 2 h

Product: naphthalene with OMe and C(OAc)(CO$_2$Et)$_2$ (40)

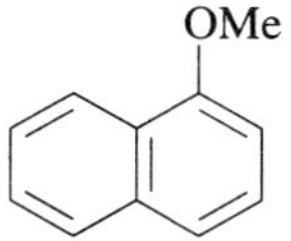

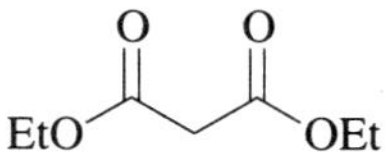

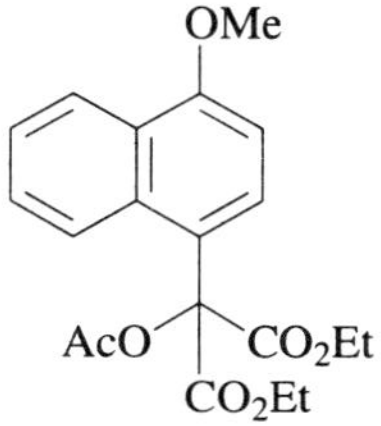

Refs.: 38

TABLE XXIA. ARENES AND CARBONYL COMPOUNDS (*Continued*)

Substrate	Reagent	Conditions	Product(s) and Yield(s) (%)	Refs.
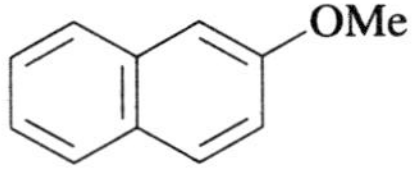	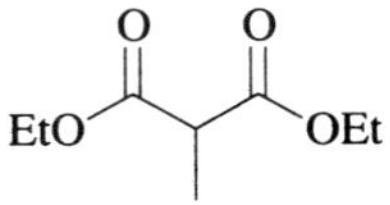	$Mn(OAc)_3$, AcOH, 80°, 3 h	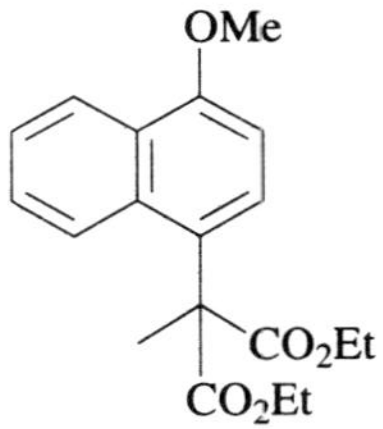 (51)	38, 165
	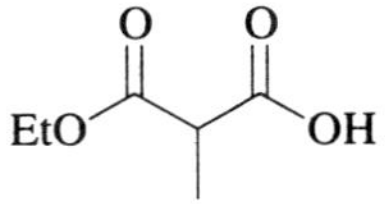	$Mn(OAc)_3$, AcOH, 80°, 9 h	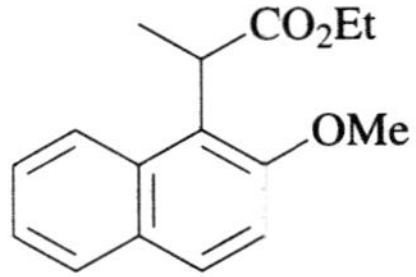 (35)	165
	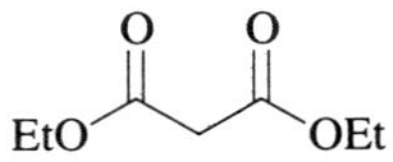	$Mn(OAc)_3$, AcOH, 80°, 2 h	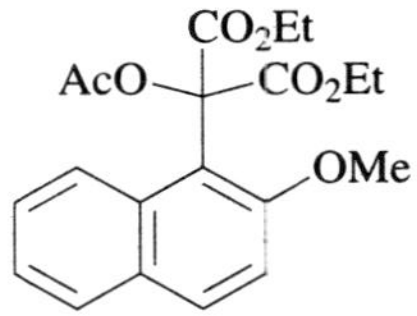 (12)	38
	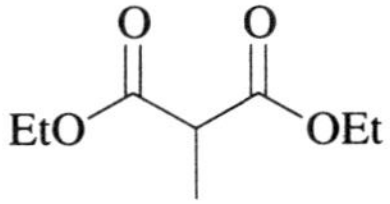	$Mn(OAc)_3$, AcOH, 80°, 4 h	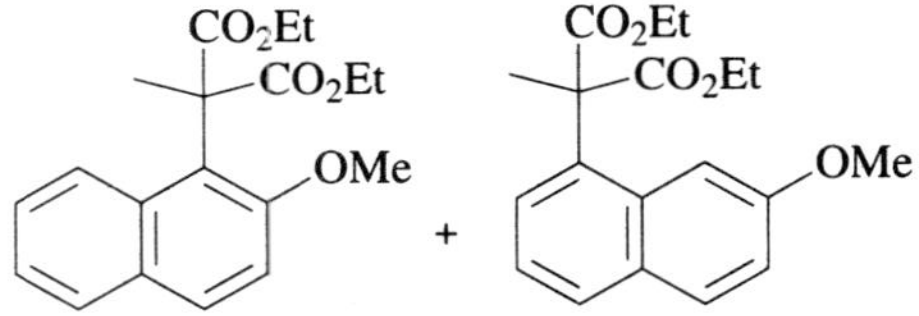 83 : 13 (52)	165

TABLE XXIA. ARENES AND CARBONYL COMPOUNDS (*Continued*)

Substrate	Reagent	Conditions	Product(s) and Yield(s) (%)	Refs.
C_{11}-C_{12} R = 2,7-(MeO)$_2$ R = 1-MeO R = 2-MeO R = 1,5-(MeO)$_2$ R = 1,3-(MeO)$_2$ R = 2,6-(MeO)$_2$ R = 1,7-(MeO)$_2$	Mn(acac)$_3$	Mn(acac)$_3$, AcOH, 80°, 2.5 h	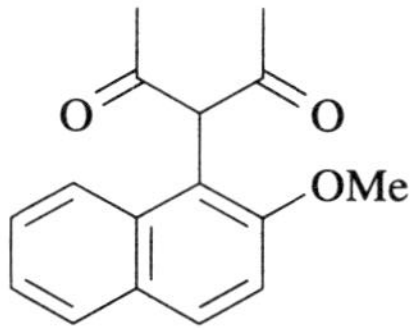 (48)	38
		Mn(OAc)$_3$, AcOH reflux, 1 min	R = 2,7-(MeO)$_2$ (55) R = 4-MeO (46) R = 2-MeO (33) R = 4,8-(MeO)$_2$ (58) R = 2,4-(MeO)$_2$ (53) R = 2,6-(MeO)$_2$ (34) R = 4,6-(MeO)$_2$ (26) R = 2,8-(MeO)$_2$ (22)	39
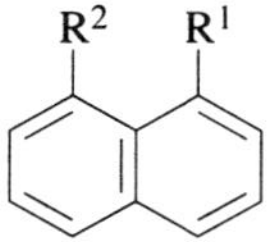	Mn(acac)$_3$	Mn(OAc)$_3$, AcOH, 100°, 2-25 min	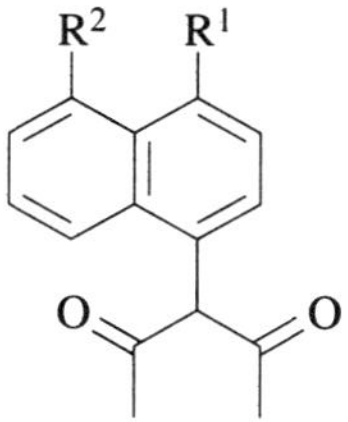	40

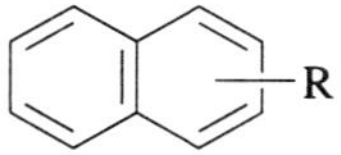

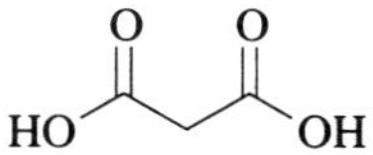

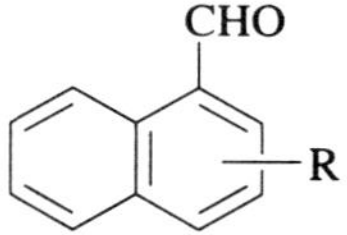

R^1	R^2	
MeO	H	(9)[c]
Me	Me	(30)
(CH$_2$)$_2$		(24)
Me	H	(15)

Substrate	Reagent	Conditions	Product(s) and Yield(s) (%)	Refs.
C_{13} 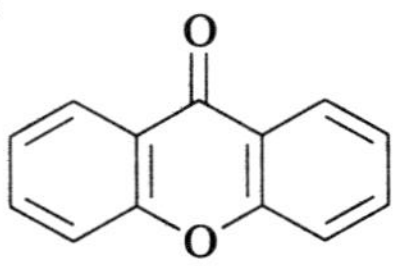	AcOH	Mn(OAc)₃, AcOH, Ac₂O, reflux, 40 min	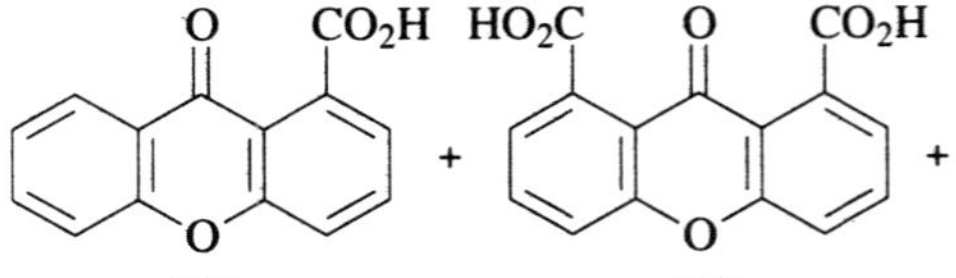 (30) + (12) 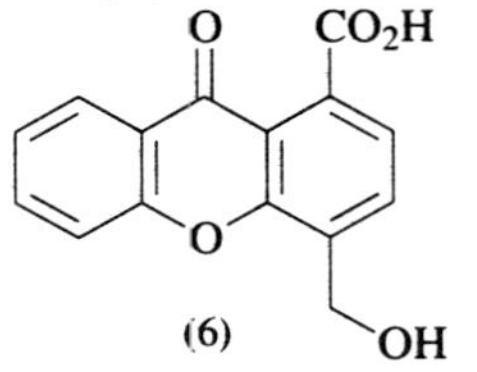	220
C_{16}-C_{23} 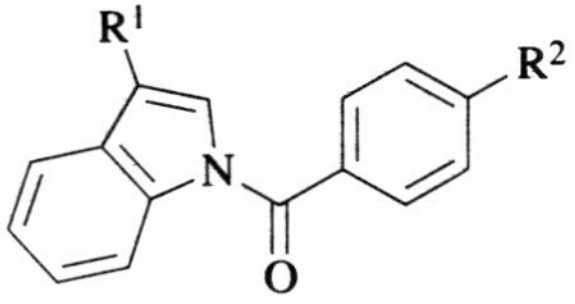	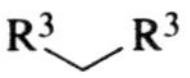 $R^3 = CO_2Me$	Mn(OAc)₃, AcOH, 80°, 24 h	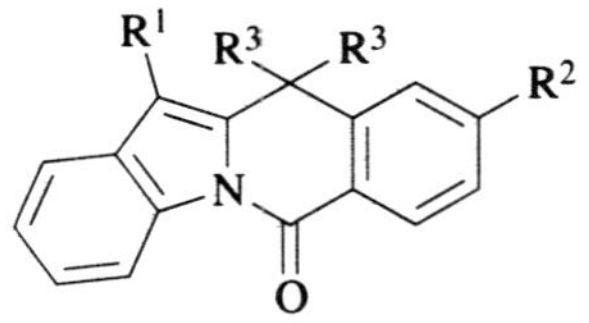	221

R^1	R^2	
CO_2Me	H	(83)
CO_2Me	H	(74)
CO_2Me	Ph	(81)
CN	H	(59)
CN	Me	(57)

R^1	R^2	
CN	OMe	(62)
COMe	H	(48)
COMe	Me	(52)
COMe	Ph	(42)

[a] Content (%) of *ortho* isomer.
[b] Ratio of *o* : *m* : *p* isomers.
[c] An α-overoxidation product was also formed (40%).

TABLE XXIIB. HETEROCYCLES AND CARBONYL COMPOUNDS

Substrate	Reagent	Conditions	Product(s) and Yield(s) (%)	Refs.
C₄				

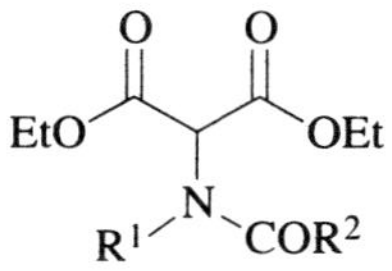

C_4

Substrate: furan

Reagent (diethyl malonate with R^1–N–COR^2):

R^1	R^2
H	Me
H	H
Me	Ph
	C_6H_4CO

Conditions: Mn(OAc)₃, 65°, 12 h

Product: furan–C(CO₂Et)(NR¹COR²)(CO₂Et)

(70)
(74)
(64)
(82)

Refs.: 107

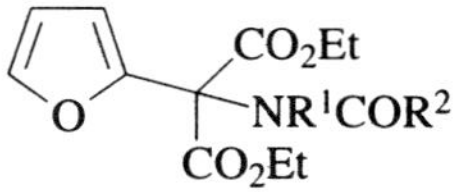

Substrate: thiophene

Reagent (diethyl malonate with R^1–N–COR^2):

R^1	R^2
H	Me
	C_6H_4CO

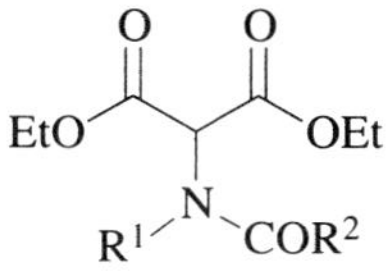

Conditions: Mn(OAc)₃, 65°, 12 h

Product: thiophene–C(CO₂Et)(NR¹COR²)(CO₂Et)

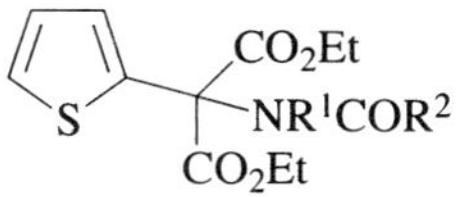

(60)
(85)

Refs.: 107

C_4–C_5

Substrate:

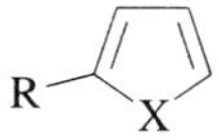

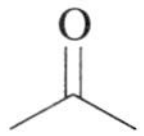

R	X
H	O
Me	O
H	S

Reagent: acetone

Conditions: Mn(OAc)₃, AcOH, 70 - 80°

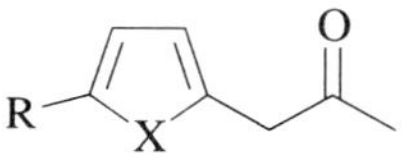

Product:

(35)
(50)
(40)

Refs.: 166

Substrate	Reagent	Conditions	Product(s) and Yield(s) (%)	Refs.

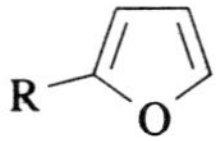

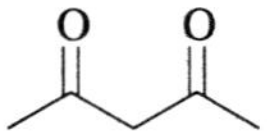

Substrate:

R = H
R = Me

Reagent: (diketone structure)

Conditions: Mn(OAc)$_3$, AcOH, 80°, 1 h

Products:

I + **II**

	I	**II**
R = H	(5)	(0)
R = Me	(14)	(15)

Refs: 166

Substrate:

C$_4$ R = H
C$_6$ R = COMe

Reagent: HC(CO$_2$Et)$_3$

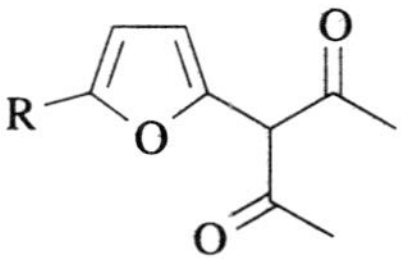

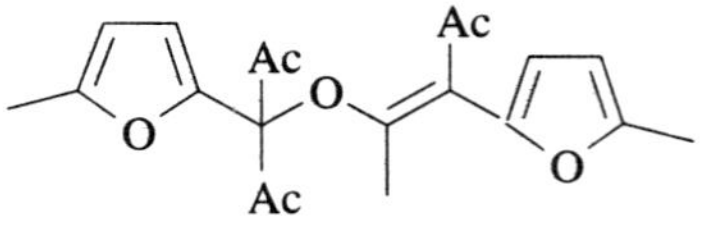

Conditions: Mn(OAc)$_3$, AcOH, AcONa, 60 - 65°, 1 d

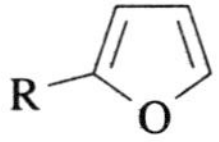

Product: R—(furan)—C(CO$_2$Et)$_3$

(92)
(48)

Refs: 37

Substrate:

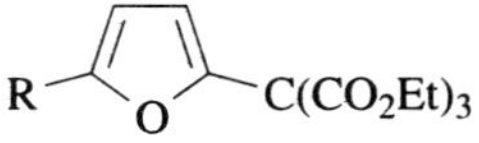

C$_4$ R = H
C$_6$ R = COMe

Reagent: HC(CO$_2$Et)$_3$

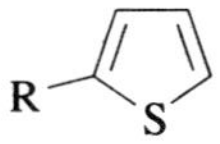

Conditions: Mn(OAc)$_3$, AcOH, AcONa, 60 - 65°, 1 d

Product: R—(thiophene)—C(CO$_2$Et)$_3$

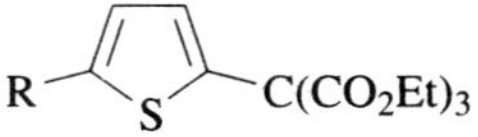

(55)
(53)

Refs: 37

TABLE XXIIB. HETEROCYCLES AND CARBONYL COMPOUNDS (Continued)

Substrate	Reagent	Conditions	Product(s) and Yield(s) (%)	Refs.
C_5-C_{13} 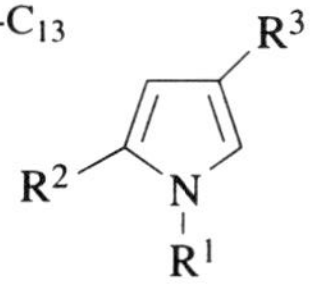	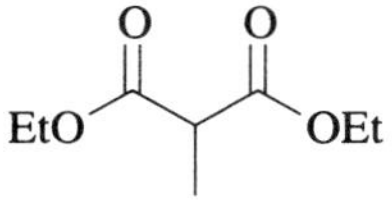	$Mn(OAc)_3$, AcOH, AcONa, 70°, 4 - 6 h	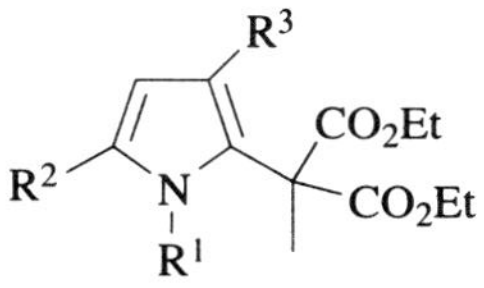	167

R^1	R^2	R^3	
H	CHO	H	(80)
Me	CO_2Me	H	(60)
Me	COMe	H	(64)
Me	$4\text{-}MeC_6H_4CO$	H	(70)
H	PhCO	H	(83)
Me	H	COMe	(19)
H	H	CO_2Me	(24)

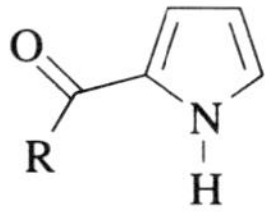	$HC(CO_2Et)_3$	$Mn(OAc)_3$, AcOH, AcONa, 60 - 65°, 1 d	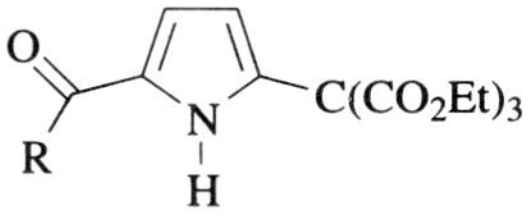	37

C_6 R = Me   (61)

C_{11} R = Ph   (86)

C_8 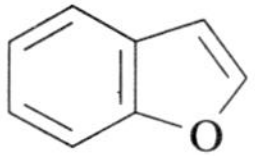	AcOH	$Mn(OAc)_3$, AcOH, Ac_2O, reflux	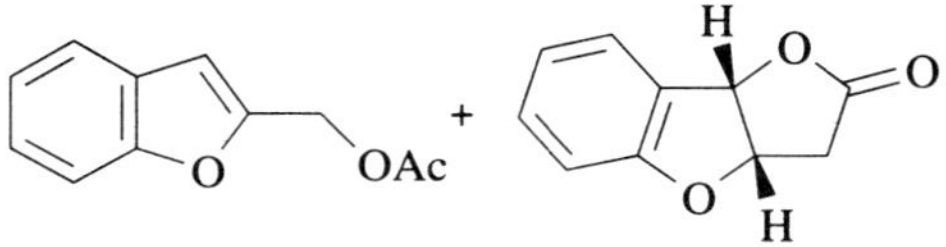	132

(21) + (42)

587

TABLE XXIB. HETEROCYCLES AND CARBONYL COMPOUNDS (*Continued*)

Substrate	Reagent	Conditions	Product(s) and Yield(s) (%)	Refs.

C$_{11}$

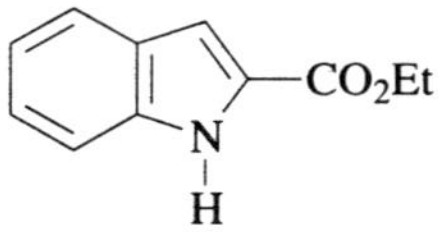

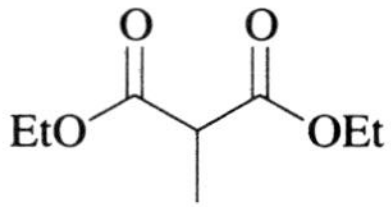

$Mn(OAc)_3$, AcOH, AcONa 70°, 4-6 h

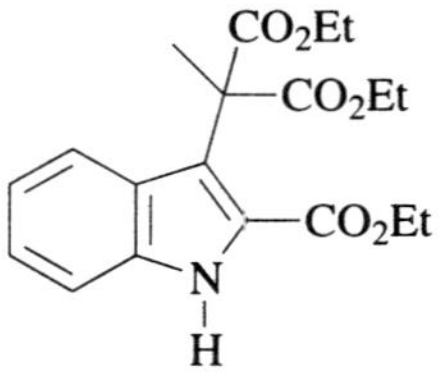

167

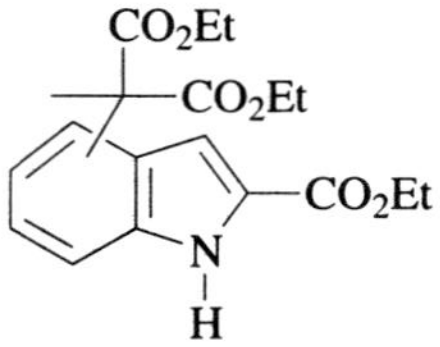

(64) + (8)

TABLE XXII. NITROALKYLATION REACTIONS

Substrate	Reagent	Conditions	Product(s) and Yield(s) (%)	Refs.
C$_5$				
cyclopentene	CH$_3$NO$_2$	Mn(OAc)$_3$, Cu(OAc)$_2$, AcOH, reflux, 1-4 h	cyclopentenyl-CH$_2$NO$_2$ (11)	79
C$_6$				
benzene	CH$_3$NO$_2$	Mn(OAc)$_2$, Pt anode, AcOH, LiBF$_4$, 83°, 3 h	C$_6$H$_5$CH$_2$NO$_2$ (73)	169
cyclohexene	CH$_3$NO$_2$	Mn(OAc)$_3$, Cu(OAc)$_2$, AcOH, reflux, 1-4 h	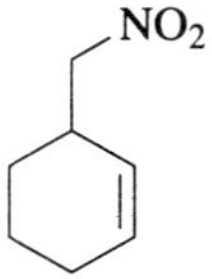 cyclohexenyl-CH$_2$NO$_2$ (38)	79
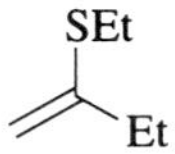 SEt / Et alkene	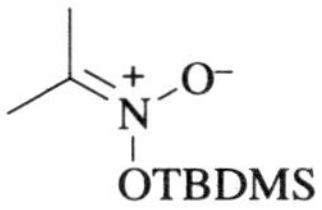 nitronate (N$^+$–O$^-$, OTBDMS)	Mn(pic)$_3$, DMF, rt	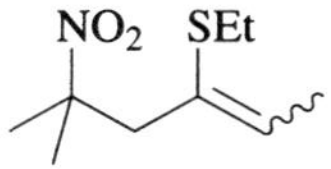 SEt / NO$_2$ product (30)	118
C$_6$-C$_7$ 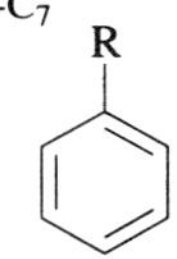			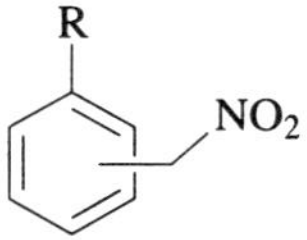	
R–C$_6$H$_4$	CH$_3$NO$_2$	Mn(OAc)$_3$, AcOH, 83°	R–C$_6$H$_4$CH$_2$NO$_2$ R = H (78)[a]; R = Me (66)[a]; R = OMe (77)[a]; R = Cl (20)[a]	36
R = H, R = Me		Mn(OAc)$_3$, AcOH, reflux	(14); (55)	168

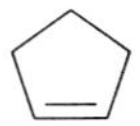

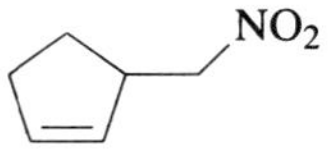

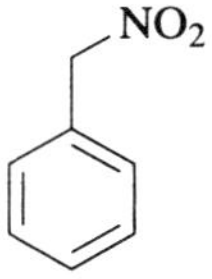

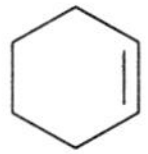

TABLE XXII. NITROALKYLATION REACTIONS (Continued)

Substrate	Reagent	Conditions	Product(s) and Yield(s) (%)	Refs.
C$_{10}$ 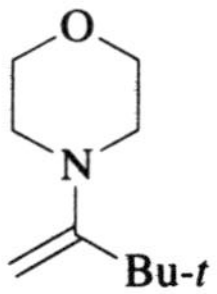 morpholine enamine, Bu-t	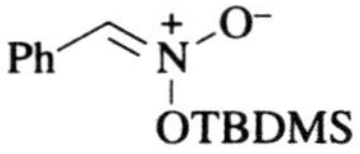 Ph–CH=$\overset{+}{N}$(O$^-$)–OTBDMS	Mn(pic)$_3$, DMF, rt	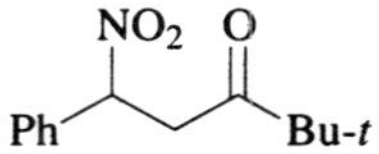 NO_2 … Bu-t, Ph (46) + morpholine–N… Bu-t, NO_2, Ph (7) + isoxazoline Ph, Bu-t, morpholine, O$^-$–$\overset{+}{N}$ (19)	118
C$_{14}$ 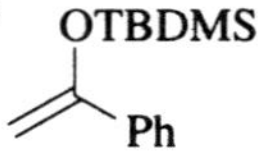 OTBDMS, Ph	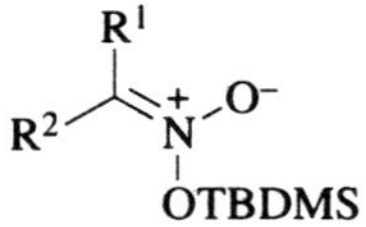 R^1, R^2 C=$\overset{+}{N}$(O$^-$)–OTBDMS	Mn(pic)$_3$, DMF, rt, 24 h	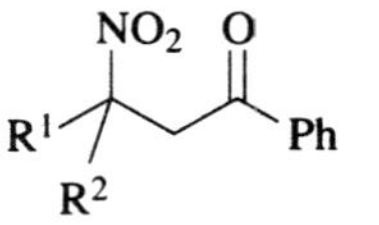 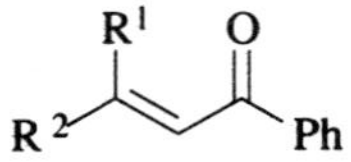 NO_2, R^1, Ph, R^2 + R^1 (O), Ph, R^2	118
C$_{15}$ OTBDMS, OBn	(CH$_3$)$_2$C=$\overset{+}{N}$(O$^-$)–OTBDMS	Mn(pic)$_3$, DMF, rt	NO_2 … OBn (46) + … OBn (12)	118

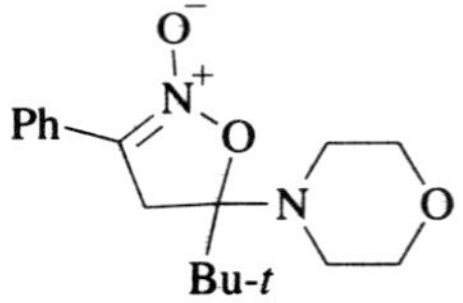

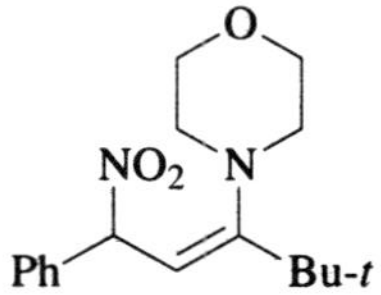

Reagent/product variants for C$_{14}$:

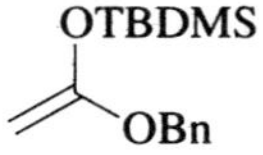

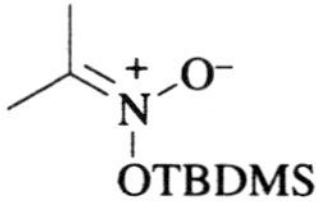

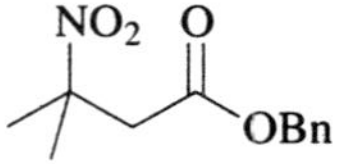

R^1	R^2	(middle product %)	(right product %)
Me	Me	(74)	(11)
Me	H	(0)	(69)
Ph	Me	(30)	(0)
Ph	H	(0)	(70)

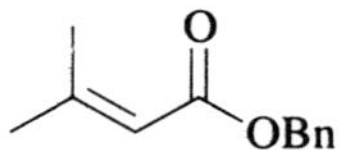

a The product was a mixture of o (60 - 70 %) and m and p (30 - 40 %) isomers.

TABLE XXIIIA. ADDITION-CYCLIZATION REACTIONS - ALKENES

Substrate	Reagent	Conditions	Product(s) and Yield(s) (%)	Refs.
C₄				
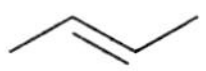 (2-butene)	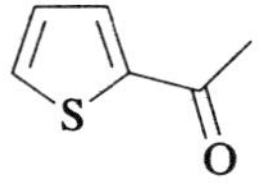 (1-(thiophen-2-yl)ethanone)	Mn(OAc)₃, AcOH, 85°	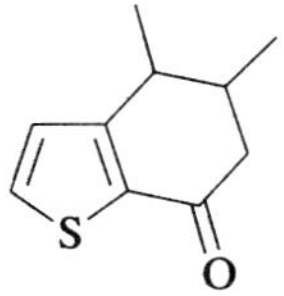 (43)	170
 (1-butene)	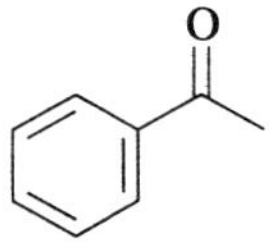 acetophenone	Mn(OAc)₃, AcOH, 85°	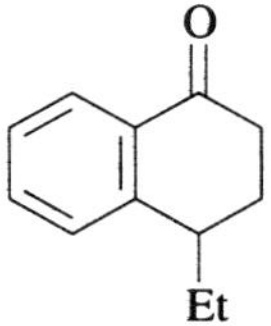 (49)	170
 (2-butene)	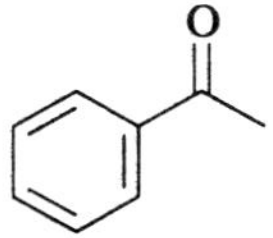 acetophenone	Mn(OAc)₃, AcOH, 85°	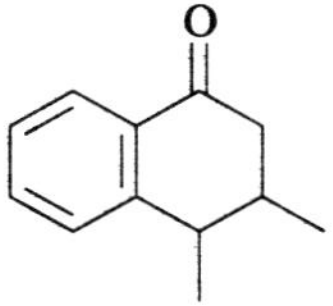 (40)	170
 (isobutylene)	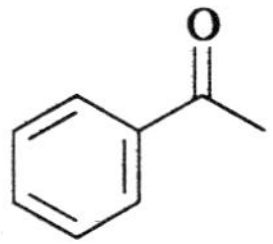 acetophenone	Mn(OAc)₃, AcOH, 85°	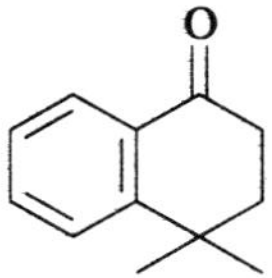 (43)	170
C₄-C₆				
$CH_2=CH-R$	$EtO_2C-CH(CO_2Et)$	Mn(OAc)₃, Cu(OAc)₂, AcOH, 75°, 24 h	R = OAc (40) R = n-Bu (35) R = CH₂TMS (89)	80

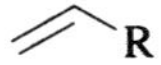

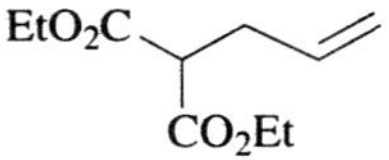

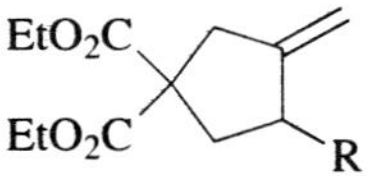

591

TABLE XXIIIA. ADDITION-CYCLIZATION REACTIONS - ALKENES (*Continued*)

Substrate	Reagent	Conditions	Product(s) and Yield(s) (%)	Refs.
C$_6$ 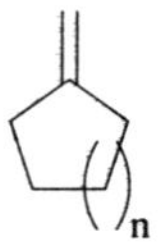	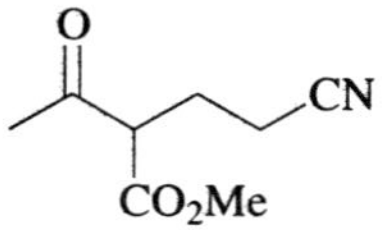		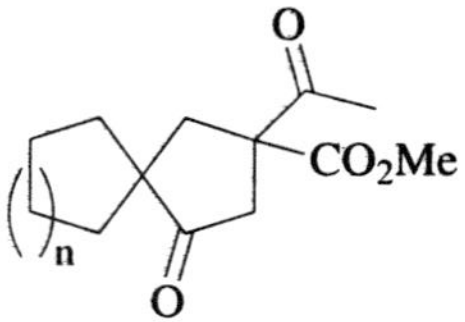	
(methylenecyclopentane)$_n$	CN, CO$_2$Me reagent		(acetyl-CO$_2$Me spiro product)	
n = 0		Mn(OAc)$_3$, EtOH, TFA, 25°, 18 h	(30)	171
n = 0		Mn(OAc)$_3$, Mn(OAc)$_2$, EtOH, TFA, 25°, 28 h	(51)	
n = 1		Mn(OAc)$_3$, EtOH, 25°, 6 h	(39)	
	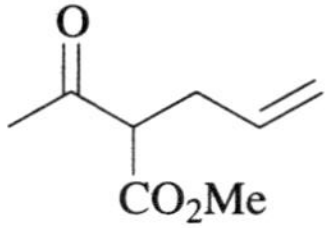 CO$_2$Me reagent (methyl ketone)	Mn(OAc)$_3$, Cu(OAc)$_2$, AcOH, 25°, 16 h	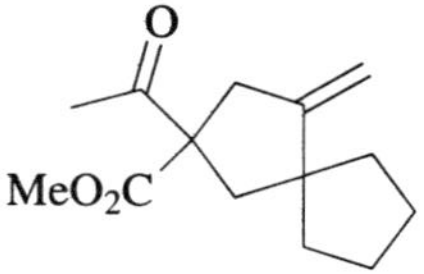 MeO$_2$C spiro product (75)	80
	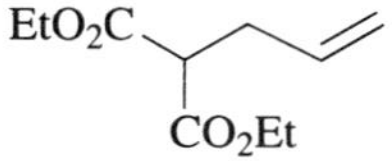 EtO$_2$C, CO$_2$Et reagent	Mn(OAc)$_3$, Cu(OAc)$_2$, AcOH, 75°, 24 h	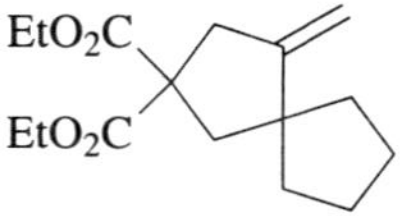 EtO$_2$C, EtO$_2$C spiro product (100)	80

TABLE XXIIIA. ADDITION-CYCLIZATION REACTIONS - ALKENES (*Continued*)

Substrate	Reagent	Conditions	Product(s) and Yield(s) (%)	Refs.
	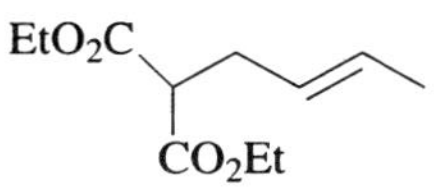 EtO$_2$C, CO$_2$Et	Mn(OAc)$_3$, Cu(OAc)$_2$, AcOH, 70°, 38 h	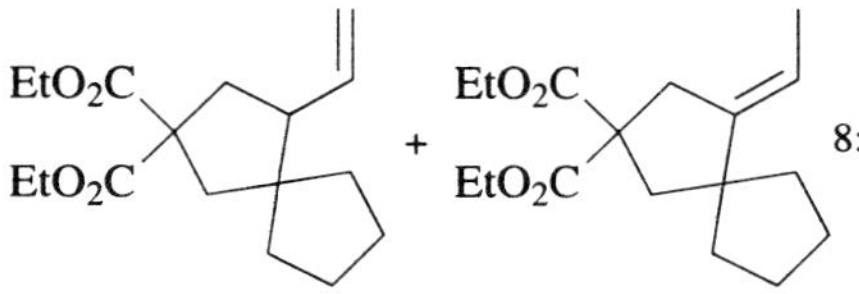 EtO$_2$C / EtO$_2$C (+) EtO$_2$C / EtO$_2$C 8:1 (49)	80
	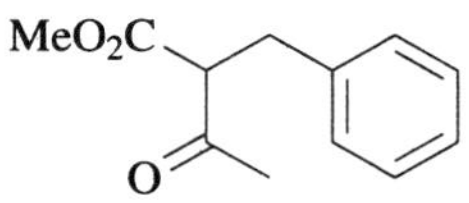 MeO$_2$C, O	Mn(OAc)$_3$, AcOH, 25°, 1 h	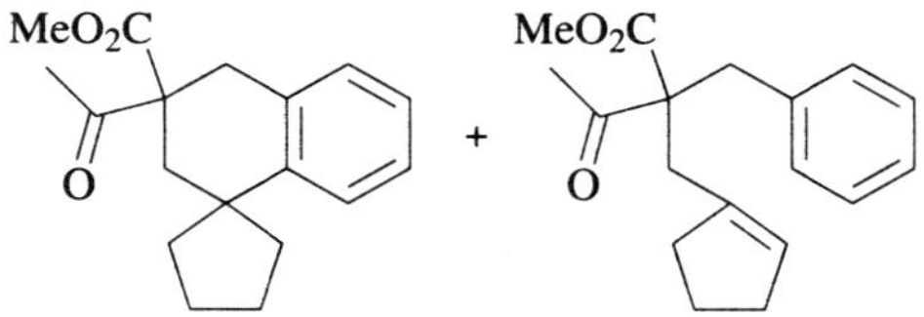 MeO$_2$C, O (+) MeO$_2$C, O	80
	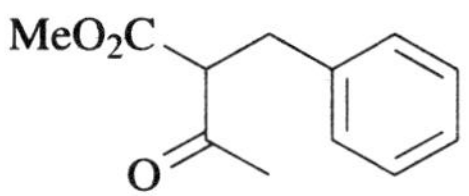 MeO$_2$C, O	Mn(OAc)$_3$, AcOH, 25°, 9 h	**I** (79) **I** 1.4:1 (77)	80
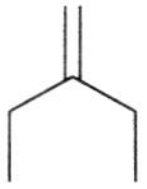	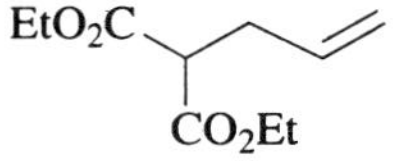 EtO$_2$C, CO$_2$Et	Mn(OAc)$_3$, Cu(OAc)$_2$, AcOH, 75°, 24 h	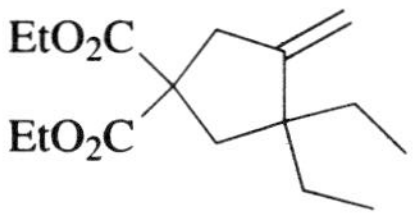 EtO$_2$C / EtO$_2$C (86)	80
C$_8$ 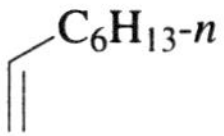 C$_6$H$_{13}$-n	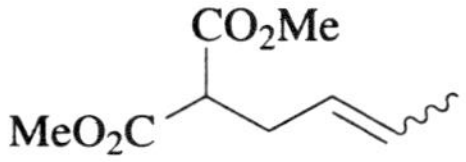 MeO$_2$C, CO$_2$Me / MeO$_2$C	Mn(OAc)$_3$, Cu(OAc)$_2$, AcOH, 90°, 6 h	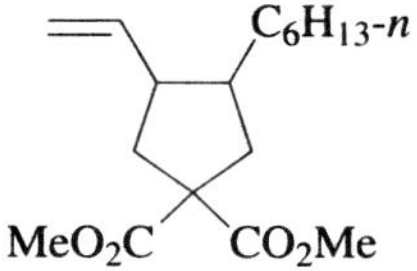 C$_6$H$_{13}$-n / MeO$_2$C, CO$_2$Me (81)	81

593

TABLE XXIIIA. ADDITION-CYCLIZATION REACTIONS - ALKENES (*Continued*)

Substrate	Reagent	Conditions	Product(s) and Yield(s) (%)	Refs.
	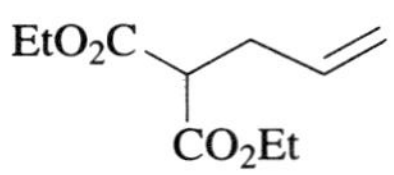 EtO$_2$C, CO$_2$Et (allyl)	Mn(OAc)$_3$, Cu(OAc)$_2$, AcOH, 90°, 6 h	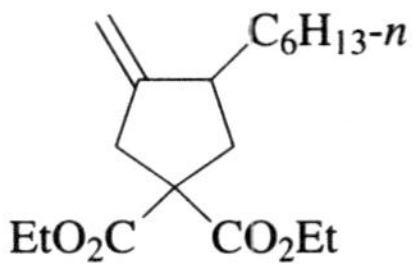 C$_6$H$_{13}$-n, EtO$_2$C, CO$_2$Et (80)	81
	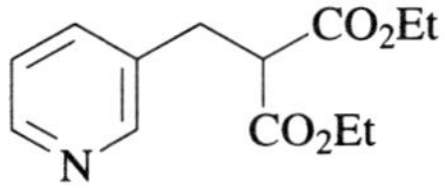 pyridine-CH$_2$CH(CO$_2$Et)CO$_2$Et	Mn(OAc)$_3$	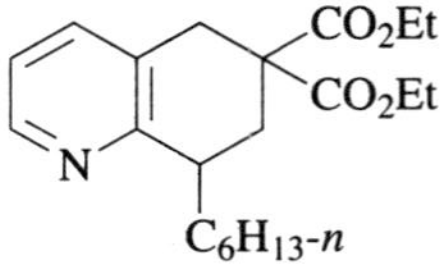 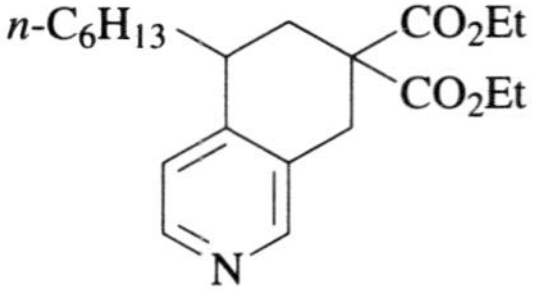 CO$_2$Et, CO$_2$Et, C$_6$H$_{13}$-n **I** + n-C$_6$H$_{13}$, CO$_2$Et, CO$_2$Et **II**	172
		AcOH, 70°, 8 h	**I + II** (95) **I:II** = 1.4:1	
		MeCN, CF$_3$CO$_2$H, 20°, 12 h	**I + II** (95) **I:II** = 1.1:1	
R^3, R^2, R^1 pyridine-CH$_2$CH(CO$_2$Et)CO$_2$Et	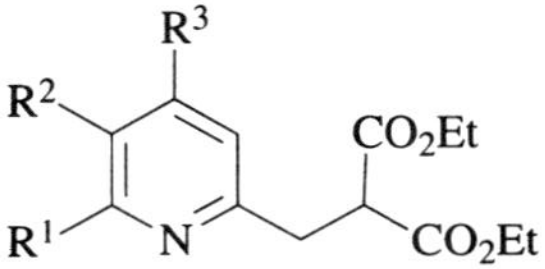	Mn(OH)$_3$, AcOH, 70°, 12 h	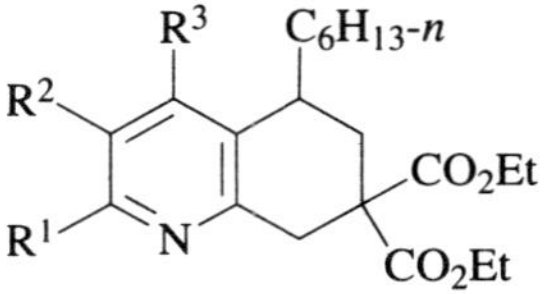 R^3, C$_6$H$_{13}$-n, CO$_2$Et, CO$_2$Et, R^2, R^1, N	172

R^1	R^2	R^3	
H	H	H	(90)
H	Et	H	(89)
Me	H	Me	(61)
CH=CHCH=CH		H	(85)

594

TABLE XXIIIA. ADDITION-CYCLIZATION REACTIONS - ALKENES (*Continued*)

Substrate	Reagent	Conditions	Product(s) and Yield(s) (%)	Refs.
	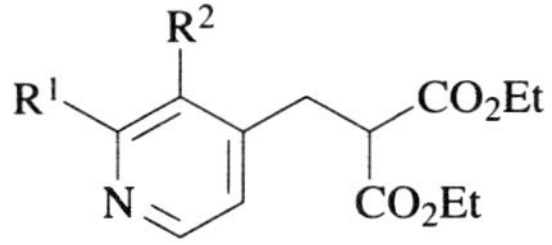	$Mn(OH)_3$, AcOH, 70°, 12 h	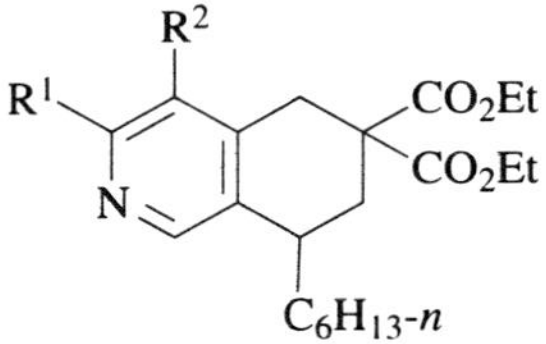	172
			(83)	
			(88)	
	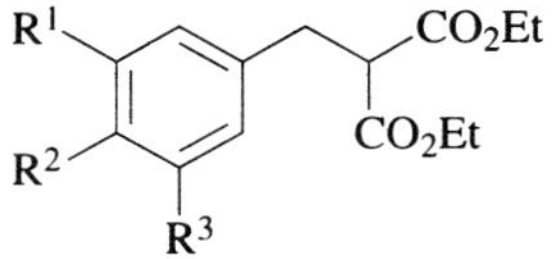	$Mn(OAc)_3$, AcOH, 60°, 12 h	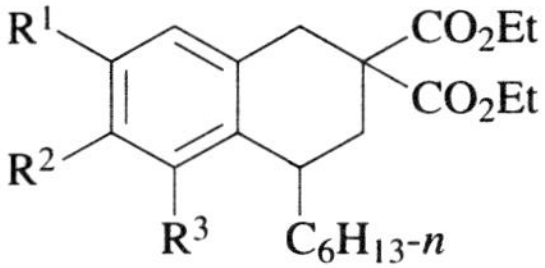	173

R^1	R^2	R^3	
H	H	H	(90)
H	Cl	H	(86)
H	F	H	(85)
H	CO_2Me	H	(85)
F	H	H	(89)
MeO	H	H	(83)
NO_2	H	H	(80)
Me	H	H	(85)

| | 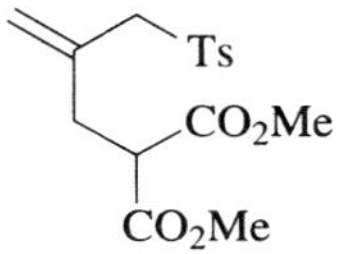 | $Mn(OAc)_3$, $Cu(OAc)_2$, AcOH, 90°, 6 h | 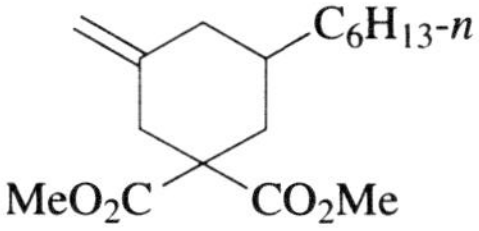 (35) | 81 |

595

TABLE XXIIIA. ADDITION-CYCLIZATION REACTIONS - ALKENES (*Continued*)

Substrate	Reagent	Conditions	Product(s) and Yield(s) (%)	Refs.
 C_8 $\quad$ R = n-C$_6$H$_{13}$ $\quad$ C_9 $\quad$ R = CH$_2$Ts	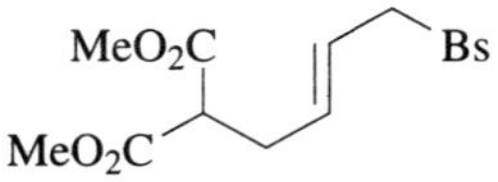 (MeO$_2$C)$_2$CH–CH$_2$–CH=CH–CH$_2$Bs	Mn(OAc)$_3$, Cu(OAc)$_2$, AcOH, 90°, 6 h	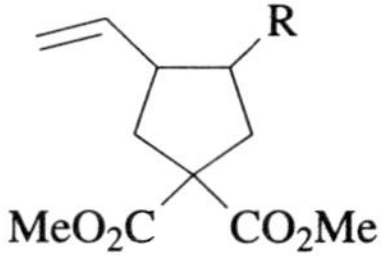 (62) (77)	81
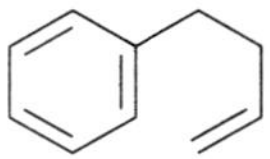 C_{10}	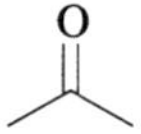 acetone (O)	Mn(OAc)$_3$, AcOH	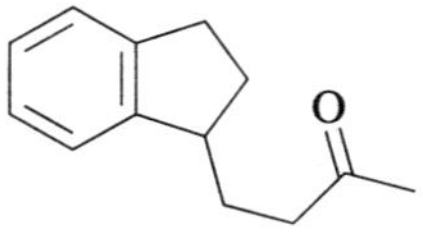 (—)	170
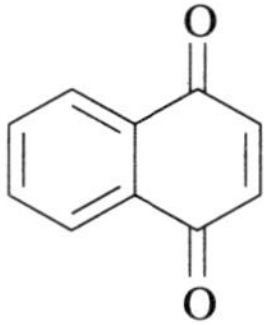	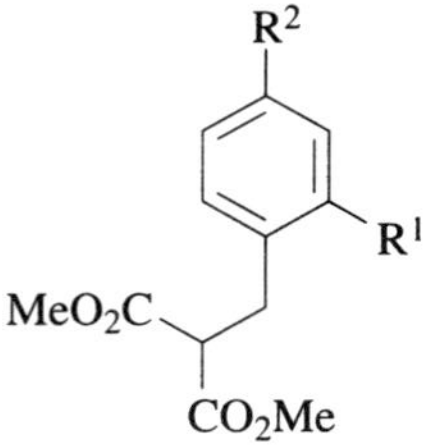	Mn(OAc)$_3$, AcOH, 80°, 8 h	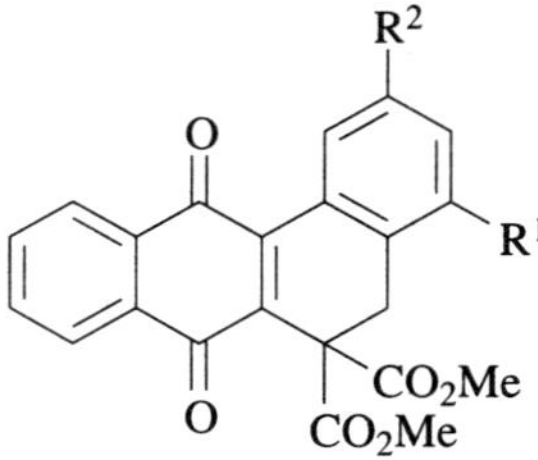	177

R^1	R^2	
H	H	(59)^a
H	Me	(51)
H	OMe	(58)
H	Br	(51)
Me	Me	(35)
Me	H	(32)

TABLE XXIIIA. ADDITION-CYCLIZATION REACTIONS - ALKENES (*Continued*)

Substrate	Reagent	Conditions	Product(s) and Yield(s) (%)	Refs.
C_{11}				
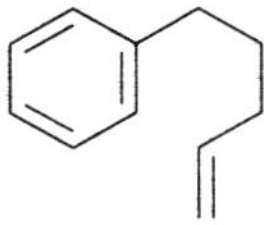	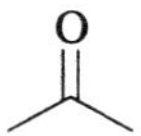	$Mn(OAc)_3$, AcOH	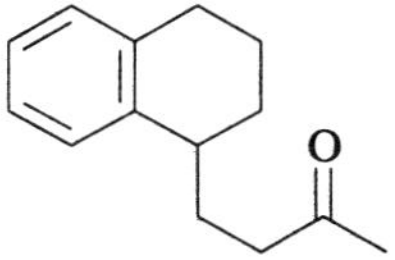 (70)	170
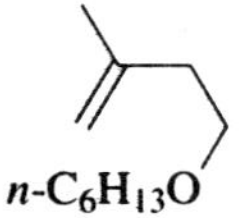	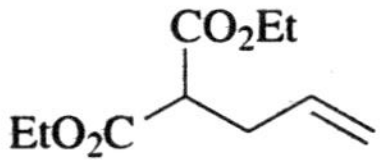	$Mn(OAc)_3$, $Cu(OAc)_2$, AcOH, 90°, 6 h	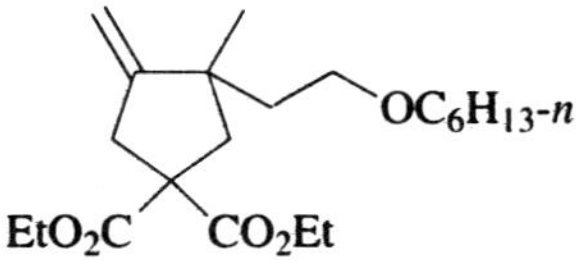 (60)	81
	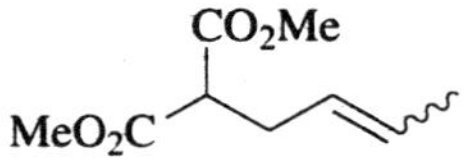	$Mn(OAc)_3$, $Cu(OAc)_2$, AcOH, 90°, 6 h	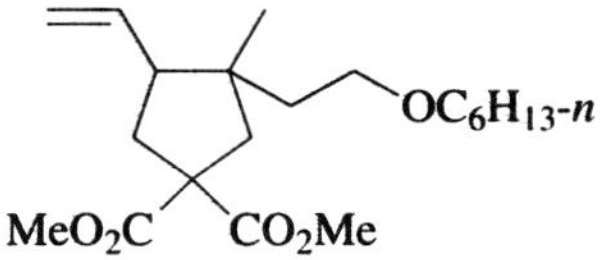 (60)	81

[a] Analogous results were obtained with 1,4-dihydroquinones as substrates.

TABLE XXIIIB. ADDITION-CYCLIZATION REACTIONS - ALKYNES

Substrate	Reagent	Conditions	Product(s) and Yield(s) (%)	Refs.
RC≡CH	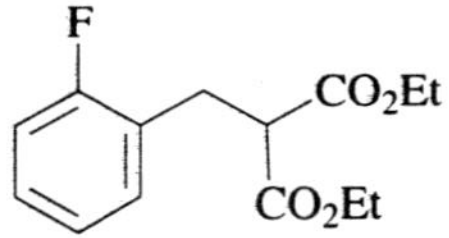	Mn(OAc)₃, AcOH	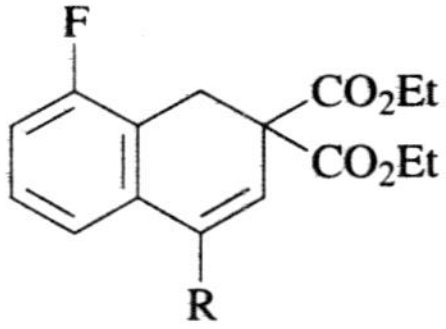 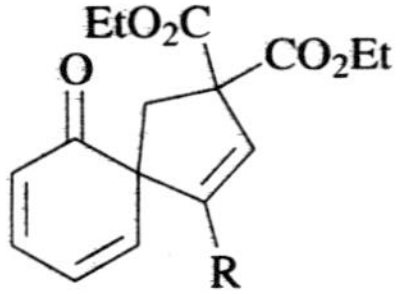	174
			C₃ R = CH₂OH (12), (65)	
			C₈ R = n-C₆H₁₃ (6), (77)	
R¹C≡CH C₃-C₈	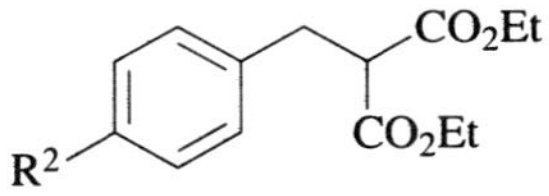	Mn(OAc)₃, AcOH	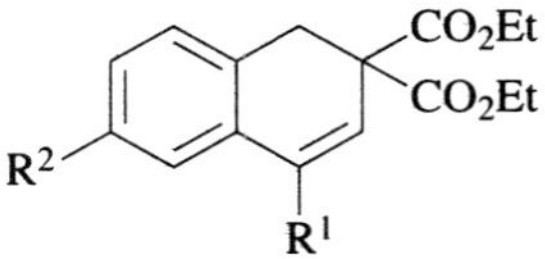 I 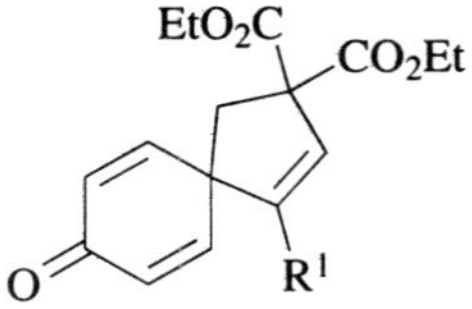 + II	174

R¹	R²	I	II
n-C₆H₁₃	OMe	(10)	(79)
CH₂OH	OMe	(13)	(48)
n-C₆H₁₃	F	(28)	(66)
TMS	F	(32)	(62)
Ph	F	(13)	(69)
Ph	i-Pr	(85)	(0)
Ph	H	(92)	(0)

TABLE XXIIIB. ADDITION-CYCLIZATION REACTIONS - ALKYNES (*Continued*)

Substrate	Reagent	Conditions	Product(s) and Yield(s) (%)	Refs.
C_3-C_{14} $R^1C{\equiv}CR^2$	(diethyl benzylmalonate with CO_2Et, CO_2Et)	$Mn(OAc)_3$, AcOH, 70°, 6-24 h	(naphthalene with CO_2Et, CO_2Et, R^2, R^1)	30

R^1	R^2	
n-C_6H_{13}	H	(92)
CO_2Et	H	(50)
TMS	H	(91)
Ph	H	(95)
CH_2OH	H	(70)
Ph	Ph	(67)
Et	Et	(19)
CO_2Et	CO_2Et	(20)
CO_2Me	Me	(12)
n-Pr	n-Pr	(23)

Substrate	Reagent	Conditions	Product(s) and Yield(s) (%)	Refs.
C_5-C_8 $R^1C{\equiv}CH$	(with CO_2Et, CO_2Et, R^2)	$Mn(OAc)_3$, AcOH	I (with CO_2Et, CO_2Et, R^1, R^2) + II (EtO_2C, CO_2Et, R^1, AcO, R^2)	174

R^1	R^2	I	II
n-C_6H_{13}	CF_3	(65)	(12)
TMS	CF_3	(41)	(6)
Ph	CF_3	(63)	(18)
Ph	CO_2Me	(60)	(31)

TABLE XXIIIB. ADDITION-CYCLIZATION REACTIONS - ALKYNES (Continued)

Substrate	Reagent	Conditions	Product(s) and Yield(s) (%)	Refs.
RC≡CH	p-tolyl-CH$_2$-CH(CO$_2$Et)(CO$_2$Et)	Mn(OAc)$_3$, AcOH	**I** (naphthalene diester) + **II** R / I / II: TMS (37) (52) Ph (23) (67) n-C$_6$H$_{13}$ (41) (55)	174
C$_8$ HC≡CR	3-pyridyl-CH$_2$-CH(CO$_2$Et)(CO$_2$Et)	Mn(OAc)$_3$, AcOH, 60°, 12 h	(quinoline diester) + (isoquinoline diester) R: n-C$_6$H$_{13}$ 1.8 : 1 (80) Ph 1.9 : 1 (74)	172

600

TABLE XXIIIC. ADDITION-CYCLIZATION REACTIONS - ALKADIENES

Substrate	Reagent	Conditions	Product(s) and Yield(s) (%)	Refs.
C$_6$ 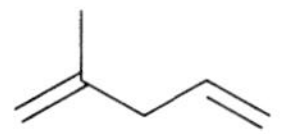	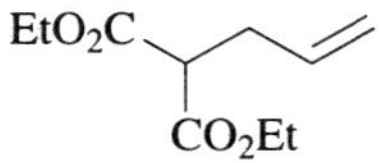 EtO$_2$C CO$_2$Et	Mn(OAc)$_3$, Cu(OAc)$_2$, AcOH, 75°, 24 h	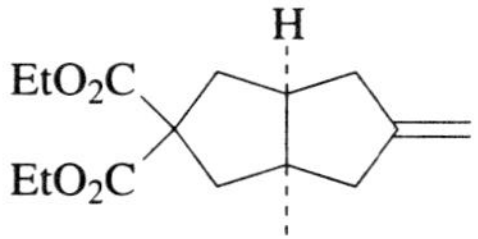 EtO$_2$C EtO$_2$C (12) + EtO$_2$C EtO$_2$C (7)	80
C$_7$ 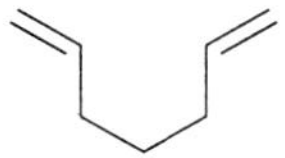	AcOH	Mn(OAc)$_3$, AcOH, AcOK, reflux	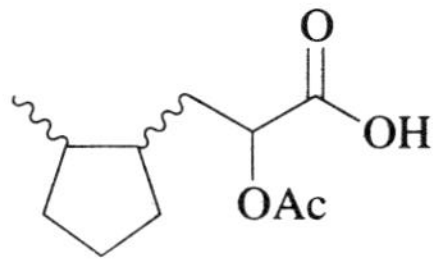 (12) + (48) 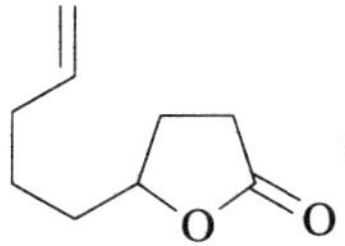 + (13)	117
C$_{10}$	O R R = Me R = Ph	Mn(OAc)$_3$ reflux, 24 h dioxane, 12 h	R (—) (—)	82

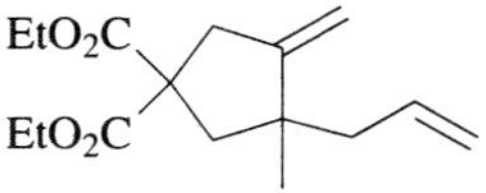

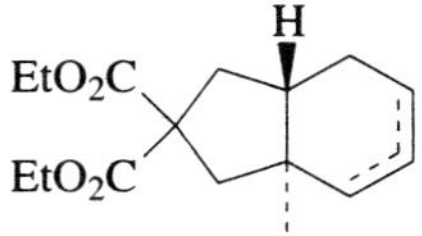

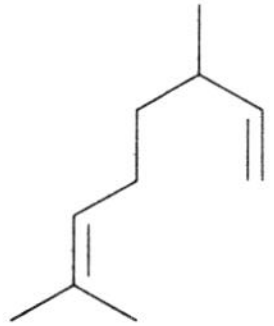

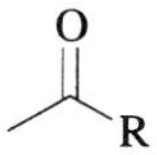

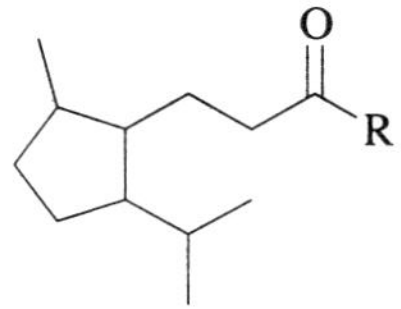

601

TABLE XXIIIC. ADDITION-CYCLIZATION REACTIONS - ALKADIENES (*Continued*)

Substrate	Reagent	Conditions	Product(s) and Yield(s) (%)	Refs.
	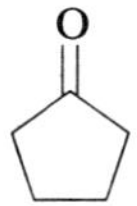	Mn(OAc)$_3$, Cu(OAc)$_2$, AcOH, 110°, 15 min	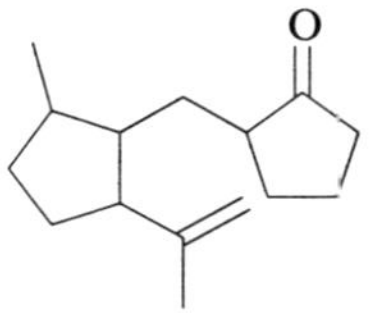 (—)	82
	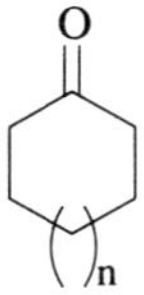	Mn(OAc)$_3$, dioxane	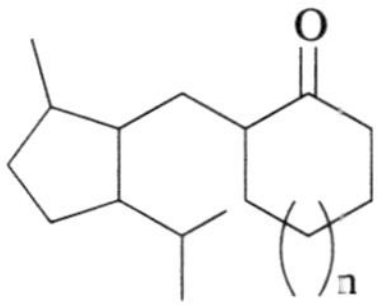	82
	n = 0	110°, 30 h,	(—)[a]	
	n = 1	12 h	(57)[a]	
	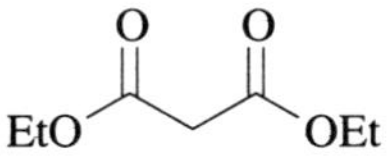	Mn(OAc)$_3$, Cu(OAc)$_2$, AcOH, reflux, 10 min	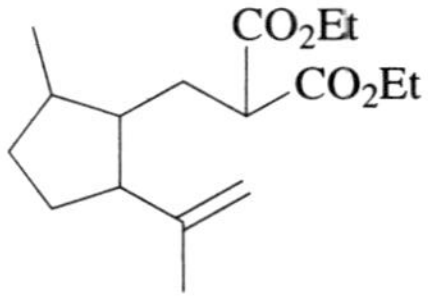 (—)	82

[a] An α,β-unsaturated enone was also formed.

TABLE XXIV. INTRAMOLECULAR CYCLIZATIONS OF 2-SUBSTITUTED 3-KETOESTERS

Substrate	Conditions	Product(s) and Yield(s) (%)	Refs.
C_{12}	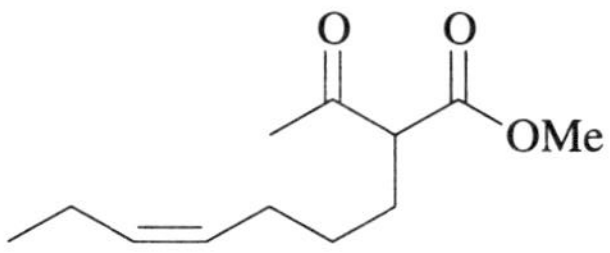 Mn(OAc)$_3$, Cu(OAc)$_2$, AcOH, 50°, 1 h	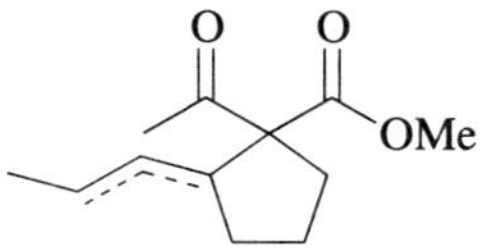 (67)	83
C_{14}	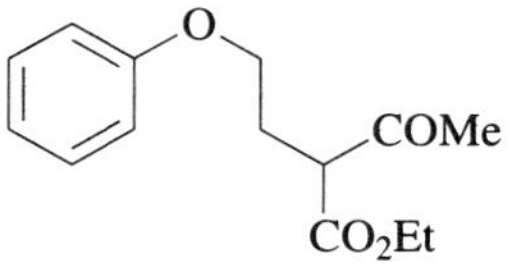 Mn(OAc)$_3$, AcOH, AcONa, 70-80°, 3-4 h	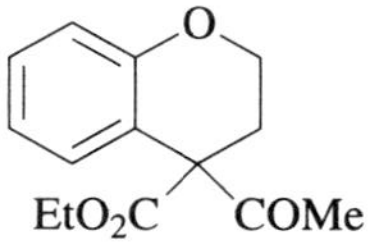 (92)	181

Substrate	Conditions	Product(s) and Yield(s) (%)	Refs.
C$_8$ 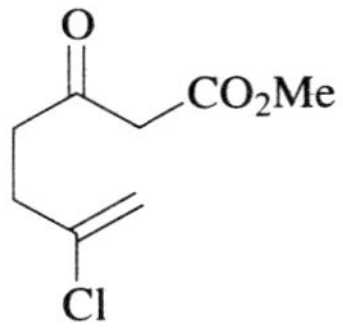 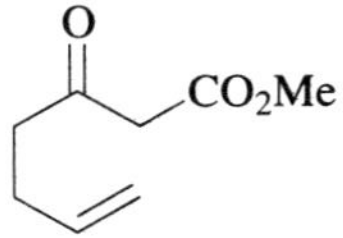	Mn(OAc)$_3$, Cu(OAc)$_2$, AcOH	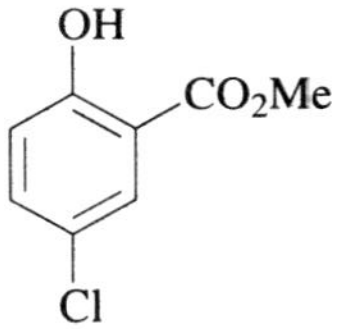 (10) 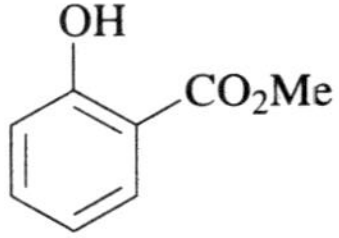	84
	Mn(OAc)$_3$, Cu(OAc)$_2$, AcOH, AcOK, 50°, 50 min	(94)	89
	Mn(OAc)$_3$, Cu(OAc)$_2$, AcOH	(78)	84, 90
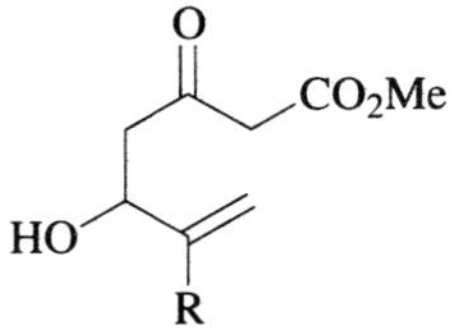 C$_8$ R = H C$_9$ R = Me	Mn(OAc)$_3$, AcOH, 25°, 3h	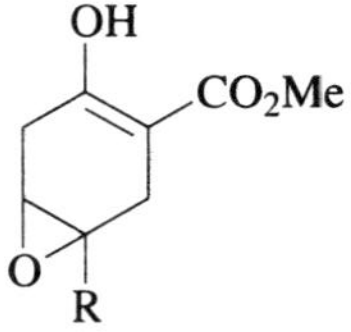 (30) (50)	84
C$_9$ 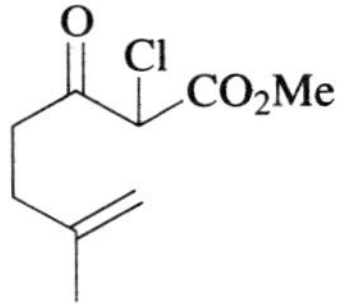	1. Mn(OAc)$_3$, Cu(OAc)$_2$, AcOH 2. AcOH, TFA, 120°, 10 h	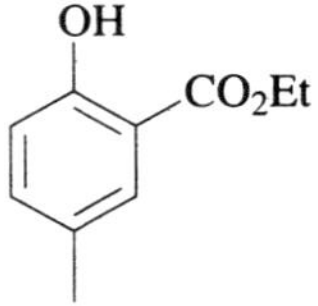 (71)	84

604

TABLE XXVA. INTRAMOLECULAR CYCLIZATIONS OF 4-SUBSTITUTED 3-KETOESTERS (D-MODE) (*Continued*)

Substrate	Conditions	Product(s) and Yield(s) (%)	Refs.
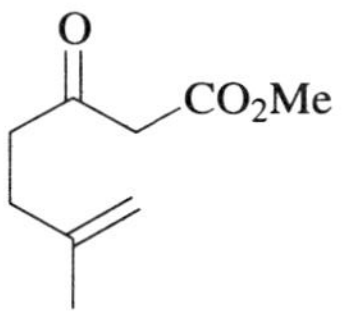	1. Mn(OAc)$_3$, LiCl, AcOH 2. AcOH, LiCl, 100°, 24 h	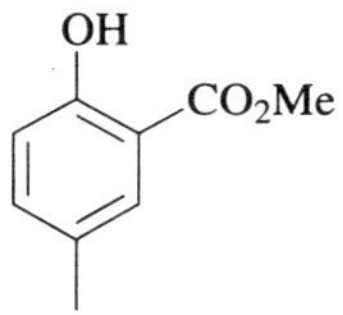 (70)	84
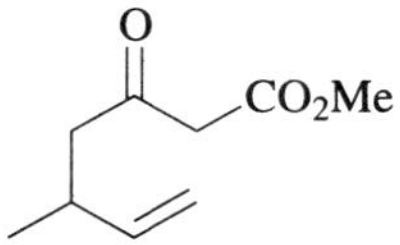	Mn(OAc)$_3$, Cu(OAc)$_2$, AcOH	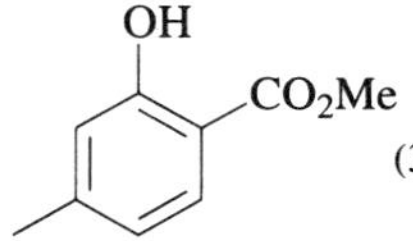 (38)	84
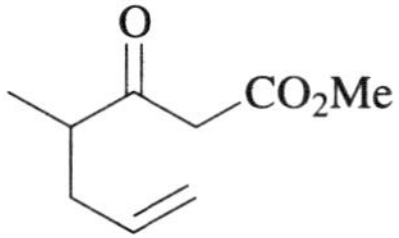	Mn(OAc)$_3$, Cu(OAc)$_2$, AcOH	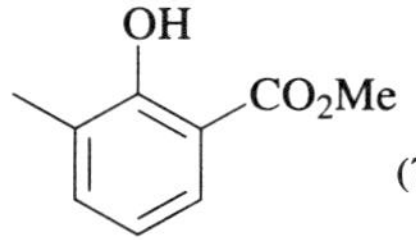 (78)	84
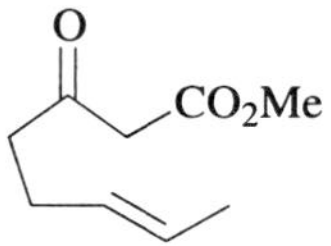	Mn(OAc)$_3$, Cu(OAc)$_2$, AcOH, 60°, 1 h	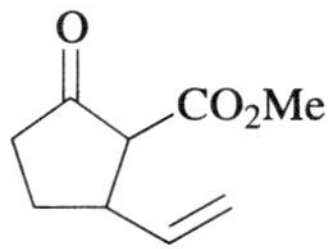 (21)	85
C$_{10}$	Mn(OAc)$_3$, Cu(OAc)$_2$, AcOH, 25°	(50) + (18)	88

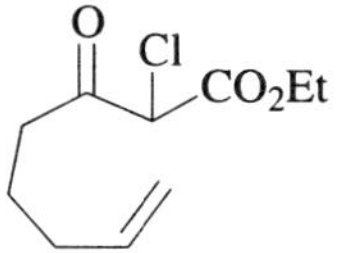

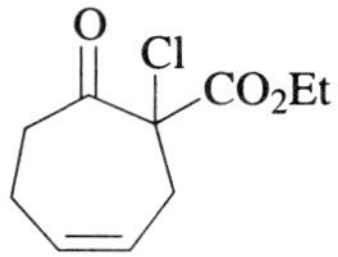

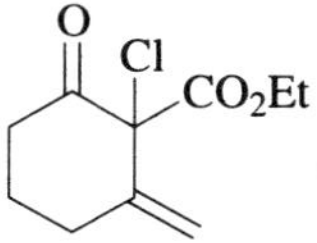

605

TABLE XXVA. INTRAMOLECULAR CYCLIZATIONS OF 4-SUBSTITUTED 3-KETOESTERS (D-MODE) (*Continued*)

Substrate	Conditions	Product(s) and Yield(s) (%)	Refs.
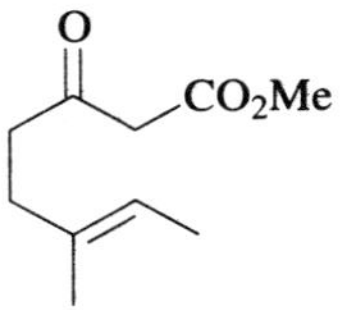 (structure, CO_2Me)	1. Mn(OAc)$_3$, LiCl, AcOH 2. AcOH, LiCl, 100°, 24 h	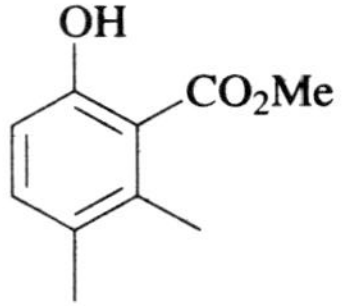 (structure, OH, CO_2Me) (40)	84
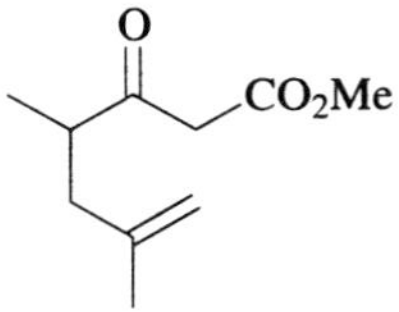 (structure, CO_2Me)	1. Mn(OAc)$_3$, LiCl, AcOH 2. AcOH, LiCl, 100°, 24 h	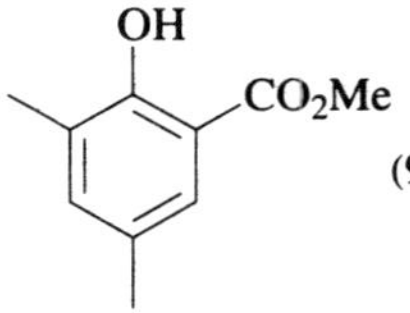 (structure, OH, CO_2Me) (91)	84
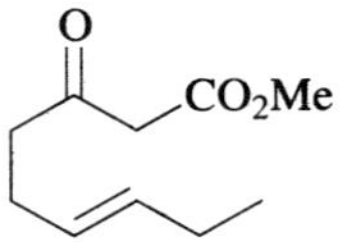 (structure, CO_2Me)	Mn(OAc)$_3$, Cu(OAc)$_2$, AcOH, 60°, 1 h	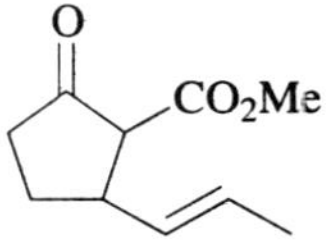 (structure, CO_2Me) (36) 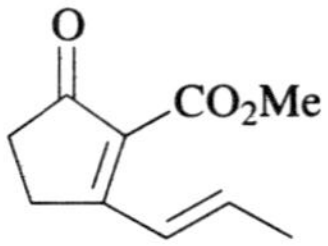 + (structure, CO_2Me) (10)	85
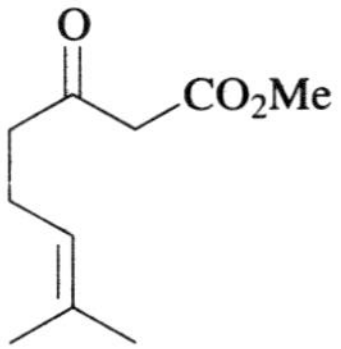 (structure, CO_2Me)	Mn(OAc)$_3$, Cu(OAc)$_2$, AcOH, 60°, 1 h	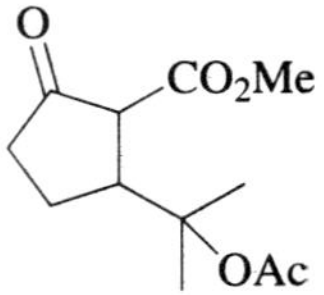 (structure, CO_2Me, OAc) (10) 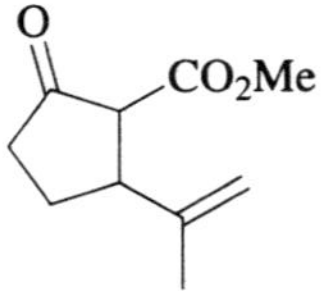 + (structure, CO_2Me) (8)	85
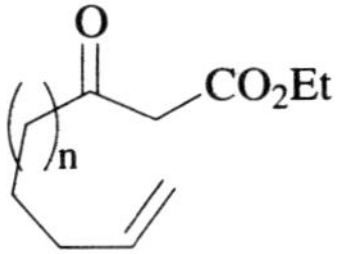 (structure, CO_2Et) C_{10} n = 1 C_{11} n = 2	Mn(OAc)$_3$, Cu(OAc)$_2$, AcOH, 25°, 44 h	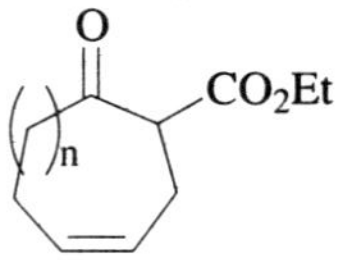 (structure, CO_2Et, n) (13) (17)	88

TABLE XXVA. INTRAMOLECULAR CYCLIZATIONS OF 4-SUBSTITUTED 3-KETOESTERS (D-MODE) (*Continued*)

Substrate	Conditions	Product(s) and Yield(s) (%)	Refs.
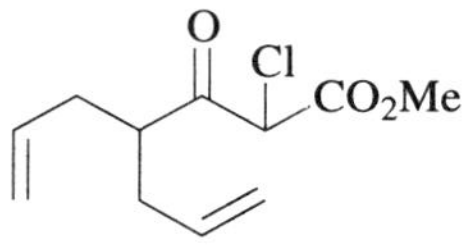 C₁₁	Mn(OAc)₃, Cu(OAc)₂, AcOH, rt, 30 h	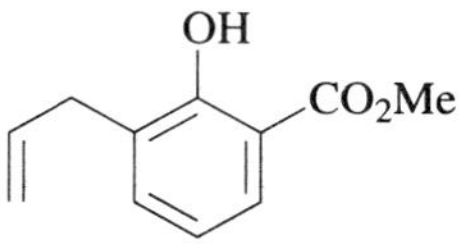 (24)	86
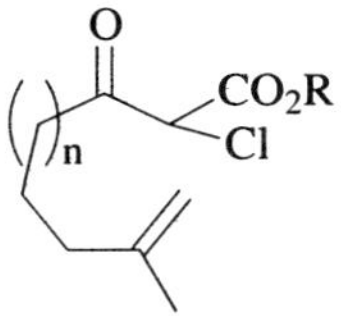 n / R 1 / Et 2 / Me	Mn(OAc)₃, Cu(OAc)₂, AcOH, 25°	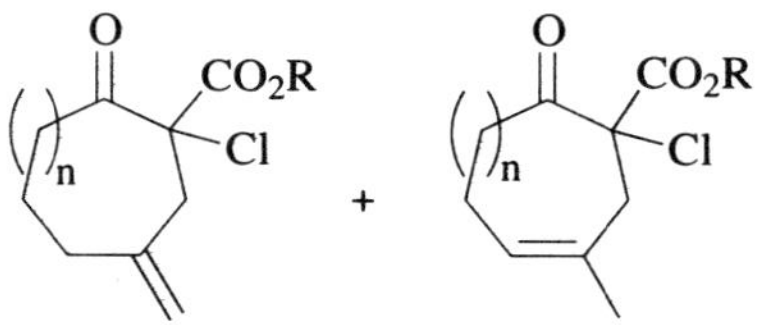 (20)/(49) + (49)/(20)	87
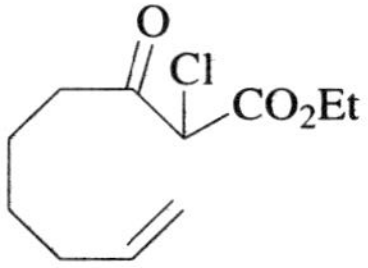	Mn(OAc)₃, Cu(OAc)₂, AcOH, 25°	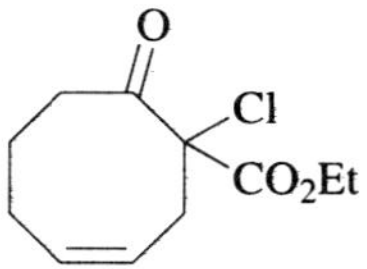 (47)	88
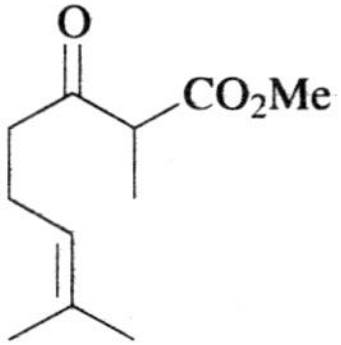	Mn(OAc)₃, Cu(OAc)₂, AcOH, 60°, 1 h	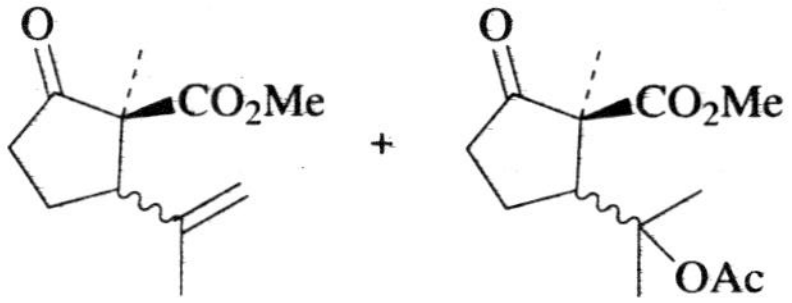 β : α 3 : 2 (20) + β : α 3 : 2 (27)	85

TABLE XXVA. INTRAMOLECULAR CYCLIZATIONS OF 4-SUBSTITUTED 3-KETOESTERS (D-MODE) (*Continued*)

Substrate	Conditions	Product(s) and Yield(s) (%)	Refs.
(structure, CO$_2$Et)	Mn(OAc)$_3$, Cu(OAc)$_2$, AcOH	(structure, CO$_2$Et) (29) + (structure, CO$_2$Et) (11)	90
(structure, CO$_2$Me, HO)	Mn(OAc)$_3$, Cu(OAc)$_2$, AcOH, 60°, 1 h	(structure, CO$_2$Me, HO) α-OH (41) β-OH (9)	85
(structure, CO$_2$Me)	Mn(OAc)$_3$, Cu(OAc)$_2$, AcOH, 60°, 1 h	(structure, CO$_2$Me) (41)	85
(structure, CO$_2$Me) Z, E	Mn(OAc)$_3$, Cu(OAc)$_2$, AcOH	(structure, CO$_2$Me) (71), (64)	83
(structure, CO$_2$Et)	Mn(OAc)$_3$, Cu(OAc)$_2$, AcOH, 25°, 4 d	(structure, CO$_2$Et) + (structure, CO$_2$Et, OAc) (7) β : α 2.5 : 1 (39)	83

50°, 1 h
25°, 26 h

TABLE XXVA. INTRAMOLECULAR CYCLIZATIONS OF 4-SUBSTITUTED 3-KETOESTERS (D-MODE) (Continued)

Substrate	Conditions	Product(s) and Yield(s) (%)	Refs.
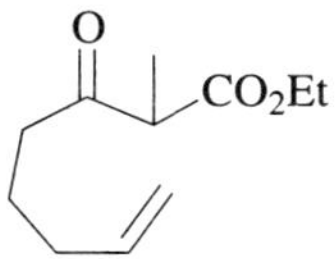 [structure: ketoester with CO₂Et]	Mn(OAc)₃, Cu(OAc)₂, AcOH, 25°	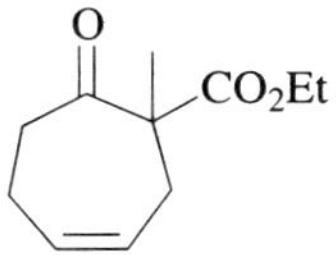 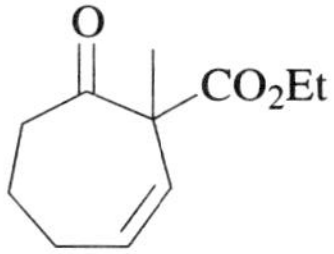 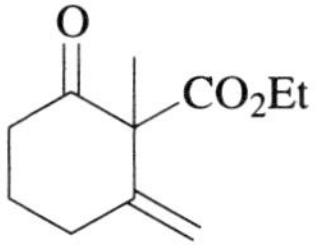 [CO₂Et product] + [CO₂Et product] + [CO₂Et cyclohexanone] + [bicyclic lactone] (5) 3 : 1 : 1 (46) 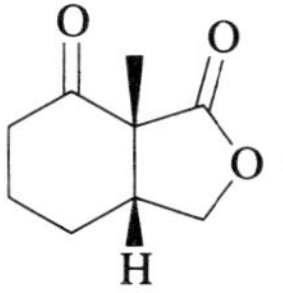 +	88, 87
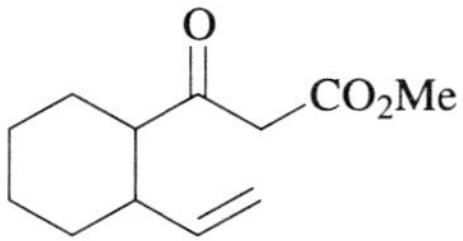 [CO₂Me vinyl cyclohexane]	Mn(OAc)₃, Cu(OAc)₂, AcOH	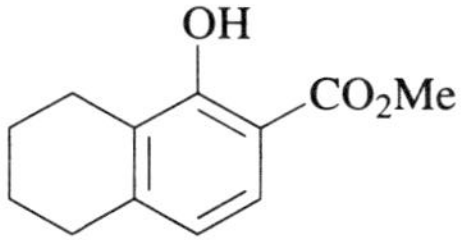 [OH CO₂Me naphthalenol] (46)	80
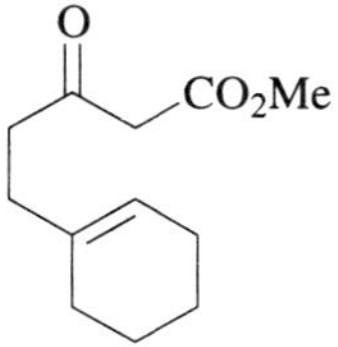 C₁₂ [CO₂Me cyclohexene]	1. Mn(OAc)₃, LiCl, AcOH 2. AcOH, LiCl, 100°, 24 h	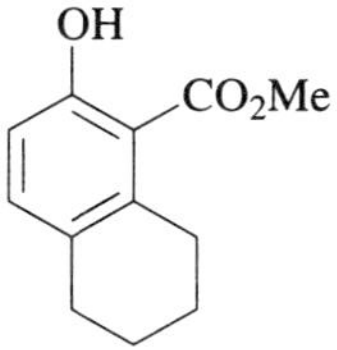 [OH CO₂Me naphthalenol] (37)	80
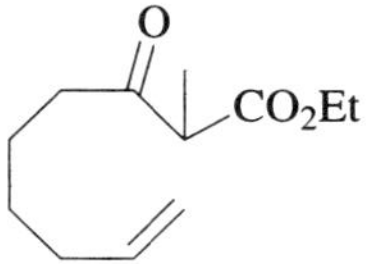 [CO₂Et ketoester]	Mn(OAc)₃, Cu(OAc)₂, AcOH, 25°	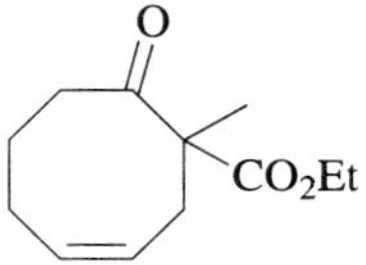 [CO₂Et cyclooctenone] (38)	88

609

Substrate	Conditions	Product(s) and Yield(s) (%)	Refs.
C$_{13}$	Mn(OAc)$_3$, Cu(OAc)$_2$, AcOH, 50°, 18 h	(18) + (18)	83
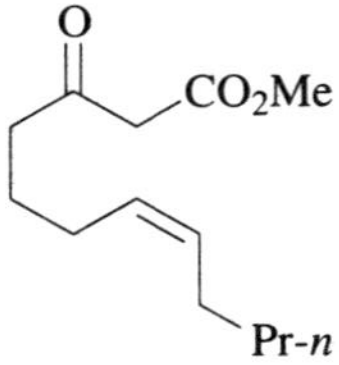	Mn(OAc)$_3$, Cu(OAc)$_2$ AcOH, 60°, 1 h	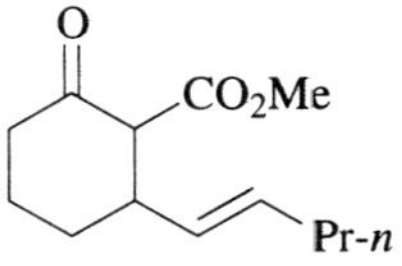 (75)	85
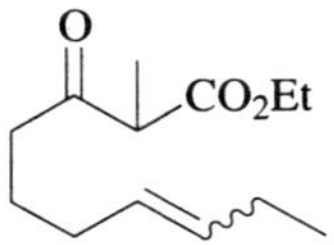 Z E	Mn(OAc)$_3$, Cu(OAc)$_2$ AcOH, rt, 17 h	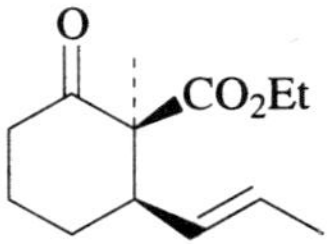 I 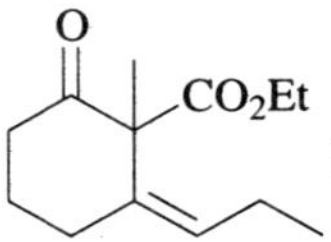 + II 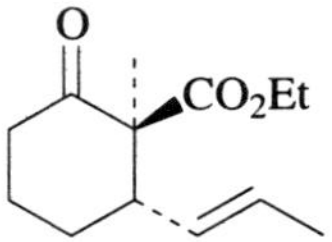 + III	83

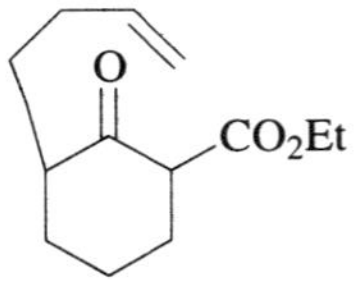

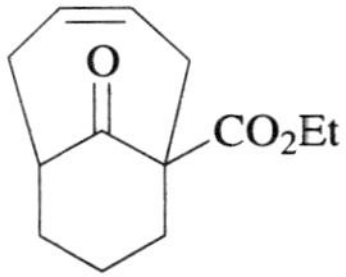

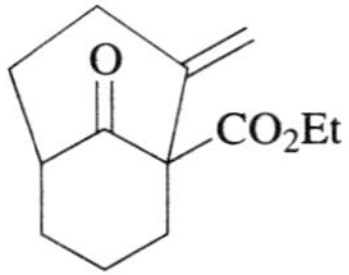

	I	II	III
Z	(56)	(14)	(3)
E	(43)	(10)	(10)

Substrate	Conditions	Product(s) and Yield(s) (%)	Refs.
C$_{14}$	Mn(OAc)$_3$, Cu(OAc)$_2$, AcOH	(20)	84

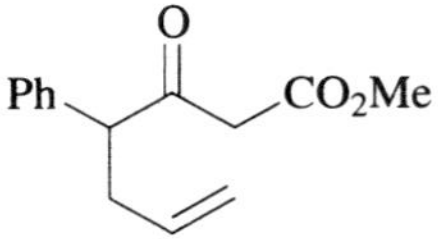

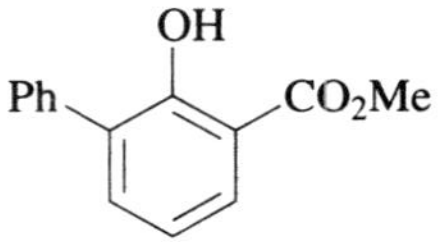

TABLE XXVA. INTRAMOLECULAR CYCLIZATIONS OF 4-SUBSTITUTED 3-KETOESTERS (D-MODE) (*Continued*)

Substrate	Conditions	Product(s) and Yield(s) (%)	Refs.
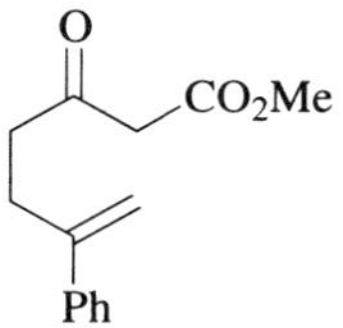	1. Mn(OAc)$_3$, LiCl, AcOH 2. AcOH, LiCl, 100°, 24 h	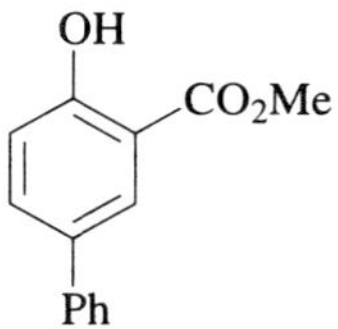 (44)	84
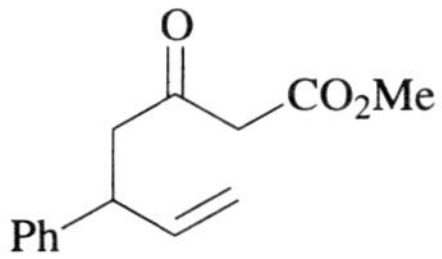	Mn(OAc)$_3$, Cu(OAc)$_2$, AcOH	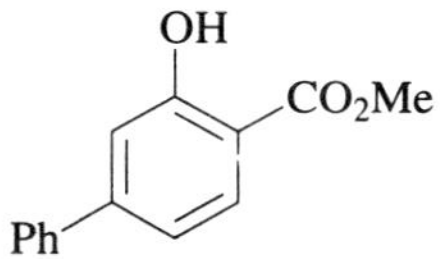 (11)	84
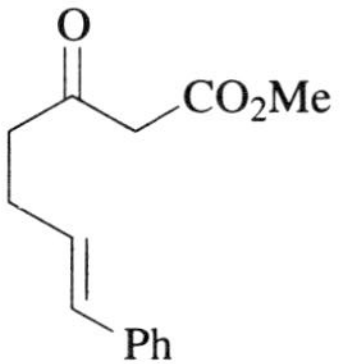	Mn(OAc)$_3$, Cu(OAc)$_2$, AcOK, AcOH, 50°, 50 min	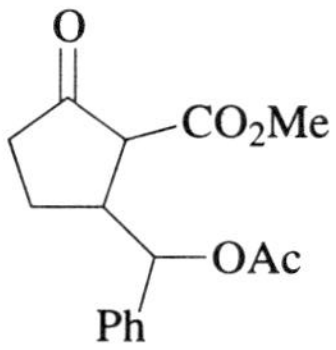 (70)	89
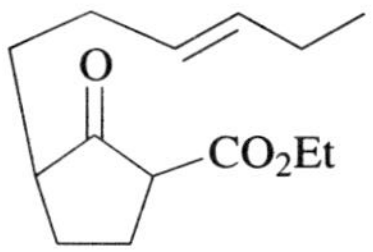	Mn(OAc)$_3$, Cu(OAc)$_2$, AcOH, 25°, 24 h	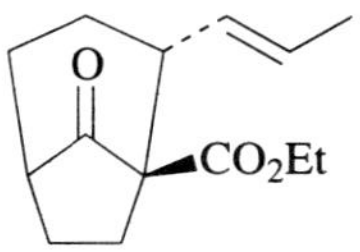 (78)	83
C$_{15}$ 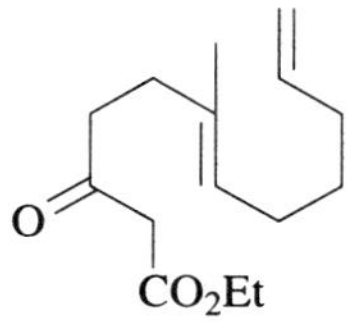	Mn(OAc)$_3$, O$_2$, AcOH, rt, 18 h	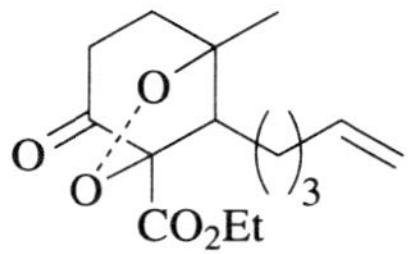 (25)	56

TABLE XXVA. INTRAMOLECULAR CYCLIZATIONS OF 4-SUBSTITUTED 3-KETOESTERS (D-MODE) (*Continued*)

Substrate	Conditions	Product(s) and Yield(s) (%)		Refs.
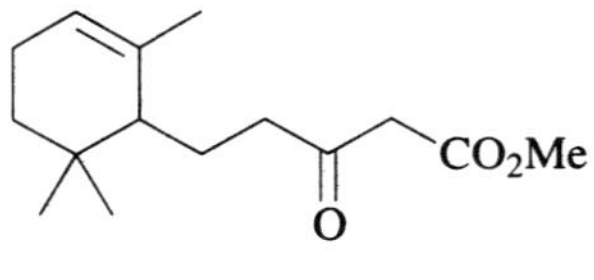	Mn(OAc)₃, AcOH, 58°, 15 min	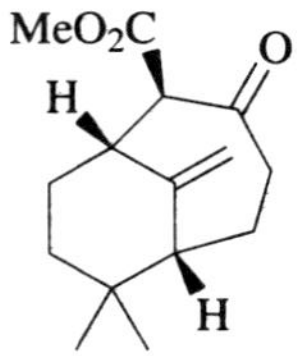	(60)	180
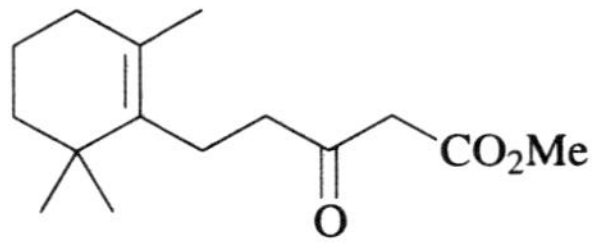	Mn(OAc)₃, AcOH, rt, 24 h	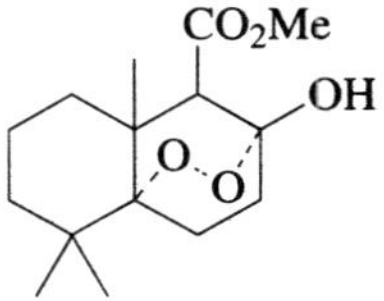	(30)	180
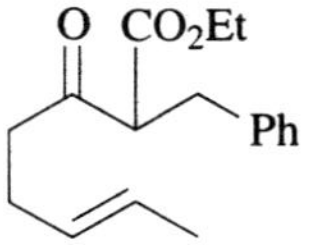	Mn(OAc)₃, Cu(OAc)₂, AcOH	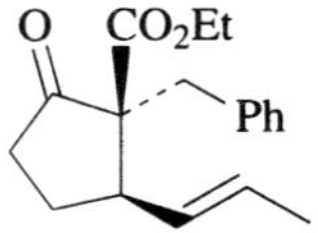	(50)	90
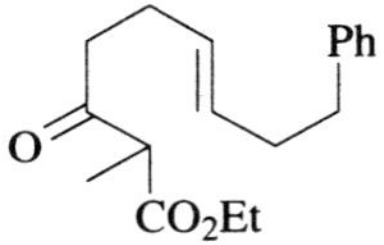	Mn(OAc)₃, Cu(OAc)₂, AcOH, 50°	α : β 3 :1 (64)		91

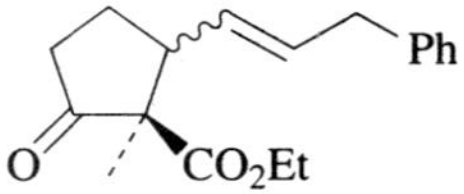

612

TABLE XXVB. INTRAMOLECULAR CYCLIZATIONS OF 4-SUBSTITUTED 3-KETOESTERS (T-MODE)

Substrate	Conditions	Product(s) and Yield(s) (%)	Refs.
C_{10}	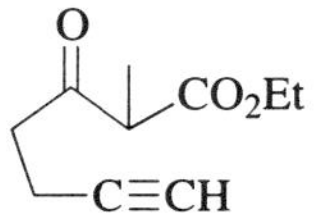 Mn(OAc)$_3$, EtOH, 25°, 21 h	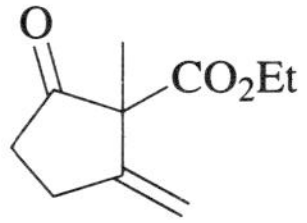 (20) 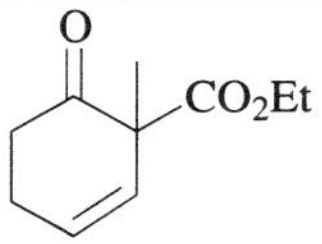 + (12)	99
C_{11}	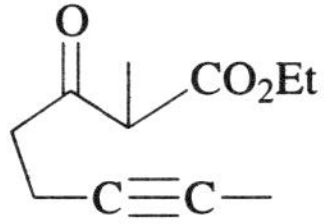 Mn(OAc)$_3$, EtOH, 25°, 4.5 h	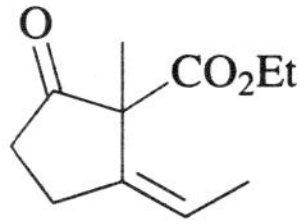 (18) 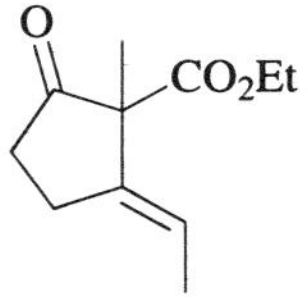 + (48)	99
C_{11} $n = 1$ C_{12} $n = 2$	Mn(OAc)$_3$-anhydr., EtOH	(35) (34)	87
C_{12}	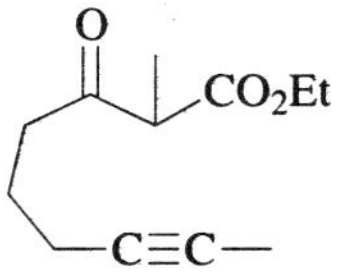 Mn(OAc)$_3$-anhydr., EtOH	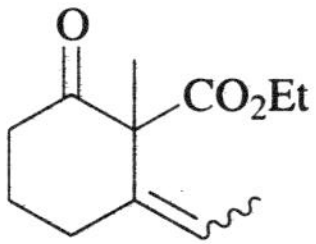 E : Z 2.5 : 1 (59)	87
C_{12} $n = 1$ C_{13} $n = 2$	Mn(OAc)$_3$-anhydr., EtOH	(6) (13)	87

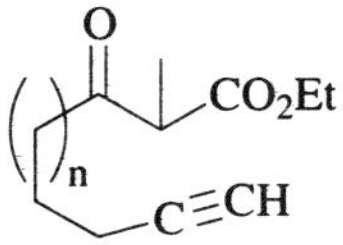

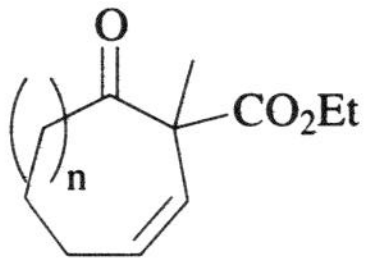

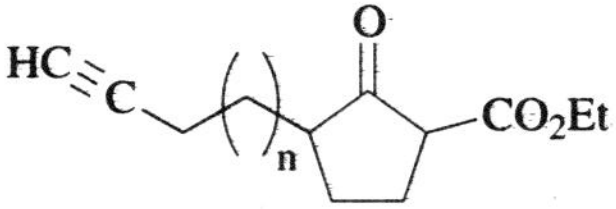

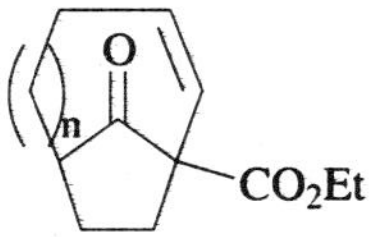

TABLE XXVC. INTRAMOLECULAR CYCLIZATIONS OF 4-SUBSTITUTED 3-KETOESTERS (B-MODE)

Substrate	Conditions	Product(s) and Yield(s) (%)	Refs.
C_{13}-C_{14}			

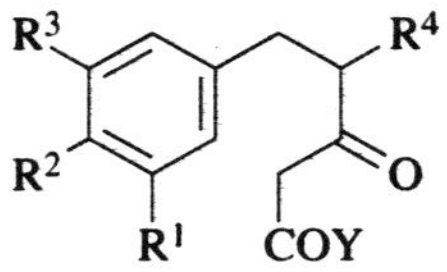

1. Mn(OAc)$_3$, AcOH, 40°–70°
2. SiO$_2$, C$_6$H$_6$, reflux

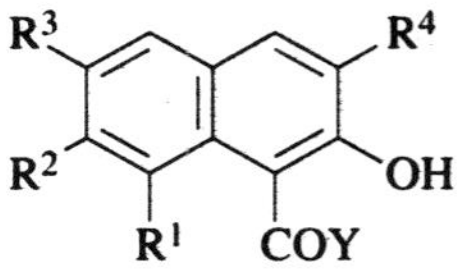

182

R^1	R^2	R^3	R^4	Y	
H	H	H	H	OEt	(24)
H	H	H	Me	OEt	(30)
H	F	H	Me	OMe	(10)
OMe	H	H	Me	OMe	(41)
H	H	OMe	Me	OMe	(36)

614

TABLE XXVI. INTRAMOLECULAR CYCLIZATIONS OF *O*-SUBSTITUTED 3-KETOESTERS

Substrate	Conditions	Product(s) and Yield(s) (%)	Refs.
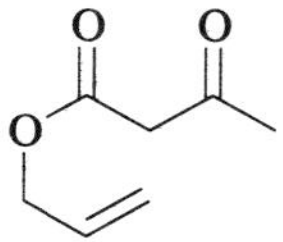 C7	Mn(OAc)$_3$, Cu(OAc)$_2$, AcOH, AcOK, 75°	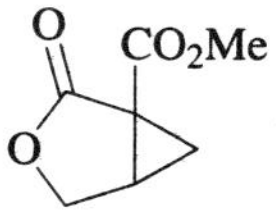 (57)	92

TABLE XXVII. INTRAMOLECULAR CYCLIZATIONS OF 4-SUBSTITUTED 1,3-DIKETONES

Substrate	Conditions	Product(s) and Yield(s) (%)	Refs.
C$_8$	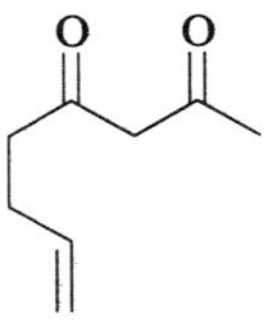 Mn(OAc)$_3$, Cu(OAc)$_2$, AcOH, AcOK, 50°, 50 min	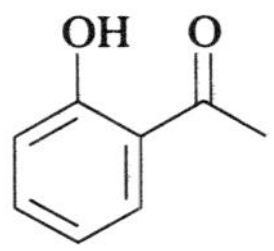 (96)	89
C$_9$	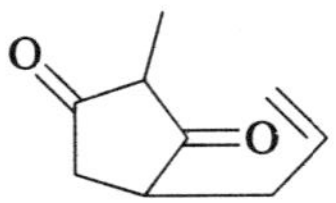 Mn(OAc)$_3$, Cu(OAc)$_2$, AcOH, 25°, 18 h	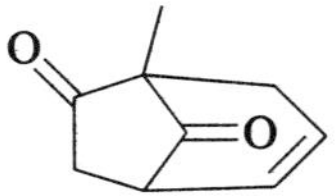 (38)	83
C$_{10}$	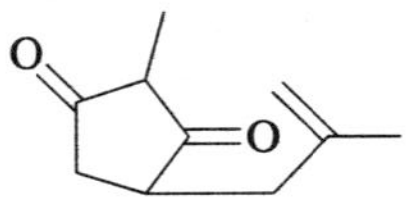 Mn(OAc)$_3$, Cu(OAc)$_2$, AcOH, 25°, 18 h	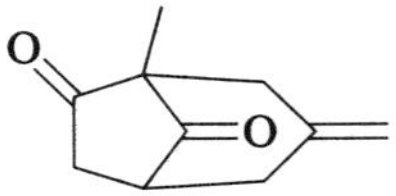 (48)	83
C$_{13}$	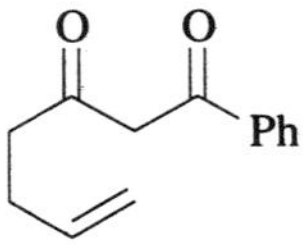 Mn(OAc)$_3$, Cu(OAc)$_2$, AcOH	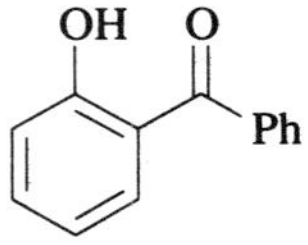 (47)	84

616

TABLE XXVIII. INTRAMOLECULAR CYCLIZATIONS OF 2-SUBSTITUTED 1,3-DIKETONES

Substrate	Conditions	Product(s) and Yield(s) (%)	Refs.
C$_{17}$ 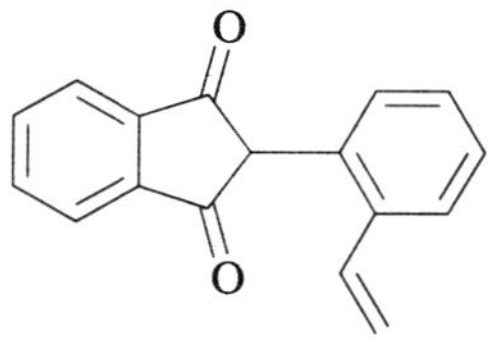	Mn(OAc)$_3$, Cu(OAc)$_2$, AcOH, rt, 30 min	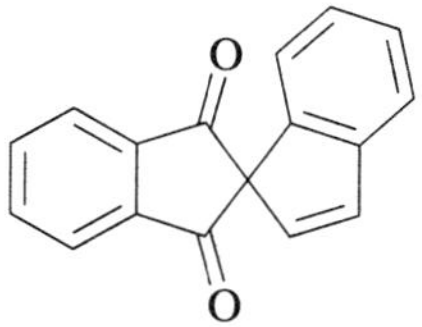 (72)	93
	Mn(OAc)$_3$, Cu(OAc)$_2$, CHCl$_3$, rt, 30 min	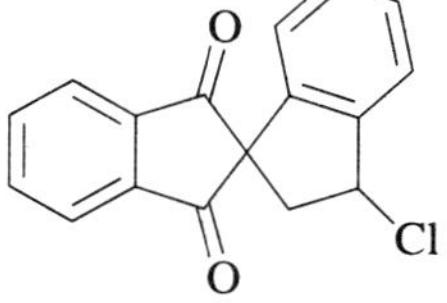 (68)	93
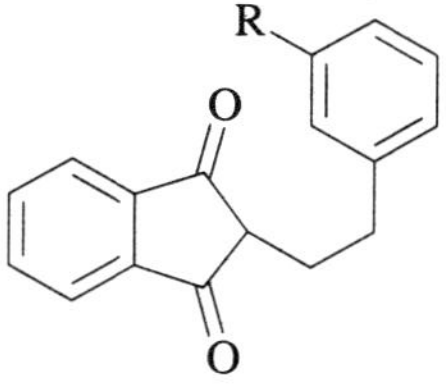 C$_{18}$ R = OMe C$_{24}$ R = OBn	Mn(OAc)$_3$, AcOH, 95-100°, 2 h	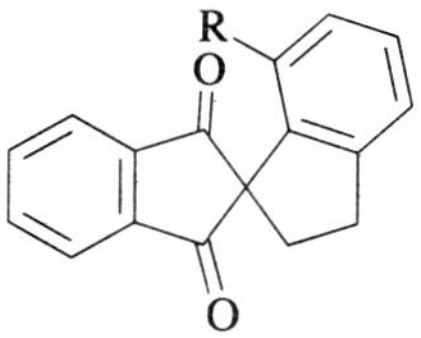 (32) (25)	29

TABLE XXIX. INTRAMOLECULAR CYCLIZATIONS OF O-SUBSTITUTED MALONIC ESTERS

Substrate	Conditions	Product(s) and Yield(s) (%)	Refs.
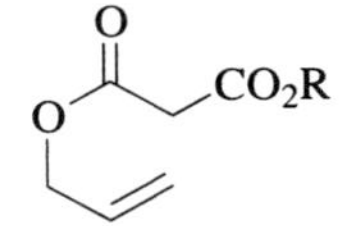 C₇ CO₂R R = Me C₉ R = CH₂CH=CH₂	Mn(OAc)₃, Cu(OAc)₂, AcOH, AcOK, reflux	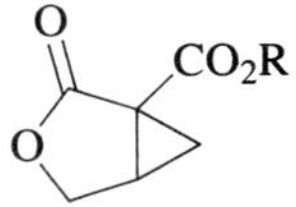 CO₂R (42) (43)	92, 96
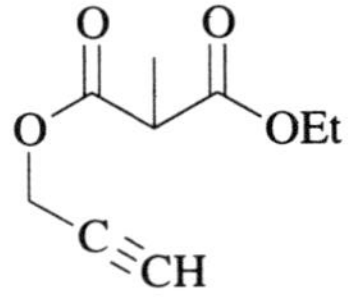 C₉ OEt C≡CH	Mn(OAc)₃, Cu(OAc)₂, AcOH, AcONa, reflux	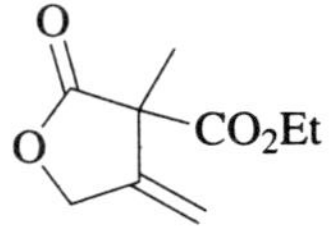 CO₂Et (38) 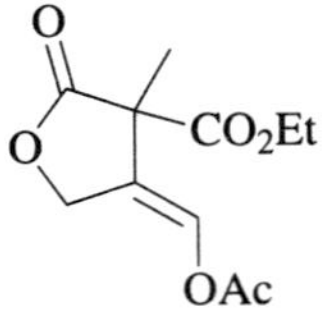 + CO₂Et (33) OAc	96
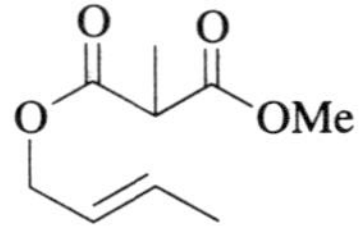 C₉ OMe	Mn(OAc)₃, Cu(OAc)₂, AcOH, AcOK, reflux	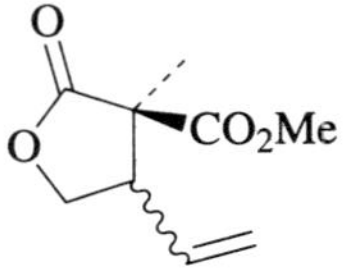 CO₂Me 70 : 30 (54)	92, 96
C₁₀ n = 1 OEt Et C₁₁ n = 2	Mn(OAc)₃, Cu(OAc)₂, AcOH, 50°, 3.5 h	OEt (18) (27)	83
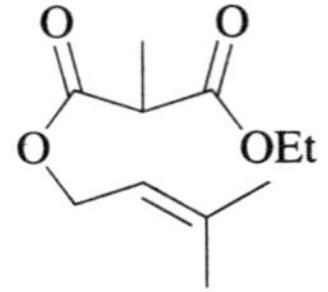 C₁₁ OEt	Mn(OAc)₃, Cu(OAc)₂, AcOH, AcONa, reflux	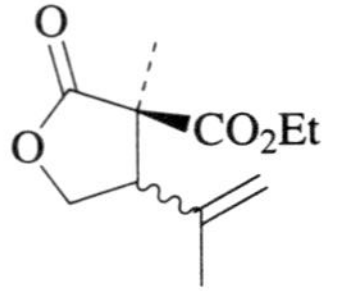 CO₂Et (30) 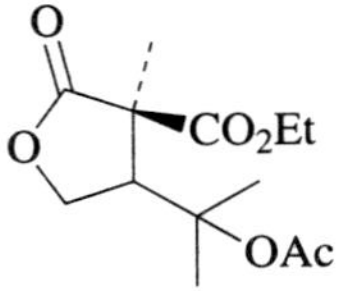 + CO₂Et (17) OAc	96

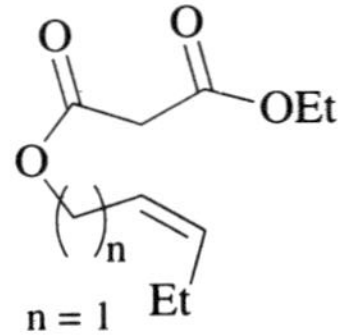

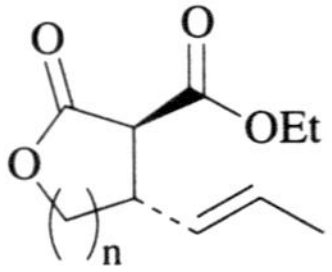

TABLE XXIX. INTRAMOLECULAR CYCLIZATIONS OF *O*-SUBSTITUTED MALONIC ESTERS (*Continued*)

Substrate	Conditions	Product(s) and Yield(s) (%)	Refs.
C$_{12}$ 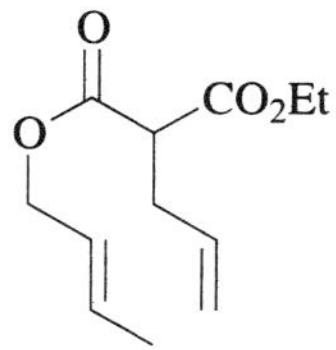	Mn(OAc)$_3$, Cu(OAc)$_2$, AcOH	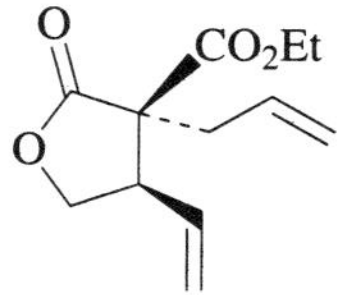 (43) 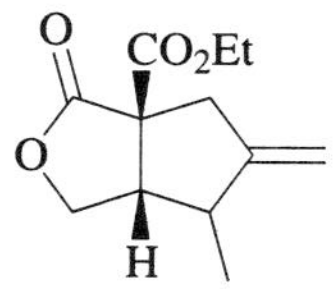 + (18)	94
C$_{13}$	Mn(OAc)$_3$, Cu(OAc)$_2$, AcOH	(31) + (9) + (26) + (11)	94
C$_{15}$ 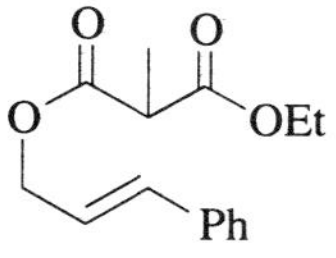	Mn(OAc)$_3$, Cu(OAc)$_2$, AcOH, AcOK, reflux	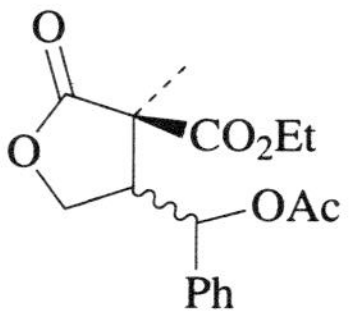 (43) 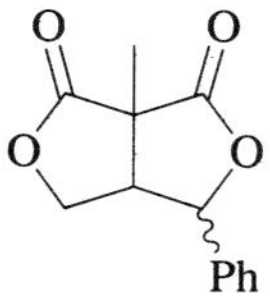 + (20)	92, 96

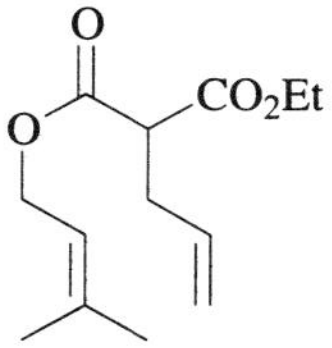

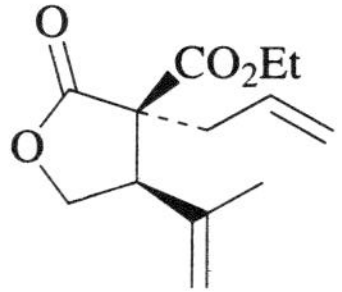

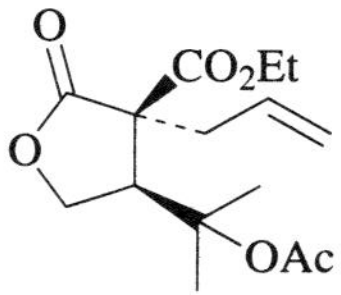

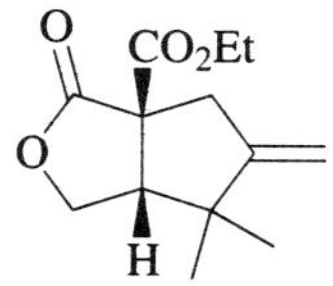

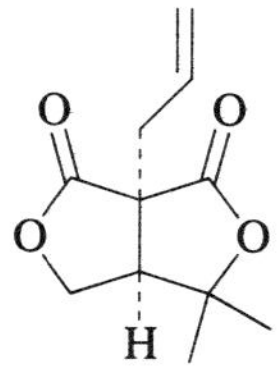

Substrate	Conditions	Product(s) and Yield(s) (%)	Refs.
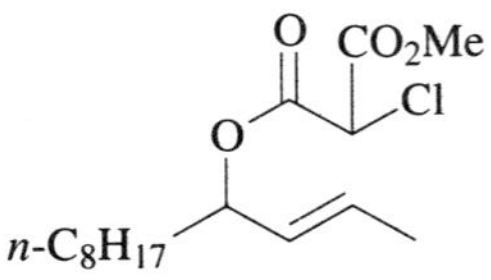	Mn(OAc)$_3$, Cu(OAc)$_2$, EtOH, 75°, 2.5 h	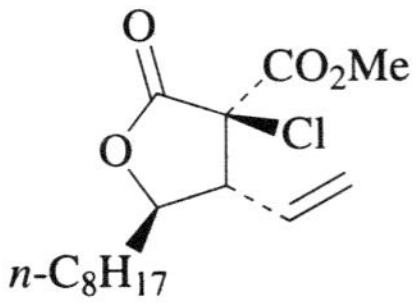 (62) 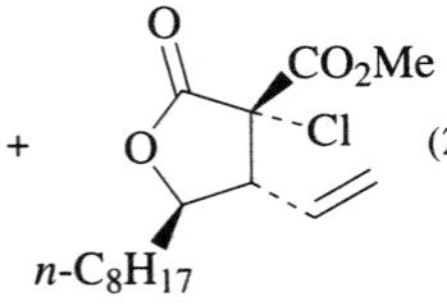 + (20)	95
C$_{16}$	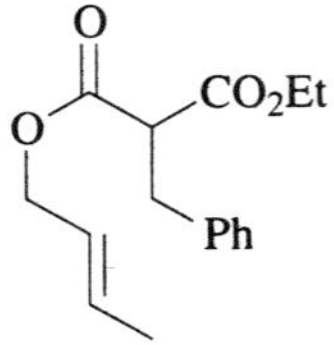 Mn(OAc)$_3$, Cu(OAc)$_2$; AcOH	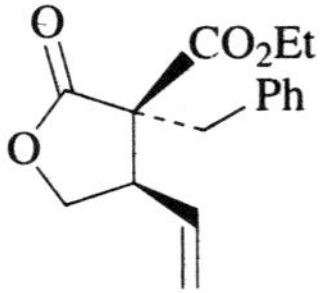 (56) 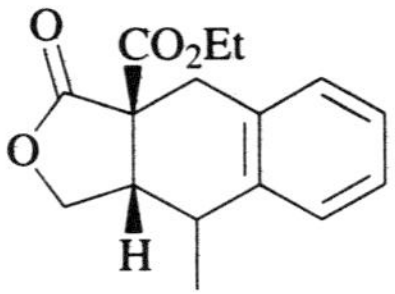 + (24) 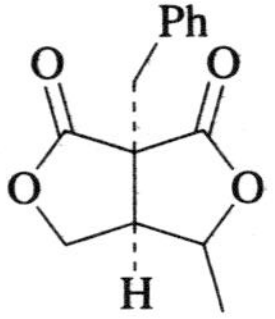 + (10)	94

TABLE XXXA. INTRAMOLECULAR CYCLIZATIONS OF C-SUBSTITUTED MALONIC ESTER DERIVATIVES (D-MODE)

Substrate	Conditions	Product(s) and Yield(s) (%)	Refs.
C$_{10}$ (CN, CO$_2$K substrate)	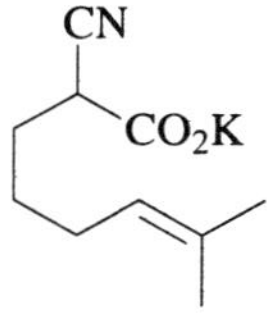 1. Mn(OAc)$_3$, AcOH, 70° 2. CH$_2$N$_2$	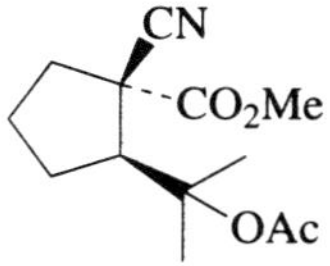 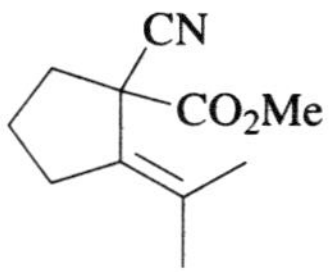 (CN, CO$_2$Me, OAc product) (17)^a + (CN, CO$_2$Me, isopropylidene product) (9)	178
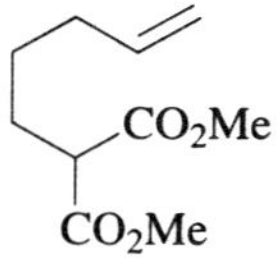 (CO$_2$Me, CO$_2$Me substrate)	Mn(OAc)$_3$, EtOH, 55°	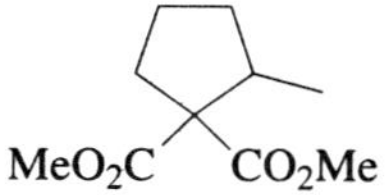 (MeO$_2$C, CO$_2$Me product) (40)	99
C$_{13}$ R = CN C$_{14}$ R = CO$_2$Me	Mn(OAc)$_3$, Cu(OAc)$_2$, AcOH, 55° 2 d 3 d	(MeO$_2$C, R product) (35) (65)	97

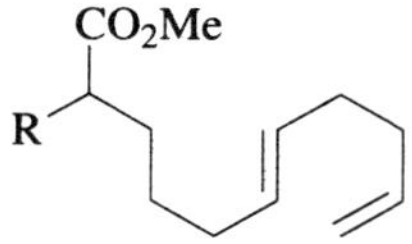

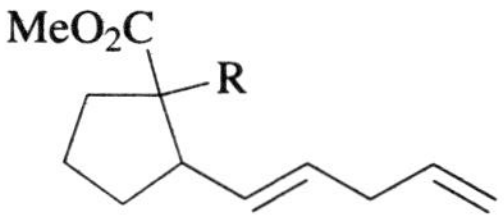

^{a} A tandem cyclization product was also formed (17%).

621

TABLE XXXB. INTRAMOLECULAR CYCLIZATIONS OF C-SUBSTITUTED MALONIC ESTERS (B-MODE)

Substrate	Conditions	Product(s) and Yield(s) (%)	Refs.
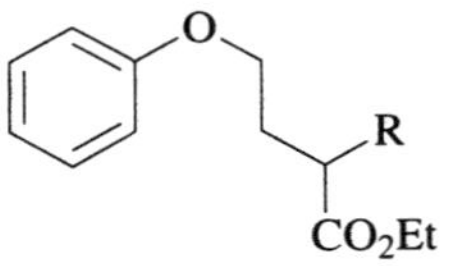 C$_{13}$ R = CN C$_{15}$ R = CO$_2$Et	Mn(OAc)$_3$, AcOH, AcONa, 70-80°, 3-4 h	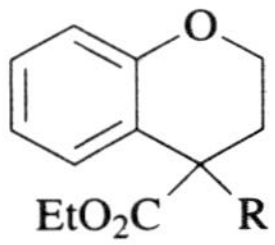 (8) (90)	181
C$_{15}$	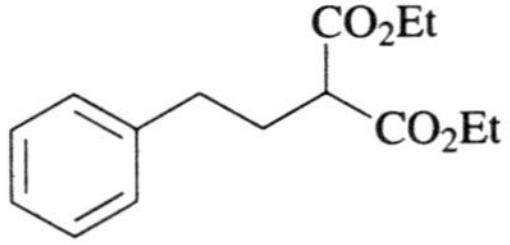 Mn(OAc)$_3$, AcOH, AcONa, 70°, 8 h	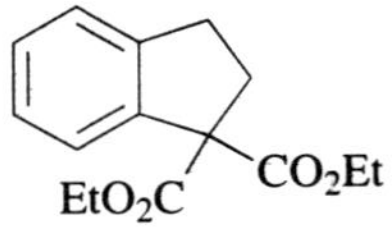 (39)	181
C$_{15}$-C$_{16}$	Mn(OAc)$_3$, AcOH, AcONa, 70-80°, 3-24 h	R^1 R^2 H H (39) OMe H (47) H OMe (30)	181
C$_{16}$	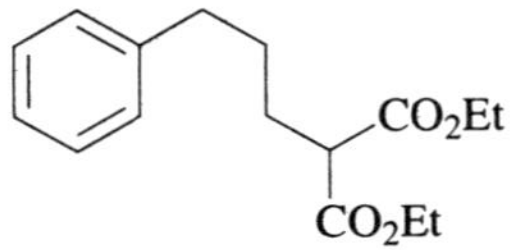 Mn(OAc)$_3$, AcOH, AcONa, 70°, 3 h	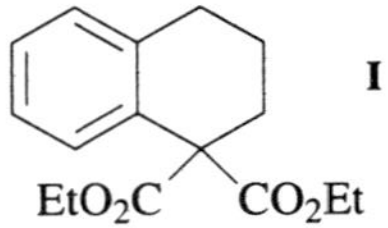 **I** (85)	181
	Mn(OAc)$_3$ anhydr., AcOH, 60°, 6 h	**I** (89)	222

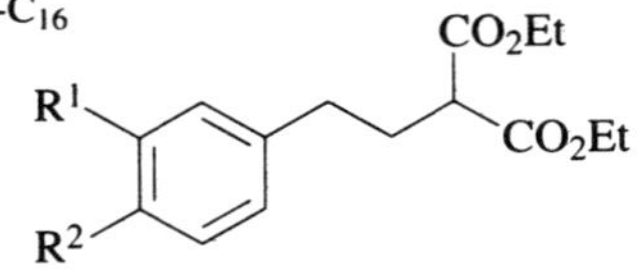

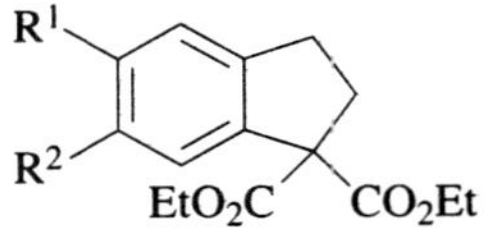

Substrate	Conditions	Product(s) and Yield(s) (%)	Refs.
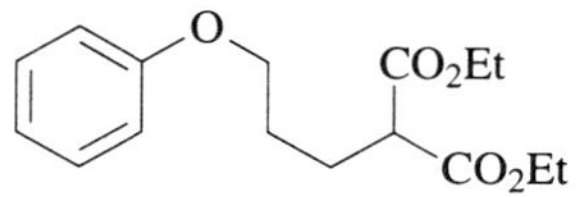	Mn(OAc)$_3$, AcOH, AcONa, 70-80°, 7 h	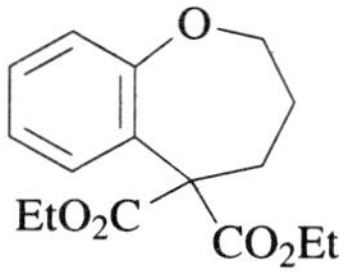 (15)	181
C_{16}–C_{18} 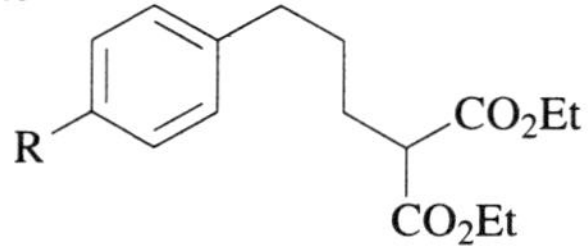 R = NO$_2$ R = OMe R = NHCOCH$_3$	Mn(OAc)$_3$, AcOH, AcONa, 70-80°, 3-10 h	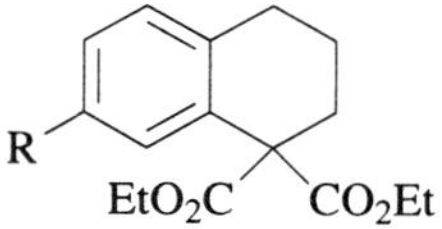 (85) (80) (88)	181
C_{18} 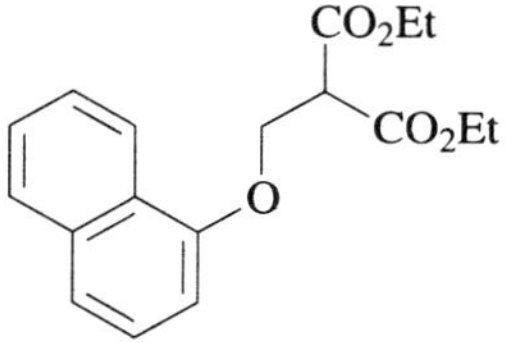	Mn(OAc)$_3$, AcOH, AcONa, 70-80°, 3 h	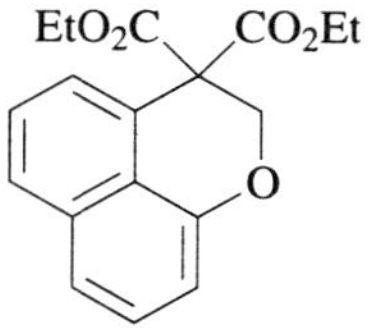 (91)	181
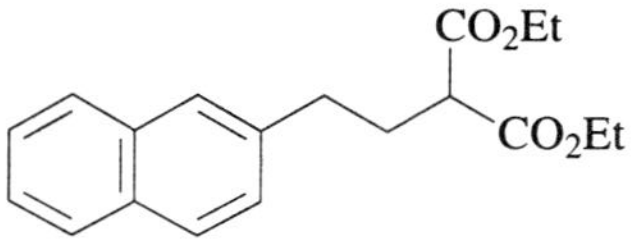	Mn(OAc)$_3$, AcOH, AcONa, 70-80°, 4 h	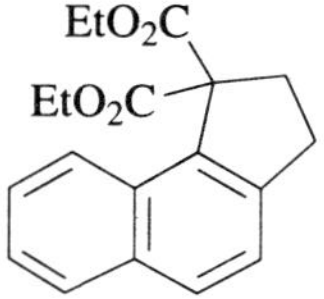 (90) 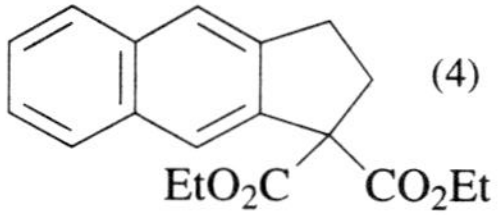 + (4)	181

623

Substrate	Conditions	Product(s) and Yield(s) (%)	Refs.
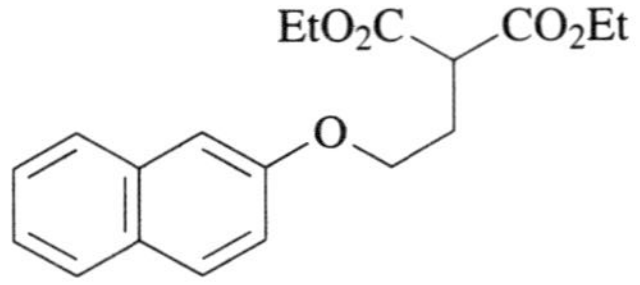	Mn(OAc)$_3$, AcOH, AcONa, 70-80°, 3 h	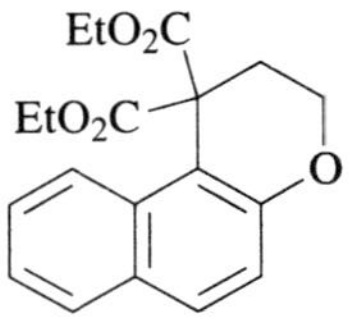 (90)	181
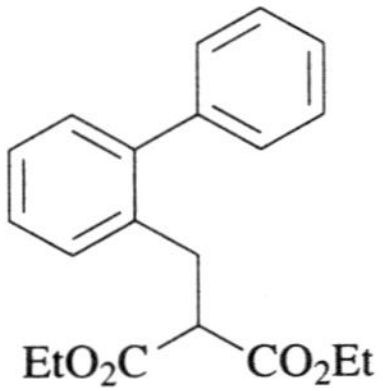	Mn(OAc)$_3$, AcOH, AcONa, 70°, 3 h	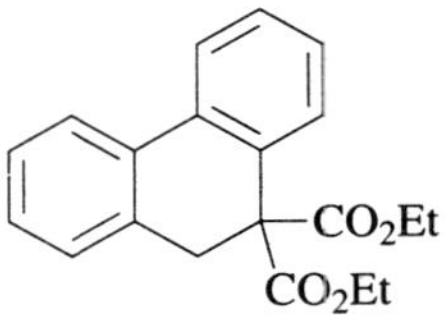 (93)	181
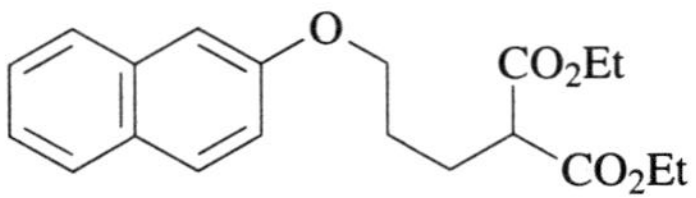	Mn(OAc)$_3$, AcOH, AcONa, 70-80°, 5 h	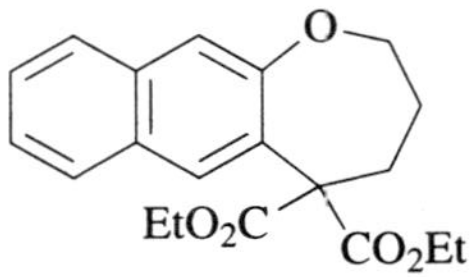 (70)	181

C$_{20}$

TABLE XXXI. INTRAMOLECULAR CYCLIZATIONS OF *N*-SUBSTITUTED AMIDES

Substrate	Conditions	Product(s) and Yield(s) (%)	Refs.
C$_9$ 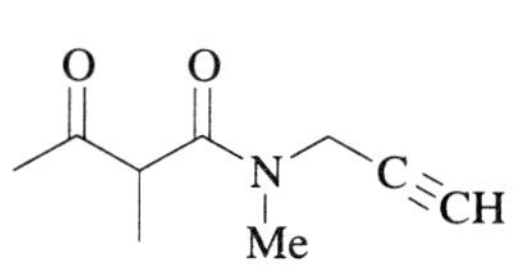	Mn(OAc)$_3$-anhydr., EtOH, rt, 1 h	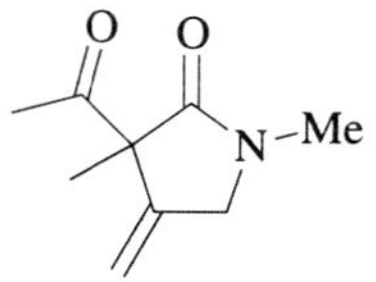 (40)	179
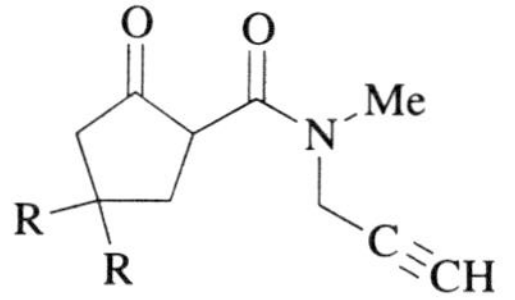 C$_{10}$ R = H C$_{12}$ R = Me	Mn(OAc)$_3$-anhydr., EtOH, rt, 1 h	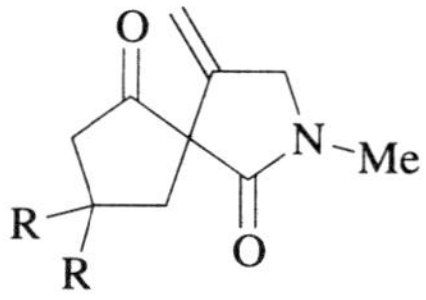 (60) (55)	179
C$_{11}$ 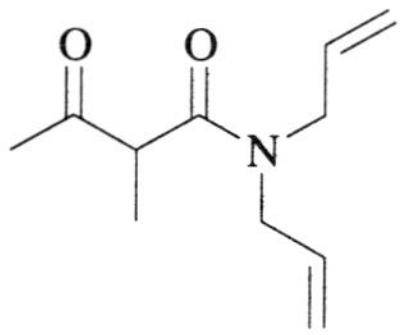	Mn(OAc)$_3$-anhydr., EtOH, rt, 1 h	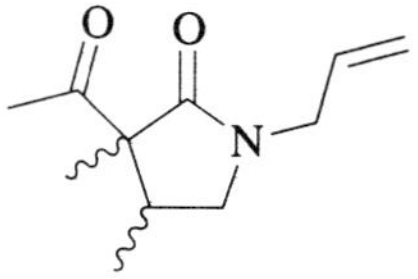 (40)	179
C$_{12}$ R = H C$_{14}$ R = Me	Mn(OAc)$_3$-anhydr., EtOH, rt, 1 h	(47) + (3) (47) (3)	179

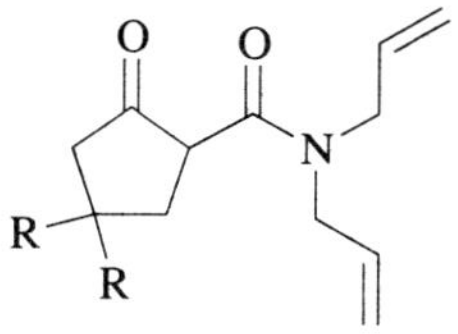

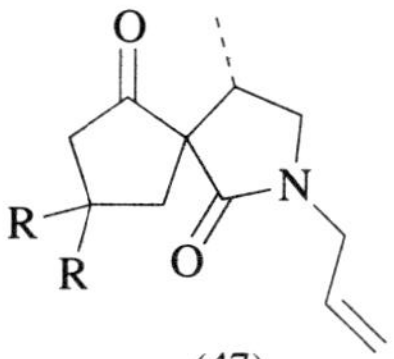

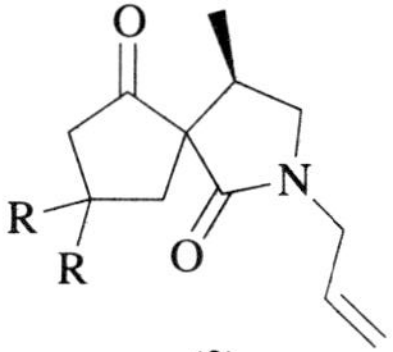

TABLE XXXI. INTRAMOLECULAR CYCLIZATIONS OF *N*-SUBSTITUTED AMIDES (*Continued*)

Substrate	Conditions	Product(s) and Yield(s) (%)	Refs.
C$_{13}$ 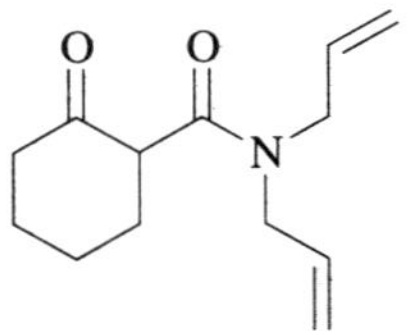	Mn(OAc)$_3$-anhydr., EtOH, rt, 1 h	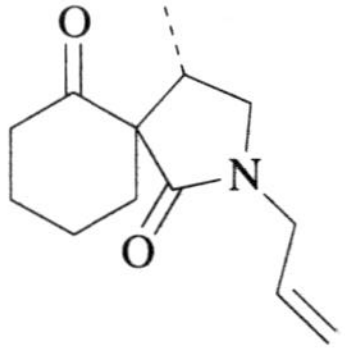 (30) 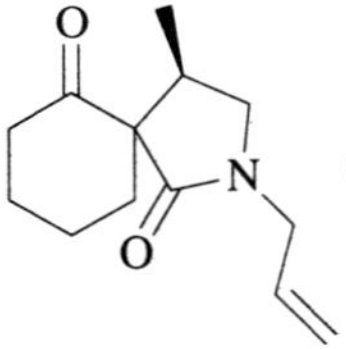 + (10)	179
C$_{19}$ 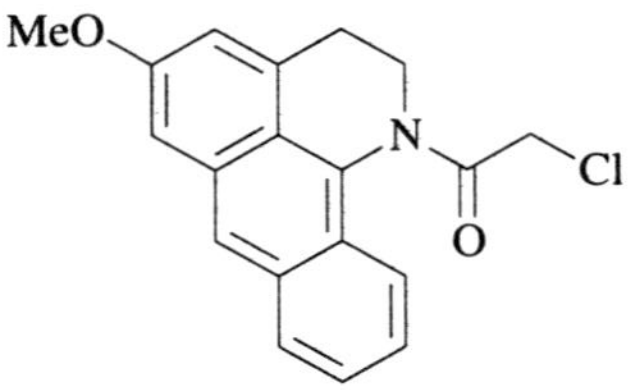	Mn(OAc)$_3$, AcOH, 50°, 5 h	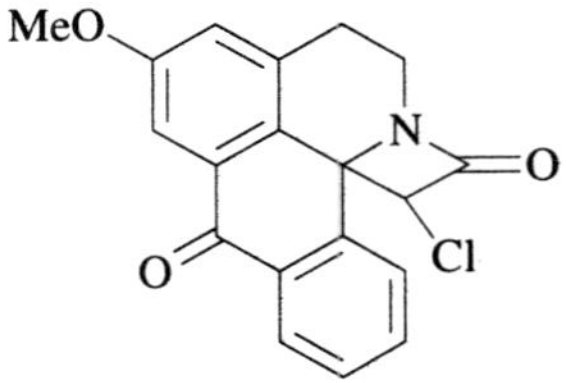 (21)	223

TABLE XXXII. TANDEM CYCLIZATIONS (DD-MODE)

Substrate	Conditions	Product(s) and Yield(s) (%)	Refs.
C₁₁	Mn(OAc)₃, Cu(OAc)₂, AcOH, rt, 20 h	1 : 1 (53)	103
	Mn(OAc)₃, Cu(OAc)₂, AcOH, AcONa, reflux	38 : 54 : 8ᵃ (66)	96
	Mn(OAc)₃, Cu(OAc)₂, AcOH, rt, 38 h	CO₂Me (11) + (8)	86
	Mn(OAc)₃, Cu(OAc)₂, AcOH, 25°, 5 h	(48) + (18)	100

627

TABLE XXXII. TANDEM CYCLIZATIONS (DD-MODE) (*Continued*)

Substrate	Conditions	Product(s) and Yield(s) (%)	Refs.
C₁₁-C₁₅	Mn(OAc)₃, Cu(OAc)₂, AcOH, rt, 26 h	R = Cl (72) R = OPO(OEt)₂ (77) R = CH₂TMS (30)	101
C₁₂	Mn(OAc)₃, Cu(OAc)₂, AcOH, 25°, 12 h	2 : 4 : 1 : 1 (66)	103
	Mn(OAc)₃, Cu(OAc)₂, AcOH, rt, 14 h	(33) + (35)	86

628

TABLE XXXII. TANDEM CYCLIZATIONS (DD-MODE) (*Continued*)

Substrate	Conditions	Product(s) and Yield(s) (%)	Refs.
	Mn(OAc)₃, Cu(OAc)₂, AcOH, rt, 10.5 h	6 : 1 (41)	86
	Mn(OAc)₃, Cu(OAc)₂, AcOH, 25°, 26 h	(86) + (0)	100, 102
	Mn(pic)₃, Cu(OAc)₂, AcOH, 25°	(0) (15)	102
	Mn(OAc)₃, Cu(pic)₂, AcOH, 25°	(76) (4)	102
	Mn(OAc)₃, AcOH, 25°	(14) (24)	102
C₁₃	Mn(OAc)₃, EtOH, 55°, 2d	4 : 1 : 1.1 : 1.2 (35)	97

629

TABLE XXXII. TANDEM CYCLIZATIONS (DD-MODE) (*Continued*)

Substrate	Conditions	Product(s) and Yield(s) (%)	Refs.

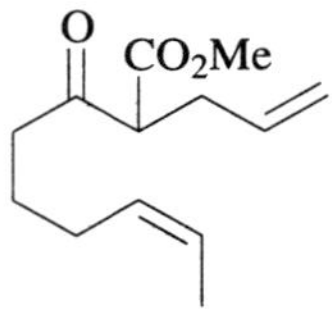

Row 1:
- Conditions: Mn(OAc)$_3$, Cu(OAc)$_2$, AcOH, 25°, 13 h
- Product: (67)
- Refs.: 100

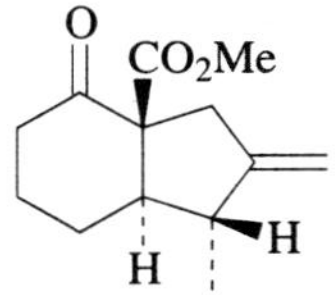

Row 2:
- Conditions: Mn(OAc)$_3$, Cu(OAc)$_2$, AcOH
- Product: 2 : 1 (65) + (5)
- Refs.: 100

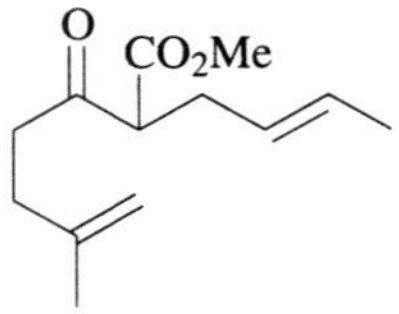

Row 3:
- Conditions: Mn(OAc)$_3$, Cu(OAc)$_2$, AcOH
- Product: (32) + α (8) β (4)
- Refs.: 88

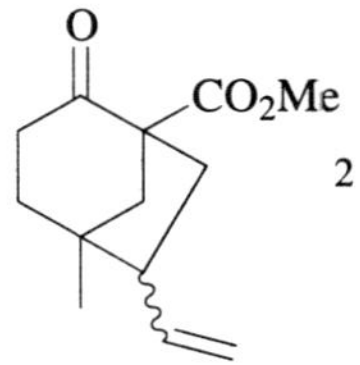

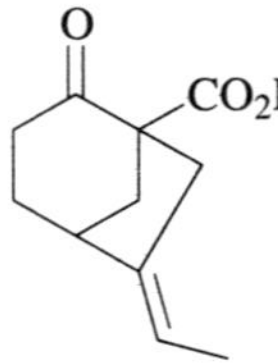

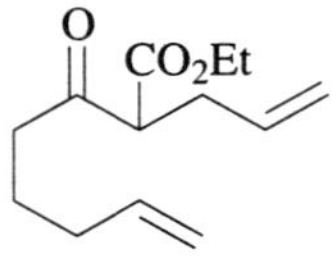

Row 4:
- Conditions: Mn(OAc)$_3$, Cu(OAc)$_2$, AcOH
- Product: (68) n = 1, (70) n = 2
- Refs.: 87

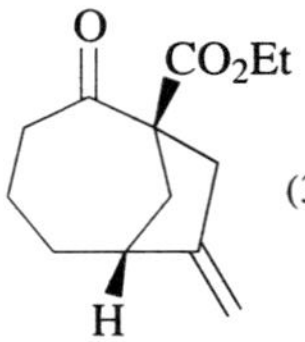

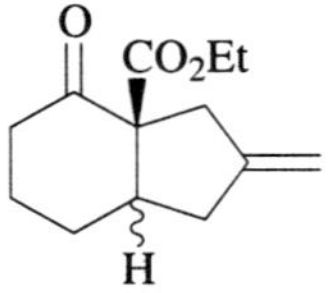

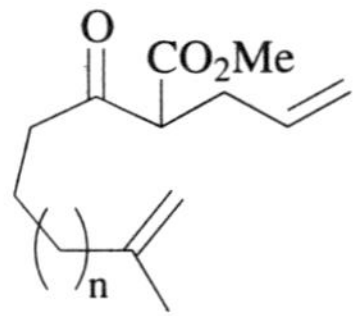

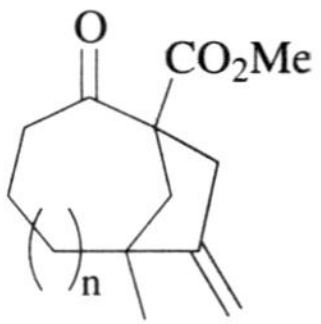

C$_{13}$ n = 1
C$_{14}$ n = 2

TABLE XXXII. TANDEM CYCLIZATIONS (DD-MODE) (Continued)

Substrate	Conditions	Product(s) and Yield(s) (%)	Refs.

C13–C16

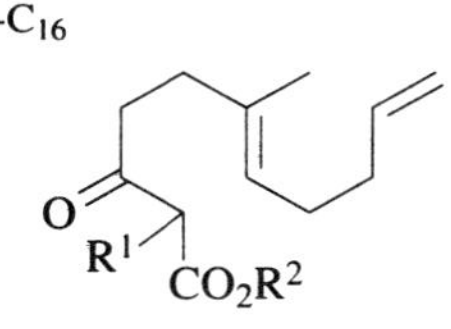

Mn(OAc)$_3$, Cu(OAc)$_2$, AcOH, 25°, 2 - 5 d

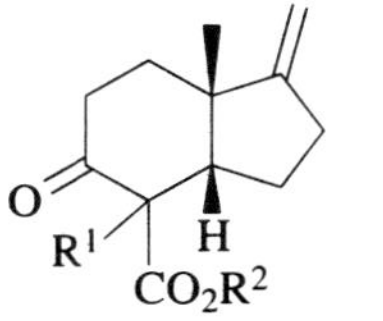

R^1	R^2	
H	Me	(44)
Me	Et	(46)
Cl	Et	(48)

101

C14

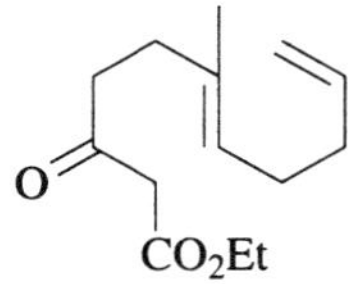

Mn(OAc)$_3$, Cu(OAc)$_2$, AcOH, rt, 4 h

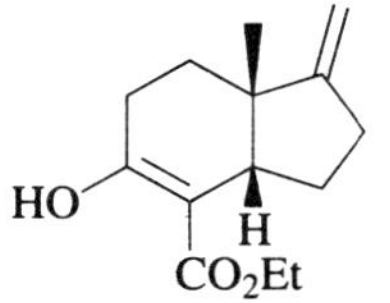

(41)

98

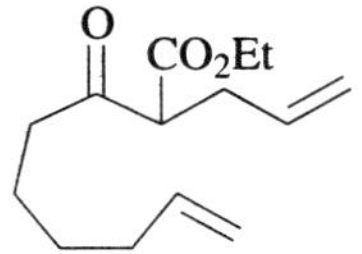

Mn(OAc)$_3$, Cu(OAc)$_2$, AcOH

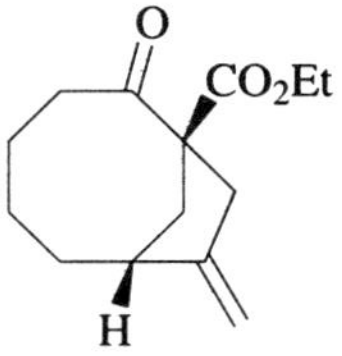

(11)^b

88

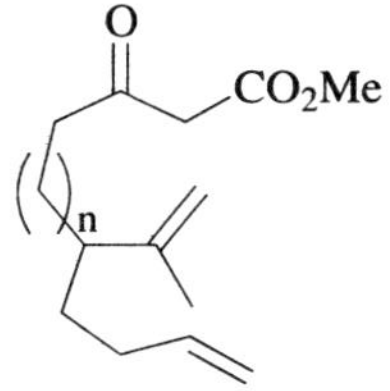

C14   n = 1
C15   n = 2

Mn(OAc)$_3$, Cu(OAc)$_2$, AcOH

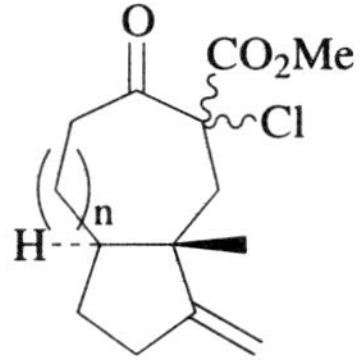

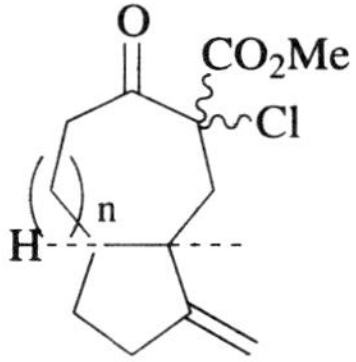

(45)
(57)

+

(20)
(19)

87

TABLE XXXII. TANDEM CYCLIZATIONS (DD-MODE) (*Continued*)

Substrate	Conditions	Product(s) and Yield(s) (%)	Refs.
C_{15} 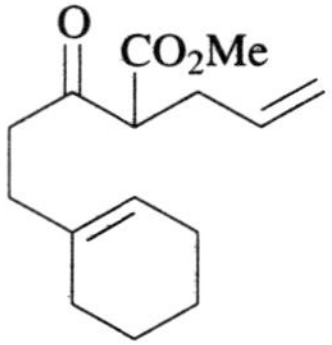	Mn(OAc)$_3$, Cu(OAc)$_2$, AcOH	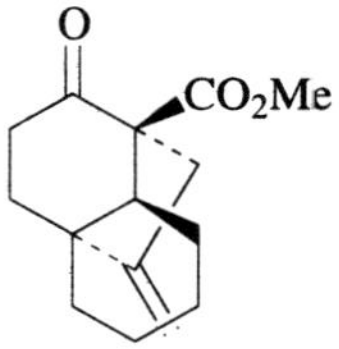 (73)	100
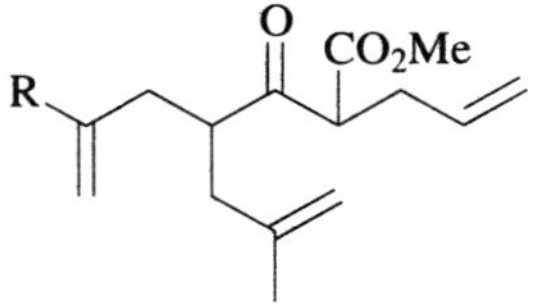	Mn(OAc)$_3$, Cu(OAc)$_2$, AcOH, rt, 12 h	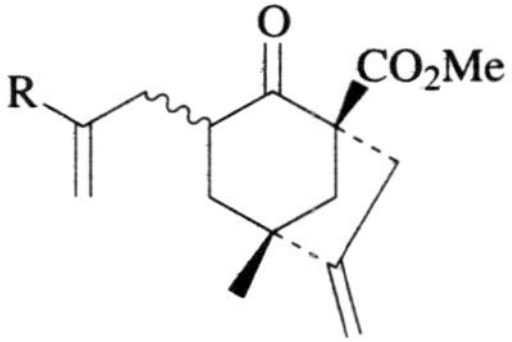 $\alpha : \beta$ 1 : 3 (52) α (55), β (36)	86
R = H R = Cl			
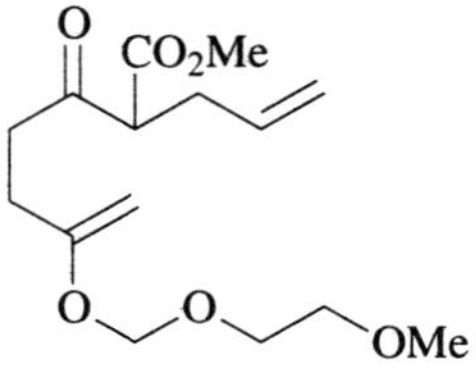	Mn(OAc)$_3$, Cu(OAc)$_2$, EtOH, 60°, 13 h	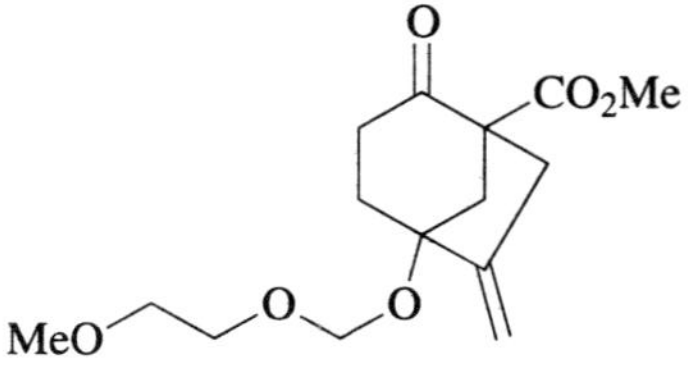 (52)	99
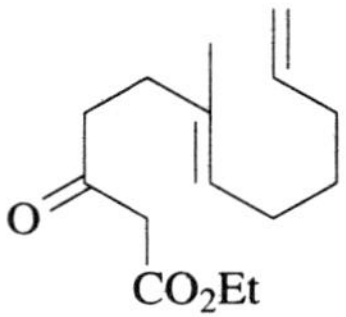	Mn(OAc)$_3$, Cu(OAc)$_2$, AcOH, rt, 7 h	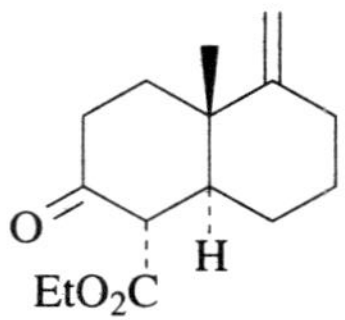 (43) 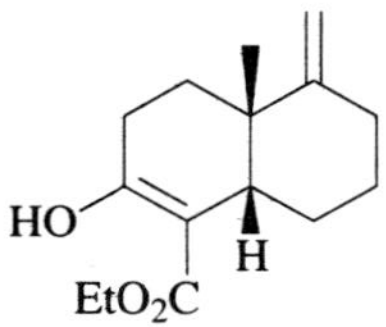 + (12)	56

632

TABLE XXXII. TANDEM CYCLIZATIONS (DD-MODE) (Continued)

Substrate	Conditions	Product(s) and Yield(s) (%)	Refs.
C$_{19}$	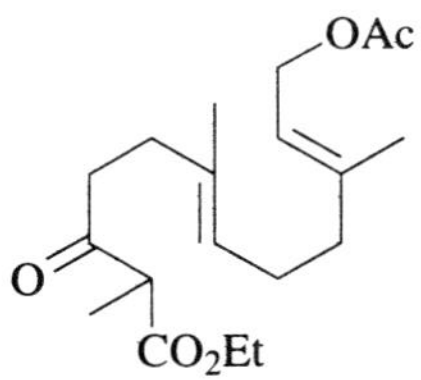 Mn(OAc)$_3$, AcOH, 25°, 30 min	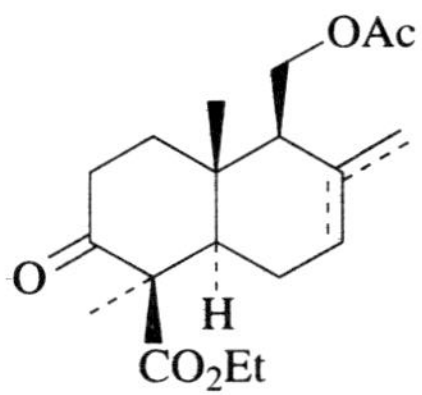 exo (48) endo (15)	91
C$_{23}$	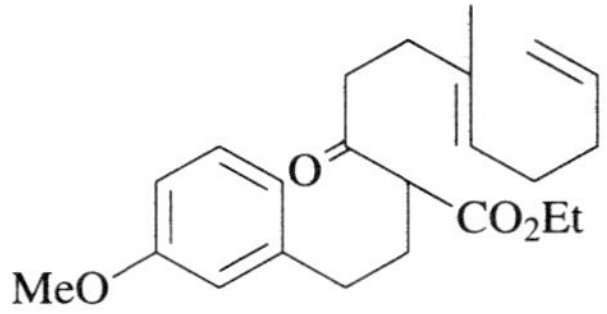 Mn(OAc)$_3$, Cu(OAc)$_2$, AcOH, rt, 7 h	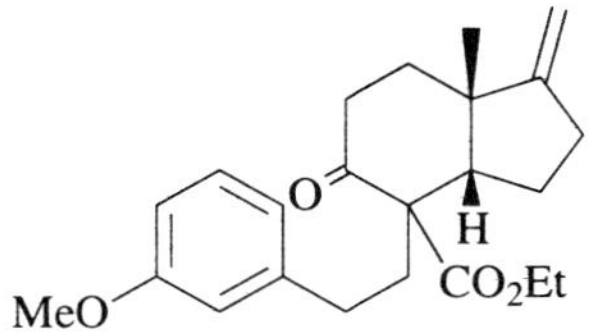 (50)	98
C$_{24}$	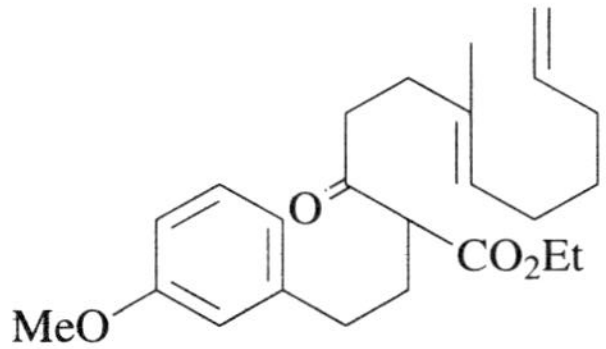 Mn(OAc)$_3$, Cu(OAc)$_2$, AcOH, rt, 4 h	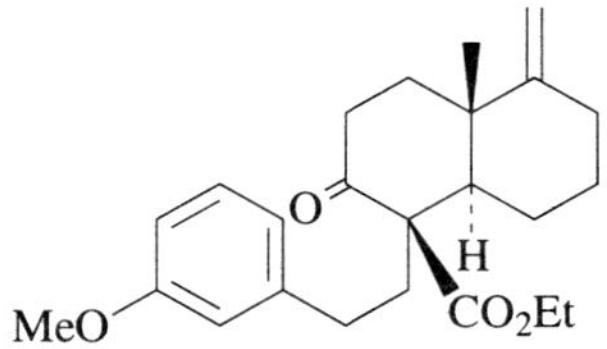 (57)	56

[a] Ratio of stereoisomers anti-symm : unsymm : syn-symm.
[b] A monocyclization product was also formed (17%).

TABLE XXXIII. TANDEM CYCLIZATIONS (DB-MODE)

Substrate	Conditions	Product(s) and Yield(s) (%)	Refs.
C_14	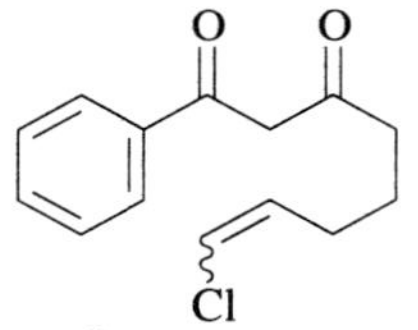 Mn(OAc)_3, AcOH, 35°, 15 h	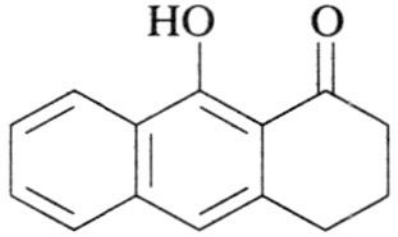 (79)	103
C_15 R = H C_16 R = Me C_16	Mn(OAc)_3, AcOH	(23) (57) + (15) (20)	94
C_16	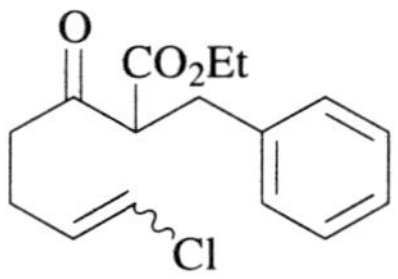 Mn(OAc)_3, AcOH, rt, 20 h	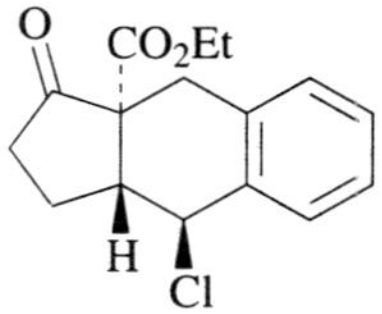 (55) 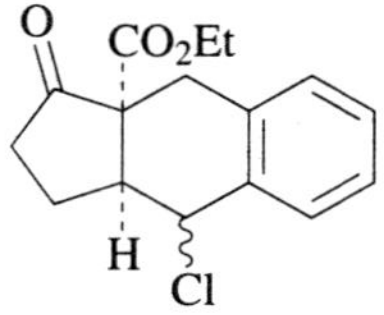 + (17)	103
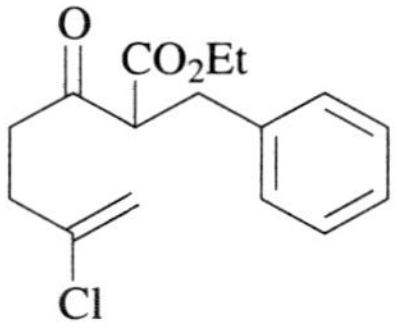	Mn(OAc)_3, AcOH, rt, 20 h	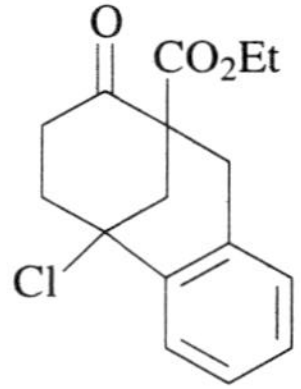 (64)	103

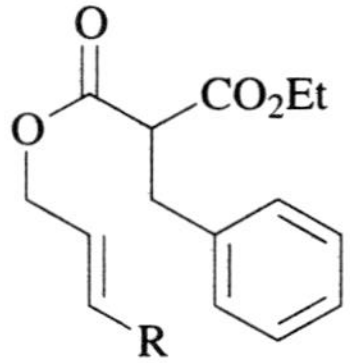

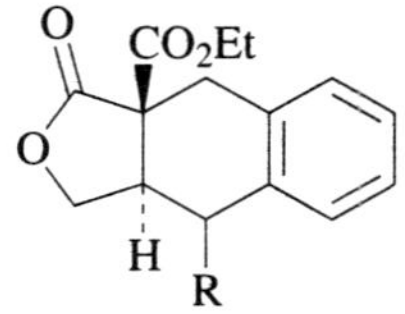

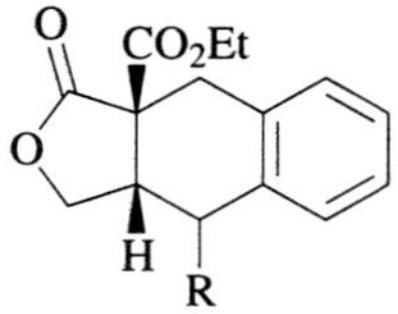

634

TABLE XXXIII. TANDEM CYCLIZATIONS (DB-MODE) (*Continued*)

Substrate	Conditions	Product(s) and Yield(s) (%)	Refs.
	1. Mn(OAc)₃, AcOH, AcOK, 35°, 17 h 2. K₂CO₃, MeOH	(43)	103
Z E	Mn(OAc)₃, AcOH	(58) (85)	90
	Mn(OAc)₃, AcOH	+ 9 : 3 (55)	90
C₁₇	Mn(OAc)₃, AcOH, 35°, 28 h	(54) + (15)	103

TABLE XXXIII. TANDEM CYCLIZATIONS (DB-MODE) (*Continued*)

Substrate	Conditions	Product(s) and Yield(s) (%)	Refs.
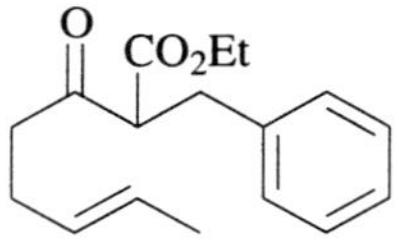 C_{18}	Mn(OAc)$_3$, AcOH, 25°, 24 h	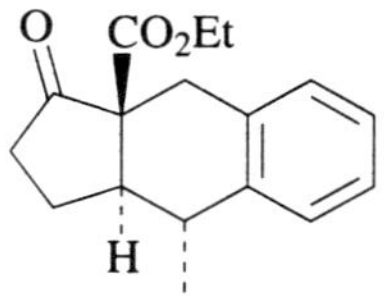 (74)	90
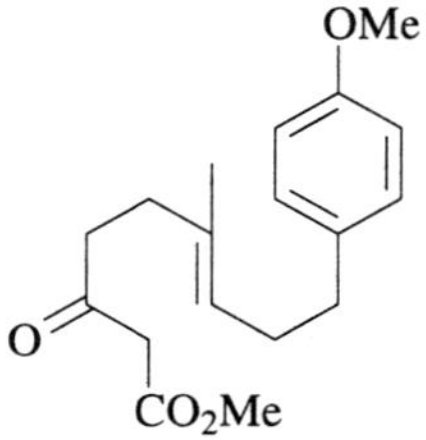	Mn(OAc)$_3$, AcOH, 20°, 1 h	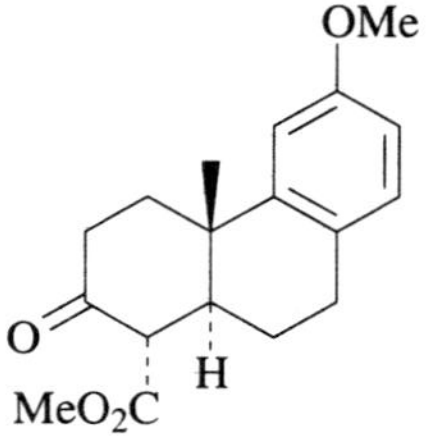 (70)	85
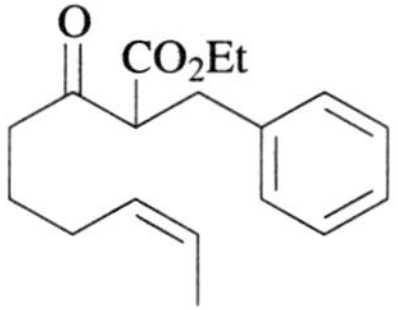 C_{19}	Mn(OAc)$_3$, AcOH, 25°, 24 h	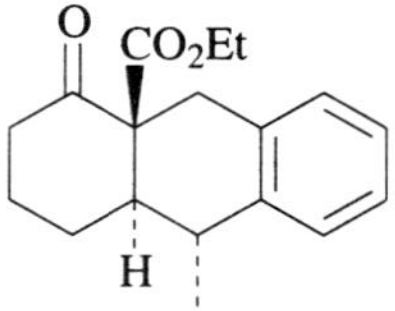 (83)	90
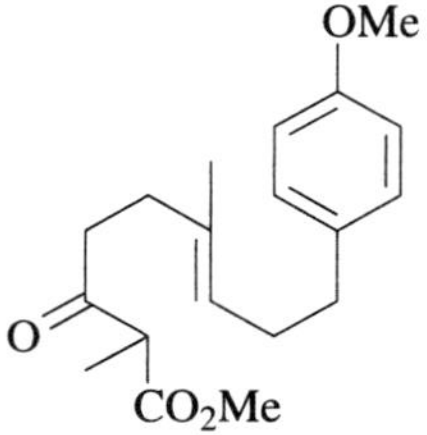	Mn(OAc)$_3$, AcOH, 20°, 1 h	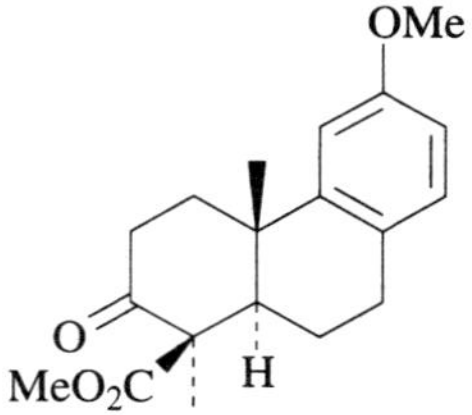 (50)	85

636

TABLE XXXIII. TANDEM CYCLIZATIONS (DB-MODE) (*Continued*)

Substrate	Conditions	Product(s) and Yield(s) (%)	Refs.
C$_{20}$	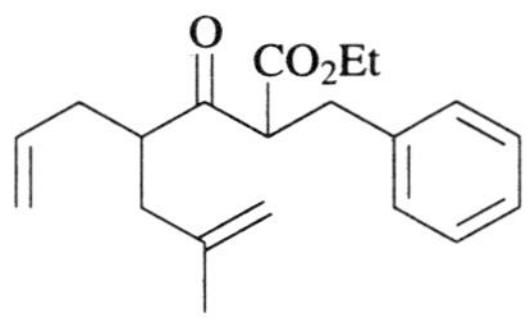 Mn(OAc)$_3$, Cu(OAc)$_2$, AcOH, rt, 14 h	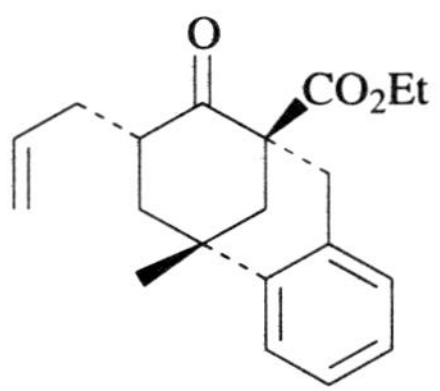 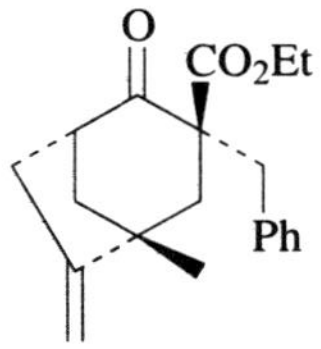 + 3 : 2 (33)	86

637

TABLE XXXIV. TANDEM CYCLIZATIONS (TD-MODE)

Substrate	Conditions	Product(s) and Yield(s) (%)	Refs.
C_{11} 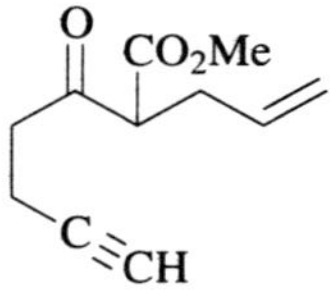	Mn(OAc)$_3$ - anhydr., EtOH, 25°, 8 h	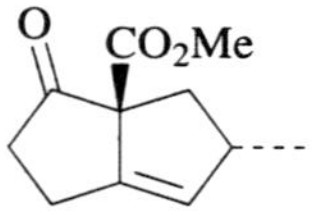 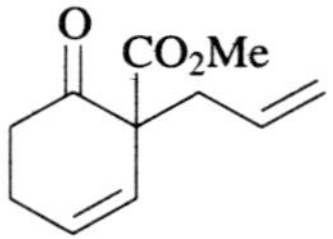 + 2 : 1 (20)	99
C_{12} 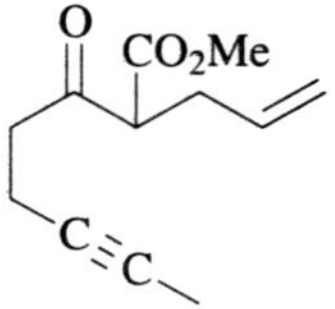	Mn(OAc)$_3$ - anhydr., EtOH, 25°, 27 h	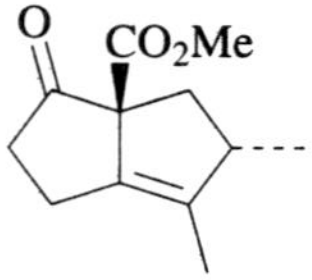 (35)	99
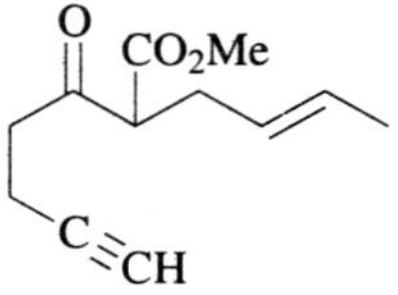	Mn(OAc)$_3$ - anhydr., Cu(OAc)$_2$, EtOH, 25°, 23 h	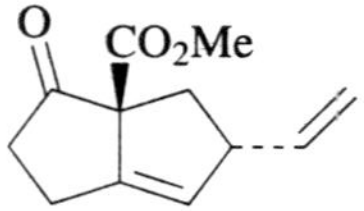 (32) 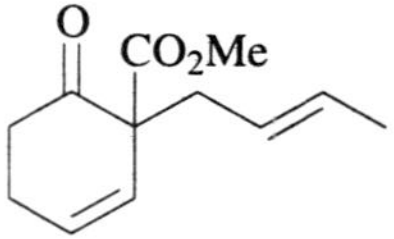 + (15)	99

TABLE XXXV. TANDEM CYCLIZATIONS (TB-MODE)

Substrate	Conditions	Product(s) and Yield(s) (%)	Refs.
C$_{15}$	Mn(OAc)$_3$, AcOH, 35°, 17 h	(81)	103
C$_{17}$	Mn(OAc)$_3$, AcOH, 35°, 17 h	(71)	103

TABLE XXXVI. TANDEM CYCLIZATIONS (DC-MODE)

Substrate	Conditions	Product(s) and Yield(s) (%)	Refs.
C9			
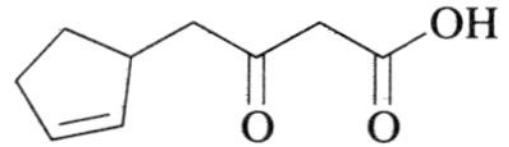	Mn(OAc)₃, AcOH, 23°, 1 h	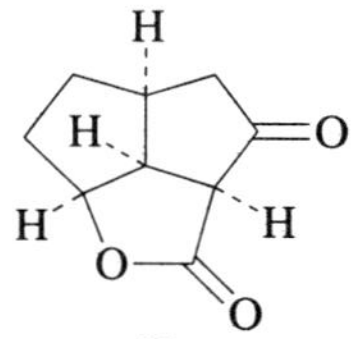 (52)	183
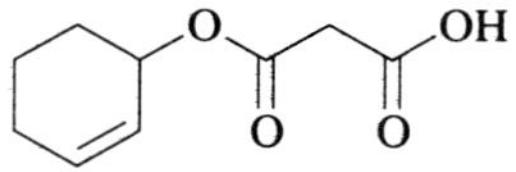	Mn(OAc)₃, AcOH, 40°, 24 h	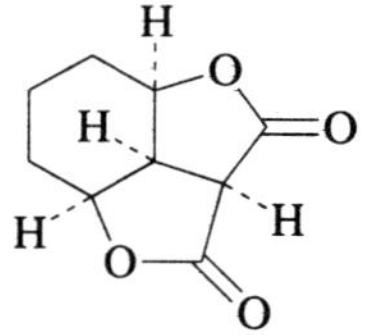 (64)	183
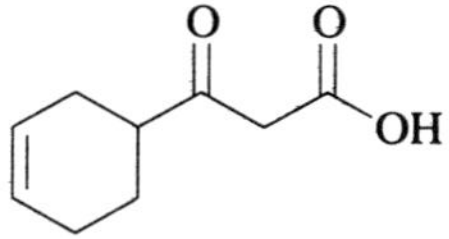	Mn(OAc)₃, AcOH, 23°, 24 h	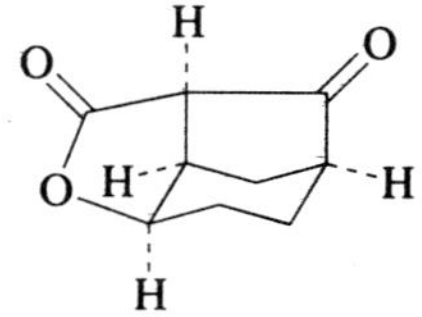 (61)	183
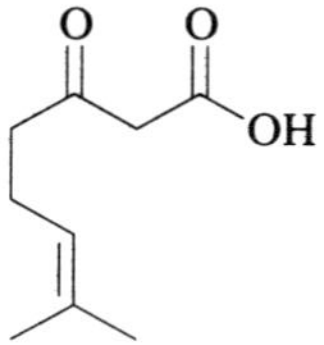	Mn(OAc)₃, AcOH, 23°, 20 min	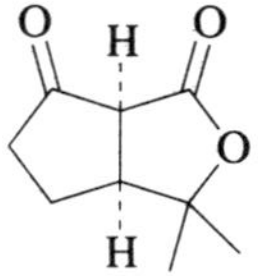 (80)	183
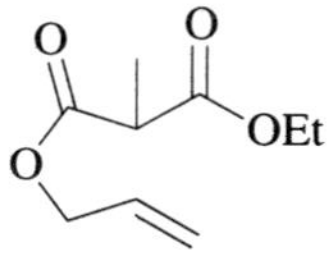	Mn(OAc)₃, Cu(OAc)₂, AcOH, AcONa, reflux	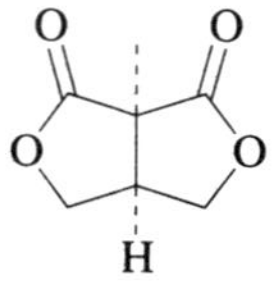 (53) 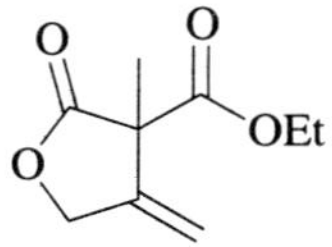 + (20)	96

TABLE XXXVI. TANDEM CYCLIZATIONS (DC-MODE) (*Continued*)

Substrate	Conditions	Product(s) and Yield(s) (%)	Refs.
C$_{10}$	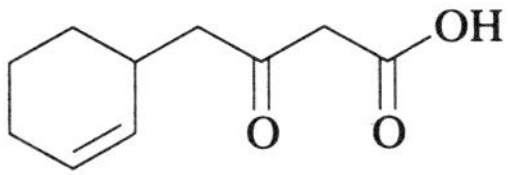 Mn(OAc)$_3$, AcOH, 23°, 20 min	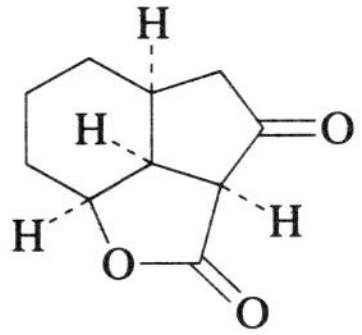 (63)	183
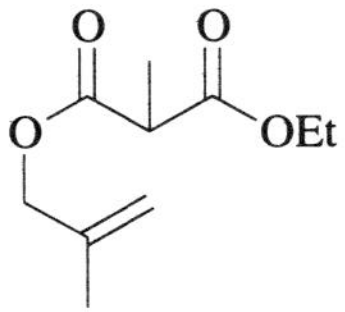	Mn(OAc)$_3$, Cu(OAc)$_2$, AcOH, AcONa, reflux	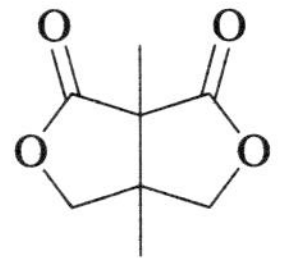 (21)	96
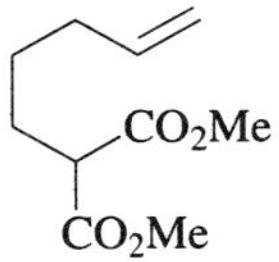	Mn(OAc)$_3$, Cu(OAc)$_2$, AcOH, 55°, 28 h	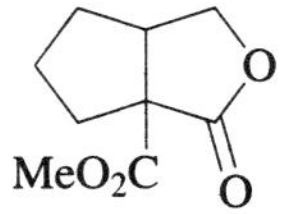 (48)	99
C$_{11}$ X = CN X = CO$_2$Me	Mn(OAc)$_3$, AcOH, 25°	(14) (41) + (0) (3)	178

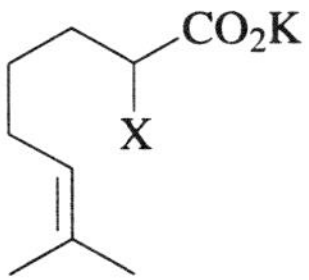

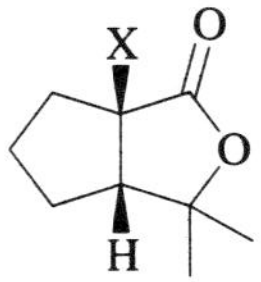

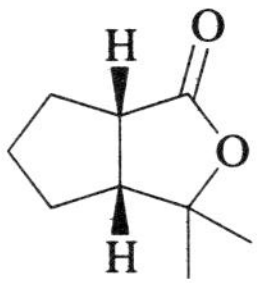

641

TABLE XXXVI. TANDEM CYCLIZATIONS (DC-MODE) (*Continued*)

Substrate	Conditions	Product(s) and Yield(s) (%)	Refs.
C_{11}	Mn(OAc)$_3$, Cu(OAc)$_2$, AcOH	(47)	94
C_{11} X = CN C_{12} X = CO$_2$Me	Mn(OAc)$_3$, AcOH, 70°	+ (0) (50) (8) (48)[a]	178
C_{15}	Mn(OAc)$_3$, Cu(OAc)$_2$, AcOH	(40) + CO$_2$Et (19)	94
C_{17}	1. Mn(OAc)$_3$, AcOH 2. MeCHN$_2$	(40) + (20)	91

642

TABLE XXXVI. TANDEM CYCLIZATIONS (DC-MODE) (*Continued*)

Substrate	Conditions	Product(s) and Yield(s) (%)	Refs.
C$_{21}$	Mn(OAc)$_3$, Cu(OAc)$_2$, AcOH	(61)	91
	Mn(OAc)$_3$, AcOH	(40)	
		(33)	
		endo : exo 4 : 1 (55)	

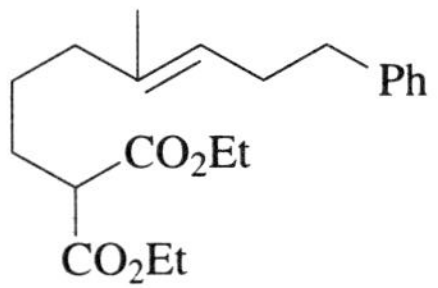

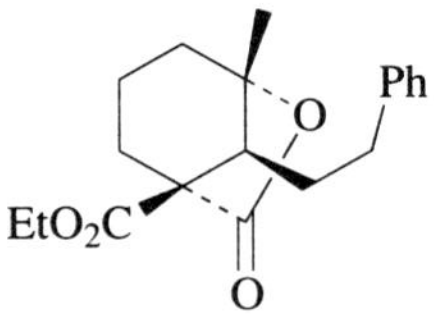

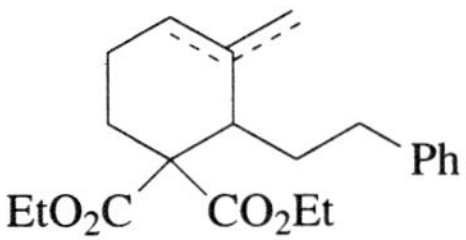

^{a} The product contained 20% of the *E* isomer.

643

TABLE XXXVII. TANDEM CYCLIZATIONS (DN-MODE)

Substrate	Conditions	Product(s) and Yield(s) (%)	Refs.
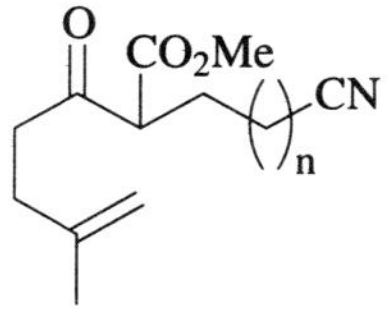 C₁₁ n = 0 C₁₂ n = 1	Mn(OAc)₃, AcOH, 25°, 24 h	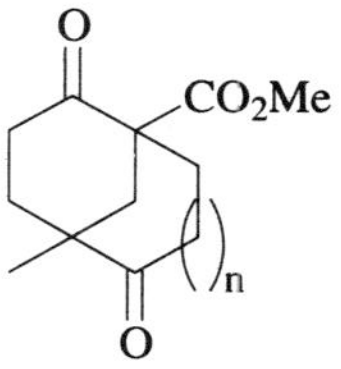 (51) (8)	171
C₁₃	Mn(OAc)₃, AcOH, 25°, 22 h Mn(OAc)₃, EtOH, TFA, 25°, 21 h	(40) (57)	171
	Mn(OAc)₃, EtOH, 25°, 18 h	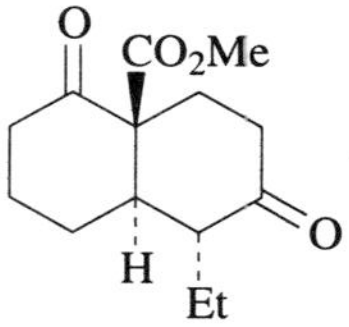 (13) 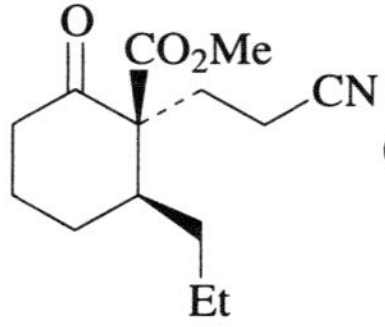 + (4)	171

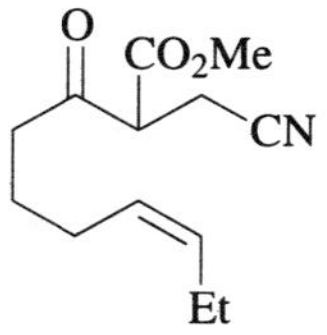

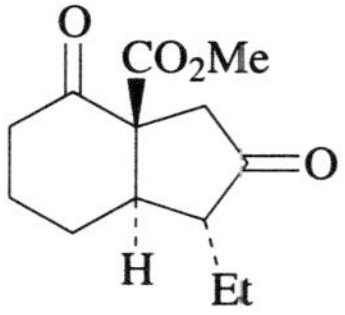

644

TABLE XXXVIII. POLYCYCLIZATION REACTIONS (DDDD-MODE)

Substrate	Conditions	Product(s) and Yield(s) (%)	Refs.
C$_{25}$	Mn(OAc)$_3$, Cu(OAc)$_2$, AcOH, rt, 20 h	(31)	104

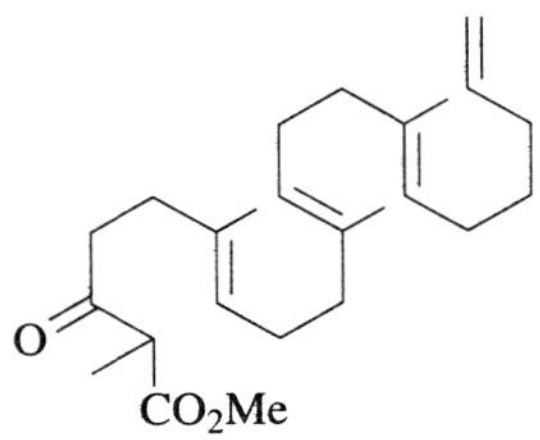

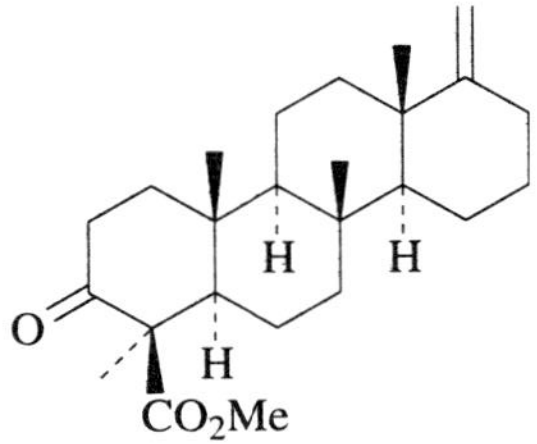

TABLE XXXIX. CARBON MONOXIDE TRAPPING REACTIONS

Substrate	Reagent	Conditions	Product(s) and Yield(s) (%)	Refs.
C_8		Mn(OAc)₃, AcOH CO, 70°, 10 h	(43)	188
C_9		Mn(OAc)₃, AcOH CO, 70°, 15 h	(37)	188
		Mn(OAc)₃, AcOH CO, 70°, 13 h	(55)	188
C_{10}		Mn(OAc)₃, AcOH CO, 70°, 10 h	(50)	188
C_{12}		Mn(OAc)₃, AcOH CO, 70°, 10 h	56 : 44 (44)	188

Substrate	Reagent	Conditions	Product(s) and Yield(s) (%)	Refs.
C$_5$				
AcO⟶⟍	CBr$_4$	Mn(OAc)$_2$, anode, AcOH, AcOK, 40°	AcO, Br, CBr$_3$ (96)	190
	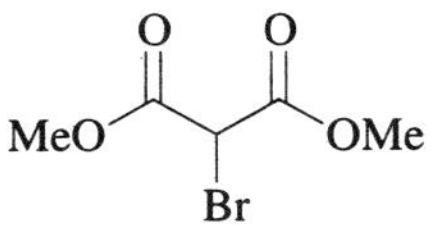 MeO–C(O)–CH(Br)–C(O)–OMe	Mn(OAc)$_2$ (cat.), anode, AcOH, AcOK, 40°	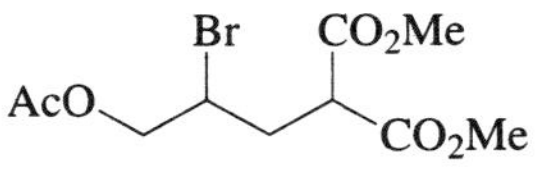 AcO, Br, CO$_2$Me, CO$_2$Me (80)	205
	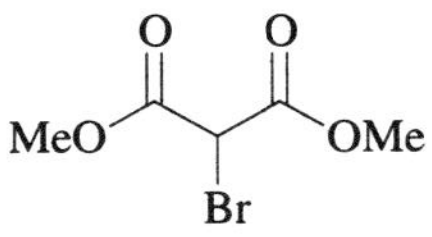 MeO–C(O)–CH(Br)–C(O)–OMe	Mn(OAc)$_2$ (cat.), anode, AcOH, AcOK, 40°	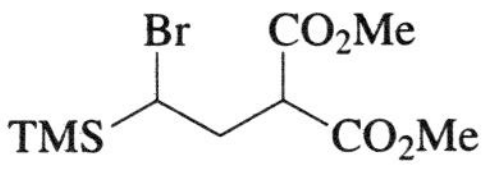 TMS, Br, CO$_2$Me, CO$_2$Me (40)	205
TMS⟶⟍	EtO–C(O)–CH$_2$–CN	Mn(OAc)$_2$ (cat.), anode, AcOH, AcONa, EtOAc, 40°	O=, OEt, CN, cyclohexene(n) (64) (60) (51) (60) (50)	203
C$_5$-C$_{12}$				

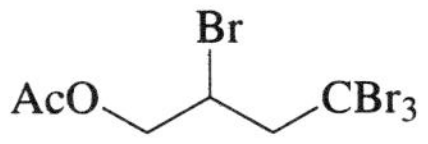

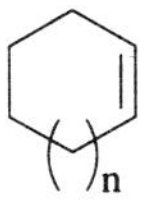

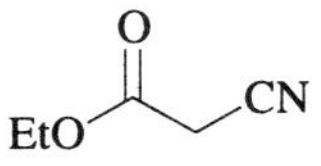

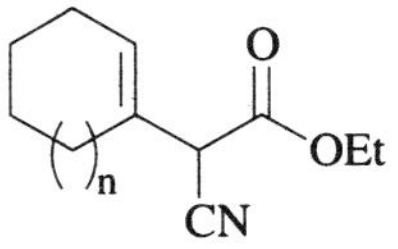

n	
0	(64)
1	(60)
2	(51)
3	(60)
7	(50)

647

Substrate	Reagent	Conditions	Product(s) and Yield(s) (%)	Refs.
C$_6$	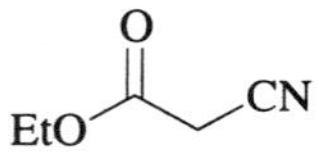 EtO—C(O)—CH$_2$CN	Mn(OAc)$_2$ (cat.), anode, Cu(OAc)$_2$, AcOH, AcONa, EtOAc, 40°	$$\begin{array}{c|c} n & \\ \hline 0 & (44) \\ 1 & (53) \\ 2 & (54) \\ 3 & (47) \\ 7 & (49) \end{array}$$	203, 204
C$_6$–C$_8$ 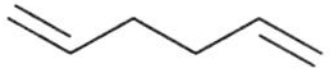	MeO$_2$C–CHBr–CO$_2$Me	Mn(OAc)$_2$ (cat.), anode, AcOH, AcOK, 40°	Br—CH(CO$_2$Me)—...—CO$_2$Me (75)	205
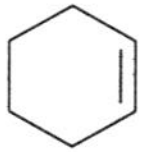	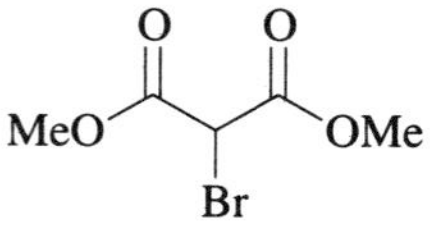 MeO$_2$C–CHBr–CO$_2$Me	Mn(OAc)$_2$ (cat.), anode, AcOH, AcOK, 40°	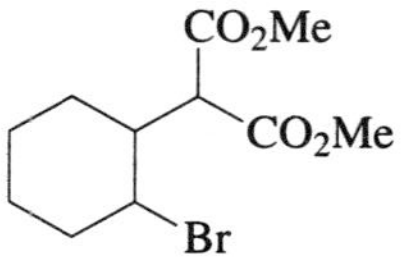 (42)	205
	EtO$_2$C–CH$_2$–C(O)–CH$_3$	Mn(OAc)$_2$ (cat.), anode, AcOH, EtOAc, AcONa, 40°	$$\begin{array}{c|c|c} R^1 & R^2 & \\ \hline n\text{-Pr} & H & (86) \\ n\text{-C}_5\text{H}_{11} & H & (80) \\ t\text{-Bu} & Me & (82) \end{array}$$	204

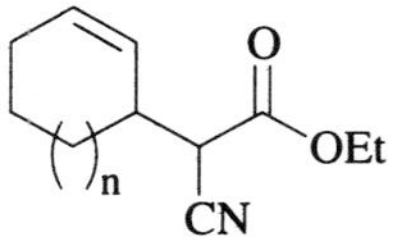

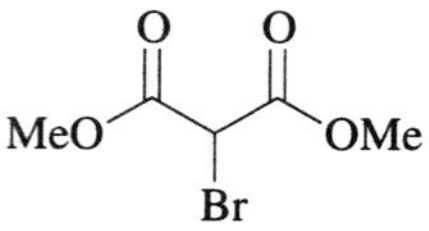

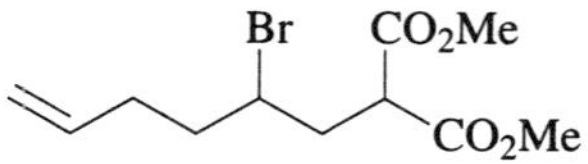

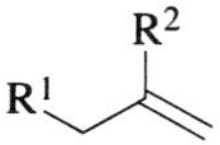

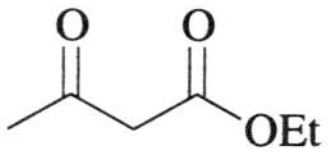

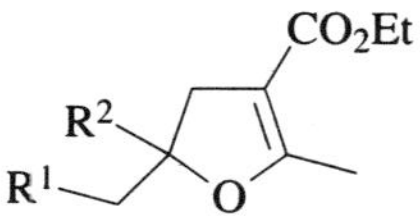

648

TABLE XLA. INTERMOLECULAR REACTIONS WITH ELECTROCHEMICALLY GENERATED Mn(OAc)$_3$ (*Continued*)

Substrate	Reagent	Conditions	Product(s) and Yield(s) (%)	Refs.
C$_6$-C$_{10}$				
R⌒⌒ (allyl)	O⧗EtO⌒CN	Mn(OAc)$_2$ (cat.), anode, AcOH, EtOAc, AcONa, 40°	R⌒⌒⌒CH(CO$_2$Et)CN	203, 204
R = *n*-Pr			(62)	
R = *n*-C$_5$H$_{11}$			(54)	
R = *n*-C$_7$H$_{15}$			(56)	
	O⧗EtO⌒CN	Mn(OAc)$_2$, anode, Cu(OAc)$_2$, AcOH, AcONa, EtOAc, 40°	R⌒⌒=⌒CH(CO$_2$Et)CN	203, 204
R = *n*-Pr			(45)	
R = *n*-C$_5$H$_{11}$			(55)	
R = *n*-C$_7$H$_{15}$			(58)	
C$_7$				
n-C$_5$H$_{11}$⌒	MeO$_2$C⌒C(Br)(R)CO$_2$Me R = Br R = H	Mn(OAc)$_2$ - anode, AcOH, AcOK, 40°	*n*-C$_5$H$_{11}$CH(Br)CH$_2$C(CO$_2$Me)(CO$_2$Me)(R) (61) (78)	190, 205
C$_8$				
n-C$_6$H$_{13}$⌒	CF$_2$Br$_2$	Mn(OAc)$_2$ - anode, AcOH, AcOK, rt	*n*-C$_6$H$_{13}$CH(Br)CH$_2$CF$_2$Br (40)	190
	BrCCl$_3$	Mn(OAc)$_2$ - anode, AcOH, AcOK, 40°	*n*-C$_6$H$_{13}$CH(Br)CH$_2$CCl$_3$ (95)	190
	Br⌒CH(Br)CO$_2$Me	Mn(OAc)$_2$ - anode, AcOH, AcOK, 40°	*n*-C$_6$H$_{13}$CH(Br)CH$_2$CH(Br)CO$_2$Me (52)	190

649

TABLE XLA. INTERMOLECULAR REACTIONS WITH ELECTROCHEMICALLY GENERATED Mn(OAc)₃ (*Continued*)

Substrate	Reagent	Conditions	Product(s) and Yield(s) (%)	Refs.
	$C_8F_{17}I$	Mn(OAc)₂ - anode, AcOH, AcOK, rt	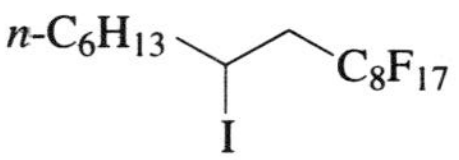 $n\text{-}C_6H_{13}$ ⌇ C_8F_{17} (80)	190
$C_8\text{-}C_{14}$				
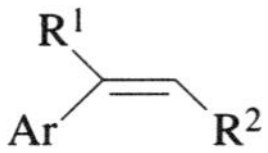	AcOH, Ac₂O	Mn(OAc)₂ (cat.), anode, Cu(OAc)₂, AcOH, Ac₂O, AcONa, 95 - 97°	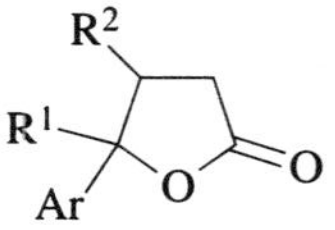	204

Ar	R¹	R²	
Ph	H	H	(80)
Ph	Me	H	(75)
Ph	H	Ph	(61)
$p\text{-}ClC_6H_4$	H	H	(58)
$p\text{-}MeC_6H_4$	H	H	(81)
Ph	H	CO₂Et	(84)
Ph	H	CH₂OH	(78)

TABLE XLB. ADDITION-CYCLIZATIONS WITH ELECTROCHEMICALLY GENERATED Mn(OAc)$_3$

Substrate	Reagent	Conditions	Product(s) and Yield(s) (%)	Refs.
C$_6$ 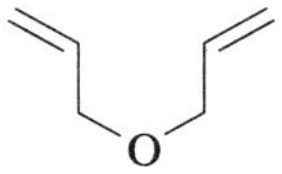 (allyl ether)	CBr$_4$	Mn(OAc)$_2$ (cat.), anode, AcOH, AcOK, 40°	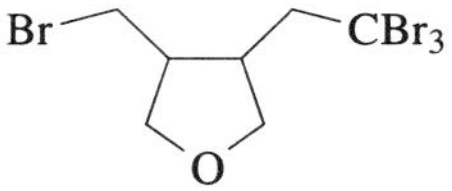 tetrahydrofuran with CH$_2$Br and CH$_2$CBr$_3$ substituents (72)	190
C$_8$ 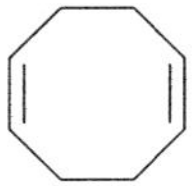 (cyclooctadiene)	BrCCl$_3$	Mn(OAc)$_2$ (cat.), anode, AcOH, AcOK, 40°	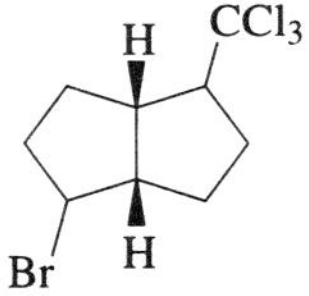 bicyclic, H/H, CCl$_3$, Br (83)	190
	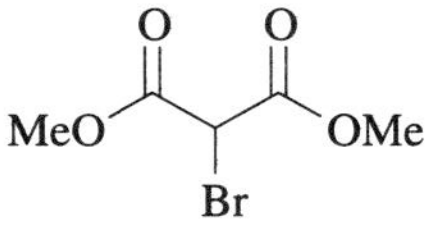 MeO$_2$C–CHBr–CO$_2$Me	Mn(OAc)$_2$ (cat.), anode, AcOH, AcOK, 40°	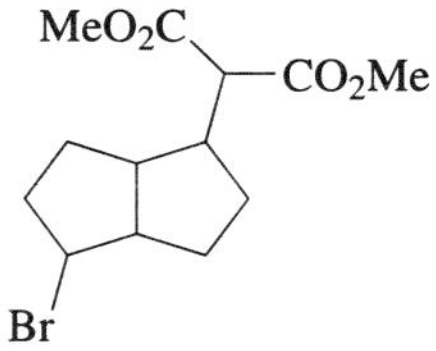 MeO$_2$C–CH(CO$_2$Me)– bicyclic, Br (85)	205
	R–CH$_2$–CO–OEt	Mn(OAc)$_2$ (cat.), anode, AcOH, EtOAc, 20 - 25°	bicyclic with R–CH–C(O)OEt; R = CN (76); R = CO$_2$Et (78)	204

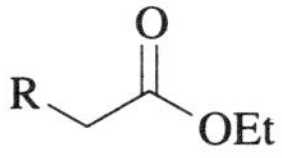

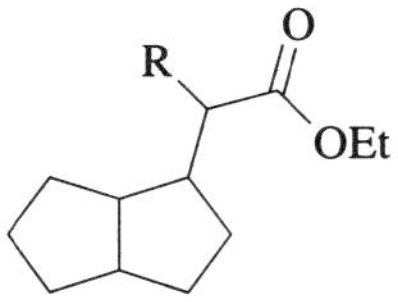

Substrate	Reagent	Conditions	Product(s) and Yield(s) (%)	Refs.
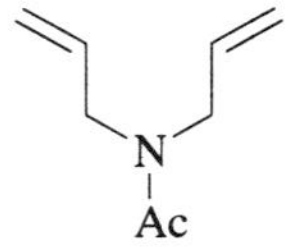 C₁₁-C₁₂	BrCCl₃	Mn(OAc)₂ (cat.), anode, AcOH, AcOK, 40°	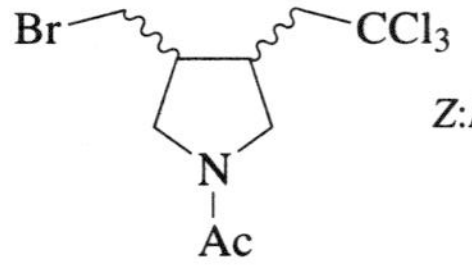 $Z{:}E$ 4.5:1 (60)	190
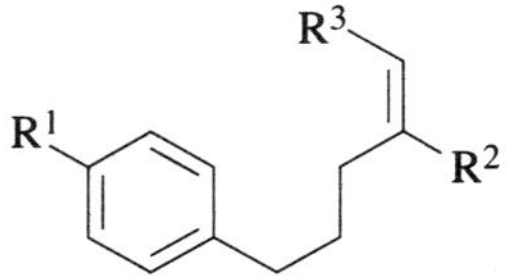	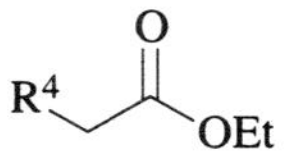	Mn(OAc)₂ (cat.), anode, AcOH, EtOAc	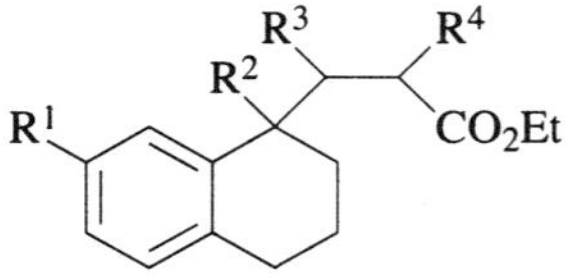	204

R^1	R^2	R^3	R^4	
H	H	H	CN	(70)
Me	H	H	CN	(58)
F	H	Me	CN	(53)
H	Me	H	CN	(60)
H	H	Me	CN	(37)
H	H	H	CO₂Et	(79)
Me	H	H	CO₂Et	(61)
F	H	H	CO₂Et	(60)
H	Me	Me	CO₂Et	(73)
H	H	Me	CO₂Et	(39)

TABLE XLI. SONOCHEMICAL REACTIONS

Substrate	Reagent	Conditions	Product(s) and Yield(s) (%)	Refs.
C$_4$  (2,5-dihydrofuran)	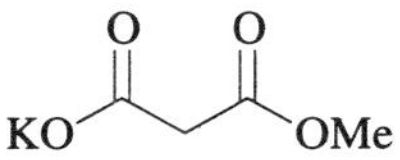 KO / OMe (potassium methyl malonate)	Mn(OAc)$_3$, ultrasound, AcOH, AcOK, 0°, 15 - 25 min	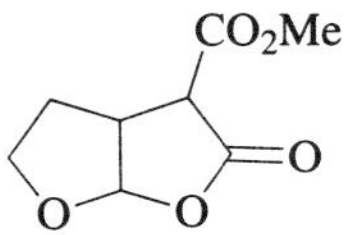 lactone, CO$_2$Me (67)	191
		Mn(OAc)$_3$ (cat.), ultrasound, AcOH, 0°, 1.5 h	(41)	
C$_5$ 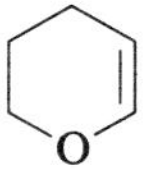 (3,4-dihydro-2H-pyran)	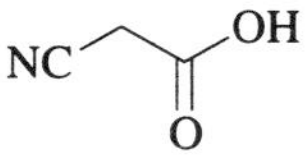 NC / OH (cyanoacetic acid)	Mn(OAc)$_3$, ultrasound, AcOH, AcOK, 0°, 15 - 120 min	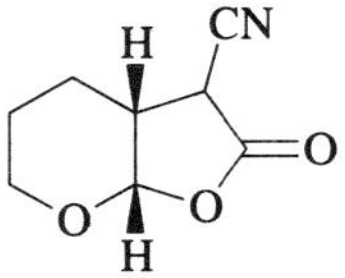 bicyclic lactone, CN, H, H (65)	191
	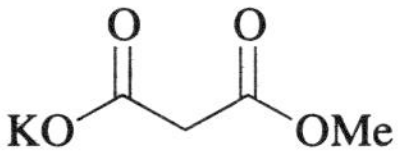 KO / OMe (potassium methyl malonate)	Mn(OAc)$_3$, ultrasound, AcOH, AcOK, 0°, 15 - 25 min	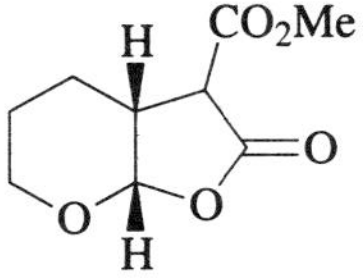 bicyclic lactone, CO$_2$Me, H, H (80)	191
		Mn(OAc)$_3$ (cat.), ultrasound, AcOH, 0°, 1.5 h	(39)	
C$_6$ 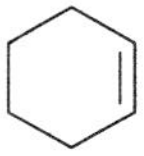 (cyclohexene)	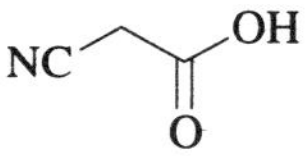 NC / OH (cyanoacetic acid)	Mn(OAc)$_3$, ultrasound, AcOH, AcOK, 0°, 15 - 120 min	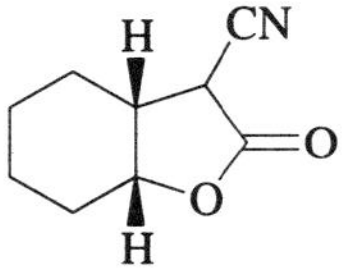 bicyclic lactone, CN, H, H (65)	191

653

TABLE XLI. SONOCHEMICAL REACTIONS (Continued)

Substrate	Reagent	Conditions	Product(s) and Yield(s) (%)	Refs.
C₇		Mn(OAc)₃, ultrasound. AcOH, AcOK, 0°, 15 - 25 min	(78)	191
		Mn(OAc)₃ (cat.), ultrasound, AcOH, 0°, 1.5 h	(22)	
		Mn(OAc)₃, ultrasound. AcOH, AcOK, 0°, 15 - 120 min	(50)	191
		Mn(OAc)₃, ultrasound. AcOH, AcOK, 0°, 15 - 25 min	(75)	191
		Mn(OAc)₃ (cat.), ultrasound, AcOH, 0°, 1.5 h	(34)	

654

TABLE XLI. SONOCHEMICAL REACTIONS (Continued)

Substrate	Reagent	Conditions	Product(s) and Yield(s) (%)	Refs.
C$_8$				
	KO, OMe (dimethyl malonate potassium salt)	Mn(OAc)$_3$, ultrasound, AcOH, AcOK, 0°, 15 - 25 min	CO$_2$Me (81)	191
Ph	NC, OH	Mn(OAc)$_3$, ultrasound, AcOH, AcOK, 0°, 15 - 120 min	CN Ph (55)	191
	NC, OH	Mn(OAc)$_3$, ultrasound, AcOH, AcOK, 0°, 15 - 120 min	CN (73)	191
	NC, OH	Mn(OAc)$_3$, ultrasound, AcOH, AcOK, 0°, 15 - 120 min	CN E : Z 9 : 1 (65)	191
Ph	KO, OMe	Mn(OAc)$_3$, ultrasound, AcOH, AcOK, 0°, 15 - 25 min	CO$_2$Me Ph (62)	191

TABLE XLI. SONOCHEMICAL REACTIONS (Continued)

Substrate	Reagent	Conditions	Product(s) and Yield(s) (%)	Refs.
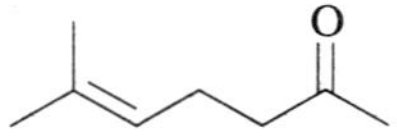 [methyl ketone with trisubstituted alkene]	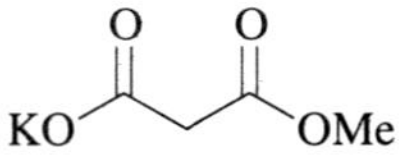 KO–CO–CH₂–CO–OMe	Mn(OAc)₃, ultrasound, AcOH, AcOK, 0°, 15 - 25 min	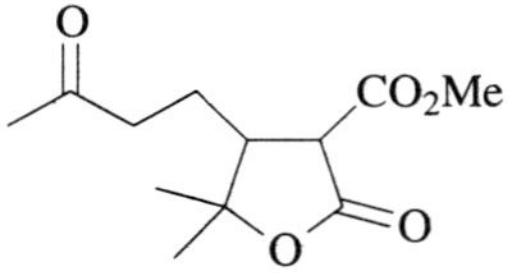 [γ-butyrolactone bearing CO₂Me and oxobutyl chain] (75)	191
[cyclooctene]	KO–CO–CH₂–CO–OMe	Mn(OAc)₃, ultrasound, AcOH, AcOK, 0°, 15 - 25 min	[bicyclic lactone with CO₂Me] **I** **I** E : Z 7 : 3 (70)	191
		Mn(OAc)₃ (cat.), ultrasound, AcOH, 0°, 1.5 h	**I** (39)	
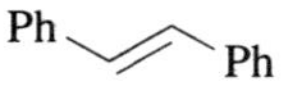 C₁₄ Ph—CH=CH—Ph	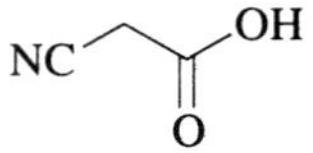 NC–CH₂–CO–OH	Mn(OAc)₃, ultrasound, AcOH, AcOK, 0°, 15 - 120 min	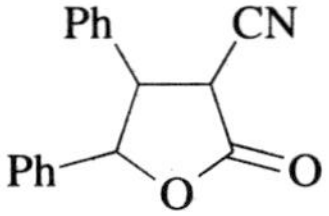 [γ-butyrolactone with CN, Ph, Ph] (45)	191
	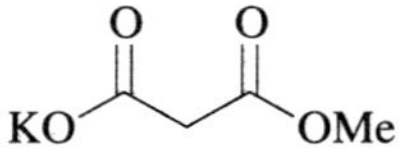 KO–CO–CH₂–CO–OMe	Mn(OAc)₃, ultrasound, AcOH, AcOK, 0°, 15 - 25 min	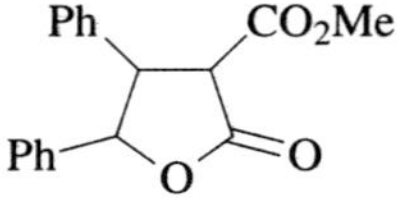 [γ-butyrolactone with CO₂Me, Ph, Ph] (71)	191

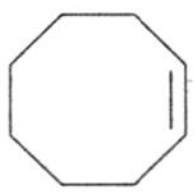

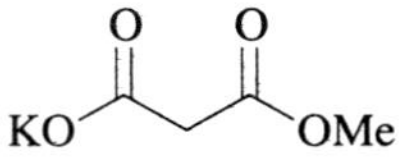

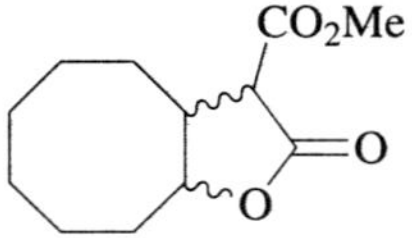

TABLE XLIIA. CYCLOPROPANOL DERIVED ALKYL RADICALS: INTERMOLECULAR ADDITIONS

Substrate	Reagent	Conditions	Product(s) and Yield(s) (%)	Refs.
C$_3$ 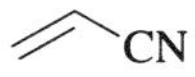 CN	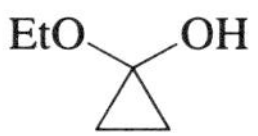 EtO⟩—OH (cyclopropyl)	Mn(pic)$_3$, DMF, 0°, 1 - 2 h	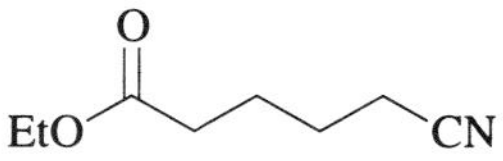 EtO-C(=O)...CN (72)	23
	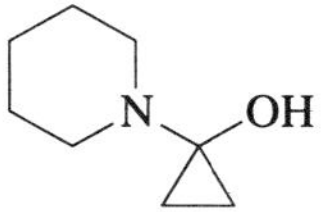 piperidine-N—OH (cyclopropyl)	"	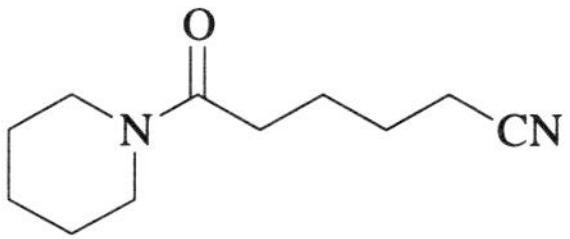 piperidine-N-C(=O)...CN (75)	23
	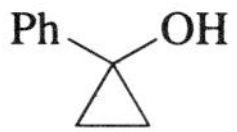 Ph—C(OH) (cyclopropyl)	"	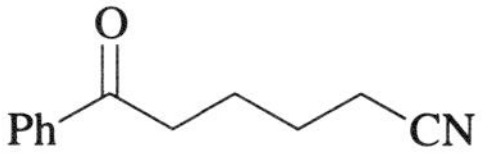 Ph-C(=O)...CN (47)	23
	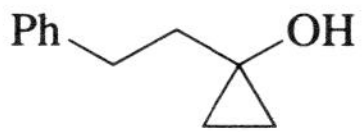 Ph-CH$_2$CH$_2$—C(OH) (cyclopropyl)	"	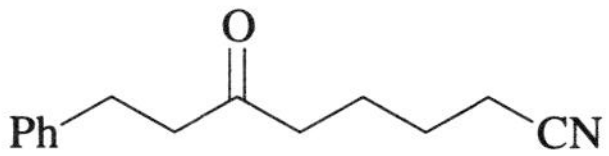 Ph...C(=O)...CN (52)	23
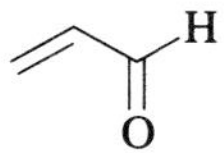 (acrolein) CHO	 EtO⟩—OH (cyclopropyl)	"	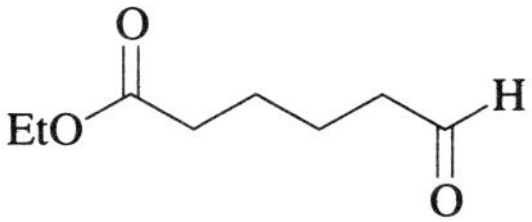 EtO-C(=O)...CHO (47)	23
	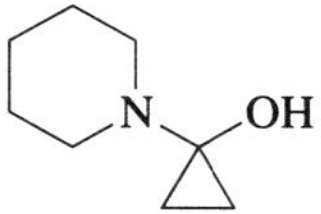 piperidine-N—OH (cyclopropyl)	"	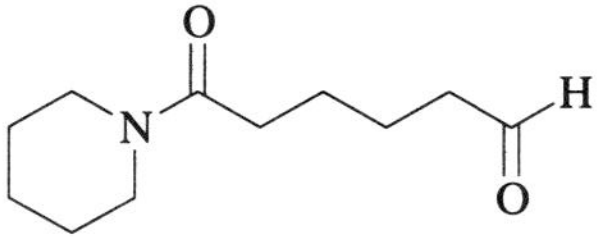 piperidine-N-C(=O)...CHO (48)	23
	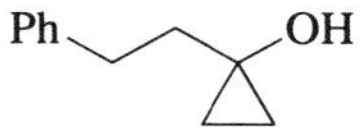 Ph-CH$_2$CH$_2$—C(OH) (cyclopropyl)	"	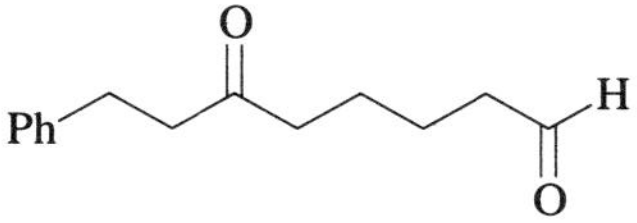 Ph...C(=O)...CHO (46)	23

657

	Substrate	Reagent	Conditions	Product(s) and Yield(s) (%)	Refs.
C₄			Mn(pic)₃, DMF, 0°, 1 - 2 h	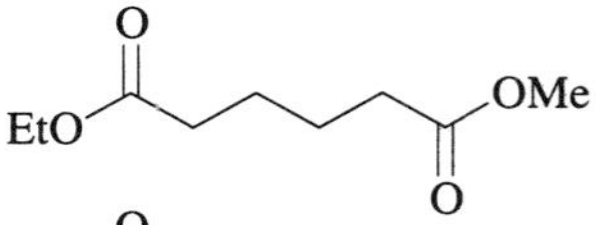 (44)	23
			"	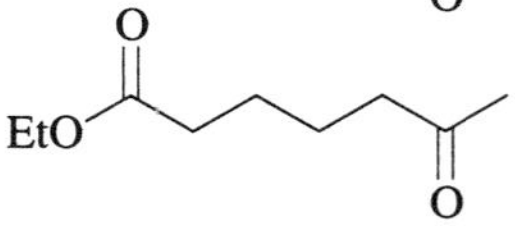 (51)	23
			"	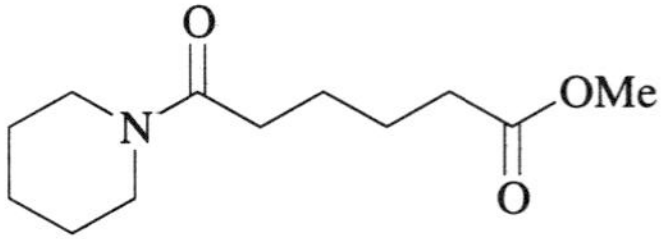 (49)	23
			"	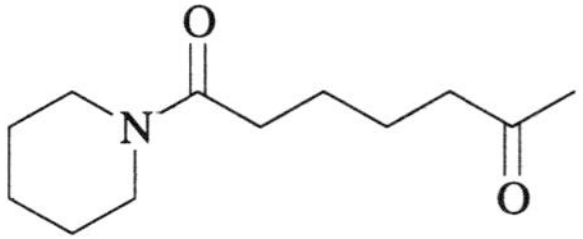 (52)	23
			"	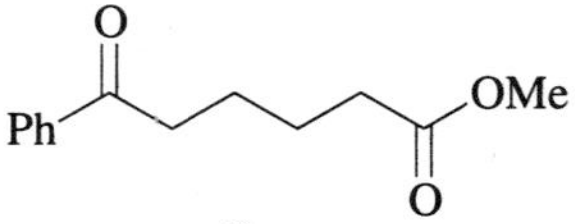 (57)	23
			"	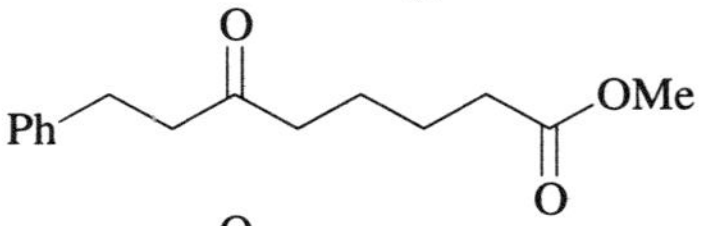 (60)	23
			"	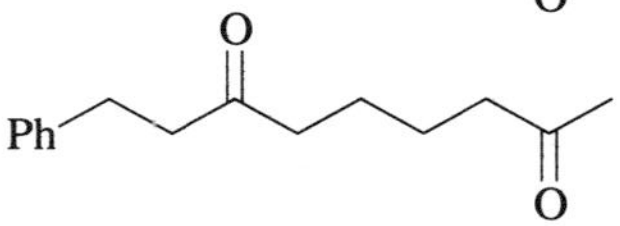 (43)	23

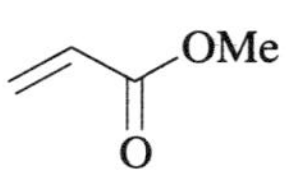

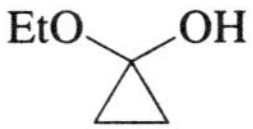

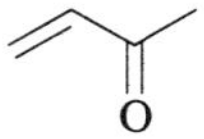

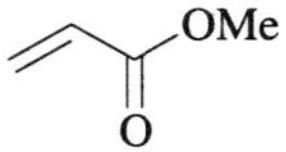

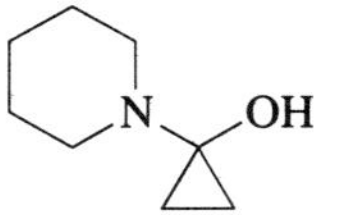

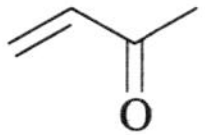

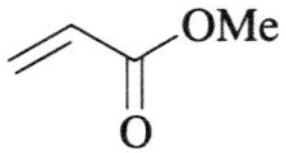

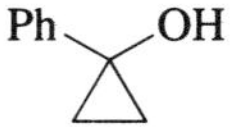

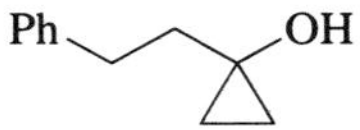

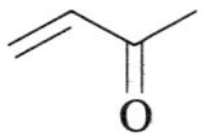

658

TABLE XLIIA. CYCLOPROPANOL DERIVED ALKYL RADICALS: INTERMOLECULAR ADDITIONS (Continued)

Substrate	Reagent	Conditions	Product(s) and Yield(s) (%)	Refs.
C$_5$ (NMe$_2$, O)	Ph—cyclopropane—OH	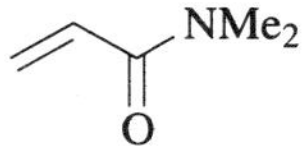 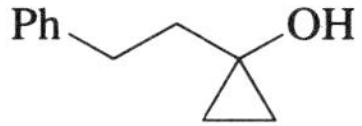 Mn(pic)$_3$, DMF, 0°, 1 - 2 h	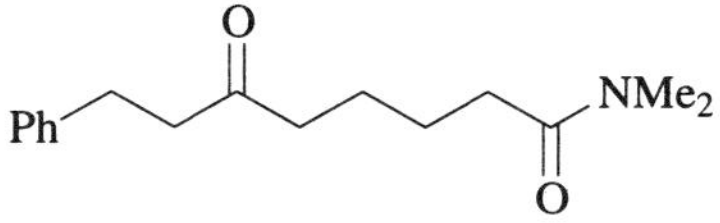 (NMe$_2$ product) (27)	23
C$_6$ (SEt, SEt)	Ph—cyclopropane—OH	Mn(pic)$_3$, DMF, 0°, 0.5 - 5 h	(SEt, SEt product) (71)	23
C$_9$ (OTMS)	"	"	(14)	22, 23
(OMe, Ph)	"	"	(72)	22, 23
C$_{12}$ (OTMS, Ph)	"	"	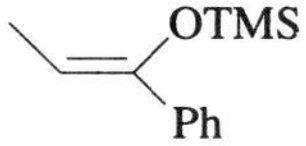 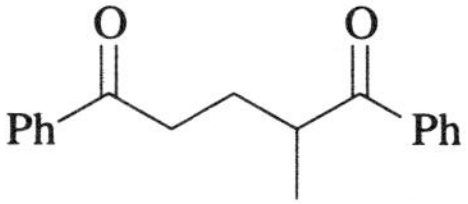 (41)	22, 23
(furan, OTBDMS)	"	"	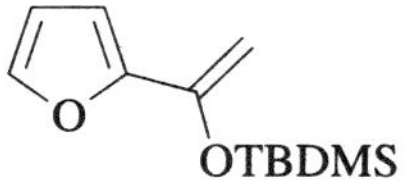 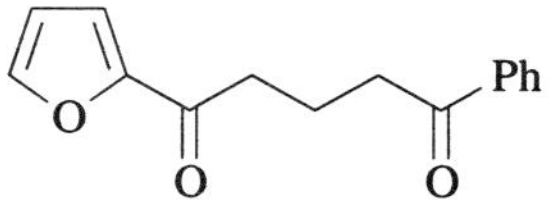 (Ph, O product) (80)	23
C$_{14}$ (OTBDMS, Ph) 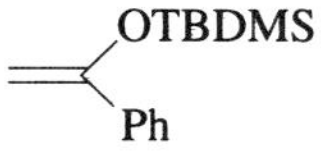 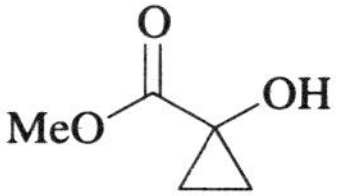 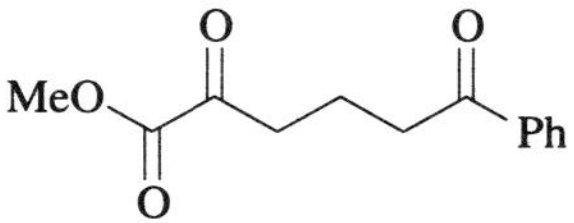	MeO(O)—cyclopropane—OH	"	(MeO, Ph product) (29)	23

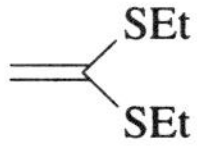

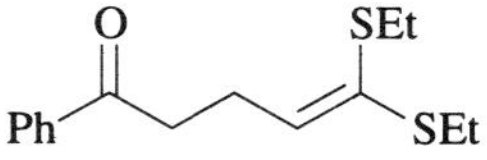

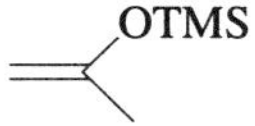

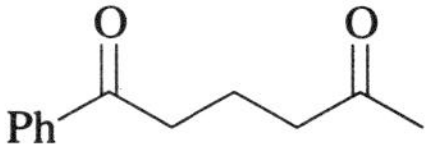

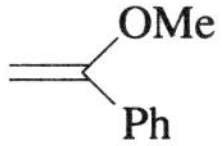

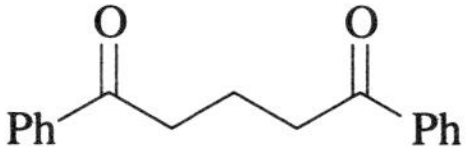

659

TABLE XLIIA. CYCLOPROPANOL DERIVED ALKYL RADICALS: INTERMOLECULAR ADDITIONS (*Continued*)

Substrate	Reagent	Conditions	Product(s) and Yield(s) (%)	Refs.
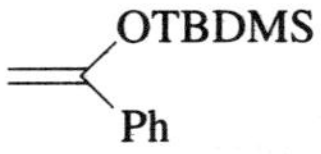 OTBDMS / Ph	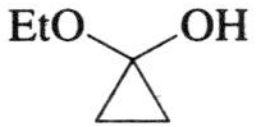 EtO—⊄—OH	Mn(Pic)₃, DMF, 0°, 0.5 - 5 h	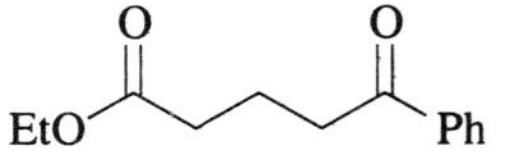 Ph (85)	22, 23
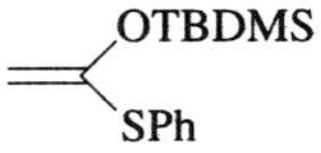 OTBDMS / SPh	"	"	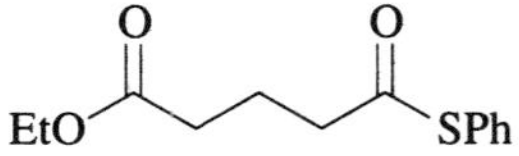 SPh (63)	22, 23
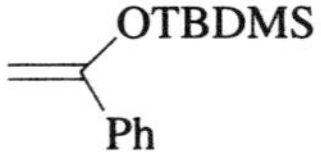 OTBDMS / Ph	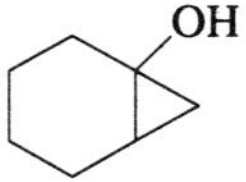 bicyclic OH	"	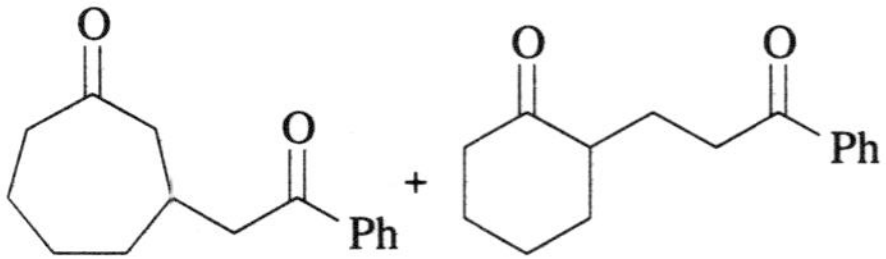 Ph (77) + Ph (5)	22, 23
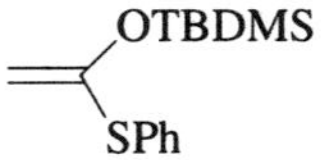 OTBDMS / SPh	"	"	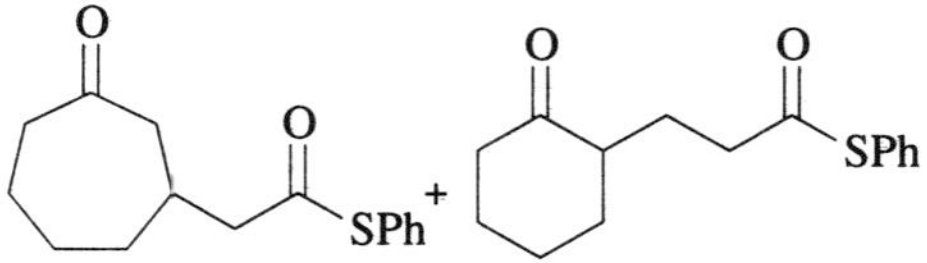 SPh (54) + SPh (10)	22, 23
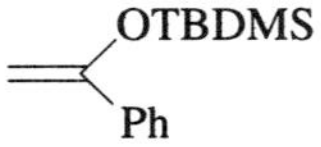 OTBDMS / Ph	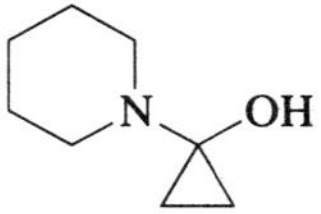 piperidine N—⊄—OH	"	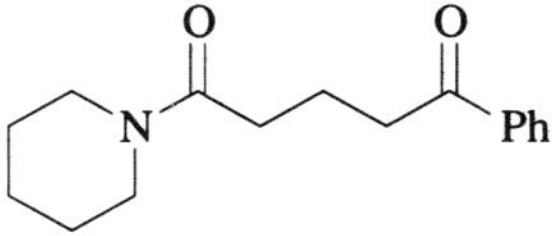 Ph / N-piperidine (65)	23
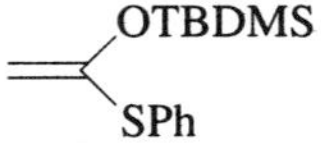 OTBDMS / SPh	"	"	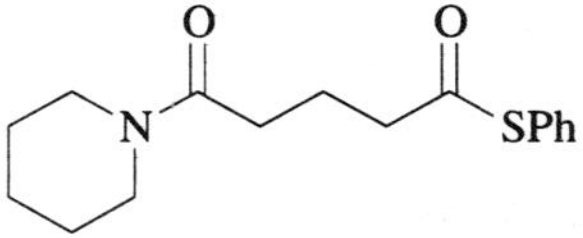 SPh / N-piperidine (46)	23

660

TABLE XLIIA. CYCLOPROPANOL DERIVED ALKYL RADICALS: INTERMOLECULAR ADDITIONS (*Continued*)

Substrate	Reagent	Conditions	Product(s) and Yield(s) (%)	Refs.
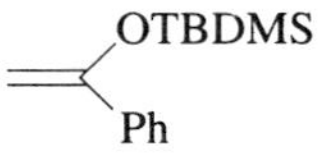 OTBDMS, Ph	 Ph—OH (cyclopropanol)	Mn(pic)$_3$, DMF, 0°, 0.5 - 5 h	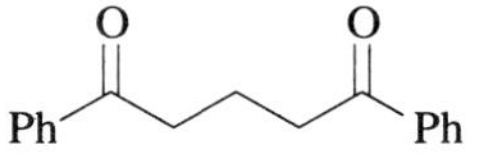 (89)	22, 23
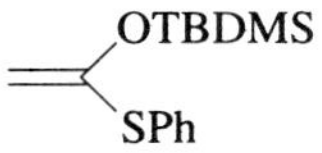 OTBDMS, SPh	"	"	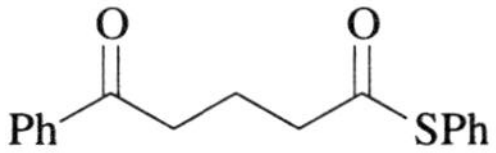 (66)	22, 23
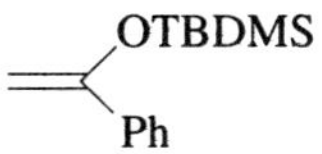 OTBDMS, Ph	 OH (cyclopropanol, Ph-CH$_2$)	"	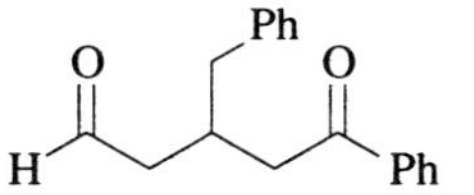 (78)	22, 23
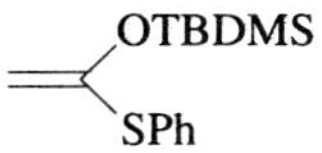 OTBDMS, SPh	"	"	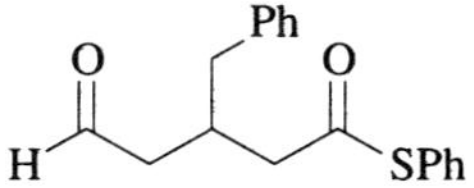 (33)	22, 23
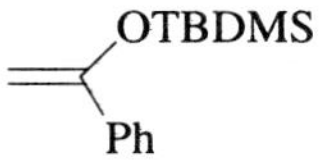 OTBDMS, Ph	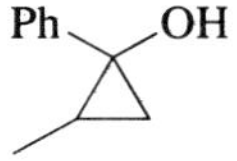 Ph—OH (methylcyclopropanol)	"	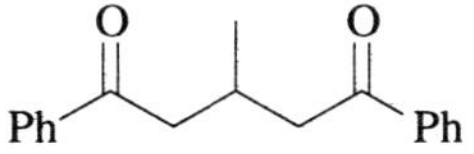 (78)	22, 23
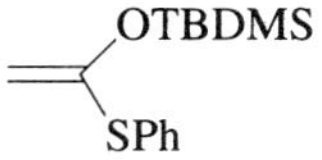 OTBDMS, SPh	"	"	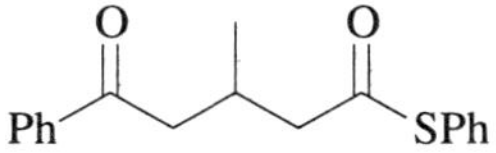 (66)	22, 23
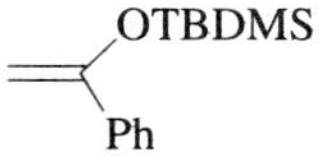 OTBDMS, Ph	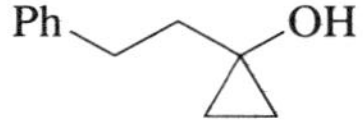 Ph-CH$_2$CH$_2$—OH (cyclopropanol)	"	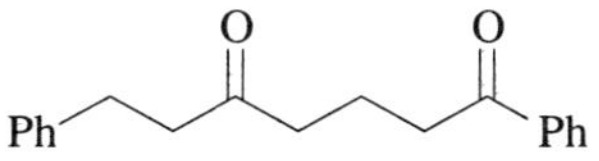 (80)	22, 23

TABLE XLIIA. CYCLOPROPANOL DERIVED ALKYL RADICALS: INTERMOLECULAR ADDITIONS (*Continued*)

Substrate	Reagent	Conditions	Product(s) and Yield(s) (%)	Refs.
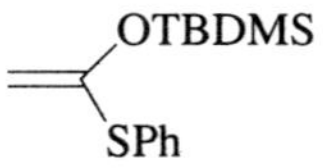 OTBDMS, SPh	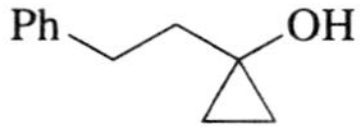 Ph—⟨OH⟩ (cyclopropanol)	Mn(pic)$_3$, DMF, 0°, 0.5 - 5 h	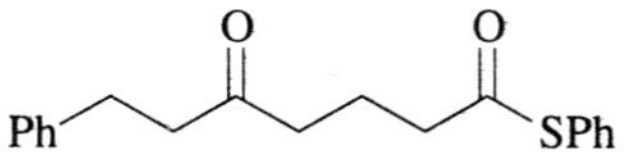 Ph—(O)(O)—SPh (59)	22, 23
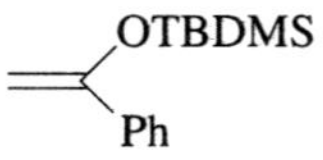 OTBDMS, Ph	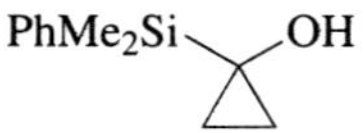 PhMe$_2$Si—⟨OH⟩ (cyclopropanol)	"	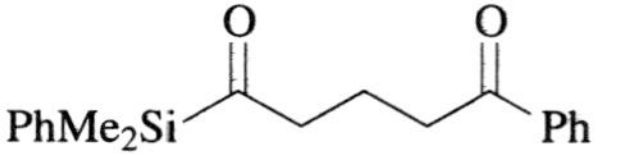 PhMe$_2$Si—(O)(O)—Ph (60)	23
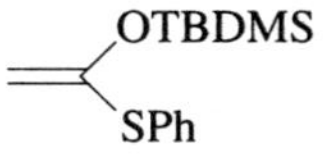 OTBDMS, SPh	"	"	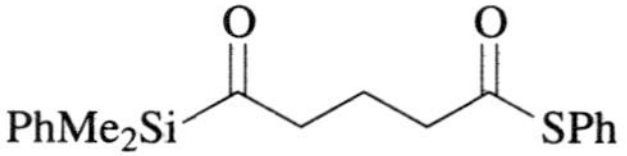 PhMe$_2$Si—(O)(O)—SPh (61)	23
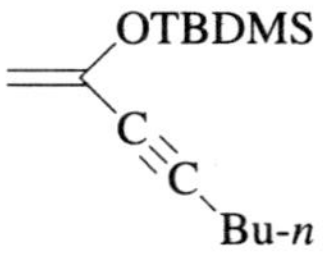 OTBDMS, C≡C—Bu-*n*	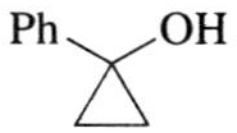 Ph—⟨OH⟩ (cyclopropanol)	"	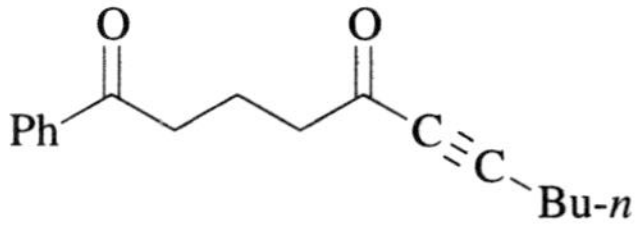 Ph—(O)(O)—C≡C—Bu-*n* (88)	22, 23

TABLE XLIIB. CYCLOPROPANOL DERIVED ALKYL RADICALS: INTRAMOLECULAR ADDITIONS

Substrate	Reagent	Conditions	Product(s) and Yield(s) (%)	Refs.
C$_{11}$ 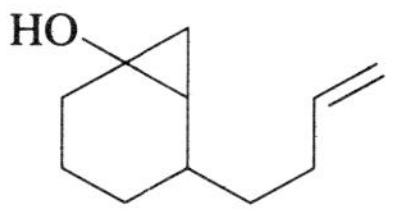	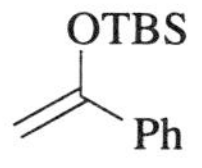 OTBS Ph	Mn(pic)$_3$, DMF, 0°	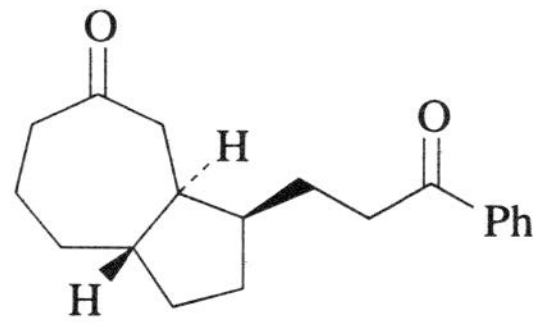 (81)	192
	n-Bu$_3$SnH	"	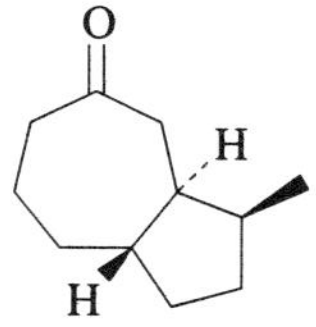 (75)	192
	PhSeSePh	"	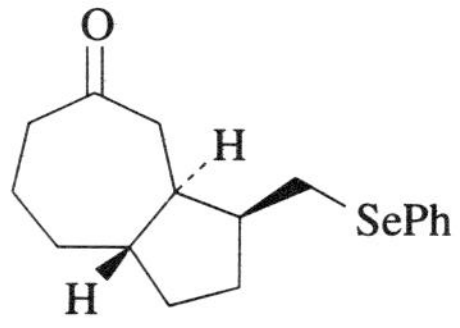 (68)	192
	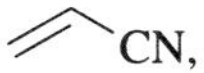 CN, n-Bu$_3$SnH	"	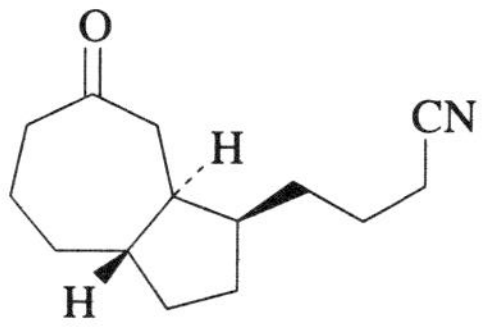 (66)	192

663

TABLE XLIIB. CYCLOPROPANOL DERIVED ALKYL RADICALS: INTRAMOLECULAR ADDITIONS (*Continued*)

Substrate	Reagent	Conditions	Product(s) and Yield(s) (%)	Refs.
C₁₂				
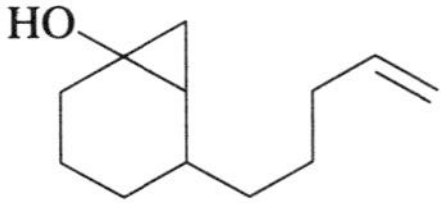	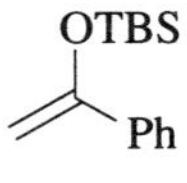	Mn(pic)$_3$, DMF, 0°	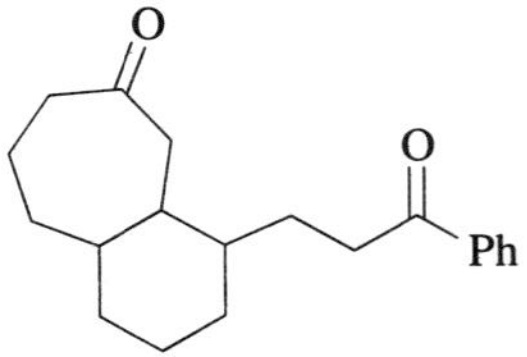 (64)	192
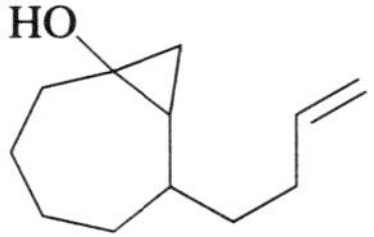	"	"	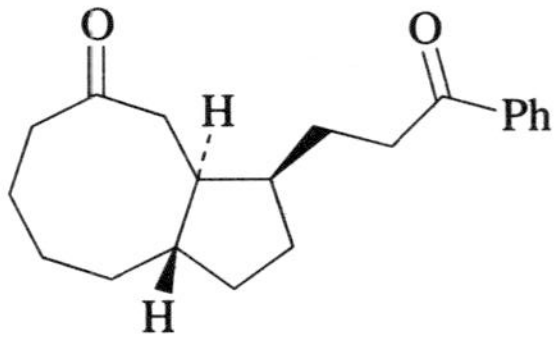 (63)	192

664

Substrate	Conditions	Product(s) and Yield(s) (%)	Refs.
C$_{11}$			
	Mn(OAc)$_3$, EtOH, 25°, 12 h	(45)	24
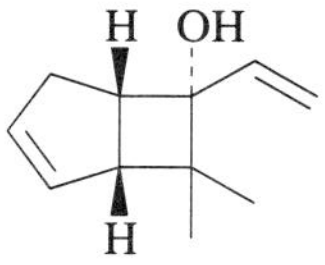	Mn(OAc)$_3$, Cu(OAc)$_2$, EtOH, reflux, 1 h	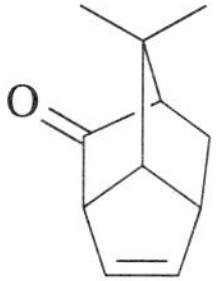 (44) 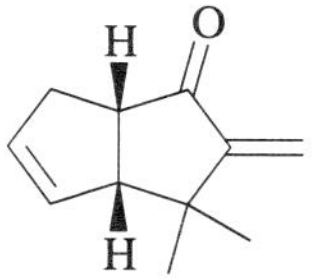 + (22)	24
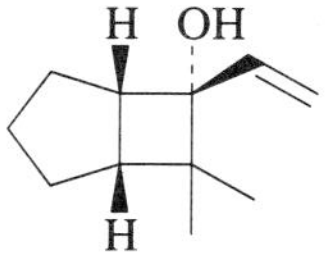	Mn(OAc)$_3$, Cu(OAc)$_2$, EtOH, 25°, 3 h	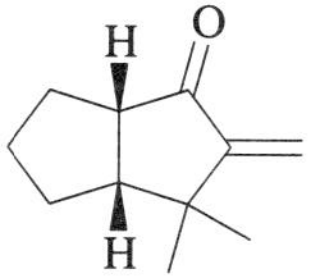 (83)	24
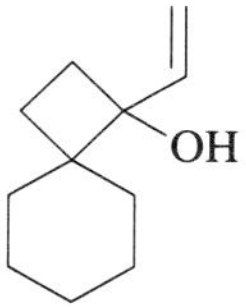	Mn(OAc)$_3$, Cu(OAc)$_2$, EtOH, reflux, 40 min	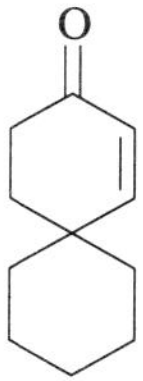 (30) 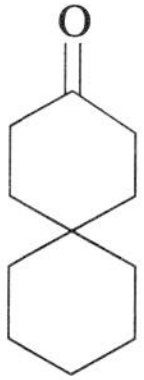 + (24) 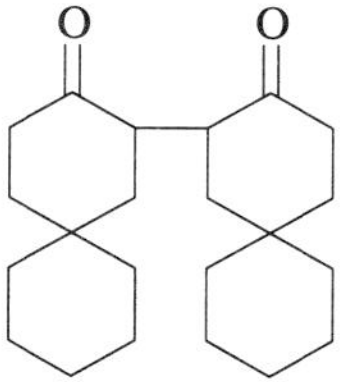 + (22) 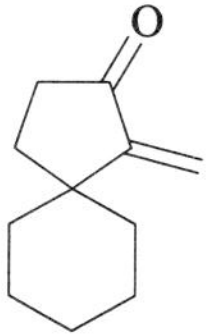 + (6)	24

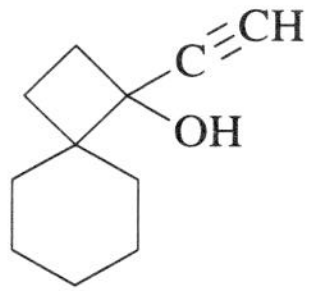

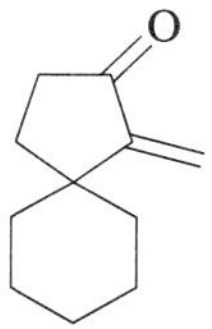

TABLE XLIIC. CYCLOBUTANOL DERIVED ALKYL RADICALS: INTRAMOLECULAR ADDITIONS (*Continued*)

Substrate	Conditions	Product(s) and Yield(s) (%)	Refs.
C$_{12}$ 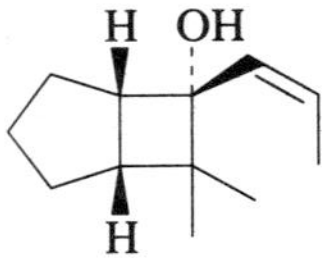	1. Mn(OAc)$_3$, Cu(OAc)$_2$, EtOH, reflux, 0.5 h 2. Et$_3$N, Et$_2$O	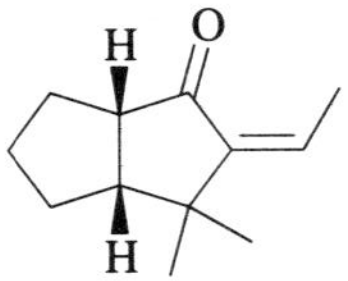 (34) 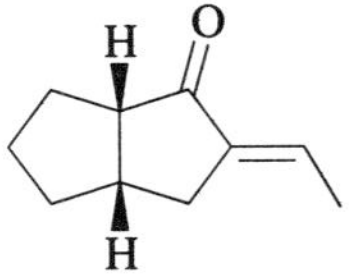 + (29) 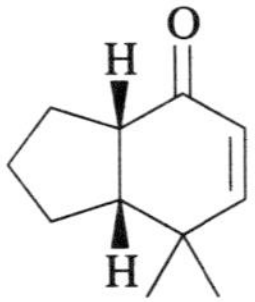 + (10)	24
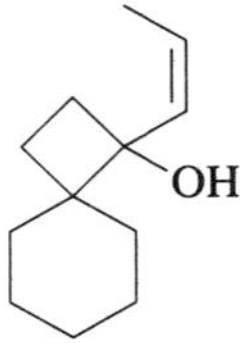 C$_{13}$	Mn(OAc)$_3$, Cu(OAc)$_2$, EtOH, reflux, 30 min	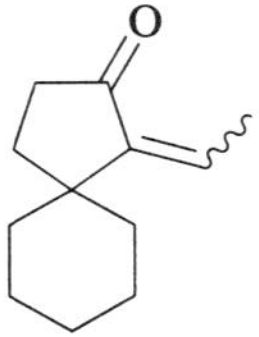 Z (8) + E (22) 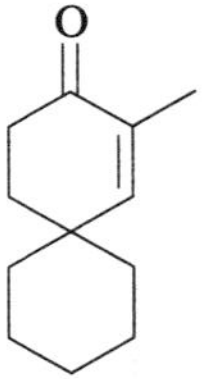 + (24)	24
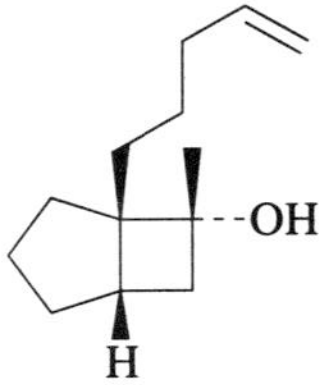	Mn(OAc)$_3$, Cu(OAc)$_2$, EtOH, 25°, 4 h	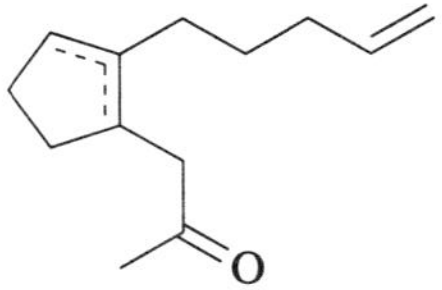 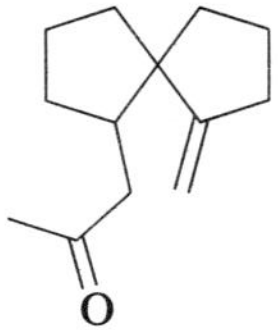 1 : 4 (69)	24
C$_{14}$ 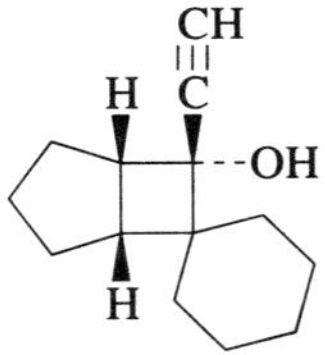	Mn(OAc)$_3$, EtOH, reflux, 20 min	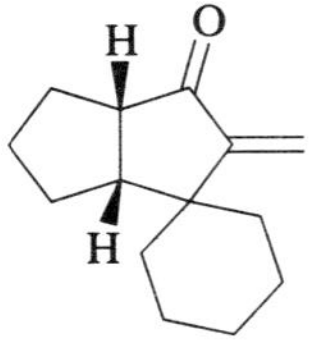 (40)	24

TABLE XLIIC. CYCLOBUTANOL DERIVED ALKYL RADICALS: INTRAMOLECULAR ADDITIONS (*Continued*)

Substrate	Conditions	Product(s) and Yield(s) (%)	Refs.
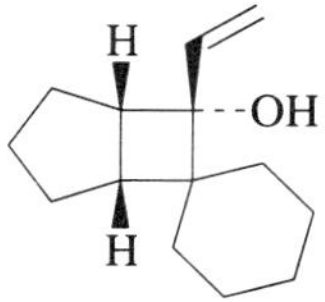 C₁₅	Mn(OAc)₃, Cu(OAc)₂, EtOH, reflux, 40 min	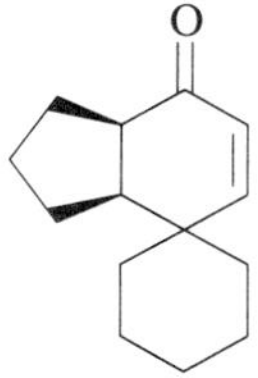 (26) 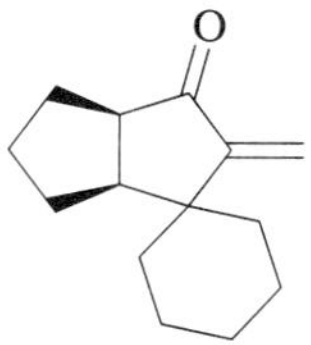 + (22) 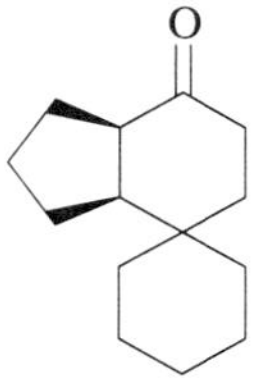 + (17)	24
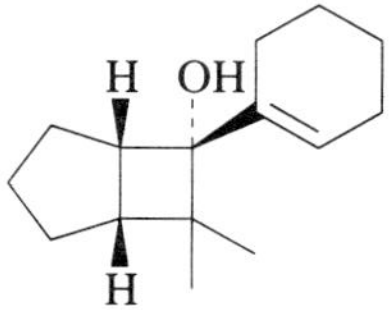	Mn(OAc)₃, Cu(OAc)₂, EtOH, 25°, 11 h	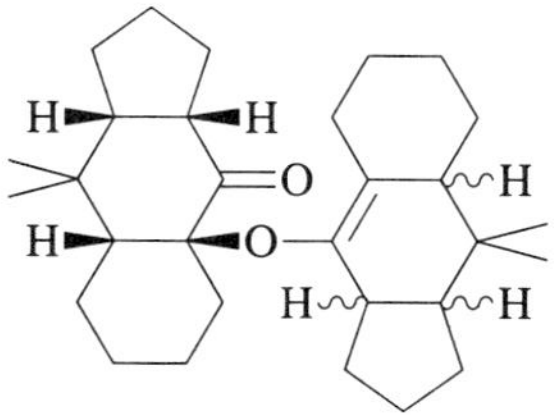 (40) 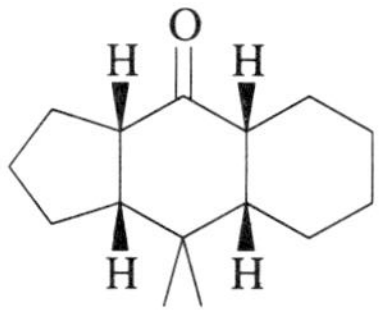 + (21) 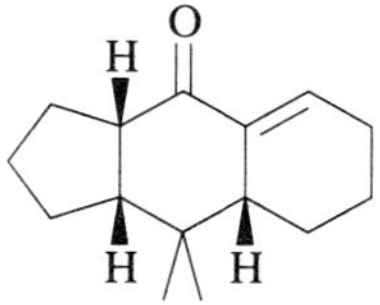 + (14)	24
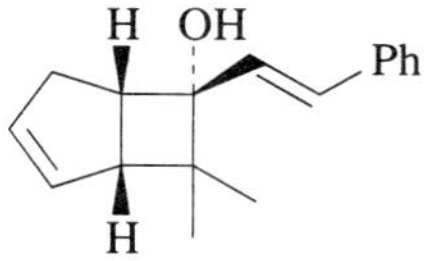 C₁₇	Mn(OAc)₃, Cu(OAc)₂, EtOH, 25°, 12 h	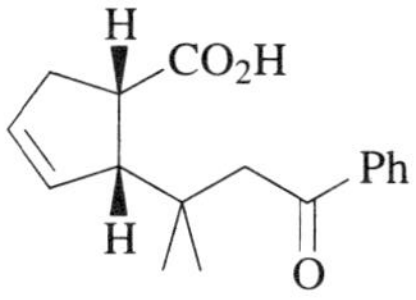 (25)	24

TABLE XLIII. Cr(0) COMPLEX-DERIVED ALKYL RADICALS

Substrate	Reagent	Conditions	Product(s) and Yield(s) (%)	Refs.
C_3 =–CN	$(CO)_5Cr{=}C(O^-\,NMe_4^+)(Ph)_3$	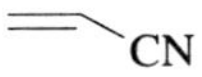 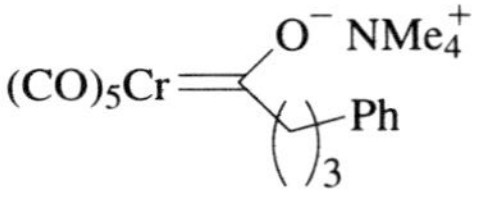 Mn(pic)$_3$, DMF, 0°, 1 h	 Ph–––CN $(24)^a$ $(28)^b$ $(68)^c$	25
C_4 =–CHO	"	"	 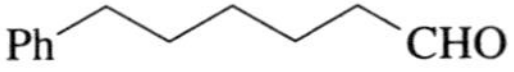 Ph–––CHO $(48)^c$	25
=–COMe	"	"	 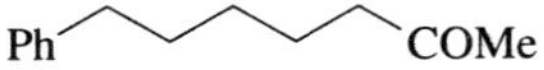 Ph–––COMe $(47)^c$	25
=–CO$_2$Me	"	"	  Ph–––CO$_2$Me $(77)^c$	25
C_6 =–C(SEt)$_2$	"	"	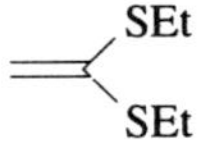 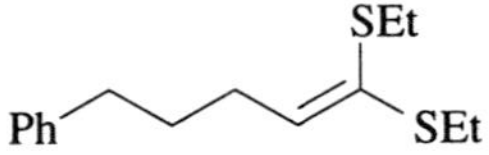 Ph–––CH=C(SEt)(SEt) (64)	25
C_{18} TBDMSO–(furan)–C(=CH$_2$)	"	"	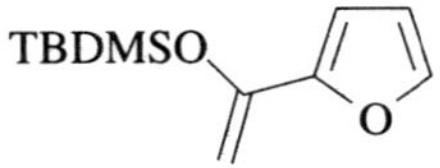 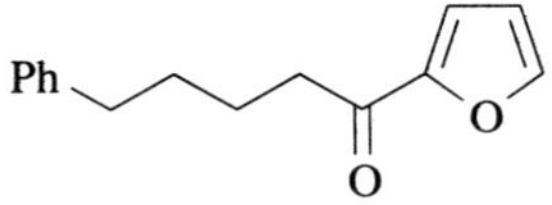 Ph–––C(=O)–(furanyl) (61)	25
C_{20} OTBDMS, Ph =–C(OTBDMS)(Ph) 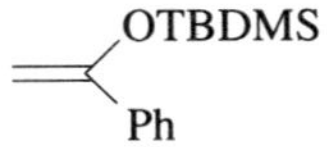 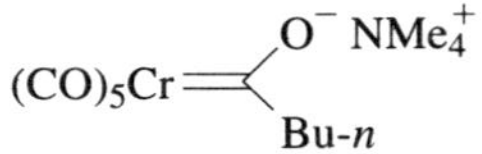 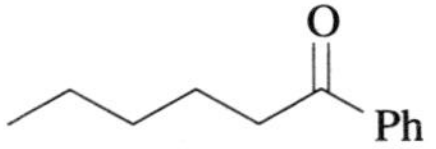	$(CO)_5Cr{=}C(O^-\,NMe_4^+)(Bu\text{-}n)$	"	–––C(=O)Ph (68)	25

668

TABLE XLIII. Cr(0) COMPLEX- DERIVED ALKYL RADICALS (*Continued*)

Substrate	Reagent	Conditions	Product(s) and Yield(s) (%)	Refs.
	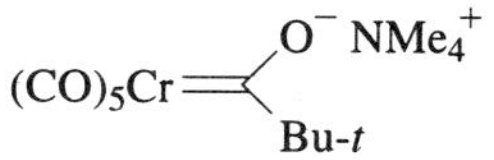 (CO)$_5$Cr=C(O$^-$ NMe$_4^+$)Bu-*t*	Mn(pic)$_3$, DMF, 0°, 1 h	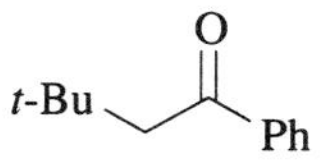 *t*-Bu—CH$_2$—C(O)—Ph (64)	25
	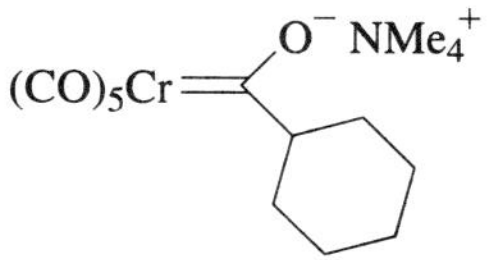 (CO)$_5$Cr=C(O$^-$ NMe$_4^+$)(cyclohexyl)	"	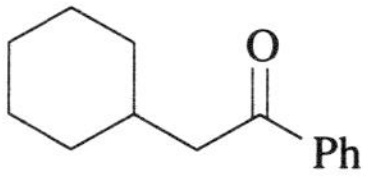 cyclohexyl—CH$_2$—C(O)—Ph (64)	25
	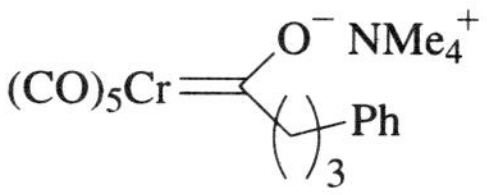 (CO)$_5$Cr=C(O$^-$ NMe$_4^+$)(CPh)$_3$	"	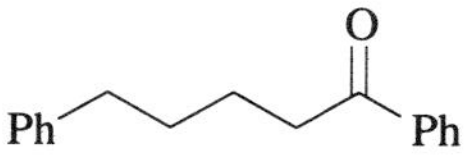 Ph—CH$_2$CH$_2$CH$_2$—C(O)—Ph (74)	25
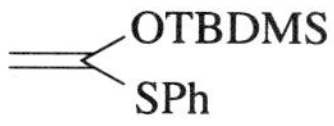 OTBDMS SPh	"	"	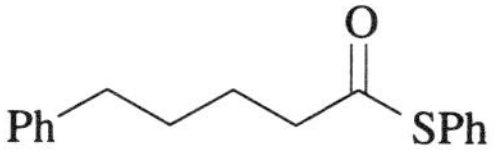 Ph—CH$_2$CH$_2$CH$_2$—C(O)—SPh (33)	25

[a] The trapping agent was 9,10-dihydroanthracene.
[b] The trapping agent was (TMS)$_3$SiH.
[c] The trapping agent was Bu$_3$SnH.

669

REFERENCES

[1] Minisci, F. A. *Acc. Chem. Res.* **1975**, *8*, 165.

[2] *Organic Syntheses by Oxidation with Metal Compounds*, Mijs, W. J.; de Jonge, R. H. I., Eds., Plenum Press, New York, 1986.

[3] Giese, B. *Radicals in Organic Synthesis: Formation of Carbon–Carbon Bonds*, Pergamon Press, Oxford, 1986.

[4] Ramaih, M. *Tetrahedron* **1987**, *43*, 3541.

[5] Curran, D. P. in *Comprehensive Organic Synthesis*, Trost, B. M., Ed., Pergamon Press, Oxford, 1992, chapters 4.1, 4.2, and references therein.

[6] Iqbal, J.; Bhatia, B.; Nayyar, N. K. *Chem. Rev.* **1994**, *94*, 519.

[7] Walling, C.; Hauser, E. S. *Org. React.* **1966**, *13*, 103.

[8] Stacey, F. W.; Harris, J. F. *Org. React.* **1966**, *13*, 170.

[9] de Klein, W. J. in *Organic Syntheses by Oxidation with Metal Compounds,* Mijs, W. J.; de Jonge, R. H. I., Eds., Plenum Press, New York, 1986, p. 261.

[10] Snider, B. B. *Chemtracts-Org. Chem.* **1991**, *4*, 403.

[11] Melikyan, G. G. *Synthesis* **1993**, 833 and references therein.

[12] Fristad, W. E.; Peterson, J. R.; Ernst, A. B.; Urbi, G. B. *Tetrahedron* **1986**, *42*, 3429.

[13] Nishino, H.; Itoh, N.; Nagashima M.; Kurosawa, K. *Bull. Chem. Soc. Jpn.* **1992**, *65*, 620.

[14] Williams, G. J.; Hunter, N. R. *Can. J. Chem.* **1976**, *54*, 3830.

[15] Demir, A. S.; Akgun, H.; Tanyeli, C.; Sayrac, T.; Watt, D. S. *Synthesis* **1991**, 719 and references therein.

[16] Kurosawa, K.; Yamaguchi, K. *Bull. Chem. Soc. Jpn.* **1981**, *54*, 1757.

[17] Tsuruta, T.; Harada, T.; Nishino, H.; Kurosawa, K. *Bull. Chem. Soc. Jpn.* **1985**, *58*, 142.

[18] Yonemura, H.; Nishino, H.; Kurosawa, K. *Bull. Chem. Soc. Jpn.* **1986**, *59*, 3153.

[19] Yonemura, H.; Nishino, H.; Kurosawa, K. *Bull. Chem. Soc. Jpn.* **1987**, *60*, 809.

[20] Kochi, J. K. in *Free Radicals*, Kochi, J. K., Ed., Vol. 1, Wiley, New York, 1973, Chapter 11.

[21] Kochi, J. K. *Acc. Chem. Res.* **1974**, *7*, 351.

[22] Iwasawa, N.; Hayakawa, S.; Isobe, K.; Narasaka, K. *Chem. Lett.* **1991**, 1193.

[23] Iwasawa, N.; Hayakawa, S.; Funahashi, M.; Isobe, K.; Narasaka, K. *Bull. Chem. Soc. Jpn.* **1993**, *66*, 819.

[24] Snider, B. B.; Vo, N.; Foxman, B. *J. Org. Chem.* **1993**, *58*, 7228.

[25] Narasaka, K.; Sakurai, H. *Chem. Lett.* **1993**, 1269.

[26] Vinogradov, M. G.; Direi, P. A.; Nikishin, G. I. *J. Org. Chem. USSR* **1976**, *12*, 518.

[27] van der Ploeg, R. E.; de Korte, R. W.; Kooyman, E. C. *J. Catal.* **1968**, *10*, 52.

[28] Bush, J.; Finkbeiner, H. *J. Am. Chem. Soc.* **1968**, *90*, 5903.

[29] Aidhen, I.; Narasimhan, N. *Tetrahedron Lett.* **1989**, *30*, 5323.

[30] Santi, R.; Bergamini, F.; Citterio, A.; Sebastiano, R.; Nikolini, M. *J. Org. Chem.* **1992**, *57*, 4250.

[31] Snider, B. B.; Patricia, J. J.; Kates, S. A. *J. Org. Chem.* **1988**, *53*, 2137.

[32] March, J. *Advanced Organic Chemistry*, 3rd ed., Wiley, New York, 1985, p. 618.

[33] Heiba, E. I.; Dessau, R. M. *Disc. Faraday Soc.* **1968**, *46*, 189.

[34] Min, R. S.; Aksenov, V. S.; Vinogradov, M. G.; Nikishin, G. I. *Bull. Acad. Sci. USSR, Div. Chem. Sci.* **1979**, 2114.

[35] Kurz, M.; Baru, V.; Nguyen, P-N. *J. Org. Chem.* **1984**, *49*, 1603.

[36] Kurz, M. E.; Chen, R. T. *J. Chem. Soc., Chem. Commun.* **1976**, 968.

[37] Cho, I.; Muchowski, J. *Synthesis* **1991**, 567.

[38] Citterio, A.; Fancelli, D.; Santi, R.; Pagani, A.; Bonsignore, S. *Gazz. Chim. Ital.* **1988**, *118*, 405.

[39] Nishino, H.; Tsunoda, K.; Kurosawa, K. *Bull. Chem. Soc. Jpn.* **1989**, *62*, 545.

[40] Nishino, H. *Bull. Chem. Soc. Jpn.* **1986**, *59*, 1733.

[41] Nikishin, G. I.; Vinogradov, M. G.; Il'ina, G. *J. Org. Chem. USSR* **1972**, *8*, 1422.

[42] Nikishin, G. I.; Vinogradov, M. G.; Verenchikov, S. P.; Kostyukov, I. N.; Kereselidze, R. V. *J. Org. Chem. USSR* **1972**, *8*, 544.

[43] Nikishin, G. I. *Bull. Acad. Sci. USSR, Div. Chem. Sci.* **1984**, 109 and references therein.

[44] Vinogradov, M. G.; Verenchikov, S. P.; Fedorova, T. M.; Nikishin, G. I. *J. Org. Chem. USSR* **1975**, *11*, 937.

[45] Vinogradov, M. G.; Fedorova, T. M.; Nikishin, G. I. *J. Org. Chem. USSR* **1975**, *11*, 1366.

[46] Dessau, R. M.; Heiba, E. I. *J. Org. Chem.* **1974**, *39*, 3457.

[46a] Chatzopoulos, M.; Montheard, J.-P. *C. R. Acad. Sci. Paris* **1977**, *284C*, 133.

[47] Midgley, G.; Thomas, C. B. *J. Chem. Soc., Perkin Trans. 2* **1987**, 1103.

[48] Okano, M.; Aratani, T. *Bull. Chem. Soc. Jpn.* **1976**, *49*, 2811.

[49] Vinogradov, M. G.; Direi, P. A.; Nikishin, G. I. *J. Org. Chem. USSR* **1977**, *13*, 2323.

[50] Vinogradov, M. G.; Dolinko, V. I.; Nikishin, G. I. *Bull. Acad. Sci. USSR, Div. Chem. Sci.* **1984**, 1884.

[51] Qian, C.; Nishino, H.; Kurosawa, K. *Bull. Chem. Soc. Jpn.* **1991**, *64*, 3557.

[52] Qian, C.; Yamada, T.; Nishino, H.; Kurosawa, K. *Bull. Chem. Soc. Jpn.* **1992**, *65*, 1371.

[53] Qian, C.; Nishino, H.; Kurosawa, K. *J. Heterocycl. Chem.* **1993**, *30*, 209.

[54] Qian, C.; Nishino, H.; Kurosawa, K.; Korp, J. *J. Org. Chem.* **1993**, *58*, 4448.

[55] Tategami, S.; Yamada, T.; Nishino, H.; Korp, J.; Kurosawa, K. *Tetrahedron Lett.* **1990**, *31*, 6371.

[56] Zoretic, P. A.; Ramchandani, M.; Caspar, M. L. *Synth. Commun.* **1991**, *21*, 915.

[57] Melikyan, G. G.; Vostrowsky, O.; Bauer, W.; Bestmann, H. J.; Khan, M.; Nicholas, K. M. *J. Org. Chem.* **1994**, *59*, 222.

[58] Heiba, E. I.; Dessau, R. M. *J. Am. Chem. Soc.* **1971**, *93*, 524.

[59] Breuilles, P.; Uguen, D. *Bull. Soc. Chim. Fr.* **1988**, 705.

[60] Midgley, G.; Thomas, C. B. *J. Chem. Soc., Perkin Trans. 2* **1984**, 1537.

[61] Vinogradov, M. G.; Pogosyan, M. S.; Shteinshneider, A. Ya.; Nikishin, G. I. *Bull. Acad. Sci. USSR, Div. Chem. Sci.* **1983**, 768.

[62] Melikyan, G. G.; Mkrtchyan, V. M.; Badanyan, Sh. O.; Vostrovsky, O.; Bestmann, H. J. *Chem. Ber.* **1991**, *124*, 2037.

[63] Vinogradov, M. G.; Dolinko, V. I.; Nikishin, G. I. *Bull. Acad. Sci. USSR, Div. Chem. Sci.* **1984**, 334.

[64] Nikishin, G. I.; Vinogradov, M. G.; Fedorova, T. M. *J. Chem. Soc., Chem. Commun.* **1973**, 693.

[65] Vinogradov, M. G.; Il'ina, G.; Nikishin, G. I. *J. Org. Chem. USSR* **1974**, *10*, 1167.

[66] Vinogradov, M. G.; Petrenko, O. N.; Verenchikov, S. P.; Nikishin, G. I. *J. Org. Chem. USSR* **1980**, *16*, 626.

[67] Vinogradov, M. G.; Petrenko, O. N.; Verenchikov, S. P.; Terent'ev, A. B.; Nikishin, G. I. *Bull. Acad. Sci. USSR, Div. Chem. Sci.* **1979**, 1333.

[68] de Klein, W. *Recl. Trav. Chim. Pays-Bas* **1975**, *94*, 151.

[69] Vinogradov, M. G.; Petrenko, O. N.; Verenchikov, S. P.; Nikishin, G. I. *Bull. Acad. Sci. USSR, Div. Chem. Sci.* **1979**, 1782.

[70] Vinogradov, M. G.; Fedorova, T. M.; Nikishin G. I. *J. Org. Chem. USSR* **1976**, *12*, 1183.

[71] Breuilles, P.; Uguen, D. *Tetrahedron Lett.* **1990**, *31*, 357.

[72] Vinogradov, M. G.; Pogosyan, M. S.; Shteinshneider, A. Ya.; Nikishin, G. I. *Bull. Acad. Sci. USSR, Div. Chem. Sci.* **1981**, 1703.

[73] Melikyan, G. G.; Mkrtchyan, D. A.; Mkrtchyan, V. M.; Badanyan, Sh. O. *Chem. Heterocycl. Compd.* **1985**, 253.

[74] Melikyan, G. G.; Mkrtchyan, D. A.; Badanyan, Sh. O. *Chem. Heterocycl. Compd.* **1981**, 678.

[75] Melikyan, G. G.; Mkrtchyan, D. A.; Badanyan, Sh. O. *Arm. Khim. Zh.* **1981**, *34*, 1011 (*Chem. Abstr.* **96**, 199443g).

[76] Melikyan, G. G.; Mkrtchyan, D. A.; Badanyan, Sh. O. *Chem. Heterocycl. Compd.* **1982**, 14.

[77] Melikyan, G. G.; Sargsyan, A. B.; Giri, V. S.; Grigoryan, R. T.; Badanyan, Sh. O. *Chem. Heterocycl. Compd.* **1988**, 258.

[78] Melikyan, G. G.; Sargsyan, A. B.; Badanyan, Sh. O. *Chem. Heterocycl. Compd.* **1989**, 606.

[79] Kurz, M.; Reif, L.; Tantrarat, T. *J. Org. Chem.* **1983**, *48*, 1373.

[80] Snider, B.; Buckman, B. *Tetrahedron* **1989**, *45*, 6969.

[81] Chuang, C. *Synlett* **1991**, 859.

[82] McQuillin, F. J.; Wood, M. *J. Chem. Soc., Perkin Trans. 1* **1976**, 1762.

[83] Kates, S. A.; Dombroski, M. A.; Snider, B. B. *J. Org. Chem.* **1990**, *55*, 2427.

[84] Snider, B. B.; Patricia, J. J. *J. Org. Chem.* **1989**, *54*, 38.

[85] Snider, B. B.; Mohan, R. M.; Kates, S. A. *J. Org. Chem.* **1985**, *50*, 3659.

[86] Dombroski, M. A.; Snider, B. B. *Tetrahedron* **1992**, *48*, 1417.

[87] Snider, B. B.; Merritt, J. E. *Tetrahedron* **1991**, *47*, 8663.

[88] Merritt, J. E.; Sasson, M.; Kates, S. A.; Snider, B. B. *Tetrahedron Lett.* **1988**, *29*, 5209.

[89] Peterson, J. R.; Egler, R. S.; Horsley, D. B.; Winter, T. J. *Tetrahedron Lett.* **1987**, *28*, 6109.

[90] Mohan, R. M.; Kates, S. A.; Dombroski, M. A.; Snider, B. B. *Tetrahedron Lett.* **1987**, *28*, 845.

[91] Snider, B. B.; Mohan, R. M.; Kates, S. A. *Tetrahedron Lett.* **1987**, *28*, 841.

[92] Oumar-Mahamat, H.; Moustrou, C.; Surzur, J.-M.; Bertrand, M. *Tetrahedron Lett.* **1989**, *30*, 331.

[93] Rama Rao, A.; Venkateswara, R. B.; Reddappa, R. D.; Singh, A. *J. Chem. Soc. Chem. Commun.* **1989**, 400.

[94] Bertrand, M. P.; Surzur, J.-M.; Oumar-Mahamat, H.; Moustrou, C. *J. Org. Chem.* **1991**, *56*, 3089.

[95] Snider, B. B.; McCarthy, B. A. *Tetrahedron* **1993**, *49*, 9447.

[96] Oumar-Mahamat, H.; Moustrou, C.; Surzur, J.-M.; Bertrand, M. P. *J. Org. Chem.* **1989**, *54*, 5684.

[97] Snider, B. B.; Armanetti, L.; Baggio, R. *Tetrahedron Lett.* **1993**, *34*, 1701.

[98] Zoretic, P. A.; Ramchandani, M.; Caspar, M. L. *Synth. Commun.* **1991**, *21*, 923.

[99] Snider, B. B.; Merritt, J. E.; Dombroski, M. A.; Buckman, B. O. *J. Org. Chem.* **1991**, *56*, 5544.

[100] Snider, B. B.; Dombroski, M. A. *J. Org. Chem.* **1987**, *52*, 5487.

[101] Dombroski, M. A.; Kates, S. A.; Snider, B. B. *J. Am. Chem. Soc.* **1990**, *112*, 2759.

[102] Snider, B. B.; McCarthy, B. A. *J. Org. Chem.* **1993**, *58*, 6217.

[103] Snider, B. B.; Zhang, Q.; Dombroski, M. A. *J. Org. Chem.* **1992**, *57*, 4195.

[104] Zoretic, P. A.; Weng, X.; Caspar, M. L.; Davis, D. G. *Tetrahedron Lett.* **1991**, *32*, 4819.

[105] Vinogradov, M. G.; Dolinko, V. I.; Nikishin, G. I. *Bull. Acad. Sci. USSR, Div. Chem. Sci.* **1983**, 2036.

[106] Heiba, E. I.; Dessau, R. M. *J. Org. Chem.* **1974**, *39*, 3456.

[107] Citterio, A.; Marion, A.; Maronati, A.; Nicolini, M. *Tetrahedron Lett.* **1993**, *34*, 7981.

[108] Lamarque, L.; Meou, A.; Brun, P. *Tetrahedron Lett.* **1994**, *35*, 2903.

[109] Nishino, H.; Yoshida, T.; Kurosawa, K. *Bull. Chem. Soc. Jpn.* **1991**, *64*, 1097.

[110] Vinogradov, M. G.; Il'ina, G. P.; Ignatenko, A. V.; Nikishin, G. I. *J. Org. Chem. USSR* **1972**, *8*, 1425.

[111] McQuillin, F. J.; Wood, M. *J. Chem. Res. S (Synopses)* **1977**, 61.

[112] Vinogradov, M. G.; Kovalev, I. P.; Nikishin, G. I. *Bull. Acad. Sci. USSR, Div. Chem. Sci.* **1984**, 342.

[113] Vinogradov, M. G.; Direi, P. A.; Nikishin, G. I. *J. Org. Chem. USSR* **1978**, *14*, 1894.

[114] Vinogradov, M. G.; Verenchikov, S. P.; Nikishin, G. I. *J. Org. Chem. USSR* **1976**, *12*, 2245.

[115] Kamiya, Y.; Kotake, M. *Bull. Chem. Soc. Jpn.* **1973**, *46*, 2780.

[116] Vinogradov, M. G.; Min, R. S.; Aksenov, V. S.; Nikishin, G. I. *Bull. Acad. Sci. USSR, Div. Chem. Sci.* **1982**, 1760.

[117] Fristad, W. E.; Peterson, J. R. *J. Org. Chem.* **1985**, *50*, 10.

[118] Narasaka, K.; Iwakura, K.; Okauchi, T. *Chem. Lett.* **1991**, 423.

[119] Curran, D. P.; Morgan, T. M.; Schwartz, E. C.; Snider, B. B.; Dombroski, M. A. *J. Am. Chem. Soc.* **1991**, *113*, 6607.

[120] Anderson, J. M.; Kochi, J. K. *J. Am. Chem Soc.* **1970**, *92*, 2450.

[121] Vinogradov, M. G.; Kovalev, I. P.; Nikishin, G. I. *Bull. Acad. Sci. USSR, Div. Chem. Sci.* **1981**, 1265.

[122] Nikishin, G. I.; Vinogradov, M. G.; Il'ina, G. P. *Synthesis* **1972**, 376.

[123] Petrenko, O. N.; Vinogradov, M. G.; Verenchikov, S. P.; Shteinshneider, A. Ya.; Terent'ev, A. B.; Nikishin, G. I. *J. Org. Chem. USSR* **1978**, *14*, 1292.

[124] Citterio, A.; Gentile, A.; Minisci, F.; Serravalle, M. *Gazz. Chim. Ital.* **1983**, *113*, 443.

[125] Heiba, E. I.; Dessau, R. M.; Koehl, W. *J. Am. Chem. Soc.* **1968**, *90*, 5905.

[126] Okano, M. *Bull. Chem. Soc. Jpn.* **1976**, *49*, 1041.

[127] Sherwani, M.; Ahmad, M.; Ahmad, I.; Osman, S. *Indian J. Chem., Sect. B* **1985**, *24B*, 629.

[128] Heiba, E. I.; Dessau, R. M.; Rodewald, P. G. *J. Am. Chem. Soc.* **1974**, *96*, 7977.

[129] Fristad, W. E.; Peterson, J. R.; Ernst, A. B. *J. Org. Chem.* **1985**, *50*, 3143.

[130] Kurosawa, K.; Harada, H. *Bull. Chem. Soc. Jpn.* **1979**, *52*, 2386.

[131] Tanimoto, S.; Ohnishi, A. *Bull. Inst. Chem. Res.* **1989**, *66*, 369.

[132] Kasahara, A.; Izumi, T.; Suzuki, A.; Takeda, T. *Bull. Chem. Soc. Jpn.* **1976**, *49*, 3711.

[133] Witkiewicz, K.; Chabudzinski, Z. *Rocz. Chem.* **1977**, *51*, 2155.

[134] Sugie, A.; Shimomura, H.; Katsube, J.; Yamamoto, H. *Tetrahedron Lett.* **1977**, 2759.

[135] Nishino, H. *Bull. Chem. Soc. Jpn.* **1985**, *58*, 1922.

[136] Yang, F. Z.; Trost, M. K.; Fristad, W. E. *Tetrahedron Lett.* **1987**, *28*, 1493.

[137] Corey, E. J.; Ghosh, A. *Chem. Lett.* **1987**, 223.

[138] Corey, E. J.; Ghosh, A. *Tetrahedron Lett.* **1987**, *28*, 175.

[139] Mellor, J. M.; Mohammed, S. *Tetrahedron Lett.* **1991**, *32*, 7107.

[140] Mellor, J. M.; Mohammed, S. *Tetrahedron Lett.* **1991**, *32*, 7111.

[141] Mellor, J. M.; Mohammed, S. *Tetrahedron* **1993**, *49*, 7547.

[142] Mellor, J. M.; Mohammed, S. *Tetrahedron* **1993**, *49*, 7557.

[143] Mellor, J. M.; Mohammed, S. *Tetrahedron* **1993**, *49*, 7567.

[144] Ito, N.; Nishino, H.; Kurosawa, K. *Bull. Chem. Soc. Jpn.* **1983**, *56*, 3527.

[145] Fristad, W. E.; Hershberger, S. S. *J. Org. Chem.* **1985**, *50*, 1026.

[146] Fujimoto, N.; Nishino, H.; Kurosawa, K. *Bull. Chem. Soc. Jpn.* **1986**, *59*, 3161.

[147] Corey, E. J.; Gross, A. W. *Tetrahedron Lett.* **1985**, *26*, 4291.

[148] Peterson, J. R.; Do, H. D.; Surjasasmita, I. B. *Synth. Comm.* **1988**, *18*, 1985.

[149] Rosario-Chow, M.; Ungwitayatorn, J.; Currie, B. L. *Tetrahedron Lett.* **1991**, *32*, 1011.

[150] Nishino, H. *Bull. Chem. Soc. Jpn.* **1985**, *58*, 217.

[151] Sato, H.; Nishino, H.; Kurosawa, K. *Bull. Chem. Soc. Jpn.* **1987**, *60*, 1753.

[152] Vinogradov, M. G.; Dolinko, V. I.; Nikishin, G. I. *Izv. Akad. Nauk SSSR, Ser. Khim.* **1981**, 700 (*Chem. Abstr.* **95**, 24673j).

[153] Melikyan, G. G.; Mkrtchyan, D. A.; Lebedeva, K. V.; Maeorg, U.; Panosyan, G. A.; Badanyan, Sh. O. *Chem. Nat. Compd.* **1984**, 94.

[154] Hosogai, T.; Omura, Y.; Mori, F.; Aihara, S.; Wada, F.; Fujita, W.; Onishi, T.; Nishida, T. *Jpn. Kokai 77 57,163* (*Chem. Abstr.* **88**, 37598u).

[155] Bader, A. *Aldrichimica Acta* **1976**, *9*, 49.

[156] Melikyan, G. G. unpublished results.

[157] Melikyan, G. G.; Mkrtchyan, V. M.; Atanesyan, K. A.; Asaryan, G. Kh.; Badanyan, Sh. O. *Chem. Nat. Compd.* **1990**, 78.

[158] Melikyan, G. G.; Aslanyan, G. Kh.; Atanesyan, K. A.; Mkrtchyan, D. A.; Badanyan, Sh. O. *Chem. Nat. Compd.* **1990**, 83.

[159] Melikyan, G. G.; Mkrtchyan, V. M.; Atanesyan, K. A.; Asaryan, G. Kh.; Badanyan, Sh. O. *Bioorg. Khim.* **1990**, *16*, 1000 (*Chem. Abstr.* **113**, 230995j).

[160] Melikyan, G. G.; Mkrtchyan, D. A.; Badanyan, Sh. O. *Arm. Khim. Zh.* **1982**, *35*, 163.

[161] Melikyan, G. G.; Sargsyan, A. B.; Badanyan, Sh. O. *Arm. Khim. Zh.* **1986**, *39*, 228 (*Chem. Abstr.* **106**, 176079d).

[162] Melikyan, G. G.; Vostrowsky, O.; Bauer, W.; Bestmann, H. J. *J. Organometal. Chem.* **1992**, *423*, C24.

[163] Melikyan, G. G.; Sargsyan, A. B.; Badanyan, Sh. O. *Khim. Geterotsikl. Soedin.* **1986**, 562 (*Chem. Abstr.* **106**, 67021h).

[164] Heiba, E. I.; Dessau, R. M.; Koehl, W. J. *J. Am. Chem. Soc.* **1969**, *91*, 138.

[165] Citterio, A.; Santi, R.; Fiorani, T.; Strologo, S. *J. Org. Chem.* **1989**, *54*, 2703.

[166] Min, R. S.; Aksenov, V. S.; Vinogradov, M. G.; Nikishin, G. I. *Bull. Acad. Sci. USSR, Div. Chem. Sci.* **1981**, 1902.

[167] Baciocchi, E.; Muraglia, E. *J. Org. Chem.* **1993**, *58*, 7610.

[168] Kurz, M.; Ngoviwatchai, P.; Tantrarant, T. *J. Org. Chem.* **1981**, *46*, 4668.

[169] Bellami, A. J. *Acta Chem. Scand.* **1979**, *B33*, 208.

[170] Heiba, E. I.; Dessau, R. M. *J. Am. Chem. Soc.* **1972**, *94*, 2888.

[171] Snider, B. B.; Buckman, B. O. *J. Org. Chem.* **1992**, *57*, 322.

[172] Citterio, A.; Sebastiano, R.; Carvayal, M. *J. Org. Chem.* **1991**, *56*, 5335.

[173] Citterio, A.; Sebastiano, R.; Marion, A.; Santi, R. *J. Org. Chem.* **1991**, *56*, 5328.

[174] Citterio, A.; Sebastiano, R.; Maronati, A.; Santi, R.; Bergamini, F. *J. Chem. Soc., Chem. Commun.* **1994**, 1517.

[175] Julia, M. *Acc. Chem. Res.* **1971**, *4*, 386.

[176] Beckwith, A. L. *J. Chem. Soc. Rev.* **1993**, 143.

[177] Chuang, C.; Wang, S. *Tetrahedron Lett.* **1994**, *35*, 4365.

[178] Ernst, A. B.; Fristad, W. E. *Tetrahedron Lett.* **1985**, *26*, 3761.

[179] Cossy, J.; Leblanc, C. *Tetrahedron Lett.* **1989**, *30*, 4531.

[180] Colombo, M. I.; Signorella, S.; Mischne, M. P.; Gonzales-Sierra, M.; Ruveda, E. A. *Tetrahedron* **1990**, *46*, 4149.

[181] Citterio, A.; Fancelli, D.; Finzi, C.; Pesce, L.; Santi, R. *J. Org. Chem.* **1989**, *54*, 2713.

[182] Citterio, A.; Pesce, L.; Sebastiano, R.; Santi, R. *Synthesis* **1990**, 142.

[183] Corey, E. J.; Kang, M. *J. Am. Chem. Soc.* **1984**, *106*, 5384.

[184] Snider, B. B.; Wan, B.; Buckman, B.; Foxman, B. *J. Org. Chem.* **1991**, *56*, 328.

[185] Snider, B. B.; Zhang, Q. *Tetrahedron Lett.* **1992**, *33*, 5921.

[186] Zhang, Q.; Mohan, R. M.; Cook, L.; Kazanis, S.; Peisach, D.; Foxman, B. M.; Snider, B. B. *J. Org. Chem.* **1993**, *58*, 7640.

[187] Zoretic, P. A.; Weng, X.; Biggers, C.; Biggers, M.; Caspar, M. L.; Davis, D. *Tetrahedron Lett.* **1992**, *33*, 2637.

[188] Ryu, I.; Alper, H. *J. Am. Chem. Soc.* **1993**, *115*, 7543.

[189] Snider, B. B.; McCarthy, B. A. *ACS Symposium Series 577* **1994**, Ch. 7, 84.

[190] Nohair, K.; Lachaise, I.; Paugam, J.; Nedelec, J. *Tetrahedron Lett.* **1992**, *33*, 213.

[191] Allegretti, M.; D'Annibale, A.; Trogolo, C. *Tetrahedron* **1993**, *49*, 10705.

[192] Iwasawa, N.; Funahashi, M.; Hayakawa, S.; Narasaka, K. *Chem. Lett.* **1993**, 545.

[193] Subramaniam, C. S.; Thomas, P. J.; Mamdapur, V. R.; Chadha, M. S. *Indian J. Chem.* **1978**, *16B*, 318.

[194] Trivedi, S. V.; Mamdapur, V. R. *Indian J. Chem., Sect. B* **1986**, *25B*, 1160.

[195] Fukamiya, N.; Oki, M.; Okano, M.; Aratani, T. *Chem. Ind.* **1981**, 96.

[196] Gardrat, C. *Synth. Commun.* **1984**, *14*, 1191.

[197] White, J.; Somers, T.; Yager, K. *Tetrahedron Lett.* **1990**, *31*, 59.

[198] Corey, E. J.; Wu, Y. *J. Am. Chem. Soc.* **1993**, *115*, 8871.

[199] Paquette, L. A.; Schaefer, A.; Springer, J. *Tetrahedron* **1987**, *43*, 5567.

[200] Coleman, J. P.; Hallcher, R. C.; McMackins, D. E.; Rogers, T. E.; Wagenknecht, J. H. *Tetrahedron* **1991**, *47*, 809.

[201] Breuilles, P.; Uguen, D. *Tetrahedron Lett.* **1984**, *25*, 5759.

[202] Brauer, G. *Handbook of Preparative Inorganic Chemistry*; Vol. 2, 2nd ed., Academic Press, New York, 1965, p. 1469.

[203] Shundo, R.; Nishiguchi, I.; Matsubara, Y.; Hirashima, T. *Chem. Lett.* **1990**, 2285.

[204] Shundo, R.; Nishiguchi, I.; Matsubara, Y.; Hirashima, T. *Tetrahedron* **1991**, *47*, 831.

[205] Nedelec, J.; Nohair, K. *Synlett* **1991**, 659.

[206] Hessel, L. W.; Romers, C. *Recl. Trav. Chim. Pays-Bas* **1969**, *88*, 545.

[207] Prabhakaran, C. P.; Sarasukutty, S. *Thermochimica Acta* **1984**, *82*, 391.

[208] de Klein, W. J. *Recl. Trav. Chim. Pays-Bas* **1977**, *96*, 22.

[209] Narasaka, K.; Miyoshi, N.; Iwakura, K.; Okauchi, T. *Chem. Lett.* **1989**, 2169.

[210] Mee, A. Patent Ger. 2,016,820 (*Chem. Abstr.* **74**, 12630w).

[211] Hoekman, M.; Fagan, G. L.; Webb, A. D.; Kepner, R. E. *J. Agric. Food Chem.* **1982**, *30*, 920.

[212] Mee, A. Patent Ger. 1,927,233 (*Chem. Abstr.* **72**, 78456j).

[213] Heiba, E. I.; Dessau, R. M.; Williams, A. L.; Rodewald, P. G. *Org. Synth.* **1983**, *61*, 22.

[214] Goudgaon, N. M.; Nayak, U. R. *Indian J. Chem., Sect. B* **1981**, *20B*, 955.

[215] Goudgaon, N. M.; Nayak, U. R. *Indian J. Chem., Sect. B* **1985**, *24B*, 493.

[216] Witkiewitcz, K.; Chabudzinski, Z. *Rocz. Chem.* **1976**, *50*, 1545.

[217] Witkiewicz, K.; Chabudzinski, Z. *Rocz. Chem.* **1977**, *51*, 825.

[218] Witkiewicz, K.; Chabudzinski, Z. *Rocz. Chem.* **1977**, *51*, 475.

[219] de Klein, W. J. *Recl. Trav. Chim. Pays-Bas* **1975**, *94*, 48.

[220] Nishino, H.; Kurosawa, K. *Bull. Chem. Soc. Jpn.* **1983**, *56*, 474.

[221] Chuang, C.; Wang, S. *Tetrahedron Lett.* **1994**, *35*, 1283.

[222] Citterio, A.; Nikolini, M.; Sebastiano, R.; Carvajal, M. C.; Cardani, S. *Gazz. Chim. Ital.* **1993**, *123*, 189.

[223] Bremner, J.; Jaturonrusmee, W. *Aust. J. Chem.* **1990**, *43*, 1461.

CUMULATIVE CHAPTER TITLES
BY VOLUME

Volume 1 (1942)

1. **The Reformatsky Reaction**: Ralph L. Shriner

2. **The Arndt-Eistert Reaction**: W. E. Bachmann and W. S. Struve

3. **Chloromethylation of Aromatic Compounds**: Reynold C. Fuson and C. H. McKeever

4. **The Amination of Heterocyclic Bases by Alkali Amides**: Marlin T. Leffler

5. **The Bucherer Reaction**: Nathan L. Drake

6. **The Elbs Reaction**: Louis F. Fieser

7. **The Clemmensen Reduction**: Elmore L. Martin

8. **The Perkin Reaction and Related Reactions**: John R. Johnson

9. **The Acetoacetic Ester Condensation and Certain Related Reactions**: Charles R. Hauser and Boyd E. Hudson, Jr.

10. **The Mannich Reaction**: F. F. Blicke

11. **The Fries Reaction**: A. H. Blatt

12. **The Jacobson Reaction**: Lee Irvin Smith

Volume 2 (1944)

1. **The Claisen Rearrangement**: D. Stanley Tarbell

2. **The Preparation of Aliphatic Fluorine Compounds**: Albert L. Henne

3. **The Cannizzaro Reaction**: T. A. Geissman

4. **The Formation of Cyclic Ketones by Intramolecular Acylation**: William S. Johnson

5. **Reduction with Aluminum Alkoxides (The Meerwein-Ponndorf-Verley Reduction)**: A. L. Wilds

5. **The Reaction of Halogens with Silver Salts of Carboxylic Acids**:   C. V. Wilson

6. **The Synthesis of β-Lactams**:   John C. Sheehan and Elias J. Corey

7. **The Pschorr Synthesis and Related Diazonium Ring Closure Reactions**:
    DeLos F. DeTar

Volume 10 (1959)

1. **The Coupling of Diazonium Salts with Aliphatic Carbon Atoms**:
    Stanley J. Parmerter

2. **The Japp-Klingemann Reaction**:   Robert R. Phillips

3. **The Michael Reaction**:   Ernst D. Bergmann, David Ginsburg, and Raphael Pappo

Volume 11 (1960)

1. **The Beckmann Rearrangement**:   L. Guy Donaruma and Walter Z. Heldt

2. **The Demjanov and Tiffeneau-Demjanov Ring Expansions**:   Peter A. S. Smith and
    Donald R. Baer

3. **Arylation of Unsaturated Compounds by Diazonium Salts**:
    Christian S. Rondestvedt, Jr.

4. **The Favorskii Rearrangement of Haloketones**:   Andrew S. Kende

5. **Olefins from Amines: The Hofmann Elimination Reaction and Amine Oxide
    Pyrolysis**:   Arthur C. Cope and Elmer R. Trumbull

Volume 12 (1962)

1. **Cyclobutane Derivatives from Thermal Cycloaddition Reactions**:   John D.
    Roberts and Clay M. Sharts

2. **The Preparation of Olefins by the Pyrolysis of Xanthates. The Chugaev
    Reaction**:   Harold R. Nace

3. **The Synthesis of Aliphatic and Alicyclic Nitro Compounds**:   Nathan Kornblum

4. **Synthesis of Peptides with Mixed Anhydrides**:   Noel F. Albertson

5. **Desulfurization with Raney Nickel**:   George R. Pettit and Eugene E. van Tamelen

Volume 13 (1963)

1. **Hydration of Olefins, Dienes, and Acetylenes via Hydroboration**:   George Zweifel
    and Herbert C. Brown

2. **Halocyclopropanes from Halocarbenes**: William E. Parham and Edward E. Schweizer

3. **Free Radical Addition to Olefins to Form Carbon-Carbon Bonds**: Cheves Walling and Earl S. Huyser

4. **Formation of Carbon-Heteroatom Bonds by Free Radical Chain Additions to Carbon-Carbon Multiple Bonds**: F. W. Stacey and J. F. Harris, Jr.

Volume 14 (1965)

1. **The Chapman Rearrangement**: J. W. Schulenberg and S. Archer

2. **α-Amidoalkylations at Carbon**: Harold E. Zaugg and William B. Martin

3. **The Wittig Reaction**: Adalbert Maercker

Volume 15 (1967)

1. **The Dieckmann Condensation**: John P. Schaefer and Jordan J. Bloomfield

2. **The Knoevenagel Condensation**: G. Jones

Volume 16 (1968)

1. **The Aldol Condensation**: Arnold T. Nielsen and William J. Houlihan

Volume 17 (1969)

1. **The Synthesis of Substituted Ferrocenes and Other π-Cyclopentadienyl-Transition Metal Compounds**: Donald E. Bublitz and Kenneth L. Rinehart, Jr.

2. **The γ-Alkylation and γ-Arylation of Dianions of β-Dicarbonyl Compounds**: Thomas M. Harris and Constance M. Harris

3. **The Ritter Reaction**: L. I. Krimen and Donald J. Cota

Volume 18 (1970)

1. **Preparation of Ketones from the Reaction of Organolithium Reagents with Carboxylic Acids**: Margaret J. Jorgenson

2. **The Smiles and Related Rearrangements of Aromatic Systems**: W. E. Truce, Eunice M. Kreider, and William W. Brand

3. **The Reactions of Diazoacetic Esters with Alkenes, Alkynes, Heterocyclic, and Aromatic Compounds**: Vinod Dave and E. W. Warnhoff

4. **The Base-Promoted Rearrangements of Quaternary Ammonium Salts**: Stanley H. Pine

AUTHOR INDEX, VOLUMES 1–49

Volume number only is designated in this index.

CHAPTER AND TOPIC INDEX, VOLUMES 1–49

Many chapters contain brief discussions of reactions and comparisons of alternative synthetic methods related to the reaction that is the subject of the chapter. These related reactions and alternative methods are not usually listed in this index. In this index, the volume number is in **boldface**, the chapter number is in ordinary type.